Current Trends
in International
Fusion Research

Current Trends in International Fusion Research

Edited by
Emilio Panarella
*Advanced Laser and Fusion Technology, Inc.
Hull, P.Q., Canada; and
University of Tennessee
Knoxville, Tennessee*

Springer Science+Business Media, LLC

Library of Congress Cataloging-in-Publication Data

Current trends in international fusion research / edited by Emilio Panarella.
    p. cm.
    "Proceedings of the 1st Symposium on Current Trends in International Fusion Research: Review and Assessment, a conference organized under the auspices of the Global Foundation, Inc., held November 14-18, 1994, in Washington, D.C., U.S.A."--Ack.p.
    Includes bibliographical references and index.
    ISBN 978-0-306-45513-1   ISBN 978-1-4615-5867-5 (eBook)
    DOI 10.1007/978-1-4615-5867-5
    1. Nuclear fusion--Congresses. 2. Pellet fusion--Congresses.
I. Panarella, Emilio. II. Symposium on Current Trends in International Fusion Research: Review and Assessment (1st : 1994 : Washington, D.C.)
QC790.95.C87 1997
539.7'64--dc21   97-11196
                                                                        CIP

Proceedings of the First International Symposium on Evaluation of Current Trends in Fusion Research, held November 14 – 18, 1994, in Washington, D.C.

ISBN 978-0-306-45513-1

© 1997 Springer Science+Business Media New York
Originally published by Plenum Press, New York in 1997

http://www.plenum.com

10 9 8 7 6 5 4 3 2 1

All rights reserved

No part of this book may be reproduced, stored in a retrieval system, or transmitted in any form or by any means, electronic, mechanical, photocopying, microfilming, recording, or otherwise, without written permission from the Publisher

1st SYMPOSIUM ON CURRENT TRENDS IN INTERNATIONAL FUSION RESEARCH

Review and Assessment

INTERNATIONAL ADVISORY BOARD

Bruno Coppi
Department of Physics
Massachusetts Institute of Technology
Rm. 26-217
Cambridge, MA 02139

George H. Miley
216 Nuclear Engineering Laboratory
University of Illinois
103 S. Goodwin Avenue
Urbana, IL 61801

Hans Persson
Fusion Plasma Physics
Alfvén Laboratory
Royal Institute of Technology
S-100 44 Stockholm
Sweden

Richard F. Post
Lawrence Livermore National Laboratory
P.O. Box 808, L-644
Livermore, CA 94550

Shenggang Liu
The University of Electronic Science
 and Technology of China
Chegdu, Sichuan Province, 610054
People's Republic of China

R. N. Sudan
Laboratory for Plasma Studies
Cornell University
369 Upson Hall
Ithaca, NY 14853

Edward Teller
Herbert Hoover Memorial Bldg.
Room 208
Stanford University
Stanford, CA 94305

Guillermo Velarde
Instituto de Fusion Nuclear
Universidad Politecnica de Madrid
José Guitiérrez Abascal, 2
Madrid 28006
Spain

Chiyoe Yamanaka
Institute for Laser Technology
Yamadaoka 2-6
Suita, Osaka 565
Japan

Masaji Yoshikawa
Japan Atomic Energy Research Institute
2-2, Uchisaiwaicho 2-chome, Chiyoda-ku
Tokyo 100
Japan

STEERING COMMITTEE

Francesco Giammanco
Department of Physics
University of Pisa
56100 Pisa
Italy

Julio Herrera
Instituto de Ciencias Nucleares
Universidad Nacional Autonoma de Mexico
Circuito Exterior C.U.
A.P. 70-543
04510 Mexico DF
Mexico

Ron Kirkpatrick
MS B229
Los Alamos National Laboratory
Los Alamos, NM 87544

Behram Kursunoglu
Global Foundation, Inc.
P.O. Box 249055
Coral Gables, FL 33124-9055

Vincenzo Molinari
Department of Nuclear Engineering
University of Bologna
Via dei Colli, 16
40136 Bologna
Italy

Sadao Nakai
Institute of Laser Engineering
Osaka University
2-6 Yamada-oka
Suita Osaka, 565
Japan

Emilio Panarella
Advanced Laser and Fusion Technology, Inc.
189 Deveault St., Unit No. 7
Hull, P.Q., J8Z 1S7 Canada, and
Department of Electrical Engineering
University of Tenneessee
Ferris Hall
Knoxville, TN 37996-2100

Francesco Pegoraro
Dipartimento di Fisica
Università di Torino
10125 Torino
Italy

Hafiz-Ur Rahman
Institute of Geophysics and Planetary Physics
University of California
Riverside, CA 92521-0412

Norman Rostoker
Department of Physics
University of California
Irvine, CA 92697-4575

J. Reece Roth
Department of Electrical Engineering
The University of Tennessee
Knoxville, TN 37996-2100

Nikos A. Salingaros
Division of Mathematics
The University of Texas
San Antonio, TX 78249-0664

Y. Chia Thio
Massey University at Albany
Mathematics Department
Private Bag 102-904
North Shore MSC, Auckland
New Zealand

Moreno Vaselli
Scientific Attache
Embassy of Italy
Uliza Jesnina, 5
121002 Moscow
Russia

PANEL OF EVALUATORS

Edward C. Creutz
P.O. Box 2757
Rancho Santa Fe, CA 92067

Arthur R. Kantrowitz
4 Downing Road
Hanover, NH 03755

Joseph E. Lannutti
515 Keen Building
Florida State University
Tallahassee, FL 32306-3016

Hans J. Schneider-Muntau
National High Magnetic Field Laboratory
Florida State University
Tallahassee, FL 32306-4005

Glenn T. Seaborg
Bldg. 70/A Rm. 3307
Lawrence Berkeley Laboratory
1 Cyclotron Road
Berkeley, CA 94720

Proceedings Committee

Frederick Seitz
The Rockfeller University
1230 York Ave.
New York, NY 10021-6399

William B. Thompson
Department of Physics
University of California at San Diego
La Jolla, CA 92093

ACKNOWLEDGMENTS

The Symposium Organizers gratefully acknowledge the financial contribution from the Institute of Laser Engineering, Osaka, Japan.

CONTENTS

1. Opening Remarks 1
   Emilio Panarella, Chairman of the Steering Committee

SECTION I

2. Controlled Fusion, Soon! 5
   Edward Teller

Magnetic Confinement

3. Comments on the Feasibility of Achieving Scientific Break-Even with a Plasma
   Focus Machine 11
   J. S. Brzosko, J. H. Degnan, N. V. Filippov, B. L. Freeman, G. F. Kiutlu, and
   J. W. Mather

4. Self-Colliding Beams as an Alternative Fusion System for $D-He^3$ Reactors 33
   Norman Rostoker and Michl Binderbauer

Inertial Confinement

5. Target Physics for Inertial Fusion Energy 43
   J. M. Martínez-Val, G. Velarde, and S. Eliezer

6. Spherical Pinch Research: Historical Background, Achievements, and
   Projections 67
   F. Giammanco, E. Panarella, N. Salingaros, D. P. Singh, and M. Vaselli

SECTION II

7. Perspectives of Advanced Confinement Programs 119
   Bruno Coppi

Magnetic Confinement

8. Present Status of Field-Reversed Configurations 121
   John Slough

9. Ignition Physics and the Ignitor Project 125
   Francesco Pegoraro

Other Confinement

10. The Inertial Electrostatic Confinement Approach to Fusion Power 135
    George H. Miley

11. The D-^{3}He Dipole Fusion Reactor 149
    Michael E. Mauel

12. Open-Ended Magnetic Confinement Systems for Fusion 153
    Richard F. Post and Dmitri D. Ryutov

13. Formation, Compression, and Acceleration of Magnetized Plasmas 179
    J. H. Degnan, D. E. Bell, A. L. Chesley, S. K. Coffey, J. L. Eddleman,
    S. E. Englert, T. J. Englert, M. H. Frese, D. G. Gale, J. D. Graham,
    J. Hammer, C. W. Hartman, J. Havranek, T. W. Hussey, G. F. Kiuttu, F. M.
    Lehr, G. J. Marklin, H. S. McLean, A. W. Molvik, C. D. Holmberg,
    C. A. Outten, R. E. Peterkin, Jr., D. W. Price, N. F. Roderick, E. L. Ruden,
    U. Shumlak, P. J. Turchi, and J. J. Watrous

14. Prospects of Magnetic Electrostatic Plasma Confinement 197
    T. J. Dolan

Tutorial Talk

15. Analysis of the Fusion Breakeven Conditions for D-T Plasmas of Prescribed
    Temperature Evolution 211
    E. Panarella

SECTION III

16. Progress in Inertial Fusion Research 245
    Chiyoe Yamanaka

Inertial Confinement

17. Heavy-Ion Driven Inertial Fusion Energy 279
    R. O. Bangerter and T. J. Fessenden

18. X-Ray Driven Implosions on the Nova Laser 295
    J. D. Kilkenny and the LANL and LLNL ICF team

19. Present Status and Future Prospects of Laser Fusion Research at Osaka 297
    Chiyoe Yamanaka

20. Magnetized Target Fusion: An Overview of the Concept 319
    Ronald C. Kirkpatrick and Irvin R. Lindemuth

21. Thermonuclear Fusion in a Staged Pinch 333
    F. J. Wessel, Norman Rostoker, H. U. Rahman, P. Ney, and E. L. Ruden

22. Novel Staged Z-Pinch Concept as Super Radiant X-Ray Source for ICF 347
    Vitaly Bystritskii, Frank J. Wessel, Norman Rostoker, and Hafiz Rahman

SECTION IV

23. Fusion, the Competition, and the Prospects for Alternative Fusion Concepts 365
    L. John Perkins, James. H. Hammer, and R. Paul Drake

Magnetic Confinement

24. Ideas for Future RFP Experiments 375
    James A. Phillips, Don A. Baker, and Robin F. Gribble

25. Dense Z-Pinches for Fusion 385
    David Scudder and Jack Shlachter

26. Assessment of Field-Reversed-Configuration Stability 387
    Richard E. Siemon

Other

27. Muon-Catalyzed Fusion in 1996 389
    Steven E. Jones

28. Experimental Investigation of Muon-Catalyzed Fusion in Mixtures of Hydrogen
    Isotopes 401
    V. M. Bystritsky

29. Magnetoelectric Toroidal Confinement 421
    J. Reece Roth

30. Ball Lightning: What Nature Is Trying to Tell the Fusion Community 459
    J. Reece Roth

31. Fusion Implications of Free-Floating Plasmak™ Magnetoplasmoids 475
    Paul M. Koloc

SECTION V

32. Alternate Fusion Concepts 489
    Norman Rostoker

Inertial Confinement

33. Inertial Fusion Driven by Intense Cluster Ion Beams 497
    C. Deutsch, A. Bret, S. Eliezer, J. M. Martinez-Val, and N. A. Tahir

34. Inertial Fusion Energy: An Approach to Low Maintenance and Cost of
    Electricity, and the Role of the National Ignition Facility Testing the
    Target Physics 541
    B. Grant Logan

35. Magnetized Target Fusion: An Ultrahigh Energy Approach in an Unexplored
    Parameter Space 543
    Ronald C. Kirkpatrick, Irvin R. Lindemuth, Robert E. Reinovsky, and
    Peter T. Sheehey

SECTION VI

36. Concluding Remarks 561
    Emilio Panarella, Chairman of the Steering Committee

SECTION VII

37. Report of the Evaluators 563
    Edward C. Creutz, Arthur R. Kantrowitz, Joseph E. Lannutti,
    Hans J. Schneider-Muntau, Glenn T. Seaborg, Frederick Seitz, and
    William B. Thompson

38. Biographies of Evaluators 577

Participants 583

Index 591

OPENING REMARKS

Emilio Panarella, Chairman of the Steering Committee

Advanced Laser and Fusion Technology, Inc.
189 Deveault St., No. 7, Hull, P.Q. J9Z 1S7 Canada, and
Department of Electrical Engineering
University of Tenneessee
Knoxville, Tennessee 37996-2100

I take great pleasure in opening this Symposium on "Current Trends in International Fusion Research". As you know, this Symposium has been organized under the auspices of the Global Foundation, "A Nonprofit Organization for Global Issues and Frontier Problems in Science". Since Prof. Behram Kursunoglu, the Chairman of the Board of that Foundation, could not make it this morning to give you his personal welcome because of early commitments, he has sent me this letter that I want to read to you, so that you can have an idea of the objectives and mandate of the Global Foundation, and why this Symposium fits very well within the range of its objectives.

Dear Emilio:

I wish to congratulate you for your success in organizing the conference "Current Trends in International Fusion Research". You have also succeeded in forming two excellent committees to help you with the conference. It appears from your preliminary program that you are bringing together quite a few luminaries in the field of fusion research. You are certainly a very persuasive organizer of a conference in an important field. Your conference is the only one on the subject matter of fusion research organized under the auspices of the Global Foundation, Inc. In the Foundation we have organized a series of seventeen conferences on the problem of energy, and only some of the meetings included a few presentations on fusion, but none of these meetings were solely devoted to fusion.

I hope all will go well with your conference and that you will be pleased with the results. With best wishes,

Sincerely yours,

Berham N. Kursunoglu
Chairman of the Board
and Professor Emeritus

As to the Global Foundation itself, and the reasons why I accepted the invitation to join its Advisory Board, I believe that I was particularly attracted by two things: 1) the high standing of the scientists that are part of the Foundation, and 2) its objectives. If I read some of the names on the Board of Trustees, I find, besides Jean Couture of the Institut Français de l'Énergie, and Henry King Stanford, President Emeritus of the University of Miami, other names such as Manfred Eigen of the Max Planck Institut in Göttingen, Robert Hofstadter of Stanford University, Willis Lamb, Jr., of the University of Arizona, Louis Néel, of the University of Grenoble, Abdus Salam, of the International Center for Theoretical Physics, Glenn T. Seaborg, of the University of California at Berkeley, and Eugene P. Wigner of Princeton University. You will have recognized from my reading of these last seven names that they are Nobel Prize Laureates.

As to the objectives of the Foundation, I will read a few paragraphs from the section entitled "Objectives", from which it will be clear that this Symposium fits precisely within the scope of the Foundation.

> The Global Foundation concentrates on major global problems (and fusion energy is certainly one of those). The real value of the Foundation to world society depends on the degree to which it succeeds in meeting its global purpose, i.e., making definitive impacts by disseminating and effectively communicating its research and incisive analyses.

Another sentence that I find within the section "Objectives" is:

> The Foundation offers its research results on strategic issues to interested governments.

And, finally, another sentence that I find of interest in relation to this Symposium has subtitle:

Problems of Energy

> The Foundation organizes conferences on the problems of energy, and recommends appropriate sources of energy which are compatible with environmental protection and economic development.

In short, I see that the stated objectives of the Foundation, namely: 1) to make definitive impact by disseminating and effectively communicating its research and incisive analyses, 2) to offer its research results on strategic issues to interested governments, and 3) to recommend appropriate sources of energy which are compatible with environmental protection and economic development, are precisely those objectives that we hope to achieve, and I personally trust that we will be able to achieve with this Symposium.

I will not go now into the history of how this Symposium came to be conceived and organized, except to say that, at the beginning, it was a history of hesitation, doubts, and question marks on the way, such as "are we doing the right thing?" or "will we succeed?", etc. I will only tell you that those hesitations and doubts quickly disappeared as we were moving along, when I found that the invitations to join me that I made to the outstanding people who are now on the International Advisory Board and Steering Committee did not require an intense effort on my part to be accepted. Quite to the contrary, after I had explained the rationale behind the Symposium, the almost standard comment that followed were the reassuring words: "This Symposium has been long overdue".

On March of this year we had the first meeting in Washington of the International Advisory Board and Steering Committee where we set the path for the present coming together here of all of us. As Chairman of the Steering Committee, I could not have had an easier job, in that there was a complete meeting of the minds, and all deliberations were approved with unanimous consent. In particular, we unanimously agreed on the Statement of the Objectives of the Symposium, on the criteria for the selection of the fusion concepts to be presented at the Symposium, on the criteria for the selection of the invited speakers, of the invited participants, and of the evaluators. The Statement of the Symposium Objectives is summarized in its opening sentence, which reads:

> The objectives of the Symposium are to assess the benefits and uncertainties involved in following the conventional and alternative approaches to fusion energy production and assess what industrial spinoffs and other applications (including space propulsion and pure science) might result from each approach in light of the uncertainties.

The Symposium is therefore centered around the words: "assessment of the risks". It is therefore an analysis of the fusion concepts that have made possible, individually and collectively, to be where we are now in fusion research and to understand the nature of the fusion problem, thus paving the way for solutions.

Since the Symposium is basically an assessment, it requires two components: 1) a presentation of results, and 2) an evaluation of the results.

For the presentation of the results, the selection of the speakers has been guided by the criterion that the leading exponents of the various fusion programs were going to be invited. But this is not enough. We wanted to have a debate or discussion at the end of each presentation, which can take place only if those people who will be asking tough questions to the speakers are competent scientists with a strong interest in fusion. And this is the reason why we selected the participants with as much care as the speakers. Their role can be clarified by saying that they are just as important as the speakers because, through their questions, they will provide useful elements of assessment for the evaluators.

As to the evaluators, their names speak by themselves, and I will not add more words to these. Only one comment, however, is in order, regarding the choice of the evaluators, and why we did not select them all among plasma physicists. The reason is that we believe that stature is more important than specific competence. And since we want that the assessment of our work be a credible assessment, we believe that this can be achieved only if the stature of the evaluators is so high that there cannot be any doubt about their objectivity and impartiality.

What are the elements that the evaluators will have to carry out their work? They are three: the first is the abstract of all presentations, which they received well before this Symposium began; the second is provided by the presentations that will be heard in the next five days, and by the question ad answer period after each presentation; and the third, and perhaps the most important element, is provided by the full manuscripts of all presentations, which will be sent to them after the Symposium. Since what we are doing now is essentially a promotional effort of our fusion work, I urge you to give the proper emphasis, in terms of quality and clarity, to your presentations of the next few days and to the follow-up manuscript.

The open-door policy that we have established in the selection of the fusion concepts represented at the Symposium is based on the rationale that there is room for innovative ideas in fusion, and that we should allow them to come forward and make their case for acceptance. Although the field of controlled fusion is a mature one, there is ample space for rejuvenation through novel concepts. One of the purposes of the Symposium is to begin a process of renewal, if the innovative ideas that we start to present now, and the others that will certainly follow in the future, will have merit. The concepts in fusion that will be presented in the next few days are therefore a mixture of classical concepts, centered around the definitions of magnetic and inertial confinement, and novel concepts, which have been generically categorized under the heading of "other concepts". I expect to see that this open-door policy will lead to a flourishing of novel concepts, which will necessarily be brought forward by the next and younger generation of plasma physicists.

In the selection of the "current trends in fusion research" we were guided by the criterion that all major programs should have adequate representation. And indeed, we extended our invitation to their major exponents in the U.S.A., Europe, Russia and Japan. If you see now that some of the major programs do not have here the adequate representation that we wanted and that they deserve, I want to assure you that it is not for lack of invitation, but for other reasons which are not clear to me now.

I would like to conclude with a few words of hope that controlled fusion, to which some if not most of us has dedicated a lifetime, will come out of this Symposium reinforced in the belief that it can be achieved not in some far away year such as 2025, but within our own life. I would like to be able to say one day that, on 14 November 1994, I was part of a team that began the intellectual effort to reach that elusive goal of controlled fusion within a time frame of much more modest dimensions. I would like to be able to say that, on that same day, we began a journey in which not only the goal of controlled fusion appeared to be near, but the heritage of knowledge built, in terms of basic science and applications of plasma science, was of high value for our future generations.

It is my sincere hope that this Symposium will provide renewed and successful directions to fusion research. On behalf of the International Advisory Board and Steering Committee, I convey to you my best wishes of good work in your promotion of the value and importance of fusion for present and future generations.

CONTROLLED FUSION, SOON!

Edward Teller

Hoover Institution
Stanford, California
Lawrence Livermore National Laboratory
Livermore, California

I will try to speak briefly and to restrict myself to a number of obvious points that I believe are known to most of you. It's not an easy job because there are many approaches to controlled fusion. I may not mention all of them. I am most anxious, however, that after I have finished, there should be comments and questions. I would suggest that the questions be completely free, not necessarily connected with what I have said. I will try to give you an answer whether I have one or not.

There has developed an interest in fusion, an interest with a peculiar direction about which I, and some of my friends, are not terribly happy about. Fusion is being supported, but the underlying motivation is that fusion is going to be *the* energy source of the future. That might be. There are at least two great difficulties. I don't mention the difficulty that fusion may not work. I am completely convinced that it will. However, how expensive it will be is a very different question. That it will take for its development decades speaks in itself for the complications into which we are running. The apparatus which in the end will result will probably be rather complicated. Furthermore, if you will subject it to strong bombardment, particularly pulses of very fast neutrons, this means that the apparatus may not last very long.

In the general discussion of energy, politicians and the general public have emphasized again and again the raw material that delivers the energy. We have plenty of it, but it is the apparatus where the great expenditures will originate.

Then there is a second point concerning the eventual feasibility, which I think is less discussed and more urgent. That is the acceptability of fusion. Fission energy actually has worked fine, including the safety of fission. If you take the actual accidents, they are relatively few and the damage is limited. In fact, we already know how to make fission safer than practically any other energy source. The public, and even many scientists do not accept this. They don't accept it for a simple reason: the exaggerated fear of radioactivity. There is, however, not the slightest doubt that, in fusion reactors, radioactive material will be generated. One cannot make fusion without neutrons, including 14 MeV neutrons, and they will introduce nuclear changes in almost anything. The problem of radioactivity will occur. My point is not that this in itself is a real difficulty. The perception of it is the difficulty. There is not the slightest doubt that, if fusion becomes feasible, then this connection of radioactivity and fusion will be emphasized, and I think it is very important that we anticipate this. The way to anticipate it is to put the whole question of radioactive danger into proper proportion. In the interest of nuclear energy production by fusion or fission, the two are not very different, and one has therefore to consider carefully how great the danger in reality is.

This is the first half of my talk. The second half will be specifically to discuss the various methods of approach, not only the one that is being favored but that and others.

About radioactivity: If you take the cases of the number of people who actually have been hurt by radioactivity, they are exceedingly few. That cannot be and is not the reason of fear.

The reason of fear is that many people have been exposed to low-level radioactivity. We know that low-level radioactivity does not cause any noticeable damage in the short run. There is, however, an assumption that the dose and the effects are proportional. If you assume this and consider the very great number of people exposed, then multiply the low probability of effects with a great number of people exposed, you do get damage to many people. This is the motivation of fear.

Is there or is there not such proportionality? To begin with, it is very difficult to find out because the main danger is supposed to be cancer, and it is absolutely clear that there is no chance of an immediate causation of cancer. The cancer that emerges is delayed by years. If you have a low probability of one in thousand or one in ten thousand and try to find it out after several years, that becomes a particularly difficult problem. Millions of people have to be observed for years if proof is to established.

What is the reason for assuming proportionality? The original reason goes back to studies of genetic changes where it has been found that there is a quantitative proportionality between mutations and the amount of radiation received. That, furthermore, appealed to everybody because radiation is apt to cause random changes, and random mutations must have negative effects.

There are at least two definite pieces of information that one has to add to what I have already said. One is that, in the very field where original studies have been made, mutations of definite properties, there is a special position for the mature male sperm. If the mature male sperm is irradiated, the effect is, indeed, proportional to the dose. If the irradiation does not go to a male sperm, but to the parent cells of the male, the spermatogonia, then the proportionality is not there. Repair mechanisms appear. Furthermore, the mature female cells, the egg, fail to show proportionality.

I want, therefore, to propose an explanation. The mature male sperm has a unique property: It does not have any cell body. It has the single strand of unpaired genetic information, and it has methods of moving the cell, but otherwise it has no cell body. Wherever there is a cell body, like in the egg, in the female mature cell, or in the spermatogonia, repair mechanisms appear. I claim that most of our cells are not like the sperm, they are like the others, they have cell bodies, and there must be repair mechanisms. Furthermore, the modern development of detailed knowledge both of proteins and of genes tells us that they each consists of chains, in the one case of amino acids, in the other case of nucleic acids. The mutations occur in the amino acids and in the nucleic acids. The new fact, which I did not realize until relatively recently, is that in each one of our cells, there are thousands of mutations each day. Mutations are not the exception, they are the rule. Furthermore, there are repair mechanisms. For instance, the famous p52 protein (52 stands for the molecular weight of 52 thousand), which is responsible for repair mechanisms.

To this has to be added that, when one looks for proportionality, one does not find it. I recommend to all of you a book by a Japanese by the name of Kondo.* The title of the book is ***The Health Effects of Low Level Radiation***. The book contains tables of observation on Hiroshima, Nagasaki, Chernobyl and other places where there is radioactivity in great amounts. The book makes an explicit statement: The proportionality is a myth. There is no evidence for it. If I look at the evidence in detail, says Kondo, I find that a little radiation is helpful rather than harmful.

I am not stating that Kondo is right in that, nor does he claim to be right. He only says: That is the trend. All of us are exposed to radiation in the amount of approximately one-third of an r per year in connection with natural radiation, radon, for instance. Groups of people have been exposed to much more radiation. Studies of this kind have gone on in the United States, in Sweden, and, peculiarly enough (I happen to know) in Hungary. The evidence seems to be that under one r, i.e., three times the natural amount, there is no ill effect whatsoever. Probably a few more years will be needed to harden that fact. Kondo himself says that he believes there is a threshold above which there are harmful affects. He is uncertain whether the threshold should be put at 10 r or 30 r. Under 1 r, there is no evidence of harmful effects. I believe that if this point could be made absolutely clear and generally accepted, then a lot of activities that are now going on, for instance the demand to clean up the nuclear national laboratories which is supposed to cost hundreds of billions of dollars, will disappear.

I believe that there also will be sooner or later real trouble with fusion. I, therefore, would like to appeal to you to give a close look to the question of low-level radiation.

Now we come to the next question: What are the methods of producing fusion? There is a leading contender, magnetic confinement in the form of a ring. The construction that is at least theoretically the best is called the Tokamak. That is where most of the money goes, to develop the Tokamak into something that will give large amounts of energy. A few years ago, I visited Culham Laboratory in England where they are working on a big Tokamak, and asked them when will they have the first working model, not yet something practical, but just a good model. They told me in the year 2010. I told them it was too late and asked: Can't they make it in the year 2008 because that will be my 100th birthday, and I want to come! I think that here are big plans which I believe are sure to succeed, except that the exaggerated environmental movement might step in and stop it when it becomes final. Otherwise, it will succeed. At what expense and when is another question.

Now I would like to give a short list of other developments. I try to concentrate on those which I think are most obvious, and briefly comment on their feasibility and usefulness. To begin with, the most obvious thing to do is the smaller Tokamak. One should try to do it at a size where it can produce energy and neutrons at an early time at a minimum of expense. That will have a great value partly in giving us ideas what the real expenditures will be in the big model. There are also scientific reasons. A small Tokamak, or any small model device that depends on magnetic confinement (including the magnetic mirror), will be a remarkable neutron source. Anything one can do with neutrons, including using neutrons to observe properties of solids, will benefit from that device. So, I claim that this more modest stage, which is parallel to what we are already going, needs careful consideration.

Let me mention one additional small point. What is the fuel? The fuel may be deuterium. That will work. An easier fuel is deuterium plus tritium. Where do we get tritium?

---

* Published in 1993 by Sohei Kondo, Kinki University Press Co., Ltd. Higashiosaka, Osaka 577 Japan. Published and distributed in North America by Medical Physics Publishing Corporation, Madison, WI 53705, USA.

Probably there is a great amount of He^3 on the surface of the Moon, deposited by the solar wind. He^3, in turn, can be relatively easily transformed into tritium by neutron bombardment. The D-T reaction as an energy source, particularly for the small Tokamaks, is by no means excluded. Neither is the D-He^3 reaction if one wants to minimize the emission of neutrons.

Magnetic confinement is only one of the methods to obtain controlled fusion. Let me mention briefly that controlled fusion is already available, only we won't use it; the opposition to it is too strong. It has been proposed to make a small hydrogen bomb in a big cavity. We know how to contain it. Make the cavity big, put in the kind of material that turns the shock into heat. One can fire small explosives, use the energy and produce more fuel. Not the least question that it could be done. Not everyone could agree that it would be economical, and that the psychological opposition could be overcome. It has been pursued for a while, then it has been stopped. I do not recommend that it be revived, but at least I want to say that it exists, and maybe somebody will find reasons why it should be resumed.

Another possibility of which all of you know, which I should briefly mention, is the use of the negative μ meson. The μ-meson, playing the role of an electron attached to a proton, gives a neutral particle of a very small radius. This proton-negative-meson combination can capture another proton and make the analogous composition to the hydrogen molecular ion: But instead of the electron you have a 100 times more heavy meson. Then the two hydrogen nuclei, the deuterons, will be close enough so that they will react and liberate the meson which then can go to work and, before it decays, it can catalyze the thermonuclear reaction a number of times. It took some energy to produce the meson, approximately 100 MeV. You have to get that paid back. In general, it looks that it will be difficult to make that work in an effective way from the point of your energy production. From a point of view of experimentation and getting neutrons generated, it does not look like the best possibility, but it might have applications.

The main alternative is none of these, but inertial confinement fusion, ICF. I am sure that it will work. Quite a bit of progress has been made. I shall remind you of the way it is supposed to work. You compress DT or just deuterium to very high density, at a density of approximately 100 times normal density it will work. The reason is very simple. You heat it, and the usual reason why heating alone will not do the trick is that most of the energy goes into radiation, and the radiation is not useful. The kinetic energy of the particles is what brings the neutrons together. If you go to high density, the amount of energy in radiation will be less compared to the amount of energy in the kinetic form and the burning can proceed. This actually we know will work. Can we do it on a laboratory scale? Well, a lot of work has been done and, fortunately, a year or two ago, work in Livermore was declassified. A DT mixture has actually been compressed and up to 10^{12}, 10^{13} neutrons were produced in one shot. That, of course, is still by far not enough to give the amount of energy that we needed to compress. I believe, however, that this can be improved, and the way to improve it is to put the deuterium or the DT into a heavy envelope and implode the heavy envelope by putting it into a cavity and bringing the cavity to a high temperature. The way to give the cavity the high temperature is to shoot into it particles: You can do it with an accelerator. Alternatively, you can deliver the energy with lasers. Both have been tried, both are apt to work.

As usual, the difficulties are the instabilities. The implosion of the heavy envelope does not under all conditions proceed smoothly, and if it does not proceed symmetrically and smoothly, then you won't get the needed one-hundredfold compression.

Livermore has received the go-ahead signal for a particular apparatus that will be constructed in approximately the next five years. It is called NIF (National Ignition Facility). The main point in NIF is the delivery of two megajoules of energy by many laser beams. That compares to lasers that we now have of approximately 1/50th the power. We believe that with the two-megajoule laser, we are going to get fusion.

I might spend a little of the remaining time to tell you about the expected uses. To begin with, it would be a very powerful neutron source. All the nice things I said in that regard about the little Tokamak would hold in this case. To develop it into a big-scale power source is possible but difficult.

With the 14-million volt neutrons that will be obtained in great abundance, it ought to be possible to cause fission in elements lower than uranium. We shall get better understanding of fission itself.

Remarkably enough, the main scientific advantage may not be in fusion at all. In order to get the fusion, there must be compression. This compression can be applied not only to the fusion fuel; it can be applied to anything else. We know that in this way we will be able to compress any metal tenfold, one-hundredfold, any other material to a similar degree of extent. We may verify the remarkable prediction by Seitz and Wigner that, at sufficiently high pressure, hydrogen will turn into a metal. The required pressures we can reach even today, but we cannot reach them without also getting high temperatures, and at high temperatures the conductivity does not prove metallic behavior. The interesting thing is that hydrogen will turn metallic at low temperatures, and for that we need a more careful, more slow compression leading to a high density. I believe that this kind of thing could be done. You could compress materials to obtain the equation of state easily at high temperatures, and at greater difficulties at low temperatures. If we get it at low temperatures, we could probably study solids at five-fold, tenfold their normal density. These studies are only moderately interesting when the temperature is high. When it is low, we can ask when will everything turn conductive. What will happen to ferromagnetism when iron or anything else is compressed. This is a big area of new science comparable in its extent to solid state physics. I believe that the development of the NIF facility is apt to open up this new science. What will come out of it, I do not know. Quite possibly, at such high pressure new alloys could be formed, which could still be there when the pressure is reduced.

An implosion along a line could be easily achieved. I have seen a very ingenious proposal of an implosion in a Tokamak-like device to a circle rather than a line. The result may be fusion in a device that is half Tokamak, half inertial confinement fusion.

Now, I have one last comment to make. I expressed doubts whether all this will have a big effect. It will be too expensive, it may be too strongly opposed. In spite of this, I feel very comfortable that fusion will be very important. I think one point that might come out of this conference is to try to estimate what its importance might be, and I will give only one statement where it might be useful. Space travel over long distances taking a long time just cannot be done without nuclear energy. If you use nuclear energy, then fission or fusion are the options. There is even today a plan called TAO, Thousand Astronomic Units. It has all kinds of interesting applications in trying to find details about the galaxy, about the distance of stars. If you get a thousand astronomic units away, that is in the neighborhood of a light week, still quite short compared to the distance of a next star. It will take 40 years to travel. Not many people will volunteer. We, therefore, need a mechanism to deliver energy reliably without human supervision. I am a little doubtful about fission because fission will give nothing but heat. The heat has to be transformed into electricity, the electricity then has to accelerate particles and all of that has to work for 40 years without human interference. If I could do this with magnetically confined fusion, the confinement that you have to develop anyway, will give you the configuration out of which with a little modification you can get a jet. If you can get the energy out of fusion at all, you can transform it into propulsive power much more easily.

If fusion can be accomplished, it might be used for long distance space propulsion and might give you a bigger reach into our neighborhood of the universe.

Thank you very much.

COMMENTS ON THE FEASIBILITY OF ACHIEVING SCIENTIFIC BREAK-EVEN WITH A PLASMA FOCUS MACHINE

J. S. Brzosko,[1] J. H. Degnan,[2] N. V. Filippov,[3] B. L. Freeman,[4] G. F. Kiutlu,[2] and J. W. Mather[5]

[1]Avegadro Energy Systems Inc.
 Staten Island, New York
[2]Phillips Laboratory
 Kirtland Air Force Base, New Mexico
[3]Kurchatov Institute
 Moscow, Russia
[4]Los Alamos National Laboratory
 Los Alamos, New Mexico
[5]University of New Mexico
 Albuquerque, New Mexico

ABSTRACT

This paper discusses some aspects of dense plasma focus (DPF) operation relevant to its extrapolation to hundreds of MJ energy storage. Experiments show, that the main fusion production mechanism is based on the plasma target - beam interaction i.e.: the pinch and the expanding column form hot and dense plasma target, that confines the hundred keV ion beams, produced during the column instability phase. Using a new compilation of neutron yield scaling with capacitor bank energies (W_o) it is expected that the scientific break-even will occur for W_o of a few hundreds of MJ, not less than W_o=50 MJ, assuming that plasma parameters will evolve in the same manner as for the existing DPF machines.

Strength of DPF research program lies in the fact that, it is based on two complementary and realistic lines of R&D actions. The first line of action considers two independent scaling tests of DPF performance at ten MJ energy level in Russia and USA. Both tests can use existing energy storages and chambers. The scaling tests will additionally demonstrate the operation of DPF in the insulator free version and the use of inductive storage. The second line of action (already implemented) is to use the DPFs, as intense and pulsed sources of neutrons, X-rays and ion beams. Selected examples of 100 kJ class DPF (and below) are shown to demonstrate the potentials of industrial applications and near-term payoffs.

1. BRIEF INTRODUCTION TO PLASMA FOCUS PHYSICS

Initially, dense plasma focus machines, DPF, were constructed independently by N.V. Filippov in Moscow [1] and J.W. Mather in Los Alamos [2] at the end of the fifties. Since then, over 30 laboratories carry vigorous DPF-programs, oriented mainly towards the studies of physics and DPF applications as intense neutron and X-ray sources.

The conceptual design of a plasma focus machine is shown in Fig. 1. The discharge chamber is usually pre-filled with deuterium or deuterium and tritium gas (D_2 or $D_2 + T_2$) at a pressure of p= 0.5 - 15 Torr, depending on the DPF-mode of operation and/or type of electrodes. For the neutron mode of operation, this pressure is larger than for the ion beam operation mode. The particular pressure is dependent on the electrode geometry and the circuit-parameters of the energy supplying system. For the X-ray production, the filling gas is usually Ne, Ar or Xe doped with H_2. The plasma focus operation can be divided into the following phases:

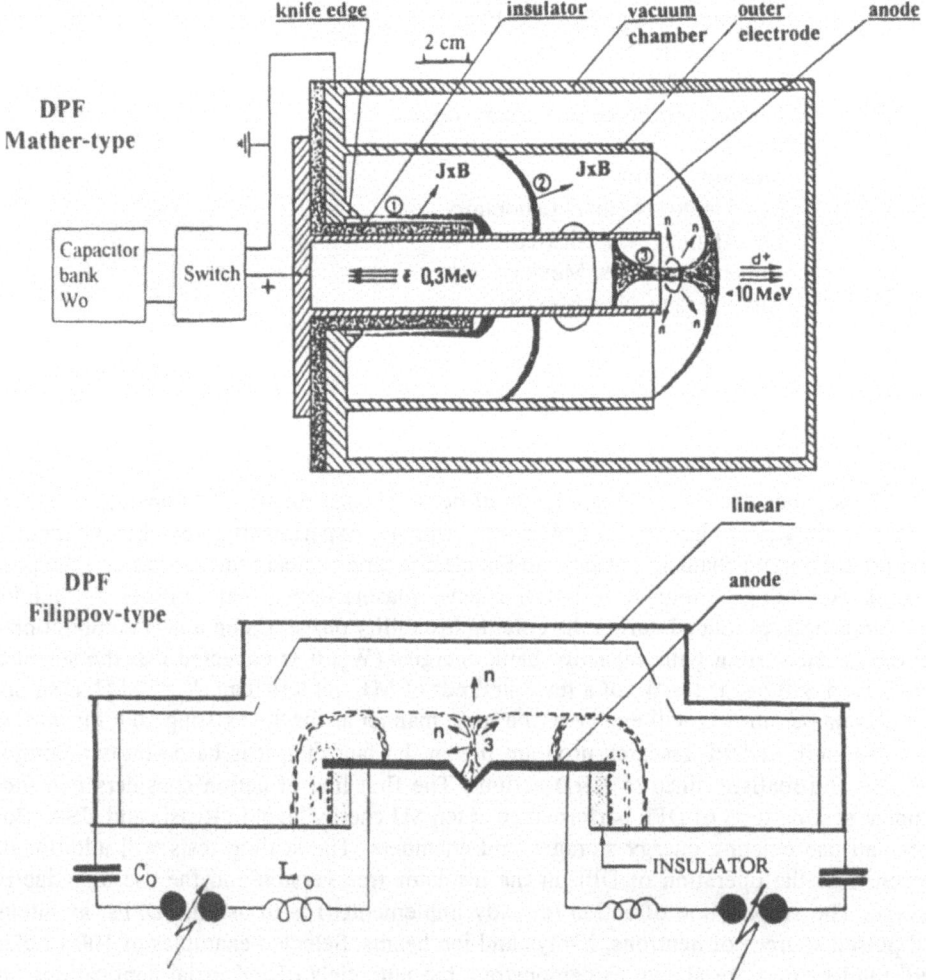

Figure 1. Schematic view of DPF machines. 1. Plasma sheath formation. 2. Run-down phase. 3. Radial compression phase. Instabilities and intense fusion occur after maximum compression.

(1) Initiation of the Discharge i.e. Breakdown, Sheath Formation and Take-Off. The breakdown occurs along a cylindrical insulator and is governed by a modified version of Paschen's law [3,4].

(2) Run-Down Phase of the Plasma Sheath (PS). This phase occurs in the Mather-type electrodes only. After its formation, the PS begins a lift-off from the insulator surface due to the **BxJ** magnetic forces, and accelerates in the direction of the electrode muzzle. The current sheath, moving with the velocity ca. 10^7 cm/s into an unperturbed deuterium gas region, generates a shock front. Its thickness is typically < 1 mm, corresponding to the free mean path of neutrals. Its Mach number at this point is of the order of 100. Deuterium molecules pass through this front in about 1 ns. This time is about two orders of magnitude smaller than the ionization relaxation time. Therefore, a non equilibrium transition region, with a temperature of several eV, exists. It gradually develops into a plasma layer, where 100% of ionization ($kT_e > 20$ eV, $n_e > 10^{18}$ cm^{-3}) and equilibrium is established. The dynamics of the run-down and the following compression phase are fairly well described by Newton's Second Law and by the circuit equation. For a well operating system, three conditions are fulfilled: (i) almost all of the momentum increase is allocated in the growth of the PS-mass, (ii) ohmic resistivities of the external circuit and of the plasma are negligible, and (iii) the inductance of the electric circuit is comparable or smaller than the plasma inductance.

(3) Radial Compression Phase. When the plasma sheath arrives at the end of the electrode muzzle, it changes its direction of motion from axial to radial. It forms an inward moving, funnel type sheath. In the Filippov type DPF, after the take-off is completed, the PS rolls-off into the radial compression. The radial compression is governed by similar rules as in the run-down phase. Here however, the driving magnetic force has a B ~ 1/r dependence, instead of the relatively position-independent **B** occurring in the axial phase. This allows for the acceleration (up to ca. 10^8 cm/s) of the funnel-shaped plasma sheath, before the funnel itself reaches a radius comparable to its thickness (r < 3 mm). For r < 3 mm, collision of sheets must be included in the description of the process. At the maximum compression (pinch instant), the plasma reaches a density of $n=10^{19}$ cm^{-3} and an electron temperature of $kT_e = 1$ keV [5,6]. At this time however, the neutron emission is negligible in the majority of DPFs as compared to the following phase i.e. plasma column formation and disruption phase.

(4) Nuclear Activity Phase. The collision of almost cylindrical plasma sheaths on the PF axis leads to plasma heating and begins the reverse movement. This motion soon (within 30 ns) stagnates, due to magnetic pressure outside of the plasma, compensating for the expansion momentum. For hundreds of kJ plasma focus, the column may be about cm in diameter and a few cm long, remaining stable for 50–100 ns [7] [8], [9], [10]. After this time, instabilities begin to develop, leading to plasma decompression. The strong neutron and hard X-ray signal accompanies this disruption of the plasma column. At that point, the plasma is characterized by $kT_e=0.3$–0.6 keV, $kT_i<1$ keV and $n=2$–$4 \cdot 10^{18}$ cm^{-3} only [5,6]. These parameters do not change significantly for the existing plasma focus machines operating in the neutron optimized mode and 5 kJ $<W_o<$ 600 kJ. Similar sequence of radiation and plasma column evolution as well as plasma parameter, characterize both Mather's and Filippov's type devices.

The neutron spectra from D(d,n)^{3}He fusion reveals immense broadening (see Fig. 2). The FWHM of the neutron spectra, Γ, spans from 400 keV to 1 MeV. If this were interpreted

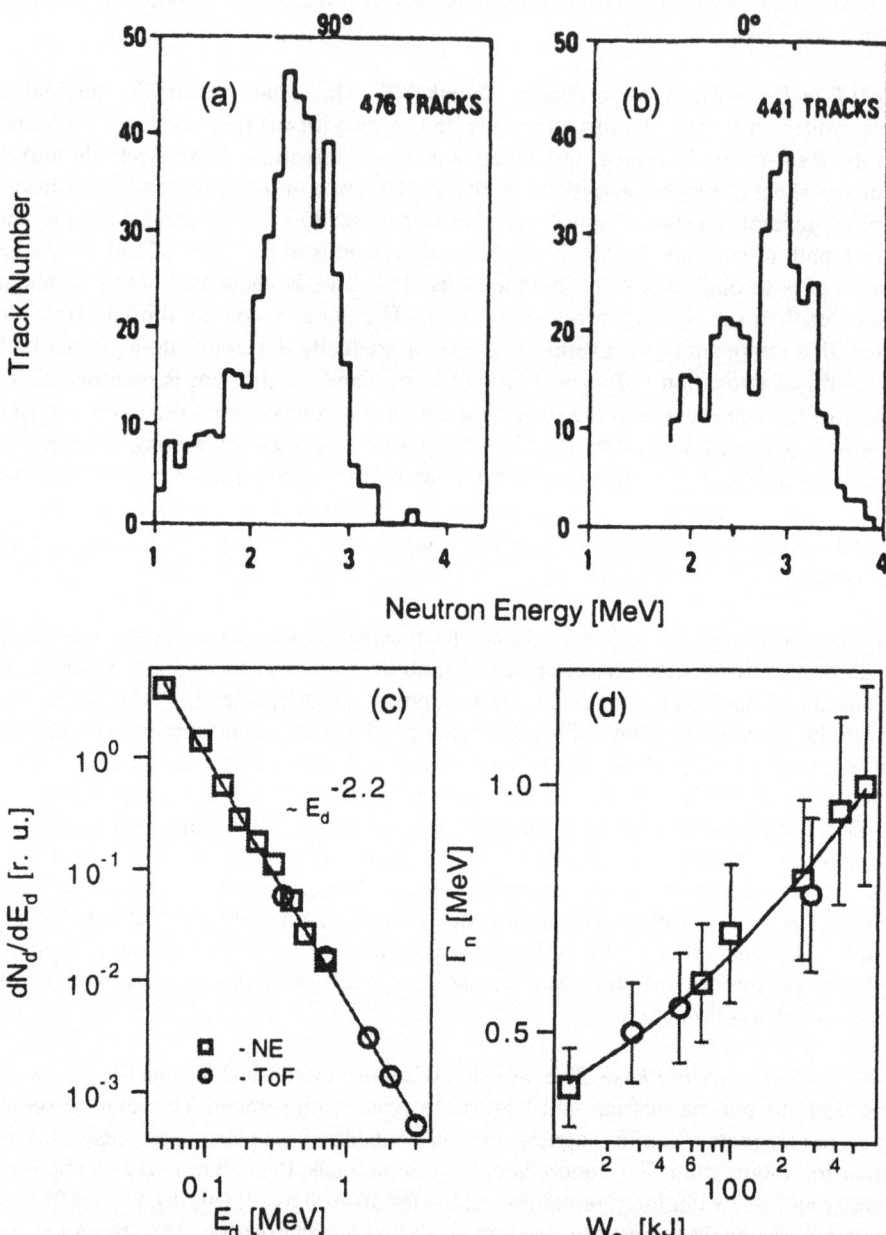

Figure 2. Neutron spectra measured with nuclear emulsions NE [7], (a) and (b), and their convolution to D⁺ spectra (c). In (c) additionally to NE data, the time of flight, ToF data [11] were used. W_o = 400 kJ, p= 6 Torr D_2; (d) shows the dependence of D-D neutron spectra width on W_o. The experimental data are for measurements at side-on detector position (90°) and for the following W_o values: 12 kJ [12]; 27 kJ [13]; 50 kJ [14]; 67 kJ [15]; 97 kJ [15]; 250 kJ [7]; 280kJ [16]; 400 kJ [7] and 560 kJ [17]. Data points marked as circles, come from NE analysis done by the same laboratory at Heidelberg University.

as the result of a high plasma temperature it would indicate an ion temperature 40 keV < kT_i < 150 keV [18]. A convolution of the time integrated neutron spectra, measured at different angles to the anode symmetry axes, allows us to estimate the D^+ ion spectra involved in the neutron production. This approach, initiated by Bernstein [19] and further developed at Stuttgart [8], reveals that there exists in the plasma a small population of fast ions with an average energy on the order of $<E_d>$ = 100 keV, responsible for the D-D fusion. An example of the neutron spectra convolution is presented in Fig. 2c. Observed D^+ spectrum, dN_d/dE_d, associated with the majority of D-D fusion, and normalized to one fast ion is well represented by:

$$dN_d/dE_d = (m-1) E_o^{(1-m)} E_d^{-m} \qquad (1)$$

where E_o (usually ca. 5 keV) is the lower limit of E_d, above which D^+ ions have an accountable participation in the produced neutron yield, Y_n. m is a fitted parameter, which is close to m=3 for W_o of 10 kJ and decreases to m=2 for a DPF operating at W_o=0.5 MJ. The same shape of the spectra, extended up to E_d<10 MeV is observed for D^+ emitted out of the plasma [20].

Independent information about the energy of D^+ ions, involved in the fusion, comes from the operation of the DPF with a deuterium - 3-helium mixture. In this case, the difference in D^+ energy dependence on fusion cross sections, for $D(d,n)^3$He and ^{3}He$(d,p)^4$He, is reflected on the measured neutron to proton yield, $\theta=Y_n(E_n<4\ MeV)/Y_p(E_p>12\ MeV)$. The compilation of experimental data, for θ is shown in Fig. 3. The increase in Γ (of the neutron spectra) and the θ decrease with the growth of W_o, both indicate a hardening of ion spectra involved in fusion reactions.

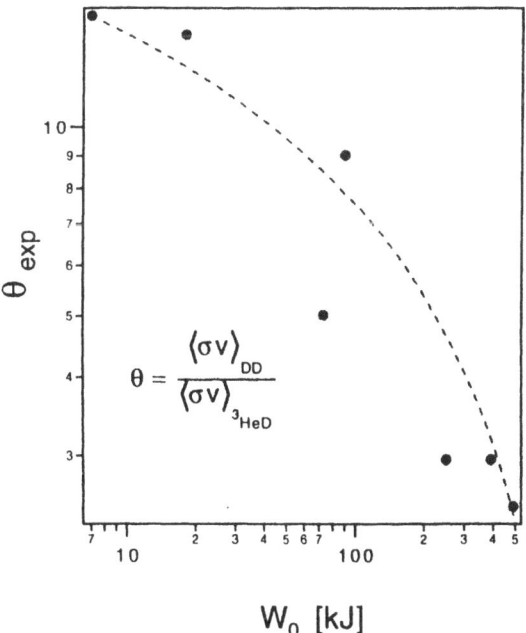

Figure 3. Compilation of measured reaction rate ratios for D-D and D-^{3}He fusion as a function of capacitor bank energy 7 kJ ≤ W_o ≤ 500 kJ. Experimental data come from the following sources: W_o = 7 kJ [21]; W_o = 18 kJ [22]; W_o =67 kJ [23], W_o =90 kJ [24] and W_o =250 - 500 kJ [25].

The above discussed information, related to D-D neutron production conditions, shows unambiguously that the source of fusion neutrons is a sort of self renewing, moderately hot, dense plasma target, with an embedded and trapped ion population of hundreds of keV controlling the fusion rate.

2. EXTRAPOLATION OF PLASMA FOCUS PARAMETERS TOWARD SCIENTIFIC BREAK EVEN

A comparison of the neutron yield, Y_n, produced by plasma focus devices at different capacitive, storage energies, W_0, indicate that for deuterium (D-D) and deuterium-tritium (D-T) plasmas, Y_n scales as W_0^2 for energies between 5 kJ and 600 kJ (see Fig. 4). It was proven experimentally that up to W_0=2 MJ the DPF of Filippov type (DPF-3, at Kurchatov) holds all discharge parameters without any obstacles for high-Z plasmas [26]. In particular: (a) pinch current is proportional to W^2, (b) K-shell X-rays yield scales as $\sim I_p^4$, (c) plasma sheath does not leave any substantial amount of residual gas behind and (d) plasma current (above 4 MA) during compression phase does not cause any feasible ablation of central electrode. Bearing in mind the above facts, one can consider the DPF scaling up to W_0=2 MJ as experimentally proven.

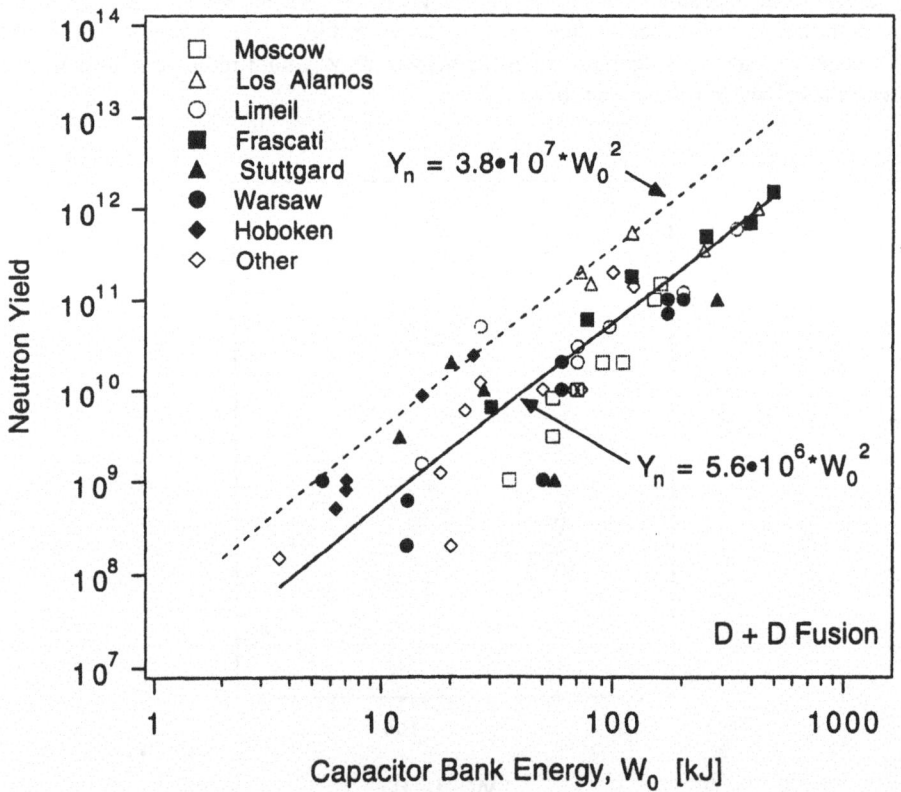

Figure 4. Neutron yield scaling for DPF operating with deuterium plasma. From all reported Y_n values the ones cited by their authors as average or typical for conditions of their experiment were chosen. Only those laboratories are identified, where experiments were done at different W_0 values. The solid line is the best fit to the shown data while the dashed line is for the best results, including the yet unpublished ones from Hoboken and Los Alamos.

For a D-T plasma, the conservative scaling law, based on typical DPF's performance is:

$$Y_n(\text{D-T}) = 5.6 \cdot 10^8 \, W_o^2;$$

$$[W_o] = [kJ] \qquad (2a)$$

The D-D neutron yield is two orders of magnitude lower. Matching the equally important scaling:

$$Y_n = 10 \cdot I_p^4;$$

$$[I_p] = [kA] \qquad (2b)$$

in terms of the plasma peak current, I_p, but relating to a deeper insight on the physics of the DPF operation. Unfortunately, reliable data for current scaling is far more scarce than the capacitor bank energy scaling. For this reason the energy scaling is used in the present context.

The scaling constants quoted in Eq. 2 are based on the compilation of all available data but are limited by requirement of Y_n being an average (or at least typical) value from many shots in the same initial DPF conditions. Research centers are identified only when data are available for at least few W_o values (not obviously the same device) otherwise data are marked as "others". This type of compilation is clearly not favoring recent research progress. If one takes into account DPFs with the best neutron efficiency, including the recent unpublished yet results, attained with the field distortion element at Stevens Institute of Technology (W_o = 15–25 kJ) and at Los Alamos National Laboratory (W_o = 70–120 kJ) then:

$$Y_n(\text{D-T}) = 3.8 \cdot 10^9 \, W_o^2;$$

$$[W_o] = [kJ] \qquad (2c)$$

Eq. 2a suggests that the scientific break-even i.e. break-even between energy stored in the capacitor bank, W_o, and the nuclear energy produced in plasma, E_{nucl}, may occur at W_b = 600 MJ (see Fig. 5). If one takes as the extrapolation base the DPF with the best fusion efficiency (Eq. 2c), then W_b = 90 MJ is estimated. In addition, W_b may be substantially reduced if pinch chaining (during one discharge) is realized [27]. For example, the scientific break-even energy might be reduced below W_b < 90 MJ by just such a mechanism. For the purpose of the present discussion, one can use conservative scaling of Eqs. 2 and assume that the scientific break-even will probably occur at a few hundred MJ, certainly below a 1 GJ. The projected W_b energy is 2 orders of magnitude higher than W_o at which energy DPF are operated. This in itself raises the question of feasibility. There are no severe problems with the know-how of hundreds of MJ energy storage. The feasibility question should be addressed to the DPF itself and to the kind of theoretical studies (based on MHD models) that were done in the seventies. Responses are encouraging and they set the W_b energy at 100 MJ [28]. In the present analysis we address the issue of trends in parameters of the fusion producing medium, using a semi-empirical approach.

Experimental data shows a two step mechanism, leading to fusion i.e. formation of the plasma target and acceleration/injection of the fast ions. Unfortunately, there is discon-

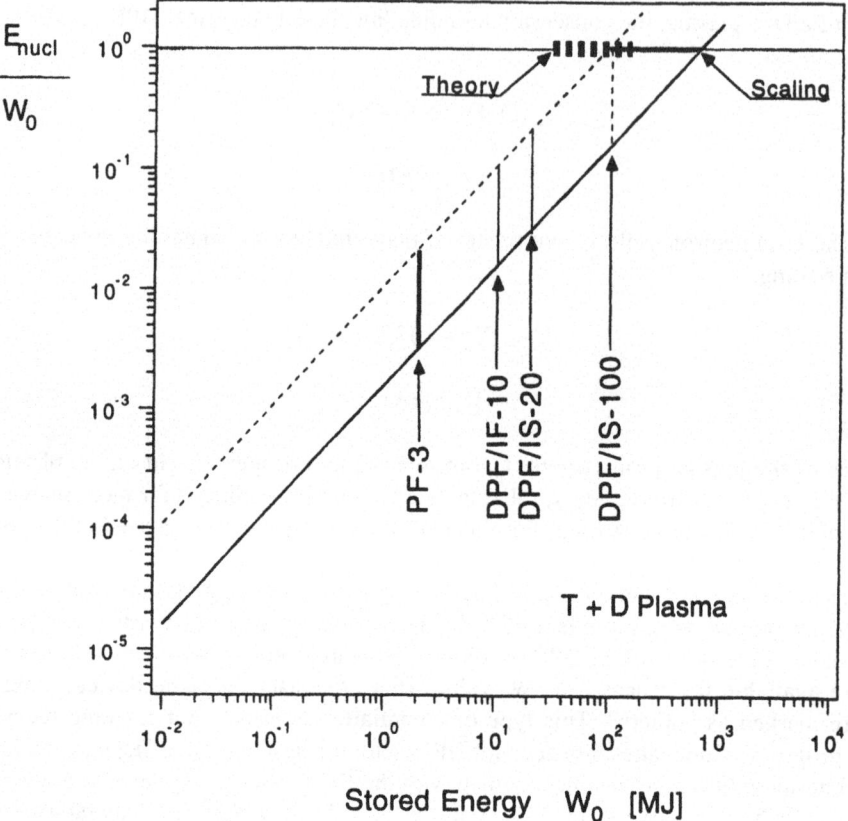

Figure 5. Scaling of DPF fusion efficiency as a function of stored energy. The solid line corresponds to the average DPF performance, Eq. 2a, (fit in Fig. 4 transferred from D-D to D-T plasma), the dashed line represents DPFs with the best fusion efficiency. | is for DPF-3 successfully operated at the largest W_0. Vertical lines mark near future devices discussed in this paper.

tinuity in the theoretical description of DPF physics at the instant of fast beam production. This does prevent from checking whether or not there exists obstacles of efficient DPF functioning at high stored energies. On the other hand, smooth evolution of plasma parameters with W_0 (in experimentally controlled range) may be used as zero-order approximation for defining the plasma in close to break-even conditions.

To examine the self-consistency of the plasma parameters, with large stored energies of many mega joules, let us use a simplified formula for the neutron yield:

$$Y_n = n \tau \langle\sigma v\rangle N_i \qquad (3)$$

where: n and τ are plasma target density and the trapping time of the fast ions, respectively. N_i is the total number of trapped fast ions with the spectra $dN_i/dE_i \sim E_i^{-m}$ (see Eq. 1), $\langle\sigma v\rangle$ is the fusion reactivity.

Plasma Density and Temperature. Experimental data does not reveal a significant dependence of n and kT on the stored energy and one can assume that the plasma target remains almost unchanged in MJ region with nominal values of kT=1 keV and n=10^{19} cm^{-3}.

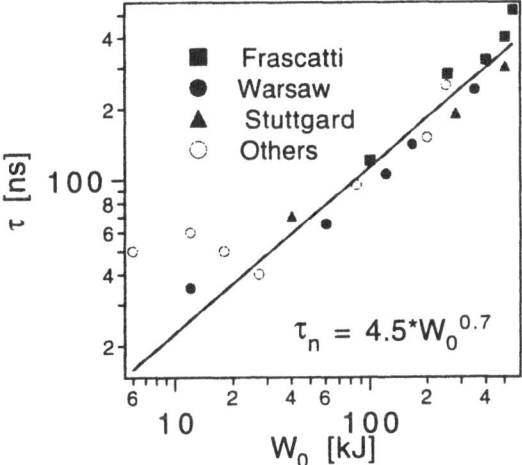

Figure 6. Dependence of neutron emission time (τ_n) on capacitor bank energy. Only those laboratories are identified where experiments were done at different W_o values. The solid line is the best fit to the shown data.

Trapping Time of Fast Ions. There is no straightforward experimental data about fast ion trapping/reacting but one can consider the width of the neutron signal τ_n as a measure of τ. The compilation of τ_n presented as function of W_o is shown in Fig. 6 and can be approximated as:

$$\tau_n \approx 4.5\, W_o^{0.7} \qquad [\tau_n]=[\text{ns}],\ [W_o]=[\text{kJ}] \qquad (4)$$

Fusion Reactivity. There are two categories of experimental data showing a hardening of the fast ion spectrum with the increase of W_o. One is the broadening of the neutron spectrum Γ_n (see Fig. 2d) and other is the decrease of the $\theta = \langle\sigma v\rangle(D\text{-}D)/\langle\sigma v\rangle(D\text{-}^3He)$ (see Fig. 3). Changes in $\theta(W_o)$ and $\Gamma_n(W_o)$ can be transferred to $m(W_o)$ and further to $\langle\sigma v\rangle_{DT}(W_o)$ because of:

$$\langle\sigma v\rangle = 9.6\cdot 10^{-16}(m-1)E_o^{(m-1)} \int_{E_o}^{\infty} E_i^{(0.5-m)}\sigma dE_i \qquad [\sigma v]=[\text{cm}^3/\text{s}];\ [\sigma]=[\text{b}] \qquad (5)$$

In the simplified form the above transformation suggests that:

$$\langle\sigma v\rangle_{DT} \sim W_o^{0.5} \qquad \text{for } W_o < 1\text{ MJ} \qquad (6)$$

Total Number of Trapped Fast Ions. The number of fast ions trapped in the plasma column region is:

$$N_i = \gamma\, n\, V \qquad (7)$$

where γ is the fraction of plasma ions that are accelerated to $E_i > E_o$. V is the neutron emitting plasma volume. Experimental data are scanty but it will be fair to say that:

$$V \sim W_o^k \quad \text{for } 0.5 < k < 1 \tag{9}$$

Combining Eqs. 3–7 and taking into consideration that $Y_n \sim W_o^2$ one can postulate that γ does not change significantly with W_o. In fact, γ estimates, from ^{3}He-D experiments, as well as from the GDR model fits to reaction proton spectra [8], [21], [25] suggest $\gamma \cong .005$. It means that for $W_o = 200$ MJ and $\gamma \ll 1$ the balance between population of trapped fast ions and plasma target population will not be affected. The energy balance between both populations for $W_o = 1$ MJ and $\gamma=0.005$ is b=1 and will further increase for break-even conditions to $b_b=2.5$ only.

As a result, the projected parameters of a near break-even experiment at $W_b= 200$ MJ are: $n=10^{19}$ cm^{-3}, $kT=1$ keV, $V=10^3$ cm^3, $\tau=200\mu s$, $Y_n=2 \cdot 10^{20}$, $\gamma=0.005$, $E_{beam}[E_i>10 \text{ keV}] / E_{plasma}=2.5$ and they do not introduce any obstacles from the feasibility point considered here.

All of the above discussions must be treated as an approximate cross-check of basic plasma parameters. If the DPF scaling saturates at $0.1 W_b$ of break-even, it will still be a very useful and worthwhile radiation source.

The same pulse power needed to drive a high energy plasma focus is applicable to several candidates for achieving break-even fusion. These include solid liner implosions, compressing magnetized plasmas. One version of this is the Russian MAGO concept. These candidates include compressed, accelerated Compact Toroids with inductive storage-opening switch pulse compression, and Z-pinches, including staged pinches. The pulse power ability to drive any of these to fusion levels is the ability to drive all of them. They all have reasonable prospect of scaling to break-even at a few hundred mega joules electrical (pulsed power) energy. The path to a commercial fusion power plant is not obvious yet. However, pulse power driven fusion is the cheapest, fastest route toward break-even. Achieving break-even by any means, or even better, by multiple means, at modest cost (10's of M$), will dispel the myth that fusion can not be achieved, and raise the credibility of reaching fusion by other means, more amenable to the use in commercial power production.

This is why the 10 MJ class experiments have to be performed, especially that the effort to do so is moderated by the existing basic hardware.

3. FEASIBLE EXPERIMENTS WITH 10 MJ-CLASS PLASMA FOCUS MACHINES

The effort to develop a break-even generation plasma focus may be optimized by utilizing 10 MJ - class energy storage systems. This will extend and prove the underlying principles at five to twenty times the historic capacitor bank energy. Presently, at least two energy storage systems with the appropriate characteristics do exist, and can be used for this purpose.

3.1. 10 MJ Dense Plasma Focus with Conical Electrodes and Insulator-Free Design (DPF/IF-10 and CT-DPF/IF-10 Modes)

In order to consider the industrial applications of DPF machines or their upgrading to operate at MJ energy storage, the one important technical task is to insure the survival of the insulator. The insulator is exposed to significant electrical, thermal and mechanical stresses. In MJ fusion devices, it will additionally be exposed to strong neutron and gamma radiation.

The aim of the discussed project is to test the performance of a Mather type geometry at MJ levels while at the same time, to prove the feasibility of the device with the insulator-free design. The project is at its conceptual level and considers the use of similar hardware as used in the MARAUDER experiment at SHIVA-STAR capacitor energy storage (Phillips Lab., KAFB, USA) [29], [30]. Advantages of the above concept are as follow:

a. Existing hardware allows for the insulator-free operation of DPF i.e. discharges are neither initiated nor can occur in the later stages of operation on the insulator.
b. Relevant parameters of the electric circuit and linear dimension of the conical electrodes are almost equivalent to those of the 1 MJ Frascati plasma focus (Mather type).
c. Successful experience gathered with MACH-2 (2.5-D MHD code) in the reconstruction and projection of MARAUDER experiments allows the development and optimization of DPF/IF-10.

The schematic design of DPF/IF-10 is shown in Fig. 7. The resulting design is supposed to fulfill the basic requirements for the efficient delivery of the plasma sheath (PS) to the muzzle and the execution of the compression phase. To insure proper conditions for the PS-formation and the following run-down, based on SHIVA-STAR discharges, three sets of gas injectors are required. They should produce the required gradients of the deuterium density. The length adjustment of the cylindrical external electrode, composed of stainless steel rods, should be chosen experimentally. This will function for the proper shaping of the plasma funnel during the compression phase. The following phases of DPF/IF-10 operation should be considered:

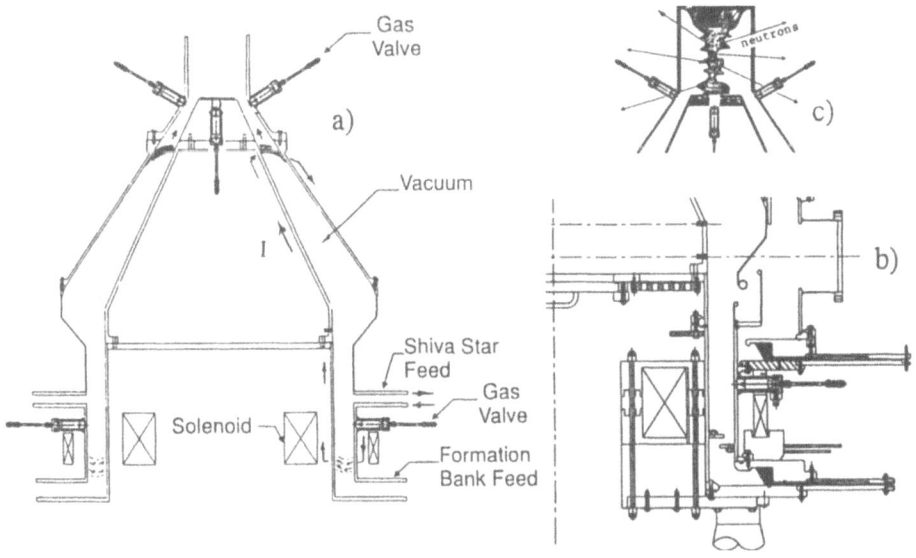

Figure 7. Conceptual design of DPF/IF-10 on the base of MARAUDER hardware. (a) The electrodes system and the main current path during the run-down phase. (b) Engineering design of breakdown area. (c) Instability of plasma column and fusion phase.

1. Injection of Neutral Gas (Xenon or Deuterium) at the Level of Coils. This is carried by 60 valves, a traditional initial gas injection in the MARAUDER operation.

2. Injection of Deuterium below the Muzzle and the Axial Injection from the Valve in the Central Electrode. These two operations, have to insure a proper z-gradient of deuterium density in the inter electrode gap, to achieve constant PS-velocity in the run-down phase, and proper radial density distribution at the muzzle as required for maximal velocity in the compression phase. To attain this goal, the mass of the gas per unit length of run down trajectory that is absorbed by the moving PS has to be constant. This requires $\rho(b^2 - a^2)$ = const. Here ρ is the gas density, **b** and **a** are the radii of the external and internal electrodes, respectively. The ρ, **b** and **a** values are z-dependent for the MARAUDER conical electrodes and additionally **b/a** = const.

3. Application of Voltage from the Formation Bank and Formation of Diffuse and Highly Conductive Plasma. Voltage from the formation bank applied between the inner and outer conductor (with controlled delay related to the opening of the MARAUDER gas valves) will initiate the diffuse discharge. The BxJ forces push the plasma upward, along the axis, while the pressure gradient works to expand the discharge in both z-directions. As a result, we expect the formation of a diffuse and highly conductive plasma. At the time when a substantial part of the discharge passes the SHIVA-STAR feed (some part still remains near the lower baffle), the injection of a slowly rising magnetic field, fed by external magnetic field coils, into the inter electrode gap will occur in the region of the lower baffle. This will cause a stagnation of the diffuse discharge for the whole duration of the DPF operation, the same as during classical MARAUDER experiments [29].

4. SHIVA-STAR Discharge and Run-Down Phase. The storage bank is fired, as part of the diffused plasma passes the bank feed. The diffused discharge is rapidly cut into two parts at the level of the feed. The plasma current flows in two different directions, between the outer and inner electrodes, closing the feed circuit. The upper part is pushed upward, forming a classical run-down phase of the plasma sheath. The lower part is concentrated and stagnated at the altitude of the external magnetic field and safeguards a low resistance connection between the electrodes. For this phase, it is important to maintain a proper z-gradient of the deuterium density to insure a thin, constant velocity PS, that will deliver plasma to the compression phase at a defined instant, before the maximum of the plasma current.

5. Compression Phase. During this phase, the plasma rolls-off from the axial movement into the axial compression and forms a funnel-like PS shape. It achieves maximum compression, with a time delay increasing with the z-coordinate. The central valve allows for the control of additional deuterium injection, before maximum compression, improving the energy pumping into the plasma. After the maximum compression, the plasma expands, forming a column with gradually developing instabilities.

6. Fusion Phase. The development of instabilities in the column coincides with the strong fusion rate. One can expect that, at the medium capacitor bank energy of SHIVA-STAR, W=5 MJ, with the neutron yield from D-D fusion $Y_n = 5 \cdot 10^{14}$ (equivalent to $5 \cdot 10^{16}$ of D-T fusion neutrons).

The other option to use the SHIVA-STAR, with almost MARAUDER regime operation is shown in Fig. 8. In accordance to this concept, Compact Toroid (CT) of high-Z

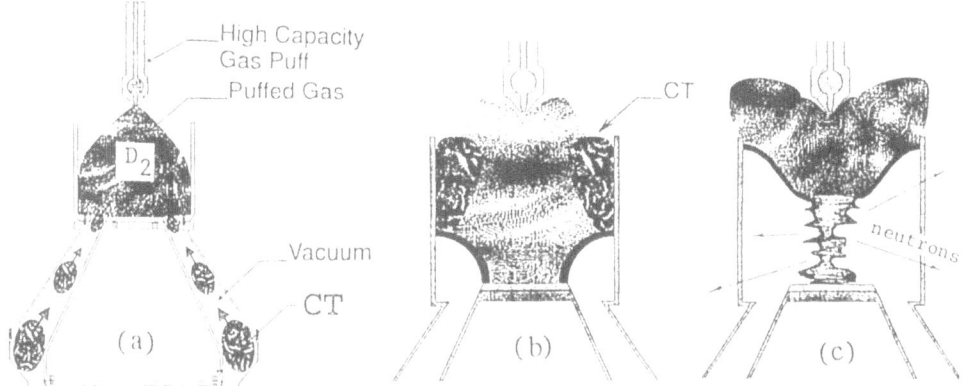

Figure 8. The main concept of the CT-DPF/IF-10 operation mode and the sequence of main events leading to fusion neutron production. Alternation of the MARAUDER external electrode (at the muzzle) shows the concept of efficient reconnection of the current-carrying plasma sheath from CT to the axially injected D_2 (or D_2+T_2) gas. (a) CT run-down (CT shown in three consecutive positions), current ramp-up phase, (b) Current transfer and beginning of radial compression phase, (c) Fusion phase initiated by instability induced acceleration of ions. For design of breakdown area see Fig. 7b.

plasma will play the role of the current opening (or transfering) switch for plasma focus action occurring in D_2 [31]. For details of MARAUDER physics and its operation, one could see the paper by J.H. Degnan, published in the materials of this conference and Ref. 29. As it is shown in Fig. 8 the CT is driven into compression stage by the magnetic piston, generated due to the compression discharge current. The CT acts like a gasket as it maintains contact with the conducting walls and keeps magnetic flux from the compression discharge behind it. When the CT leaves the muzzle of the central electrode, the current sheath switches from the inner electrode to the gas column and goes through radial compression in a conventional plasma focus fashion. Although the CT-DPF concept is clear, a good theoretical and experimental work should be done to optimize the current switching from high-Z plasma sheath to the deuterium one together with handling the CT that will fly-off the inter electrode gap.

3.2. 20 MJ Dense Plasma Focus with Filippov Electrodes and Powered from Inductive GJ Storage (DPF/IS-20)

In the past, the majority of plasma focus studies were made with capacitive energy storage. There are experimental data showing that the use of inductive storage instead of the capacitive one can increase the fusion yield 2-4 times with the same stored energy [32]. The DPF/IS program considers to utilize the inductive energy storage at TRINITY (Troisk, Russia). Inductive storage will power the Filippov-type DPF with large anode (ϕ_a = 2.5 m). It is expected that from 100 MJ of energy stored, as much as W=20 MJ will be transferred into the plasma. The plasma current can thus reach I_p=10 MA. This project is considered to be a proof-of-principles test, for the second step, larger scale effort, that will use the full capacity of T-14 (W = 1 GJ) inductive storage to deliver ca. 100 MJ (I_p=50 MA) into the plasma.

The conceptual design of the DPF/IS-20 device is shown in Fig. 9. To insure proper condition matching, the plasma sheath is produced, and initially accelerated in argon. It

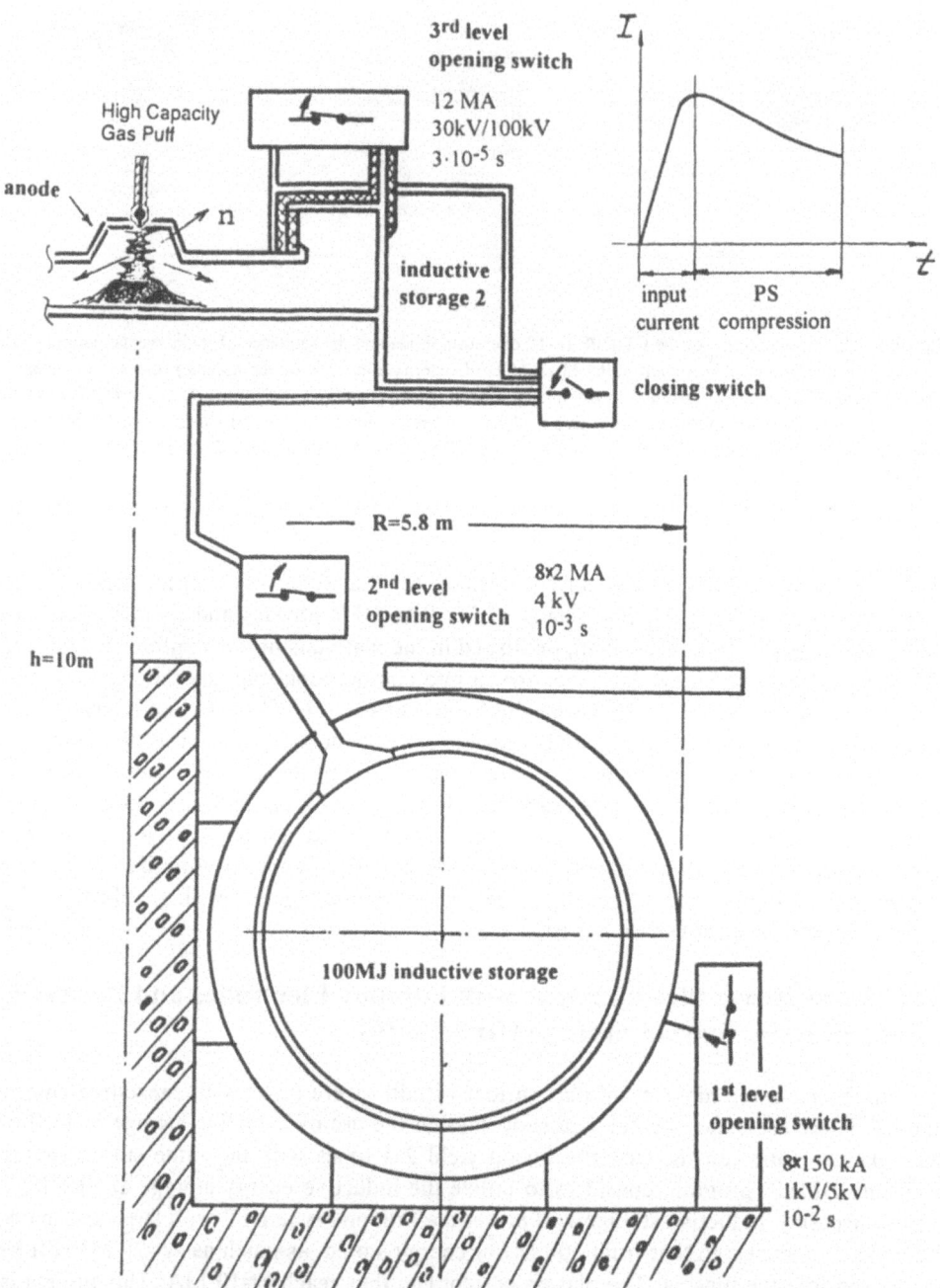

Figure 9. Conceptual design of DPF/IS-20 powered from the 100 MJ inductive storage at TRINITY. The insert shows the current delivered to plasma.

later compresses on axially injected deuterium. Due to the specific scheme of energy transfer from 100 MJ inductive storage into the DPF/IS-20 plasma feed (developed at TRINITY [33]) the current wave form will have improved (as compared to capacitive storage) characteristics, see Fig. 9. This project is the continuation of—and will profit from—experiences gathered in the last decade on PF-3 at Kurchatov, Moscow (capacitive storage, W=3 MJ, Filippov-type, ϕ_a=0.9 m, operation with Ar and Ne) [26]. At the present stage of the project, the inductive storages are operational together with the first level opening switch. The following phases of DPF/IS-20 operation should be considered:

1. Static Gas-Filing of the Chamber. The chamber will be filled with argon gas (ca. 3 Torr), since the discharge initiation, PS formation, roll-off and beginning of acceleration, has to be done in a high-Z gas to insure the proper matching conditions.

2. Energy Transfer from Main Inductive Storage, to Storage Feeding DPF Electrodes. The energy transfer from the main inductive storage to the DPF collector is based on a multistage switch developed at TRINITY. In fact, the proposed system disentangles the original energy storage from intermediate inductive storage connected to the DPF feed and allows for power amplification, control of matching between the storage and DPF as well as controlled modification of the current wave form.

3. Breakdown and Roll-Off Phase. Formation of the PS near the insulator, its take-off and roll-down to the radial compression, will occur in the similar conditions as known from experiments at PF-3 device [26]. Indeed MHD simulations of PF-3 performance, extended to DPF/IS-20 show no drastic changes in the plasma sheath behavior.

4. Deuterium Injection and Compression. At a proper instant of Ar- plasma sheath compression/acceleration, the deuterium will be injected axially, to form a deuterium cylinder coaxial with the anode. The arrangement of the valves is not yet designed. However, it will comply with the requirements defined by MHD - optimized transfer of current from argon to deuterium plasma sheath and the optimum of deuterium heating /compression.

5. Fusion Phase. Once a dense/hot pinch is achieved, the plasma column, with instabilities, is formed and fusion occurs. For plasma current I_p=10 MA, one can expect the neutron yield from D-D fusion of ca. Y_n=$3·10^{15}$ (equivalent to $3·10^{17}$ of D-T fusion neutrons). For the second step of the project, where I_p=50 MJ is expected, the D-D neutron yield will be Y_n= $8·10^{17}$ (equivalent to $8·10^{19}$ for T-D fusion neutrons).

4. SELECTED EXAMPLES OF DPFs FOR INDUSTRIAL APPLICATIONS AND NEAR-TERM PAYOFFS

First and second generation DPFs are relatively compact neutron and/or X-ray sources. Additionally, machines of this class have extraordinary capabilities for ion beam production and radioisotope breeding. The foundation for future applications of plasma focus machines lay in their outstanding (and scaling with ~W^2) efficiency in the transformation of stored energy into radiation. As an example, let us mention that for W_o=0.5 MJ, $2·10^8$ 14 MeV neutrons/J are produced [34], while for W_o= 1 MJ 10% of the energy is converted to 1 keV X-ray photons [26]. These numbers are a few orders of magnitude higher than for any known accelerators. With modern technology DPFs can be engineered to be reliable in operation and straightforward in maintenance. To illustrate the diversity of DPF applications, some (from among many) examples are selected.

1. X-Ray Lithography. A X-ray source of DPF origin ($E_x=1$ keV) is point like in the axial direction with very stable (from shot to shot) axial position. This allows for its application in the multi-shot mode, with a few kJ capacitor bank [35], [36], or in the single shot mode for a bank of W=1 MJ. In the second case, the mask can be evaporated without affecting the overall print since the X-ray pulse is shorter than 100 ns. An example of a high quality inexpensive X-ray source produced with the DPF ($W_o=4.5$ kJ, $\nu = 2$ Hz) is shown in Fig. 10.

2. Nondestructive Testing with 14 MeV Neutrons. Use of DPF as a point-like, pulsed neutron source, allows for the imaging of extended and bulk objects. Medium and high-Z matter is well transparent to 14 MeV neutrons while X-rays and thermal neutrons inefficiently examine solid objects of thickness above 5 cm. Additionally, the short pulse of neutron emission allows for a cut-off of parasitic radiation coming from multi scattered neutrons. An example of expected images simulated for prototype design for this purpose

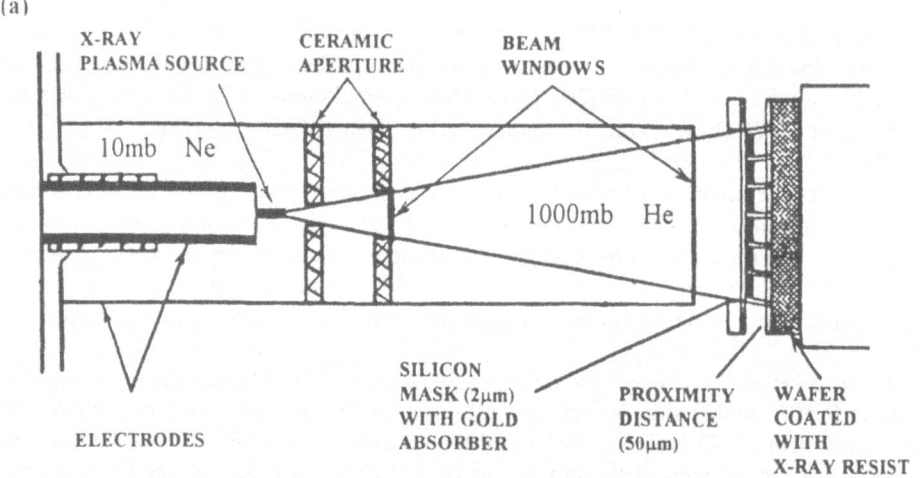

Figure 10. (a) Set-up for X-ray lithography with plasma focus. (b) examples of scanning electron micro graphs of sub micron pattern in a 60 mJ resist, obtained with $W_o=4.5$ kJ, $\nu = 2$ Hz at the Fraunhofer Inst. fur Lasertechnik, Aachen, Germany [35].

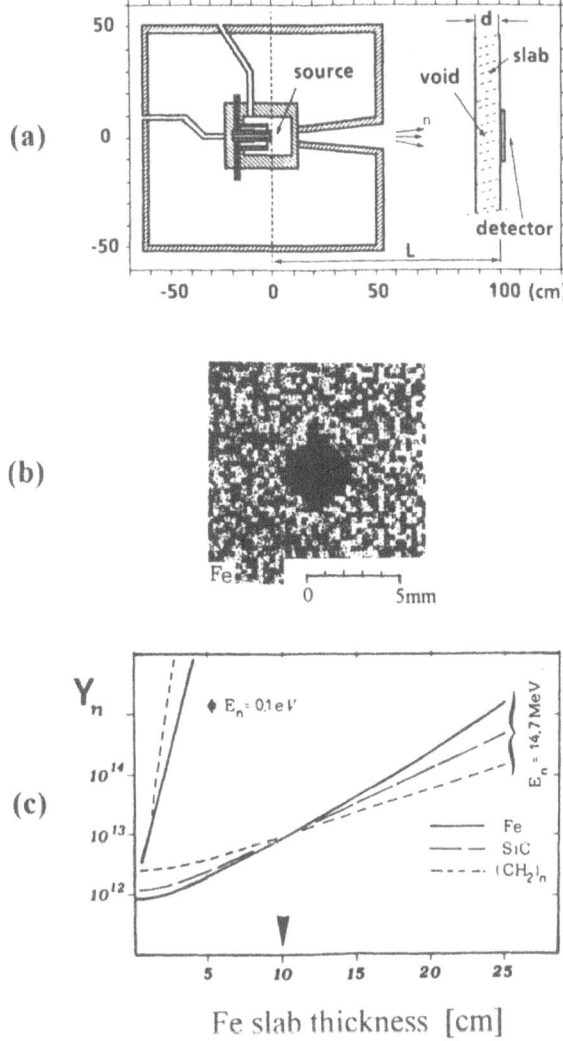

Figure 11. Fast neutron radiography, FNR, with DPF source. (a) is the main segment of FNR system. (b) is the numerically generated image of a void in 10 cm Fe slab with one DPF shot at 100 kJ. (c) is the FNR scaling, to produce image of 3×3×3mm^3 void as shown in (b).

are shown in Fig. 11 from Ref. [37]. Results indicate that DPFs with capacitor bank energies of 20 kJ < W_o < 150 kJ should satisfy the general demands of automobile, aviation and building industries.

3. Elemental Analysis with 14 MeV Neutrons. DPF service for elemental analysis can be considered in two ways.

The first way is to profit only from the outstanding neutron yield not achievable with present compact accelerators. In this class of applications, a good example would be the control of Al-ore enrichment by fast neutron activation analysis carried with ^{241}Am/Be source [38]. If one will consider the substitution of presently used ^{241}Am/Be-20Ci or elec-

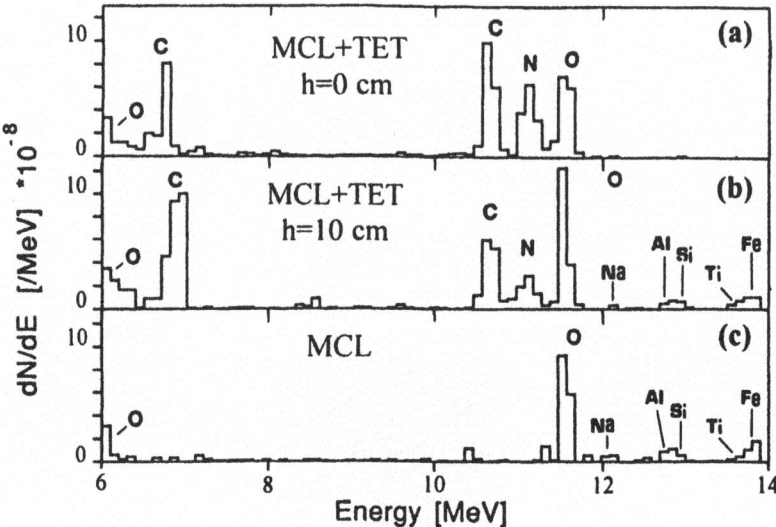

Figure 12. Spectrum of back scattered D-T neutrons from TNT (10 kg) mine. (a) surface mine, (b) 10 cm below ground level, (c) MCL (Matatula Clay Loam) soil only. Neutron response is normalized to one neutron, emitted by DPF (1 m above ground) and 1 cm² of detector surface (1.5 m above ground level).

trostatic accelerator with a DPF (W_o=50 kJ) then almost the same apparatus will examine more samples per unit time improving the quality of the test (decreasing statistical error) by a factor of 3. The second approach will combine high neutron yield with high power neutron radiation for identification of the elemental composition of a substance. The demand for this kind of service comes from the urgent need for the identification of plastic explosives and narcotics. Although this method is at an introductory level of development, due to its importance, an example of its potential is shown in Fig. 12 [39].

4. Short-Life Radioisotope Production in DPF. It is known that the plasma focus accelerates and traps light, energetic ions ($\leq$ 5 MeV/nucleon) in the hot plasma. They can interact with the doped high-Z elements leading to the production of radioisotopes [40]. The special interest is in the availability of ^{13}N, ^{15}O, ^{18}F....elements used for medical testing. Those isotopes, however have short life times (~ minutes) and should be produced at the hospital facilities. This is accomplished usually by accelerating hydrogen or helium ions (ca. 10 MeV) with a cyclotron and impinging them on targets with 11<A<16. Experience with the operation of a plasma focus plasmas, composed of light (Z $\leq$ 2) and heavy (6 $\leq$ Z $\leq$ 8) isotopes, reveal that the quality of DPF operation is not sensitive to the gas mixture [40] and well predicted by model of high nuclear reactivity plasma domains [40],[41], see Fig 13. DPF is therefore an efficient breeder of the required isotopes. It is very important (for further chemical processing) that the produced radioisotopes appear in a gaseous form. Projection of experimental data shows that, a DPF (W_o=150 kJ) working 1 min. at a discharge frequency of 0.2 Hz will produce 70 mCi of ^{15}O i.e. the amount typically used for medical diagnostics. Simplicity and low cost DPF maintenance, simplification of chemical reprocessing and lack of parasitic radiation makes the DPF a very competitive option as compared to the accelerator.

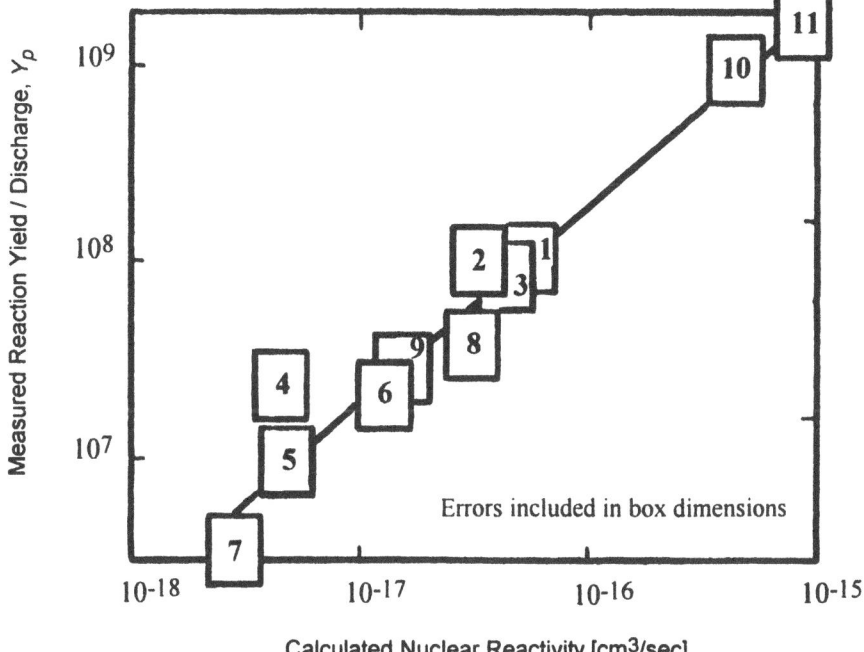

Figure 13. Comparison of experimental (W_o =7 kJ) and predicted values of nuclear reaction yield for different plasma composition [41]. Each box (except #10) represents plasma composed of two isotopes in equal atomic densities. Projected DPF radioisotope breeder should be $W_o \cong 150$ kJ and operating at 0.2 Hz frequency. [1 – $^{12}C(d,n)$ ^{13}N; 2 – $^{14}N(d,n)$ ^{15}O; 3 – $^{16}O(d,n)$ ^{17}F; 4 – $^{12}C(^3He,n)$ ^{14}O; 5 – $^{12}C(^3He,d)$ ^{13}N; 6 – $^{14}N(^3He,d)$ ^{15}O; 7 – $^{14}N(^3He,\alpha)$ ^{13}N; 8 – $^{16}O(^3He,p)$ ^{18}F; 9 – $^{16}O(^3He,\alpha)$ ^{15}O; 10 – $D(d,n)$ ^3He; 11 – $^3He(d,n)$ ^4He].

5. MJs DPF as driver for Safe Fission Reactor. One can consider the application of a sub critical fission reactor with an external neutron source that does compensate for the negative reactivity of a sub critical system [42]. The neutron balance necessary for energy production (at the level of a critical system) is achieved with a high flux of "external" neutrons. As an example, this mode of operation requires the injection of $5\ 10^{17}$ neutrons/s in order to drive a 1 GW fission reactor with a negative criticality $\Delta k/k \cong 0.06\%$, to insure a power half-life of about 100 ns. Here k is the effective multiplication factor of neutrons, set below unity. A safe reactor should not have the possibility to become critical due to human error or hardware failure. Its functional nature should allow for a rapid shut-down in case of reactor failure. Conceptual analysis [34] shows that, the second generation of DPF machines can be compatible with this task. In particular four DPF-2 MJ machines operating in multipinch mode (10 pinch train) at a discharge frequency of $\nu=1$ Hz will provide the required balance of neutron flux to insure criticality. This concept can be tested by $W_o= 0.5$ MJ units and a 1 MW reactor. The conceptual design for the proof of the principle experiment is shown in Fig. 14 [34].

5. SUMMARY

Presented material briefly reviews the current status, near future developments and the potentials of plasma focus machines as future sources of fusion energy. In evaluating

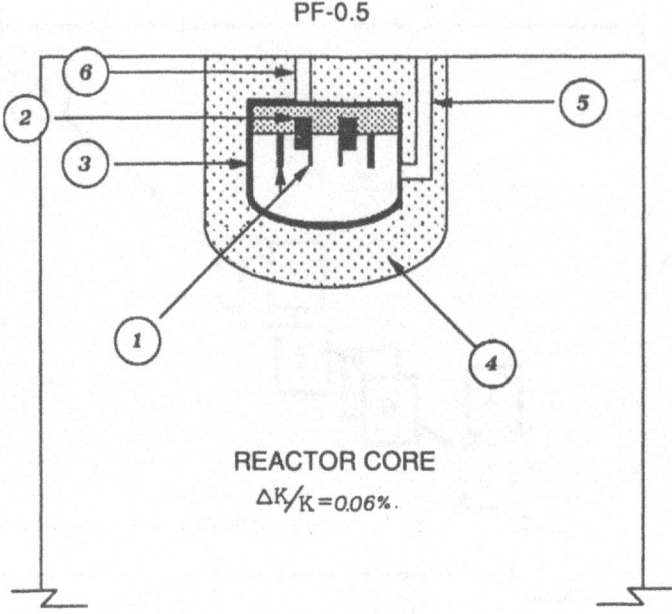

Figure 14. Conceptual design of DPF -0.5 MJ, working with 1 MW sub critical fission reactor. 1- electrodes; 2- insulators; 3- chamber; 4- multiplier/moderator; 5- gas circulation duct; 6- power delivery duct.

the DPF potentials, one has to take into account that the source of fusion energy is the sort of self renewing, moderately hot, dense plasma target with embedded and trapped in it a population of hundreds of keV ions controlling the fusion rate. Some relevant conclusions are:

1. The present state of art in DPF understanding and technology is a solid foundation for the immediate and widespread industrial applications of the 100 kJ class units and assures a minimum risk in designing the second generation units (10 MJ class).
2. Construction of the second generation DPFs can be based on the existing high energy storages making the above discussed options very attractive and economically feasible .
3. The analysis of the scaling laws does not reveal any obstacles in approaching scientific break-even. Future fusion reactors based on the DPF concept have two unique potentials:

   (a) The DPF has compact structure and does not require any supporting plasma heating (rf, or external beams), and
   (b) The DPF operation based on self acceleration and trapping of hundred keV ions has potentials to be used for the neutron-lean fusion variant.

ACKNOWLEDGMENTS

Authors would like to express their thanks to the students of Stevens Institute of Technology: Jan R. Brzosko and David B. Goldstein for their help in preparing the above text.

REFERENCES

1. Filippov, N.V., Filippova, T.I., and Vinogradov, V.P., 1962, Dense, High Temperature Plasma in a Non-Cylindrical Z-Pinch Compression, Nuclear Fusion Suppl. (1962) Part 2, p.577.
2. Mather, J.W., 1965, Formation of a High-Density Deuterium Plasma Focus, Physics Fluids 8: 336; see also J. Marshall and I. Henins (1965), Fast Plasma from a Coaxial Gan, Nuclear Fusion Supplement, IAEA-CN-21 (1966), Vol. 2, p. 449.
3. Kies, W., 1986, Power Limits for Dynamical Pinch Discharges, Plasma Phys. Controlled Fusion 28: 1645.
4. Brzosko, J.B., and Klobukowska, J., 1983, Self-Sustaining Partial Discharges in Argon at the Dielectric Surface, IEEE Trans. EI-18: 420.
5. Herold, H., et al., 1985, Two Phases of Neutron Production in the POSEIDON Plasma Focus, Nuclear Fusion Supplement (IAEA-CN-44) p.579.
6. Erhardt, J., Kirchesch, P., Hubner, K., and Rager, J.P., 1982, Light-Scattering in the Frascati Plasma Focus, Physics Lett., 89A: 285.
7. Rager, J.P., et al., 1981, Experiments on Neutron Production Phase of Frascati 1-MJ Plasma Focus, Nuclear Fusion Supplement (IAEA-CN-38) p.579.
8. Jager, U., and Herold, H., 1978, Fast Ions Kinetics and Fusion Reaction Mechanism in the Plasma Focus, Nuclear Fusion 27: 407.
9. Ivanov, V.D., et al., 1981, Experimental Studies of Plasma Focus, Nuclear Fusion Supplement (IAEA-CN-38) p.161.
10. Vinogradov, V.P., Golubtshikov, D.G., Filippov, N.V., and Filippova, T.I., 1967, Toepler Device with Pulsed Ruby Laser for the Plasma Sheath Study, Teplofizika Vysokih Temperatur 5: 343.
11. Brzosko, J.S., Klobukowska, J., and Robouch, B.V., 1987, A Macroscopic Study of the Neutron, Gamma- and X-Ray Emissivity in the Frascati Plasma Focus, Fusion Technology 12: 71.
12. Decker, G., et al., 1979, Neutron Emission Parameters in Plasma Focus Devices, Nuclear Fusion Supplement (IAEA-CN-37) p.135.
13. Bernard, A., Coudeville, A., Jolas, A., Launspach, J., and de Mascureau, J., 1975, Experimental studies of the Plasma Focus and Evidence for Nonthermal Processes, Physics of Fluids, 18: 180.
14. Conrads, H., Cloth, P., Demmeler, M., and Hecker, R., 1972, Velocity Distribution of the Ions Producing Neutrons in a Plasma Focus, Physics of Fluids 15: 209.
15. Gratreau, P., Luzzi,G., Maisonnier, Ch., Pecorella, F., Rager, J.P., Robouch, B.V., and Samuelli, M., 1971, Structure of the Dense Plasma Focus, Part II: Neutron Measurements and Phenomenological Description, Nuclear Fusion Supplement (IAEA-CN-28) p. 523.
16. Schmidt, R., and Herold, H., 1986, A Method for time resolved Neutron Spectroscopy on Short Pulsed Fusion Neutron Sources, Plasma Phys. Controlled Fusion, 29: 523.
17. Gentilini, A., et. al., 1980, Comparison of Four Calibration Techniques of a Silver Activated Geiger Counter for the Determination of the Neutron Yield on the Frascati Plasma Focus, Nuclear Instr. Meth., 172: 541.
18. Lehner, G., 1970, Reaction Rates and Energy Spectra for Nuclear Reactions in High Energy Plasmas, Z.Physik 232: 174.
19. Bernstein ,M.J., and Comisar, G.G., 1972, Neutron Energy and Flux Distribution from a Crossed-Field Acceleration Model of Plasma Focus and Z-Pinch Discharges, Physics of Fluids 13: 700.
20. Nardi, V., Bortolotti, A., Brzosko, J.S., Esper, M., Luo, C.M., Pedrielli, F., Powell, C., and Zeng, D., 1988, Simulated Acceleration and Confinement of Deuterons in Focused Discharges - Part I, IEEE Trans. PS-16: 368.
21. Brzosko, J.S., Nardi, V., Brzosko, J.R., and Goldstein, D., 1994, Observation of Plasma Domains with Fast Ions and Enhanced Fusion in Plasma-Focus Discharges, Physics Lett. A192: 250.
22. Yamada, Y., Kitagawa, Y., Yokoyama, M., 1985, High-Energy Deuteron Beam Generated by a Plasma Focus Device, J. Applied Phys. 58: 188.
23. Gullickson, R.L., Luce, J.S., Sahlin, H.L., 1977, Operation of a Plasma-Focus Device with D2 and 3He, J. Applied Phys. 46: 3718.
24. Filippov, N.V., Belaieva, I.F., Filippova, T.I., 1978, Measurements of the Effective Temperature of Ions in Plasma Focus by Measurements of Reaction Rate in the D2–3He Mixture, Fizika Plasmy (Sov.) 4: 1364.
25. Brzosko, J.S., Rager, J.P., Robouch, B.V., Bahr, A.H., Klapdor, H.V., Anderson, E., and Herges, P., 1983, Measurements of the D(d,n) and 3He(d,p) Reaction Yields for the 1-MJ Plasma Focus Device Operating with a D2–3He Gas Mixture, Nuclear Technology/Fusion 4: 263.
26. Filippov, N.V., Filippova, T.I., Khutoretskaia, I.V., Krivtsov, V.A., Mialton, V.V., Vinogradov, V.P., 1996, Megajule Plasma Focus as Efficient X-Ray Source, Physics Lett. 211: 168.

27. Conrads, H., Salge, J., 1983, On Repetitive Operation of a Plasma Focus, Energy Storage, Compression, and Switching (Plenum Press, NY), p.87.
28. Imshennik, V.S., Filippov, N.V., and Filippova, T.I., 1973, Similarity Theory and Increased Neutron Yield in a Plasma Focus, Nuclear Fusion 13: 929.
29. Degnan, J.H., et. al., 1993, Compact Toroid Formation, Compression and Acceleration, Physics of Fluids, B5: 2938.
30. Degnan, J.H., 1995, Formation, Compression and Acceleration of magnetized Plasmas, in Evaluation of Current Trends in Fusion Research (Publ. Plenum Publ. Corp (published in the same volume)
31. Peterkin Jr, R.E., Degnan, J.H., Hussey, T.H., Roderick, N.F., and Turchi, P.J., 1993, A Long Conduction Time Compact Torus Plasma Opening Switch, IEEE Trans. Plasma Science 21: 522.
32. Chernyshev, V.K., at al., 1986, Change in the Parameters of a Plasma Focus when the Capacitive Energy source is Replaced by an Inductor, Sov. Phys. Tech. Phys. 31: 558.
33. Aziziv, E.A., Lototsky, A.P., Nastoyashchy, A.F., Folippova, T.I., and Filippov, N.V., 1994, Pulsed - High - Power Neutron Source Based on a magnetic Energy Storage, private communication from State Centre TRINITY (Russia); see also Dense Z-Pinches (Ed. Haines, M., and Knight, A., Publ. American Inst. Phys., New York, 1994) p.271.
34. Nardi, V., Brzosko, J.S., and Powell, Ch., 1992, Neutron Sources of the Advanced Plasma Focus Type for Demonstration Tests of Safe Reactors, Los Alamos Nat. Lab. Raport LA-UR-92–3552.
35. Neff, W., Eberle, J., Holz, R., Lebert, R., and Richter, F., 1989, The Plasma Focus as a Soft X-Ray Source for Microscopy and Lithography, Proc. SPIE 1140: 13.
36. Kato, Y., and Be, S.H., 1986, Generation of Soft X-Rays Using a Rare Gas-Hydrogen Plasma Focus and Its Application to X-Ray Lithography, Appl. Physics Lett. 48: 686.
37. Brzosko, J.S., Robouch, B.V., Ingrosso, L., Bortolotti, A., and Nardi, V., 1992, Advantages and Limits of 14-MeV Neutron Radiography, Nuclear Instr. Meth., B72: 119.
38. Holmes, R.J., Phillips, P.L., Roczniok, A.F., and Waddington, P.J., 1987, A Prototype Bauxite Analyser Based on Fast-Neutron-Activation Analysis, Nuclear Geophisics 1: 41.
39. Nardi V., Brzosko, J.S., Powell, Ch., and Bortolotti, A., 1993, High Fluence Pulsed Neutron Source for Neutron Interrogation, Proc. Int. Symp. on Substance Identification Technologies (Innsbruck, 1993) in press (publ. by IAEA, Vienna).
40. Brzosko, J.S., and Nardi, V., 1991, High Yield of 12C(d,n)13N and 14N(d,n)15O Reactions in the Plasma Focus Pinch, Physics Lett. A155: 162; see also Physics Lett. A192: 250 (1994).
41. Brzosko, J.S., Nardi, V., Goldstein, D.B., and Brzosko, J.R., 1993, High Z - Low Z Nuclear Reactions in the Plasma Focus Pinch, IEEE Conf. Record (1993 ICOPS) 93CH3334–0: 188.
42. Teller, E., 1991, Adress at Int. Conf. of Plasma Focus Applications (Hoboken, 1991); see also Los Alamos Nat. Lab. Raport LA-UR-92–3552 (1992).

SELF-COLLIDING BEAMS AS AN ALTERNATIVE FUSION SYSTEM FOR D–He^3 REACTORS

Norman Rostoker and Michl Binderbauer

Department of Physics
University of California
Irvine, CA 92697-4575

ABSTRACT

A plasma consisting of large orbit non-adiabatic ions and adiabatic electrons is considered. For such a plasma it is possible that the anomalous transport characteristic of Tokamaks can be avoided provided that long wavelength low frequency instabilities are controlled. A specific example is investigated that is a Field Reversed Configuration (F.R.C.) in which the current is carried by the energetic ions D and He^3. The analysis begins with self-consistent equilibrium solutions of the Vlasov–Maxwell equations. The classical Fokker–Planck collision operator is used to evaluate Coulomb collisions, transport, etc. The sustained operation of a D–He^3 reactor involves the injection of fuel ions and the transfer of energy and momentum to fuel ions by interaction with the fusion products, 14.7 MeV protons and 3.67 MeV He^4.

1. INTRODUCTION

Magnetic confinement systems in plasma physics involve particles where the orbit radius and orbit period are small compared to the characteristic scales of length and time. In particle accelerators they are of the same order. There is a considerable difference between the physics of adiabatic and non-adiabatic particles; for example, strong focusing is non-existent for adiabatic particles and even weak focusing is quite different. It is an experimental fact that the non-adiabatic particles are much better confined than the adiabatic particles of plasmas. High energy particles have been studied and employed in fusion plasmas as a minority particle for heating and because the reaction products must be considered. Fusion plasmas where non-adiabatic ions are the majority particle have been studied many years ago — *DCX* and *OGRA* (1958–1968)[1] and *MIGMA*[2] (1975–1985). From 1988–1992 experiments were carried out with high energy large orbit ions in Tokamaks (*TFTR*, D-III-D and *JET*).[3] These experiments proved that large orbit ions slow down and diffuse classically in the presence of anomalous fluctuations and transport of the majority plasma particles which were of much lower energy. The possibility of classical confinement for a dense plasma of large orbit ions is suggested by these experiments.[4]

The evaluation of this possibility requires many calculations for a specific reactor configuration; a considerable effort with large scale computational tools and personnel. In this paper a preliminary analytical investigation is carried out for a high beta F.R.C. Many approximations are employed; the F.R.C. is considered to be infinitely long in the axial direction; some important finite length effects are evaluated. Coulomb collisions are described by the Fokker–Planck collision operator for a plasma.[5] The test-particle method is employed to simplify the calculations of transport, slowing down, etc. The present methods illuminate the physical processes and provide the guidelines for a detailed numerical investigation.

2. EQUILIBRIUM DISTRIBUTIONS OF A D–He^3 PLASMA

Consider distribution functions of the form

$$f_j(\mathbf{x},v) = \left(\frac{m_j}{2\pi T_j}\right)^{3/2} n_j(r,z) \exp\left\{-\frac{m_j}{2T_j}\left[(v_x - \omega_j y)^2 + (v_y + \omega_j x)^2\right]\right\} \tag{1}$$

where ω_j and T_j are the same for all ions. $\langle v_\theta \rangle = -\omega_j r$; this is a rigid rotor distribution. The density is of the form

$$n_j = n_{0j} \exp\left[\frac{m_j(\omega_j r)^2}{2T_j} - \frac{e_j \Phi}{T_j} - \frac{e_j A_\theta \omega_j r}{cT_j}\right]. \tag{2}$$

Φ and $A_\theta = \Psi/r$ are potentials. The electric and magnetic fields are $\mathbf{E} = -\nabla\Phi$ and $\mathbf{B} = \nabla \times (A_\theta \hat{\theta})$. The equilibrium solution of the Vlasov–Maxwell equations is obtained by solving simultaneously

$$\nabla \times \mathbf{B} = \frac{4\pi}{c} \sum n_j e_j r \omega_j \hat{\theta} \tag{3}$$

and

$$\sum n_j e_j \cong 0. \tag{4}$$

If n_j depends on r and z numerical methods are required.[4] If it depends only on r analytical solutions can be obtained. For distribution functions like Eq. (1), the Vlasov equation can be replaced by the fluid equations for conservation of momentum.[6]

$$-n_j m_j r \omega_j^2 = n_j e_j \left(E_r - \frac{r\omega_j}{c} B_z\right) - T_j \frac{dn_j}{dr} \tag{5}$$

$$n_e = \sum_i n_i Z_i \tag{6}$$

$$\frac{dB_z}{dr} = \frac{4\pi e r}{c} \left[\sum_i n_i Z_i \omega_i - n_e \omega_e\right]. \tag{7}$$

In Fig. 1 a two-dimensional solution for the magnetic fields is illustrated. A one-dimensional solution for the particle densities is a reasonable approximation except at the ends. The

Self-Colliding Beams as an Alternative Fusion System for D–He³ Reactors

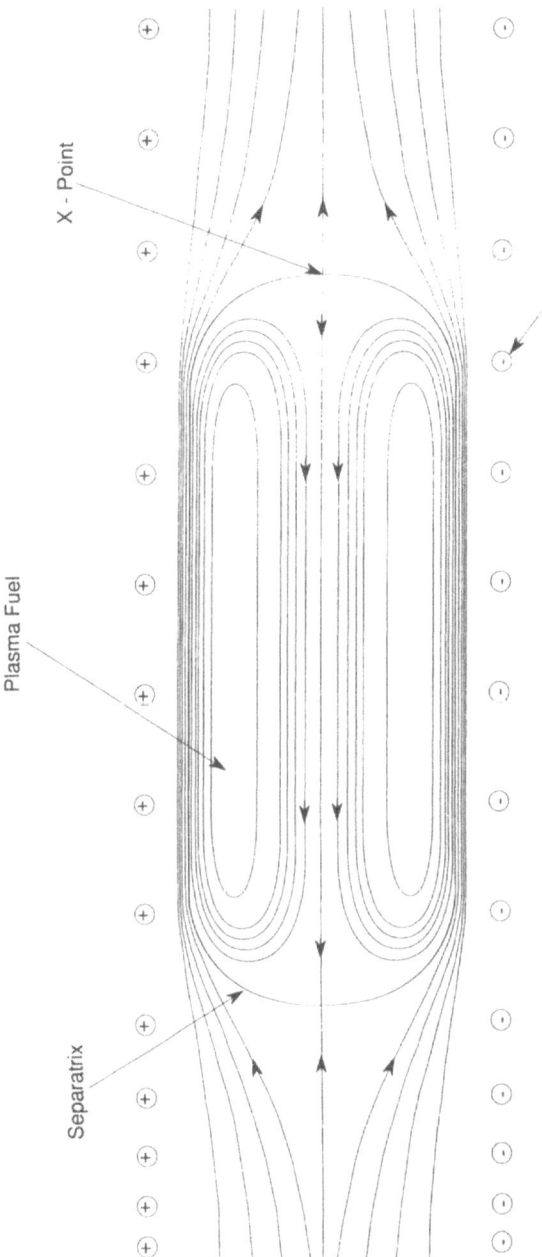

Figure 1. Magnetic flux surfaces for the F.R.C.

balance of this paper is will be a one-dimensional analysis. An essential feature of the density distributions is that the particle density at the walls is reduced by a factor of 10^6 compared to the peak density. This should make electron transport to the wall negligible. The following numerical data are assumed:

- peak ion densities of (D) $n_1 = .143 \times 10^{15}$ cm^{-3}, (He^3) $n_2 = .429 \times 10^{15}$ cm^{-3};
- peak electron density $n_e = 10^{15}$ cm^{-3};
- coherent ion energy $W_i = (1/2) m_i (r\omega_i)^2$ with $W_1 = 200$ keV, $W_2 = 300$ keV;
- radius of peak ion density $r_o = 30$ cm;
- wall radius $r_B = 2^{1/2} r_o$;
- reactivity $\langle \sigma v \rangle = 2 \times 10^{-16}$ cm^3/sec;
- energy per reaction 14.7 MeV protons, 3.67 MeV He^4;
- temperatures $T_1 = T_2 = T_e = 200$ keV.

The following equilibrium properties are calculated:

- externally applied magnetic field $B_o = 2.35$ kG;
- maximum magnetic field $B_{max} = 118$ kG
- $B_{min} = -113$ kG
- $B_z(r) = B_o[1 + \beta^{1/2} \tanh(r^2 - r_o^2)/r_o \Delta r]$;
- $\beta = 2430$;
- $\Delta r = 2.60$ cm.

Line density (electrons)	$N_e = 2\pi r_o \Delta r n_e = 4.88 \times 10^{17}$ cm^{-1}
Current	$I_\theta = 1.82 \times 10^5$ A/cm
Inductance	$L = 16.2$ μH/cm
Magnetic energy	$(1/2) L I_\theta^2 = 268$ k-Joules/cm
Particle energy	$\sum_i N_i [W_i + (3T_i/2)] = 25.6$ kJ/cm.

3. STABILITY CONSIDERATIONS

Large orbit ions average fluctuations so that transport is produced only by fluctuations of wavelength larger than the gyro-radius. This has been observed with two dimensional models and computer simulation.[7] It explains the results with large orbit ions in tokamaks.[3] For the plasmas under consideration where essentially all ions are non-adiabatic, micro-instabilities would not be important. Long wavelength stability is required but it should be noted that the expected magnetohydrodynamic (MHD) instabilities do not occur experimentally[8]:

- Alfvèn wave instabilities that are driven by beams in tokamaks are absent (as are Alfvèn waves).
- The tilt mode is usually not observed.

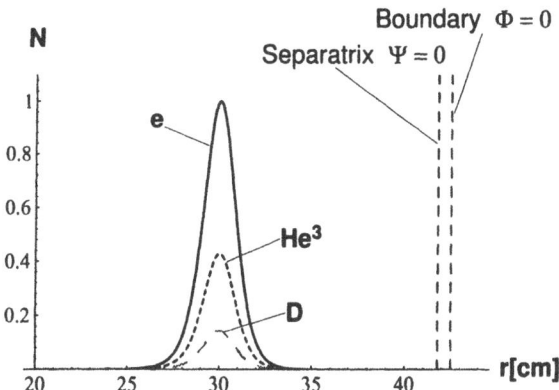

Figure 2. Density profiles of electrons, D-ions and He^3-ions in the one-dimensional model.

- The rotational kink mode requires stabilizing multipole windings, but only during formation.
- The tearing mode is not observed.

The FRC is evidently a plasma to which MHD does not apply. A new treatment has been initiated for a high β non-adiabatic plasma.[9]

4. CLASSICAL SCATTERING

The test particle method to evaluate slowing down and diffusion is based on the Fokker–Planck collision operator where one particle is singled out and the remainder of the particles have Maxwell Distributions

$$\left(\frac{\partial f_i}{\partial t}\right)_c = -\frac{\partial}{\partial v}\cdot f_i \langle \Delta v \rangle_i + \frac{1}{2}\frac{\partial}{\partial v}\cdot \frac{\partial}{\partial v}\cdot f_i \langle \Delta v \Delta v \rangle_i \qquad (8)$$

$$\langle \Delta v \rangle_i = \langle \Delta v_\parallel \rangle_i (v/v)$$

$$\langle \Delta v \Delta v \rangle_i = \langle \Delta v_\parallel^2 \rangle_i (vv/v^2) + \langle \Delta v_\perp^2 \rangle_i [1-(vv/v^2)] \qquad (9)$$

$$\langle \Delta v_\parallel \rangle_i = \frac{4\pi(Z_i e^2)^2}{m_i} \ln\Lambda \sum_j n_j Z_j^2 \left(\frac{1}{m_j}+\frac{1}{m_i}\right) \frac{\partial}{\partial v}\left(\frac{1}{v}\operatorname{erf}\frac{v}{\sqrt{2}v_j}\right) \qquad (10)$$

$$\langle \Delta v_\perp^2 \rangle_i = \frac{4\pi(Z_i e^2)^2}{m_i^2 v} \ln\Lambda \sum_j n_j Z_j^2 \left[\operatorname{erf}\frac{v}{\sqrt{2}v_j} + \frac{\partial}{\partial v}\left(\frac{v_j^2}{v}\operatorname{erf}\frac{v}{\sqrt{2}v_j}\right)\right] \qquad (11)$$

$m_j v_j^2 \doteq T_j$, $\ln\Lambda \cong 20$, the test-particle is denoted by i and the summation is over all types of field particles. It is convenient to separate the contributions from each type of field particle. For example, the time for scattering through a large angle of particle i of energy W_i by electrons is

$$\tau_{ie} = \frac{W_i}{(\partial W_\perp/\partial t)_{ie}} \qquad \text{where} \qquad \left(\frac{\partial W_\perp}{\partial t}\right)_{ie} = m_i \langle \Delta v_\perp^2 \rangle_{ie} \qquad (12)$$

and $\langle \Delta v_\perp^2 \rangle_{ie}$ means the term in the sum in Eq. (21) due to electrons is the only term retained The inequalities usually satisfied are $v_e > v$, v_i where v is the velocity of an ion test particle The scattering times are as follows

$$\tau_{ee} = \frac{\sqrt{m}\, T_e^{3/2}}{2\pi n_{eo} e^4 \ln \Lambda} = 10^{-4} \text{ sec.} \tag{13}$$

This is the time for the establishment of a Maxwell distribution for electrons. It is the shortest time scale compared to diffusion time, slowing down time, etc. Therefore, the electron distribution function must be close to Maxwellian at all times. The ion-ion collision time is $\tau_{ii} \sim 10^{-2}$ sec and should also be close to Maxwellian.

The equilibrium calculations require values for T_e and T_i, but the Vlasov/Maxwell equations do not provide any way to determine these quantities. Fusion reactions and collisions must be considered. The fusion power density can be expressed as

$$P_F = 1.6 \times 10^{-19} n_1 n_2 \varepsilon_F \langle \sigma v \rangle \text{ kW/cm}^3 \tag{14}$$

$\varepsilon_F = 18.4$ MeV is the energy per fusion reaction. The main energy loss is Bremsstrahlung which is

$$P_B = 4.71 \times 10^{-34} n_e \sum_i n_i Z_i^2 T_e^{1/2} \text{ kW/cm}^3$$

$$= \sum_i \frac{n_i W_i}{t_{ie}} \tag{15}$$

where $t_{ie} = (3/8\sqrt{2\pi}) m_i T_e^{3/2}/n_{eo} Z_i^2 e^4 m^{1/2} \ln \Lambda$. Eq. (15) can be solved for $T_e = 185$ keV which is close to the previous guess. The slowing down times are thus $t_{1e} = 4.93$ sec; $t_{2e} = 1.85$ sec. The quantity $Q = P_F/P_B = 3.04$. We have selected a fuel ratio of D:T=1:3 in order to limit the neutron energy produced by D–D side reactions to 1% of the fusion power.[10] The helium rich fuel mixture then radiates more and reduces Q compared to a 1:1 fuel ratio, which is the price that must be paid for an aneutronic reactor. These calculations assume that the fuel ion energy remains constant as well as the particle densities and the current that maintains the FRC configuration.

5. SUSTAINED OPERATION OF A REACTOR

Considering the selective confinement of the fusion products it is possible that sustained operation requires no external current drive. This possibility has been studied[11] for a conventional FRC. Typical particle orbits in an FRC are illustrated in Fig. 3. Particles with betatron orbits and average azimuthal velocity in the diamagnetic direction ($v_\theta < 0$) would find the Lorenz-force at the ends focusing. The radial magnetic field would be defocusing for $v_\theta > 0$ and the particles would be promptly expelled. All of the confined fuel ions have $v_\theta < 0$. There is a loss cone for $v_\theta > 0$, but since only electrons have confinement for $v_\theta > 0$ the time for scattering into the loss cone is quite long, i.e.:

$$\tau_{ie} = (3/2\pi^{1/2}) W_i T_e^{1/2}/(Z_i^2 n_e e^4 \ln \Lambda)$$
$$= 208 \text{ sec (D); } 118 \text{ sec (He}^3\text{).}$$

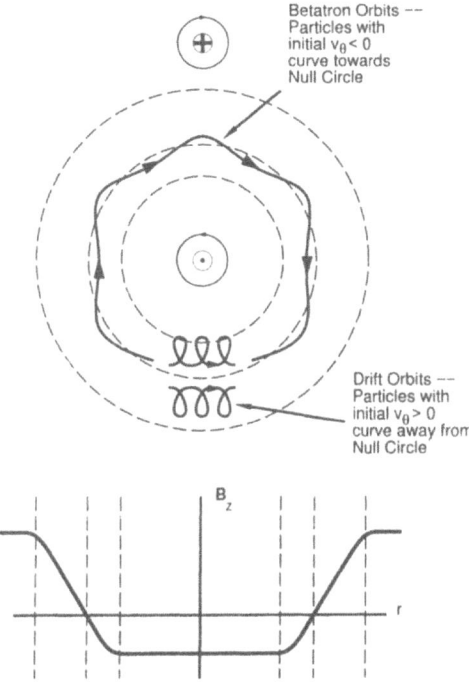

Figure 3. Typical Particle Orbits (for $v_\theta < 0$ and $v_\theta > 0$) and Magnetic Field Configuration.

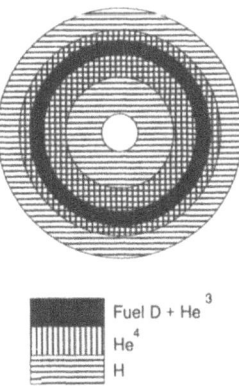

Figure 4. Distribution of Fusion Fuel and Reaction Products.

The fusion products are created in the center of mass system with an isotropic distribution and a definite energy or velocity: protons with $|V_3| = 5.37 \times 10^9$ cm/sec in a coordinate system rotating with $|V_e| = |V_1| = |V_2| = .44 \times 10^9$ cm/sec. After the protons rotating with $v_\theta > 0$ are expelled, the remaining protons have an average rotation velocity

$$|V_3| = |V_c| + \frac{1}{2}(|V_o| - |V_c|) = 2.91 \times 10^9 \text{ cm/sec.}$$

The fraction of protons that escape promptly is

$$\frac{|V_o| - |V_c|}{2|V_o|} = .459$$

For He^4, $|V_o| = 1.34 \times 10^9$ cm/sec; $|V_4| = .89 \times 10^9$ cm/sec and the fraction that escapes promptly is .336. The selective confinement of protons and α-particles produces a contribution to the current I_θ. In addition the fusion products have a net angular momentum so that collisions with the fuel ions can maintain their angular momentum. In order to treat these effects we consider the Vlasov equation with the addition of the collision terms of Eq. (8). Fluid equations can be derived by taking moments

$$n_j m_j \frac{d\mathbf{V}_j}{dt} = n_j e_j \left[\mathbf{E} + \frac{1}{c}\mathbf{V} \times \mathbf{B} \right] - \nabla(n_j T_j) + n_j m_j \sum_{k \neq j} \frac{(\mathbf{V}_k - \mathbf{V}_j)}{t_{jk}} \quad (16)$$

$$\frac{\partial n_j}{\partial t} + \nabla \cdot n_j \mathbf{V}_j = 0 \quad (17)$$

$j = e, i$ are the particle type indices. The fluid velocity is

$$n_j \mathbf{V}_j = \int f_j v dv$$

and t_{jk} are momentum transfer collision times determined by $\langle \Delta v_\parallel \rangle_j$ of Eq. (10). Consider only the angular velocity θ-component of Eq (16), multiply by $2\pi r$ and integrate from $r = 0$ to $r = 2^{1/2} r_o$;

$$\frac{dV_{e\theta}}{dt} = \sum_i \frac{V_{i\theta} - V_{e\theta}}{t_{ei}} - \frac{2\pi}{N_e} \int_{x=-r_o^2/2}^{x=r_o^2/2} dx n_e(x) V_{er}(x) \Omega_e(x) \quad (18)$$

$$\Omega_e(x) = \frac{e(B_z - B_o)}{mc} \quad N_e = 2\pi \int_{-r_o^2/2}^{r_o^2/2} n_e(x) dx$$

(19)

$n_e, v_{er}\Omega_e$ are functions of $x = (r^2 - r_o^2)/2$; n_e is even, V_{er} and Ω_e are odd. The magnetic field has no effect unless V_{er} is finite. V_{er} is finite because of diffusion. If burning is not considered there would only be diffusion from electron–ion collisions which are negligible. However, if burning is considered there would be ion diffusion due to collisions of fuel and fusion product ions and also between different fuel ions because the distribution functions do not remain Maxwellian (a fact that the present treatment does not cover). If there is ion diffusion electrons must follow. There is an additional radial velocity V_{er} because when fusion takes place D and He^3 ions disappear and are replaced by H and He^4 ions which have a much larger energy. They leave the fuel region and take electrons with them; they subsequently scatter until they leave confinement with accompanying electrons. The following is an estimate of this effect

$$\langle V_{er}\Omega_e\rangle = \frac{\int dx\, n_e(x) V_{er}(x) \Omega_e(x)}{\int dx\, n_e(x)}$$

$$\simeq \Omega_m r_o \langle \sigma v\rangle \frac{n_{1o}n_{2o}}{n_{eo}} \simeq \Omega_m \langle V_{er}\rangle$$

Ω_m involves $B_{\max} - B_o$, n_{1o} etc. are maximum densities; $\Omega_m \simeq 1.8 \times 10^{12}$ sec^{-1} and $\langle V_{er}\rangle = .3$ cm/sec, $V_{i\theta} = .45 \times 10^9$ cm/sec and $(1/t_{e1}) + (1/t_{e2}) = 1.35 \times 10^3$ so that $dV_{e\theta}/dt \simeq 0$.

The equation for fuel ions is

$$\frac{dV_1}{dt} = -\frac{(V_1 - V_e)}{t_{1e}} + \frac{2\pi}{N_1}\int dx\, n_1(x) V_{1r}(x) \Omega_1(x) + .54\frac{(V_p - V_1)}{t_{F1}}\frac{m_p}{m_1} \qquad (20)$$

for D with a similar equation for He^3.

$$\Omega_1(x) = \frac{e(B_z - B_o)}{m_1 c} \qquad n_1 V_{1r}(x) = -D_1 \frac{\partial n_1}{\partial x}$$

and D_1 is the classical diffusion coefficient. The last term describes the transfer of momentum from the fusion product protons to the Deuterium fuel ions. $t_{F1} = [n_{2o}\langle\sigma v\rangle]^{-1} = 11.6$ sec is the fusion time for D-ions. Estimates similar to those for Eq. (18) indicate that the first and second terms are of similar magnitude and the last term is the smallest. The diffusion comes mainly from D–He^3 collisions which would not produce diffusion if Maxwell distributions were applicable. Collisions with the fusion products change the distributions so that these collisions would produce diffusion; however, the methods employed in this paper preclude accurate estimates. They indicate the relative magnitude and importance of different terms. The result is that a self-sustained steady state of a burning plasma is plausible. A more sophisticated analysis with computer simulations is necessary for quantitative results.

ACKNOWLEDGMENTS

This work was supported in part by a grant from the Plasma Physics Research Institute of the Lawrence Livermore National Laboratory.

References

[1] S. Glasstone and R. H. Lovberg, *Controlled Thermonuclear Reactions*, D. Van Nostrand Co., New York (1960), Chapter 9.
[2] D. Al-Salameh et al., *Phys. Rev. Lett.* **54**, 796 (1985); B. C. Maglich, *Nucl. Instrum. Methods Phys. Res., Sect. A* **271**, 13 (1988).
[3] W. Heidbrink, J. Kim, and R. J. Groebner, *Nucl. Fusion* **28**, 2097 (1988); W. Heidbrink, *Phys. Fluids B* **2**, 4 (1990); W. Heidbrink et al., *Phys. Fluids B* **3**, 3167 (1991); S. J. Zweben et al., *Nucl. Fusion* **31**, 2219 (1991).
[4] N. Rostoker, F. Wessel, H. Rahman, B. C. Maglich, B. Spivey and A. Fisher, *Phys. Rev. Lett.* **70**, 1818 (1993).
[5] M. N. Rosenbluth, W. M. MacDonald and D. L. Judd, *Phys. Rev.* **107**, 1 (1957).
[6] N. Rostoker and H. Rahman, "Large Orbit Confinement for Aneutronic Systems," *Nonlinear and Relativistic Effects in Plasmas*, Ed. V. Stefan, Am. Inst. of Phys., New York, p. 116 (1992).
[7] H. Naitou, T. Kamimura, and J. Dawson, *J. Phys. Soc. Jpn.* **46**, 258 (1979).
[8] M. Tuszewski, *Nucl. Fusion* **28**, 2033 (1988).
[9] H. V. Wong, H. L. Berk, R. V. Lovelace, and N. Rostoker, *Phys. Fluids B* **3**, 2973 (1991).
[10] L. J. Wittenberg et al., *Fusion Technology* **10**, 167, (1986).
[11] H. L. Berk, H. Momota and T. Tajima, *Phys. Fluids* **30**, 3548, (1987).

TARGET PHYSICS FOR INERTIAL FUSION ENERGY

J. M. Martínez-Val, G. Velarde, and S. Eliezer[*]

Institute of Nuclear Fusion, DENIM
E.T.S.I.I., Madrid Polytechnic University
J. Gutiérrez Abascal, 2/28006 Madrid, Spain

ABSTRACT

A short review of Inertial Fusion Targets is presented according to a systematics based on the four consecutive phases of the evolution of a target: beam illumination; implosion hydrodynamics; fuel deceleration and compression; and ignition and burn propagation. Main critical points of each phase are pointed outed. Besides that, the paper deals with three novel topics: volume ignition; externally guided implosions; and solid DT fuels at room temperatures. All of them are good examples of the richness of ideas that must not be disregarded in the framework of Inertial Fusion.

INTRODUCTION AND BACKGROUND

Inertial Fusion Energy is aimed at producing high-gain microexplosions in the range of few hundreds MJ by compressing and heating tiny capsules of fusionable fuel.

After three decades of IFE research, the field is still too young and remains too open before a final choice can be made in any of the three major areas in which IFE can by divided: drivers, targets and reactor chambers (Velarde, Ronen, Martínez-Val 1993).

Mainstream ideas on Target design can be classified into two lines: direct drive and indirect (or radiation) drive (Caruso 1989). In the former, the target is directly illuminated by a set of beams (laser photons, heavy ions, light ions, clusters...) that deposit their energy into the target coating so producing the mechanical reaction that induces the target core (fuel) to implode (Yabe 1993). Main drawbacks of this scheme arise from the nonuniformities in the illumination. If the implosion is not perfectly spherical, the fuel can not reach the high densities to selfsustain a huge fusion burst, even if keV temperatures are

---

[*] Permanent address: Soreq N.R.C. (Israel)

achieved in some spots (which is also questionnable if the implosion is not very perfect) (Mc Crory and Verdon 1989).

A new scheme has been proposed in this framework to try to overcome this problem (Eliezer, Honrubia and Velarde 1992) (Desselberger et al. 1992). The target is driven by a double (or hybrid) pulse. The first pulse is made of thermal X-rays (which are produced in the interaction of the first driver pulse with an outer casing which is burned-through and emits a strong X-ray burst almost isotropically). The second (or main) pulse, which passes through the casing rarefied plasma, directly impinges on the target coating, around which a plasma corona has been created. Within the corona, nonuniformities of the main pulse disappear to a large extent by thermal smoothing.

Indirect drive is based on the creation of a radiation field within an oven (so called hohlraum) where the driver beams energy is deposited (Lindl, Mc Crory and Campbell 1992). The radiation field becomes isotrope in the middle of the oven by the multiple absorption-reemission of X-ray photons from the wall, and drives the implosion of the target core by mechanical reaction to the sudden evaporation of the target coating. The main drawback in this scheme is the poor energy conversion efficiency from the driver beam into fuel implosion energy, because of the intermediate conversion steps (Murakami and Meyer-ter-Vehn 1991). In the case of laser indirect drive, the net energy gain could be negative because of the low efficiency of the lasers. Gain could be positive for heavy-ion indirect drive because of the much higher efficiency of particle accelerators (Bangerter 1993). Nevertheless, the latter presents specific problems, as the need of very high current (of the order of kA and beyond) and the photoionization of the ion beams by the radiation field (that produces a strong coulombian expansion of the beams, which become severely unfocused from the target) (Velarde et al. 1992).

New alternatives are emerging in Target Physics and may contribute to solve the problem of finding a target robust enough to be very unsensitive to non-uniformities. One proposal in this field is based on the non-linear acceleration of undercritical foams of fuel illuminated by ultrashort high power laser beams (Hora et al. 1994).

Another proposal is based on externally guided plasma acceleration, collision and implosion (Martínez-Val and Piera 1994). The external guidance is provided by a dense and thick tube made of iron or lead (high density material). The principle is similar to the acceleration of a bullet inside a cannon. In this case, two opposite cannons (only two driver beams) are needed, and they converge into a central small cavity where the fuel pellet collision takes place. The tube provides an interim tamper effect for the fuel to become compressed and hot, so undergoing a fusion burst. Although the gain of this scheme seems to be smaller than the theoretical energy gain of a spherical implosion by a factor of 2, the advantages of needing only two beams and the robustness of the design deserve further investigation. This proposal will be further exposed in section 4 of this paper.

Similarly, new fuel materials as Li_2DT and CDT must be taken into account for the development of IFE Targets in the future. The advantage of these fuels is that they can not be degraded by mixing (because they are homogeneous). This possibility will also be examined in deeper detail in this paper (section 5). On the contrary, in multilayered shells targets, the fuel can be displaced from the core by mixing with the surrounding coating. As the coating can have a bigger mass than the fuel, this one becomes very diluted during the implosion.

Besides the difficulties of finding the best drive method, IFE targets present another area of research (Bruckner and Jorna 1974) (Bodner 1974) (Meyer-ter-Vehn 1982) where the field is also very open: the ignition and triggering of the fusion burst. A school of research consider that the best (and only) way to ignite a target is by the creation of a central hot spark at the stagnation phase (when the fuel has been compressed to the minimum radius and some shock waves go back and forth from the target center due to the overpressure of the pusher and to the high pressure of the central fuel). In this scheme, fusion conditions are initially reached in the hot spark only (which is 10% of the fuel mass). Ignition is then propagated to the surrounding fuel, which is denser but much cooler (Lindl 1989).

Another ignition model is the so-called Volume Ignition (Martínez-Val, Eliezer and Piera 1994) (Hora and Ray 1978) (Martínez-Val et al. 1993) (Martínez-Val and Piera 1993), where a central hot spark is not clearly formed (although the peak temperature is in the center). This configuration can be created by stagnation-free implosions (Mima, Takabe and Nakai 1989), typically produced by pusherless deceleration (the pusher becomes almost fully ablated by the end of the implosion). Both density and areal density can reach very high values in this scheme, what is needed for alpha-particle and neutron energy deposition. The main advantage of this scheme arises from its hydrodynamic stability. Several experiments on spark-ignition have found that the stagnation phase is very unstable and the fuel confinement is rapidly destroyed (without time enough for the fusion burst to take place) (Bayer et al. 1984) (Yamanaka 1994). Section 3 of this paper will be devoted to this ignition configuration. Before presenting this analysis, a general overview of target physics will be outlined in section 2.

TARGET PHYSICS SYSTEMATICS

The evolution of an ICF target can be divided into four consecutive phases: target illumination by the driver beams; payload (fuel) implosion towards the center of the hollow target; deceleration of the fuel as the inner void closes because of the fuel central collision, so producing high temperatures and high densities; and ignition and burn propagation if the conditions achieved at the end of the compression are good enough to trigger a fusion burst.

The first phase depends on the type of driver and the illumination method, but in all the cases there is a fundamental requirement that must be satisfied: the effect of the illumination (measured as ablation pressure, or as energy deposition or as any other way) must be very uniform all over the external surface of the target. Otherwise, the implosion will not be spherical and some sectors of the fuel will be imploding much faster than others, with a final negative consequence: the target core will be broken into pieces (jets) without reaching any significant compression. The degree of uniformity has to be better than 2% for the most critical modes, which are typically close to the Legendre order l=16 (Mc Crory 1989).

The uniformity issue has to be solved by each type of driver according to their capabilities. In Laser Direct Drive, there are already a number of smoothing techniques (Induced Spatial Incoherence, Random Phase Plates, Fiber Optics Band Broadning...) that has successfully been used to drive spherical implosions. In Japan, very high compressed densities, up to 600 g/cm^3 have been reported (Yamanaka 1993).

For Particle Beams, there are not yet any sound porposal to get uniform implosions with Direct Drive, although a possibility must be mentioned: to create convective currents into the outer part of the coating (where the beams deposit their energy) so that a uniformization of the imploding pressure can be achieved. This is a proposal under research by the authors of this paper, but there are not conclusive results so far.

As already said in section 1, Indirect Drive is the method selected by mainstream research to produce uniform spherical implosions. The physical reasons supporting this idea can be understood with the help of fig. 1.

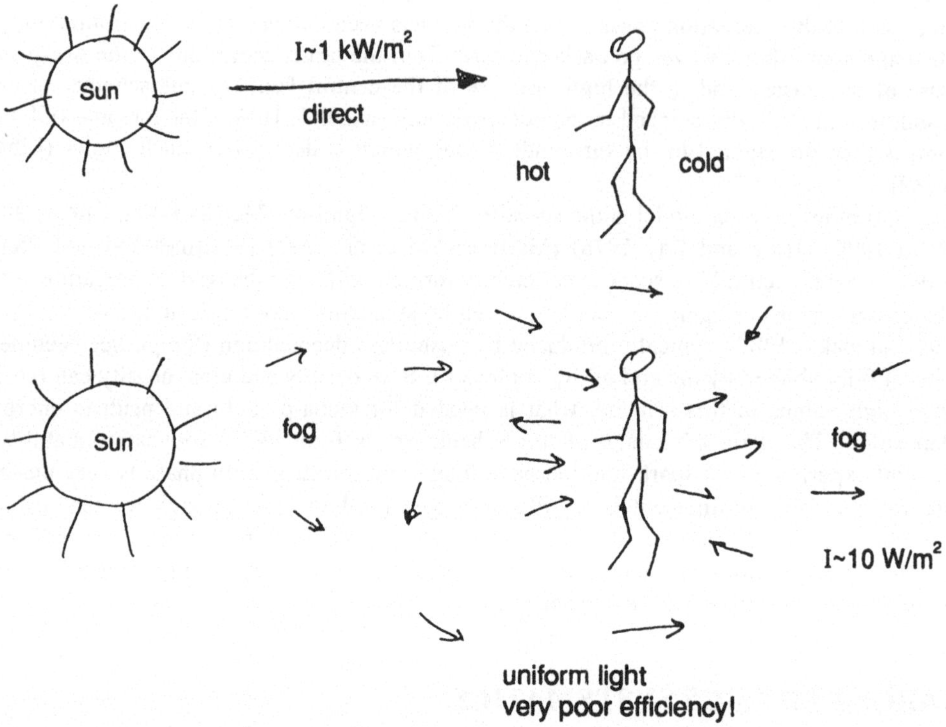

Figure 1. A sketch of direct (a) and indirect (b) illumination from the sun onto a sunbather. In the upper figure, the direct illumination is sketched, and it can be seen how a high intensity beam (through thin air and clear sky) impinges in the front side of the person (at 1kW/m² about). However, the rear side of the person does not receive any (direct) photon, and therefore the illumination is rather asymmetric. On the contrary, for a person inside a thick fog, the sunbeams do not directly reach upon him. He sees an isotropic flux of photons coming from everywhere, and thus the illumination is very uniform. However, the intensity on his skin is rather low, about 10 W/m² what points out that Indirect illumination is not energy effective.

Indirect Drive is based on the conversion of the driver beam energy into thermal radiation at high T ($\simeq$ 300 eV). Subsequently, the radiation field undergoes a multiple process of absorptions and reemissions in order to isotropize the field (as in the fog of the sketch). In order to reduce radiation losses, the system is confined inside a casing (so called hohlraum) which acts as a microwave oven (although in this case the electromagnetic waves are of much shorter wavelengths). As there are several papers on this topic in this conference, the hohlraum approach will not be described or commented any longer in this one.

The second main issue, related to the implosion phase, is the problem of hydrodynamic instabilities. The deleterious effect of these phenomena can be described simply: different layers of material become mixed up and, what is even worse, a heavy material acting as implosion pusher can reach the center of the target and to displace the fusion fuel from there, so avoiding any possibility of a fusion burst.

Hydrodynamic instabilities are well known in ICF target physics (Gamaly 1993). What is not known is how to get rid of them or, at least, how to refrain them from growing too much. There are two phases of the implosion where instabilities can grow quite a lot. First, the ablation or pushing phase, because there is a layer of target where $\nabla P \cdot \nabla \rho < 0$.

Nevertheless, if the illumination is very uniform and the target finishing is good (without long-wavelength important perturbations) the effect of the instabilities remains acceptable. Moreoever, the effect can really be neglegible if there are not density jumps between different materials. If the scale-length of the density gradient is much longer than the width of the unstable region ($\nabla P.\nabla \rho <0$), implosion will proceed without instabilities growth.

Of course, the former requirement is very difficult to meet if two very different materials are put together because of the target design. For instance, if a plastic ablator of 1 g/cm^3 is trying to push lead (11 g/cm^3).

In order to avoid (or to minimize) the problem of instabilities during the acceleration phase, we have proposed (Martínez-Val and Piera 1993) the use of very simple targets, namely, a DT shell (cryogenic, as usual in reactor scenarios) coated by a lithium layer (0.2 and 0.5 g/cm^3 respectively). No heavy material is placed in the middle to act as a preheating shield. We have shown that the compression performance of these targets is extremely good if the implosion is suitably driven (Velarde et al. 1992) as will be explained later on.

There is still another phase where instabilities can produce a total disaster: the deceleration phase.

It must be remembered that, before void closure, the fuel will be flying inwards at a very high speed (~4 x 10^7 cm/s) and around it the pusher will be flying not so fast (~2 x 10^7 cm/s) with a much higher density. When the void closes, the fuel kinetic energy begins to be converted to internal energy along a process that takes 1 ns (or more) because all the fuel is not stopped at the same time. Because of that, a maximum of pressure appears at the center, its gradient ∇P pointing inwards. However, in the fuel-coating interface the density gradient $\nabla \rho$ will be pointing outwards, because the coating is denser. It is well known that a light material can not stop a heavier one. This is called Rayleigh-Taylor instability, and it is similar to have a layer of mercurium floating over the water in a glass. The mercurium will fall down to the very bottom of the glass. Similarly, the heavier pusher will reach the very center of the target and the fusionable fuel will be displaced to the outside, where it can not ignite.

The Japanese team of Osaka made a very good series of experiments on what is called pusherless targets, i.e., targets in which the fuel becomes almost naked at the end of the implosion because the coating is fully evaporated just before void closure (Mima, Takabe and Nakai, 1989) (Nakai 1989). Those experiments achieved very high densities and were not affected by mixing problems. The main difficulty of this type of targets is that they must be imploded in a finely tuned way, because the fuel has to become uncoated in the proper time (neither sooner nor later).

We have shown (Martínez-Val et al. 1994) that pusherless targets can be imploded by particle beams direct drive (presuming uniform illumination). As in the laser case (where pulse time shape is used to make the time multiplexing of shocks leading to pusherless implosions) in particle beam drive is also necessary to tune the beam specifications to the target description. However, such a tuning is not difficult to determine by numerical simulation (Martínez-Val et al. 1994).

Figure 2 shows the temperature and density profiles of a pusherless target, and it can be seen that the lithium density (coating, meshpoints from 1 to through 60) is very low. Hence, the density gradient never points out during the deceleration phase, and the lithium remains outside while the fuel is in the central part.

A consequence of this type of driving is that "central-spark" ignition configurations can not be achieved, because there is not a stagnation phase along which a train of shockwaves sweep the fuel back and forth as they are reflected in the center and in the fuel-pusher interface. Neither the temperature can be very high in the central 10% of the fuel

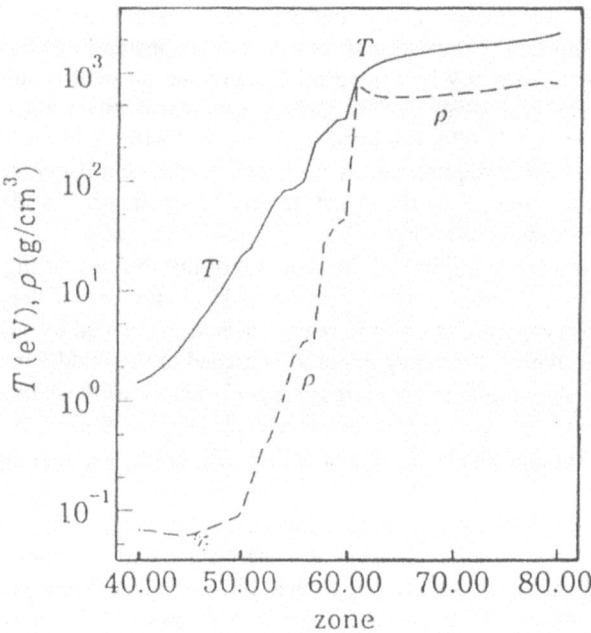

Figure 2. Temperature and density profile at the end of the implosion of a pusherless target (from Martínez-Val et al., 1994).

and very low outside, nor the density can be higher in the outer fuel than in the center, because the pressure drops to zero close to the fuel border and so does the density and the temperature. As a matter of fact, the pusherless scheme leads to the so called Volume Ignition configuration, that will be treated in next section.

Another issue of paramount importance (mainly for pusherless targets) is hydrodynamic decoupling. This term means that, along the acceleration, and particularly at the end of the implosion, the forefront (innermost) fuel is flying towards the center at a much higher speed that the bulk of the fuel. If this happens, the fuel is internally hydrodynamically decoupled, what has very negative effects. Some numerical calculations (Velarde et al. 1992) have shown that the compressions reached in hydro decoupled targets are very poor as compared with those achieved in the cases where decoupling has been avoided, i.e., where the bulk of the fuel is flying at a high speed, very close to the forefront speed.

The former problem can be understood better if it is formulated in the following way: How the fuel has to be accelerated in order to have the maximum T and the maximum ρ when it is decelerated after void closure? If the fuel is hydro decoupled, when the forefront is stopped (void closure) the bulk of the fuel is still very far from the center. The central pressure rises very rapidly because of the conversion of the kinetic energy into internal energy. When the pressure shock moving outwards clashes against the bulk of the fuel, still flying inwards, the fuel is gradually stopped and can not reach positions closer to the center. Hence, the outer fuel radius remais very big, and therefore both the density and the areal density do not achieve very high values.

It has already been said that the way to avoid hydro decoupling is to produce a tuned acceleration, in such a way that the outer fuel tries to catch up with the forefront. In laser direct drive, it can be done by shock multiplexing in such a way that (almost) of the shockwaves converge into the center at the same time. For Indirect Drive, the radiation field temperature has to be tailored in time for producing such a type of coupled implosion. In this case, a strong problem is found in relation to fuel preheating. It is obvious that if the fuel becomes very hot (over 30 eV) before the end of the implosion, it will be very difficult to reach high densities. If it is taken into account that radiation temperature in the hohlraum are about 300 eV, the preheating problem can dominate unless there is an optically thick coating around the fuel by the end of the implosion. Unfortunatelly, radiation shields have to be very heavy (for instance, lead) what rises again the problem of instabilities in the deceleration phase (see the paper by C. Yamanaka in these Proceedings). The radiation drive regime has to be fully studied under this perspective. So far, the maximum density reported by Lawrence Livermore National Laboratory in hohlraum experiments is 20 g/cm^3, much less than the D$_2$ maximum density obtained by Osaka ILE with pusherless implosions by Direct Drive (120 g/cm^3, i.e., a compression factor of 600).

As a summary of this section, it can be said that stagnation free targets initially coated by light ablators (as lithium) lead to very high compression and are, at the same time, very insensitive to instabilities all along the target evolution. Those targets have a feature that has already been pointed out: they lead to Volume Ignition configurations. Next section is devoted to this topic.

VOLUME IGNITION

Before quoting and commenting some results of numerical simulation of Volume Ignition, it is mandatory to clarify a misperception about it: initial simplistic analysis considered that the driver has to produce an implosion at the end of which the temperatures of the fuel would be 5 keV. This number came from the ignition temperature criterion established in the following way: the plasma reheating by alpha-particles must exceed the radiation losses by bremmstrahlung.

This requirement is applicable to Magnetic Confinement, where the bremmstrahlung photons m.f.p. is much longer than the size of the plasma. In other words, the plasma is optically thin and radiation escapes freely.

On the contrary, this is not the case in ICF compressed targets. We will put it into the right perspective by giving some relevant numbers characterizing an ICF compressed target in a reactor scenario. For instance, a DT sphere of 1 mg compressed up to a final radius of 100 μm will have an electron density of 10^{26} cm^{-3}. If a temperature of a few keV is assumed, the bremmstrahlung photons will have a broad spectrum with a sudden cut at about 0.3 nm wavelength. The maximum frequency will thus be about 10^{18} Hz but most of the photons will have much longer wavelengths. The maximum m.f.p. (for 0.3 nm) will be 10 μm. As it varies with the frequency as ν^3, it is found that most of the photons will have a m.f.p. much shorter than 0.1 μm. It must be remembered that a photon being born in the center of the sphere will undergo about $(R/2\lambda)^2$ collisions (of absorption and reemission) before reaching the outer border. This means that the compressed fuel acts (at the beginning of ignition) as a black body. With an outer temperature of 1 keV, it will be radiating at 6 x 10^{16} W/cm^2. For a 100 μm sphere radius that would be 6 x 10^{13} W, i.e. 2 x 10^4 J in a 0.33 ns compression phase. This is a very low value as compared to the internal energy of a fuel compressed to such temperatures and densities, which is about 2 x 10^5 J.

Moreoever, the confinement time of such a compressed sphere can be estimated as R/c$_s$, where R=100 μm and c$_s$ is the sound speed, that can be expressed as follows by using ideal gas EOS:

$$c_s = (\gamma T / 2.5\, m_p)^{1/2} \qquad (1)$$

This Eq., as the rest of the paper, is written in c.g.s. units plus keV for the temperature (with the only exception of the fuel mass, which is given in mg in some Eqs. as is noted).

For 2 keV plasmas, this means about 4.0×10^7 cm/s (400 μm/ns) what in turn represents about 250 ps of confinement. The total energy radiated will be about 15 kJ. If we compare this value with the internal fuel energy of the fuel at that moment, it is seen that the radiated energy is of the order of 6% this value (at 2 keV, the internal energy of 1 mg DT is 240 kJ). It is thus important to realize that there is not any significant cooling mechanism of an ICF target at maximum compression. If the compressed matter is not a fusionable fuel, the microsphere will explode and most of its internal energy will be converted into kinetic energy of the mechanical disassambly (the radiation losses being almost neglegible). In figure 3, it can be seen the areal density evolution of a target. In the pure hydro calculation without computing the reheating effect by fusion product, an almost symmetric curve is observed, with an effective width of 200 ps, in accordance with the previous estimates by simple but realistic physical considerations.

In the same figure it can be seen the evolution when the reheat is computed (Martínez-Val et al. 1994). The confinement time shortens because the ion temperature rises to more than 200 keV, but there is time enough to burn an important fraction of the fuel (>30%). In order to explain the results of the numerical simulations we can proceed as follows:

First, the reheating power can be estimated in terms of the fusion reactivity $\langle\sigma v\rangle$ and the energy deposition both by alpha-particles and neutrons. In the range of interest to

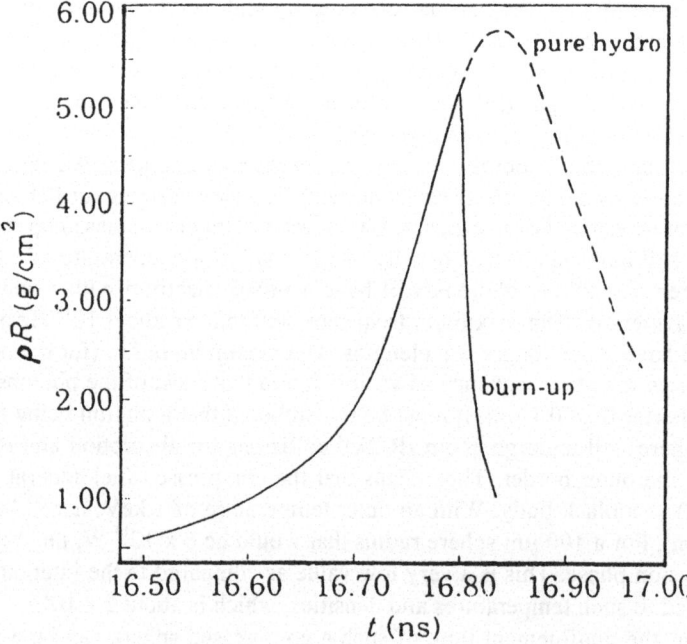

Figure 3. Time evolution of the areal density of a target with and without computing the reheating effect of fusion products.

analyze if fusion is ignited or not, i.e., between 1 keV and 10 keV, the reactivity cm³/s can be estimated as

$$<\sigma v> = 10^{-20} T^4 \quad (T \text{ in } keV) \qquad (2)$$

As the fuel will have an areal density (see fig. 3) higher than 4 g/cm², the 14.1 MeV neutrons will be scattered by the D and T nuclei and they will transfer to the plasma a fraction of their energy according to their Kerma factors (Martínez-Val 1990). Credit will not be given to the effect of suprathermal fusions because it will be smaller than 10% of the thermonuclear fusion effect (Martínez-Val 1990). The optical m.f.p. of 14.1 MeV neutrons in stechiometric DT is 4.7 g/cm² and the Kerma factor is about 3 MeV per collision. Hence the reheating power can be expressed by

$$P_h\left(\frac{keV}{cm^3 s}\right) = 10^3 <\sigma v> n_i^2 \left(3.5 + 3\frac{\rho R}{4.7}\right) \qquad (3)$$

The former Eq. is only valid for $\rho R < 22$ g/cm². For areal densities higher than this value, it must be taken into account that the neutron reheat can not be bigger than 14.1 MeV (if suprathermal fusions are not taken into account (Martinez-Val 1990)). In this case of very high areal densities, the heating power can be written as

$$P_h\left(\frac{keV}{cm^3.s}\right) = 17.6 \times 10^3 <\sigma v> n_i^2 \qquad (4)$$

It is obvious that for triggering the fusion burst in this configuration of Volume Ignition, the time required to heat the plasma from the initial temperature of the hydro implosion T_i up to 10 keV must be smaller than the disassembly time. (From 10 keV on, the fusion burst proceeds so quickly that the it takes less than 30 ps to finish the process by fuel depletion and mechanical disassembly. Remember that it is not a generation-type chain reaction, but a pure reaction process stimulated by thermal motion).

The heating time can be computed by

$$t_h = \int_{T_i}^{10} \frac{6 n_i dT}{P_h(T)} \qquad (5)$$

By replacing in this Eq. the values formerly given, it is found that

$$t_h = a(\rho_i, \rho R_i) T_i^{-3} \qquad (6)$$

where a $(\rho_i, \rho R_i)$ is a function of the density and the areal mass reached at the end of the implosion. If the neutron reheating is not taken into account and "f" stands for the compression factor (ρ/ρ_0), the former Eq. can be written as

$$t_h = 0.23 \times 10^{-5} f^{-1} T_i^{-3} \qquad (7)$$

Values of t_h are reported in Table 1 for different values of ρ_i and T_i for 1 mg DT targets.

Table 1. Heating times values (in ns) for 1 mg DT target as a function of the density and the temperature reached at the end of the implosion

T_i(keV)	Density (g/cm^3)		
	f=210 g/cc	f=420 g/cc	f=630 g/cc
1	2.30	1.15	0.77
1.1	1.73	0.86	0.58
1.2	1.33	0.67	0.44
1.3	1.04	0.52	0.35
1.4	0.84	0.42	0.28
1.5	0.68	0.39	0.23
1.6	0.56	0.28	0.19
1.7	0.47	0.24	0.16
1.8	0.39	0.20	0.13
1.9	0.33	0.17	0.11
2.0	0.29	0.15	0.10
3.0	0.084	0.042	0.028

Of course, when the heating time becomes very small (<10 ps) the former calculation is not applicable because we have considered an instantaneous slow-own of knock-on ions and alpha-particles, what is not true (it takes between 2 and 5 ps to thermalize those particles). Nevertheless, if t_h becomes so small, there is no doubt of triggering a fusion burst. In general, the burst would be ignited if the confinement time t_c is at least two or three times as long as the heating time t_h. It was already said that t_c is given by

$$t_c = R/c_s$$

where c_s was estimated in Eq. 1. In turn, the fuel compressed radius R can be defined in terms of the mass of fuel M and the compression factor f (ρ_c/ρ_0). For pure DT it holds

$$R(cm) = 0.1 \left(\frac{M(mg)}{f} \right)^{1/3} \quad (8)$$

Hence, we can write

$$t_c = 0.42 \times 10^{-7} M^{1/3} f^{-1/3} T_i^{-1/2} \quad (9)$$

By comparing the results of Eq. (5) for the heating time with those of the disassembly time, it is found that for a 1 mg DT pellet compressed to a density of 300 g/cm$_3$, the compression temperature T_i must be higher than 1.5 keV. If the compression is still higher, for instance 600 g/cm^3, the ignition drops to 1.2 approximately. Of course, these values are an indication of the average values needed at the end of the implosion. Accurate numerical simulations are needed to have a finer representation of the target performance, but we can say that the results in the literature agree quite well with the estimates of this theory.

An interesting scaling law can be obtained to complete it. The volume ignition criterion can be expressed as

$$t_c > t_h \tag{10}$$

If we take into account the corresponding equations and their dependence on the initial mass M and the compression factor f, it can be written

$$0.42 \times 10^{-7} M^{1/3} f^{-1/3} T_i^{-1/2} > 0.23 \times 10^{-5} f^{-1} T_i^3 \tag{11}$$

And therefore, for the case of not computing neutron reheat

$$M^{1/3} f^{2/3} > 55 T_i^{5/2} \tag{12}$$

This means that in order to trigger volume ignition with a given material (for instance, DT) the former product has to be high enough. The criterion can thus be satisfied by increasing the mass M but it is still more effective to increase the compression factor $f=(=\rho_i/\rho_0)$. The foregoing analysis justifies the results obtained by some numerical calculations (G. Velarde et al. 1992) (Martínez-Val, Eliezer and Piera 1994) that show how it is possible to obtain high energy gain by volume ignition. In fact, if T_i can be of the order of 1.25, a gain close to 100 could be obtained with 1 mg DT targets driven by implosions where the hydro efficiency is about 10% and the fusion burnup fraction is around 40%.

Moreoever, it is worth remembering that Volume Ignition configurations are the natural hydrodynamic outcome of pusherless implosions aimed at avoiding destructive instabilities at the final phase of the implosion-compression.

EXTERNALLY GUIDED ICF TARGETS

A new proposal (Martínez-Val and Piera 1994) to drive and ignite ICF targets will briefly be commented in this section, as an example of new ideas that can emerge in this field. Figure 4 depicts a conceptual design for a reactor working under this new concept. The sketch shows a liquid lithium free-falling blanket that surrounds the cavity where the target explodes. Two driver beams are only needed to implode the target, a section view of which is seen is Fig. 5.

The evolution of the target will proceed along the following steps. First, as in any other target, the beams will interact with the fuel-pellet coatings. It goes without saying that material and thickness of the coatings have to match the type of driver beam. Heavy ion beams will need a much thicker coating than laser beams.

The hydro-mechanical reaction to the coating evaporation (ablation) will accelerate the fuel pellets inwards along each barrel. A fundamental objective in this phase is to produce an internally coupled acceleration, i.e., to avoid the generation of an ultrafast forefront of fuel (with a very low density) that leaves behind itself the bulk of the fuel. In section 2, the problem of hydro decoupling was analyzed in a general context.

Our early numerical results show that an almost planar acceleration can be achieved along a cylindrical barrel, but the problem of avoiding the hydrocoupling is not negligible. In the laser drive case, a sharp time shape is needed to produce the suitable time-multiplexing of small shocks. If the laser can really give such a pulse (with an initial prepulse

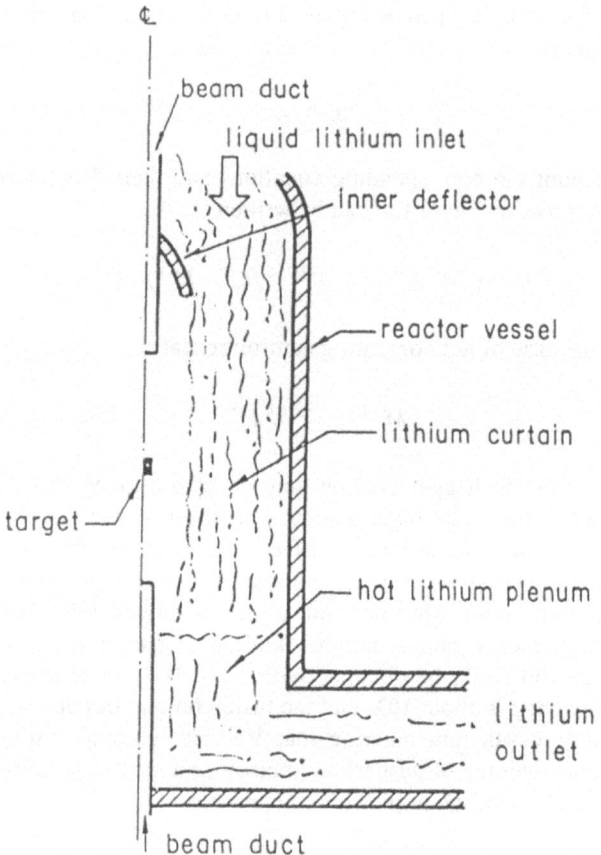

Figure 4. A conceptual design of externally guided ICF reactors.

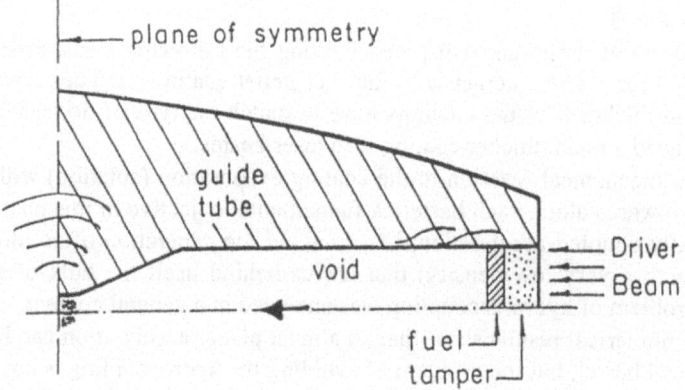

Figure 5. A sketch of an externally guided ICF target.

lower than 10^{13} W/cm^2 and a maximum peak of 10^{15} W/cm^2) the fuel pellet will fly inwards at a speed close to 4 x 10^7 cm/s or even higher. Then, the central collision will happen and very high compressions and temperatures are obtained. Two points deserve to be commented further: First, the nozzle-like central cavity of the tube guide provides a very good geometry for the forefront to slow-down just a little before the collision (after passing through the throat) what is important for the bulk of the fuel to fill the cavity without finding a very big central overpressure too early. Moreoever, the compressed fuel filling the cavity will be a sphere where the fusion products will produce the fuel reheat needed to trigger ignition. Of course, enough density and areal density are essential for it, as was discussed in the preceding section. In any case, a sphere is much better than a thin cylinder as central cavity shape not only because hydrodynamic reasons in the deceleration phase, but also to provide a good geometry to catch the fusion products (in a fine cylinder, the transversal escape will reduce the reheat). The second point to be underlined is that the confinement and the density will decrease because of the expansion of the central cavity. It must be taken into account that a DT plasma with a density ρ(g/cm^3) and temperature T (keV) will have a pressure of the order of

$$P(dy/cm^2) = 0.8 \times 10^{15} \rho T \qquad (13)$$

for non-degenerate plasmas (which is the applicable case for this type of driving). This means that for 500 g/cm^3 and 1 keV, the pressure is about 4 x 10^{17} (dy/cm^2) i.e. 400 Gbar. If this pressure is applied inside a sphere, its radius will be expanding at a speed

$$v_t = (0.8 \times 10^{15} \rho T / \beta \rho_t)^{1/2} \qquad (14)$$

where ρ_t is the initial density of the guide tube (11 g/cm^3 for lead) and β is a parameter connected to the Hugoniot (Eliezer et al. 1986) of this material (typically 1.1<β<1.5). For the former case, it means an expansion speed close to 2 x 10^8 cm/s. Because of the foregoing values, the confinement time of the plasma inside the central cavity will be much lower than the confinement time predicted for spherically compressed targets (R/c_s) given by Eq. (9). In fact, if the central cavity has an initial radius R_0, a criterion for the confinement time can be defined as

$$t_c = \frac{0.2 R_0}{v_t} \qquad (15)$$

When the radius R(t) has increased up to 1.2 R_0, the volume of the central cavity will be 1.8 time the initial volume. This figure has been chosen for the definition of the confinement time. It leads to the expresion

$$t_c = 0.455 \alpha\, M^{1/3} f^{-5/6} T^{-1/2} \qquad (16)$$

where α is a parameter depending on the density and EOS of the tube material.

$$\alpha = (\beta \rho_t / 0.8 \times 10^{15})^{1/2} \qquad (17)$$

For lead, $\alpha \approx 0.14 \times 10^{-6}$ (in c.g.s. units).

Confinement times given by the former Eq. are much shorter than those given by Eq. (9) for spherical implosion. For instance, for 1 mg DT with a compression factor f=1000 and T=1 keV, the spherical compression t_c is 390 ps, while the externally guided confinement gives only 20 ps. This value is almost of the order of the slowing down time of the charged particles in DT plasmas ($t_\alpha < 5$ ps). Nevertheless, it will be found in the following analysis that there is a sound possibility of triggering ignition in externally guided plasmas by properly choosing the specifications of the target. In order to do that, the former confinement time will be compared with the heating time. An index of quality i is defined as

$$i = \frac{t_c}{t_h} \qquad (18)$$

and this index has to be larger than 1 for ignition be possible.

The heating time, including the neutron reheat (for the standard case of an areal density $2<\rho R<22$) can be written as

$$t_h = 1.65 \times 10^{-6} T^{-3} (0.735 f + 0.03 M^{1/3} f^{5/3})^{-1} \qquad (19)$$

where the first term within the bracket takes into account the alpha particle reheat and the second one is the neutron reheat (with the limitation of 14 MeV, which is not written in the equation but can easily be taken into account in the calculations).

Hence the index of quality is

$$i = 0.04 T^{5/2} (0.735 M^{1/3} f^{1/6} + 0.03 M^{2/3} f^{5/6}) \qquad (20)$$

This Eq. is a very useful scaling law to understand the dependence of the ignition feasibility i on the three implosion parameters: the fuel mass M, the compression factor f and the temperature at the end of the implosion T. The index fundamentally depends on T, what is obvious if the values of reactivity $<\sigma v>$ are remembered. It also depends on f, although not very significantly. For low areal densities, in which case the neutron reheat is almost negligible, it only depends as $f^{1/6}$. For high areal densities, the dependence is higher, but always smaller than linear. Something similar happens with the dependence on mass: it lies between $M^{1/3}$ for small targets and $M^{2/3}$ for large ones.

Figure 6 depicts the index value for some ranges of the independent variables. It is seen that i becomes larger than 1 (i.e. ignition conditions) for a broad field of values. However, it is not so for small targets. For M=1 mg, i is only larger than 1 for very high T (T>4 keV) and high compression factors (f>5000). Furthermore, the heating time is so small in these cases ($5<t_h<10$ ps) that Eq. (19) is not applicable because it presumes instantaneous energy deposition of the charged particles (which need a few ps to stop). However, for M=5 mg the situation is much better. The range of i>1 is much wider and the heating times are bigger. For instance, a good choice for final implosion conditions would be f=1000 (ρ=210 g/cm^3) and T=4.5 keV. For this case, t_h=14 ps. For cases with higher compressions, the ignition temperature is smaller. For example, for f=5000, T=3.5 keV, but the confinement time is very short (5 ps) what is questionnable to sustain a fusion burst. Nevertheless, it must be noted that most of the fusions take place in any ICF target when the fuel is aready in the disassembly phase with speeds higher than 10^8 cm/s.

From the previous examples it is obvious that the fuel mass has to be close to the largest size compatible with a reactor scenario. Besides that, the ignition temperature in

Target Physics for Inertial Fusion Energy

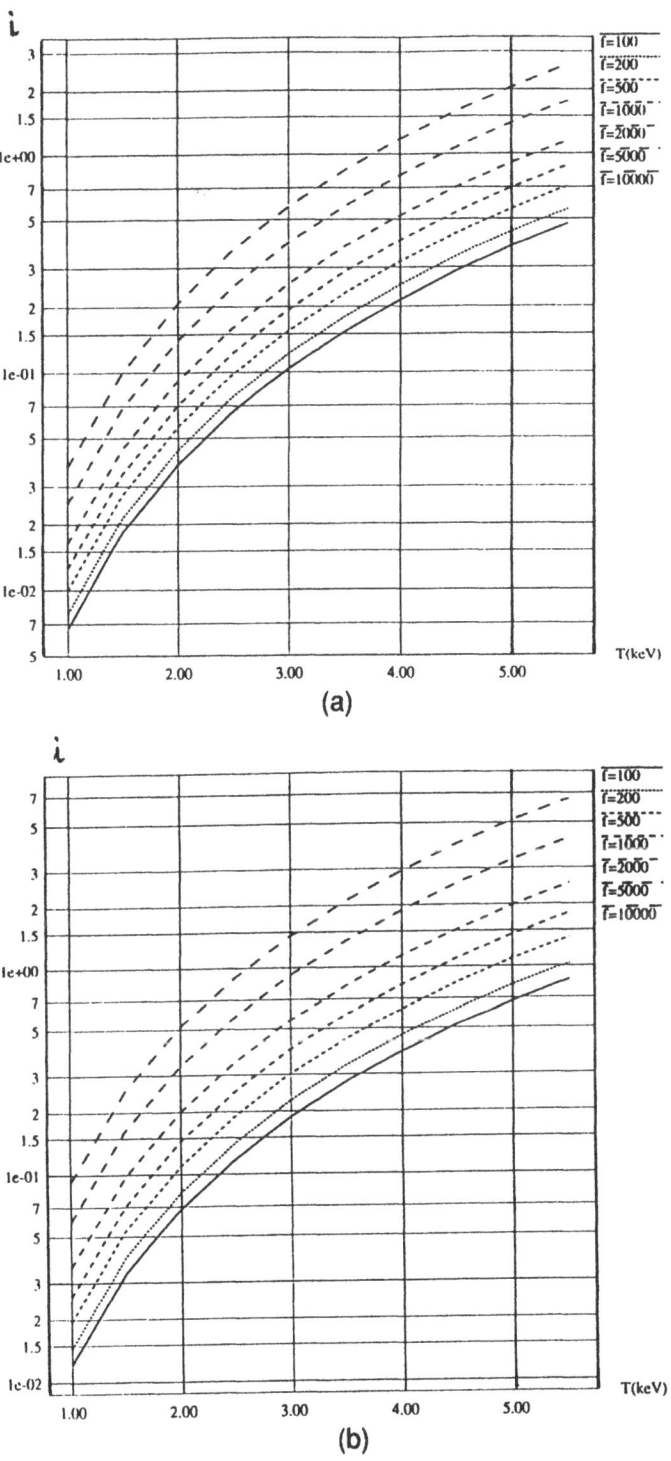

Figure 6. Ignition index i (t_c/t_h) for targets with 1, 5 and 10 mg DT as a function of the compression factor f (ρ/ρ_0) and ignition temperature T (figures (a) .001 gr; (b) .005 gr; and (c) .01 gr; respectively).

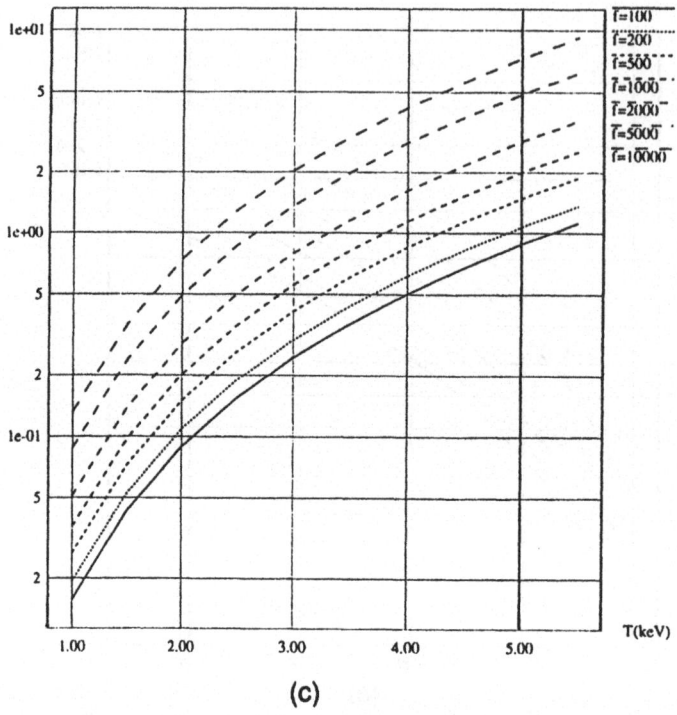

Figure 6c.

this concept is higher than the values given for spherical implosions. A consequence of this requirement is a reduction in energy gain. For instance, if a burnup fraction could be 40% and the hydrodynamic efficiency of the implosion could be 15%, the maximum gain would be about 40. Although this value could seem very low as compared to other designs, it is worth pointing out that it would be suitable for a driver with an energy conversion efficiency of 20% and an energy system efficiency of 40%. In this case, the recirculating fraction would be about 1/3, an acceptable figure for a fusion reactor.

In figure 4, an outline was depicted of a fusion reactor based on externally guided targets. The driver beams (only two) enter the cavity through the vertical ends in order to simplify the mechanical configuration of the cooling system, which is composed on a thick blanket (>0.5 m) of free-falling liquid lithium. Other materials as $Pb_{83}Li_{17}$ euthectic can also be used for this purpose. The target debris and the fusion products are captured in the liquid blanket. The hot plenum in the lower part of the vessel is connected to an intermediate heat exchanger as in any energy conversion system involving liquid alkalis.

The vessel heigth can be moderate, about 1 m or less. The inner blanket radius can be 0.3 m and the outer radius 0.8 m. The total volume of falling lithium will be about $2m^3$. The falling time would be 0.4 s. The total flow would thus be 5 m^3/s, what corresponds to 2.5 Mg/s (i.e. 150 tons per minute). The pumping power can be small because the main ducts are of large diameter and the speed is small. The main pressure loss would happen inside the heat exchanger and can be estimated in 2 bars. Hence, the pumping power for 5 m^3/s would be 1 MW (a pump of about 2 MW would be required, because of the friction internal losses and actual efficiency).

On the basis of exploding a 5 mg DT target per second, with a burnup fraction of 33%, a power of 500 MW would be obtained. By taking into account the energy multiplication in the blanket by neutron-induced reactions, the total thermal power would be about 600 MW. The coolant energy balance can be established by taking into account the lithium specific heat (at high temperature) that is 3.5 J/gK. Hence, the temperature jump in the falling lithium would be 600 MW/(2.5 x 3.5) (Mg . J/gK) what gives 68°C, an acceptable value for a reactor. If a smaller value is sought to avoid thermal stresses, a bigger mass flow can be provided by increasing the cavity volume.

SINGLE-MATERIAL TARGETS FOR ICF

One of the critical points in target performance is the destructive effects of hydrodynamic instabilities. They are particularly deleterous at the early acceleration phase and at the deceleration phase (if a tampered compression is used) because the fuel can become mixed with the pusher.

A way to avoid mixing problems is to use pellets made of a single material. Unfortunately, DT (hydrogen) can not be used as the only material because of several reasons related to its hydrodynamic and thermal properties (Velarde et al. 1992). In particular, the thermal diffusivity ($k/\rho c_p$) is so high that all the fuel becomes isothermalized during the acceleration phase and it reaches very high temperature (>100 eV) before void closure. The attempt to make a cold acceleration has failed so far (in theoretical and numerical analysis) because of the ablation pressures needed to reach relevant speeds (>4 x 10^7 cm/s). There are two alternatives to pure H_2 as single materials for ICF targets: an organic polymer (of CDT as standard molecular formula) and lithium hydryde Li_2DT. The latter has the advantage of having a very high melting point, and remains solid at reactor cavity temperatures. The polymers will melt down as they are placed in the cavity, unless a strong cooling mechanism is provided in the target holder.

Two main problems arise from the use of compounds as fuel material. First, more energy is needed to heat its components up to the ignition temperature. Second, the inner pressure of the plasma is bigger because of the higher electron pressure.

If an analysis of Volume Ignition is made for Li_2DT, we find the following scaling laws. First, the sound speed must take into account the total number of particles,

$$c_s^2 = \frac{12\gamma T}{19 m_0} \tag{21}$$

where m_0 is the proton mass. The confinement time for spherical implosions is thus

$$t_c = \frac{R}{c_s} = \frac{0.26 M^{1/3} m_0^{1/2}}{3.46 f^{1/3} (\gamma T)^{1/2}} \tag{22}$$

The heating time must also take into account the total number of particles, hence

$$t_h = \int_{T_i}^{10} \frac{18n}{n^2 <\sigma v>} \frac{dT}{E_f} \tag{23}$$

where the energy reheat (per fusion) can be estimated as

$$E_f(keV) = 10^3(3.5+0.33\rho R) \tag{24}$$

and the heating time is then

$$t_h = 6 \times 10^{20} n^{-1} E_f^{-1} T_i^3 \tag{25}$$

where n is the deuterium density ($n=n_D=n_T$). If a sound speed of 5 x 10^7 cm/s is assumed for characteristic value at ignition, the ignition criterion $t_c=t_h$ will lead to

$$0.12 \times 10^{-8} M^{1/3} f^{1/3} = 0.02 f^{-1} E_f^{-1} T_i^3 \tag{26}$$

In the case of neglecting the neutron reheat, E_f=3500, and the criterion can be rewritten as

$$M^{1/3} f^{2/3} T^3 \geq 5000 \tag{27}$$

where M is in mg and T in keV. For instance, for a 10 mg Li$_2$DT with a compression factor of 1000, the ignition temperature would have to be higher than 2.85 keV (As the neutron reheat has been neglected, this is a conservative, upper estimate). If the compression increases to 3000, the temperature decreases to 2.23 keV. On the other hand, if M is risen to 30 mg (maximum DT output of 2.7 GJ in 100% burnup) the ignition temperature would be 2.5 keV. Thus, high compressions are very important for Li$_2$DT in order to get low ignition temperatures (i.e., in order to get high gains).

Externally Guided Targets with a Single Fuel Material

This idea of using CDT or Li$_2$DT as the only material for making ICF targets can also be applied to Externally Guided implosions. In this case, as already known, the confinement time is expected to be

$$t_c = \frac{0.2R}{v_t} \tag{28}$$

where R is the cavity radius and v_t is the exploding velocity of the tube. The radius in this case is related to the amount of Li$_2$DT that can be compressed inside it by the formula (c.g.s.)

$$R = 0.6 M^{1/3} f^{-1/3} \tag{29}$$

where M must be in g. The ideal EOS pressure is

$$P = NT \tag{30}$$

where N must account for all the species densities

$$N = 12n = 12f n_0 = 3.8 \times 10^{23} f \tag{31}$$

n_0 being the n_D density at regular Li$_2$DT (~ 1 g/cm^3). Thus the estimate of the pressure in c.g.s. and keV for T is

$$P = 0.6 \times 10^{15} fT \tag{32}$$

The former equation gives a pressure four times as high as the corresponding pressure for DT with the same f and T (for DT, P=0.16 x 10^{15} fT). This means that the confinement time will be shorter than in the DT case. If now is taken into account that the tube radial speed will be

$$v_t = \sqrt{\frac{P}{\rho_t \beta}} \tag{33}$$

where β is an Hugoniot-related parameter (~1.4), the confinement time is

$$t_c = 2 \times 10^{-8} M^{1/3} f^{-5/6} T^{-1/2} \tag{34}$$

This confinement time is shorter than the DT confinement time (with the same M, f and T by a factor ~1/3).

On the other hand, the Li$_2$DT heating time is

$$t_h = \int_T^{10} \frac{18n}{n^2 <\sigma v>} \frac{dT}{e_f} \tag{35}$$

where the fusion products reheat (per DT reaction alone) can be estimated as

$$E_f(keV) = 10^3 (3.5 + 0.2 M^{1/3} f^{2/3}) \tag{36}$$

where the use has been made of the relation $\rho R = 0.6\ M^{1/3}f^{2/3}$, for this material. By integrating the former Eq. we obtain

$$t_h = 1.9 \times 10^{-5} T^{-3} (3.5f + 0.2 M^{1/3} f^{5/3})^{-1} \tag{37}$$

Because of the electrons and the lithium ions, the Li^{2}DT heating time is about 2.7 times higher than the corresponding heating time for pure DT with the same M, f and T. Although it must be taken into account that the mass of Li$_2$DT will be higher than the mass of pure DT for the same amount of DT (the same maximum explosion yield) by a factor close to 4, the effect both in t_c and t_h is very small.

The index of quality for ignition in Externally Guided Li$_2$DT targets in then expressed by

$$i = \frac{t_c}{t_h} = 0.0035\ T^{5/2} (M^{1/3} f^{1/6} + 0.06\ M^{2/3} f^{5/6}) \tag{38}$$

with M in grams and T in keV.

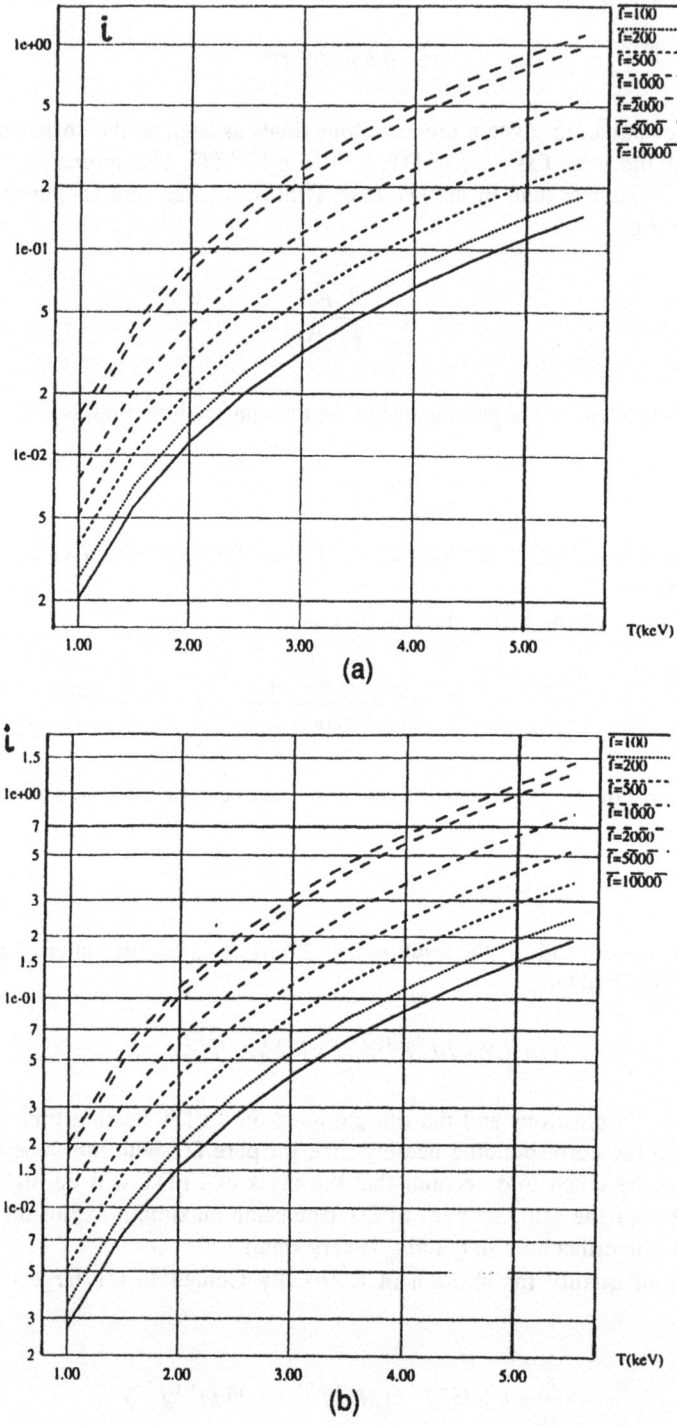

Figure 7. Quality index for igntion of Externally Guided Li_2DT targets for 10 mg (a) 20 mg (b) and 5 mg (c).

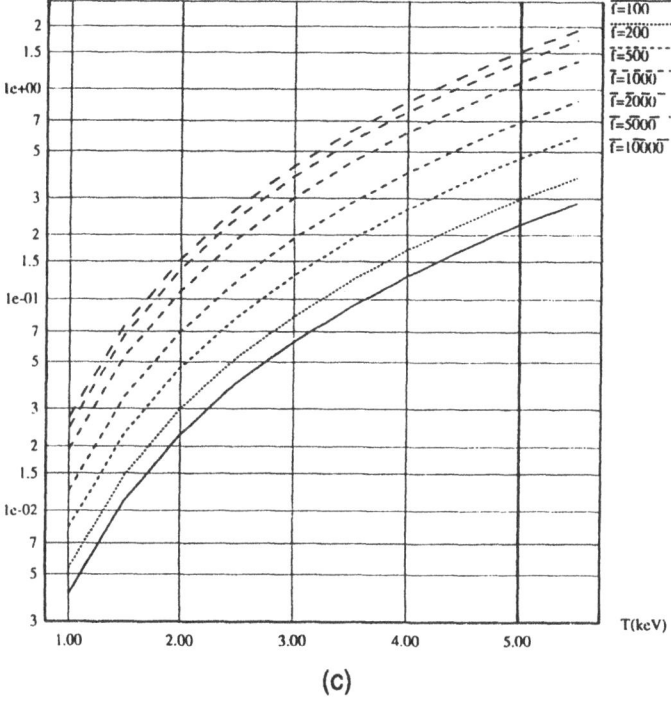

Figure 7c.

For the same values of M, f and T, this Eq. gives values smaller than the index of DT, by a factor ~1/8. Besides that, the driver energy to produce a given implosion with the same amount of DT fuel, f and T is higher for Li_2DT than for pure DT, what gives an energy gain of the former that is smaller than the gain of the latter.

Nevertheless, a range of variables is found in which Externally Guided Li_2DT targets can undergo ignition and burn.

CONCLUSIONS

Both the theoretical predictions and, above all, the experimental confirmations have demonstrated that is physically possible to implode microcapsules of light material.

However, it is not evident that those implosions will produce high energy gain because of the existence of assymmetries and hydroinstabilities that degrade the quality of the confinement. In order to avoid these problems, an alternative is radiation drive, but this means a loss of beam-target coupling efficiency. Another alternative, without the inherent loss of efficiency, is pusherless implosions leading to stagnation-free volume ignition. In this case, the maximum energy gain is bound by the minimum temperature capable to trigger ignition. A maximum gain of 100 could be achieved, but not higher (unlike in spark-ignition, where the gain can theroretically be higher if the saprk mass is much smaller than the total fuel mass). Nevertheless, a gain of 100 (and even a gain of 50) is more than enough for reactor scenarios if the driver efficiency is not very low. Because of these reasons, this paper has been particularly devoted to stagnation-free targets, including the new concepts of externally guided targets and the single-material targets (and the use of both ideas in the same scheme).

As a general conclusion, it can be said that more experimental work is needed to obtain more accurate data (in atomic physics, EOS and so forth) and to check integral performance of different types of targets and illuminations schemes.

In any case, the field of ICF must be considered as still to open to make final decisions on the best choice for driver, target specification, type of illumination and related items. It is compulsory to make an in-depth scientific analysis of all the relevant alternatives in ICF because its intelectual richness deserves it.

ACKNOWLEDGMENTS

This work was partially supported by the European Union Network "High Energy Density Matter" and by the "Iberdrola" Visiting Professor Chair in ETSII.

REFERENCES

Bangerter, R.O., 1993, "The Induction Approach to Heavy-Ion Inertial Fusion: Accelerator and Target Consideration", *Il Nuovo Cimento*, **106A**, 1445.
Bayer, C., et al., 1984, "Laser Implosion of Microballoons: Study of the Transition from Exploding-Pusher to Ablative Regime", *Nucl. Fusion*, **24**, 573.
Bodner, S., 1974, *Phys. Rev. Letters*, **33**, 761.
Bruckner, K.A., and Jorna, S., 1974, *Rev. Modern Physics*, **46**, 325.
Caruso, A., 1989, *Inertial Confinement Fusion*, pp. 139–162, Società Italiana de Fisica, Editrice Compositori (Bologna) Italia.
Desselberger, M. et al., 1992, *Phys. Review Letters*, **68**, 1539.
Eliezer, S., Ghatak, G., and Hora, H., 1986, "An Introduction to Equations of State: Theory and Applications". Cambridge University Press, U.K.
Eliezer, S., Honrubia, J.J., Velarde, G., 1992, *Phys. Letters A*, **166**, 249.
Gamaly, E., 1993, "Hydrodynamic Instabilities of Target Implosion in ICF", Chapter 13 of *Nuclear Fusion by Inertial Confinement*, Velarde, G., Ronen, Y., and Martínez-Val, (eds.). CRC Press Inc., Boca Raton, Florida (USA).
Hora, H., and Ray, P.S., 1978, *Naturforsch.*, **33A**, 890.
Hora, H., Eliezer, S., Hopfl, R., Martínez-Val, J.M., Velarde, G., Honrubia, J.J., Piera, M., 1994, "ICF fast ignitor with high hydroefficiency, nonlinear force drive", IAEA Conference of Plasma Physics and Fusion Research, Seville 1994, paper IAEA-CN-60-B-P-2.
Lindl, J.D., 1989, pp. 617–632 in *Inertial Confinement Fusion*, Società Italiana de Fisica, Ed. Compositore (Bologna) Italia.
Lindl, J.D., Mc Crory, R.M., and Campbell, E.M., 1992, *Physics Today*, September 1992, 32.
Martínez-Val, J.M., 1990, "Neutronic effects in ICF targets", *Fusion Technology*, **17**, 476.
Martínez-Val, J.M. et al., 1993, "Numerical and Theoretical Studies on the Ignition of ICF Plasmas Driven by Ions Beams", *Il Nuovo Cimento*, Vol. 106A, N. 12.
Martínez-Val, J.M., and Piera, M., 1993, *Fusion Technology*, **23**, 218.
Martínez-Val, J.M., Eliezer, S., and Piera, M., 1994, "Volume Ignition for Heavy Ion Inertial Fusion", *Laser and Particle Beams*, Vol. 12, N. 4.
Martínez-Val, J.M., and Piera, M., 1994, "An Externally Guided Target for ICF", submitted to *Fusion Technology*.
Meyer-ter-Vehn, J., 1982, *Nucl. Fusion*, **22**, 561.
Mc Crory, R.L., and Verdon, C.P., 1989, pp. 83–98, in *Inertial Confinement Fusion*, Societá Italiana de Fisica, Ed. Compositori (Bologna) Italia.
Mima, K., Takabe, H., and Nakai, S., 1989, "Pusherless Implosion, Pulse Tailoring and Ignition Scaling Law for Laser Fusion", *Laser Particle Beams*, **7**, 249.
Murakami, M., and Meyer-ter-Vehn, J., 1991, "Indirectly Driven Targets for ICF", *Nucl. Fusion*, **31**, 1315.
Nakai, S., 1989, "Laser Fusion Experiement", *Laser Particle Beams*, **7**, 467.

Velarde, G., Martínez-Val, J.M., Piera, M., Aragonés, J.M., Honrubia, J.J., Mínguez, E., Perlado, J.M., and Velarde, P.M., 1992, "An Analysis of the Implosion of Heavy Ion Directly Driven Simple Targets", *Particle Accelerators*, 37, 537.

Velarde, G., Ronen, Y. and Martínez-Val, J.M. (eds.), 1993, *Nuclear Fusion by Inertial Confinement*, CRC Press Inc. Boca Raton, Florida (USA).

Yabe, T., 1993, "The Compression Phase in ICF Targets", Chapter 11 of *Nuclear Fusion by Inertial Confinement*, Velarde, G., Ronen, Y. and Martínez-Val, J.M., (eds.). CRC Ress Inc., Boca Raton, Florida (USA).

Yamanaka, C., 1993, "Diagnostics of Laser-Imploded Plasma", Chapter 20 of *Nuclear Fusion by Inertial Confinement*, Velarde, G., Ronen, Y., and Martínez-Val, (eds.). CRC Press Inc., Boca Raton, Florida (USA).

Yamanaka, C., 1994, *IAEA Conference on Controlled Thermonuclear Fusion and Plasma Physics*, Seville, September 1994 (España).

6

SPHERICAL PINCH RESEARCH[*]

Historical Background, Achievements, and Projections

F. Giammanco,[1] E. Panarella,[2†] N. Salingaros,[3] D. P. Singh,[4] and M. Vaselli[4]

[1]Department of Physics
University of Pisa
56100 Pisa, Italy

[2]Advanced Laser and Fusion Technology, Inc.
189 Deveault St., Unit No. 7
Hull, P.Q., J8Z 1S7 Canada

[3]Division of Mathematics
University of Texas
San Antonio, Texas 78249

[4]Institute of Atomic and Molecular Physics
National Research Council
56127 Pisa, Italy

ABSTRACT

Several years of research in a particular plasma configuration designated by the name of Spherical Pinch has led to a device that has the potential for long-term fusion energy benefits while providing immediate industrial benefits. The former benefits derive from its ability to approach fusion breakeven conditions in a spherical configuration, which is an improved version over the classical inertial confinement fusion configuration. The latter benefits derive from its exploitation as a high flux broadband radiation source for uses that can go from soft-X-ray microlithography, to neutron radiography for nondestructive detection of hidden defects in metallic structures. This paper highlights the milestones of the Spherical Pinch research program carried out over the past several years, up to its present status as a fusion device and an industrial tool. The international collaboration which is being established on spherical pinch research is also highlighted. A clear path of future work is then projected for research and development, in both the fusion area and in industrial spinoffs.

---

[*] Parts of this work are abstracted from previous works by the same authors.
[†] Also with the Department of Electrical and Computer Engineering, University of Tennessee, Knoxville, Tennessee, and the National Research Council, Ottawa, Canada.

1. INTRODUCTION

The only known controlled fusion reaction takes place in the sun and stars. Its principal features are:

1. Both the fusion reactor and the reacting material are spherically symmetric;
2. The reaction is confined by the spherical-symmetric radially-inward gravitational field;
3. The solar reactor is thermally and elastically stable: it accommodates surface oscillations, and allows disruptions such as flares and prominences without affecting its overall integrity;
4. The temperature of the reactor is very high throughout, and the reactor is enveloped in a blanket of gas (solar corona), which has distinct physical properties from both the reacting material and the surroundings.

The hydrogen bomb, a source of uncontrolled reactions, is also spherical in shape. The investigation of fusion in weapons, based on laser-driven inertial fusion, deals with spherical symmetry. The spherical geometry seems therefore the most logical choice to reproduce controlled fusion on Earth.

Since the late 70s, an experimental and theoretical program has been carried out at the National Research Council of Canada in Ottawa on an evolution of a well-known plasma physics concept, the theta pinch, from the cylindrical geometry to the spherical. The motivation for such a study was given by the ability of the spherical configuration to overcome the serious problem of plasma end losses inherent in the cylindrical geometry. If this scheme were supplemented by the presence of a hot plasma in the centre of the spherical vessel, which could act as a target for the imploding shock waves generated in the Spherical Pinch, then these shock waves would contain and compress such a central plasma, and further raise its temperature. The spherical pinch concept would then become a device capable of seriously competing in the fusion race.

The conversion from cylindrical to spherical geometry along the lines indicated above was successful. A number of experiments were carried out and copious X-ray and neutron emission was observed from modest scale plasmas.[1] This in turn led, at the beginning of the 80s, to analyze the experimental conditions required for a spherical pinch to satisfy the Lawson criteria for fusion breakeven. Following a study and modelling of the phenomenon, the scaling laws for spherical pinch devices were derived in 1983.[2] Although obtained under simplifying conditions, they nevertheless indicated that the spherical pinch concept had the potential of reaching breakeven. A series of pilot experiments designed to approach the conditions required by the scaling laws were then carried out, which verified the stability of the spherically pinched plasma under those conditions. Most important, neutron emission, a signature of fusion reactions, was again observed.[3,4]

While this work was going on in Canada, a series of independent investigations began in Italy on a configuration similar to the spherical pinch. It uses a variation of this concept by exploiting the reflected shock wave from a spherical surface to increase the temperature and density of a laser plasma produced in the center of a vessel.[5] In the U.S.A. another investigation was initiated on the merit comparison between spherical and toroidal geometries in their approach towards fusion breakeven conditions.[6] A similar scheme was independently investigated in Los Alamos, designated by the name of Magnetized Target Fusion, where the central plasma is magnetized so that conduction losses are largely reduced.[7]

The experimental achievements encouraged an industrial interest in the Spherical Pinch. In 1987 a Company (Advanced Laser and Fusion Technology, Inc. - ALFT) became operational in Canada and a particular target industrial spinoff, soft X-ray production for microlithography,[8–9] was selected as initial application of the device. This project has now completed the final stage of machine prototyping, and the production stage is beginning.

Another industrial spinoff of the Spherical Pinch is also being pursued. It is neutron generation for nondestructive testing of materials. A Feasibility Study has indicated that the Spherical Pinch can become a transportable neutron generator for a neutron radiography system. An experimental proof-of-principle is now being planned, in which a 1 MJ condenser bank facility will be used for the project.[10]

The knowledge acquired through these industrial spinoffs on the basic operation and characteristics of the Spherical Pinch has proven fundamental in understanding the direction to follow to reach fusion ignition conditions. In particular, a numerical study carried out with realistic energy input parameters has revealed that the Spherical Pinch can become a serious contender for fusion.[10–11] The advantages of pursuing this line of research are therefore threefold: 1) the laws that govern the phenomenon are known and understood; 2) experiments aimed at proving conditions near or at ignition are, by all standards, modest scale experiments, and therefore economical; and 3) even if ignition conditions are not achieved in the short term, the Spherical Pinch is already providing economical returns through its industrial spinoffs.

In the following sections a broad overview of past achievements in the Spherical Pinch program will be provided, and future directions of research will be outlined.

2. THE PLASMA COMPRESSION CONCEPT

This is an evolution of concepts that began in the late 60's with a study of the classical cylindrical implosion process, and continued with the demonstration that a central plasma in a spherical vessel can be compressed by imploding shock waves.[1]

The way these imploding shock waves are generated leads to a classification of the Spherical Pinch either as an inductive or a resistive heating device.[4] Fig. 1 shows the structure of an inductive device. In a sphere whose radius can range from 2 to 10 cm, current I enters along the equatorial plane QQ', and is then forced by a set of slots aligned along the meridians to proceed towards the north pole, return to the equator, cross the south pole, and then exit the sphere as I on the plane of the equator. This structure assures that current flows always along the meridians of the sphere, which in turn assures a three-dimensional plasma implosion within the sphere. Because the discharge is inductively generated, the working gas pressure cannot exceed 10 Torr. Beyond this pressure, the gas does not break down. The great advantage of this geometry is its simplicity, in that energy is transferred into the plasma by a simple mechanism, inductive transfer. The disadvantage is the low efficiency of transfer, normally only a few percent.

If one wants to improve the efficiency of energy transfer, resistive discharges, i.e., electrical sparks in a high pressure gas, are the preferred route. In this way the energy transfer efficiency can reach 30% in a sphere of reasonable size, of the order of a few centimetres (2 - 10 cm).

Figure 2 shows a resistive heating device in schematic form. In the interior of a spherical metal vessel, a spherical cavity is produced. A hot plasma is created in the center of the cavity by means of an electrical discharge. Simultaneously, a number of im-

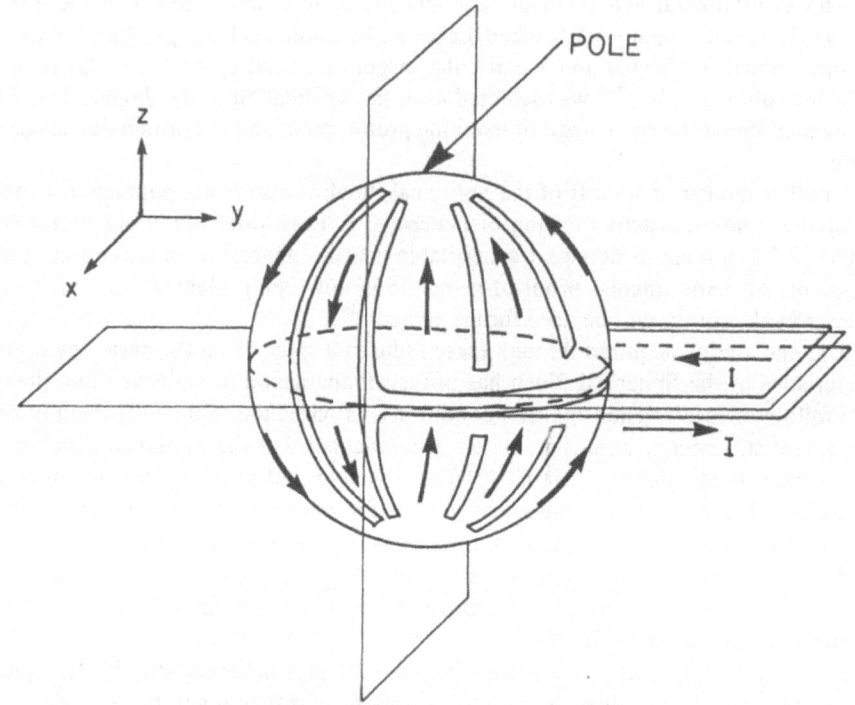

Figure 1. Inductive generation of implosive shock waves.

ploding shock waves is launched from the periphery of the cavity by discharging the energy of a condenser bank through a set of electrodes positioned at the boundary of the cavity. The discharges in this case are resistively generated by sparks. The shocks which break loose from the expansion of the discharge plasmas at the periphery of the vessel coalesce. During their convergence towards the center of the spherical vessel, they merge into a single shock front and, because of area convergence, experience a large pressure amplification effect. Because the pressure of the imploding shock wave far exceeds the pressure of the central plasma, when the spherical shock front reaches the central plasma, two phenomena take place: a) a strong transmitted shock penetrates the plasma and converges towards the centre with high velocity, thus raising the plasma temperature further, and b) the central plasma is compressed. The combination of shock heating and plasma compression makes the central plasma temperature rise rapidly, and fusion reactions can take place.

3. SCHEMATIC SEQUENCE OF SPHERICAL PINCH OPERATIONS

Fig. 3 is a diagram of the time evolution of the shock fronts (dashed lines) and contact surfaces (solid lines) when a plasma is first created in the centre of the vessel (R = 0), and is later compressed by the imploding shock waves launched from the periphery of the spherical vessel (R = R). We observe the following sequence of operations:[2]

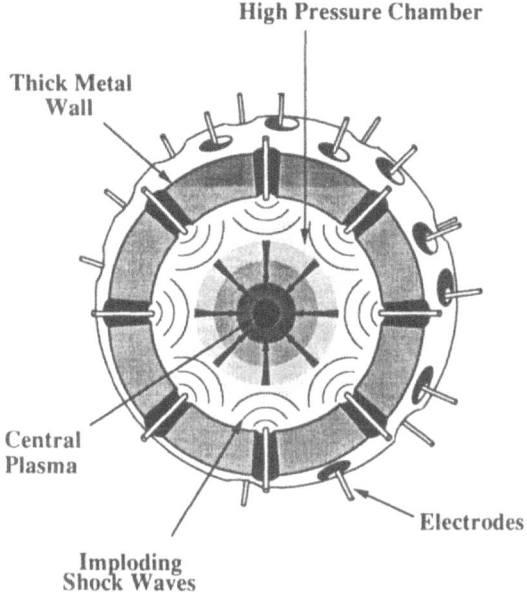

Figure 2. Resistive generation of imploding shock waves.

1. A central plasma is formed, as hot as possible. This is necessarily a rarefied plasma, i.e., of density lower than the original density of the gas in which it is formed. A diverging shock wave S_e breaks loose from this exploding plasma;
2. Later on in time, a peripheral plasma is exploded, which gives rise to an imploding shock wave S_i that collides with the central plasma and starts compressing it;
3. During compression, the central plasma temperature increases and significant nuclear reactions begin to take place;
4. The reaction rates increase as the central plasma is being compressed. Simultaneously, a transmitted shock wave TS inside the central plasma raises its temperature further;
5. Reflected shock waves RS_e and RE bounce back and forth between the contact surfaces of the central plasma and peripheral plasma;
6. Finally, when the central plasma pressure equals the pressure of the imploding shock waves, equilibrium is reached, and the plasma begins to be disassembled. The phenomenon ends at this time.

4. APPROXIMATE SCALING LAW FOR SPHERICAL PINCH EXPERIMENTS

The Spherical Pinch phenomenon is governed by well-known fluid dynamics and shock wave equations. In order to derive the scaling law for breakeven fusion conditions, it is assumed that the reaction rates in the central plasma begin to be significant during the compression phase, when T = 2.58 KeV, which is the minimum temperature for breakeven as required by the Lawson conditions. From this time on and until the plasma is disassembled, one wants the plasma to obey the Lawson criterion for breakeven in terms of plasma density and containment time. In a planar geometry, the scaling law is:[2,4]

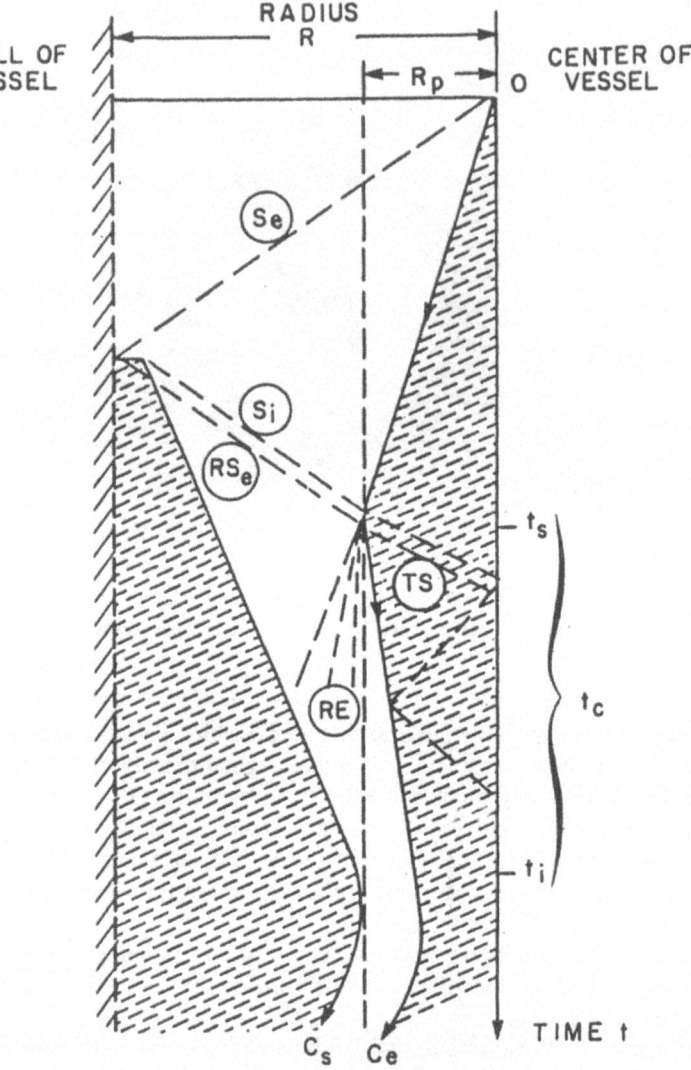

Figure 3. Time evolution of Spherical Pinch operations.

$$\rho R \left(\frac{E_s}{M_s}\right)^{\frac{1}{2}} \geq 1.96 \times 10^2 \qquad cm^2 \, g^{\frac{1}{2}} \, J^{\frac{1}{2}}$$

where:

ρ is the initial gas density;
R is the radius of the vessel; and
E_s/M_s is the energy density deposited at the periphery of the vessel.

If one plots the above expression in Fig. 4, one sees that, in order to keep the radius R of the vessel to an acceptable value (2 - 10 cm) and the energy density E_s/M_s within a

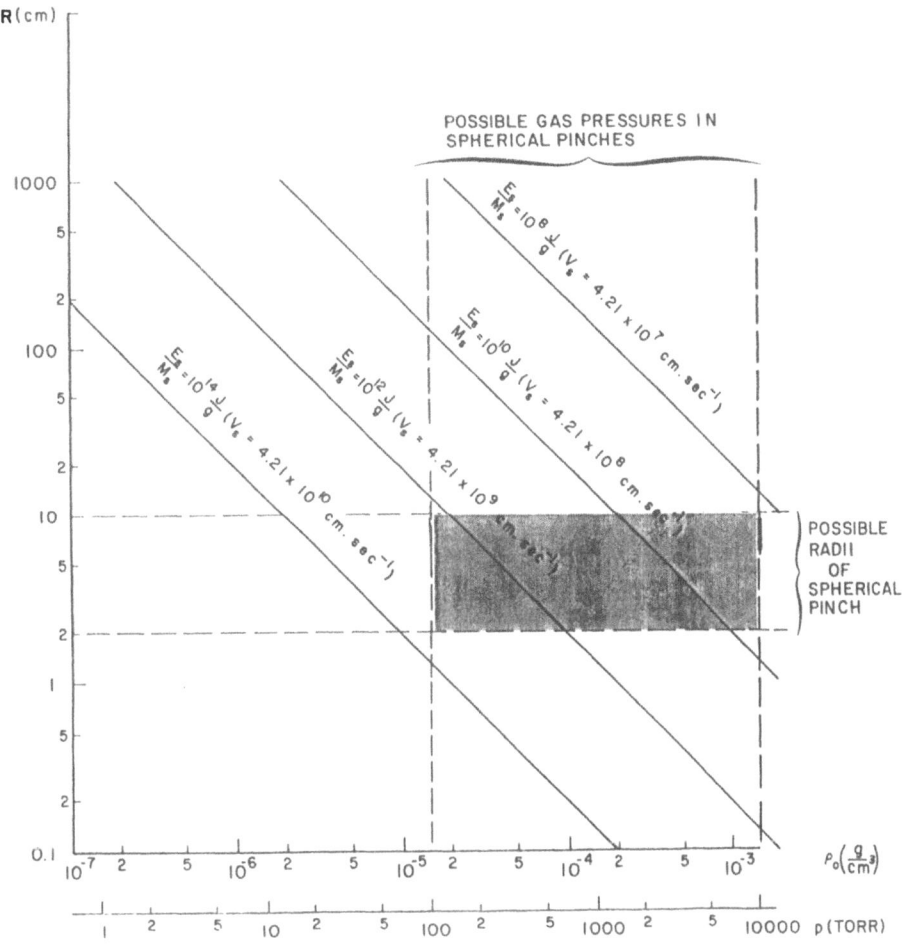

Figure 4. Plot of the scaling law for spherical pinch devices.

realistic range (~10^8 J/g), breakeven conditions can be achieved only at high pressures of the working gas.

5. FULL ANALYTICAL STUDY[12]

In the following sections, a full analytical treatment of the spherical pinch concept will be given, starting with a study of the propagation of diverging shock waves caused by the explosion of the central gas in a spherical vessel, to their interaction with the imploding shock waves generated by the peripheral explosion. The scaling laws of neutron production from deuterium gas are derived in a self-consistent manner for the cases when the converging shock wave interacts directly with the expanding central plasma, or with the explosive shock wave detached from the central fireball. The agreement of the analytical results with the experimental results reported in the literature provides an indication that the analytical treatment is correct.

5.1. Premise

The generation and propagation of spherical shock waves is of great importance in view of achieving plasma compression and heating relevant to fusion.[13–17] Many experimental and numerical studies have been devoted to analyzing diverging and converging shock waves in different geometries. The interaction of two colliding shocks, which, under certain conditions, leads to the formation of a third Mach wave during the irregular reflections, has been discussed widely in the literature.[18] The formation of shock waves during plasma expansion has opened up new regimes of states of matter, where the equation of state is relevant to dense and hot material.[19] In this context, the interaction of spherical converging shock waves with an expanding plasma/diverging shock wave occupies a distinctive place. It applies in analyzing the feasibility of the spherical pinch plasma fusion scheme,[15] in which a hot central plasma is confined and compressed by imploding shock waves.

We are now concerned with the plasma evolution from the time of formation of the diverging shock wave up to its interaction with the converging shock wave. Under appropriate conditions, this leads to compression of the central plasma and an enhancement of the plasma temperature to very large values during the collapse phase of the converging shock wave. The generation of neutrons under these conditions is also studied, and the relevant numerical estimates are compared with other data. Section 5.2 deals with the plasma expansion phase just after the plasma formation at the center of the spherical cavity and the formation of the explosive shock wave. The evolution of the converging shock wave launched from the periphery of the vessel and related interactions are studied in Sec. 5.3. The formulas for neutron generation in the appropriate cases are derived in Sec. 5.4, and a brief discussion of the numerical results along with their comparison with experimental data are given in Sec. 5.5.

5.2. Fireball Evolution and Explosive Shock Wave

5.2.1. Explosive Stage after Breakdown. When sufficient energy is in a small region of a gas in a short time, gas breakdown occurs. The plasma subsequently expands in all directions. Several mechanisms of plasma evolution have been proposed in the literature to understand the plasma behaviour during the energy deposition time.

Following Ref. 20, we assume that hydrodynamic motion is negligible during the deposition time in an explosive plasma. Therefore, the plasma evolution can be described by a self-similar electron thermal wave,[21] whose front position R_{fL} is given by:

$$R_{fL} = (\chi_f \Delta \tau_L)^{1/2},$$

where

$\chi_f = aT_{fL}^{5/2}/N_0 =$ electron thermal diffusion coefficient
$T_{fL} =$ temperature behind the shock thermal front, which is assumed uniform[21]
$\Delta \tau_L =$ pulse duration of energy deposition
$N_0 =$ gas density
$a = \dfrac{1.3 \times 10^{-5}}{k_B m Z \ln \Lambda}$ (cgs)

Z =   charge of ions
m =   mean degree of ionization of the gas
k_B =   Boltzmann constant
$\ln \Lambda$ =   Coulomb logarithm
$\ln \Lambda \approx$ $\ln \dfrac{k_B T_{fL}}{(4\pi)^{1/2} Z^2 e^2 N_0^{1/3}}$ (Ref. 21).

The absorbed energy is used in ionizing the gas, as radiation loss, and in plasma heating. From the energy conservation law, the temperature of the plasma may be written as:

$$k_B T_{fL} = \frac{E_L - E_I - E_{Rad}}{2\pi m N_0 R_{fL}^3}, \tag{1}$$

where E_L, E_I, and E_{Rad} represent the absorbed energy, the energy used to ionize the gas, and the radiation losses, respectively. The expressions for E_I and E_{Rad} may be further given, respectively, by:

$$E_I = \frac{4}{3}\pi m I_Z N_0 R_{fL}^3, \tag{2}$$

where I_Z is the ionization potential of the Z ion, and:

$$E_{Rad} = \left(\frac{D_{Ric}}{T_{fL}^{1/2}} + D_B T_{fL}^{1/2}\right)\frac{4}{3}\pi (Zm)^2 N_0^2 R_{fL}^3 \Delta \tau_L. \tag{3}$$

The first and second terms on the right-hand side of Eq. (3) represent the particle recombination and bremsstrahlung radiation, where $D_{Ric} = 3 \times 10^{-22}$ cgs, $D_B = 10^{-27}$ cgs (Ref. 21), and T_{fL} is expressed in degrees Kelvin.

A relative numerical estimate of energy distribution in the aforementioned different processes reveals that, for typical plasma discharge experiments, the energies used in ionizing the gas and in radiation are quite small, and that the major part of the energy couples with plasma motion. The radius and the temperature of the expanding plasma, in practical units, can be written as:

$$R_{fL} = 0.052 \left(\frac{E_{L_u}^5 \Delta \tau_{L_u}^2}{N_{0_u}^7}\right)^{1/19} \text{(cm)} \tag{4}$$

and

$$T_{fL} = 3 \times 10^2 \frac{E_{L_u}}{N_{0_u} R_{fL}^3} (^\circ K), \tag{5}$$

where E_L and $\Delta \tau_L$ are given in Joules and nanoseconds, respectively. For a typical case of fully ionized deuterium gas, $E_{L_u} = 20$ J, $\Delta \tau_{L_u} = 10$, $N_{0_u} = 1$, $N_{0_u} = N_0/3 \times 10^{19}$, $R_{fL} \approx 0.15$ cm,

$T_{fL} = 1.8 \times 10^6 \, °K$, the ionization energy (assuming an ionization potential ~24 eV) is $E_I \approx 1.22$ J, and $E_{Rad} \approx 0.02$J. Henceforth, we will neglect the contributions of ionization and radiation energies in the following analysis.

5.2.2. Explosive Shock Wave Formation. Following Ref. 21, the equation governing the shock position R_{sE} may be written as:

$$\frac{R_{sE}\ddot{R}_{sE}}{\left(\dot{R}_{sE}\right)^2} = \frac{\alpha - 1}{\alpha},$$

where α is the self-similar exponent. Assuming the initial conditions as $R_{sE}(0) = R_{fL}$ and $\dot{R}_{sE}(0) = V_{s0}$, and the exponent of self-similar motion as $\alpha = 2/5$, for a spherical shock wave we find the solution of the preceding equation as:

$$R_{sE} = R_{fL}\left(1 + \frac{5V_{s0}}{2R_{fL}}t\right)^{2/5}, \tag{6}$$

and the velocity of the shock front as:

$$V_{sE} = V_{s0}\left(\frac{R_{fL}}{R_{sE}}\right)^{3/2}. \tag{7}$$

A comparison of Eq. (7) with the relation obtained from the usual shock wave expansion shows that

$$R_{sE} = R_{fL}\left[1 + \left(\frac{E_L}{MN_0}\right)^{1/2}\frac{t}{R_{fL}^{5/2}}\right]^{2/5}, \tag{6a}$$

where M is the ion mass. The condition:

$$\left(\frac{E_L}{MN_0}\right)^{1/2}\frac{t}{R_{fL}^{5/2}} \gg 1$$

leads to the usual relation for the explosive shock wave, and henceforth, the shock wave formation time may be defined as

$$\tau_{SW} = \left(\frac{MN_0}{E_L}\right)^{1/2} R_{fL}^{5/2}.$$

Therefore, a well formed shock wave occurs at a time τ when the condition $\tau \gg \tau_{SW}$ is fulfilled. Actually, τ_{SW} represents the energy transfer time from the thermal wave due *to electrons to mass motion* and then to shock wave due to ions.

5.2.3. Fireball Expansion. Assuming that, after the gas breakdown, the plasma expansion is spherically symmetric and radially outward, the boundary of the thermal conduction zone is found by integrating the continuity equation from the center of the gas breakdown to the thermal wavefront. In a spherical geometry, we have:[22-23]

$$\frac{\partial}{\partial t} \int_0^{R_f} N(r, t) r^2 dr = -N_f R_f^2 \frac{dR_f}{dt}. \tag{8}$$

The plasma density profile (for $\alpha = 2/5$) may be written as:[23]

$$N(r, t) = \beta_E(\gamma) N_0 \left(\frac{r}{R_{sE}}\right)^{9/2},$$

where $\beta_E(\gamma) = (\gamma + 1)/(\gamma - 1)$ is the plasma density jump for a strong shock wave. Inserting this relation into Eq. (8) gives:

$$R_f = R_{fL} \left(\frac{R_{sE}}{R_{fL}}\right)^{3/10}, \tag{9}$$

where R_f is the thermal wavefront radius at the end of the energy deposition pulse.

In general, for a self-similar exponent α, the plasma density profile is:[23]

$$N(r, t) = \beta_E(\gamma) N_0 \left(\frac{r}{R_{sE}}\right)^\varepsilon,$$

where

$$\varepsilon = \frac{2\frac{1-\alpha}{\alpha}}{\gamma - \frac{2}{3}\left(\frac{1-\alpha}{\alpha}\right)}. \tag{10}$$

Therefore, the time evolution of the fireball radius is given by

$$R_f = R_{f0} \left(\frac{R_s}{R_{f0}}\right)^{\varepsilon/2(\varepsilon+3)} \tag{9a}$$

By assuming a constant pressure behind the shock front,[23] the temperature is given by

$$T(r, t) = \delta(\gamma) \frac{M}{k_B} V_{sE}^2 \left(\frac{R_{sE}}{r}\right)^\varepsilon, \tag{11}$$

where a strong shock wave has $\delta(\gamma) = (\gamma - 1)/(\gamma + 1)^2$, and V_{sE} is the velocity of the shock front.

In the case of a spherical first-order shock wave,[21] i.e., $\alpha = 2/5$, Eqs. (10) and (11) lead to

$$N(r, t) = \frac{\gamma+1}{\gamma-1} N_0 \left(\frac{r}{R_{sE}}\right)^{9/2} \tag{12}$$

and

$$T(r,t) = 2 \frac{(\gamma-1)}{(\gamma+1)^2} \frac{M}{k_B} V_{sE}^2 \left(\frac{R_{sE}}{r}\right)^{9/2} \tag{13}$$

or, substituting V_{sE} from Eq. (7) gives:

$$T(r, t) = \frac{8}{25} \frac{(\gamma-1)}{(\gamma+1)^2} \left(\frac{E_L}{k_B N_0}\right)\left(\frac{R_s}{r^3}\right)^{3/2}. \tag{13a}$$

This concludes our derivation for an exploding shock wave.

5.3. Converging Shock Wave

We assume that the converging shock wave is generated by the breakdown of a thin spherical shell at a radius R_0, which propagates in a radially inward direction. During the converging motion, the shock wave meets the preformed central plasma or diverging shock wave detached from it. Therefore, the subsequent interaction physics depends on two different physical situations. In the first case, the converging shock wave reaches the vicinity of the center of the cavity before the formation of the explosive shock wave, and the central plasma is then compressed if the pressure at the front of the converging shock wave is greater than the fireball pressure. This situation is of particular interest due to its applicability. The maximum central plasma compression obtainable is $(\gamma + 1)/(\gamma - 1)$, but the central plasma temperature may reach enormous values in comparison to those obtained in the classical inertial confinement scheme which does not have a central plasma. This scheme is the Spherical Pinch.[4] Actually, the experimental results of Ref. 4 show that compression and heating of the central plasma are greatly enhanced if the implosive shock wave meets the central explosive shock wave formed already, before the collapse takes place. Again, similar to the preceding case, if the pressure at the converging front exceeds the pressure at the diverging shock front, the latter wave is reflected back and heats the central plasma. This may further lead to multiple reflections. In both situations, it is important to analyze and understand the propagation of a converging shock wave.

The motion of the converging shock wave is determined by

$$R_{sC} = R_0 \left(1 - \frac{V_{sOC}}{\alpha_{R0}} t\right)^\alpha, \tag{14}$$

where R_0 is the initial radius and α is the self-similarity exponent, with $\alpha \approx 0.687$ for $\gamma = 5/3$ (Refs. 13 and 21). The index C refers to the peripheral converging shock wave. If the convergent shock wave propagates in the unperturbed gas of density N_0, the front pressure is:

$$P_{sC} = \frac{8}{25} \frac{1}{(\gamma+1)} M N_0 V_{s0C}^2 \left(\frac{R_0}{R_{sC}}\right)^{2[(1-\alpha)/\alpha]}, \tag{15}$$

and the condition to compress the central plasma is expressed as:

$$\frac{8}{25} \frac{1}{(\gamma+1)} M V_{s0C}^2 \left(\frac{R_0}{R_{fL}}\right)^{2[(1-\alpha)/\alpha]} > k_B T_{fL}. \tag{15a}$$

Because the implosive shock wave belongs to the class of second-order self-similar motions, in principle we cannot apply the procedure of the preceding section to relate the initial velocity to the initial energy. Nevertheless, we can maintain a similar approach, although approximate, taking into account that the value of the exponent of a second-order self-similar motion $\alpha \approx 0.687$ is close to the self-similar exponent of a first order plane shock wave, i.e., $\alpha = 2/3$ (Ref. 21). Therefore, the plasma temperature and the shell width, at the end of the energy deposition, are given by:

$$T_{fLC} = \left(\frac{E_{LC}}{2\pi m k_B}\right)^{4/9} \left(\frac{1}{N_0 R_0^4 a \Delta \tau_{LC}}\right)^{2/9} \tag{16}$$

and

$$R_{fLC} = \left(\frac{E_{LC}}{2\pi m k_B}\right)^{5/9} \left[\frac{(a \Delta \tau_{LC})^2}{R_0^{10} N_0^7}\right]^{1/9}, \tag{17}$$

where E_{LC} is the corresponding absorbed energy in a time $\Delta \tau_{LC}$ and in a shell whose width is R_{fLC}.

The formation time and the initial velocity for the converging shock wave are, therefore:

$$\tau_{CSW} = \left(\frac{4\pi M N_0}{E_{LC}}\right)^{1/2} R_0 R_{fLC}^{3/2} \tag{18}$$

and

$$V_{0sC} = \frac{2}{3} \left(\frac{E_{LC}}{4\pi M N_0}\right)^{1/2} \frac{1}{R_0 R_{fLC}^{1/2}}. \tag{19}$$

Equation (19), in practical units, may further be written as

$$V_{OsC_u} = 0.12 \left(\frac{E_{LC_u}^2}{R_0^4 N_{0_u} \Delta \tau_{LC_u}} \right)^{1/9} \tag{19a}$$

and

$$R_{n,c} = 2 \times 10^{-3} \left(\frac{E_{LC_u}^5 \Delta \tau_{LC_u}^2}{R_0^{10} N_{0_u}^7} \right)^{1/9} \quad (cm),$$

where $\Delta \tau_{LC}$ is expressed in nanoseconds.

For the case of the reflection of a diverging shock wave from the converging wave, the pressure at the front of the latter wave should exceed that of the former. The velocity of the reflected shock front in the frame of the converging wave is:[22–23]

$$V_{rel} = \frac{4}{5} \frac{\gamma-1}{\gamma+1} \left(\frac{E_L}{M N_0} \right)^{1/2} R_1^{-3/2},$$

whereas relative to a frame at rest and to the central plasma, it is:

$$V_{0I} = \frac{4}{5} \frac{\gamma-1}{\gamma+1} \left(\frac{E_L}{M N_0} \right)^{1/2} R_1^{-3/2} + V_{OsC} \left(\frac{R_0}{R_1} \right)^{(1-\alpha)/\alpha}, \tag{20}$$

where R_1 is the distance from the fireball center where reflection of the explosive shock wave occurs.

After central reflection, the re-explosive shock wave can meet the converging shock wave in a new position R_2. After second reflection at the center, the velocity increases by a factor of $\{2\gamma/(\gamma+1)\}^{1/2}$ and the self-similar exponent is the same for the converging and re-explosive shock waves, so total reflection cannot occur in R_2 (Ref. 23). Actually, the re-explosive shock wave velocity overcomes the converging shock wave by the same factor of $\{2\gamma/(\gamma+1)\}^{1/2} \approx 1.1$. The difference between velocities being small, a weak reflected shock wave can be expected. In the spherical pinch experiment, a converging shock wave, generated in the first part of the capacitor discharge time, can be followed by a piston whose pressure depends on a different mechanism, say, a $(\mathbf{j} \times \mathbf{B})$ pusher. In such a case, a piston operates as a rigid wall, and multiple reflections can occur.[4]

The converging shock wave may implode the plasma up to the minimum radius, which depends on the ion mean free path in the shock front:[22–23]

$$R_{lim} \approx \frac{u_i^2}{\pi Z^2 e^4 \ln\Lambda\, N_{s\,lim}}, \tag{21}$$

where

Spherical Pinch Research

$U_i =$ total ion energy per particle $= K T_{sC\,lim}/(\gamma -1) + M v_{sC\,lim}^2/2$

$N_{slim} =$ front density related to the plasma density at the implosion N_{lim} by $N_{slim} = N_{lim}(\gamma+1)/(\gamma-1)$

$T_{sClim} =$ function of $v_{sC\,lim}$, making use of the relation for a strong shock wave.

Taking into account that:

$$v_{sC\,lim} = v_{0sC}\left(\frac{R_0}{R_{lim}}\right)^{(1-\alpha)/\alpha},$$

after some rearrangement, we obtain:

$$R_{lim} \approx \left\{\frac{M^2}{2\pi Z^2 e^4 \ln\Lambda}\left[\frac{4}{(\gamma+1)^2}+1\right]\left(\frac{\gamma-1}{\gamma+1}\right)\left[\frac{v_{0sC}^4 R_0^{4[(1-\alpha)/\alpha]}}{N_{lim}}\right]\right\}^{\alpha/(4-3\alpha)} \quad (21a)$$

This concludes our discussion of converging shock waves.

5.4. Neutron Yields

Large numbers of nuclear reactions occur during the re-explosive phase just after the collapse of a converging shock wave. As demonstrated in Ref. 22, the velocity, temperature, and density are related to the values reached at R_{lim} by

$$v_{lim1} = \left(\frac{2\gamma}{\gamma+1}\right)^{1/2} v_{lim}, \quad (22)$$

$$T_{lim1} - \left(\frac{3\gamma-1}{\gamma}\right) T_{lim}, \quad (23)$$

and

$$N_{lim1} = \frac{\gamma(\gamma+1)}{(\gamma-1)^2} N_{lim}. \quad (24)$$

Moreover, we assume that the main contribution to the generation of neutrons comes from the central fireball, after the re-explosion, whose time evolution is given by Eq. (9a), and where T and N can be assumed to be spatially uniform. By changing the variable $dt = dR_f/v_f$, we get the total number of neutrons as

$$N_n = \frac{4\pi}{3}\int_{Rlim}^{\infty} R_{fl}^3 N_{fl}^2 \frac{a}{v_{fl} T_{fl}^{2/3}} \exp\left(-\frac{b}{T_{fl}^{1/3}}\right) dR_{fl} \quad (25)$$

where the index f1 refers to fireballs quantities after re-explosion, a ≈ 10^{-7} cgs, and b ≈ 4.3 × 10^3 cgs for deuterium-tritium (D-T) plasma.

All quantities can be expressed as a function of their values at the limiting radius (T_{lim}, N_{lim}, and υ_{lim} for r = R_{lim}) as given in Ref. 23,

$$N_{f1} = \frac{\gamma(\gamma+1)}{(\gamma-1)^2} N_{lim} \left(\frac{R_{lim}}{R_s}\right)^{\delta}, \tag{26}$$

where

$$\delta = \frac{\varepsilon(\varepsilon+6)}{2(\varepsilon+3)},$$

$$T_{f1} = \left(\frac{3\gamma-1}{\gamma}\right) T_{lim} \left(\frac{R_{lim}}{R_s}\right)^{\beta}, \tag{27}$$

where

$$\beta = 2\frac{1-\alpha}{\alpha} - \frac{\varepsilon(\varepsilon+6)}{2(\varepsilon+3)},$$

$$\upsilon_{f1} = \left(\frac{2\gamma}{\gamma+1}\right)^{1/2} \upsilon_{lim} \left(\frac{R_{lim}}{R_s}\right)^{\xi}, \tag{28}$$

where

$$\xi = \frac{1-\alpha}{\alpha} - \frac{(\varepsilon+6)}{2(\varepsilon+3)},$$

and

$$R_{f1} = R_{lim} \left(\frac{R_s}{R_{lim}}\right)^{\varepsilon/[2(\varepsilon+3)]} \tag{29}$$

After some rearrangement, and introducing a new variable $\chi = R_s/R_{lim}$, Eq. (25) becomes:

$$N_n = A \frac{R_{lim}^4 N_{lim}^2}{\upsilon_{lim} T_{lim}^{2/3}} \int_1^{\infty} \chi^{\psi} \exp\left(-\frac{B}{T_{lim}^{1/3}} \chi^{\beta/3}\right) d\chi, \tag{30}$$

where

$$\psi = \frac{7}{3}\left(\frac{1-\alpha}{\alpha}\right) - \frac{\varepsilon(5\varepsilon + 21)}{6(\varepsilon + 3)} \approx 0.324,$$

$$\beta/3 \approx 0.202,$$

$$A = \frac{4\pi\varepsilon}{6(\varepsilon+3)}\left(\frac{\gamma}{3\gamma-1}\right)^{2/3}\left(\frac{\gamma+1}{2\gamma}\right)^{1/2}\left[\frac{\gamma(\gamma+1)}{(\gamma-1)^2}\right]^2 a,$$

and

$$B = \left(\frac{\gamma}{3\gamma-1}\right)^{1/3} b.$$

By the variable change $x^{0.2} = \xi$, we can derive a useful formula for neutron yield by approximating $\xi^{5.5}$ with ξ^5 in the new integral, namely:

$$N_n = 5\frac{A}{B}\frac{R_{lim}^4 N_{lim}^2}{\upsilon_{lim} T_{lim}^{1/3}} P\left(\frac{T_{lim}^{1/3}}{B}\right)\exp\left(-\frac{B}{T_{lim}^{1/3}}\right), \tag{31}$$

where

$$P\left(\frac{T_{lim}^{1/3}}{B}\right) = 1 + 5\frac{T_{lim}^{1/3}}{B} + 20\left(\frac{T_{lim}^{1/3}}{B}\right)^2 + 60\left(\frac{T_{lim}^{1/3}}{B}\right)^3 + 120\left(\frac{T_{lim}^{1/3}}{B}\right)^4 + 120\left(\frac{T_{lim}^{1/3}}{B}\right).$$

Equation (31) describes the general scaling law for both of the preceding specific cases.

First, we express T_{lim}, R_{lim}, and υ_{lim} as a function of the initial radius R_0 and velocity V_{0sC} of implosion. The value of N_{lim} depends on the properties of the central plasma. Note that R_{lim} is related to N_{lim} as well [see Eq. (21a)]. Then Eq. (31) becomes:

$$N_n = A_1 V_{0sC}^{43\alpha/3(4-3\alpha)} R_0^{43(1-\alpha)/3(4-3\alpha)} N_{lim}^{(19-25\alpha)/3(4-3\alpha)}$$

$$\cdot \exp\left[-\frac{B_1}{\upsilon_{0sC}^{2\alpha/(12-9\alpha)}(R_0 N_{lim})^{\frac{2(1-\alpha)}{12-9\alpha}}}\right], \tag{32}$$

where

$$A_1 = 5\frac{A}{B}\frac{C_2^{(7\alpha-5)/3\alpha}}{C_1^{1/3}},$$

$$B_1 = B \frac{C_2^{2(1-\alpha)/3\alpha}}{C_1^{1/3}},$$

$$C_1 = \frac{2(\gamma-1)M}{(\gamma+1)^{2K}},$$

and

$$C_2 = \left\{ \frac{M^2}{2\pi Z^2 e^4 \ln\Lambda} \left| \frac{4}{(\gamma+1)^2} + 1 \right| \frac{\gamma-1}{\gamma+1} \right\}^{\alpha/(4-3\alpha)}.$$

For deuterium mass with $\gamma = 5/3$, $A = 1.3 \times 10^{-16}$ a/b (cgs), and $B_1 = 22$ b (cgs). It is convenient to express V in units of 10^7 cm/s and n in units of 3×10^{19} atm/cm^3.

Note that:

$$N_n = 10^{26} \frac{a}{b} v_{0sC_u}^{5.1} R_0^{2.3} N_{lim_u}^{0.3} \exp\left[-\frac{3.9 \times 10^{-3} b}{v_{0sC_u}^{0.24} \left(R_0 N_{lim_u}\right)^{0.11}} \right], \quad (33)$$

where the index u refers to the practical units. For the sake of simplicity, we assume in Eq. (33) that the exponent is >1, and then $P\left(T_{lim}^{1/3}/B\right) \approx 1$.

5.4.1. Shock Implosion and Re-Explosion in the Central Plasma. By introducing Eq. (19a) into Eq. (33), we obtain:

$$N_n = 2 \times 10^{21} \frac{a}{b} \frac{E_{LC_u}^{1.13} N_{lim_u}^{0.3}}{\left(N_{0_u} \Delta \tau_{LC_u}\right)^{0.57}} \exp\left[-\frac{6.4 \times 10^{-3} b \left(N_{0_u} \Delta \tau_{LC_u}\right)^{0.026}}{E_{LC_u}^{0.05} \left(N_{lim_u}\right)^{0.11}} \right]. \quad (34)$$

In this scheme, the central plasma is produced in the unperturbed gas, and the implosion occurs before the expansion takes place. Hence, in Eq. (34), N_{lim} should be equal to N_0. Nevertheless, during the strong implosion, an adiabatic compression of the plasma occurs ahead of the converging front, as observed in Ref. 4. Maximum compression is given by

$$\zeta_{lim} = \frac{N_{lim}}{N_0} = \left(\frac{R_{fl}}{R_{lim}}\right)^3,$$

where R_{fl} and R_{lim} are given by Eqs. (4) and (21a), respectively. Note that N_{lim} in Eq. (21a) corresponds to the density N_0 ahead of the shock front at the instant of collapse. In practical units,

$$\zeta_{lim} = 5.3 \times 10^{-3} \frac{E_{LC_u}^{0.79}}{E_{LC_u}^{0.95}} \Delta \tau_{L_u}^{0.32} \Delta \tau_{LC_u}^{0.47} N_{0_u}^{0.43}. \tag{35}$$

The final scaling law then reads

$$N_n = 2 \times 10^{21} \frac{a}{b} \frac{E_{LC_u}^{1.13} \zeta_{lim}^{0.3}}{N_{0_u}^{0.25} \Delta \tau_{LC_u}^{0.57}} \exp\left[-\frac{6.4 \times 10^{-3} b \left(\Delta \tau_{LC_u}\right)^{0.026}}{E_{LC_u}^{0.05} \left(\zeta_{lim}\right)^{0.11} N_{0_u}^{0.08}}\right]. \tag{36}$$

5.4.2. Reflection of Explosive Shock Wave at Converging Shock Front. Neutron yield is obtained from Eq. (33) by replacing V_{0sC} and R_0 with V_{0I} and R_1, respectively. From Eq. (31), the neutron yield formula remains the same as in the previous scheme, except for N_{lim}, which is the fireball density at the total time of shock wave implosion t_I (time to reach R_1 and then to implode). It is evident that reflection must occur quite close to the radius of the central fireball, which corresponds to the first instance of a well-formed explosive shock wave, to have a small decrease in the central density. Also, $N(t_I)_{lim}$ is given by Eq. (14) with $R_{sE}(t_I)$. Actually, if the implosion time after reflection is very short compared with the time to reach R_1, by using Eqs. (9) and (12), $N(t_I)_{lim}$ simplifies into:

$$N_{lim}(\tau_I) = \frac{\gamma+1}{\gamma-1} N_0 \left(\frac{R_{fL}}{R_1}\right)^{63/20}. \tag{37}$$

Assuming a negligible variation of N_{lim}, the plasma is further compressed by a converging shock wave by a factor of $\gamma(\gamma+1)/(\gamma-1)^2$. Moreover, the adiabatic compression ratio now becomes:

$$\zeta_{lim} = \frac{N(\tau_I)_{lim_u}}{N_0} = 4\left(\frac{R_1}{R_{lim}}\right)^3 \left(\frac{R_{fL}}{R_1}\right)^{63/20} \approx 4\left(\frac{R_{fL}}{R_{lim}}\right)^3,$$

which leads to an additional factor of 4. We then find:

$$N_n = 6.4 \times 10^{21} \frac{a}{b} \frac{E_{LC_u}^{1.13} \zeta_{lim}^{0.3}}{N_{0_u}^{0.25} \Delta \tau_{LC_u}^{0.57}} \exp\left[-\frac{4.3 \times 10^{-3} b \left(\Delta \tau_{LC_u}\right)^{0.026}}{E_{LC_u}^{0.05} N_{0_u}^{0.08} \left(\zeta_{lim}\right)^{0.11}}\right]. \tag{38}$$

Additional compression gives a noticeable increase in nuclear yields especially because of a reduction of the exponential argument. Referring to Eq. (31), one sees that as the exponential converges to 1, the value of $P(T_{lim}^{1/3}/B)$ approaches 326. Therefore, the shock wave reflection by the implosive front can increase the neutron production by several orders of magnitude compared with the simple compression of the central plasma.

5.5. Results and Comparison with Experiment

We tested the model against the experimental results given in Ref. 4. Central and peripheral shock waves are produced by capacitor bank discharges whose duration is relatively long. Thus, in principle, an electron thermal wave does not constitute the main mechanism of plasma expansion during the energy deposition time. Nevertheless, the approach of Sec. 5.2 can be retained by taking into account that the detachment of well-formed shock waves occurs at a time much longer than the energy deposition time. In such a case, the energy transferred to the shock waves could not be equal to the energy stored in the capacitor banks.

In the experiment of Ref. 4, the central plasma energy E_{Lu} =30 J is deposited in a sphere of radius R_0 = 2 cm filled with deuterium at 8.2 atm. An energy E_{LC} =1 kJ is deposited at the periphery of the gas. The imploding shock wave meets the central plasma after ~ 0.8 μs. During the collapse, ~10^7 neutrons were emitted by the central plasma, whose maximum temperature was ≈0.7 keV. The central plasma time evolution was measured by recording the plasma luminosity with an image converter camera. After collapse, the luminous front of the central plasma had a roughly constant radius of ~ 8 x 10^{-2} cm for ~ 5 μs. An average density of ~ 3 x 10^{19} ion/cm^3 was inferred from total emitted neutrons, measured temperature, volume, and confinement time of the central plasma.

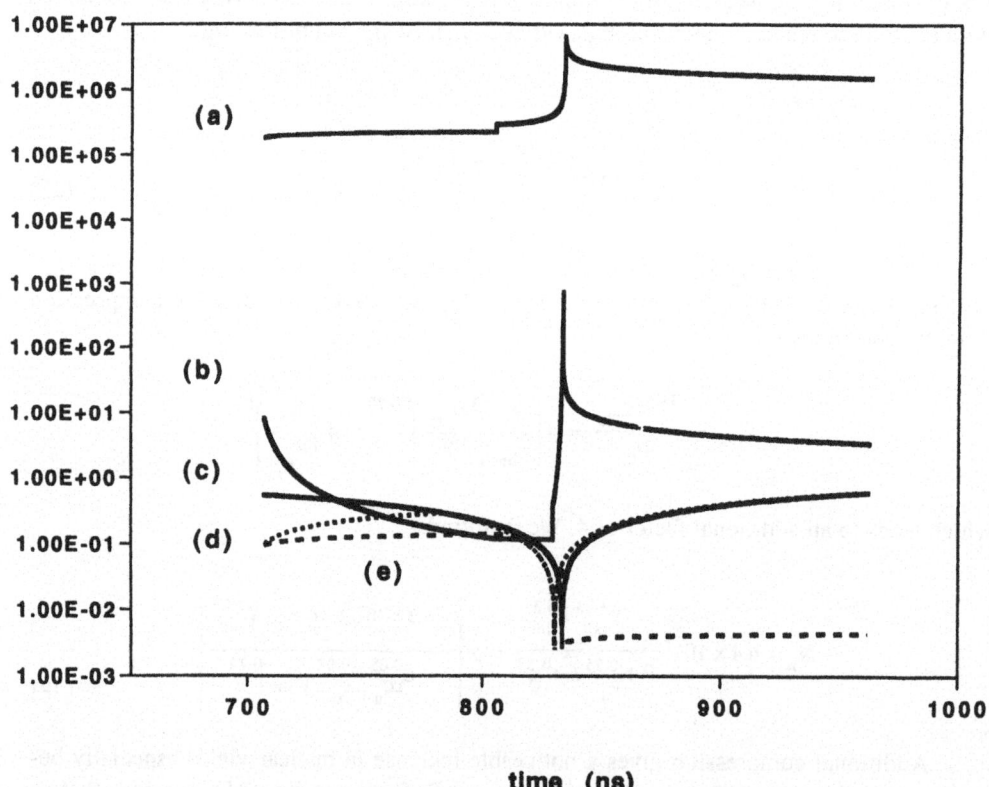

Figure 5. Central shock wave reflection by peripheral converging shock wave under Ref. 4 conditions (see text): (a) central plasma temperature (K), (b) central plasma density (atm), (c) radius of converging shock front (cm), (d) radius of central shock front (cm), and (e) radius of central fireball (cm).

Figure 5 shows the time evolution of the relevant quantities - namely, density, temperature, position of the implosive shock front, central shock wave, and central fireball - in a time interval around the collapse, obtained by the model under the conditions of Ref. 4. A vertical logarithmic scale has been chosen to display all quantities simultaneously. According to measurements in Ref. 4, the energy transferred to the central shock wave is ~ 3 J. At collapse, the temperature and density reach the values of 0.8 keV and 10^3 atm, respectively, for a minimum radius $R_{lim} = 2.5 \times 10^{-3}$ cm. After the collapse, the central plasma is quite stable for ~ 3 μs, neglecting effects due to further implosions. The average temperature and density in this phase are ~ 0.4 keV and 1 atm, respectively. The total neutron number is $N_n = 1.5 \times 10^7$.

The radius of the central plasma, measured by the position of the luminous plasma front, does not coincide with the fireball front. Actually, according to Eqs. (10) and (11), the temperature decreases as the radius increases during the re-explosive phase. Then, the luminous front can be quite close to the re-explosive shock front for several hundreds of nanoseconds. Therefore, we can conclude that the model gives a satisfactory description of the experiment in Ref. 4. Moreover, we find that simple compression of the central plasma does not lead to neutron production.[4] In fact, Eq. (36) leads to $N_n = 10^3$.

From the previous example, the scheme based on the central shock wave reflected by a strong converging shock wave introduces several advantages. First, the scheme makes it easier to compress the central plasma because of the reduction of internal pressure due to the shock wave expansion. Moreover, the reflected shock wave provides a first heating and compression (by a factor of 10) of the central plasma, before the final implosion takes place.

To carry out breakeven conditions, we find it worthwhile to introduce the compression factor in Eq. (38), which, for a D-T mixture, leads to

$$N_n = 2 \times 10^{12} \frac{E_{LC_u}^{0.85} E_{LC_u}^{0.24} \Delta \tau_L^{0.1}}{N_{0_u}^{0.13} \Delta \tau_{LC_u}^{0.43}} P\left(\frac{T_{lim}^{1/3}}{B}\right) \exp\left(-\frac{7.3 E_{LC_u}^{0.05}}{E_{L_u}^{0.09} N_{0_u}^{0.13} \Delta \tau_{L_u}^{0.04} \Delta \tau_{LC_u}^{0.025}}\right). \quad (39)$$

Equation (39) yields maxima as a function of E_{LC_u}, N_{0_u}, and $\Delta \tau_{LC_u}$ very far from achievable values on the laboratory scale. Then, neutron yields increase as E_{LC_u}, N_{0_u}, and $\Delta \tau_{LC_u}$ increase. Equation (39) emphasizes the role of the central plasma. In fact, N_n increases as E_{L_u} and $\Delta \tau_{LC_u}$ increase, provided that the pressure of converging shock wave overcomes the explosive pressure at the instant of reflection; i.e.,

$$V_{OsC}\left(\frac{R_0}{R_1}\right)^{(1-\alpha)/\alpha} > \frac{4}{5}\frac{\gamma-1}{\gamma+1}\left(\frac{E_L}{MN_0}\right)^{1/2} R_1^{-3/2}.$$

As an example, we assume that an energy $E_{LC_u} = 300$ kJ is deposited in 0.5 μs at the periphery of a sphere of radius $R_0 = 2$ cm, filled with D - T at 700 atm. An energy E_{LC_u} is absorbed by the central plasma in 0.1 μs. Fig. 6 shows the behaviour of the relevant parameters during the collapse. The maximum density and temperature are 8×10^5 atm and 7 keV, respectively. The total neutron number is $N_n = 10^{16}$. Simple compression of the central plasma, which is similar to the implosion of pellets, should give $N_n = 5 \times 10^{14}$ [Eq. (36)].

To emphasize the role of the central shock wave, by assuming $E_{L_u} = 10$ kJ and deposition time 0.5 μs in the previous example, we find $N_n = 6 \times 10^{17}$, corresponding to a fu-

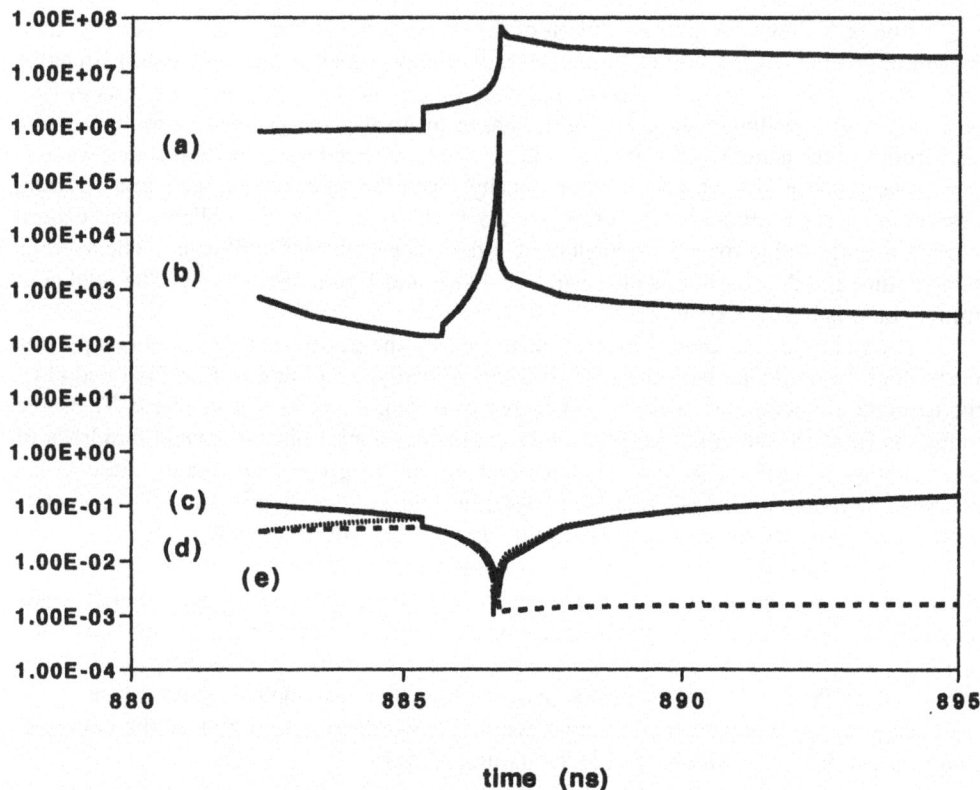

Figure 6. Central shock wave reflection by peripheral converging shock wave under Ref. 23 conditions (see text): (a) central plasma temperature (K), (b) central plasma density (atm), (c) radius of converging shock front (cm), (d) radius of central shock front (cm), and (e) radius of central fireball (cm).

sion energy of ~1.2 MJ. Note that the reflection condition for the central shock wave leads to $E_{L_u} \to 10^9$ J with the current data.

Because of the good agreement between the results of our model and the experimental data reported in Ref. 4, where plasma compression occurs in a quasi-spherical geometry, we can conclude that geometrical instabilities and dissipative phenomena related to the electron thermal conduction[23] play a secondary role in the spherical pinch experiment. Moreover, the scaling law [Eq. (39)] seems promising to achieve breakeven conditions at relatively low energy values.

6. NUMERICAL COMPARISON BETWEEN THE TWO MODELS: INERTIAL CONFINEMENT FUSION (ICF) AND THE SPHERICAL PINCH (SP)

The spherical pinch is clearly a variation of the classical inertial confinement fusion scheme, and therefore it is important to study the differences and the advantages, if any, of one scheme over the other.

In the following sections, a detailed computational comparison of the inertial confinement model and the spherical pinch is provided in terms of density, pressure, temperature, confinement time, total accumulated number of neutrons, and time-resolved neutron flux from reactions in deuterium-tritium mixture. It is shown that temperature, confinement time, and total accumulated number of neutrons for the spherical pinch improve upon the classical inertial confinement fusion scheme.

6.1. Premise

Controlled fusion research falls into two general categories: magnetic confinement[24] and inertial confinement[25-26]. Magnetic confinement fusion uses high-intensity magnetic fields to confine a relatively low-density (10^{14} cm^{-3}) plasma for times on the order of seconds. Inertial confinement fusion depends on rapid heating and compression of fusion fuel contained in a spherical target (such as a polymer or glass shell) of a few millimetres diameter to ultrahigh densities. There are four intercorrelated processes in inertial confinement fusion: a) energy deposition (driver); b) energy transport; c) high compression; and d) nuclear fusion. The key requirements for inertial fusion are the simultaneous achievement of high density and high temperature and the energy confinement time. Temperatures of the order of 8 keV and neutron yields of the order 10^{13} have been obtained with inertial confinement fusion in a deuterium-tritium mixture. However, the plasma density has reached only about 0.2 g/cm^3 for initial gas fill densities of about 0.001 g/cm^3, much below the densities required, which are of the order of 200 g/cm^3 for efficient thermonuclear burn ($\rho R \approx 3$ g/cm^2) in small targets [17a, 27-28].

Attempts to improve the system have led to the idea of the spherical pinch. This is therefore an outgrowth of the inertial confinement model[2,4]. The salient feature of the spherical pinch is the creation of a hot plasma in the center of a sphere. This plasma is then compressed by strong imploding shock waves launched from the periphery of the vessel, which are the main feature of all inertial confinement schemes.

The spherical pinch belongs to the inertial confinement fusion category for the following reasons: a) the geometry (spherical); b) the rapid supersonic shock wave phenomena; c) the imploding shock, which here plays the role of a driver as in the inertial confinement in the above mentioned four intercorrelated processes.[27] The central plasma in the spherical pinch performs the same role as the laser-imploded plasma or beam-imploded plasma, except that it is preformed, and is not a product of the shock wave collapse at the center of the vessel. In order to distinguish the spherical pinch from the magnetic confinement pinch model, the term ICF-Spherical Pinch is sometime used.

The physical models of inertial confinement fusion and the spherical pinch can be described in schematic form. Figs. 7a and b illustrate the salient features of the ICF and SP, respectively. One notices that the inertial confinement fusion scheme exploits the pressure amplification due to shock area convergence effect in order to raise the final plasma temperature to very high values. By contrast, in the spherical pinch, the large pressure amplification due to shock area convergence is now used to act upon a preformed rarefied hot plasma, which is compressed by the imploding shock waves, thus raising its temperature further. The containment time can now be adjusted according to a defined scaling law (see Sec. 4 and 5), for breakeven fusion conditions.

The present work is intended to give a detailed comparison between the ICF and the SP schemes. In Sec. 6.2, numerical modelling and simulation of the ICF and ICF-SP will be given. The details of numerical experiments to compare the two physical models will be given in section 6.3. Some conclusions will be reported in section 6.4.

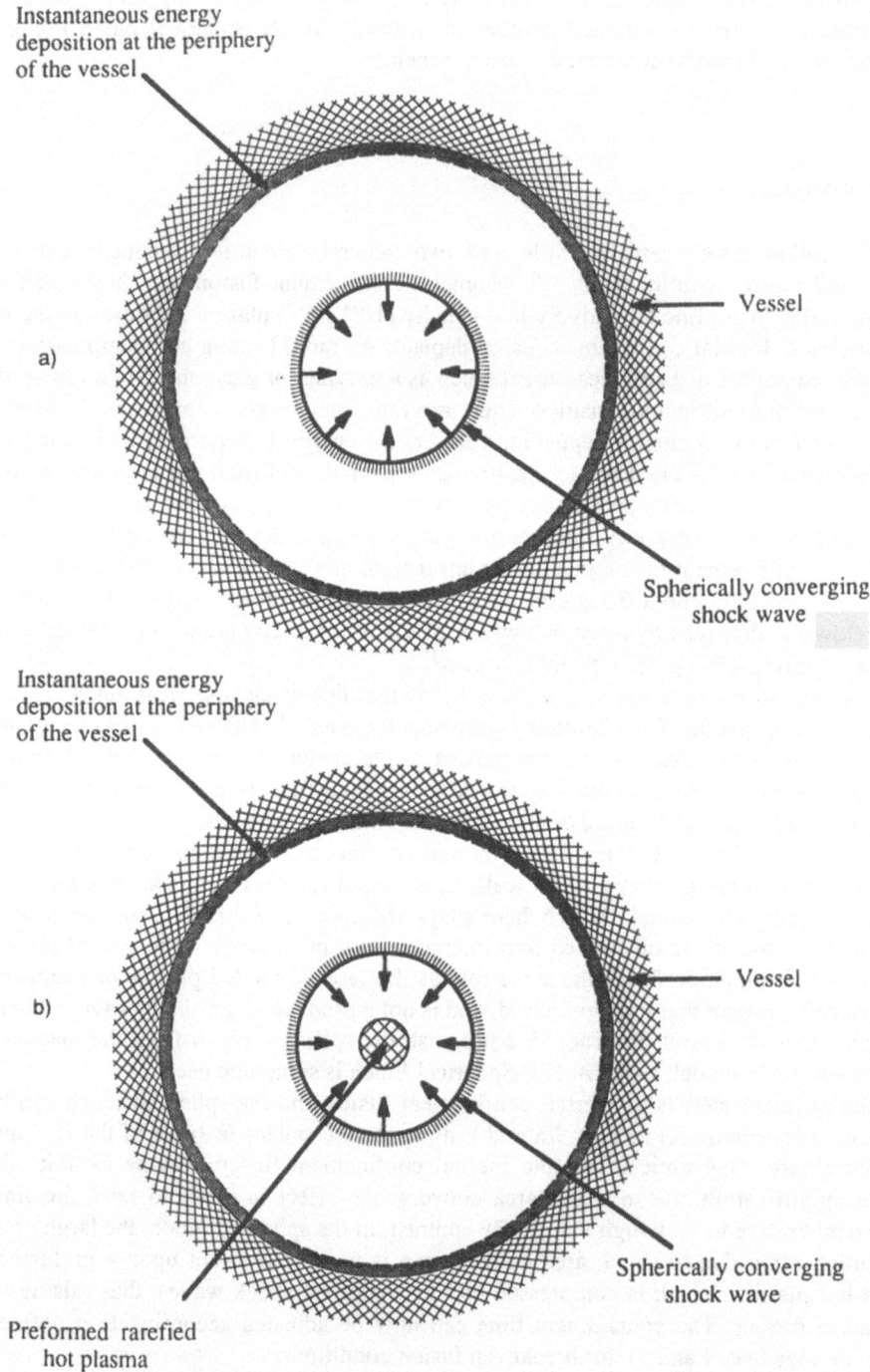

Figure 7. (a) Schematic configuration of ICF; (b) Schematic configuration of ICF- SP.

Table 1. Physical process of the ICF-SP

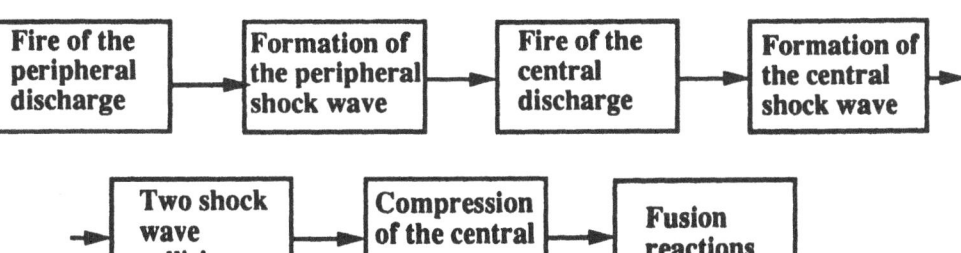

6.2. Numerical Modelling and Simulation of the ICF-SP

6.2.1. Physical Phenomena and Model. The physical process and model of the ICF-Spherical Pinch is described in schematic form in Table 1, and the phases of numerical modelling are described in schematic form in Table 2. The process is described in the next few sections.

6.2.2. Governing Equations. For completeness, the governing equations of the physical process of the spherical pinch are listed in this section.

If we define the Lagrangian mass coordinate in spherical geometry by

$$dm = 4\pi r^2 \rho dr,$$

Table 2. Phases of numerical modeling

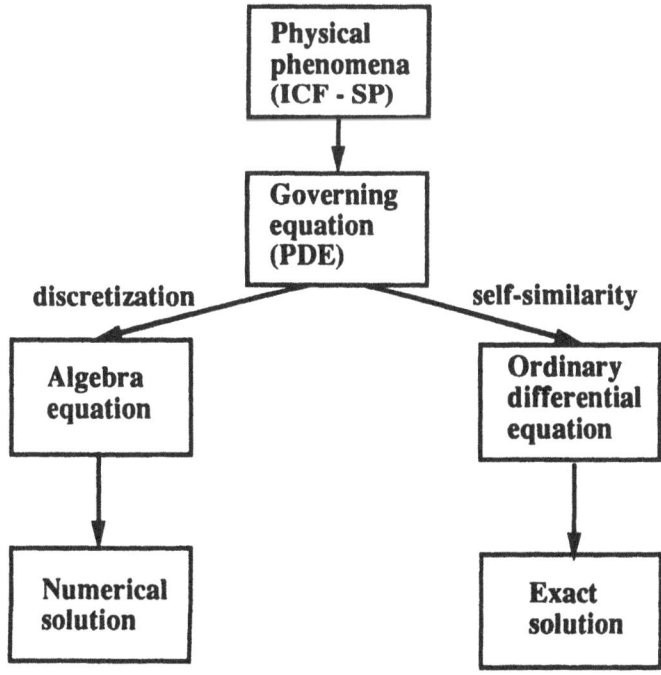

then the Lagrangian form of the equation of continuity can be written as

$$\frac{\partial V}{\partial t} = \frac{\partial}{\partial m}(4\pi r^2 u), \qquad (40)$$

where $V = 1/\rho$ is the specific volume, u is the r component of the velocity of the mass, and the equation of motion becomes

$$\frac{\partial u}{\partial t} = -4\pi r^2 \frac{\partial p}{\partial m}, \qquad (41)$$

where p is the pressure.

Usually, we omit the 4π factor and define the dependent variables per steradian. Then the Lagrangian form of the hydrodynamic equations in a spherical geometry become

$$\frac{\partial V}{\partial t} = \frac{\partial}{m}(r^2 u) \qquad (42)$$

$$\partial \frac{\partial u}{\partial t} = -r^2 \frac{\partial p}{\partial m}. \qquad (43)$$

The temperature equation is simply the one dimensional planar heat conduction equation

$$\frac{\partial T}{\partial t} = \frac{\partial}{\partial x}\left(k(T)\frac{\partial T}{\partial x}\right) \qquad (44)$$

and the nonlinear electron thermal conduction is $k(T) = K_{sp}T^{5/2}$, where K_{sp} is a constant given by Spitzer[29].

Equations (42), (43), and (44) are the governing equations in Lagrangian form of one dimensional (spherical symmetry), hydrodynamic, heat transfer, and one fluid plasma model. The artificial viscosity q (Von Neunman) is implemented to deal with the shock discontinuity

$$q = \begin{cases} \frac{1}{v}\left[2^{1/2}\ m\frac{\partial u}{\partial m}\right]^2, & \frac{\partial V}{\partial t} < 0 \text{ (compression)} \\ 0 & \frac{\partial V}{\partial t} \geq 0 \text{ (expansion)} \end{cases}$$

The importance of the Von Neunman q is based on the fact that it dissipates energy in a shock to a few surrounding finite difference zones while preserving the Rankine-Hugoniot relations across the shock. The essential features of the shock are preserved while the gradients across the shock are reduced to values that allow their treatment by finite difference methods.

6.2.3. Computational Tools and Numerical Methods. Originally developed to simulate laser driven ICF pellet implosion, a one-dimensional, spherical symmetry, Lagrangian hydrodynamic and heat transfer code is implemented for the spherical pinch configuration. The external wall is treated as adiabatic, but bremsstrahlung cooling of the plasma is accounted for (though in the assumption that the plasma is optically thin, i.e., all emitted light escapes and none is reabsorbed). Since we are only concerned with the plasma at the time when it is hot (fractions of keV and above) and at such temperature hydrogen is essentially stripped, it is felt that under the circumstances neglect of line radiation is fully justified. Computational domains consist of 3 regions (Fig. 8). Region 1 is the central part of the sphere, region 2 is the outer periphery of the sphere, while region 3 is the region between region 1 and region 2. Region 1 is designed for depositing the central energy, region 2 is designed for depositing the peripheral energy. The sizes of the region 1 and region 2 are set by a priori energy densities,

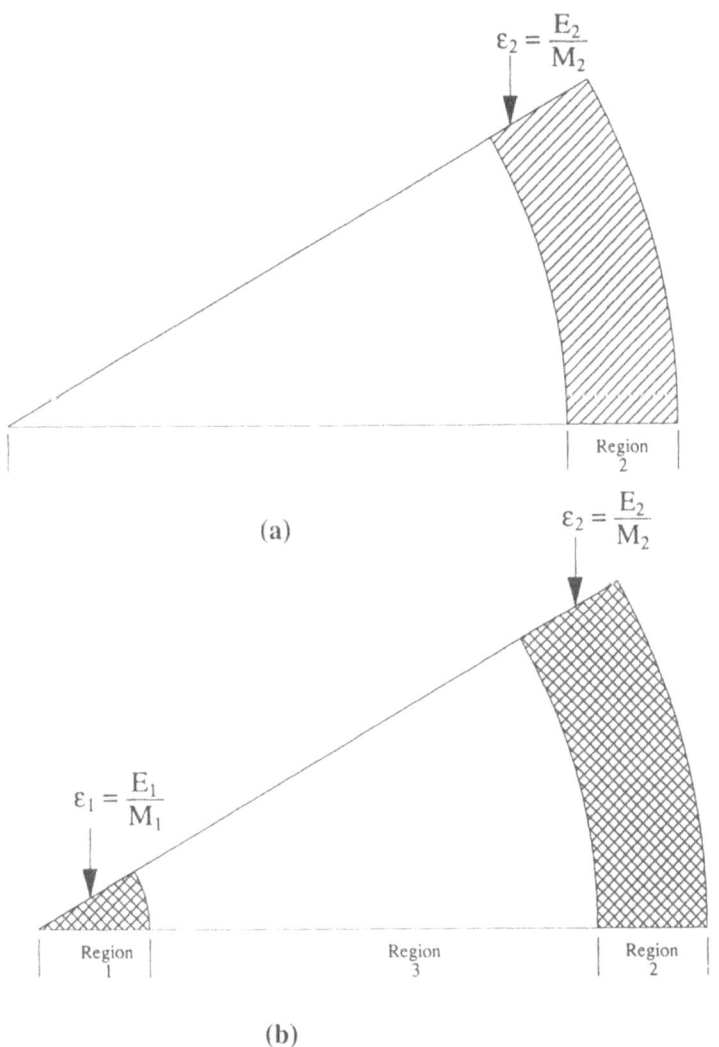

Figure 8. (a) Computational domain of the ICF; (b) Computational domain of the ICF-SP.

$$\varepsilon_1 = \frac{E_1}{M_1}$$

and

$$\varepsilon_2 = \frac{E_2}{M_2},$$

and it is assumed that $\varepsilon_1 = \varepsilon_2 = 10^9$ J/g. This energy density is necessary to have numerically a shock velocity, both in the exploding and imploding case, of the order of 10^7 cm/sec, as found experimentally. The time profile of the energy deposition is taken to be as a squared sinusoid.

6.2.4. Features and Definition of ICF and ICF-SP in the Modelling. The main features of ICF and ICF-SP and the definition of the terms in the present work are given as follows.

As mentioned in the Premise, the main feature of the ICF is the imploding shock wave launched from the periphery of the vessel. The salient feature of the spherical pinch is that, besides the imploding shock waves launched from the periphery, a hot plasma in the center of the sphere is created by a strong explosion. If we let E_1 and E_2 represent the amount of energy being deposited in the center and in the periphery of the vessel, respectively, then in our computations we define the ICF as being the case $E_1 = 0$, $E_2 \neq 0$, and the ICF-SP as being the case $E_1 \neq 0$, $E_2 \neq 0$. For the ICF-SP case, most of the results presented are for the case with $E_1 = 500$ J, and $E_2 = 3 \times 10^5$ J.

A term Tdlay is used to describe the time difference between the central and peripheral discharges. If Tdlay is 0, it means that the central and the peripheral energies are deposited simultaneously. If the sign of the Tdlay is positive, it means that the central energy is deposited first, the peripheral energy is deposited next. If the sign of Tdlay is negative, it means that the peripheral energy is deposited first, the central energy is deposited next.

6.3. Numerical Results

6.3.1. The Structure of the Shock Waves: Explosion within Implosion. The fluid mechanics and high-temperature hydrodynamic phenomena of the spherical pinch can be described briefly as follows: A point explosion shock wave takes place within an implosion of a spherical shock wave. In Oppenheim's apt words,[30–31] the spherical pinch can be simply described as: *Explosion within Implosion.*

A point explosion is described by Zel'dovich and Raizer.[32] They give a similarity solution for a strong point explosion. For implosion of a spherical shock wave, both Landau[33] and Zel'dovich and Raizer[32] give a detailed similarity solution of the phenomena. Both assume an imaginary "spherical piston" as the origin of the wave. The spherical pinch brings this classic imagination into reality—the spherical piston is the imploded spherical plasma acting as a driver. Numerical solutions provide us now with some insight into high-temperature hydrodynamic phenomena of *an explosion within an implosion.*

Numerical computations have been conducted for the above mentioned configurations and parameters. We note that all the unknowns are functions of time t and space r. In order to gain physical insight of the spherical pinch, we plot the solutions of interest as a function of time and space. Figs. 9 to 11 show the structures of the shock wave as a func-

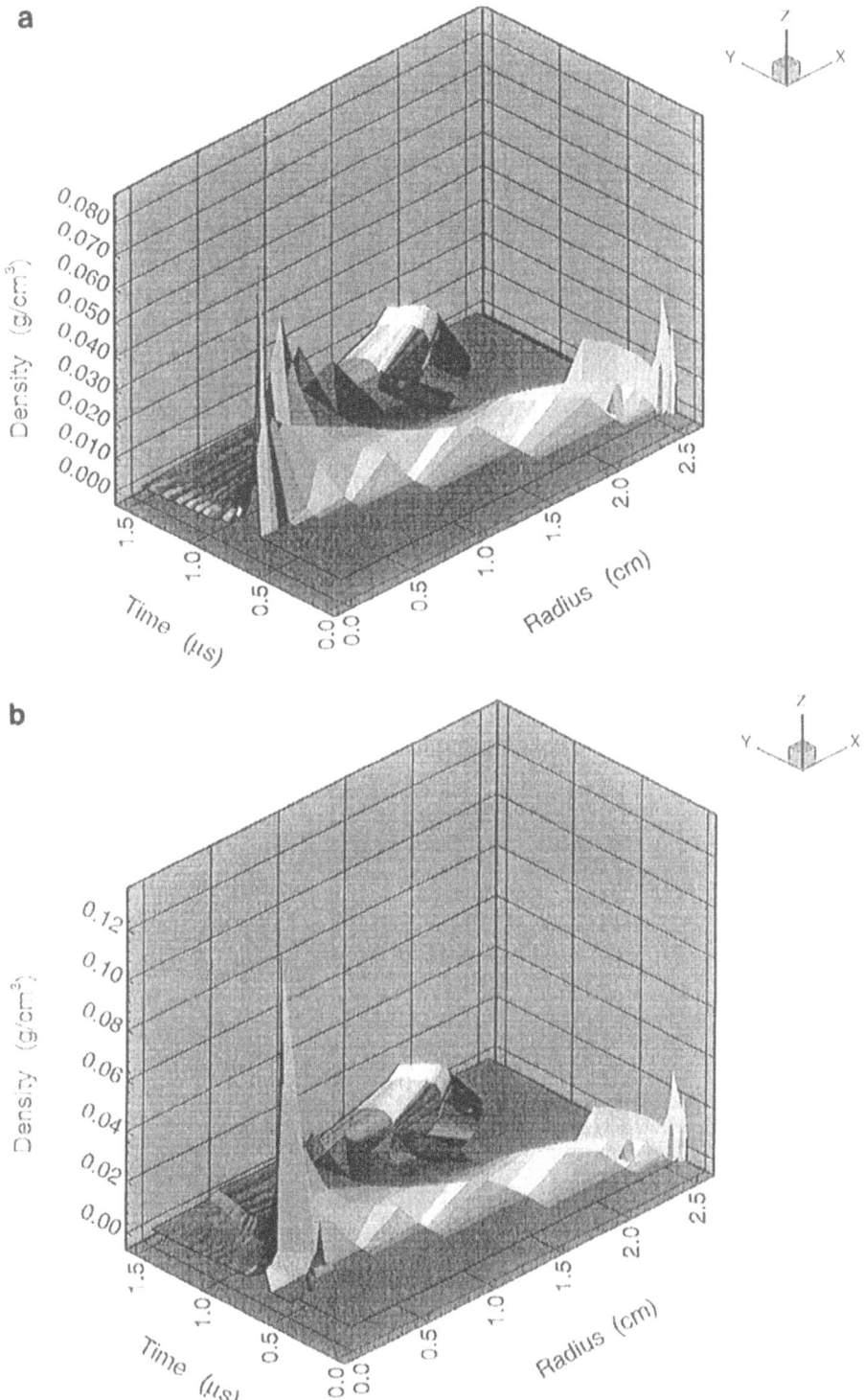

Figure 9. (a) Density profile vs. space and time for the ICF case (r = 2.5 cm, $\rho_0 = 3 \times 10^{-3}$ g/cm3, $E_1 = 0$, $E_2 = 300$ kJ); (b) Density profile vs. space and time for the ICF-SP case (r = 2.5 cm, $\rho_0 = 3 \times 10^{-3}$ g/cm^{-3}, $E_1 = 500$ J, $E_2 = 300$ kJ).

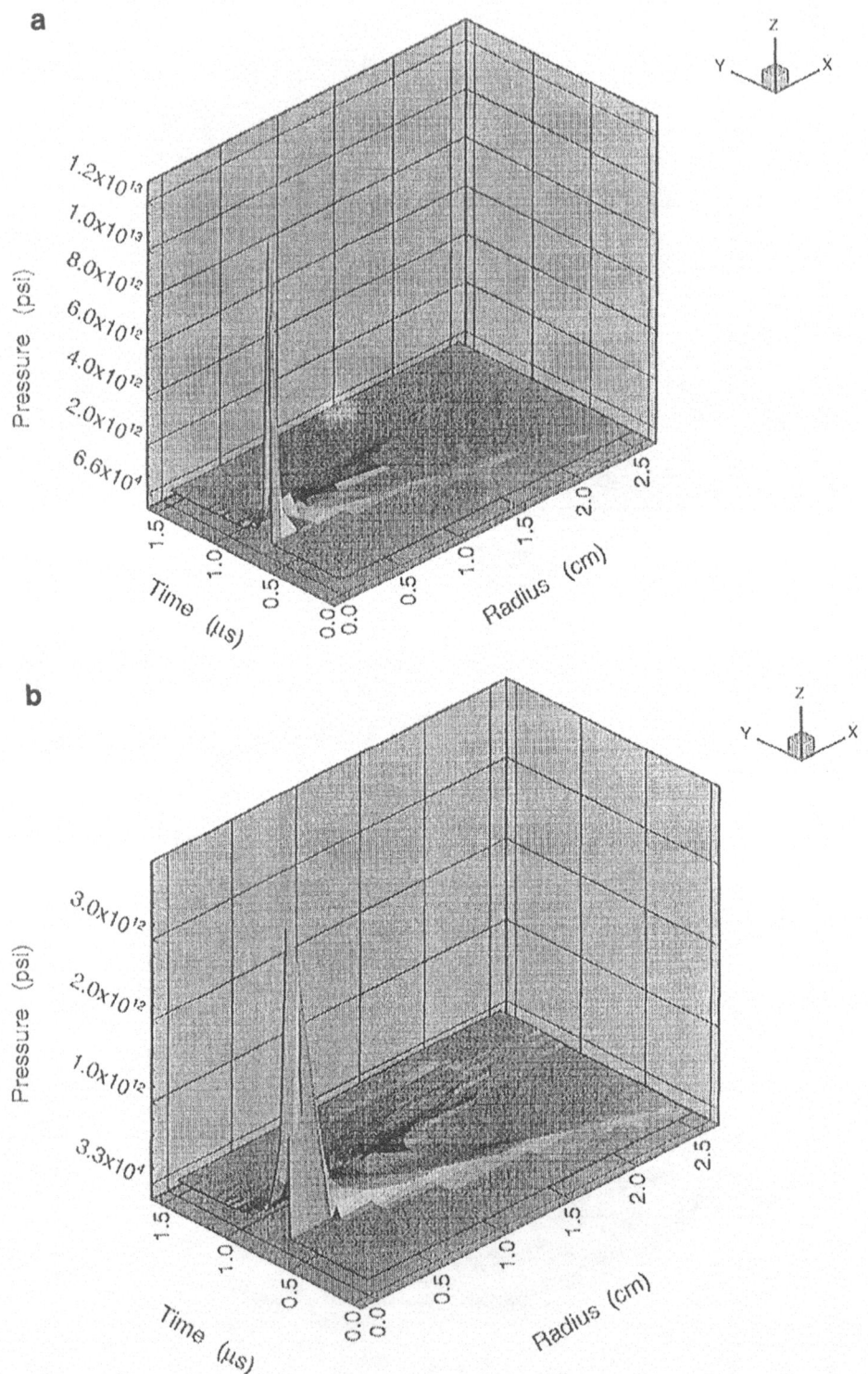

Figure 10. (a) Pressure profile vs. space and time for the ICF case (r = 2.5 cm, $\rho_0 = 3 \times 10^{-3}$ g/cm^3, $E_1 = 0$, $E_2 = 300$ kJ); (b) Pressure profile vs. space and time for the ICF-SP case (r = 2.5 cm, $\rho_0 = 3 \times 10^{-3}$, $E_1 = 500$ J, $E_2 = 300$ kJ).

tion of radius and time in terms of density, pressure and temperature for the ICF (Figs. 9a, 10a, and 11a) and for the ICF-SP (Figs. 9b, 10b, and 11b). Fig. 9a shows the density profiles as a function of radius and time for the ICF. The radius of the vessel is assumed to be r = 2.5 cm and the original gas density $\rho_o = 3 \times 10^{-3}$ g/cm³. The peripheral shock wave starts from r = 2.5 at t = 0. A spherical piston pushes the gas inward[32]. Due to the high speed and high energy, a shock wave front forms (far right corner of Fig. 9a) and travels towards the center radially as time increases. At time $t = 6.5 \times 10^{-7}$ sec, it reaches the center. At this time the density reaches its maximum $\rho = 0.08$ g/cm³ and then decreases as the shock is reflected back towards the periphery. Similar phenomena are observed in the SP case (Fig. 9b) before $t = 4 \times 10^{-7}$ sec. After $t = 4 \times 10^{-7}$ sec, the central discharge (E_2) is fired and a "tiny" shock wave forms. This explosion shock wave collides with the implosion shock wave. From the figures it can be seen that both the density (Fig. 9b) and temperature (Fig. 11b) are higher than in ICF case. Few peculiar facts are to be noted here. The peak density for ICF takes place at the center of the sphere, while the peak density for the ICF-SP takes place away from the center. Moreover, due to the form of the central shock wave, the peak density and peak temperature for ICF-SP are higher than those of the ICF. It is shown that, due to the appearance of the central shock wave, the density ratio increases from

$$\frac{\rho_{peak}}{\rho_0} = \frac{8 \times 10^{-2}}{3 \times 10^{-3}} = 26$$

to

$$\frac{\rho_{peak}}{\rho_0} = \frac{13.5 \times 10^{-2}}{3 \times 10^{-3}} = 45,$$

the peak temperature increases from 500 eV to 650 eV. It was found that the term

$$\int_{t_1}^{t_2} T(r_c, t) dt$$

is much larger for the ICF-SP than the case of ICF, where $t_1 = 0$, $t_2 = 1.5$ μsec and $r_c = 0.034$ cm is the edge of the central plasma zone (i.e., the zone edge for region 1 prior to initiating energy input to the central zone; see Fig. 8). This implies that the temperature is

Table 3. Effects of the delay (Tdlay) between discharges

Tdlay (μsec)	Peak Temp. (ev)	A = # T dt (eV.μsec)	Δ Neutrons	Total Number of Neutrons
-0.2	516	101.34	3.95×10^9	6.0×10^9
-0.3	567	89.64	6.5×10^9	8.6×10^9
-0.35	618	84.90	1.0×10^{10}	1.22×10^{10}
-0.4	699	82.40	5.8×10^9	8.0×10^9
-0.5	608	72.10	2.95×10^9	5.0×10^9
-0.6	724	57.41	2.8×10^8	2.35×10^9

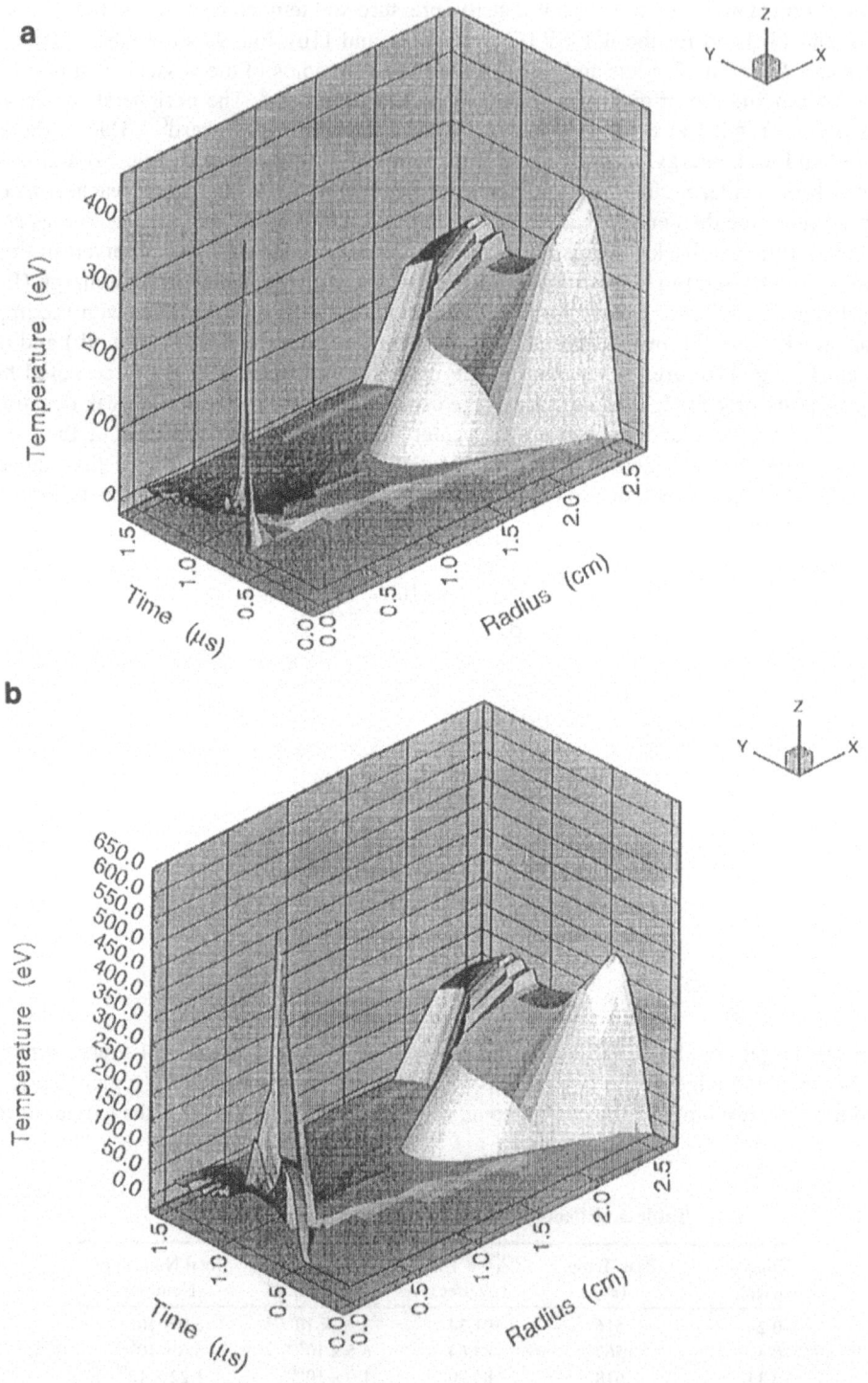

Figure 11. (a) Temperature profile vs. space and time for the ICF case (r = 2.5 cm, $\rho_0 = 3 \times 10^{-3}$ g/cm^3 $E_1 = 0$, $E_2 = 300$ kJ); (b) Temperature profile vs. space and time for the ICF-SP case (r = 2.5 cm, $\rho_0 = 3 \times 10^{-3}$ g/cm^3, $E_1 = 500$ J, $E_2 = 300$ kJ).

maintained in the ICF-SP longer than in the ICF. Since the goal of fusion is to compress the target to high density and to maintain it to high temperatures for as long as possible, it seems that the ICF-SP is an improvement over the ICF scheme.

6.3.2. Detailed Comparison of the Central Plasma Behaviour in the ICF and ICF-SP Cases. The following numerical results refer to a 2.5 cm radius sphere filled with deuterium-tritium at initial density 3×10^{-3} g/cm^3. In both the ICF case and the SP case, $E_2 = 300$ kJ of energy are deposited in the periphery of the spherical vessel in 0.5 μsec with time profile of the energy deposition to be as a squared sinusoid (power $\propto \sin^2(\pi t/\tau_2)$, where τ_2 is the energy deposition time). As defined above, for the ICF case, the central energy deposition is zero ($E_1 = 0$), while for the SP case, $E_1 = 500$ J of energy are deposited in the central plasma also in 0.5 μsec with the same squared sinusoid pulse shape.

After extensive numerical experiments, it was found that the best final plasma conditions in the SP case occur when the exploding and imploding shocks collide at the time when the central plasma reaches its maximum temperature. For this reason, different values for Tdlay time have been used. The results are shown in Table 3.

The neutron burst, Δ Neutrons, is defined as the incremental number of neutrons due to the compression of the central plasma (pinch effect). In Fig. 18, for example, Δ Neutrons can be defined as the sharp jump in total neutrons at t = 0.68 μsec.

Fig. 12 shows the radius of the central plasma (the radius of the edge of the central zone) as a function of time. The radius of this zone is assumed to correspond to the plasma boundary. The dotted curve refers to the ICF case, and the solid line to the SP case. With-

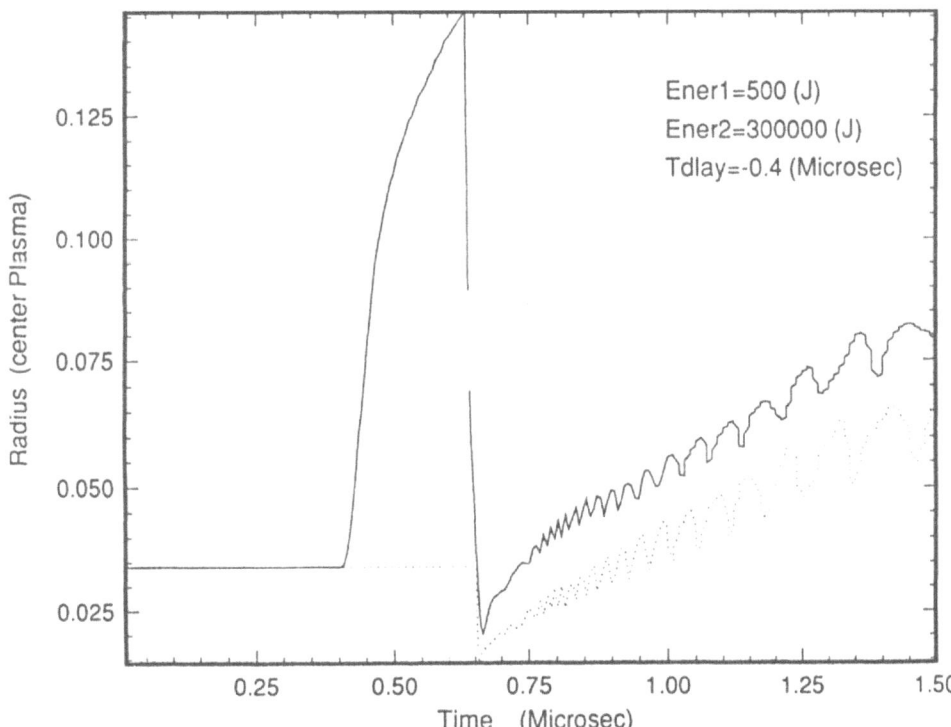

Figure 12. Radius of the central plasma vs. time (dotted line: ICF; solid line: ICF- SP; r = 2.5 cm, r_c = 0.034 cm).

out a central discharge, the radius of central mass remains constant until the shock wave launched from the periphery reach the center. At this time, a compressed central plasma is formed whose radius decreases with time. As the shock wave bounces back, the radius increases. The solid line of Fig. 12 shows that, with the central discharge, the radius of the central plasma increases until the shock wave launched from the periphery collides with the central plasma, which is subsequently compressed. Here we are only interested in the first shock, and not in the time history of successive reflected shocks.

Fig. 13 shows how the velocity of the central plasma changes with time. The positive sign of the velocity represents expansion, i.e., outward movement. The negative sign of the velocity represents compression of the central plasma (inward motion). For the ICF case (dotted line), the velocity of the central mass remains zero until the peripheral shock reaches the center. At this time the central plasma is formed and begins to move inwards. As the shock wave is reflected back, the velocity of the central plasma bounces back and oscillates. In the SP case, unlike the ICF case, the velocity of the central plasma is directed outwards (because the shock wave is generated by the central discharge), until the peripheral shock wave collides with the central region, and the velocity of the central plasma is directed inwards (negative sign). Afterwards, the same oscillation phenomena for the velocity take place as in the ICF case.

Fig. 14 plots the density of the central plasma as a function of time. The peak density is now higher in the ICF case than in the SP case at the center of the sphere. The reason for this is that the central discharge causes the density to fall, and therefore the peripheral shock wave has to compress a lower density plasma to begin with. However,

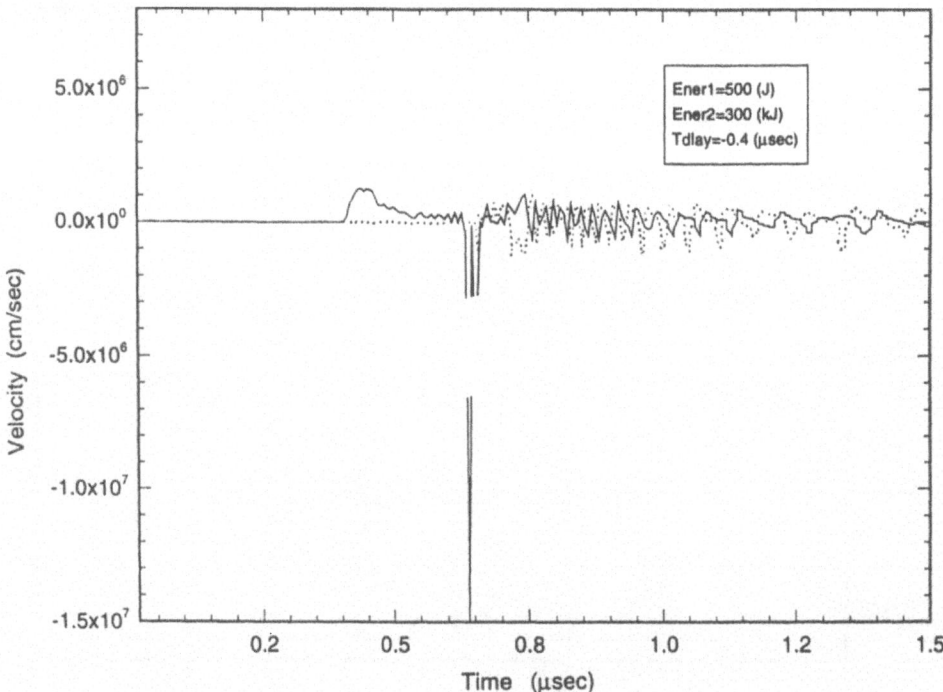

Figure 13. Velocity of the central plasma vs. time (dotted line: ICF; solid line: ICF-SP; r = 2.5 cm, r_c = 0.034 cm).

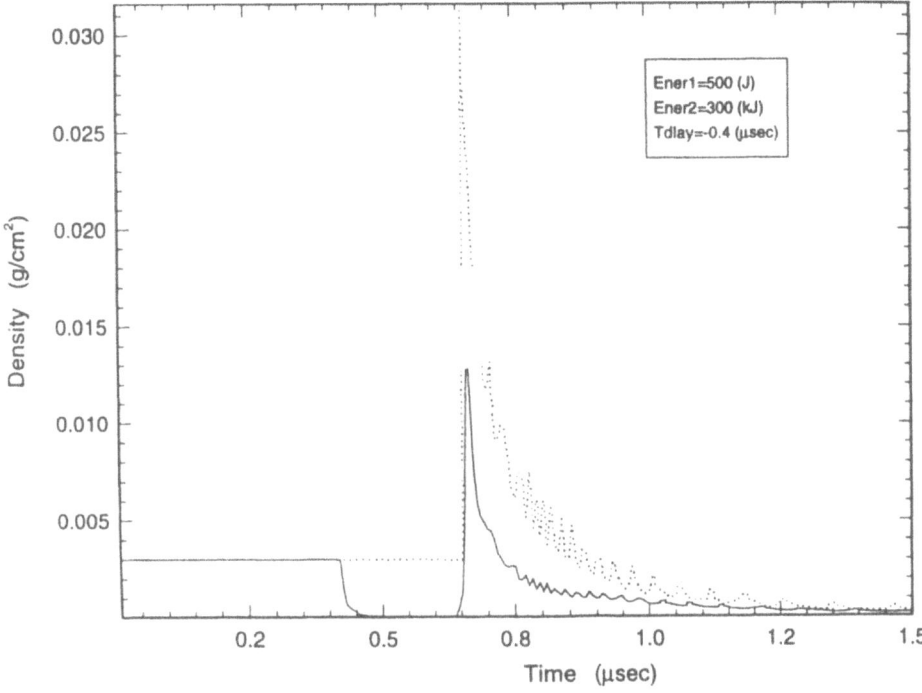

Figure 14. Density of the central plasma vs. time (dotted line: ICF; solid line: ICF- SP; r = 2.5 cm, r_c = 0.034 cm).

we will show later that the central discharge temperature reaches higher values than in the ICF case.

A careful observation shows that the peak density does not occur in the central region of the sphere for the SP case, because the central discharge lowers the density of the central region. This has been demonstrated in Fig. 9 where it was shown that the density of the central plasma for the ICF-SP case is higher than that of the ICF case away from the center of the sphere.

Fig. 15 shows the volume of the central plasma as a function of time for the ICF (dotted line) and the SP (solid line). In the SP case the volume of the central plasma increases due to the central discharge. The expanded volume will be dramatically reduced due to compression by the peripheral shock wave.

Fig. 16 shows the pressure of the central plasma as a function of time. Again, because of the central discharge, the pressure of the central plasma for the SP case is lower than that of the ICF.

The lower density and lower pressure in the centre, however, are compensated by higher temperature and longer containment time. Fig. 17 shows the temperature profile of the central plasma as a function of time for both the ICF (dotted line) and the SP case (solid line). The term

$$\int_0^{1.5} T(r_c, t) dt$$

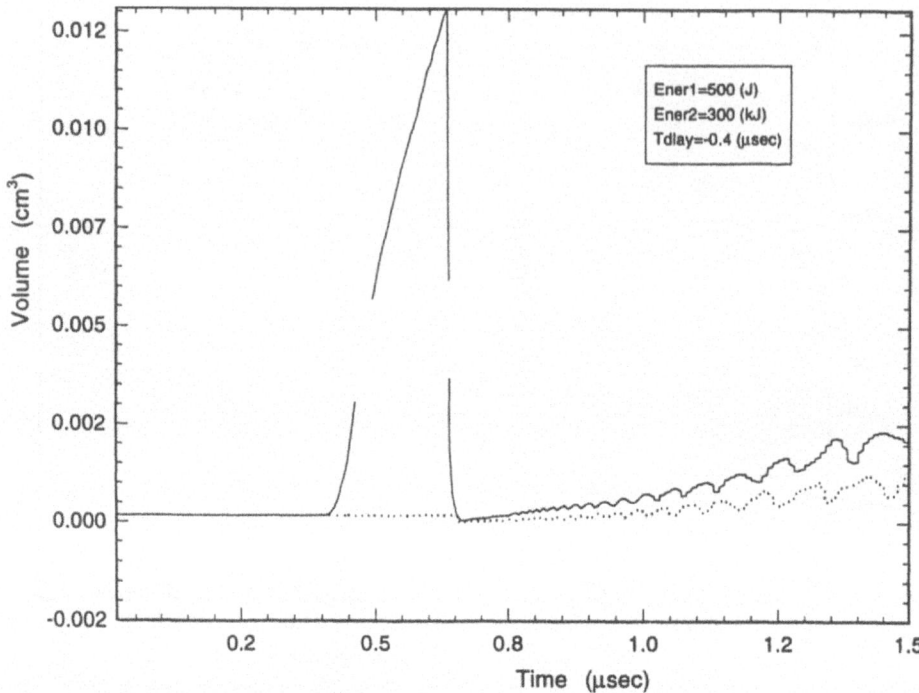

Figure 15. Volume of the central plasma vs. time (dotted line: ICF; solid line: ICF-SP; r = 2.5 cm, r_c = 0.034 cm).

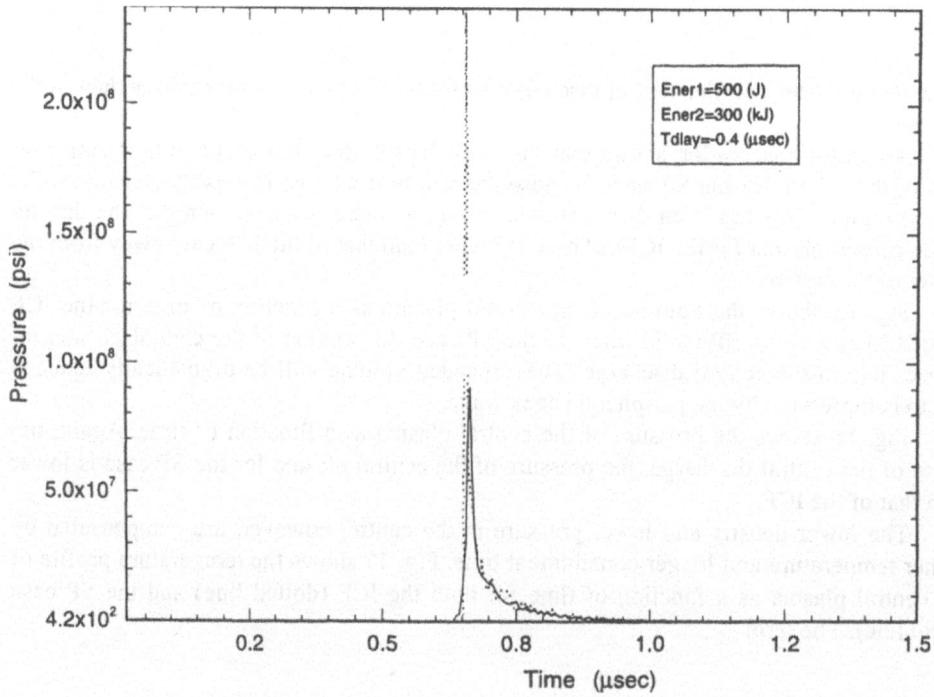

Figure 16. Pressure of the central plasma vs. time (dotted line: ICF; solid line: ICF-SP; r = 2.5 cm, r_c = 0.034 cm).

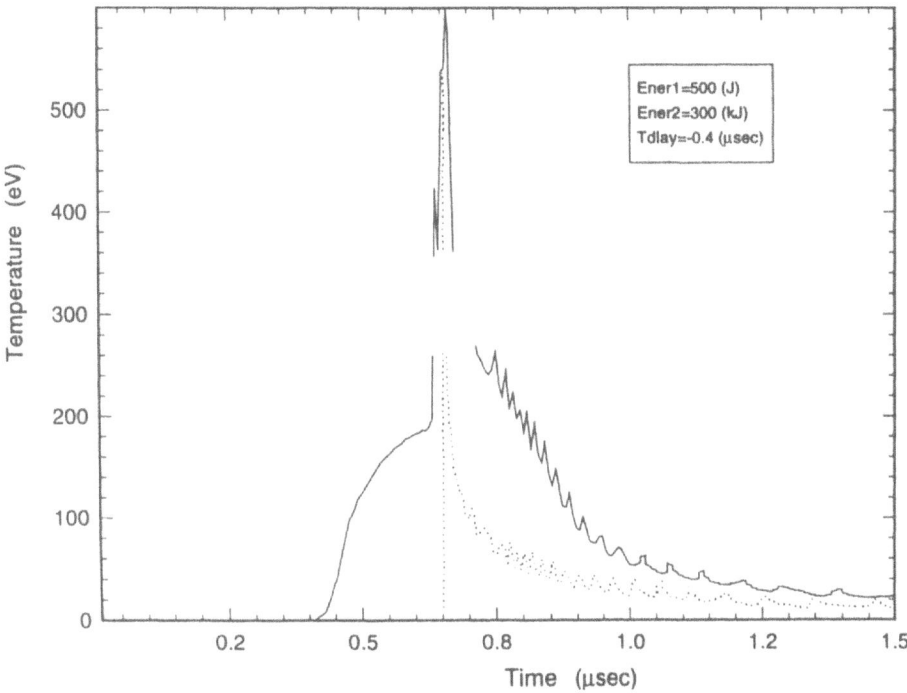

Figure 17. Temperature of the central plasma vs. time (dotted line: ICF; solid line: ICF - SP; r = 2.5 cm, r_c = 0.034 cm).

is now much larger for the SP case than for the ICF case. The introduction of the central discharge indeed leads to an elongation of the sustainment time for the temperature, and also to an increase of the magnitude of the temperature itself. This is the key feature of the spherical pinch scheme.

Neutron production is a deciding factor in the evaluation of a model for fusion. Fig. 18 shows the total number of neutrons emitted as a function of time, as derived from the numerical experiment, for both the ICF (dotted line) and the SP case (solid line). The neutrons emitted before t = 0.68 μsec are due to the periphery reactions or background heating, whereas the neutron burst, Δ Neutrons, is due to the compression of the central plasma (pinch effect). The increase of the total number of neutrons emitted is significant in the spherical pinch case, which demonstrates the advantage of the SP through the simple addition of a central plasma. Fig. 19 gives further evidence of this in terms of neutron flux for the ICF-SP case (solid line) and the ICF case (dotted line) (PWNeut = Neutron flux in neutrons/sec). If we increase the central discharge energy to 2000 J and let r = 5.5 cm and Tdlay = -2.393 μsec, the advantage of the ICF-SP over the ICF in terms of

$$\int_0^3 T(r_c, t) dt$$

(Fig. 20) and Δ Neutron (Fig. 21) is significant. These figures represent the situation where the imploding shock wave begins compressing the central plasma right at the time when the latter, in its formative stage, reaches its maximum temperature. The total accu-

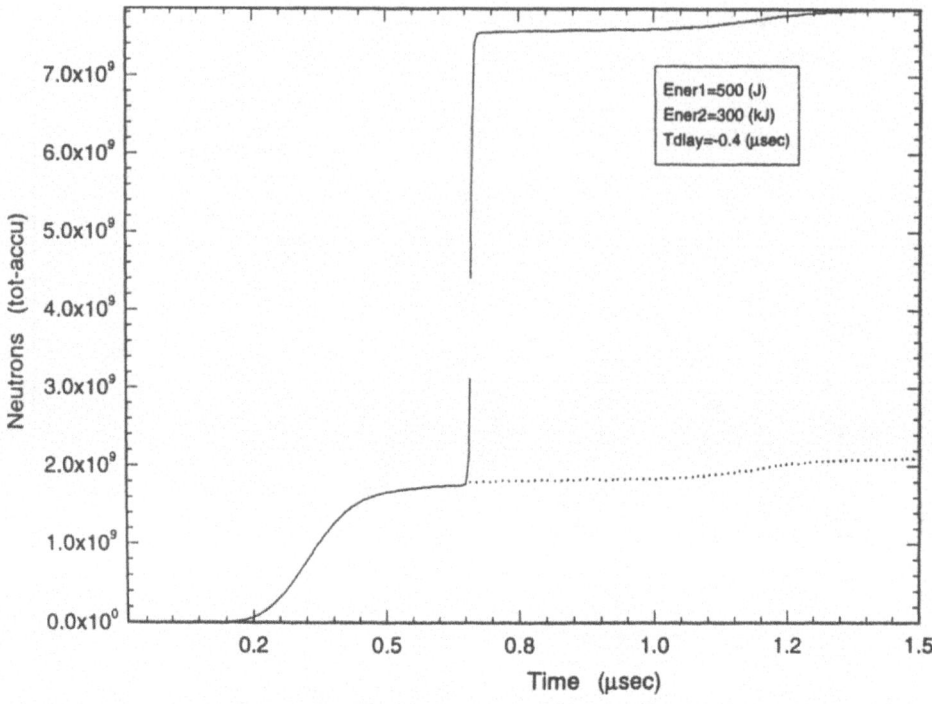

Figure 18. Total numbers of neutrons vs. time (dotted line: ICF; solid line: ICF - SP; r = 2.5 cm).

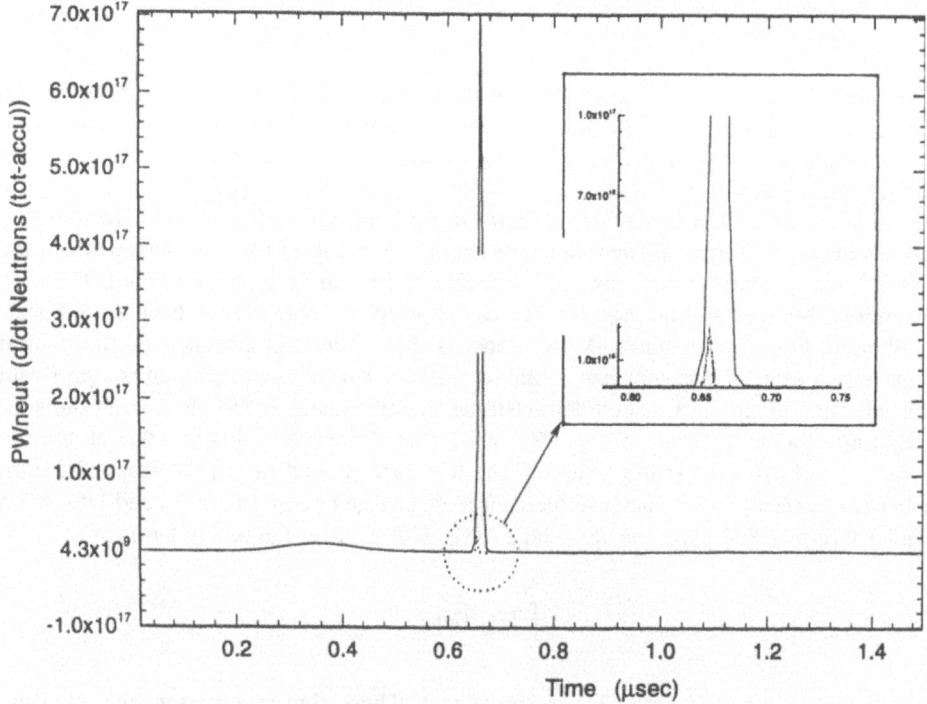

Figure 19. Flux of neutrons vs. time (dotted line: ICF; solid line: ICF - SP;). The insert is a magnified view of the neutron flux (PWNeut) around t = - 0.65 µsec (r = 2.5 cm, r_c = 0.034 cm).

Spherical Pinch Research

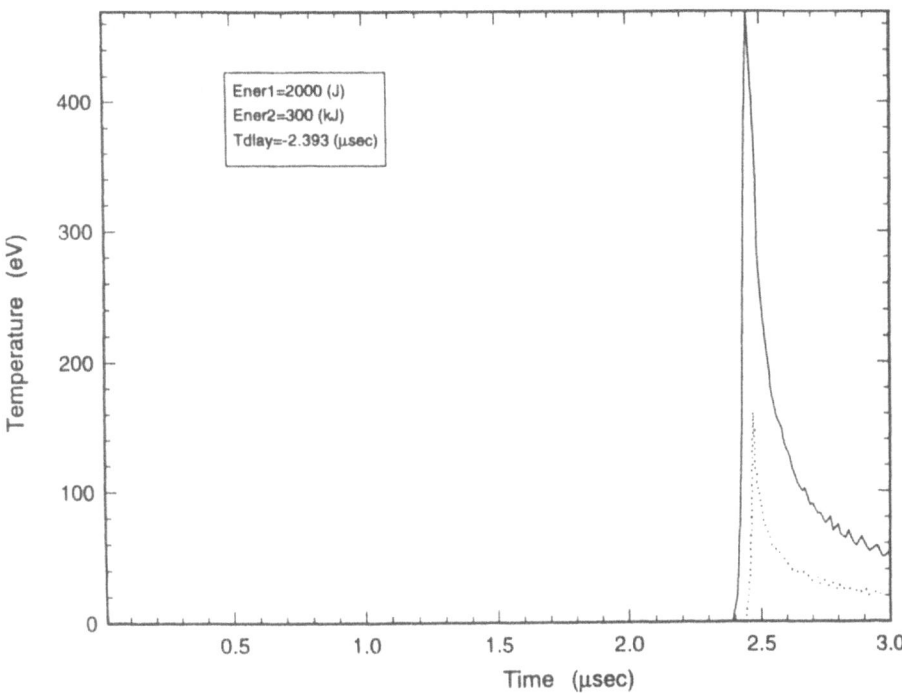

Figure 20. Temperature of the central plasma vs. time (dotted line: ICF; solid line: ICF-SP; r = 5.5 cm, r_c = 0.054 cm).

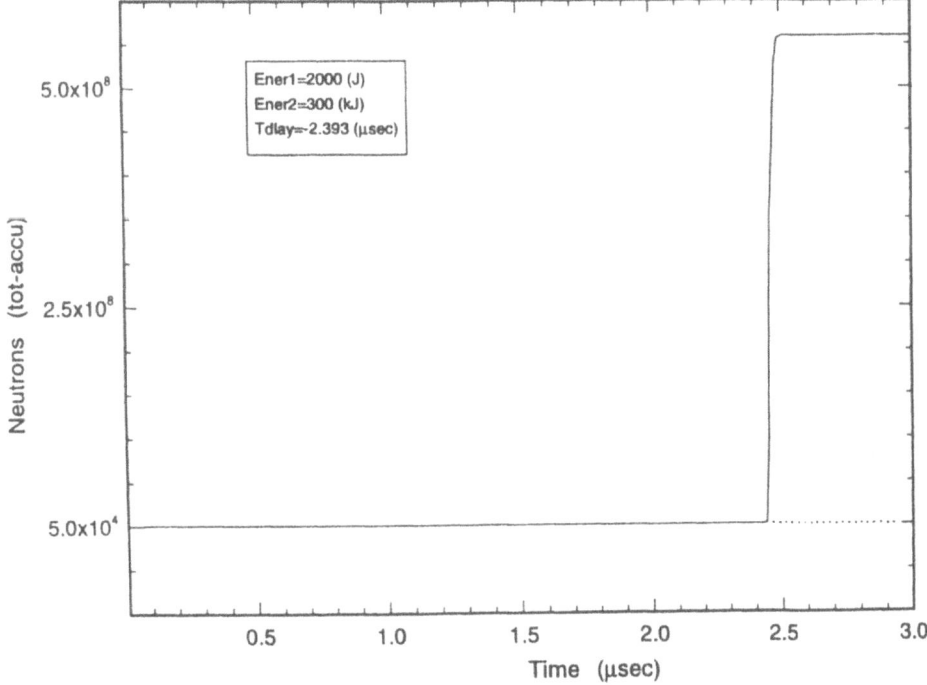

Figure 21. Total number of neutrons vs. time (dotted line: ICF; solid line: ICF-SP; r = 5.5 cm).

mulated number of neutrons is now a few orders of magnitude higher in the SP than in the ICF case. The spherical pinch obviously enhances the mechanism of inertial confinement.

This concludes our modelling of both the ICF and SP, and the report of the results obtained from our numerical experiments.

6.4. Conclusion

We have explained why the spherical pinch should be classified as an inertial confinement fusion scheme, and have given the reason for its acronym: ICF-SP. This scheme has been defined as *an explosion within an implosion,* in order to describe both the fluid mechanics and the high-temperature phenomena inside the ICF-SP vessel. A computational comparison between the inertial confinement model and the spherical pinch model as a neutron generator was undertaken in terms of density, temperature, confinement time, total accumulated number of neutrons, and neutron flux. A detailed investigation of the fluid mechanics and high-temperature phenomena relevant to the implosion alone (ICF), and for an explosion within an implosion (ICF-SP) has been carried out. This analysis revealed the structure and motion of the shock waves inside the vessel. It has been shown that this represents an improvement over the inertial confinement fusion model, because temperature, confinement time, and neutron production are favourable to the spherical pinch.

7. THE SPHERICAL PINCH AS AN INDUSTRIAL TOOL[9]

The foregoing analysis of the Spherical Pinch has indicated its potential as a fusion device. In this section we outline the exploitation of this concept as an industrial tool, with particular emphasis on its potential as a radiation source for applications to microlithography.

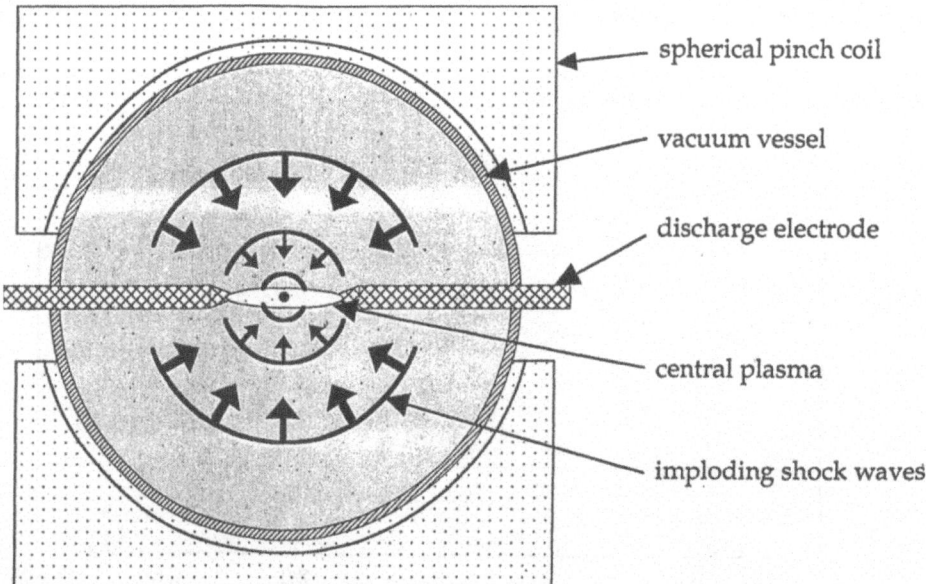

Figure 22. Schematic of the spherical pinch concept showing how the imploding shock waves collapse onto the central plasma.

Table 4. Range of the spherical pinch (SP) operating parameters against the microlithography industry requirements

Parameters	Industry Requirement*	SPX-Ray Generator Output
Source size (mm)	1	1 – 2
Source-to-wafer spacing (cm)	20	20
X-ray energy (keV)	1 – 2	1 – 3
X-ray energy/pulse (J)	1 – 5	2 – 4
Dosage on wafer/pulse (mJ/cm^2)	.2 – 1	.4 – .8
Position reproducibility (mm)	<1	<1
Repetition (Hz)	10 – 100	1 – 10
Reliability (over 10^6 shots)	1–100	?

*IEEE 6th Pulsed Power Conference Proceeding, 29 June – 1 July 1987, Washington, D.C., U.S.A.

7.1. Premise

As mentioned previously, the spherical pinch concept of plasma heating employs strong imploding shock waves to compress a preformed plasma in the center of a spherical vessel. One of the ways to generate imploding waves is by inductively discharging a condenser bank into a coil of spherical shape. As far as the central plasma is concerned, this is generated by discharging a small condenser in the center of the vessel through two suitably shaped electrodes. During compression by the imploding shock waves, this central plasma can be brought to a high temperature, which results in the emission of radiation from visible through UV, through deep UV, up to the soft X-ray region of the spectrum. The amount of radiant emission depends upon the energy imparted to the plasma through the inductive discharge, and the radiation emission is a function of the final temperature of the compressed central plasma.

The schematic configuration of the inductive spherical pinch concept is reported in Sec. 7.2, and its radiation emission over the spectral range in the soft X-ray region is reported in the same section against the microlithography requirement. In Sec. 7.3 the operating parameters of a prototype machine (SPX II) are reported. In Sec. 7.4 this machine is described. Detailed measurements of soft X-ray output from the SPX II and of the physical picture of the imploding waves are reported in Secs. 7.5 and 7.6, respectively. Specifically, in Sec. 7.5 the soft X-ray (0.8 to 10 keV) output is described as a function of the discharge voltage, the gas used (Helium, Neon, and Argon), and the neutral gas pressure. In sec. 7.6, theoretical modeling and measurements of the imploding wave velocities are reported. The comparison of the theoretical predictions and the experimental results in a wide range of the operating parameters is discussed in Sec. 7.7.

7.2. Principle of Operation and Experimental Parameters

Fig. 22 illustrates the basic concept behind the Spherical Pinch. In a thick-walled metal sphere a central plasma is generated as a linear spark by the discharge of electrical energy from a small condenser through two suitably shaped electrodes. Simultaneously, energy stored in a larger condenser bank is inductively transferred into the periphery of the gas in the same spherical vessel. The resultant imploding shock waves travel towards the center of the sphere and, when they collide with the central plasma, they compress and heat it. As a result, the temperature of the central plasma is raised to such values that soft X-rays are emitted in several pulses of about 1 μsec duration each.

Table 4 reports the preliminary operating parameters of the prototype machine (SPX II), and compares these to the microlithography industry requirements. With the exception of the reliability parameter, for which only modest data are available at present, it is apparent that the industry requirements can be satisfied by the Spherical Pinch X-ray generator.

Finally, Fig. 23 shows a pinhole photograph of the X-ray source in Argon, which proves that it is < 1 mm in diameter. As far as the reproducibility of position of the X-ray source is concerned, it has been invariably found that the variation is well within 1 mm from shot to shot. This is due to the symmetrical geometry of the imploding shocks and their consistent convergence towards the center of the vessel.

7.3. SPX II – Spherical Pinch X-Ray Generation

The spherical pinch concept has been utilized for a prototype X-ray generator that demonstrates the engineering feasibility of this approach. Design goals of the Spherical Pinch X-ray generator (SPX II) were:

- high X-ray flux in a few keV range;
- high energy transfer efficiency;
- high repetition rate;
- ease of operation;
- low electromagnetic noise, and
- operational flexibility.

Krypton gas has been used for optimum X-ray radiation in this preferred spectrum range. Other types of gas such as Argon can be used for different situations and require-

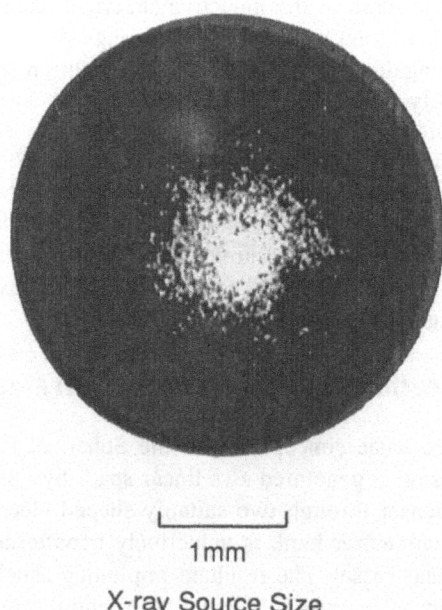

Figure 23. X-ray pinhole photograph of the spherical pinch plasma indicating X-ray source size.

ments. The gas is automatically filled to a specified pressure (typically 1 Torr) and is renewed after each discharge. This machine is essentially a pulsed high power device. The energy is stored in a capacitor bank (30 kJ at 20 kV charging voltage), then discharged in a low inductance circuit through spark-gap switches. About 20% of the stored energy is deposited into the plasma. A fraction of this absorbed energy is then converted into X-ray radiation. At present, the machine provides a few joules of X-ray energy. It is expected that at least 50 J of X-ray radiation will be emitted from each discharge, which is equivalent to an X-ray flux of 10 mJ/cm^2 at a distance of 20 cm from the source.

Since X-rays are produced in pulses of a few microseconds duration, the machine is designed to perform multiple discharges up to a number of pre-specified cycles. The repetition rate is currently up to 1/10 Hz, but it can be increased to 1 Hz and higher, limited only by the capacity of the high voltage power supply. This machine can be operated either manually or automatically. In the automatic mode, one needs only to enter the desired number of discharges into the controls, and then to initiate the discharge. The manual operation is reserved for machine optimization or diagnosis. In the actual production model, the manual operation mode will be used only for testing. Because this machine utilizes pulsed high voltages, shielding any transient electromagnetic noise is of paramount importance.

A schematic of the engineering spherical pinch coil and X-ray port is given in Fig. 24. The X-ray port is oriented vertically under the pinch as shown. A metal coated Mylar window of 25 μm thickness is used to isolate the discharge gas and to maintain the proper pressure in the discharge vessel. Another port can be added to monitor the X-ray flux from each discharge and to control the total dosage over multiple discharges.

Table 5 summarizes the machine parameters of the SPX II. The X-ray parameters of the SPX II are projected figures.

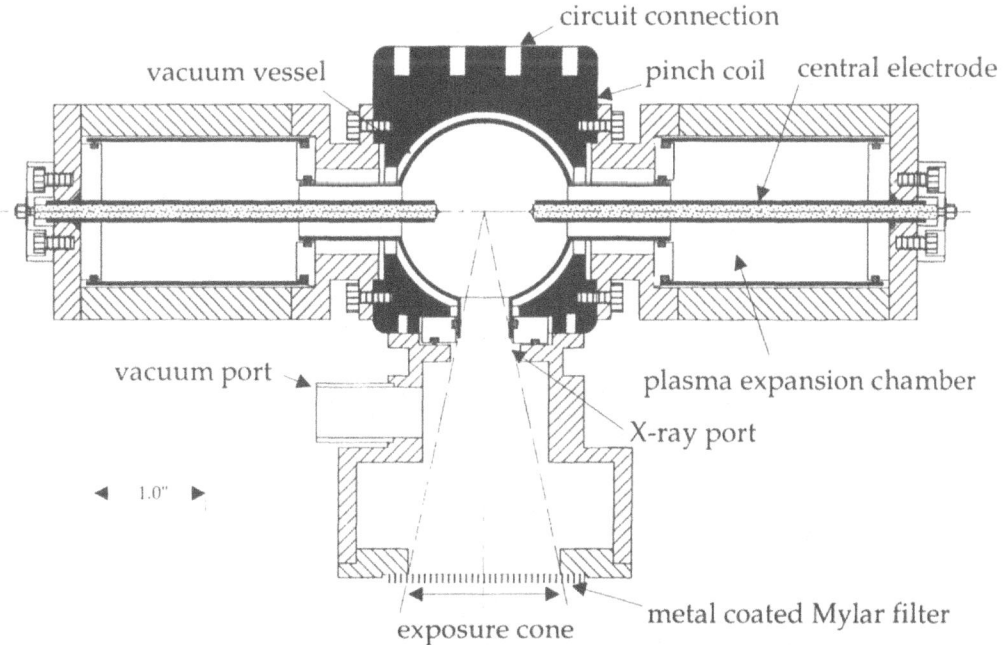

Figure 24. SPX II spherical pinch design showing pinch coil and X-ray port.

Table 5. Machine parameters of spherical pinch X-ray device

Parameters	SPX II
Operating voltage (kV)	25
Stored energy (kJ)	20
Energy transfer to plasma (%)	20
Peak current (kA)	1000
Repetition rate (Hz)	1/10
Working gas	Krypton + Oxygen
Gas pressure (Torr)	≥ 1
Gas volume (ml)	250
Dimensions	5.5' × 6.5' × 4'
Floor space (m^2)	~3
X-ray output/pulse (mJ/cm^2)	(10)
X-ray energy (keV)	(1-3)
Pulse duration (μsec)	(1)

7.4. SPX II Description and Specifications

The SPX II is a prototype machine that is being used to demonstrate its applicability as a radiation source for lithography. Figure 25 shows the overall configuration of the prototype, which consists of a discharge unit and a control console. The SPX II is completely modular in construction. All the components such as condenser banks, pinch head, switches, and trigger units can be quickly removed, serviced, and replaced if necessary.

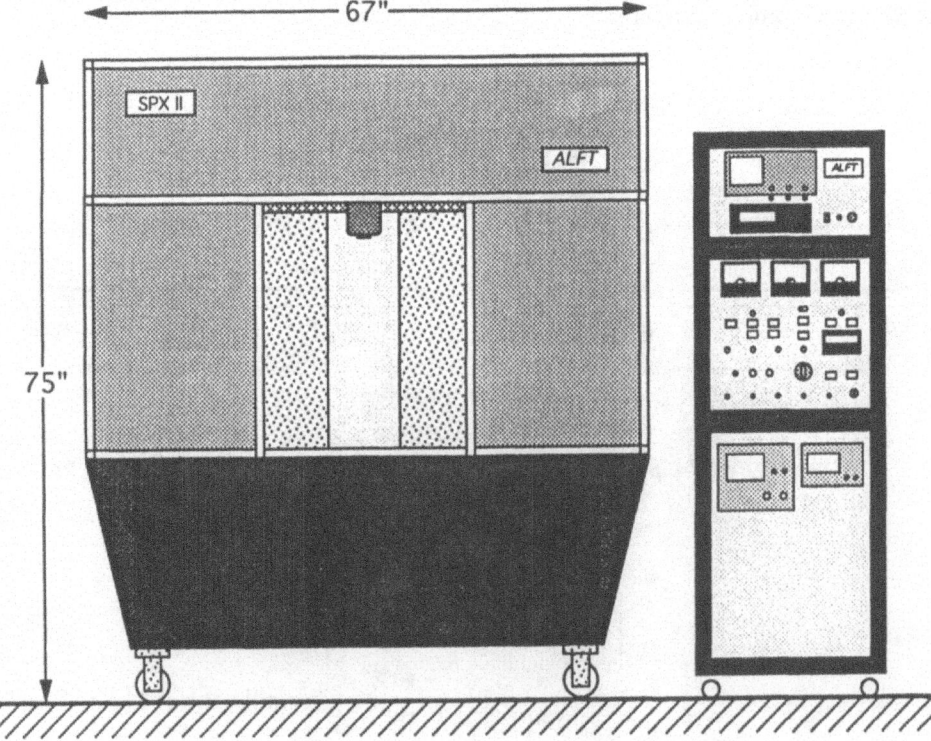

Figure 25. SPX II prototype X-ray generator, overall configuration.

Table 6. Electric specifications of the SPX II

	Spherical Discharge	Linear Discharge
Bank Energy @ 30 kV	34 kJ	2.5 kJ
Operating Voltage	15-30 kV	15-30 kV
Peak Current	500-1000 kA	25-50 kA
Ringing Frequency	86 kHz	116 kHz

Most of the components with critical lifetime are commercial, off-the-shelf parts with expected lifetimes between 10^5 and 10^6 discharges.

Two circuits for the spherical and linear discharges of the SPX II are operated through a remote control console. This is a fully automated source system with an option for manual operation. This electrical specifications are listed in Table 6.

Inert gases, such as Helium, Argon, Neon and Krypton are normally used as the working gases. The gas pressure in the vessel varies between 0.5 to 8 Torr. A commercial type machine (SPX III) of 60 kJ bank energy is now being developed. The SPX III can be discharged at the repetition rate of 1 Hz to deliver the required X-ray radiation dose to the wafer in a few seconds.

7.5. Soft X-Ray Output from the SPX II

In the SPX II, the spherical and linear circuits can be discharged with a preset delay time, which is very important in order to maximize the X-ray yield. The results reported here, however, are for zero time delay between discharges.

To eliminate the visible and UV lights from reaching the detector, an aluminum foil (> 1.9 mg/cm^2), which transmits only radiation above 800 eV, was used as window.

A vacuum compatible and light tight X-ray camera has been designed. High-sensitivity double-emulsion polyester-base Kodak Direct Exposure Film (DEF 392) was used to measure the radiation after the window. Kodak DEF has been the choice of plasma laboratories for recording 1–10 keV radiation for many years. Its high sensitivity and ease of handling have readily justified its use in spite of its high background noise level. This film is made of 185 μm thickness polyester as a base which is coated with emulsion on both surfaces. The optical transmission properties of polyester indicate that the radiation beyond 8 keV can be transmitted through the base to expose the other side of the film. Since we have not observed exposures on both sides of the film, it can be claimed that the radiation above 10 keV is negligible. On the other hand, the aluminum window blocks the radiation below 0.8 keV. It follows that the radiation observed on the film is between 0.8 and 10 keV.

The film was calibrated at five photon energies between 0.93 and 6.93 keV by Rockett[34]. Here we present only the relative and not the absolute X-ray dose generated from the source, because of lack of information about the radiation energy spectrum of the source and detailed film calibration in a range of 0.8 to 10 kV.

The relative soft X-ray output vs. discharge voltage is shown in Fig. 26. Argon was the working gas and the background pressure was 2 Torr. The soft X-ray output increases with condenser bank discharge voltage, as expected. Higher discharge voltages may generate hotter plasmas at the center and consequently larger radiation outputs. We have yet not pushed the machine to operate at voltages higher than 20 kV.

Figure 27 shows the soft X-ray output as a function of the background gas pressures for Helium, Argon and Neon. The machine was discharged at 20 kV. The results indicate

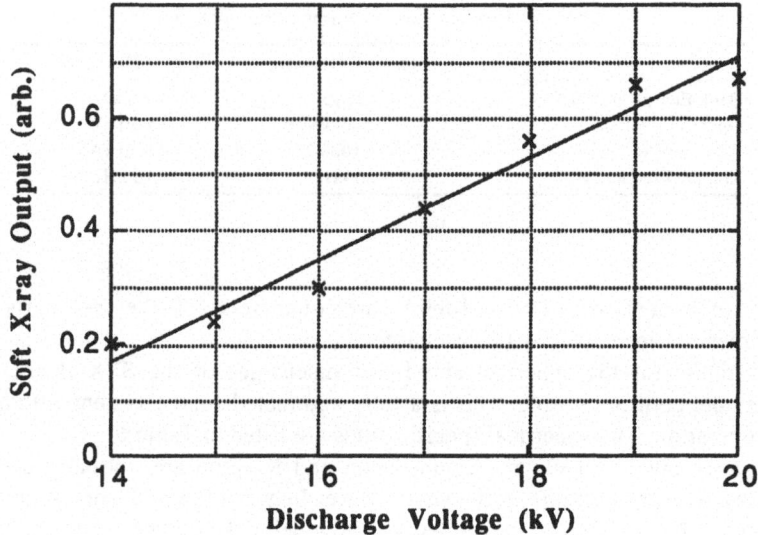

Figure 26. Relative soft X-ray output vs. discharge voltage.

that, for all gases, the optimum pressure lies between 1 and 3 Torr. This could be because the plasma temperature reaches a maximum within this pressure range, or because the competing soft X-ray absorption through the neutral gas before the window is predominant over any increase in soft X-ray radiation output.

7.6. Imploding Waves

As mentioned earlier, the background gas is first ionized by an electrical discharge at the center of the spherical chamber. This is accomplished by generating a low-tempera-

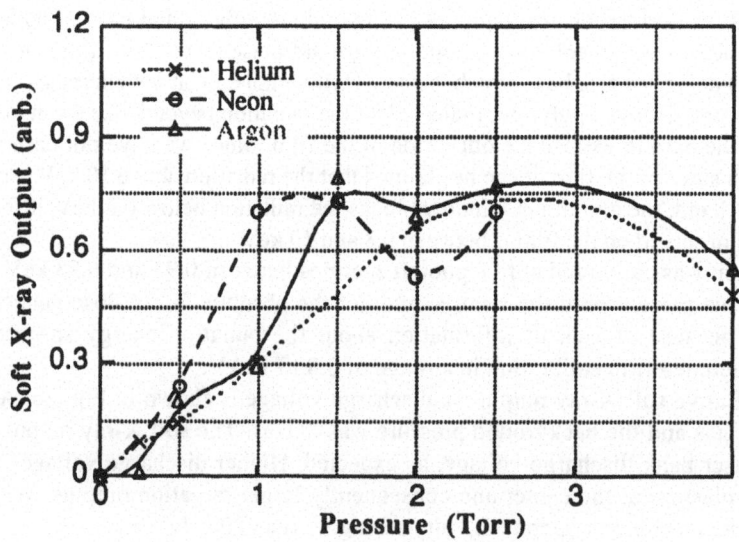

Figure 27. Relative soft X-ray output vs. gas pressure.

Spherical Pinch Research

ture plasma from a spark across two electrodes. Simultaneously, electrical energy stored in a 30 kJ capacitor banks is inductively transferred to the gas surrounding the center plasma. This discharge launches imploding waves that converge at the center of the pinch vessel. When the imploding waves collide with the center spark-ionized plasma, they compress it, causing its temperature to reach a level sufficient for soft X-ray emission. The soft X-ray yield depends not only on the total electrical energy stored in the bank, but also on the imploding velocity of the waves and the time when the imploding waves collide with the center spark-ionized plasma. This has motivated us to develop an analytic model to predict the imploding velocity and direct us to select operating parameters.

It is assumed that all plasma that has been swept up by the magnetic piston generated by the inductive discharge is piled up in a very thin layer at the piston front, and travels with it. It is also assumed that the plasma has infinite conductivity. We have an equation of motion for the imploding wave front as follows,

$$ \tag{45}$$

Here, ρ_0, P_0, and R_0 are the background gas density, pressure and the vessel radius, respectively. L is the characteristic length of the induction coil and r is the front of the imploding wave. If we know the discharge current I in the coil, the magnetic field can be derived as

$$B^2 = \frac{\mu_0^2 I^2}{4L^2}. \tag{46}$$

Since the time for an imploding wave to travel from the edge to the center of the vessel is much shorter than 1/4 of the ringing period, we can reasonably assume, that

$$I \approx \alpha t. \tag{47}$$

Also, the gas pressure is much smaller than the magnetic pressure, i.e., $P_0 \ll B^2/2\mu_0$. Hence:

$$\frac{d}{dt}\left[(R_0^2 - r^2)\frac{dr}{dt}\right] = -Art^2, \tag{48}$$

where $A = \mu_0 \alpha^2/(4\rho_0 L^2)$. The solution of this ordinary differential equation with the initial conditions $r = R_0$ and $dr/dt = 0$ is,

$$r = R_0\left(1 - \frac{\sqrt{A}}{\sqrt{11R_0}}t^2\right) \approx R_0 - 0.15\frac{\sqrt{\mu_0}\alpha}{\sqrt{\rho_0}L}t^2. \tag{49}$$

This simple analytical model shows that the plasma front accelerates with constant acceleration $0.15\sqrt{\mu_0}\alpha/\sqrt{\rho_0}L$. The lighter the gas and the lower the gas pressure, the faster the wave front travels. Higher peak discharge current, or higher discharge voltage, generates a stronger implosion, as expected.

7.7. Experimental Results and Comparison with Theory

In order to verify the applicability of the above model to the spherical pinch, we have carried out a series of experiments to detect the passage of the imploding wave front. These were monitored by a magnetic probe located at r = 2.0 cm from the center of the vessel. Simultaneously, a pinhole optical probe was used to view a narrow area at the center. An imploding wave forms at the periphery of the vessel, then moves towards to the center. It will pass the magnetic probe at r = 2.0 cm and then collapse at the center. The time difference between the front detected by the magnetic probe and the luminosity peak detected with the optical probe indicates how fast an imploding wave propagates. Helium and Krypton were used and the background gas pressure was changed from 1 to 8 Torr. The peak discharge currents varied from 400 to 700 kA.

To compare the experimental results with the model's prediction, we calculate the time required for a wave front to travel from $r = 0.02$ m to the center, using Eq. (49) with $R_0 = 0.03$ m and $L = 0.035$ m,

$$t_{r=0.02m} = 1.44 \frac{\sqrt[4]{\rho_0}}{\sqrt{\alpha}}, \tag{50}$$

and

$$t_{r=0.0m} = 2.50 \frac{\sqrt[4]{\rho_0}}{\sqrt{\alpha}}. \tag{51}$$

The time difference between $r = 0.02$ m and $r = 0$ is what we can detect in the experiments, and this can be expressed as,

$$\Delta t = t_{r=0.02\,m} - t_{r=0.0\,m} = 1.06 \frac{\sqrt[4]{\rho_0}}{\sqrt{\alpha}}. \tag{52}$$

Here, ρ_0 is a function of gas and pressure, and α can be determined from the discharge current. Figure 28 shows the time difference Δt versus $1.06 \sqrt[4]{\rho_0} / \sqrt{\alpha}$. The straight line is the model prediction and the crosses are the measured results. Good agreement is observed between the theoretical prediction and the experimental results, which indicates that the imploding wave behavior follows the snow-plow model. It should be noted here that this model is only able to describe the transit phenomenon of the imploding waves. It cannot give any information about the compression of the central plasma. Therefore, we are not able to predict the radiation emission directly from this model.

7.8. Concluding Remarks

The adaptation of the spherical pinch concept as an industrial tool for radiation generation, particularly soft X-rays, and its applicability for soft X-ray microlithography has been demonstrated, both theoretically ad experimentally. Measurements of soft X-ray output from the SPX II prototype were reported. The output increases with the discharge voltage. The optimum gas pressure for the soft X-ray yield is between 1 and 3 Torr. A simple

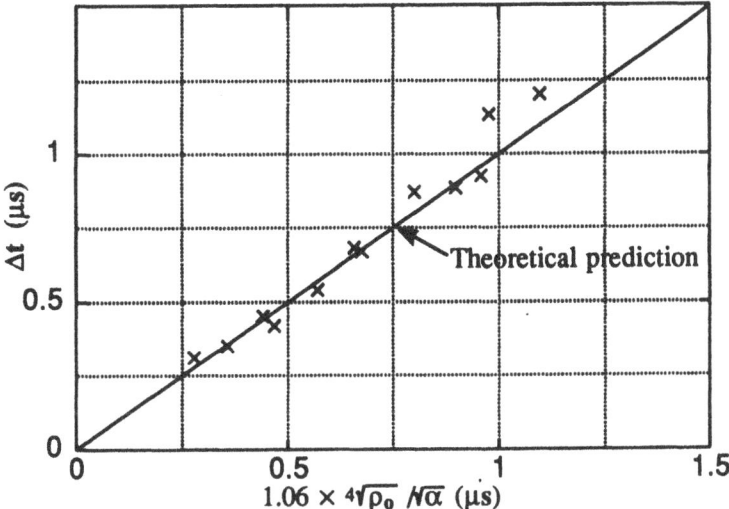

Figure 28. Time difference of the imploding wave traveling from r = 2.0 cm to the center.

analytic model has been developed based on the snow-plow assumption. It predicts the imploding wave velocity with respect to the operating parameters. The imploding wave velocities were measured in a wide range of the parameter space using both a magnetic probe located at r =2.0 cm, and an optical probe viewing the center of the vessel. The prediction of the model agrees quite well with the experimental results.

8. CONCLUSIONS

8.1. Historical Survey of the Research on the Spherical Pinch

Research on the Spherical Pinch spans a time frame of three decades, in a path that has maintained a steady pace, albeit a slow one, given the limited funding available to carry it out at an accelerated pace. In table form, the history of the research on the SP is given by the following scheme:

1964 – 1966:	Study of the cylindrical theta-pinch. Experimental observation of X-ray emission from the pinched plasma;
1967 – 1975:	Conversion from cylindrical to spherical geometry;
1976 – 1980:	Study of the spherical geometry in the inductive configuration. Experimental observation of neutron emission from the pinched plasma;
1981 – 1983:	Derivation of the scaling laws for spherical pinch devices;
1984 – 1987:	Study of the spherical geometry in the resistive configuration. Experimental observation of the stability of central plasma;
1988 – 1994:	Numerical simulation of the phenomenon and demonstration of the advantages of the spherical pinch model over the inertial confinement fusion model;

Table 7. Groups working on the Spherical Pinch

Country	Institution	Investigators	Subject
U.S.A.	University of Texas San Antonio	Dr. N. Salingaros	Theoretical analysis of spherical configurations with the use of fiber theory.
	University of Miami	Dr. C. Thio	Experiments on use of railgun plasmas to compress central plasma.
	University of Tennessee, Knoxville	Dr. M. Laroussi Dr. E. Panarella	Establishment of a research program on the spherical pinch for microlithography applications.
Canada	Advanced Laser and Fusion Technology, Inc., Hull, P.Q.	Dr. E. Panarella Dr. L. Zhang	Experimental investigation on the spherical pinch for microlithography applications. Computational analysis for simulation of implosion phenomena. Theoretical analysis of fusion breakeven conditions. Experimental investigation of neutron emission from the spherical pinch for neutron radiography and compression of a central plasma.
Italy	National Research Council, Pisa	Dr. M. Vaselli Dr. D. P. Singh	Study of the stability of imploding shock waves, both experimental and theoretical.
	University of Pisa	Dr. F. Giammanco	Theoretical analysis of spherical imploding shock waves and compression of a central plasma.
France	Centre d'Etudes de Limeil-Valenton, Villeneuve-St-Georges	Dr. M. de Peretti	Simulation work on spherical imploding shock waves and compression of the central plasma.

1995 — present: Analytical work demonstrating the advantages of the spherical pinch over the inertial confinement fusion model for achieving breakeven.

The reason why research on the Spherical Pinch has never lost its momentum during the past several years is because this concept offers distinct advantages over other fusion concepts. The most notable is that its scaling laws for breakeven conditions are well known and understood, rather than empirically derived. This allows an experimental approach to fusion through modest scale experiments, rather than more expensive large-scale experiments. Moreover, the Spherical Pinch is providing immediate industrial benefits, such as soft X-ray generation for microlithography, which are not possible when the approach to fusion is in a scale incompatible with the capability of the market to pay for it.

8.2. International Collaboration

The Spherical Pinch is gaining acceptance among the international plasma and fusion physics community, and research in this field is being carried out in various countries through collaborative work between different groups. As a result, this paper represents a joint effort between four groups in the U.S.A., Canada, and Italy. Table 7 highlights the location of these groups, and the respective work they are carrying out. The collaboration among various groups is rapidly expanding. For instance, discussions are underway with the Los Alamos group working on Magnetized Target Fusion (MTF)[7] concept in order to establish a connection between that group and those working on the Spherical Pinch.

In conclusion, it is not inconceivable that, since the essentially similar concepts of the Spherical Pinch and Magnetized Target Fusion seem to be emerging as viable and economic fusion approaches, they might become the approaches of choice by the international community for a successful quest of fusion energy production.

REFERENCES

1. E. Panarella, *Can. J. Phys. 58*, 983 (1980).
2. E. Panarella, and P. Savic, *J. Fusion Energy 3*, 199 (1983).
3. E. Panarella, and V. Guty, *Proc. 1982 IEEE Intern. Conf. on Plasma Science*, Ottawa, Canada, 17–19 May 1982, p. 135.
4. E. Panarella, *J. Fusion Energy 6*, 285 (1987).
5. F. Cornolti, F. Giammanco, and A. Giulietti, *Lett. Nuovo Cim.*, **19**, 165 (1977); A. Giulietti, M. Vaselli, and F. Giammanco, *Opt. Comm.*, **33**, 257 (1980); A. Giulietti, D. Giulietti, M. Lucchesi, and M. Vaselli, *Opt. Comm.*, **47**, 131 (1983); F. Cornolti, A. Giulietti, D. Giulietti, M. Lucchesi, and M. Vaselli, *Opt. Comm.*, **57**, 249 (1986).
6. N. A. Salingaros, *Fusion Technology 27*, 230 (1995).
7. R. C. Kirkpatrick, I. R. Lindemuth, and M. S. Ward, *Fusion Technology 27*, 201 (1995).
8. P. Savic and E. Panarella, *J. Appl. Phys. 59*, 3990 (1986).
9. K. Kawai, E. Panarella, and D. Mostacci, *SPIE Proceedings 1465* (1991), p. 308; S. Aithal, K. Kawai, M. Lamari, and E. Panarella, *SPIE Proceedings 1671* (1992), p. 383; S. Aithal, E. Panarella, M. Lamari, B. Hilko, and R. McIntoch, *SPIE Proceedings, 1924* (1993), p. 3362; J. Chen, E. Panarella, B. Hilko, and H. Chen, *SPIE Proceedings 2194* (1994), p. 231; L. Zhang, E. Panarella, B. Hilko, and H. Chen, *SPIE Proceedings 2437* (1995), p. 356.
10. DND (Department of National Defence) Final Report SSC 003SV.W2207–0-AF17, 31 December 1991.
11. D. Mostacci, K. Kawai, E. Panarella, and N. Salingaros, *Bull. Amer. Phys. Soc. 35*, 1963 (1990).
12. F. Giammanco, S. Del Tredici, D.P. Singh, and M. Vaselli, *Fusion Technology 27*, 221 (1995).
13. Y. Fujimoto and E. A. Mishikin, *Phys. Fluids, 21*, 1993 (1978).
14. J. Nuckolls, L. Wood, A. Thiessen, and G. Zimmerman, *Nature 239*, 139 (1972).
15. Reference 4 and references therein.
16. M. A. Harith, V. Palleschi, D. P. Singh, G. Tropiano, and M. Vaselli, *Optics Commun. 71*, 76 (1989).
17. a) B. Ahlborn and M. H. Key, *Plasma Phys. 23*, 435 (1981); see also b) B. Ahlborn, M. H. Key, and A. R. Bell, *Phys. Fluids 25*, 541 (1982).
18. G. Ben-Dor, *Prog. Aerospace Sci. 25*, 329 (1988).
19. D. Salzmann, S. Eliezer, A. D. Krumbein, and L. Gitter, *Phys. Rev. A 28*, 1738 (1983).
20. F. Giammanco, A. Giulietti, M. Vaselli, and G. Einaudi, *Il Nuovo Cimento B. 52*, 187 (1979).
21. Ya. B. Zel'Dovich and Yu. P. Raizer, *Physics of Shock waves and High Temperature Hydrodynamic Phenomena*, Academic Press, New York (1966).
22. G. Einaudi, F. Giammanco, A. Giulietti, and M. Vaselli, *Il Nuovo Cimento B 51*, 280 (1979).
23. F. Giammanco, *Il Nuovo Cimento B 54*, 297 (1979).
24. J. G. Cordey, R. J. Goldston, and R. R. Parker, *Phys. Today,* January (1992), p. 22.
25. J. D. Lindl, R. L. McCrory, and E. M. Campbell, *Phys. Today*, September (1992), p. 32.
26. W. J. Hogan, R. Bangerter, and G. L. Kulcinski, *Phys. Today*, September (1992), p. 42.
27. G. Velarde, Y. Ronen, and J. M. Martinez-Val, *Nuclear Fusion by Inertial Confinement: A Comprehensive Treatise*, CRC Press (1992), p. 46.
28. H. Motz, *The Physics of Laser Fusion* (Academic Press 1979).
29. Lyman Spitzer, JR., *"Physics of Fully Ionized Gases,"*, Second Edition, (Interscience Publishers 1962).
30. Ya. B. Zel'dovich, G. I. Barenblatt, V. B. Librovich, and G. M. Makhviladze, *"The Mathematical Theory of Combustion and Explosions"*, (Consultants Bureau 1985), p. 550.

31. A.K. Oppenheim, *Annual Review of Fluid Mechanics*, 5, (1973), pp. 31–58.
32. Ref. 20, p. 794.
33. L. D. Landau, and E. M. Lifshitz, *"Fluid Mechanics"*, (Pergamon Press 1959), Academic Press, (1966), p. 392.
34. P.D. Rockett, C.R. Bird, C.J. Hailey, D. Sullivan, D.B. Brown, and P.G. Burkhalter, *Applied Optics*, 24, 2536 (1985).

PERSPECTIVES OF ADVANCED CONFINEMENT PROGRAMS*

Bruno Coppi

Department of Physics
Massachusetts Institute of Technology
Cambridge, Massachusetts 02139

One of the main motivations to construct devices producing an axisymmetric toroidal confinement configuration, in which an important component (poloidal) of the confining magnetic field is produced by the current induced in the plasma, is their relative simplicity. The series of Tokamak experiments developed in Russia originally had this motivation. The Alcator program, that was proposed when the Tokamak experiments were still viewed with almost universal skepticism in the West, had adopted this configuration in view of its versatility in order to explore new high and low density plasma regimes.

In fact the technology solutions devised for the Alcator machine made it quite different from the typical Russian Tokamaks and induced L. Artsimovich to predict that because of these solutions the Russian Tokamaks would have been rapidly outdistanced. The most notable among these innovations was the aircore poloidal field system that lends itself to produce high plasma currents in compact and low aspect ratio configurations.

By following this approach of maximum simplicity and on the basis of the favourable confinement properties of high density plasma discovered by the Alcator program it has become possible since 1975 to propose and design the first ignition experiments. As of the present date there is no other magnetic confinement configuration that could be employed to attain fusion ignition conditions in the near term. Thus it would be an error to conceive of a fusion program where a preeminent role could not be given to this kind of approach.

Some of the main results of general scientific value that have been attained are described. These include, for instance, the so-called "slideaway regime" involving the formation of non-thermal electron distributions and the transverse acceleration of high energy nuclei that has been employed to explain the observation of "conic" ion distributions in space physics, the excitation and suppression of the reconnection events in high temperature regimes that are involved in the so-called sawtooth oscillations, the "anomalous" inward (in the direction of the density gradient) particle transport, the "isotopic effect" on plasma confinement, the favourable dependence of the degree of plasma purity

---

* Supported in part by the U.S. Department of Energy.

and confinement on the plasma density, the rapid achievement of record values of the confinement parameter $n\tau_E$, the attainment of collisionless regimes with the highest electron and ion temperatures, the discovery of transport barriers, etc.

It is argued that high density plasmas that are confined by high magnetic fields have the necessary degree of macroscopic stability to reach ignition conditions. This feature coupled with others (strong ohmic heating that minimizes the need for injected heating systems which deteriorate confinement) gives a window in parameter space that can be exploited to accelerate fusion research by a sequence of relatively small scale, advanced experiments. The possibility to explore by this avenue the burning conditions of attractive tritium-poor plasma mixtures is pointed out.

The urgent need for meaningful experiments to explore D-T burn conditions, where the α-particle (reaction product) slowing down time is considerably shorter than the electron energy confinement time and the electron temperature exceeds the ion temperature, is stressed. These are, in fact, the conditions that are needed to transfer energy from the reaction products to the fusing nuclei and are not met in the existing generation of operating experiments.

8

PRESENT STATUS OF FIELD-REVERSED CONFIGURATIONS

John Slough

University of Washington
Seattle, Washington 98105

A Field Reversed Configuration (FRC) is an elongated toroidal plasma having no material object linking the hole in the torus (i.e., an elongated compact toroid) and having ideally no toroidal field[1]. The plasma is confined within a separatrix by poloidal magnetic field. This field is generated by plasma diamagnetic currents that flow exclusively in the toroidal direction. Externally imposed vacuum magnetic flux insulates the FRC from the chamber walls.

Several techniques have been used to form FRCs with the most successful to date being the field reversed theta pinch (FRTP). Axial and radial pressure balance with the confining external magnetic field produce a volume averaged β in a constant radius FRTP of $\langle\beta\rangle \sim 0.5 x_s^2$, where x_s is the ratio of the FRC separatrix radius, r_s to the coil radius r_c. β is defined as the ratio between the local plasma pressure and the external magnetic field pressure, thus the equilibrium β is always large ($0.5 \leq \beta \leq 1$).

Research in FRCs is motivated by the fact that fusion reactors based on these configurations would have several important advantages over reactors with tokamak geometry. First, the cylindrical geometry of the FRC vacuum vessel allows for simpler and smaller reactors. The FRC can also be translated along guiding magnetic fields into different specialized chambers, such as a formation (source) chamber, a compression chamber, and a burn chamber. Second, the high plasma β results in more economically attractive reactor. Furthermore, the high β (and the resultant lower confining field) make FRC based reactors strong candidates to burn advanced fuels such as D-^{3}He. Third, the natural diverter formed by the external vacuum magnetic field allows the controlled deposition of exhaust particles (escaping plasma and/or charged fusion reaction products). In this way, divertor related technological problems are minimized and direct energy conversion may be possible with D-^{3}He operation.

Despite the extreme reactor attractiveness of FRCs, the concept has not been pursued more vigorously for two primary reasons, one technological and other theoretical. Firstly, FRCs are difficult to form, with standard techniques leading to relatively high densities, and thus requiring pulsed technology to overcome initial radiation losses. Secondly, FRCs have been thought to be grossly unstable, in spite of experimental evidence to the contrary.

Progress has been made over the last decade in both improving the formation technology and understanding the observed experimental stability, especially to the internal tilt mode. Kinetic effects were shown to provide tilt stability at small s, where s is the average number of ion gyroradii between the magnetic field null and the FRC edge (separatrix). The s parameter is related to the kinetic/fluid properties of the FRC plasma. Low values (s < 1) indicate strong kinetic effects while high values (s >>1) indicate more MHD-like characteristics. The largest FRC device built to date, the Large s Experiment (LSX)[2] was built specifically to explore FRC physics in the less kinetic, more MHD-like regime (s >> 1). At any s however there is a region near the magnetic null where kinetic effects will always be important. In this interior region ion orbital excursions can be comparable to the FRC radius, even for moderate or large s. MHD calculations incorporating ion orbit effects have, for example shown stability to the internal tilt mode. Assuming that the equilibrium profiles do not change, the stabilizing effect of the meandering orbits is calculated to diminish as the average number of ion gyroradii between the magnetic null and the separatrix, or s increases. However, calculations made in the fluid limit, which examined the influence of ion gyroviscosity equilibrium shape, and radial current profile found stable equilibria even in the purely MHD limit.[3]

During the brief operating period of LSX, FRCs were formed with total temperatures greater than 1 KeV, electron densities of the order of 10^{21} m^{-3}, plasma volumes of up to 0.5 m^3. The confining magnetic fields were typically around 0.5 T. Positive confinement scaling with s was observed for s values up to 5. In its one year of operation LSX was successful in forming FRCs with s up to 5 which were completely tilt stable. Empirical confinement scaling laws were extended to cover two orders of magnitude, clearly identifying diffusive scaling, with particle diffusion parameters within an order of magnitude of tokamaks despite the high beta and absence of any toroidal field.

The particle lifetime scaling observed on LSX had the following scaling:

$$\tau_N = 10\sqrt{2/A_i}\, x_s r_s^{2.1}\, (m) n_m^{0.5} (10^{20} m^{-3})\ msec$$

Examination of the energy balance in present FRCs shows that the principal energy losses are convective, so that τ_N is as good a parameter as any to use for scaling extrapolations.

The reactor prospects for the FRC depend on three requirements. The first, and most important, is continued stability as s is increased to at least 20. It appears likely that the internal tilt can be made stable by a combination of profile shaping and remaining kinetic effects from either burn products, high energy beams, or current profile shaping.

The second requirement is a slight improvement over the present empirical scaling. This might be expected to occur if the anomalous resistivity is due to drift wave turbulence, which will saturate at a lower value as the scale lengths increase. Evidence of this is seen in the measured x_s dependence of particle, flux, and energy lifetimes in LSX.

The third requirement is to be able to efficiently generate the FRCs initially. The difficulty of forming FRCs in FRTPs is related to the amount of the poloidal flux. Previously this had been ~2 mWb, and was increased to ~10 mWb in LSX. However, LSX probably achieved within a factor of 10 of what is technologically feasible. Since over 1 Wb is required for a reactor, some method for increasing the FRC poloidal flux is required.

The most straightforward methods for increasing the flux would use either neutral beams or RF to increase the toroidal current. Reactor designs based on neutral beam flux build-up have been proposed (4), but have not been demonstrated on even a small scale.

The other method, which has been demonstrated in small, low energy plasmas (5), employs the application of a rotating magnetic field with a frequency that entrains electrons but not ions.

One attractive feature of developing a flux build-up technique is that it provides a methodology for steady state operation. Rotating magnetic field frequencies of about 50 kHz, with field strengths of only about 20 G would be sufficient. It may be desirable to drive ion currents with neutral beams in addition, since it has been shown to also provide tilt stability.

REFERENCES

1. M. Tuszewski, *Nucl. Fusion 28*, 2033 (1988).
2. J. T. Slough, A. L. Hoffman, R. D. Milroy, *et al.*, *Phys. Rev. Letters 69*, 2212 (1992).
3. L. C. Steinhauer, A. Ishida, and R. Kanno, *Phys. Plasma 1*, 1523 (1994).
4. H. Momota, A. Ishida, Y. Kohzaki, *et al., Fusion Techology 21*, 2307 (1992).
5. W. N. Hugrass, I. R. Jones, M.G.R. Phillips, *J. Plasms Physics B26*

IGNITION PHYSICS AND THE IGNITOR PROJECT

Francesco Pegoraro

The Ignitor Project Group
Dipartimento di Fisica Teorica
Università di Torino
Via P. Giuria 1, 10125 Torino, Italy

ABSTRACT

Ignitor is a compact high magnetic field machine with tight aspect ratio and shaped cross section, where large plasma currents can be produced. High plasma densities, strong Ohmic heating, good confinement of particles and energy, adequate plasma purity, and stability against MHD modes are expected on the basis of the machine characteristics.

The main goals of the machine are the confinement of the fusion produced α-particles and the Ohmic heating of D–T plasmas up to ignition at relatively low peak temperatures ($T_{eo} \simeq T_{io} \lesssim 15$ keV) with the production of significant amounts of fusion power (≈ 100 MW). An ICRH system with $P_{RF} \lesssim 18$ MW is adopted to accelerate the approach to D–T burn conditions and to control at the same time the evolution of the current density, to produce significant fusion power in D–^{3}He plasmas, and to investigate the access to the high-β second stability region.

1. INTRODUCTION

The main point to be addressed in the program of magnetic thermonuclear fusion research is the objective of plasma ignition which has not yet been tackled. Indeed recent experiments on JET and TFTR have yielded sizable amounts of fusion power, but in regimes that are not directly relevant to ignition in terms of the ratio between the ion and the electron temperature, of the α-particle confinement and/or of the ratio between the slowing down time of the α-particles and the plasma energy confinement time, of the confinement parameter $n\tau_E$ and of the plasma purity and thermalization. This implies that the most important physics issues of burning magnetically confined D–T plasmas are still to be explored. These issues are:

1. whether the extrapolation to the regimes where the plasma temperature is self-sustained of presently known or explored processes is reliable. These processes include plasma stability (in particular MHD stability), plasma purity (plasma wall interaction and wall loading) and particle and energy transport.

2. what will the effect be of a large population of α-particles with an energetic and non-thermal distribution on MHD modes, on energy and particle transport, etc., and how accurately have we been able to predict it.

Understanding these physics issues is, in our opinion, the main pre-requisite in order to be able to assess which is the most promising line for a real, energy producing, magnetic fusion reactor and then to address questions such as steady state operation, current induction etc., and to study the technological problems involved in detail. A rapid and effective exploration of the physics of magnetically burning plasmas is becoming even more urgent as the feeling is growing, both in the public and in the scientific community, that despite the good results so far obtained, no sufficiently direct action is being taken in order to make fusion energy a reality on a reasonable time-scale. This feeling is also often shared by members of the fusion community outside the line of thermal, magnetic plasmas.

The Ignitor experiment was designed to reach D–T burn conditions by Ohmic heating alone at high peak densities, $n_0 \approx 10^{21}$ m^{-3}, and relatively low peak temperatures, $11 \lesssim T_o \lesssim 15$ keV. At $T_o \approx 11$ keV, ignition is achieved with values of the fusion α-particle power $P_\alpha < 20$ MW.

In this paper we shall briefly illustrate how the ignition physics issues mentioned above have been addressed in the Ignitor project by a careful choice of the machine and operation parameters. As far as MHD stability is concerned, the strategy adopted is based on the possibility of reaching ignition conditions at relatively low values of the poloidal-beta parameter and small values of the $q \leq 1$-volume (possibly by controlling the current penetration and its density profile) which are the main parameters for determining the onset and for assessing the effect of the "internal $m = 1$ modes." As far as the energy transport is concerned, the choice of Ohmic-heating as the primary heating source serves the purpose of limiting the degradation of the energy confinement that is typically observed to occur in the transition from Ohmic to auxiliary heating, whereas the combination of a large plasma current and a large poloidal magnetic field, which is necessary in order to obtain high temperatures by Ohmic heating, is a proven favorable confinement factor. Moreover, in view of the high plasma density, good plasma purity is expected.

2. REFERENCE DESIGN PARAMETERS

The Ignitor experiment (see Fig. 1) was conceived on the basis of the favorable properties of high density plasmas, in terms of good confinement and high degree of purity, that had been discovered by the high field Alcator machines at MIT[1] and confirmed by the FT experiments in Frascati and by other advanced experiments. High field high density experiments were the first to be proposed[2] in order to achieve fusion ignition conditions on the basis of existing technology and of known properties of magnetically confined plasmas.[3,4,5] Thus the reference parameters of the Ignitor Ult machine[6] given in Table 1, have been chosen in order to obtain[7]:

- a peak plasma density n_0 about 10^{21} m^{-3};

Ignition Physics and the Ignitor Project

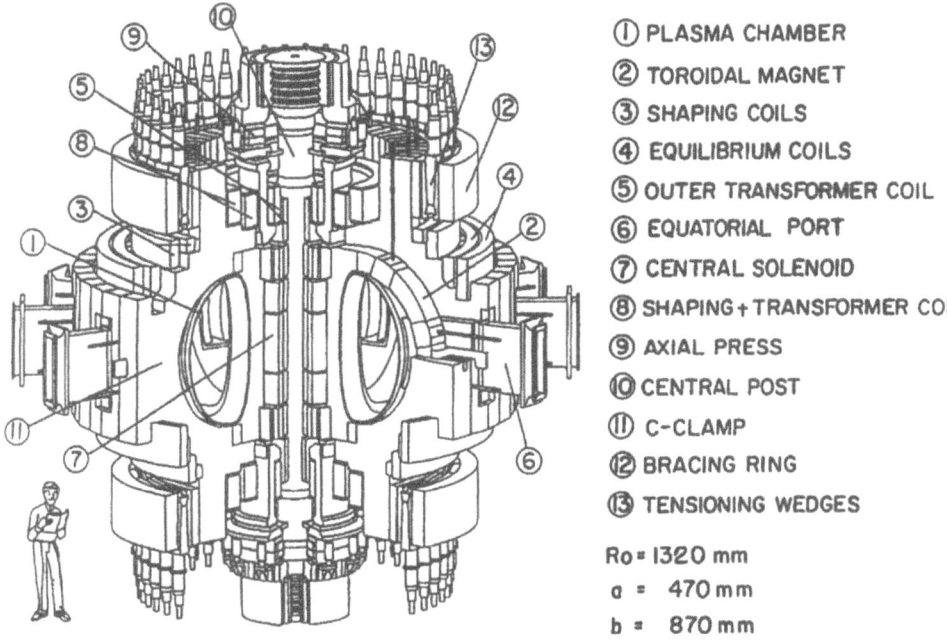

① PLASMA CHAMBER
② TOROIDAL MAGNET
③ SHAPING COILS
④ EQUILIBRIUM COILS
⑤ OUTER TRANSFORMER COIL
⑥ EQUATORIAL PORT
⑦ CENTRAL SOLENOID
⑧ SHAPING + TRANSFORMER COIL
⑨ AXIAL PRESS
⑩ CENTRAL POST
⑪ C-CLAMP
⑫ BRACING RING
⑬ TENSIONING WEDGES

Ro = 1320 mm
a = 470 mm
b = 870 mm

Figure 1. Main components of the Ignitor Ult machine.

- high mean poloidal fields ($\bar{B}_p \lesssim 3.75$ T)

- high toroidal plasma current ($I_p \lesssim 12$ MA) with relatively small volumes of the region where the "unwinding" parameter q is less than unity;

- ignition at low value of the poloidal beta parameter $\beta_p = 8\pi \langle p \rangle / \bar{B}_p^2 \lesssim 0.15$.

Consequently we can expect[7]:

- a strong rate of Ohmic heating up to ignition. This is accomplished by programming the initial rise of I_p and n_o while gradually increasing the cross section of the plasma column. During this relatively long ($t_r \gtrsim 3$ to 4 s) transient phase, the electric field is strongly inhomogeneous, is small at the center of the plasma column, where T_0 can achieve relatively high values, and is maximum at the edge of the plasma column (corresponding to loop voltages $V_\phi \simeq 1$ V);

- ignition at low average temperatures, corresponding to peak temperatures $11 \lesssim T_o \lesssim 15$ keV;

- that the energy confinement time degradation, which has been observed when a form of injected heating is applied at discrete points around the torus and when the relevant power is much larger than that of Ohmic heating, will be limited. The Ignitor strategy is to sustain a strong rate of Ohmic heating up to relatively high T_0, where fusion (α–particle) heating becomes significant;

Table 1. Reference Design Parameters of the Ignitor Ult Machine

$R_o \simeq 1.32$ m	Major radius
$a \times b \simeq 0.47 \times 0.87 \text{m}^2$	Minor radii
$\delta_G \simeq 0.4$	Triangularity
$I_p \lesssim 12$ MA	Plasma toroidal current
$I_\theta \lesssim 9$ MA	Plasma poloidal current
$B_T \lesssim 13$ T	Vacuum toroidal field at R_o
$\Delta B_T \lesssim 1.5$ T	Paramagnetic field
$\langle J_\phi \rangle \lesssim 9.3$ MA/m^2	Average toroidal current density
$\bar{B}_p \lesssim 3.75$ T	Mean poloidal field
$I_p \bar{B}_p \lesssim 45$ MN/m	Confinement strength parameter
$q_\psi \simeq 3.3$	Edge safety factor at $I_p = 12$ MA
$V_o \simeq 10$ m^3	Plasma volume
$S_o \simeq 36$ m^2	Plasma surface area
$P_{RF} \lesssim 18$ MW	Injected heating power
$I_M \lesssim 88.2$ MA $\times$ turn	Toroidal magnet current

- a relatively high degree of purity that excludes dilution of the reacting nuclei and excessive loss of thermal energy by radiation. The adopted criterion is that the effective charge number Z_{eff} should not be in excess of about 1.6 in Ignitor. The high values of the plasma density and of the toroidal magnetic field B_T, and the low thermal loads on the first wall, which are expected in Ignitor under low temperature ignition conditions are well proven factors to obtain low values of Z_{eff};

- relatively high plasma edge densities that help to confine impurities to the scrape off layer, where the induced radiation distributes the thermal wall loading on the first wall more uniformly;

- good confinement of the plasma thermal energy and of the high energy α-particles in the central part of the plasma column: the large currents that Ignitor can produce cause the α-particles to deposit their energy in the central region, where the diffusion coefficient for the plasma thermal energy is consistently found to be minimal;

- a paramagnetic plasma current I_θ (up to 9 MA), flowing in the poloidal direction, that can increase B_T at $R = R_o$ up to 14.5 T;

- a bootstrap current $I_{BS} \gtrsim 10\%$ of I_p in ignited regimes. This current slightly reduces the required magnetic flux variation to be produced by the poloidal field magnetic system;

- a good margin for stability, against the onset of ideal and resistive MHD $m^o = 1$ modes[8] that could hamper the attainment of ignition.[9] This is ensured by the low aspect ratio and triangularity of the equilibrium configuration combined with the low values of β_p at which Ignitor can operate.

- a good stability margin against the onset of the so called Toroidal Alfvèn modes and of the high frequency Fish-bone modes that can be driven unstable by the α-particles.[10] This is ensured by the relatively low value of the beta parameter of the α-particles, $\beta_\alpha \approx 3 \times 10^{-3}$, which is a consequence of the low ignition temperature.

3. DESCRIPTION OF THE MACHINE

Ignitor was conceived starting from the same high field magnet technology developed for the Alcator program, that involves cryogenically cooled normal conductors. The minimum temperature (about 30 K) in Ignitor is attained by helium gas cooling of the copper coils. The initial low temperature and the low current densities in the toroidal magnet allowed by the Ignitor design (> 100 MA/m^2) make the machine suitable for plasma current pulses that can exceed 10 τ_E.

The criteria that have guided the engineering design are based on the following aims:

- to create and control the desired variety of plasma configurations;

- to induce the toroidal plasma current up to the highest considered value and maintain the plasma discharge for an adequate number of expected τ_E;

- to operate with an acceptable thermal loading on the first wall;

- to withstand the relevant static, dynamic, thermal, electromagnetic, and disruptive loads on all the affected machine components;

- to provide adequate access for diagnostics and auxiliary systems;

- to have the shortest cooling time between discharges;

- to minimize the electrical power consumption;

- to be feasible within a reasonable time scale and to limit the total cost of the project.

The injected heating system (ICRH) is being designed to provide a power $P_{RF} \lesssim 18$ MW by means of 6 antennae and to cover a broad range of frequencies e.g.: 100 to 210 MHz.[11]

4. STRATEGY OF OPERATION

Free boundary numerical simulations,[5,7,12–15] using the Tokamak Simulation Code (TSC),[16] and JETTO code,[17] have shown that ignition is most effectively achieved soon after the end of the current rise. Favorable conditions are obtained in this phase of the discharge in terms of broad toroidal current density profile and, therefore, the region where $q < 1$ is limited to a small fraction (less than 1/10) of the total plasma volume. This strategy, combined with the low value of β_p that should prevent the onset of $m^o = 1$ modes, minimizes the effects of potential sawtooth oscillations.

For the maximum Ignitor parameters given in Table 2 (reference discharge), ignition can be reached (see Fig. 2), after a 3 s current ramp, at $t \simeq 4.3$ s, $T_o \simeq 11$ keV, $\tau_E \simeq 0.66$ s, $n_{eo} \simeq 1.1 \times 10^{21}$ m^{-3} and $n_{eo}/\langle n_e \rangle \simeq 2.2$, $\langle n_e \rangle$ being the volume average density, corresponding to a thermal stored energy $W \simeq 12$ MJ. In this case the peaking of the temperature profile $T_o/\langle T \rangle$ is about 3 and $Z_{eff} \simeq 1.2$. We notice also that the α-heating power P_α, which is by definition equal to the total power losses P_L at ignition, is about 18 MW, while $P_{OH} \simeq 9.5$ MW. Thus the thermal loading of the first wall is relatively mild. During the current ramp, P_{OH} increases continuously while n_e is also being increased.

The maximum Ohmic heating power density is generated in a region off the magnetic axis (typically, around $r \simeq a/2$), where the temperature is relatively low. After the current

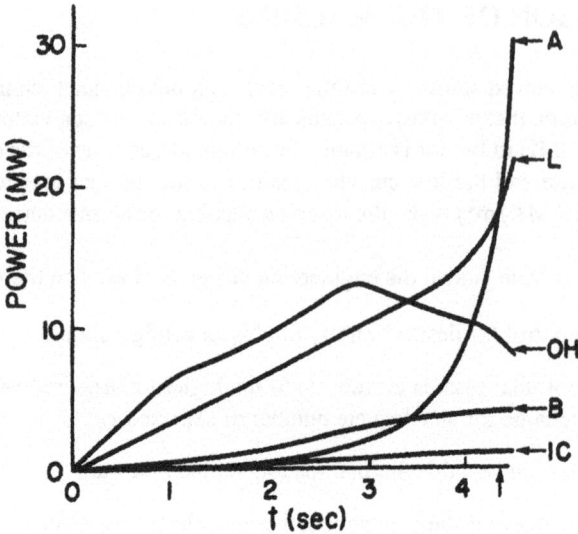

Figure 2. Time evolution of the volume integrated powers for the reference case of Table 2. The powers are labeled as: OH, ohmic heating; A, α-heating; L, total losses; B, bremsstrahlung radiation; IC, cyclotron and carbon impurity radiation ($Zeff = 1.2$).

ramp ends, P_{OH} falls gradually as the temperature rises, although not as quickly as the central voltage $V_{\phi o} \sim T_o^{-3/2}$. The relatively large values of P_{OH} are due to the fact that, under non-stationary conditions, V_ϕ is a strong function of the plasma radius and is considerably higher at the edge than in the center.

We notice that low temperature ignition is thermally unstable, since the plasma temperature tends to run away (in principle), given the temperature dependence of the fusion reactivity. At the same time, it is not difficult to envision intrinsic plasma processes that may limit the temperature excursion. Among the external means we intend to employ are the injection of pellets, the compositions and sizes of which can be chosen so that the plasma temperature is not depressed to the point where fusion burning is quenched irreparably.

The maximum magnetic field $B_T \simeq 13$ T has been taken to have a flat top of 4 s while the maximum current $I_p \simeq 12$ MA has been assumed, for the reference design, to have 3–4 s ramp and a flat top $\simeq 1$ s, followed by a gradual reduction of the current to 8 MA over 3 s. Our analyses indicate that greater values of I_p and B_T are not necessary to sustain the ignited state because the fusion α-power takes care of the power balance. Moreover, during the ramp down of the plasma current the current density is not reduced in the central plasma region ($r \leq 2/3\,a$). Thus the confinement properties of the discharge are not expected to deteriorate relative to the case where the current is not decreased. By operating at lower I_p and B_T, after ignition conditions are reached, it is possible to extend the time over which burning conditions can be sustained.

A few megawatts (e.g. 5 MW) of injected heating, started during the initial current ramp, can substantially shorten the time to ignition by increasing the central heating. It also effectively controls the size of the $q \leq 1$ region by raising the temperature in the outer half of the plasma column and slowing the rate of current penetration. High temperatures in the central region, $T_o \simeq 10$ to 15 keV, act to "freeze in" the central current density. In all the

Table 2. Reference Discharge for the Ignitor Ult Machine. ST = start of the simulation; FT = beginning of the current flat top; IG = ignition

	ST	FT	IG	
t	0.2	3.0	4.3	time (s)
R_o	1.0	1.32	1.32	major radius (m)
a	0.26	0.47	0.47	minor radius (m)
κ	1.07	1.87	1.87	elongation
δ_G	0.08	0.42	0.43	triangularity
$\ell_i/2$	0.47	0.32	0.4	internal inductance
β_p	/	0.08	0.13	poloidal beta
β	0.06	0.8	1.26	toroidal beta (%)
n_{eo}	2.5	11.0	11.0	peak electron density (10^{20} m^{-3})
$n_{\alpha o}$	/	1.5	12.0	peak α-particle density (10^{17} m^{-3})
q_o	1.75	0.83	0.71	central magnetic safety factor
q_ψ	4.2	3.3	3.6	edge magnetic safety factor
$Vol_{q=1}$	/	1.4	5.8	volume inside the $q=1$ surface (% of the total volume)
I_p	0.95	12.0	11.8	toroidal plasma current (MA)
W	0.07	7.5	11.7	internal energy (MJ)
T_o	1.1	4.0	11.0	peak electron temperature (keV)
τ_E	0.13	0.71	0.66	energy replacement time (s)
P_{OH}	1.5	13.0	9.5	Ohmic power (MW)
P_α	/	2.0	17.8	α-power (MW)
P_B	0.02	3.2	4.1	bremsstrahlung radiation power (MW)
P_{IC}	0.01	0.4	0.5	cyclotron and impurity radiation power (MW)
I_{BS}	/	0.6	1.0	bootstrap current (MA)

Ignitor cases with injected heating, the $q \leq 1$ region can be kept very small until well beyond ignition.

5. THERMAL TRANSPORT COEFFICIENTS

A model that combines the effects of transport processes originating in the main body of the plasma with those occurring in the periphery, has been adopted to identify the optimal approach to fusion burn conditions by compact ignition experiments. The observed (canonical) electron temperature profiles, both in the case where Ohmic heating is dominant and where other forms of heating are prevalent, are reproduced.

The adopted model electron heat diffusion coefficient combines the Coppi–Mazzucato–Gruber (CMG) coefficient[18] previously used to simulate Ohmic discharges[19] with an additional component χ_e^{inj} arising when injected heating is applied.[20,5] In particular, we write $\chi_e = \chi_e^{OH} + [(P_{HEAT} - P_{OH})/P_{HEAT}]\chi_e^{inj}$, where P_{HEAT} is the total input power, P_{OH} the Ohmic heating power, and $\chi_e^{OH} = \chi_e^{CMG} \times \left(V_\phi^*/V_\phi\right)^{2/3}$ extends χ_e^{CMG} to regimes in which Ohmic heating is no longer dominant. Here V_ϕ is a loop voltage and V_ϕ^* is its reference value for Ohmic discharges.

The corresponding coefficient χ_i for the ions is taken to be of the form: $\chi_i = \chi_i^{NEO} + \gamma_i[(P_{HEAT} - P_{OH})/P_{HEAT}]\chi_e^{inj}$ where γ_i is a numerical coefficient that, given the nature of the trapped particle ubiquitous mode on which the coefficient χ_e^{inj} is based, we may assume to be less than unity.

In the numerical simulations, the plasma current density diffuses according to a neoclassical electrical resistivity. Specified density profiles of the main ion species and impurities are imposed and recycling from the walls and energy losses due to ionization and charge exchange are neglected.

Simulations with different operational strategies (maximum plasma currents of 8, 10 and 12 MA, different levels of ICRH power, different values and time evolutions of the plasma density, different locations of the magnetic null point at the breakdown, etc.,) and with degraded confinement (Z_{eff} increased from 1.2 to 1.6, the anomalous part of the ion thermal diffusion coefficient increased by a factor 3, etc.,) have also been run[6] in order to find the conditions under which ignition can be obtained or to find the maximum value of the ratio between the α-particle power and the losses that can be reached under these conditions.

6. CONSTRUCTION OF KEY COMPONENTS

Since the feasibility of the machine core has been extensively analyzed, the construction of the key components has been undertaken and is close to completion. This has had the aim of validating the feasibility studies and, at the same time, of developing the tooling and the processes necessary to fabricate multiples of the following components:

- a complete module (1/24th) of the toroidal magnet that includes 10 copper plates (turns) and their insulating structure,

- a stainless steel C-clamp that constitutes the major structural element of the machine,

- a 30° sector of the thick (17 to 35 mm, Inconel 625) plasma chamber including one of the large equatorial access ports, and a 30° sector without a port. A 60° sector was then obtained by welding, with acceptable levels of deformation, and tested under vacuum and cryogenic conditions,

- a segment of the tensioning wedge system. Successful tests of this system have been performed,

- the innermost component of the central solenoid (air core transformer) consisting of a set of four tightly wound copper coils, their cooling channels and insulation system, for which construction is still underway,

- two stainless steel cylindrical segments of the Central Post, to verify the feasibility of electron beam welding over the 350 mm radius.

7. ADVANCED PLASMA REGIME EXPERIMENTS

Given the wide range of plasma currents and particle densities that Ignitor is designed to produce, and its powerful ICRH system, a significant variety of advanced plasma regimes can be explored.[21]

Since the current required to contain the 14.7 MeV protons produced by D–^{3}He reactions is about 6 MA, their effects on the thermal plasma can be studied by generating an appropriate high energy ^{3}He population with the ICRH system. Thus about 1 MW of D–^{3}He fusion power can be produced.

This may open the way to the study of the conditions required for the ignition of a D–^{3}He plasma, and is of interest in view of the appeal of a "neutron-poor" fusion reactor.

Access to the Second Stability Region, by operating at values of $q_o > 1$, can be explored by using the ICRH system for plasma heating, taking advantage of the favorable aspect ratio of Ignitor, to produce interesting plasma parameters. When operating at relatively low magnetic field, long lasting plasmas can be maintained, taking into account the fact that the magnets are cooled initially to 30 K.

In this Ignitor compares favorably to existing and proposed near term superconducting machines. The appropriate pellet injection system that has been tested successfully can be used in combination with ICRH heating to broaden the variety of regimes to be investigated.

8. CONCLUSIONS

Experience with substantial α-particle heating in reactor relevant regimes (pure plasmas, $T_i \approx T_e$, confined α-particles with a slowing-down time shorter than the energy confinement time, etc.,) is fundamental and urgently required in order to assess the potential of Magnetic Confinement Fusion as an energy source.

At present, the only option which can achieve this goal within the constraints of available technology and materials is offered by high field compact experiments (which are also of interest for Advanced Reactors).

Ignitor embodies this concept and has reached an advanced state of development and can thus be realized on a fast time scale.

Although Ignitor is designed to achieve its objective well within the explored tokamak stability domain, the $P_{RF} \lesssim 18$ MW RF system offers possibilities of advanced mode operation.

ACKNOWLEDGMENTS

This work is sponsored in part by CNR, ENEA and MURST of Italy and by the US Department of Energy.

References

[1] G. J. Boxman, B. Coppi, L. C. De Kock et al., "Low and High Density Operation of Alcator," *Proc. 7th Eur. Conf. on Plasma Physics*, 1975, Ecole Polytechnique Fédérale de Lausanne, Switzerland, **2**, 14 (1976).
[2] B. Coppi, "High Current Density Tritium Burner," *Comm. Plasma Phys. Cont. Fusion* **3**, 2 (1977).
[3] B. Coppi, *Vuoto* **18**, 153 (1988).
[4] B. Coppi and F. Pegoraro, *Il Nuovo Cimento* **9 D**, 691 (1987).
[5] B. Coppi, R. Englade, M. Nassi et al., "Current Density Transport, Confinement and Fusion Burn Conditions," *Proc. 13th Int. Conf. on Plasma Physics and Controlled Nuclear Fusion Research*, Washington D. C., USA, 1990, I. A. E. A., Vienna, **2**, 337 (1991).
[6] B. Coppi, M. Nassi and the Ignitor Project Group, *"Physics Criteria and Design Solutions for an Advanced Ignition Experiment,"* M. I. T. Report PTP-92/16 R. L. E., Cambridge, MA (1992).
[7] B. Coppi, M. Nassi and L. E. Sugiyama, *Physica Scripta* **45**, 112 (1992).
[8] A. C. Coppi and B. Coppi, *Nucl. Fusion* **32**, 205 (1992).
[9] B. Coppi, A. Taroni and G. Cenacchi, in Plasma Physics and Controlled Nuclear Fusion Research 1978, *Proc. of the 7th Int. Conf. Innsbruck*, **1**, 487 (1977).
[10] F. Romanelli, G. Vlad and F. Zonca, *ENEA Report* (in preparation).

[11] F. Carpignano, B. Coppi, P. Detragiache et al., "ICRF System and Plasma Performance of the Ignitor Experiments," *10th Topical Conference on Radio Frequency Power in Plasma*, Boston, MA, 1993.
[12] B. Coppi, M. Nassi and L. E. Sugiyama, *Fusion Technology* 21, 1612 (1992).
[13] L. E. Sugiyama and M. Nassi, *Nucl. Fusion* 32, 387 (1992).
[14] B. Coppi, P. Detragiache, S. Migliuolo et al., "Reconnection and Transport in High Temperature Regimes," *Proc. 14th Int. Conf. on Plasma Physics and Controlled Nuclear Fusion Research*, Wurzburg, Germany, 1992, paper I.A.E.A.-CN-56/D-3-1(C), Vol. 2 (1993).
[15] A. Airoldi and G. Cenacchi, *Fusion Technology* 25, 278 (1994).
[16] S. C. Jardin, N. Pomphrey and J. Delucia, *J. Comp. Phys.* 66, 481 (1986).
[17] G. Cenacchi and A. Taroni, JET-IR(88)03 (1988).
[18] B. Coppi and E. Mazzucato, *Phys. Rev. Lett.* 71A, 337 (1979).
[19] R. C. Englade, *Nucl. Fusion* 29, 999 (1989).
[20] B. Coppi, *Comm. Plasma Phys. Cont. Fusion* 12, 319 (1989).
[21] B. Coppi, P. Detragiache, S. Migliuolo et al., *Fusion Technology*, 25, 353 (1994).

10

THE INERTIAL ELECTROSTATIC CONFINEMENT APPROACH TO FUSION POWER

George H. Miley

Fusion Studies Laboratory
University of Illinois
103 South Goodwin Avenue, Urbana, Illinois 61801-2984

ABSTRACT

Inertial electrostatic confinement (IEC) of a non-Maxwellian beam-dominated plasma for fusion was originally proposed in the 1950s, but since then, only sporadic work has been devoted to the subject. Nevertheless, recent experiments have shown that small IEC devices are well-suited for commercial applications as a portable low-level neutron source for activation analysis. However, the scaling to a high- power fusion reactor is uncertain, due to the lack of experimental data with higher input currents. Three key issues need to be resolved: the stability of multiple-potential-well structures, the confinement time of energetic ions trapped in such wells, and the protection of grid structures during high-power operation. Conceptual design studies that assume a positive resolution of these issues show, however, that the resulting reactor would be economically attractive and very versatile.

INTRODUCTION

Inertial electrostatic confinement (IEC), as an approach to fusion power, was proposed originally in the late 1950s.[1,2] Despite some promising modest experiments in the 1960s, little work was done on this concept until recently, when a series of small-scale "pure" (i.e. the fusion plasma is confined solely by inertial electrostatic forces) IEC-device studies were initiated at the University of Illinois (UI),[3,4] and the High Energy Power System (HEPS) experiment, employing a hybrid magnetic IEC device, was begun by Directed Technologies, Inc. and the University of Wisconsin.[5] Ample data exists to show that the UI IEC devices are ideally suited for immediate application as low-level neutron sources.[3,6,7] However, the physical/technological feasibility of extending the concept to a high-power fusion source remains uncertain, due to the lack of experimental data needed to establish a plasma confinement scaling law. In addition, protection of the grid structure

Current Trends in International Fusion Research
edited by Panarella, Plenum Press, New York, 1997

poses a major technological problem. If a power source can be developed, however, the reactor embodiment appears to be extremely competitive and flexible. Recent conceptual design studies have examined applications ranging from a conventional electric power plant[8,9] to a hydrogen generator,[10] a space propulsion unit,[11,12] and a boron neutron capture therapy (BNCT) cancer treatment facility.[13]

IEC CONCEPTS

In the UI "pure" IEC devices, ion beams are injected into a spherical vacuum vessel containing one or more sets of spherical wire grids. A high fusion rate is generated by intersecting ion beams and the associated potential well structure in a dense plasma region created in the center of a volume circumscribed by the innermost wire cathode grid. (Fig. 1) Here, we describe a unique configuration, termed the IEC-Gaseous Discharge (IECGD), where a gaseous discharge in a single-gridded device serves as the ion source. The "Star" discharge operation mode, discovered at UI, maximizes the effective grid

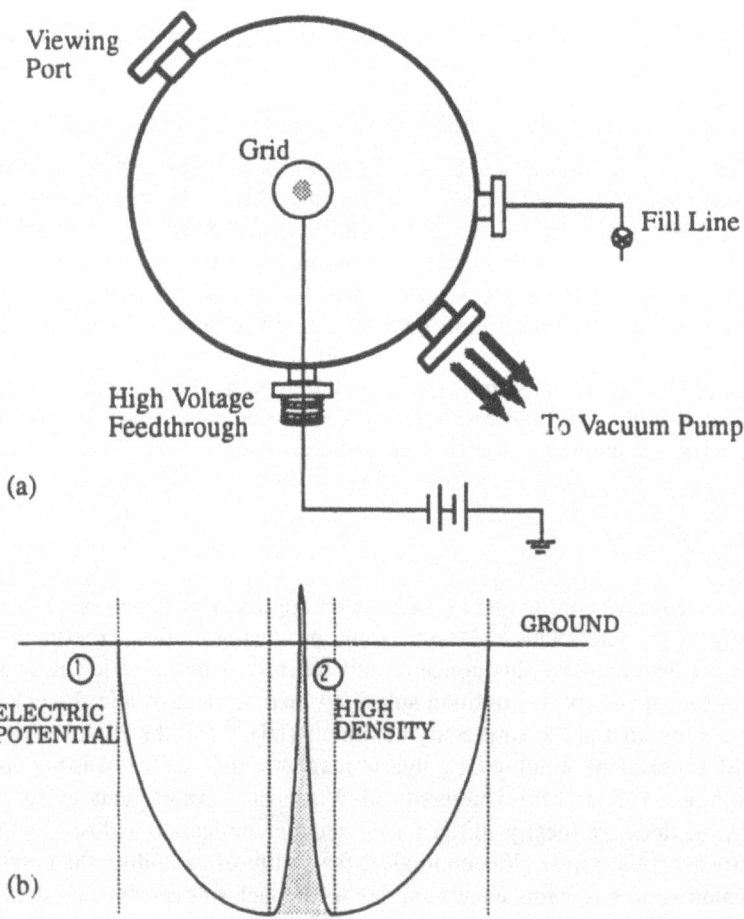

Figure 1. a) Basic components of an IECGD. b) The associated electric potential.

transmission factor for ions. This configuration, then, provides a simple, rugged, low-cost fusion neutron source, operating in the 10^6 D-D n/s or 10^8 D-T n/s range. The extension to an intense MeV proton source using D-^{3}He reactions is also possible.

The feature which distinguishes the present IEC design from earlier devices of this type is the use of a single grid to produce a gaseous discharge for ion production. The grid simultaneously serves to extract high energy ions from the discharge. This is important, in that IEC designs using other ion sources, such as external ion guns (see Hirsch's design[2]) operate at lower vacuum pressures, thus in a markedly different plasma regime. The IECGD avoids the complication of the ion guns—the goal being the simplest and cheapest design for low-level neutron production. Such a neutron source has a variety of applications involving neutron activation techniques and neutron diagnostics calibration.

PRESENT IEC NEUTRON SOURCE EXPERIMENT

Previous experimental IEC device studies employing ion-gun injectors demonstrated the ability to generate $\sim 10^9$ D-T n/s at currents and voltages within the limits set by grid heating tolerances and voltage breakdown limits.[2] However, the goal of present experiments was to develop a simpler, rugged IEC that could dependably generate steady-state yields of $\sim 10^6$–10^7 D-D n/s.

Description of Device

Two different IECGDs, designated "A" and "B," were employed. Both use a unique grid-discharge design (Fig. 1), as opposed to the Hirsch-type ion gun-injected IEC.[2] IEC-A has a 30-cm dia. vacuum vessel fabricated from 0.48-cm 304 stainless steel. The other vessel, IEC-B, has a 61-cm dia. IEC-A is vacuum-pumped with an 80-liters/second (l/s) turbo-pump, backed by a mechanical roughing pump. IEC-B is initially vacuum-pumped by a sorption pump and then switched to a 1000 l/s cryopump. Vacuum base pressures of 10^{-7} Torr are achievable in both devices without baking the chambers. The results reported in this work are for IEC-A, which incorporates nearly the optimum size and design elements for a portable neutron/proton source. IEC-B is used primarily for studying IEC scaling laws and IEC diagnostics.

Different-sized cathode grids were installed to study the effects of grid size. Typically, cathode grids of 7.6-cm and 3.75-cm dia. were used in IEC-A. The grids were made from various sizes of T302/304 stainless steel wire: 0.80 mm, 1.04 mm, and 1.30 mm in diameter. All of the grids had a geometric transparency of ~ 80–97%, with an estimated <3% deviation from exact sphericity.

Prior to operation, the IECGD is conditioned to remove absorbed gas impurities, using extended glow discharge operation. Then the vessel is pumped down, back-filled to ~ 5–20 mTorr, and a 10- to 80-kV electric potential is applied to the cathode to initiate the glow discharge. The voltage and pressure are generally related by the traditional Paschen voltage breakdown relation,[14] where the voltage is a function of a pressure-length product. [For the IECGD, the "length" in this relation is the distance from the grid to the vessel wall (vs. the grid diameter).[15]]

The diagnostics employed included pressure sensors and current and voltage meters on the cathode power supply. A BF_3 proportional counter, placed 40 cm from the IEC chamber and surrounded by a 9 cm-thick polyethylene cylinder (for thermalization of neutrons), was used to measure neutron source strength.

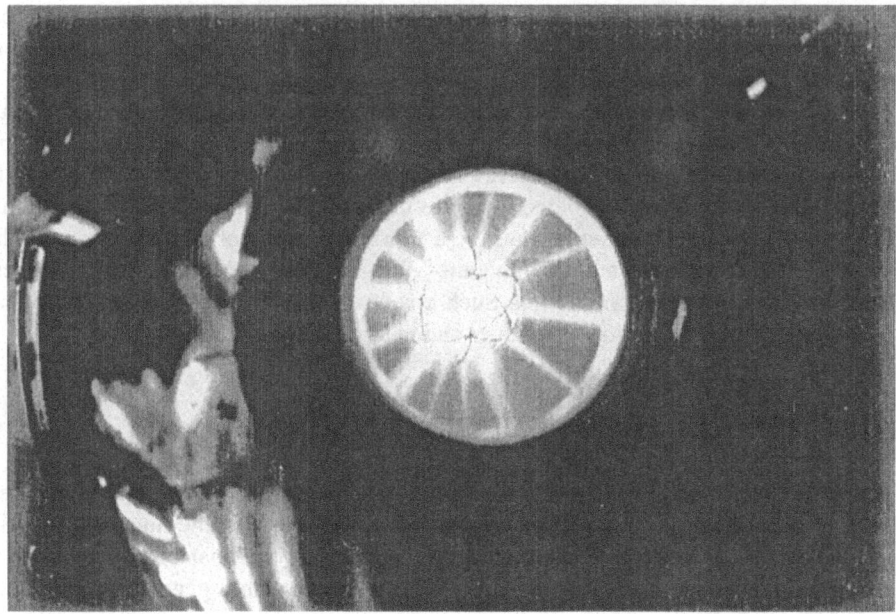

Figure 2. Photograph of the Star discharge operational mode. Beamlets of ions converge in the center of the volume circumscribed by the wire cathode grid, creating high fusion power densities in the central core region.

Modes of Operation

Glow discharge operation of the IECGD is categorized by three distinct discharge "modes": Star, Central Spot, and Halo. These names are quite descriptive of the visual appearances of the light emitted from the discharges. All three modes are reproducible and stable; each is associated with a different potential well structure, hence neutron production rate. The Star mode was used extensively in recent experiments. It is distinguished by microchannels or "spokes" radiating outward from a bright center spot (Fig. 2). As verified by magnetic deflection experiments, the spokes are primarily composed of ion beams, aligned so that they pass through the center of the openings delineated by the grid-wires. This mode is very efficient for neutron production, since the large effective grid transparency allows numerous passes of ions through the center spot before being intercepted by the grid. The Star mode is typically obtained in both IEC-A and IEC-B at lower operating pressures (<10 mTorr) and higher voltages (>30 kV), using a carefully formed grid with good sphericity and high transparency (>95%).

Measured Neutron Source Strengths

Typical plots of measured neutron source strength vs. cathode current in IEC-A are shown in Fig. 3 for different cathode voltages and currents. The neutron yield increases linearly with current and scales strongly with voltage, in agreement with the modified beam-background model described in Ref. 16. The scaling with voltage roughly corresponds to the variation of the fusion cross section with energy, i.e. approaches an exponential increase in this range of voltages. For the practical applications of interest here, operation is limited to ~70 kV, avoiding use of more complex high-voltage handling

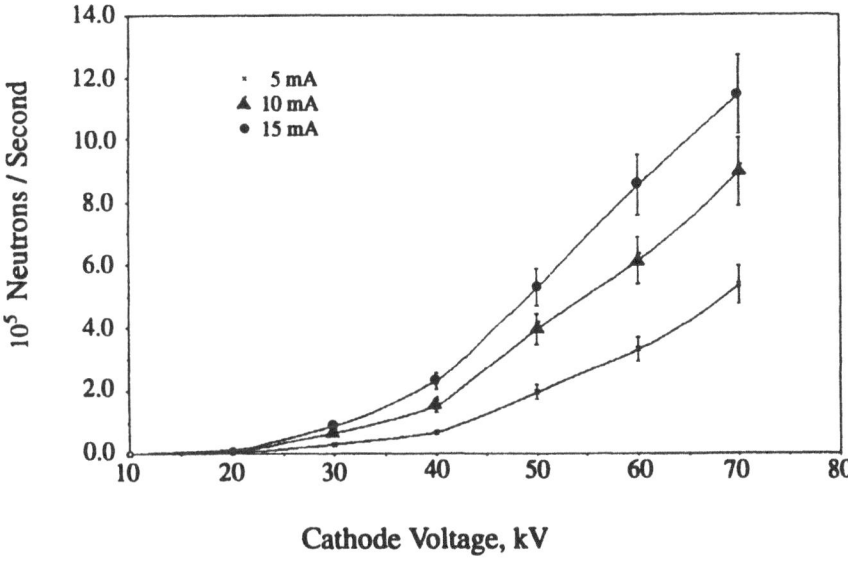

Figure 3. Observed variation of neutron output for IEC-A with voltage for several ion currents.

equipment. With this voltage limit, the maximum current is then set by the power heating limits for the grid. As seen from Fig. 3, the target source rate of 10^6 D-D n/s is achieved at 70 kV with a current of ~15 mA.

The variation of neutron production with grid design is significant, but represents a complicated issue which is not addressed here. All experiments shown in Fig. 3 used the standard IEC-A grid described earlier. Results from experiments with five different grids are presented in Ref. 16. In general, the best grid design should not only maximize the transparency without disturbing microchannel (Star) formation, but also have sufficient structural strength to maintain a near-spherical shape under high-power conditions.

Based on this data, parameters for an optimized device, designed for production of ~10^6 D-D (or 10^8 D-T) n/s, are summarized in Table I. The unit shown is restricted to 70 kV; higher neutron yields would be obtained if higher voltages were used.

Selection of Grid Materials

Present grids were constructed from stainless steel wire (vs. high-melting-point materials such as tungsten or molybdenum, which pose more difficult fabrication problems). The current limit can be illustrated by assuming that, as an upper limit, all of the power from the power supply is deposited into the grid. This represents a conservative estimate, since only the ions strike the cathode. The electrons, on the other hand, strike and deposit their energy on the anode or vessel wall. Heat is assumed to be dissipated from the grid entirely by radiation. Steel melts at 1800 K. In the present grid design, if the maximum temperature before deformation occurs is assumed to be 1500 K, the current is limited to ~35 mA at a 70-kV cathode potential. For safety, the actual current used is generally set at about one-half this theoretical limit.

If tungsten grids were used, they could operate at twice the temperature limit for steel grids. The radiated power varies as absolute temperature to the fourth power, so the

Table I. Operating parameters for a steady-state 10^6 sec^{-1} neutron source

Parameter	Value
Cathode Voltage	70 kV
Cathode Current	15 mA
IEC Diameter	30 cm
Grid Cathode Diameter	3.75 cm
Geometric Grid Transparency	95%
Background D_2 Pressure	8 mTorr

grid could operate at sixteen times the current limit for steel, or ~0.4 amps. Thus, tungsten grids hold promise as a means to further increase the neutron output in future devices, but are more difficult to fabricate. Active cooling of the grids, using a fluid coolant circulating in tubular grid wires, is also possible but adds complexity to the design and fabrication of the grid.

CORRESPONDING COMPUTER SIMULATION STUDIES

Two computer codes are employed to analyze IEC experiments. One, the IXL code, uses a one-dimensional Vlasov description coupled to Poisson's equation.[17,18] While the many assumptions used in this treatment make its degree of accuracy indeterminate, IXL has the advantage of running on a personal computer. Thus, its main utility has been for survey studies designed to identify trends. For example, IXL has been used to study ion and electron current requirements for initiating a double potential well. Some typical results are given in Refs. 19–22. As these results indicate, the threshold current predictions for double potential well formation obtained from IXL are higher than those obtained experimentally. Still, the trends predicted appear to be consistent with experimental results.

The PDS1 particle-in-cell code allows detailed one-dimensional phase space and three-dimensional velocity space studies, including a variety of atomic physics and secondary electron interactions. This code is a modification of the original PD code, developed at UC-Berkeley for use in plasma chemistry studies.[23] It was recently modified by R. Nebel et al.[19] at Los Alamos National Laboratory for use in IECGD studies by adding grid- and fusion-relevant atomic physics effects. The code has several advantages over the IXL code: 1) the PDS1 code also includes the physics of electrons and ions in the outer core region; and 2) the input parameters to the code are based on known experimental parameters, such as pressure, voltage, etc. The code also simulates self-consistent discharge processes, including external electrical circuits. However, the PDS1 run-time, even on a CRAY computer, is protracted. As a result, its main use thus far has been restricted to studies of local conditions, e.g. plasma conditions around grid wires and the associated effect of secondary electron production.[4]

SCALE-UP ISSUES

The physics issues of potential well structure and magnetic ion confinement and thermalization times are fundamental to the ultimate extension of the concept to a power-producing device. The physics involved is too complex to discuss in detail here. Further, there is considerable disagreement about the feasibility of effective beam trapping in such wells. Various theoretical studies, such as Refs. 5, 9, and 24–27, provide encouraging results, while others (Refs. 28 and 29) predict a very low Q-value, restricting the IEC to low-level neutron source applications. These issues must be resolved by further theoretical study (a fully self-consistent theoretical model for ions, electrons, plus the potential well is needed), and considerably more experimental study at higher currents is essential. A first-step objective should be the experimental measurement of the potential well structure to provide a benchmark for theoretical simulation studies. Despite the uncertainties in physics, it is still instructive to consider the present experiments and the scaling law/power flow analysis needed to project ahead to reactor parameters.

Power Flow Analysis

The power flow analysis of the IEC reactor components requires an assumption of how the fusion rate scales with injected ion current, I. Experiments have been carried out at low currents, and these results, combined with a range of theoretical scaling laws, are summarized in Fig. 4. Scaling of the fusion reaction rate with I^3 is predicted theoretically for beam-beam reactions in a deep double well. [This scaling includes a factor of I^2 from beam-beam fusion, plus an added factor of I associated with an increasing trapped ion lifetime due to increasing potential well effectiveness with current over a reasonable range of currents.] There are, however, several theories that support even higher scaling rates, due to the greater compression of the core beam-plasma associated with electrostatic wave effects. [First proposed by R.W. Bussard,[5] this is termed the "inertial collisional compression effect."] Thus, a very optimistic I^5 scaling, a median I^4 scaling, and the "conservative" I^3 scaling are included in Fig. 4 to illustrate the range of possibilities.

The ramifications of these scalings can be illustrated by considering power flows in a hypothetical 25-MWe power plant, such as the design study evaluated in Ref. 8. [In Ref.

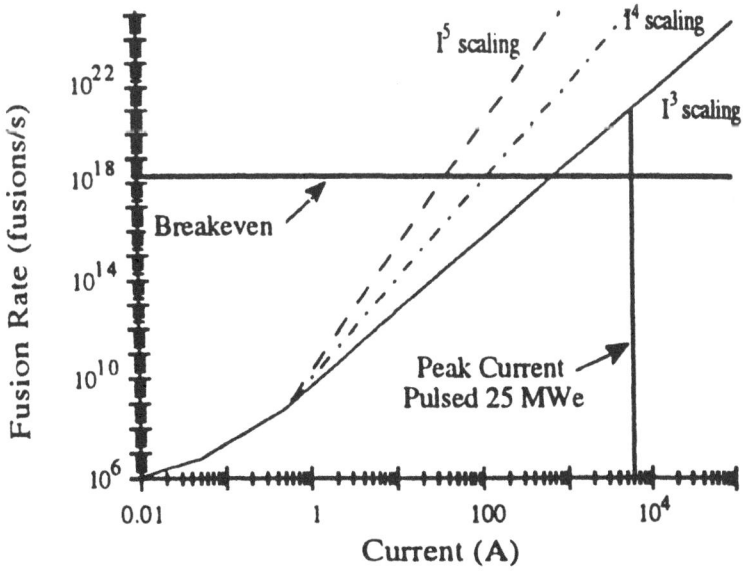

Figure 4. Range of scaling law projections for the IEC. With the higher current possible with pulsed operation, the fusion rate per time-averaged power input, i.e. the fusion energy gain, increases.

9, R.W. Bussard has also examined in some detail power flow and various technology issues for scale-up to power plants. He considers a variety of fuels, ranging from D-D to p-^{11}B. Bussard has proposed the following nomenclature, which may become standard in future papers: Inertial Electrostatic Fusion (IEF); "pure" IEC systems with spherically convergent ion flow are termed IXL systems, while the hybrid magnetic version is termed an EXL system.]

Conservative I^3 Scaling for Design Study

The more modest I^3 scaling law was chosen for a conceptual design study of a D-^{3}He IEC pilot plant.[8] With this scaling, the size of the power supply and the reactor itself become unreasonably large if steady-state operation is used. However, pulsed operation looks encouraging. The high peak currents required to take advantage of the strong Q-value scaling with current can be obtained with conventional pulsed power technology. A summary of the 25-MWe pilot power plant design is given in Table II. The size represents the radius of the spherical vacuum chamber wall, exclusive of cooling systems. Weights are for the IEC unit alone and for the entire reactor system. Most of the weight is accounted for by the large vacuum chamber wall. The relatively low weight of the 25-MWe unit, including the direct converter, is one of the attractive features of the IEC power plant, implying competitive costs for materials and construction. This is a general characteristic of the IEC concept. Thus, IEC reactors have a higher mass power density than conventional fusion systems, suggesting costs much more competitive with a light water reactor than those projected for other fusion reactor designs.[8,9,24]

The scaling law of the fusion rate determines the input power required for a 25-MWe output; however, the physical size of the reactor is determined by the limitations associated with grid heating. The small inner grid of the IEC is the primary concern—it receives a large power loading, which generally necessitates forced convection cooling in addition to radiative cooling. The other grids are larger, and their greater surface area allows sufficient heat to be dissipated by radiative cooling alone. The grids are made of tungsten; the inner grid consists of tungsten tubing, through which coolant is pumped. The heat radiated from the two IEC grids, as well as the heat from the direct energy converter (DEC) collector plates, would eventually be absorbed by the chamber wall. In the present case, 4.8 MW is received by a wall surface area of 940 m^2. Thus, this wall also employs a forced convection cooling system.

The injected current pulse width is held in the ms range, so that the potential well structure remains in quasi-equilibrium. A peak pulsed current of 6 kA is used, producing a peak fusion rate of 10^{21} fusions/sec, or a peak output power of 2550 MWe. With a 1/100 duty cycle, an output of 25 MWe is obtained with 4.8-MW injected. The pulsed power technology employed falls well within the capabilities of present state-of-the-art systems, e.g. those used for light ion beam fusion experiments.[30]

Table II. Results of IEC power plant analysis for an attractive 25-MWe pilot plant utilizing pulsed power operation and assuming an I^3 fusion rate scaling law

Input power (MW)	4.80
Size/radius (m)	6.49
IEC weight (tonnes)	2.84
Total reactor estimated weight (tonnes)	210

The energy output from the D-^{3}He fusion reaction is in the form of high-energy charged particles (14.7-MeV protons and 3.54-MeV alpha particles). In addition, there are secondary reactions of D-D and D-T fusion that produce the following charged particles: 0.8-MeV ^{3}He ions, 1.01-MeV tritium ions, 1.5-MeV alpha particles, 3.1-MeV protons, and 14.1-MeV and 1.45-MeV neutrons. However, the energy fraction carried by neutrons can be held to <10%, so their effect on structural components is minimized.

A "Venetian blind" DEC is employed to collect the energy of the charged particles and convert it to useful electrical power.[8] This type of DEC offers a high efficiency (≈80%) in the case of the nearly monoenergetic fusion product ions. Two sets of collector plates are employed, each being held at a potential slightly below the average energy of the particle that it is designed to collect. The inner plates at 3-MV collect the lower-energy particles, while the 14-MV outer plates collect the D-^{3}He protons. Electron grid suppressors minimize unwanted leakage current arising from secondary electrons.

HYBRID MAGNETIC IEC

In an effort to solve the grid problem, R.W. Bussard recently proposed an alternate configuration, whereby a quasi-spherical magnetic field be used to contain electrons in the

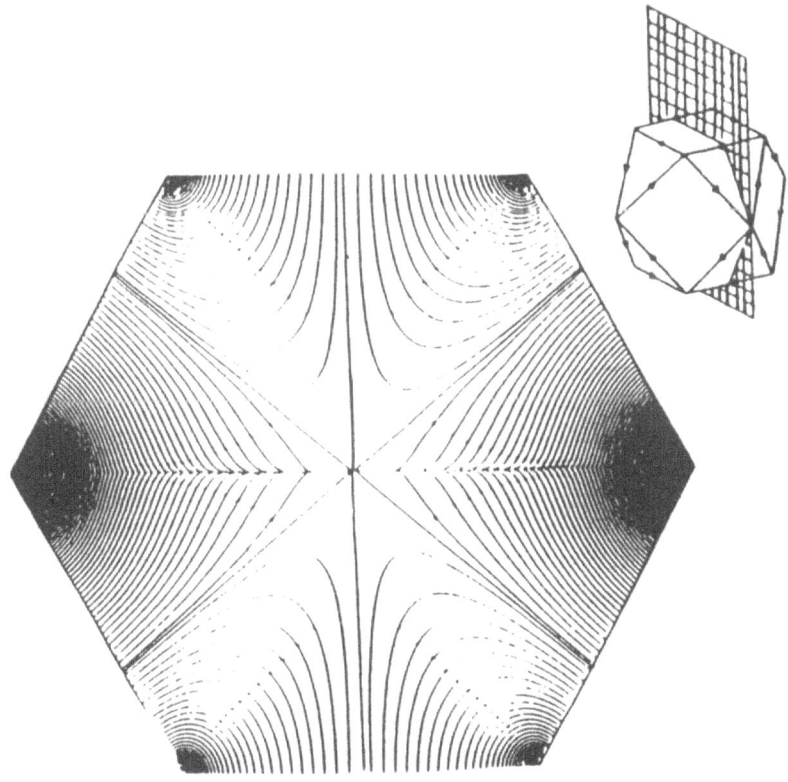

Lines of Force

Figure 5. IEC magnetic trapping well formed by energetic electron injection into cusps of polyhedral magnetic fields. (From Ref. 26).

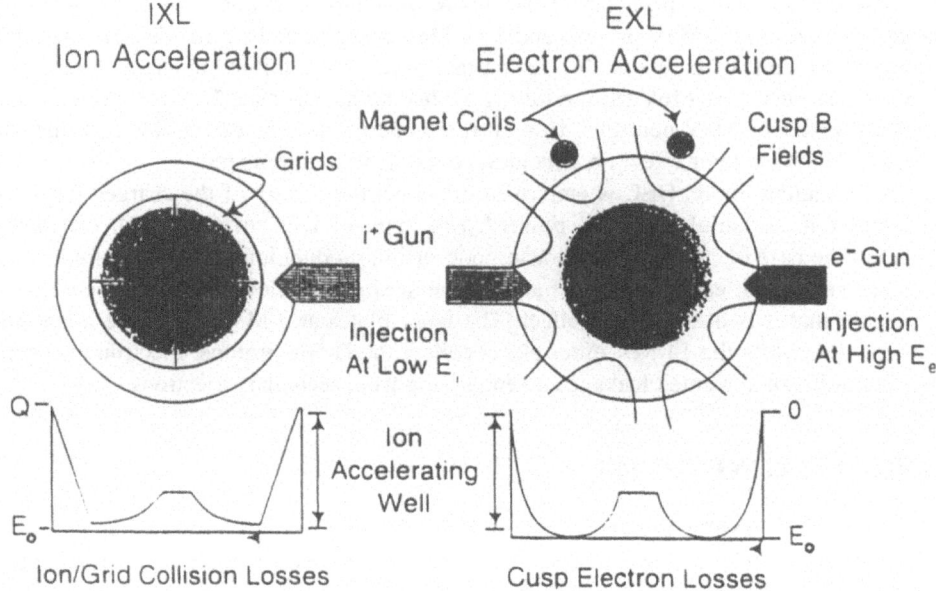

Figure 6. Two basic approaches to IEC fusion: "pure" IEC (IXL) and hybrid magnetic (EXL) systems. (From Ref. 24).

IEC configuration.[9] (Fig. 5.) In this concept, termed a Polywell™, magnetic geometry is used (Fig. 6), and electrons are first injected between cusps in the magnetic field. This creates a trapping well, such that injected ions fall into the well. Thus, in this configuration, the final potential structure is much the same as that for the pure IEC, however, the grid structure is eliminated. This configuration has been examined fairly extensively from a theoretical point of view by R.W. Bussard, N.A. Krall, et al.[5,9,24,25,26,27] This work concludes that despite the highly non-Maxwellian distribution, no show-stopping instabilities are predicted. Electron losses through the cusps in the polyhedral magnetic field could cause a fatal energy drain; however, it is thought that operation at high-β ("wiffle-ball" mode) would cause the plasma to distort field lines at the cusps, partially closing the loss cones. In that case, the loss rates would be reduced to acceptable levels. More recently, concern has been raised about thermalization of trapped ions and upscattering out of the well.[28,29] This is a generic problem that must also be faced in the conventional IEC, and the extent of the effect is still under study. As noted earlier, a comprehensive theoretical study employing a fully self-consistent plasma-potential-field-structure has not been developed, but is essential to fully evaluate the physics of high-current IECs.

Some experimental work was carried out on the Polywell™ configuration on the HEPS device in 1992. It was shown that a good electron well could be created, but lack of funding prevented additional experiments.[31] HEPS has since been moved to the University of Wisconsin-Madison (UW), where future experiments using it are being planned. In preparation for these experiments, UW staff have carried out some conventional IEC experiments, using a small device similar to the UI device. They have reported core convergence measurements, consistent with the UI studies, which show that the core radius is almost one order of magnitude smaller than the ion ballistic radius, confirming good convergence.[32] These experiments differ from UI studies, however, in that a lower pressure, non-Star mode operation is used. [The UI/LANL triple-grid experiments[4] also operate at a lower pressure, but make use of the Star discharge mode.] Also, a hydrogen plasma is employed, so that neutron production is not studied directly. Diagnostics rely primarily on Langmuir probe techniques, recordings of emitted light profiles, and foil vaporization measurements; laser-fluorescent measurements are planned.

There are several other related confinement configurations under study. As T. Dolan (Ref. 33) has pointed out, following the initial work by O.A. Lavrent'ev in Russia, a wide variety of confinement experiments involving electrostatic fields have been performed there. However, none of these studies have strictly followed the IEC route—they have typically focused on electrostatic stoppering of magnetic field configurations, and none have involved three-dimensional core convergence, which is so important to obtain high densities, hence high reaction rates, in the IEC.

The MIGMA configuration, pioneered by B. Maglich[34] uses a magnetic field to focus and collide ions at a single point in the center of the "migma." It shares, then, the colliding beam feature with the IEC. However, it differs from the IEC in two important ways: three-dimensional convergence is not used in MIGMA, and there is not a self-consistent field structure like the potential well structure created by electron and ion flow in the IEC. Thus, external electron injection into the migma becomes necessary to prevent positive space charge build-up. For these reasons, the present embodiment of the MIGMA appears to be limited to much lower power densities than the IEC.

As pointed out by D. Barnes,[35] the well-known Penning trap configuration has a mathematical similarity to the IEC. This line of reasoning has led to a concept to achieve high ion densities in a Penning trap configuration, composed of a large number of small-diameter cylindrical traps. If successful, such an approach could lead to very attractive small modular power units.

IEC APPLICATIONS IN DESIGN STUDIES

R.W. Bussard[12] carried out pioneering conceptual design studies for use of the IEC to drive an electric propulsion unit for space travel. In this design, (Fig. 7) the high-voltage output from a direct collector on a p-^{11}B IEC power unit is fed to a high-voltage diode, driving an energetic electron beam. This beam is directed into a liquid hydrogen fluid

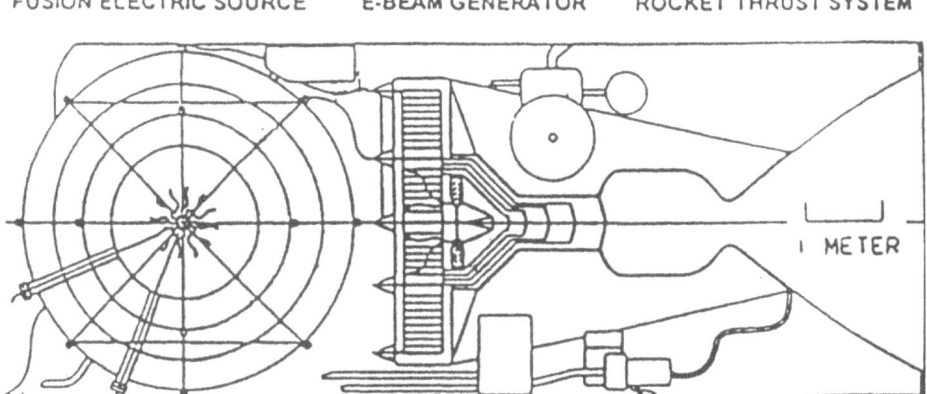

Figure 7. Schematic of Bussard's QED engine. (From Ref. 12).

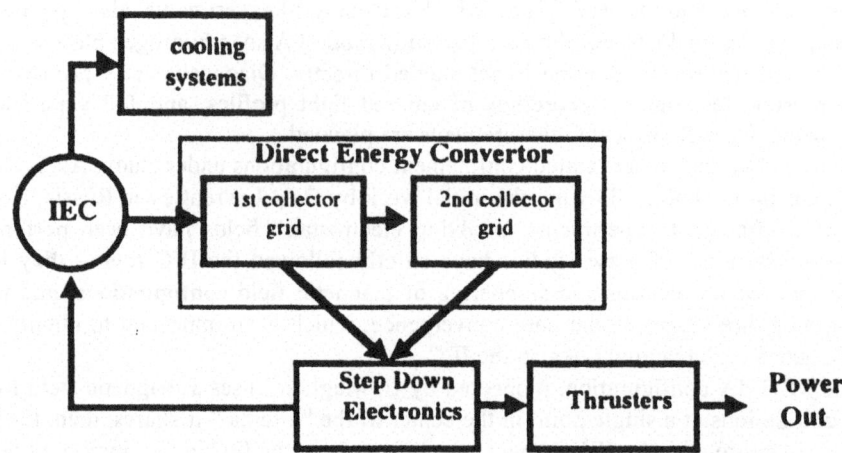

Figure 8. Block diagram of the power flow in an IEC fusion propulsion system.

layer in the nozzle of a gas-expanding thruster. The vaporized hydrogen is expanded through the nozzle, providing a high specific impulse rocket. Bussard terms this concept the QED engine and concludes that such systems can "outperform all other advanced concepts for controlled fusion propulsion by 2–3 orders of magnitude."

IEC propulsion applications were recently considered in a conceptual design study carried out at UI. This unit burned D-^{3}He, and instead of using an electron-beam diode, employed a voltage step-down converter to couple the output to a electrically driven ion thruster.[11] (Fig. 8) With five such thrusters on the spacecraft (Fig. 9), a total thrust of 6 MW was achieved, providing the capability of a fast roundtrip Mars mission.

The propulsion studies noted above capitalize on the high power/weight ratio possible with an IEC power unit. Indeed, as noted in the earlier section on scaling laws, a similar power unit would be attractive for conventional electric power production, especially in cases where smaller units are desirable.

Other applications for the IEC are possible, employing either the high-energy neutron flux generated by D-T or D-D versions, or the 14-MeV protons from a D-^{3}He version. As noted earlier, the present UI devices produce neutrons at a level of $\approx 10^6$ n/s, making

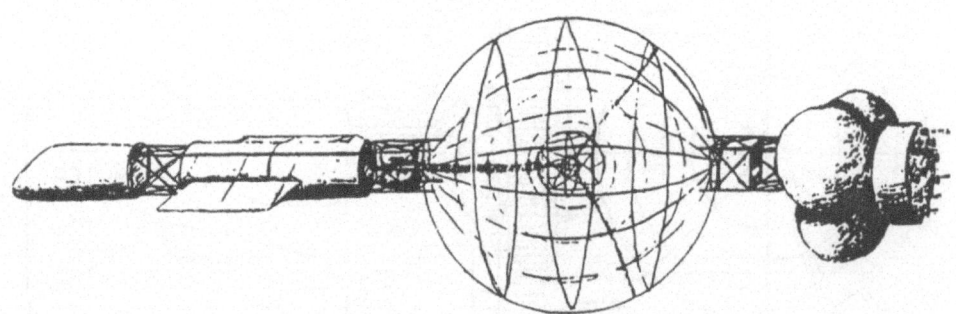

Figure 9. Illustration of conceptual design of the IEC fusion propelled rocket. The thrusters and propellant tanks are to the right, the multi-gridded IEC and DEC in the center, and the control cabin to the left of the diagram.

them attractive for portable activation analysis applications. If these levels can be scaled up 3–4 orders of magnitude, a number of additional important applications become feasible, e.g. boron neutron capture therapy (BNCT), medical radioisotope production, and positron emission tomography (PET). A preliminary design study for application in BNCT treatment facilities has been carried out.[13] If the IEC can be developed to this level, the ability to locate treatment units in local clinics and hospitals would completely revolutionize the present approach, which currently relies on use of centralized facilities at large fission reactor sites.

SUMMARY AND CONCLUSIONS

The validity of the basic IEC concept has been demonstrated at low injected currents (<0.1 A), and such devices have achieved a role as a low-level, portable neutron source for various neutron activation analysis applications. At these fusion reaction levels, however, the IEC is orders of magnitude away from a power-producing device; e.g. the 25-MWe conceptual power plant discussed earlier would employ peak pulsed currents of 6 kA. The scale-up to those levels requires the solution of a variety of physics issues involving the formation of a stable, deep, double potential well and adequate lifetime of energetic ions trapped in the well. A variety of technology issues, mainly related to grid protection or the substitution of magnetic fields for the grid, must also be studied in detail. Nevertheless, conceptual reactor studies using the IEC concept appear so attractive that an aggressive R&D program devoted to these issues appears well warranted. However, in view of the present funding limitation on alternative approaches by DOE's Office of Fusion Energy, such basic programs are difficult to start.

Fortunately, commercial applications for the IEC as a neutron or proton source may help drive this development. There are a variety of added neutron applications, e.g. medical isotope production and neutron tomography, that could employ a scaled-up version with higher neutron yields, but still be well below energy breakeven. At yet higher yields, BNCT becomes feasible. A corresponding D-^{3}He version would have proton intensities suitable for applications such as PET. Finally, near energy breakeven levels, the neutron flux from a D-T version would become useful for fusion materials studies. Thus, it may be that these various applications at intermediate steps along the route to a fusion power reactor can provide the incremental funding necessary for ultimate power development. If so, this commercial development path would be a unique one for the fusion community.

ACKNOWLEDGMENTS

This work was partially supported by Contract DOE/LANL 9-XG2–45958 and a grant from the Electric Power Research Institute. Contributions by many collaborators and colleagues listed in the references are gratefully acknowledged. Inspiration gained from a brief chance conversation with P. Farnsworth, the "father" of television, and later discussions and encouragement by R. Hirsch and R.W. Bussard represent especially memorable moments for the author.

REFERENCES

1. P. Farnsworth, U.S. Patent #3,258,402, June 28, 1966; U.S. Patent #3,386,883, June 4, 1968.

2. R. Hirsch, J. Appl. Phys., 38, 4522 (1967).
3. G.H. Miley, et al., Dense Z-Pinches, AIP Conf. Proc. #299, eds. M. Haines and A. Knight, AIP Press, New York, 675–688 (1994).
4. G.H. Miley, et al., Experimental and Theoretical Studies of Inertial Electrostatic Confinement, Final Report, DOE/LANL 9-XG2–45958 (1995).
5. R.W. Bussard, Fusion Tech., 19, 2, 273–293 (1991).
6. J.H. Nadler, G.H. Miley, et al., Fusion Tech., 21, 1639–1643 (1992).
7. Y. Gu, J.B. Javedani, G.H. Miley, Fusion Tech., 26, 929–932 (1994).
8. G.H. Miley, A.J. Satsangi, Y. Yamamoto, H. Nakashima, and J.B. Javedani, 15th Symposium on Fusion Engineering, IEEE No. 93CH3348–0, 1, 161–164 (1993).
9. R.W. Bussard, Final Report, EPRI-RP-8016–12 (1993).
10. G.H. Miley, A. Satsangi, J. Javedani, and Y. Yamamoto, Seventh Intern. Conf. on Emerging Nucl. Energy Systems (ICENES '93), ed. H. Yasuda, World Scientific, Singapore, 66–70 (1994).
11. A.J. Satsangi, G.H. Miley, J.B. Javedani, H. Nakashima, and Y. Yamamoto, AIP Conf. Proc. 301: 11th Symp. on Space Nucl. Power and Propulsion, eds. M.S. El-Genk and M.D. Hoover, Conf. 940101, AIP Press, 1297–1302 (1994). Also see G.H. Miley, et al. 1st BNCT Workshop.
12. R.W. Bussard and L.W. Jameson, AIP Conf. Proc. 301: 11th Symp. on Space Nucl. Power and Propulsion, eds. M.S. El-Genk and M.D. Hoover, Conf. 940101, AIP Press, 1289–1296 (1994).
13. Mora, M. Hulin, L. He, A. Sainz, C. Butzow, D. Worthington, E. Schaefer, T. Moschetti, and G.H. Miley, Trans. ANS, (Philadelphia, PA, 23–29 June 1995). In Press.
14. A. VonEngel, Electric Plasmas: Their Nature and Uses, Taylor & Francis, Ltd, New York, 121–127 (1983).
15. T.A. Hochberg, "Characterization and Modeling of the Gas Discharge in an SFID Neutron Generator," M.S. Thesis, Department of Nuclear Engineering, Univ. of Illinois, Urbana, IL (1992).
16. J.H. Nadler, G.H. Miley, Y. Gu, and T. Hochberg, Fusion Tech., 21, 1639 (1992).
17. D. Smithe, Mission Research Corporation Report, MRC/WDC-R-226, Mission Research Corporation, Washington, DC (1990).
18. K. King and R.W. Bussard, Energy/Matter Conversion Corporation Report, EMC2–1191–03, EMC2, Manassas, VA (1991).
19. R.A. Nebel, L. Turner, R.W. Bussard, J. Bates, H.R. Lewis, and G.H. Miley, Bult. APS, 1582 (1992).
20. T.N. Tiouririne, R.A. Nebel, L. Turner, W.D. Nystrom, R.W. Bussard, G.H. Miley, Y. Yamamoto, J. Bates, and H.R. Lewis, "Inertial-Electrostatic Confinement Studies, 1993 Intern. Sherwood Fusion Theory Conf., Newport, RI (28–31 March 1993).
21. I.V. Tzonev, "Light Intensity Measurement: Mathematical Modeling," 1st Specialist Workshop on IEC Fusion," FSL Report #513, Urbana, IL (21–24 July 1993).
22. I.V. Tzonev, G.H. Miley, and R.A. Nebel, ICPIG Abstract.
23. C.K. Birdsall, et al., Reference Manual Version 2.0, Plasma Theory and Simulation Group, Electronics Research Laboratory, Univ. of California, Berkeley, CA (1990).
24. R.W. Bussard and N.A. Krall, Fusion Tech., 26, 1326–1336 (1994).
25. N.A. Krall, Fusion Tech., 22, 42–49 (1992).
26. R.W. Bussard, Final Report, NASA/NAS 3–26711, 2 vols. (1993).
27. R.W. Bussard, IEC Study Report, DOE/LANL 9-XG2-Y5957–1 (1993).
28. W.M. Nevins, "Can Inertial Electrostatic Confinement Work Beyond the Ion-Ion Collisional Time Scale?", Physics of SCIF Workshop, Santa Fe, NM (12–14 January 1995).
29. T.H. Rider, "Collisional Electron Effects and Power Balance," Physics of SCIF Workshop, Santa Fe, NM (12–14 January 1995).
30. J.P. VanDevender and D.L. Cook, Science, 232, 831 (1986).
31. N.A. Krall, "Polywell™ Experimental Results and Interpretation," Physics of SCIF Workshop, Santa Fe, NM (12–14 January 1995).
32. L.P. Wainwright, R.D. Durst, R.J. Fonck, and T.A. Thorson, Bult. APS, 39, 7, 1740 (1994). See also T.A. Thorson, R.A. Buckles, R.D. Durst, R.J. Fonck, and L.P. Wainwright, Bult. APS, 39, 7, 1740 (1994).
33. O.A. Lavrent'ev, "Electrostatic and Electromagnetic High-Temperature Plasma Traps," trans. T.J. Dolan, in Electromagnetic Confinement of Plasmas and the Phenomenology of Relativistic Electron Beams, L.C. Marshall and H. Sahlin, eds., Ann. New York Acad. Sci., 251, 322 (1975). See also the paper by T.J. Dolan, this proceedings.
34. B.C. Maglich, T.-F. Chuang, C. Powell, J. Nering, and A. Wilmerding, Report #SAFE-94–104, Advanced Physics Corporation, Irvine, CA (1994).
35. D.C. Barnes, "Penning Trap Concept and Scaling," Physics of SCIF Workshop, Santa Fe, NM (12–14 January 1995).

11

THE D-^{3}HE DIPOLE FUSION REACTOR

Michael E. Mauel

Department of Applied Physics
Columbia University
New York, New York 10027

The dipole fusion concept was first proposed by Hasegawa[1] who was motivated by observations of high beta, energetic plasma within planetary magnetospheres[2,3]. Plasma trapped within nature's magnetic dipoles often diffuses *inward* to regions of higher magnetic field and higher plasma pressure due to the presence of low-frequency perturbations of geomagnetic and electric fields. Furthermore, active magnetospheres, such as that surrounding Jupiter, have plasma pressures exceeding the magnetic pressure, $\beta > 1$. The dipole reactor takes advantage of these properties by operating with plasma profiles which are fundamentally "stationary" and "marginally stable" at high beta. By "stationary" we mean that low-frequency fluctuations do not lead to anomalous transport, and, by "marginally stable", we refer to the stability properties of these "stationary" profiles to interchange and ballooning modes when $\beta > 1$[4]. The dipole fusion reactor is different from other fusion concepts since it is the first reactor conceived to eliminate *both* anomalous plasma transport and fast MHD plasma instabilities.

A dipole fusion reactor would consist of a single levitated circular magnet within a large vacuum chamber. The hot plasma core would encircle the levitated dipole coil forming a toroidal annulus. A large expansion region of cooler plasma extends outward from the dipole where the plasma pressure decreases with radius, L, approximately as $L^{-20/3}$ characteristic of the "stationary", "marginally stable" profiles found in magnetospheres. Although the overall dimensions of the dipole fusion reactor may be large, the size of superconducting dipole magnet is very small. Indeed, in the dipole reactor conceptual designs discussed below, the volume of the hot plasma core (40 m^3) exceeds the volume of the levitated ring (20 m^3). This feature of the dipole reactor (*i.e.*, a larger plasma volume than the volume of the high-technology superconducting magnets, shield and structure) is in sharp contrast to the tokamak where the volume of the plasma is usually less than the volume of the surrounding fusion island. (For example, in ITER, the plasma volume is 2500 m^3 and the volume of the magnets, shield and structure exceeds 5000 m^3). The dipole reactor concept also differs from the spherator[5] since the plasma profiles of the spherator are steep (*i.e.*, they cannot be made "stationary") and low-frequency fluctuations or convection cells significantly degrade confinement.

Conceptual dipole reactors designs have been reported[6,7], and the use of a dipole fusion reactor for space propulsion has been proposed[8]. In each of these designs, D-^{3}He fuel was used instead of the more highly reactive D-T fuel in order to reduce the neutron flux to the levitated dipole. Also reflectors were used to reduce synchrotron losses from the high-pressure and lower β plasma on the inside of the levitated dipole. The high β capability of the dipole reactor makes possible the use of advanced and possibly aneutronic fuels, but the high temperatures required to burn these fuels necessitate steps to reduce synchrotron emission losses. The designs reported in Refs. 6 and 7 described compact, and relatively low-power dipole reactors with large plasma expansion regions. A 20 MA dipole radius of 1.8 m confined a plasma with peak β~3 and generated 100 MW of fusion power. A higher field, 40 MA dipole with a denser plasma at the same β could generate 1000 MW. The plasma is heated to ignition with direct heating of the plasma core (using, for example, neutral beam injection) and the cooler plasma in the expansion region is populated with low-frequency drift-resonant instabilities. Axisymmetry insures confinement of energetic fusion products until after ignition when $m=1$ magnetic perturbations can be applied to drift-pump protons and alphas into direct or thermal conversion sites at natural divertors. In Ref. 8, a much larger dipole reactor was considered with a 54 MA dipole having a radius of 6 m and producing 2000 MW of fusion power. A large plasma expansion region was not used since a relatively hot plasma was diverted to an annular gas-neutralizer to generate thrust. In both Refs. 7 and 8, thermoelectric converters were located within the levitated dipole, and they provided the power to drive refrigerators for the superconducting magnets. Designs of the superconducting magnets and shields in Refs. 6 and 8 illustrate the feasibility of reactor-sized dipole magnets using present-day multi-filamentary Nb_3Sn conductors.

Although no direct financial support has been provided for the dipole fusion reactor concept, recent theoretical and experimental efforts are supportive. Patsukhov and Sokolov have shown that the intensity of drift instabilities driven by the inward temperature gradients near the surface of the levitated dipole is severely limited due to particle recycling[9,10]. Indeed, because the surface of dipole is completely surrounded by a dense plasma, the net particle flux to the ring must vanish. A cool, high-density sheath forms at the dipole surface which completely transforms the thermal flux to bremsstrahlung radiation. Finally, a laboratory terrella built at Columbia University has studied the detailed phase-space evolution of dipole-trapped energetic plasma in the presence of intense drift-resonant fluctuations[11,12]. In this experiment, an "artificial radiation belt" (a population of 5 - 50 keV energetic electrons) is produced with microwave heating which excites hot-electron interchange instabilities when the pressure gradient sufficiently exceeds the "stationary", "marginally stable" profiles envisioned for the dipole reactor. Fluctuations leading to global chaotic transport as well as thin, localized regions of stochastic drifts are observed, and these observations have been used to verify models of collisionless radial transport in dipole magnetic fields.

REFERENCES

1. A. Hasegawa, *Comments Plasma Physics and Controlled Fusion* **11**, 147 (1987).
2. See, for example, the recent reviews of ring current transport in the Earth's magnetosphere, R. B. Sheldon and D. C. Hamilton, *J. Geophys. Res.* **98**, 13491 (1993), and R. B. Sheldon, *J. Geophys. Res.* **99**, 5705 (1994).
3. L. Chen and A. Hasegawa, *J. Geophys. Res.* **96**, 1503 (1991).
4. S. M. Krimigis, *et al.*, *Science* **206**, 977 (1979).

5. R. Freeman, *et al.*, "Confinement of plasmas in the spherator," in *Proc. 4th Conf. Plasma Physics and Controlled Nuclear Fusion*, Madison, Wisconsin, June 17–23 (1971), IAEA-CN-28/A-3, Vol. 1, p. 27.
6. A. Hasegawa, L. Chen, M. E. Mauel, *Nucl. Fusion* **30**, 2405 (1991).
7. A. Hasegawa, *et al.*, *Fusion Technology* **22** (1192) 27.
8. E. Teller, *et al.*, *Fusion Technology* **22**, 82 (1992).
9. V. Pastukhov and A. Yu. Sokolov, *Nucl. Fusion* **32**, 1725 (1992).
10. A. Yu. Sokolov, *Plasma Phys. Rep.* **19**, 1454 (1993).
11. H. P. Warren and M. E. Mauel, to be published in *Phys. Rev. Letters*.
12. M. E. Mauel, H. P. Warren, A. Hasegawa, *IEEE Trans. of Plasma Science*, **20**, 626 (1992)

12

OPEN-ENDED MAGNETIC CONFINEMENT SYSTEMS FOR FUSION*

Richard F. Post and Dmitri D. Ryutov

Lawrence Livermore National Laboratory
Livermore, California 94550

ABSTRACT

Magnetic confinement systems that use externally generated magnetic fields can be divided topologically into two classes: "closed" and "open". The tokamak, the stellarator, and the reversed-field-pinch approaches are representatives of the first category, while mirror-based systems and their variants are of the second category. While the recent thrust of magnetic fusion research, with its emphasis on the tokamak, has been concentrated on closed geometry, there are significant reasons for the continued pursuit of research into open-ended systems. The paper discusses these reasons, reviews the history and the present status of open-ended systems, and suggests some future directions for the research.

I. INTRODUCTION

One obvious way to differentiate between classes of magnetic confinement systems that use externally generated magnetic fields is via the topology of their fields. Here by "the topology of the field" we mean whether the field lines of the externally applied field are "closed", i.e., are "endless" in that they are contained within a toroidal confinement region (as in a tokamak), or "open", that is, leave the plasma confinement zone at either end of an essentially linear system (for example, as in a magnetic mirror system).

It should be noted that the distinction that we have adopted here between "open" and "closed" systems, since it relates to the topology of magnetic fields generated by external coils, includes in the class of open systems not only mirror machines but FRC's (Field-Reversed Configurations), and RFM's (Reversed-Field Mirrors), both of which share the same topology of the externally generated field. Thus, in the discussion to follow, many of the points made apply to these other types of open systems.

---

* Work performed under the auspices of the United States Department of Energy by the Lawrence Livermore National Laboratory, under Contract No. W-7405-ENG-48.

The seemingly innocuous topological distinction between closed and open systems actually has profound consequences, both in the plasma physics issues involved and in the paths available toward economic fusion power systems. As an example, in closed systems particle losses, occurring only across the magnetic field lines, result necessarily in major radial variations in the plasma parameters. These include, for example, negative temperature gradients as the plasma approaches the chamber wall. Such temperature gradients, communicated back into the plasma, have been shown to be a source of plasma instabilities [1]. Such instabilities can lead to cross-field diffusion rates far in excess of those expected for a quiescent plasma. By contrast, in open-ended systems where the particle losses are out the ends, there is no requirement for contact of the plasma with the walls of its confining chamber, thus no enforced radial temperature gradient need be present. As a result the radial confinement properties of open-ended systems can in principle be superior to those of closed systems.

The comments in this paper are aimed at a discussion of some of these topologically related differences between the two systems. It is hoped that one of the byproducts of such a discussion will be to encourage fusion researchers to take a broader view of the magnetic fusion problem than the more usual one, the one that contends that the main criterion by which to compare approaches is "energy confinement time". By extension, energy confinement time is related to the attainable fusion energy-multiplication factor, Q, (ratio of fusion power to power required to heat and sustain the plasma) i.e., it determines the Q value that can be achieved. In the final analysis, capability for attaining a high Q is only one of several criteria for judging the economic potential of a magnetic confinement approach. High attainable Q may, in fact, turn out to be secondary to other criteria that are more significant in the context of comparative economic viability.

The thesis will be advanced that there are many probable economic and other advantages of open systems over closed ones that at the very least justify a continued search for ways to exploit these advantages. This line of reasoning is intended to apply especially to those ways that put minimal demands on the weakest feature of some open-ended systems, namely their in-some-cases low Q values, as exemplified by the simple mirror machine with its well-known end losses. We will, therefore, pay special attention to low-Q systems, their properties and their virtues. By relaxing the (sometimes myopic) concentration on achieving high Q there is opened up a much wider variety of possibilities for the confinement system itself. It is entirely possible, we would say even likely, that within this wider range of possibilities there exist systems that are better suited to developing into economically viable fusion power systems than is the tokamak.

After all, the ultimate goal of fusion is not the simplistic one, $Q = \infty$, but rather it is the pragmatic one, "to generate power at a competitive cost, and with superior environmental and safety characteristics as compared to other energy sources". As the search for practical fusion power proceeds towards its goal, we must pay increasing attention to the pragmatic goal, which may or may not involve a quest for ever-higher Q values.

Specifically, our intent here is to state the case for broadening the attention of the fusion community to issues beyond a search for ever-higher Q values, and to encourage a look at low-Q systems for their potential advantages. In a later section we will sketch out a hypothetical "fusion power system" where the goals are, first, to use the D-^{3}He fusion cycle with direct conversion of the fusion energy to electricity, and second, to operate in a plasma regime that permits the near-suppression of satellite neutron-producing D-D reactions. To accomplish both of these goals it will be seen that it is essential to operate at as *low* a Q value as is practical. High-Q operation is counter-productive of the stated goals.

II. BRIEF REVIEW OF THE HISTORY OF OPEN-ENDED SYSTEMS

In surveying the long history of mirror research up to the present time one is struck by the evident fact that, as contrasted to closed systems, there has always existed a close relationship between theory and experiment. Mirror systems have not only been found to be tractable to analysis, but this analysis has also pointed the way to the achievement of situations that have rarely if ever been seen in closed systems: We refer to plasmas that are so near to the quiescent state that their confinement can be predicted on the basis of classical processes, i.e., collision-related phenomena, as opposed to the turbulence-related effects that dominate the confinement in closed systems.

As we have noted earlier, open-ended confinement systems must face a problem not encountered in closed-field systems, namely, plasma losses along the field lines, out the ends of the system. The mirror machine represents one way of attacking this problem in that it utilizes the magnetic mirror effect, arising when a particle spiraling along field lines encounters, and is reflected back, by an increasing magnetic field. We can therefore trace the historical origin of mirror systems to the study of particle mirroring in connection with cosmic ray physics, analyzed early on by Störmer [2]. His studies concerned the reflection of incoming cosmic rays (protons, etc.) by their encounter with the earth's magnetic field. Many years later, in the discovery of the "Van Allen belts" [3] of charged particles magnetically trapped between the earth's north and south magnetic poles mirror confinement of charged particles was seen on a grand scale, some years after it was first studied in the laboratory (in the early 1950's).

The theory of particle confinement in mirrors, as controlled by so-called "adiabatic invariants" was elegantly derived by Teller and Northrop [4]. It was proved out in the laboratory and in a spectacular space-related experiment. In the laboratory experiments, for example one in 1962 by Gibson, Jordan, and Lauer [5], positrons from radioactive neon were trapped between mirrors and contained for times of order of seconds, corresponding to millions of reflections. Their loss occurred only as the result of collisions with background gas atoms, not as a result of some failure of the mirror process. In the "Argus" experiment in space, proposed by Christofilos and carried out in 1958, a small atomic bomb, sent into the earth's magnetosphere by rocket, was detonated, producing a cloud of energetic electrons from beta decay. These electrons, trapped between the earth's magnetic poles, spread out into a thin shell surrounding the earth. A detectable fraction of these electrons was still present decades after the detonation!

The basic principle behind both of these experiments is that particle confinement in a mirror machine is governed by adiabatic invariants, the constancy of which assures that charged particles, once trapped between mirrors, will remain trapped essentially "forever" unless some random process, for example inter-particle collisions, perturbs their motion. There are only two conditions that the particles must satisfy in order to remain trapped. The first requirement is that the size of their gyro orbit around the field lines must be "small" (i.e., less than of order 10 percent) compared to the characteristic gradient lengths of the confining field. This requirement is not difficult to satisfy practically as it simply puts lower bounds on the size of the apparatus and/or the intensity of the confining field.

The second requirement for particles to be mirror confined is that the pitch angle the particles' spiralling orbits make with the field line direction at the midplane between the end mirrors must be larger than the "loss cone" angle, θ_{LC}, defined through the relationship

$$\sin(\theta_{LC}) = R^{-1/2}, \qquad (1)$$

where $R = B_{max}/B_{min}$ is the so-called "mirror ratio", i.e., the ratio of the magnetic field intensity at the mirrors to that at the minimum point (the midplane).

The two adiabatic invariants which, together with the law of energy conservation E = constant, govern mirror confinement are: (1) the magnetic moment, $\mu = w_\perp/B$ (given by the energy associated with the cyclotronic motion of the particle around the field line on which they are moving divided by the local value of the magnetic field), and, (2) the action integral, J, carried out over the longitudinal motion along the field lines between the mirroring points. The first of these determines whether or not the particle is trapped between the mirrors; too small a value of $w_\perp$ relative to the energy of motion parallel to the field line will put the particle into the loss cone so that it is immediately lost upon its next encounter with the mirror. As will be described, the second invariant, J, determines the transverse dimension of the drift surface on which the particle moves, thus whether or not it will hit the chamber wall in the course of bouncing between the mirrors.

As a consequence of these two invariants trapped particles move back and forth between the mirroring points, while at the same time drifting transversely, but in such a way that, for a properly chosen class of magnetic geometries, the "guiding center" of their spiraling motion generates a closed drift surface surrounding the centermost field line. Then, once trapped between the mirrors, particles remain trapped until some time-varying process, such as collisional interactions, sufficiently perturb their energy and one or the other of the two invariants to destroy trapping. Perturbations of E, μ, and J lead to leakage out the mirrors and to diffusion between drift surfaces.

It is to be noted that these invariants operate just as effectively in non-vacuum magnetic fields, that is, magnetic fields that are the resultant of external coils and of diamagnetic currents flowing in the plasma, provided only that the non-vacuum field also satisfies the stated requirements for adiabaticity.

Thus, in seeking to understand the differences between open and closed systems it is important to recognize that mirror systems start with an external field that guarantees almost indefinitely long particle confinement in the absence of perturbing effects. Furthermore, confinement will still be effective provided the "perturbing effects", such as steady diamagnetic currents in the plasma, do not vitiate the conditions for adiabaticity. By contrast, a closed system such as the tokamak starts with an external field that cannot contain particles in long term (because of transverse drifts associated with the toroidal field). In order to achieve long-time confinement it must rely on currents flowing in the plasma to overcome these drifts.

Returning to the early history of mirror confinement, it began with close contact with theory, in the form of the theory of adiabatic invariants, and then proceeded toward its encounter with plasma-induced effects. The first of these, a sword of Damocles that would hang over it throughout its progress, is the effect of particle collisions on the confinement, and the particle leakage out the ends that these collisions can cause.

Since collisions can affect the magnetic moment and particle energy, they can have an immediate effect on mirror trapping, i.e., on containment. That is to say, it will take only about one "effective collision", i.e., a deflection of the pitch angle through 90 degrees, to cause a trapped particle to enter the loss cone, and thus be lost upon its next encounter with an end mirror. This situation is to be contrasted with the one associated with closed systems, such as the tokamak. In closed systems, where the only escape route for the particles is across the field lines to the chamber wall, collisions lead to a random walk process in which many collisions may be required before the particle reaches the wall.

Furthermore, as long as this process is a diffusion-like one, as opposed to a convective loss, the confinement time will be found to scale roughly as the square of the radius of the chamber, so that even in the presence of turbulent processes long confinement times can be achieved, though at the price of making the chamber dimensions very large.

Early on calculations were made, utilizing the Fokker-Planck (F-P) equation, of the confinement time against mirror losses of conventional (two-mirror) mirror machines. The first such calculations did not include an important effect: the electric potential (ambipolar potential) that arises naturally because of the disparity in scattering rates of electrons and ions (electrons scatter more rapidly). Since on the average the electron and ion loss rates must balance in steady state, a positive potential, of the plasma relative to the outside world, appears as soon as the loss rate reaches steady-state. This positive potential, trapping for the electrons and detrapping for the ions, somewhat accelerates the rate of loss of the ions over the value calculated in the absence of the potential. Thus, in the earliest calculations, the confinement time, though not robustly long, seemed adequate to achieve net fusion power, if the ion temperature was high enough. The first F-P code calculations were made in the 1950's, and by 1959 the codes had been refined to the point that all important effects were being included. As an example, the work of Kuo-Petravic, Petravic, and Watson [6] gave for the $n\tau$ containment parameter (for a D-T-electron plasma) the scaling relationship:

$$n\tau = 2.33 \times 10^{10} E_0^{3/2} \text{Log}_{10}(R) \text{ cm}^{-3}\text{sec.}, \qquad (2)$$

where E_o is the mean ion energy in keV. As can be seen immediately by putting in the numbers, to achieve $n\tau$ values leading to Q values in excess of 1.0 would require mean ion energies of order 100 keV and mirror ratios of 3 or more. This result put the simple mirror machine in the category of low Q approaches ab initio. It thereby sharpened the need for innovative improvements in order to make it a viable candidate for a fusion power system. We shall later discuss some of these improvements, following the mention of two other problems for mirrors, ones that were addressed and solved with great success.

The first of these problems, pointed out early on, during the classified days of fusion research (circa 1954) by Edward Teller, was that of MHD stability of the equilibrium. In a simple mirror machine, as shown in Fig. 1, the field lines between the mirrors are concave inward, implying a negative field gradient in the radial direction. This "bad curvature" region is one that is therefore potentially subject to MHD instabilities, i.e., to unstable transverse drifts of the plasma. In the earliest mirror experiments no such drifts were seen, for reasons that were not then understood, and the problem was put out of mind. In fact, in some experiments at Livermore where a "hot electron" plasma was created by magnetic compression in a pulsed mirror field, [7] not only was a stable plasma column observed, but radial diffusion rates were measured that were five orders of magnitude slower than the rate predicted for "Bohm" diffusion as observed in some closed-geometry systems. However, with progress in generating hot ion plasmas, once better isolated from plasmas outside the mirrors, unstable transverse drifts of the plasma were seen, in agreement with the MHD predictions. There was indeed a problem, one that demanded a solution.

In 1961, in a classic experiment, Ioffe of the Soviet Union showed one important way to suppress this instability. In his experiment Ioffe added around his plasma chamber six longitudinal conductors, carrying currents in opposite directions, pairwise, to the mirror coils. These "Ioffe bars" created a non-axially symmetric magnetic field which was characterized near the chamber wall by nested closed surfaces of magnetic field intensity,

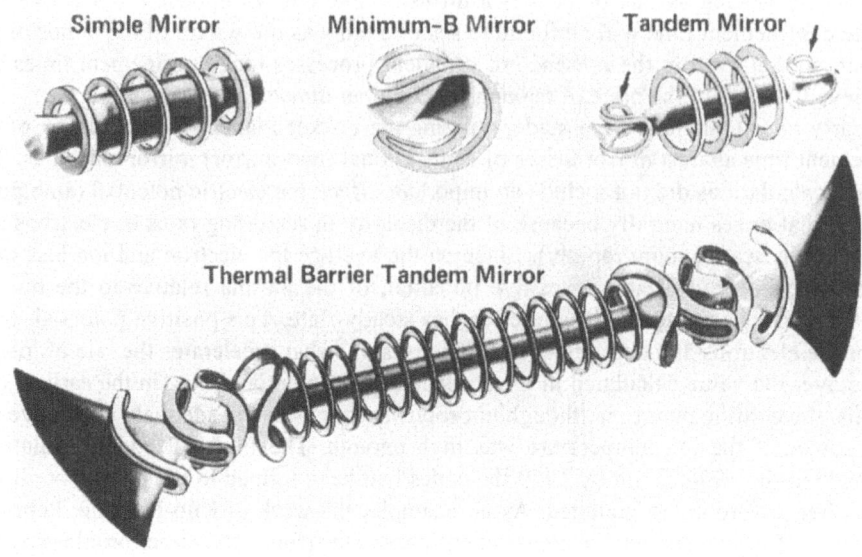

Figure 1. Schematic drawings showing the historical evolution of the mirror fusion confinement concept. In the upper left corner is shown the simple axi-symmetric mirror. In the upper middle a quadrupole magnetic well, as produced by a "baseball" coil is shown. At the upper left is the simple tandem mirror, and at the bottom a tandem mirror with thermal barriers is shown.

the field increasing from one surface to the next. In this way all MHD instabilities were suppressed, since in attempting to expand in any direction the plasma would have to move "uphill" on a magnetic pressure gradient. This principle, i.e., of the "magnetic well" would thereafter become a part of almost all mirror research, and would be adopted, where possible, in other branches of magnetic fusion research. A quadrupole version of the mirror magnetic well is shown in one of the drawings in Fig. 1.

Since the time of the Ioffe experiment additional methods have been found to suppress MHD instabilities, but the magnetic well still remains the most powerful one. These other methods include: taking advantage of so-called finite-orbit effects [8] (the difference in size of ion and electron orbits in the plasma, leading to drift motions that help to stabilize), and, stabilizing an otherwise unstable plasma in an axially symmetric mirror field by electrical conduction, along the field lines, to stabilizer cells at the ends. In these ways it has been found possible to do something not yet fully possible in toroidal systems, i.e., to eliminate the entire class of MHD instabilities, even at high plasma beta (plasma pressure approaching the confining magnetic pressure in magnitude).

In the historical development of mirror systems there was one more hurdle that was encountered, and overcome, in the march toward the fusion goal. This was the problem of the so-called "micro-instabilities", high frequency instabilities of the wave-particle type that fed from the departures of mirror-confined plasma velocity-space distributions from an isotropic maxwellian distribution. As they applied to the mirror machine these instabilities were analyzed in the 1960's and methods of controlling them were suggested. When these methods were finally experimentally tested, several years later, they proved to be successful to the point that virtually quiescent plasmas, stable to both MHD and microinstabilities, were produced. These plasmas had loss times that agreed well with the "classical" values predicted by the Fokker-Planck equation, scaling with ion energy in the way predicted by theory. Still, the confinement values obtained were marginal if projected to fusion power conditions, leaving still open the question of how to increase the Q of conventional mirror machines, where scaling-with-size does not occur as it does in toroidal systems.

At this point in mirror research (mid 1970's) one of the most spectacular successes of the mirror program occurred, in the 2XIIB experiment at the Lawrence Livermore National Laboratory. In this experiment, performed by F. Coensgen and co-workers [9,10], a hot plasma was created in a quadrupole mirror cell by means of a large array of neutral beam sources impacting on a "seed plasma" that ionized them. The combination, a low density plasma stream and the trapped high energy deuterons, was found to be stable against both MHD and microinstability modes, and to have a beta value that approached the limiting magnetic pressure of the confining field. Furthermore, the confinement time was found to agree well with the theoretically predicted values obtained from F-P code calculations, further proof of the low levels of turbulence present in the plasma. The ion temperature, 20 keV and plasma density, greater than 10^{14} cm^{-3}, together represented a record achievement at the time.

The successes in 2XIIB had the effect of focussing the attention of the fusion community on the mirror approach, with the attendant programmatic pressures that that attention implied. It meant that mirror researchers had to intensify their search for ways to improve the marginal Q of simple mirror systems, while at the same time paying attention to projected costs for mirror-based fusion systems. At the time there were at least three avenues to the goal visualized.

One of these ways, one that had been suggested and studied many years earlier [11,12], was to employ many mirror cells in series, arranging matters so that particles escaping from the innermost cell would be successively trapped in the outer cells, thereby achieving a confinement time that increased as the square of the number of cells, in this way regaining a scaling analogous to the quadratic scaling with size of closed systems. The difficulty, at that time, with the multiple-mirror idea was that it seemed to require such high plasma densities, and/or such long cells (to achieve collisional retrapping in each cell) as to make the entire system impractical. Later ideas [13] were aimed at alleviating this problem, but have not been experimentally tested.

Another approach, suggested in 1989 [14] and investigated at Livermore, came close to an approach we will later be discussing in this article. This was to devise a direct conversion system that could recover, at high efficiency, the energy of ions, both reaction products and of unburned fusion fuel, escaping through the mirrors. If coupled to a high efficiency injection system the combination might lead to a practical fusion plant. This approach was also deemed at the time to be marginal economically.

In 1976 there arose, independently, and at two different sites, a new approach to the problem—the tandem mirror idea. Dimov and co-workers at Novosibirsk [15], and Fowler and Logan [16] at Livermore came up with an improvement on an older idea of Kelly at Oak Ridge. The idea, a novel variant on the multiple mirror idea, was to have a long central cell in which fusion took place, flanked on either end with short mirror cells within which electrostatic potentials existed that electrostatically "plugged" the mirrors, reflecting back ions that would otherwise have escaped through the innermost mirrors. The overall ambipolar potential of the system would confine the electrons as usual, so that good confinement of both ions and electrons could be achieved. By making the central cell long compared to the end cells it was believed that the power required to sustain the plasmas in the end cells could be made small compared to the fusion power, so that good overall Q values could be achieved. In this, first, version of the tandem mirror the increased positive potential of the plugging cells (relative to the central cell) was to be achieved by the simple expedient of maintaining a higher plasma density in those cells than that in the central cell. Mirror theory then predicted, and experiment confirmed, that the plugging potentials would appear.

The tandem mirror idea gave a new lease on life to the mirror approach. In a relatively short time tandem mirror experiments were built, at Livermore (the TMX experiment), in Novosibirsk (the AMBAL experiment), in Tsukuba, Japan (the Gamma experiment), and elsewhere. The basic principles involved were proved out, in plasmas of modest temperature and density, and a data base was begun to be assembled. However, despite the initial success of TMX in demonstrating mirror plugging, the pressure was still on for improvements that could lead to better economic projections for mirror systems. It was thought that the need for higher plasma densities in the plugging cells as compared to the central would represent too high a cost penalty in a power plant scenario. These pressures led to the invention, by Baldwin and Logan [17], of the "thermal barrier" concept, a technique employing microwave heating of electrons in the plug cells that resulted in plugging potentials generated in a plasma whose density was *lower* than that in the central cell, with important predicted economic advantages.

Although the thermal barrier idea was successfully tested, both in TMX-Upgrade and in the Gamma 10 experiment at Tsukuba, and a very large tandem mirror experiment at Livermore, MFTF, was authorized and constructed, the budget axe fell on work on "alternate concepts" in the United States, with the virtual termination of all magnetic fusion efforts except those related to the tokamak. At the present time the main centers for mirror research are at Tsukuba in Japan and at Novosibirsk in Russia, where active programs on mirror-related concepts continues. Figure 2 is a schematic drawing of the large Gamma 10 experimental facility at Tsukuba University. In recent experiments radial and axial particle loss rates consistent with confinement times of a fraction of a second have been obtained, at multi-kilovolt plasma temperatures, in good agreement with theoretical predictions. As can be seen from the figure the central magnetic field in these experiment is only 0.4 Tesla, and the central plasma column is only a few centimeters in radius. Both the field and the plasma radius are much smaller than those required in a tokamak achieving comparable confinement time constants. The improved confinement properties can be attributed to Gamma 10's axi-symmetric central-cell magnetic fields, and to the demonstrated near-complete suppression of plasma turbulence that has been achieved, accomplished through control of the radial distribution of the plasma potential.

From the more than three decades of theoretical and experimental work on open-ended systems worldwide there now exists an unusually high degree of understanding, and of correlation between experiment and theory, of plasmas confined in open-ended systems. This understanding provides a solid base for the continued investigation of such systems. The possible reward: smaller, simpler, and less expensive approaches to practical fusion power than those presently foreseen for closed confinement systems.

This understanding can also be employed to make reliable predictions about the performance of a variety of mirror-related concepts. As a result of this work we can discern the special properties of open systems vis á vis closed systems and thereby better assess their potentialities for fusion power purposes. Below we give a list of these distinguishing characteristics.

III. DISTINGUISHING CHARACTERISTICS OF OPEN SYSTEMS

We begin this discussion by listing those characteristics of open systems of topological origin that differentiate them operationally from closed systems. The list here is closely similar to one given in a previous paper [18]. Not every embodiment (Mirror, Tan-

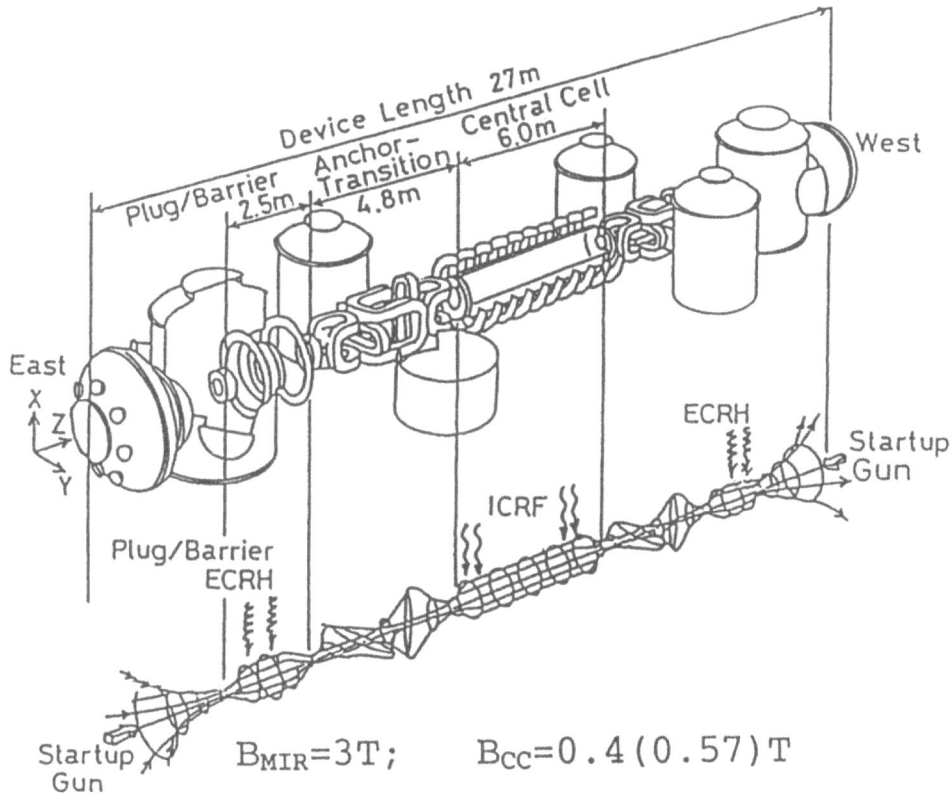

Figure 2. A schematic drawing of the Gamma 10 tandem mirror experiment at Tsukuba, Japan. Note that in its present configuration all heating and potential generation is accomplished with RF and/or microwave power. Note also that the central cell and its mirrors are axi-symmetric, leading to improved confinement as compared to non-axi-symmetric cells as were used in some earlier tandem mirror experiments.

dem Mirror, FRC, RFM, etc.) possesses all of the listed characteristics, but each possesses some of them.

1. Open field topology permits the formation of deep magnetic wells in which the vacuum field possesses a non-zero minimum within the confinement zone. In such fields, as shown both by theory [19] and experiment [20], plasma beta values approaching unity can be attained and all low-frequency instabilities of MHD origin are suppressed (for non-field-reversed configurations).

2. Plasma losses can be preferentially directed to occur along the field lines, out the ends (natural "diverter" action). As noted previously, in such cases there is no requirement for the radial boundary of the plasma to be near the plasma chamber wall, thus no requirement that there should exist a radial temperature gradient at the outer boundary of the plasma. This circumstance has at least two important consequences: from the physics standpoint, as noted, the possibility of eliminating (or minimizing) radial temperature gradients in the plasma can lead to the suppression of an important class of plasma instabilities, ones that are probably playing a significant role in enhancing the rates of radial transport in closed systems, such as the tokamak. From the engineering and economic stand-

point, the elimination of physical contact of the plasma with the chamber wall (or with a nearby diverter surface) greatly relieves severe heat-transfer and sputtering-life problems that are endemic to closed systems. In an open system the end-exiting fields lines can be extended and spread out so that they only encounter physical surfaces that are both far-removed from the main body of the plasma and are of large area. A concomitant, and important, result of having the plasma exit at the ends is that under the right circumstances it would permit the use of direct converters of the high-efficiency electrostatic type [21], both to recover energy from charged fusion reaction products and from unburned fuel ions.

3. Open field line topology permits the introduction, conditioning, and translation of plasma entities along the field lines, in at one end and out the other. This property, first exploited in the multi-stage magnetic compression mirror experiments at Livermore [22], has many practical consequences. First, it would permit the separation of the regions of plasma formation, fusion reaction, and energy extraction (by direct conversion). Second, open line topology could facilitate both the fueling and the ash removal problems (heated fuel in at one end, unburned fuel ions and fusion ash out the other end).

4. Open systems can employ linear, i.e., solenoidal, magnetic fields. Such fields are highly advantageous from an engineering standpoint when it comes to achieving high magnetic fields. It has been estimated that expected technological advances in high-field superconductors and in high-strength unidirectional fiber composites (to resist purely radial forces) should, in time, permit the design of solenoidal fields with strengths in the range of 20 to 50 T. Particularly at the high end of this range it seems not possible to envisage toroidal-geometry field coils of usual aspect ratios with fields as high as this, owing to the intractability of the 3-D force problems involved.

5. By exploiting mirror-related properties of plasma confinement it is possible both to establish and to control electric potentials within the plasma, including the possibility of maintaining electric fields parallel to the magnetic field lines in cases where it is advantageous to do so. These potentials and electric fields have various practical applications, including enhancing confinement (as in a tandem mirror) and/or controlling the direction of exit of the plasma ions or electrons to facilitate ash removal and direct conversion.

6. By contrast with toroidal systems, where particles can only exit after diffusion across the confining field, open systems can much more readily operate near to a "collisionless" state, i.e., one where collisional randomization is not complete and where the ion (or electron) distribution functions can deviate substantially from a maxwellian. There are circumstances (for example, fusion systems involving colliding ion beams) where it may be highly advantageous to be able to employ non-maxwellian distribution functions. For example, in the case of the use of the D-^{3}He reaction, such distribution functions facilitate the minimization of parasitic D-D reactions, thus minimize neutron fluxes and thus suppress induced radioactivity in the chamber walls. Another consequence of the ability to operate with ion distribution functions that are not collision-dominated is that such distribution functions are better suited than maxwellian distributions for high-efficiency direct conversion. (The narrower the energy distribution function of the ions is, the higher the efficiency of a given electrostatic direct converter will be.)

7. Linear systems of open geometry may lend themselves to a "modular" design approach, something more difficult, if not impossible, to achieve with closed geometry at usual aspect ratios. For example, the central section of a TM (tandem mirror) could consist of many identical linear modules, resulting in superior accessibility and ease of construction as compared to a toroidal system.
8. Open-ended systems of the mirror or the tandem mirror type, as has been demonstrated experimentally [23,24], have the ability, through their access to the emerging field lines at the ends, to control the potential of the plasma relative to the chamber walls, including control over the radial distribution of that potential. This control can be used to enhance confinement either through limiting or through inducing plasma rotation, and for the stabilization of certain classes of plasma instabilities. This situation is to be contrasted with that prevailing in closed systems where the inherent inaccessibility of interior field lines makes it difficult if not impossible to control either the magnitude or the interior gradients of the plasma potential.
9. A qualitative, non-quantifiable, characteristic of open systems, exemplified in part by the preceding list of attributes, is their *flexibility*. By "flexibility" we are here implying both their adaptability to a wide variety of plasma regimes (high beta, high plasma temperature, non-maxwellian distributions, etc.) and their suitability for innovative improvements (e.g., direct conversion). By contrast, for closed systems the requirement for closure within the confinement chamber of all field lines on which containment is to be effective introduces severe constraints on the flexibility of such systems with respect to accessible plasma regimes and with respect to innovative improvements.

IV. DEALING WITH THE ISSUE OF END LOSSES

Against the above-listed attributes and advantages of open system is the main disadvantage we have already cited: the need to control end losses. This one issue has dominated the research into open-ended systems almost from its initiation. It is the issue addressed by the tandem mirror idea, but that approach is not the only one that has been considered. There are, in fact, a continuum of variants of mirror-based open systems that have fusion potential, some of which we will discuss.

While the tandem mirror concept is aimed at achieving a high plasma Q value, in concert with a similar goal in the tokamak, in this paper we are addressing the possibilities inherent in low Q systems, such as the simple mirror.

A key question then for simple mirror systems becomes whether it is possible to create a viable fusion power system in which the confinement time is limited to the ion-ion relaxation time. In the language of fusion power balance, this question is whether it is feasible to operate at low Q values, as contrasted with those in so-called "ignited" states postulated for the tokamak ($Q = \infty$). Under the best conditions simple mirror systems would operate at Q values that are close to unity, thus would appear to be less desirable than systems with more robust Q values.

There exists a modern-day example of a highly practical power generator which operates very successfully at what fusion researchers would call a very low Q value. That is, the amount of power recirculated internally is large compared to the net power output. The economic practicality of this device was not achieved by the route of increasing the equivalent Q value, but by increasing the efficiency of its "injector" and "direct converter" (as they would be called by analogy with a fusion power system).

The device in question is the gas turbine. Its main components are:

1. A compressor and fuel injector (the "injector").
2. A combustion chamber (the "reaction chamber").
3. A turbine (the "direct converter"), from which power is recirculated to drive the compressor through a connecting shaft.

It is a historical fact that the first attempts to get net power from a gas turbine failed, owing to the circumstance that the efficiencies of the compressor and the turbine were too low to permit a net power output. Those first tests had to be carried out using an electric motor connected to the power shaft in order to make up for the (negative) power output of the gas turbine. Only through continued improvement in the efficiency of the compressor/turbine system was it possible to make the gas turbine into a practical device. Even so, modern gas turbines typically operate with recirculated powers that are 3 to 4 times greater than their net shaft output power. Yet for many applications they are eminently satisfactory and useful power generators.

We here suggest that there exists a possible development path for fusion that parallels that of the gas turbine. That is, one where the emphasis is on the twin goals of achieving an "adequate" Q value (which actually in principle might be smaller than unity), while at the same time the technology of injection and direct recovery is developed (always with an eye to cost) to the point where useful net power can be obtained from the fusion system.

It is at this point that we believe that the earlier-cited "flexibility" of open-ended systems could come to the fore. Because, as also noted earlier, open systems are uniquely well adapted for ease of injection and for ease of coupling to a high-efficiency direct converter, they are prime candidates for the pursuit of the low-Q option.

It must be admitted that there is a certain element of *dé-jà vu* in what has just been said. In a paper given in the 1976 I.A.E.A. Fusion Conference, Moir et al. [25], describe a reactor-study example of a simple mirror system using exactly the approach described above. Though one would not claim that that particular design would represent a useful power system, the general combination (some form of open-ended system fed by high-efficiency injectors and coupled to a high-efficiency direct converter) could form an economically attractive power system, even if the fusion Q value was not far different from unity.

In the sections to follow we will cite some examples of open-ended systems that exploit the low Q concept. That is, they minimize the plasma confinement time requirements through shifting the developmental burden to plasma injection and energy recovery efficiency issues.

V. A LOW-Q EXAMPLE: A D-^{3}He "LINEAR COLLIDER"

As a hypothetical example of a D-^{3}He fusion power system where the intent is to suppress, to the largest degree possible, the rate of satellite D-D reactions we offer some elementary calculations to define what could be called a fusion "linear collider". The present example is based on extensions of a concept described in a previous publication [26].

We use the designation "collider" to denote a linear, colliding-beam, fusion power plant where the plasma and system parameters are such as to yield net power from the one-pass collision between inter-penetrating beams. Stripped to essentials this system would consist of a long, graded-field, solenoid with coaxial injectors and direct converters. Time-synchronized plasma columns consisting of monoenergetic ions (helium-3 ions and deuterons) are repetitively launched so as to collide with each other near the midpoint in the solenoid. During the interpenetration of the two beams D-^{3}He reactions would occur, the energy from which (together with that from unreacted beam ions) would be recovered by the direct converters. At the same time the "temperature" of the the ions of each beam would begin to grow as they scatter against the ions of the other beam and with each other. However, since the energy release per D-^{3}He fusion reaction is much higher than the total kinetic energy of the beam ions, the mutual interaction time of the two beams can be limited to a time much shorter than for complete burn-up, provided that the efficiency of the injectors and the direct converters is high enough.

By maximizing the efficiency of injection and energy recovery the burn-up fraction required for net energy production could be lowered to the point that collisional heating of the deuterons would be minimal. This result would in turn insure minimal production of D-D neutrons per megawatt-year of fusion power.

A schematic power flow diagram for the linear collider is shown in Fig. 3. Taking into account the beam kinetic energies, W_1 and W_2, the fusion energy release (18.3 MeV), W_{fus}, the burn-up fraction, f, and the injector and direct converter efficiencies, η_i and η_R, there results an equation for the "enhanced Q" of the system, Q_E, where the condition $Q_E > 1$ is required for net power:

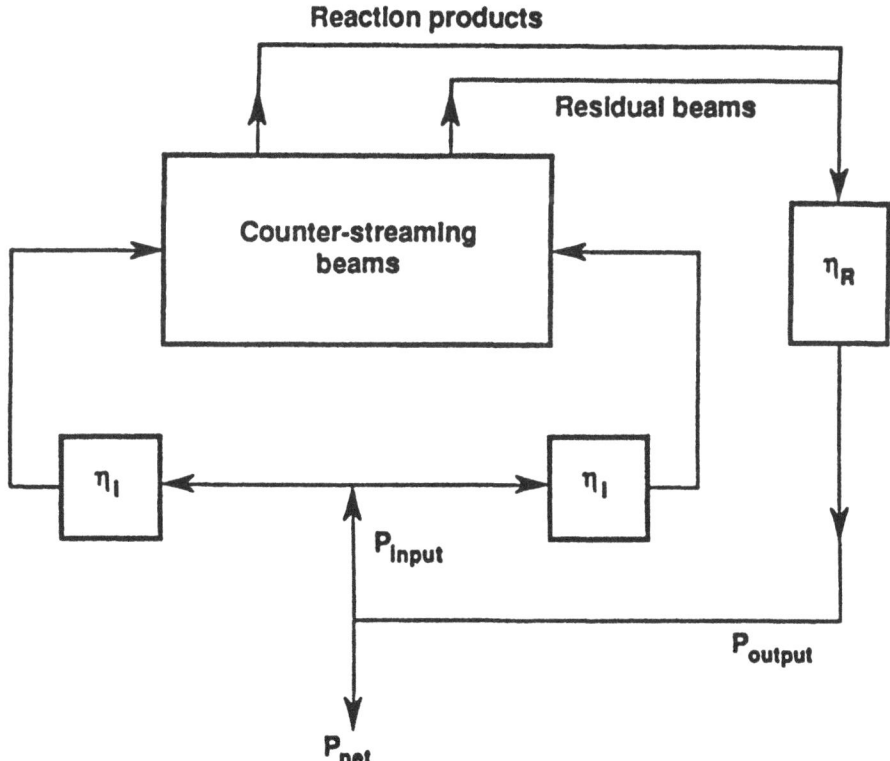

Figure 3. Block diagram of a counter-streaming "linear-collider" fusion power system. An important feature is the recirculation of unreacted ions and the direct recovery of their energy at high efficiency.

$$Q_E = \left\{\frac{fW_{fus}}{W_1 + W_2}\right\}\left\{\frac{\eta_i\eta_R}{1-\eta_i\eta_R}\right\} > 1 \text{ for net power} \tag{3}$$

At this point in the discussion there exists a branching point with respect to the approach we may follow. One such approach would be to inject high density plasma streams from both ends of the system, timed to collide in the central, highest field, region of the solenoid. The second approach, which we will also briefly discuss, would be to inject, from one end, co-moving velocity-modulated ion beams whose relative energy in the moving frame corresponded to the maximum of the fusion cross-section. Here one counts both on the converging magnetic field and the klystron-like bunching effect to achieve a high density in the interaction region, followed by decompression and energy recovery (at low plasma density) in a direct converter.

Example A: Simple Linear Collider

We will first consider the former of the two approaches, high density colliding beams, injected from opposite ends of the solenoid.

The maximum point in the D-^{3}He reaction probability occurs at an energy of the D relative to the ^{3}He ion of about 500 keV. From this fact it follows that for colliding beams the energy of each beam should be about 125 keV to yield that value of relative energy. This circumstance gives a factor-of-two advantage of colliding beams over a reacting plasma in calculating the fusion power balance. Another factor of two advantage for colliding beams over a reacting plasma comes from the fact that every primary ion-ion collision carries with it the possibility of a D-^{3}He reaction, whereas in the usual plasma case (with a 50–50 D-^{3}He fuel mixture) only half of the ion-ion collisions are of the "right" kind. These factors of two plus the circumstance that by design the ion-ion collisions between D and ^{3}He occur at the peak of the fusion cross-section help to alleviate the formidable task of achieving net power from a single-pass colliding beam situation.

The burn-up fractions that are required can be evaluated from Eq. (1) for an assumed parameter set as follows: Take $W_1 = W_2 = 125$ keV, $\eta_i = \eta_R = 0.9$ (direct conversion efficiencies approaching this value have been demonstrated in the laboratory [27]).

When these values are inserted in Eq. (1) there results for Q_E as a function of the burn-up fraction, f:

$$Q_E = 312f,$$

so that if the burn-up fraction is taken to be 0.01 the value of Q_E is 3.12. As we shall see this small a value of the burn-up fraction will allow the achievement of the objective of strongly suppressing parasitic D-D reactions, provided other quantitative requirements can be met. These requirements can be estimated by considering some plasma-related issues.

The physics points here can be illuminated by estimating two elementary quantities for the colliding beams, (1) a "mutual-scattering mean free path," and (2) a "fusion energy burn-up fraction mean free path." Comparing these two mean free paths will provide an index of the degree to which the goal of minimizing the heating of the deuterons while still providing net fusion power can be approached. Since it ignores the effect of electron drag on the deuterons, the results obtained below overestimate the ion heating resulting from ion-ion collisions between the two beams.

The effect of ion-ion collisions in causing the growth of the "temperature" of the deuteron beam ions can be estimated from collision theory as given, for example, by Spitzer [28] for a test particle (here a deuteron) moving through a field of ^{3}He ions of low temperature (in their frame of reference). We define the ratio of the perpendicular component (effectively the temperature of the test –"type 1"– ion) to initial relative kinetic energy as

$$\varepsilon = \left[\frac{<(\Delta v_\perp)^2>}{V_{12}^2}\right] \quad (4)$$

The collisional mean free path associated with a given value of ε is then given by an expression involving the Spitzer characteristic scattering time t_e and cross-section, σ_s:

$$\lambda_\varepsilon = v_1 t_\varepsilon = \left[\frac{\varepsilon}{n_2 \sigma_s}\right], \sigma_s = \left\{\frac{2\pi e^4 Z_2^2 \ell n \Lambda}{W_{12}^2}\right\} \quad (5)$$

The corresponding equation for the "burn-up fraction mean free path" is:

$$\lambda_f = \left[\frac{f}{n_2 \sigma_{12}}\right] \quad (6)$$

By setting the two mean free paths equal to each other we may find an expression for the fractional heating, ε, as a function of the burn-up fraction, given as

$$\varepsilon = 2.2 \times 10^{-12} \left[\frac{Z_2^2 f}{W_{12}^2 \sigma_{12}}\right] \quad (7)$$

Putting in the values for the relative energy, W_{12}, and the fusion reaction cross-section, σ_{12} (cm^2), we find the value

$$\varepsilon = 35f$$

The implication of this value is that if f is small, say .01 as in the example given earlier, then ε can be substantially smaller than unity, i.e., the heating of the deuterium ions by ion-ion collisions can be minimized. We should also recognize that as the two beam bunches collide the heating of the deuterium ions will start from a very low thermal temperature, rising to its maximum value only at the end of the interaction. However, throughout the encounter the colliding beams will be causing D-^{3}He fusion reactions at essentially the same rate throughout the process. This means that satellite D-D reactions will only occur at an appreciable rate during the latter part of the interaction. In code calculations of the beam-beam interactions [29] it was found that the fraction of neutron-producing reactions in the special case considered was only 2×10^{-3} of the average rate of D-^{3}He reactions.

Electron Physics Considerations

As with all other approaches to fusion, the role of the electrons in the plasma physics and the energetics of the system must be taken into account. In the colliding beam system we are here considering, when the two streams of ions interpenetrate, their accompanying electrons become intermixed. The relative motion of two oppositely directed ion streams through the resultant cloud of electrons (essentially at rest in the laboratory frame) gives rise to energy exchange between the ions and the electrons via electron drag force. Through this process the electrons may get heated to a substantial temperature, one which is relatively insensitive to the fractional burn-up, as is shown below. Furthermore, owing to their high collision rate, the electron distribution function will be Maxwellian. This circumstance will make it more difficult to recover their kinetic energy via direct conversion than it is for the more nearly monoenergetic ions. We will later discuss some possible ways to alleviate this situation. Nevertheless it is clear that electron heating is an issue that must be considered carefully in assessing the present scheme.

To assess the electron heating issue in a more quantitative way one can employ the following electron energy balance equation [30]:

$$\frac{dT_e}{dt} = \frac{176\sqrt{2\pi}}{9} \left[\frac{\sqrt{m_e}}{M_D} \right] \left[\frac{e^2 \ln(\Lambda)}{T_e^{3/2}} \right] n_D W_1, \tag{8}$$

giving the rate of heating of the electrons when exposed to the deuteron beam at density n_D. The numerical factor on the right hand side of the equation takes into account electron heating by both ion species, provided the latter have the same energy and density. Assuming that they are exposed for a time $\tau = L/v_D$ (where v_D is the deuteron velocity), and neglecting any decrease in ion energy caused by the energy transfer, with the interaction length, L, defined through the burn-up fraction, f, one obtains the following expression for the ratio of the final electron temperature to the ion energy:

$$\frac{T_e}{W_1} = \left[\frac{440}{9\sqrt{2\pi}} \frac{\sigma_s}{\sigma_{12}} f \right]^{2/5} \tag{9}$$

The energy transferred to electrons *per ion* is equal to $(9/4)T_e$. The numerical factor, 9/4, takes into account the fact that the average electron carries an energy equal to $(3/2)T_e$, and that there are 3/2 electrons per ion (owing to the presence of the ^{3}He ions). Denoting the energy conversion efficiency for electrons by η_e, one readily finds the following generalization of the equation for Q_e:

$$Q_E = \frac{fW_{fus}}{(W_1 + W_2)} \frac{\eta_i \eta_R}{1 - \eta_i \eta_R + \xi f^{2/5} \eta_i (\eta_R - \eta_e)}, \tag{10}$$

where

$$\xi \approx (9/4)(20\sigma_s/\sigma_{12})^{2/5}$$

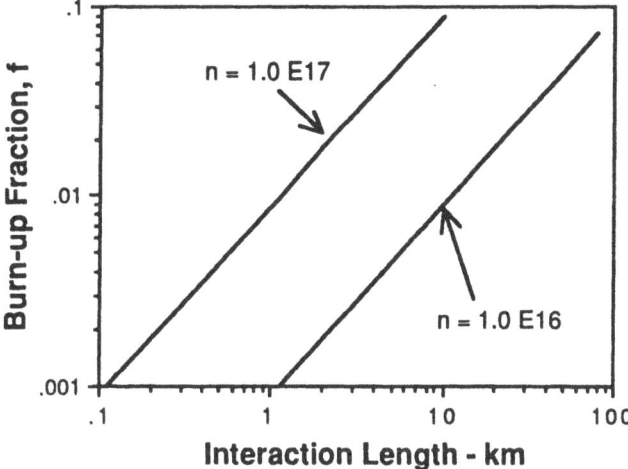

Figure 4. The burn-up fraction for colliding beams, plotted as a function of the beam-beam interaction length, with plasma density as a parameter.

To maintain the energy balance in the presence of electron heating it would be necessary to achieve good efficiency in the recovery of energy from the electrons. A preliminary estimate indicates that values of η_e of 0.7 to 0.8 would probably be sufficient to satisfy this requirement.

Plasma Column Length

We now consider a very demanding aspect of the linear collider idea - the plasma densities and associated plasma column lengths that would be required if "single-pass" operation is demanded. It is a straightforward matter to estimate the plasma density/length requirement as a function of burn-up fraction. Figure 4 presents a plot of this relationship for two different primary beam densities (the same for each beam).

The mental picture of the linear collider that emerges from the numbers on the plots in Figure 4 is that of a very long, small bore, high field solenoid, with injectors and direct converters located at each end, the whole assembly being located in a kilometer-length tunnel below the surface of the earth. There immediately comes to mind the particle colliders and TeV-energy ion accelerators in use or under construction by particle-physics researchers. Their devices also consist of high-field superconducting magnets located in deep-lying tunnels. One such facility is the completed LEP at CERN, with a 27 kilometer long tunnel. Another could be the (now abandoned) project of the SSC accelerator in Texas, which was to have a 80-kilometer tunnel. It would not therefore seem unreasonable to visualize D-^{3}He ion colliders with lengths of, say, 10 kilometers, assuming that the cost of a small-bore, superconducting, solenoidal magnet this long is economically acceptable.

Since the fusion power density for ion beams colliding with energies corresponding to the maximum point of the D-^{3}He cross-section is approximately 200 kW/cm^3 at a beam density of 10^{16} cm^{-3}, the diameter of the plasma column within the collider could be quite small, of order a centimeter or two, even assuming operation at a fairly low duty cycle. However, in order to suppress "firehose" and related streaming instabilities in the propagating and colliding beams the magnetic field must be high. A rough estimate of this requirement is that the energy density of the confining field should exceed the kinetic energy density of the beams, implying magnetic fields in excess of 20 T for the cases we have considered here. As noted earlier in this report, if one is dealing with simple solenoids it is not out of the question to consider such fields for future systems, considering developments now underway.

Example B: Velocity-Modulated Co-Injected Beams

To alleviate the technical requirements on the injector system implied in Example A an alternate approach has been considered [31,32]. Collinearly injected ion beams are to be introduced at one end of the solenoid, where the field is very low compared to its value at the midplane. These beams, derived from ion sources with accelerating grids, are to be velocity modulated so as to bunch at the point of highest field. In this way both radial and axial compression are accomplished transiently, up to densities sufficient to produce the required (small) fractional burn-ups needed for net power. Since the ion energies produced by an accelerator-grid ion source are precisely known (within a thermal spread of order 1 eV from the source plasma) it may be possible to take advantage of this fact to achieve net power at even lower fractional burn-up values than those assumed in Example A. The reason for this circumstance is that the theoretical efficiency of the electrostatic direct converter improves directly with reduction in the spread of energy of the ions that are collected. That is to say one can express the efficiency of this type of converter by the expression:

$$\eta_{DC} = 1 - \varepsilon \left\{ \frac{\Delta W}{W_1 + W_2} \right\} \quad \varepsilon \approx \frac{1}{N} \quad N = \text{Number of stages} \tag{11}$$

However, if the initial energy spread of the beam is small, then the incremental energy spread (caused by collisions with other ions in passing through the reaction region) is given by the density-dependent expression:

$$\Delta W = [n_1 n_2 / n_1 + n_2][\sigma_s v_{12} t_0 W_{12}] \tag{12}$$

The mutual scattering cross-section, σ_s, may again be estimated from Spitzer theory:

$$\sigma_s = \left\{ \frac{2\pi e^4 (Z_1)^2 (Z_2)^2 \ln(\lambda)}{[W_{12}]^2} \right\} \tag{13}$$

Note now that both the energy spread and the fractional fusion burn-up depend in the same way on the plasma density, so that in the limit $\eta_i \to 1$ and $\varepsilon\Delta W \ll W_0$, Q_E becomes independent of the beam-beam interaction time, t_0:

$$Q_E = \left\{ \frac{\sigma_{12} W_{12} W_{fus}}{[2\pi e^4 (Z_1)^2 (Z_2)^2 \ln(\lambda)]\varepsilon} \right\} \left\{ \frac{1 - \frac{\varepsilon\Delta W}{W_0}}{1 + \left[\frac{1-\eta_i}{\eta_i}\right] \frac{W_0}{\varepsilon\Delta W}} \right\} \tag{14}$$

Putting in numerical values for the constants gives:

$$Q_E = \left\{ 3.8 \times 10^{17} \frac{\sigma_{12} W_{12} W_{fus}}{(Z_1)^2 (Z_2)^2 \varepsilon} \right\} \left\{ \frac{1 - \frac{\varepsilon \Delta W}{W_0}}{1 + \left[\frac{1-\eta_i}{\eta_i}\right] \frac{W_0}{\varepsilon \Delta W}} \right\} \quad (15)$$

(all energies are in keV). (We have neglected here the effect of electron heating, assuming that the efficiency of the recovery of the energy of the electrons is sufficiently high.)

The maximum value of Q_E occurs at the maximum value of the product of the fusion cross-section and the beam-beam relative energy:

For D-T: $\sigma_{12} W_{12} = 550$ keV-barns

Take $W_{fus} = 4.5$ MeV $+ (.35 \times 14$ MeV$) = 9.4$ MeV

For D-^{3}He: $\sigma_{12} W_{12}$ (max) $= 420$ keV-barns

Inserting these results into eq. [15] one finds for D-T the value Q_E (max) $= [2.0/\varepsilon]$, and for D-^{3}He the value Q_E (max) $= [0.8/\varepsilon]$. It follows that for a 10 stage converter ($\varepsilon = 0.1$) the Q_E value for D-T would be of order 20, and for D-^{3}He it would be of order 8.0. Of course, in order to achieve these favorable values it would be necessary to find a combination of ion sources, direct converter and power circuitry, and plasma density and column length that would meet the stringent requirements stated above.

Based on some previous work [33], an analytical solution to the bunching equations has been made, in order to estimate what degree of density compression might be achieved. The calculation started with a 1-D distribution function that is a solution to the Vlasov equation for an accelerated maxwellian distribution of the ions:

$$\begin{aligned} f &= f_0 \exp\left\{-\alpha\left[(v_x)^2 - (v_0)^2\right]\right\}, & v_x &> v_0(t) \\ &= 0, & v_x &< v_0(t) \end{aligned} \quad (16)$$

The acceleration potential is to be modulated so that the launching velocity is given by:

$$v_0(t) = \frac{v_1}{1 - t/t_0} \quad t < t_{pulse}$$
$$t_0 = \frac{x_0}{v_1} \quad \alpha = M_1 / 2kT_i \quad (17)$$

To evaluate the density as a function of position and time, we can utilize the fact that the distribution function is a constant along the trajectory of the ions:

$$f(0,t) = f[x, (t-x/v_x)] \quad (18)$$

From this requirement the density as a function of position and time can be evaluated:

$$n(t,x) = \int_{v_{min}}^{v_{max}} f_0 \exp\left\{-\alpha\left[(v_x)^2 - \frac{(v_I)^2}{\left[1 - \frac{(t-x/v_x)}{t_0}\right]^2}\right]\right\} dv_x \quad (19)$$

for $t > t_p$, $v_{max} = \frac{x}{t-t_p}$

for $t < x/v_I$, $v_{min} = x/t$

for $x/v_I < t < t_I$, $v_{min} = \frac{x_0 - x}{t_0 - t}$

Fig. 5 is a plot, calculated using eq. (19), of the axial compression factor vs (x/x_0) for the case when $t/t_0 = 0.9$. Fig. 6 shows the same quantity when $t/t_0 = .9999$, i.e., at a time very nearly equal to the maximum compression. The final density will, of course, be given by the product of the axial compression and the magnetic compression factor. The latter factor also has limits, ones imposed by stability conditions. An estimate of such limits can be obtained by imposing the limit that the kinetic energy density of the field should always be less than the local energy density of the magnetic field (the elementary criterion for the avoidance of the so-called "firehose" instability). If this is done we can use the density calculations to define a "trajectory" of intensity that the magnetic field must obey as a function of x/x_0. Fig. 7 shows such a trajectory, giving the magnetic field intensity (relative to its initial value) as a function of position.

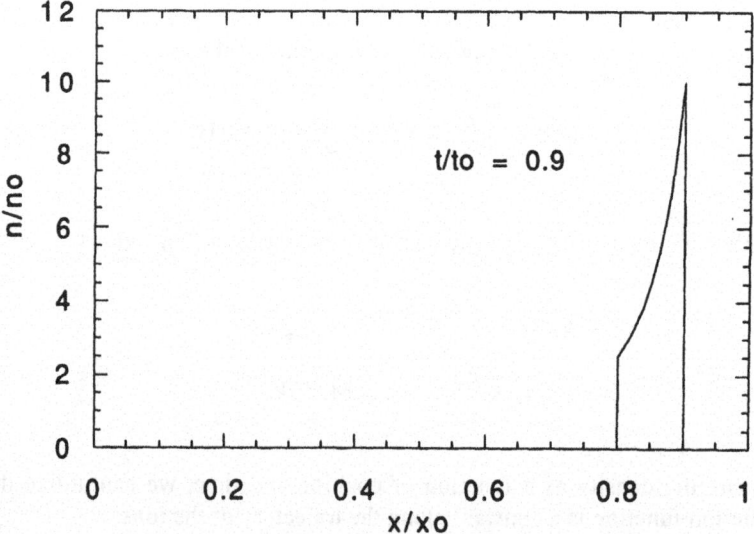

Figure 5. An in-flight "snapshot", at a normalized time of 0.9, of the beam plasma density (relative to its value at launching), as calculated by equation (19). Initial deuteron temperature: 1 eV; initial energy: 100 keV; pulse fractional time: 0.5.

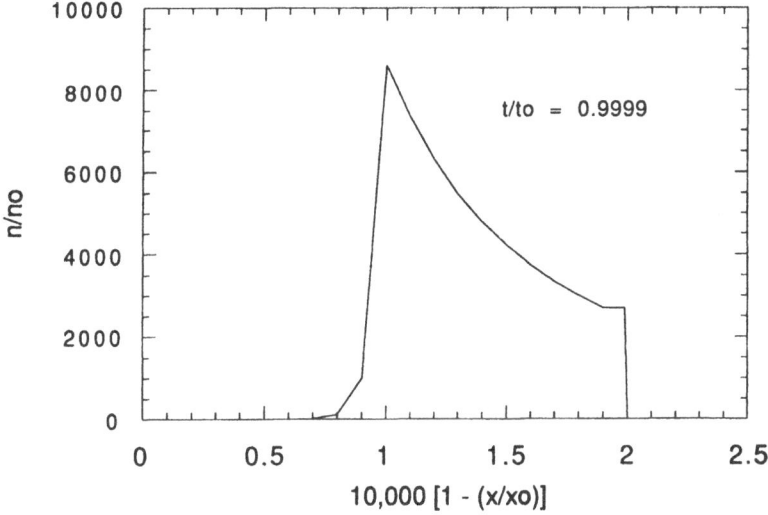

Figure 6. In-flight snapshot at normalized time 0.9999, all other parameter the same as in Fig. (5).

Given the magnetic intensity and the axial compression as a function of position we can then calculate a "stable trajectory" of the plasma density as a function of position along the plasma column. The results of such a calculation are shown in Fig. 8. As can be seen, density amplifications of several orders of magnitude would seem to be feasible. The issue then becomes the technical and economic one of how to design sources, direct converters and a solenoidal field that can satisfy practical constraints. Although such a design is far beyond the intent and scope of this paper, the extrapolations involved do not seem to be unattainable.

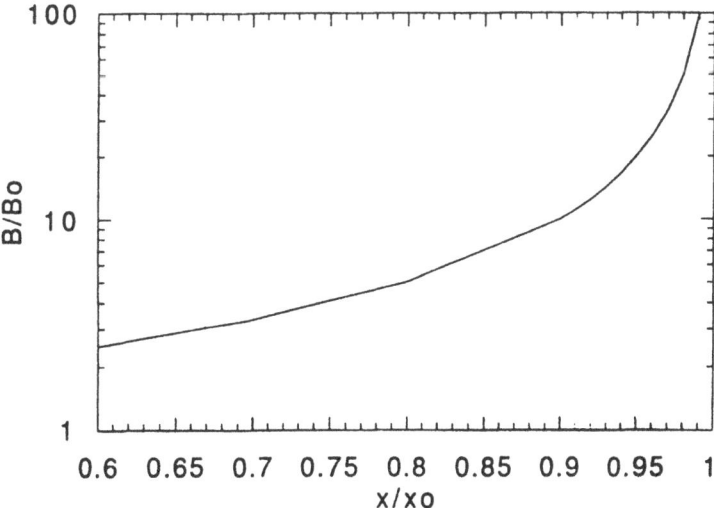

Figure 7. Relative variation of solenoid magnetic field along beam trajectory, as required to satisfy simple criterion for avoidance of the firehose instability.

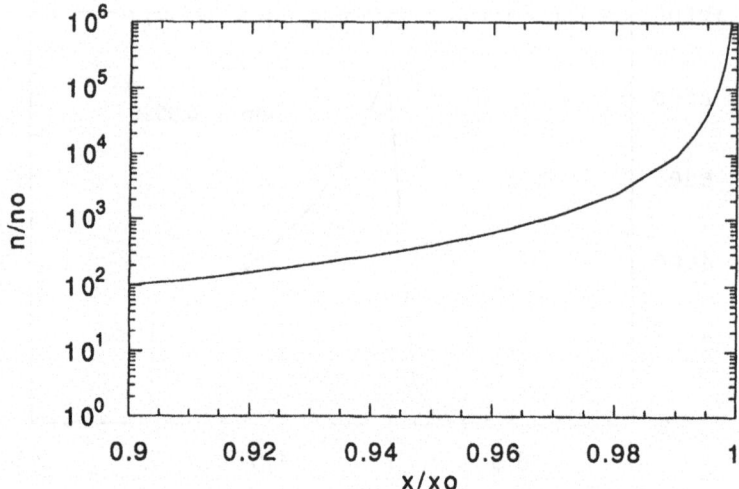

Figure 8. Firehose-stable density "trajectory" resulting from satisfaction of requirement on magnetic field shown in Fig. 7.

VI. MIRROR-BASED APPROACHES WITH REDUCED NEUTRON YIELDS

In the realm of more traditional open-ended systems, i.e., ones based on the mirror confinement of the plasma, there may also exist other schemes that allow net energy generation from D-^{3}He fuel at low Q and at low yield of parasitic D-D neutrons. One such idea was considered in a paper by Knyazev and Ryutov [34]. In this paper it was noted that the physics of open-ended systems makes it possible to maintain the temperature of the deuterons well below the energy of the ^{3}He ions, both maintaining a high rate of D-^{3}He reactions, and at the same time exponentially suppressing the rate of the parasitic D-D reactions. By using the special technique of "drift pumping" [35] one can keep the deuteron temperature as low as 5 keV, thereby reducing the neutron yield to as low a level as 10^{-5} of the fusion power [36]. Since the plasma Q value is low, the practicality of this virtually "non-radioactive" version of a fusion power plant also depends on the development of high efficiency direct converters and injectors.

VII. THE GAS DYNAMIC TRAP—ANOTHER EXAMPLE OF THE FLEXIBILITY OF MIRROR SYSTEMS

As another example of the flexibility of mirror-based systems, we describe here briefly a device that belongs to an opposite, that is, a strongly collisional, end of the spectrum as compared to the previous examples. This is a so-called "gas dynamic trap" [37], a mirror system with a very high mirror ratio ($R \approx 50$) and a length L satisfying the condition

$$L \geq \lambda_{ii} \left[\frac{\ln(R)}{R} \right] \tag{20}$$

where λ_{ii} is the mean free path with respect to scattering by 90 degrees. The significance of this condition is that implies that the length of the mirror cell is greater than the ion scattering length over a mean angle equal to the loss-cone angle, θ_{LC}, defined through the relationship, equation (1):

$$\sin(\theta_{LC}) = R^{-1/2}$$

Accordingly, the ion distribution function is nearly Maxwellian at all points, with a possible exception of the region near the mirror throat. In this situation the plasma lifetime is defined by analogy with the time taken for an ordinary collisional gas to escape from a vessel with a small orifice. That is, it is given by the ratio of the total number of particles nV contained in a vessel of volume, V, to the gas flux $nv_T S$ escaping through an orifice of area S. In the present case this circumstance yields the result for the confinement time as

$$\tau \approx RL/v_{Ti} \qquad (21)$$

The origin of this expression from gas-dynamic processes explains the origin of the name.

It is to be noted that the lifetime increases linearly with the mirror ratio (and not logarithmically as it does in simple mirror systems - see equation 2). Accordingly, it is desirable to increase the mirror ratio to the maximum value that is practically attainable. In this context it is important to note that the gas-dynamic trap can be made to be MHD-stable even with axi-symmetric fields. This possibility is based on the fact that the plasma density and pressure in the mirror and immediately beyond is approximately the same as in the central cell. Thus it is that this outer region, where the field lines are concave toward the axis (field increases radially) gives an appreciable positive contribution to MHD stability. By making this line curvature large enough one can assure MHD stability of the entire plasma. This particular approach has been confirmed experimentally on the GDT device at Novosibirsk [38]. As noted earlier, the axial symmetry of the mirror coils allows the possibility of reaching much higher field strengths than that that is possible with non-axi-symmetric coils. As the mirror coils can be made quite short, and since the plasma diameter in the mirror throat will be relatively small (10–15 cm.), it is quite conceivable that magnetic fields up to 45–50 Tesla will prove to be feasible to attain in practical systems.

A fusion power system based on the GDT would operate in a steady-state regime. Plasma and heat losses through the mirrors would be compensated by the injection of neutral beams and/or by pellet injection, in combination with RF heating. On the whole, this system is attractive, both in its simplicity and in the reliability of the confinement means (tested experimentally on a moderate-scale device [38]). Unfortunately, as with the linear colliders, this simplicity has to be paid for by the fact that a GDT fusion power plant would be relatively long, of order 2 to 3 kilometers. However, we should note that practically the entire length of the confinement chamber has only a weak field (approx. 1.5 Tesla). In the final analysis then, economic (and not doctrinaire) arguments will determine the potential of the GDT for fusion power systems.

VIII. CONCLUSION

To restate the premises on which this article was written, they were as follows: First, in the search for approaches to fusion that can lead in long run to economic and environmentally attractive fusion power systems the class of open-ended magnetic fusion systems has some special advantages. Second, that low-Q systems have advantages that have not been fully appreciated in the past. One such advantage is that such systems put far less demand on the confinement properties of the magnetic field, instead shifting the emphasis to the solution of some well-defined technological problems.

Another consequence of pursuing the low-Q option is to allow regimes in which the plasma behavior is more nearly governed by classical processes and less by turbulence-related ones, thereby introducing a higher degree of predictability on the basis of theory and simulation. This contention has been borne out, for example, in the pursuit of the mirror approach, where plasmas not dominated by turbulent transport processes have been achieved in a wide variety of experiments.

Finally, the examples of possible fusion systems given above have been given, not in the spirit of "proposals", but rather in the spirit of pointing out the wide spectrum of opportunities that exists in open-ended systems for novel approaches to the fusion problem, particularly as regards ones with what would be regarded as very low Q values relative to the ignited tokamak, for example. The effect of taking the low Q route, as noted, both shifts the emphasis from containment to technology issues, and at the same time opens up regimes, for example, ones with very low parasitic neutron-producing reactions, that are completely inaccessible to approaches that require a high Q value.

REFERENCES

1. P. W. Terry, et. al., in: *Plasma Physics and Controlled Fusion 1992*, (I.A.E.A. 1993) Vol 2., p. 313
2. C. Störmer, Z. f. Astrophys., **3**, 31, 227 (1931)
3. J. A. Van Allen, C. E. Mc Ilwain, G. H. Ludwig, J. Geophys. Res. **64**, 271 (1959)
4. T. G. Northrop, E. Teller, Phys. Rev. **117**, 215 (1960)
5. G. Gibson, W. C. Jordan, E. J. Lauer, Phys. Fluids **6**, 116 (1963)
6. L. G. Kuo-Petravic, M. Petravic, C. J. H. Watson, in *Nuclear Fusion Reactors* (Proc. Int. Conf. Culham,1969), British Nuclear Energy Society, London, 144 (1969)
7. R. F. Post, R. E. Ellis, F. C. Ford, M. N. Rosenbluth, Phys. Rev. Lett. **4**, 166 (1960)
8. M. N. Rosenbluth, N. A. Krall, N. Rostoker, Nucl Fus. Suppl., Part I, 143 (1962)
9. A. W. Molvik, F. H. Coensgen, W. F. Cummins, et. al., Phys. Rev. Lett. **32**, 1107 (1974)
10. B. G. Logan, J. F. Clauser, F. H. Coensgen, et. al., Phys. Rev. Lett. **36**, 1468 (1976)
11. R. F. Post, Phys. Rev. Lett. **18**, 232 (1967)
12. G. I. Budker, V. V. Mirnov, D. D. Ryutov, JETP Lett. **14**, 214 (1971)
13. R. F. Post, X. Z. Li, Nucl Fusion **21**, 135 (1981)
14. R. F. Post, in *Nuclear Fusion Reactors* (Proceedings British Nucl. Energy Soc. Conf. London, 1969), Culham Laboratory, Abingdon, Oxfordshire (1969) p.88
15. G. I. Dimov, V. V. Zakaidakov, M. E. Kishinevskiia, Sov. J. Plasma Phys. **2**, 326 (1976)
16. T. K. Fowler, B. G. Logan, Comments Plasma Phys. Controlled Fusion **2**, 167 (1977)
17. D. E. Baldwin, B. G. Logan, Phys. Rev. Lett. **43**, 1318 (1979)
18. R. F. Post in Proceedings of Course and Workshop: "Physics of Mirrors, Reversed Field Pinches and Compact Tori", Varenna, Italy, Vol. I, 51 (1987)
19. R. J. Hastie, J. B. Taylor, Phys. Fluids **8**, 323 (1965)
20. D. L. Correll, J. F. Clauser, F. H. Coensgen, et. al., Nucl. Fusion **20**, 655 (1980)
21. R.W. Moir, W. L. Barr, G. A. Carlson, in *Plasma Physics and Controlled Fusion 1974* (I.A.E.A. Vienna 1975) Vol. 3, p. 583
22. F. H. Coensgen, W. F. Cummins, W. E. Nexsen, Jr., A.E. Sherman, Nucl. Fusion Suppl., Part I, 125 (1962)
23. G.F. Abdrashitov, A.A.Bekhtenev, V.V. Kubarev, V.E. Pal'chikov, V.I.Volosov, Yu.N. Yudin. In: "Mirror Based and Field Reversed Approaches to Magnetic Fusion", Proc. of the 1983 Varenna School of Plasma Physics, Monotypia Franchi, v.1, p.335, 1984.
24. A. Mase et al., Nuclear Fusion, **31**, 1725 (1991)
25. R. W. Moir, W. L Barr, D. J. Bender, et al., in *Plasma Physics and Controlled Fusion 1976*, Vol. III, 237, IAEA, Vienna (1976)

26. R. F. Post, J. F. Santarius, Fusion Technology, **22**, 13 (1992)
27. See Ref. 21
28. L. Spitzer, Jr., *Physics of Fully Ionized Gases*, Interscience (1962)
29. See Ref. 26
30. NRL Plasma Formulary, NRL/PU/6790–94–265, Naval Research Laboratory, Washington DC, 1994.
31. R. F. Post, "Exploring the Limits of the 'Low Q' Fusion Power Regime", Paper 1D18, in : Proceedings of the Sherwood International Fusion Theory Conference," 14–16 March 1994.
32. R. F. Post, "Some Theoretical Aspects of 'Linear Collider' Open-Ended Fusion Power System", in: Proceedings of the Sherwood International Fusion Theory Conference," 3–5 April 1995
33. K. D. Marx, C. J. Eggens, R. F. Post, JAP **48**, 4215 (1977)
34. B. A. Knyazev, D. D. Ryutov, in: "Tritium and Advanced Fuels in Fusion Reactors," Proc. of the 1988 International School of Plasma Physics "Piero Caldirola," p. 693, Editrice Copositori, 1989
35. D. D. Ryutov, Proc. of the 2nd Wisconsin Symposium on Helium-3 and Fusion Power, (J. F. Santarius, compiler), WSCAR-TR-AR3–9307–3, University of Wisconsin, Madison, p. 121, (1993)
36. D. E. Baldwin, in: "Mirror-Based and Field-Reversed Approaches to Magnetic Fusion," Proc. of the 1983 Varenna School of Plasma Physics, Monotypia Franchi, Vol. 1, 109 (1984)
37. V. V. Mirnov, D. D. Ryutov, Sov. Technical Physics Lett., **5**, 279 (1978)
38. A. S. Ivanov, et al., Physics of Plasmas, **1**, 1529 (1994)

13

FORMATION, COMPRESSION, AND ACCELERATION OF MAGNETIZED PLASMAS

J. H. Degnan, D. E. Bell, A. L. Chesley, S. K. Coffey, J. L. Eddleman,
S. E. Englert, T. J. Englert, M. H. Frese, D. G. Gale, J. D. Graham,
J. Hammer, C. W. Hartman, J. Havranek, T. W. Hussey, G. F. Kiuttu,
F. M. Lehr, G. J. Marklin, H. S. McLean, A. W. Molvik, C. D. Holmberg,
C. A. Outten, R. E. Peterkin, Jr., D. W. Price, N. F. Roderick, E. L. Ruden,
U. Shumlak, P. J. Turchi, and J. J. Watrous

High Energy Plasma Division*
Phillips Laboratory
Kirtland AFB, New Mexico 87117–5776

INTRODUCTION

Compression and acceleration of magnetized plasmas is relevant to fusion for two reasons. Certain types of magnetized plasmas can be compressed and accelerated without fluid instability growth. These are magnetized plasmas rings or compact toroids[1,2]. Because of their stability, they can be compressed and accelerated over meters of distance and several microseconds of time, enabling economic scaling to much higher energy operation. Other types of implosions and compressions, e.g., Z-pinches, are limited by instability growth[3] to much shorter acceleration distances (a few cm) and times (less than 100 nanoseconds), making it very expensive to scale their operating energies to the fusion regime. An important second advantage of magnetized plasmas is that discussed by Lindemuth and Kirkpatrick in their magnetized target fusion (MTF) concept[4]. Reduced electron thermal conduction losses and increased alpha energy deposition result in reduced requirements of fuel density-radius product for achieving fusion ignition. In this paper, we discuss two experimental efforts at the Phillips Laboratory relevant to this topic. These are our Compact Toroid[2,5,6,7] and Solid Liner/Working Fluid[8,9,10,11] efforts. Though these efforts have potential fusion application, their present support is for the applications of intense X-ray generation and achieving high density and pressure in the laboratory, respectively.

---

* Present affiliations: Coffey, T. J. Englert, Frese, Watrous, NumerEx, Albuquerque, NM; Eddleman(deceased), Hammer, Hartman, McLean, Molvik—Lawrence Livermore National Laboratories; S. E. Englert, Roderick—University of New Mexico; Gale, Graham—Maxwell Laboratories, Inc.; Marklin—Consultant; Shumlak—University of Washington; Turchi—Ohio State University.

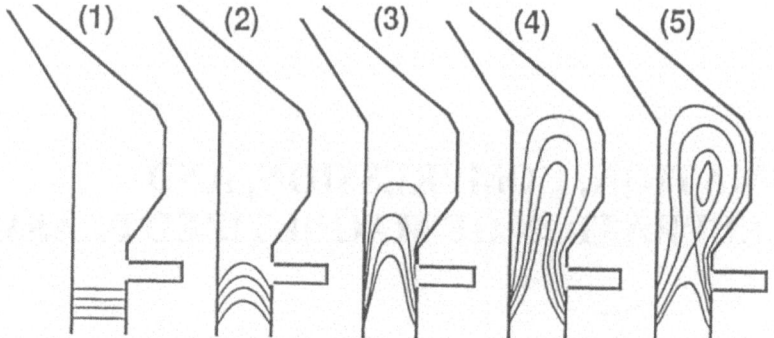

Figure 1. Compact Toroid formation process. See text.

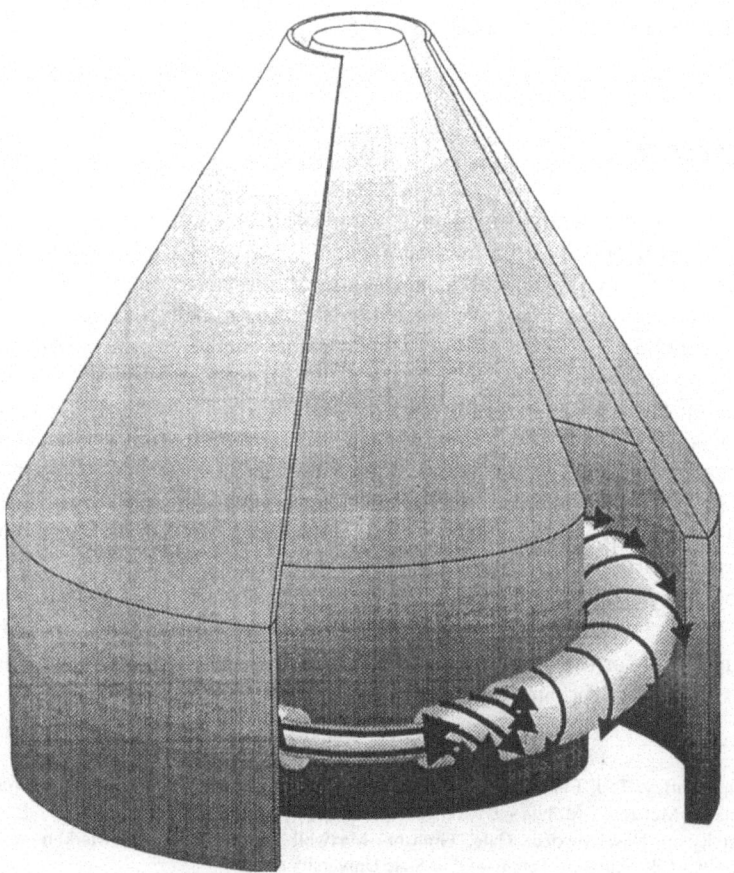

Figure 2. Compact Toroid in conical-coaxial electrode gap.

COMPACT TOROIDS

Compact Toroids are magnetized plasma rings, with comparable poloidal and toroidal magnetic fields, which are in or close to Woltjer-Taylor equilibrium[12] when in an electrode gap. The Phillips Laboratory Compact Toroid effort is a larger size and energy variant of the Lawrence Livermore National Laboratory RACE effort, which inspired it. In the present experiments, one meter major diameter, 18 cm minor diameter, one milligram mass compact toroids are formed with a magnetized, puffed gas coaxial plasma gun. They are subsequently compressed and accelerated using a second higher energy coaxial gun discharge.

This formation process occurs in an expanded gap region of the coaxial plasma gun geometry. As Fig. 1 illustrates, an initial bias magnetic field and annular gas puff are produced in the coaxial electrode gap (1). A first-stage coaxial gun discharge ionizes the annular gas puff and pushes the bias magnetic field embedded plasma annulus toward the downstream expansion region (2,3). The bias field is stretched into a toroidal bubble (4). Reconnection closes the stretched bias field lines into closed poloidal field lines around the minor circumference of the plasma ring. There are also toroidal field lines around the major circumference, arising from the first-stage plasma gun discharge current. The reconnection occurs, when energetically favorable, when the diffusion time across the stretched bias field reversal scale length becomes approximately as short as the Alfven time for that scale length. The resulting magnetized plasma ring relaxes toward the Woltjer-Taylor minimum free energy state[2,12], provided the ring is trapped in the coaxial electrode gap expansion region. A higher energy second stage coaxial gun discharge then pushes the compact toroid beyond the expansion region for subsequent acceleration and compression. An illustration of such a compact toroid in a conical-coaxial electrode gap is shown in Fig. 2. The initial magnetic field is established prior to the formation discharge, by means of a 10 millisecond rise time discharge through two solenoids (in series), located inside the inner electrode and outside the outer electrode of the coaxial gun. A two milligram annular gas puff is injected using an array of 60 fast gas valves, in a region where the preinjected magnetic field is predominately radial, about 0.2 Tesla. The formation (capacitor) discharge parameters are typically 2 megamps, 4 microsecond rise time, 270 kilojoules. The acceleration-compression discharge is driven by part or all of the 1300 microfarad, up to 120 kilovolt, up to 9.4 megajoule Shiva Star capacitor bank.[13,14]

Evidence that the magnetized plasma rings thus formed actually relax to a near Woltjer-Taylor state is shown in Figs. 3 and 4. In Fig. 3 we see the normalized sum of squares of differences between experimental and calculated poloidal field. The sum is over experimental measurement points. The two plots are for equilibrium theory and for 2D-MHD predicted field distribution. We see that for this choice of formation parameters, the approach to Woltjer-Taylor equilibrium is closest around 15 µsec into the formation discharge, or about two Alfven times for the minor diameter. Empirically, the approach to equilibrium is sufficiently close for intact subsequent acceleration at 7 µsec into the formation discharge. This data is for Argon toroids, but results are similar for other gases. In Fig. 4, we see the similarity of the vector plots of the poloidal field from experimental data and from 2D-MHD simulations at 15 µsec into the formation discharge, as well as equilibrium theory predictions.

Evidence that we have accelerated compact toroids intact is shown in Fig. 5. The magnetic probe signals have the shapes characteristic of a compact toroid traveling past the probes, including reversal of the radial magnetic field for significantly inserted probes. The probe signals characteristic of traveling compact toroids are also correlated in space

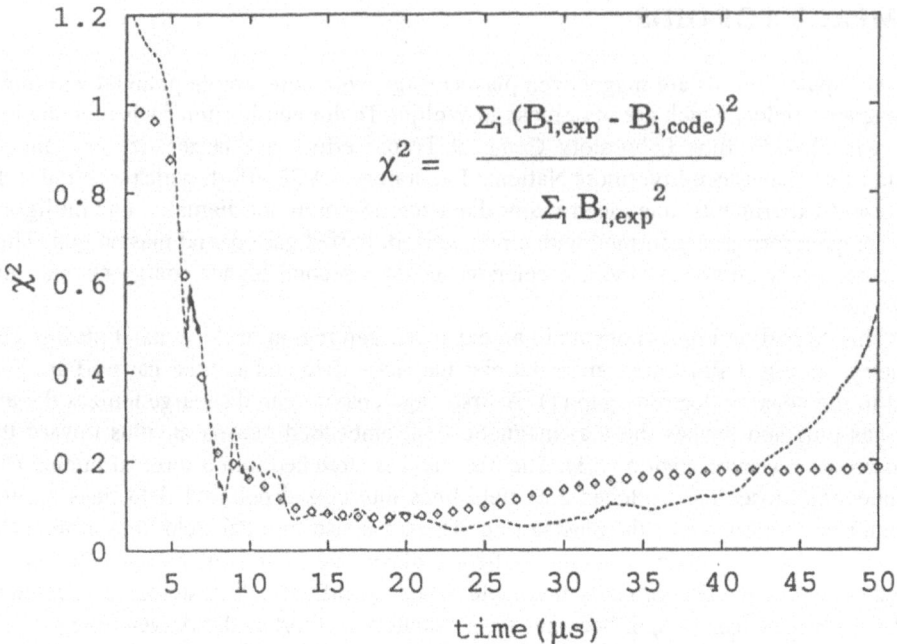

Figure 3. Poloidal field data on dynamic approach to equilibrium. See text.

and time with interferometer evidence of an electron density pulse consistent with several experimental and calculated measures of the compact toroid mass (all around one milligram)[2]. Optical spectroscopy data, such as shown in Fig. 6, indicates that the compact toroid material species is the same as the injected gas species, and that the compact toroid plasma precedes electrode plasma be several μsec, at least for modest acceleration energy discharges.

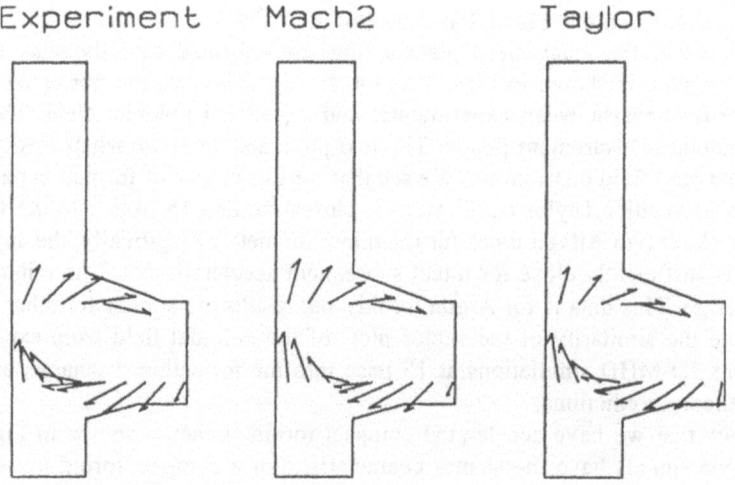

Figure 4. Poloidal field vector plots at closest approach to equilibrium.

Formation, Compression, and Acceleration of Magnetized Plasmas

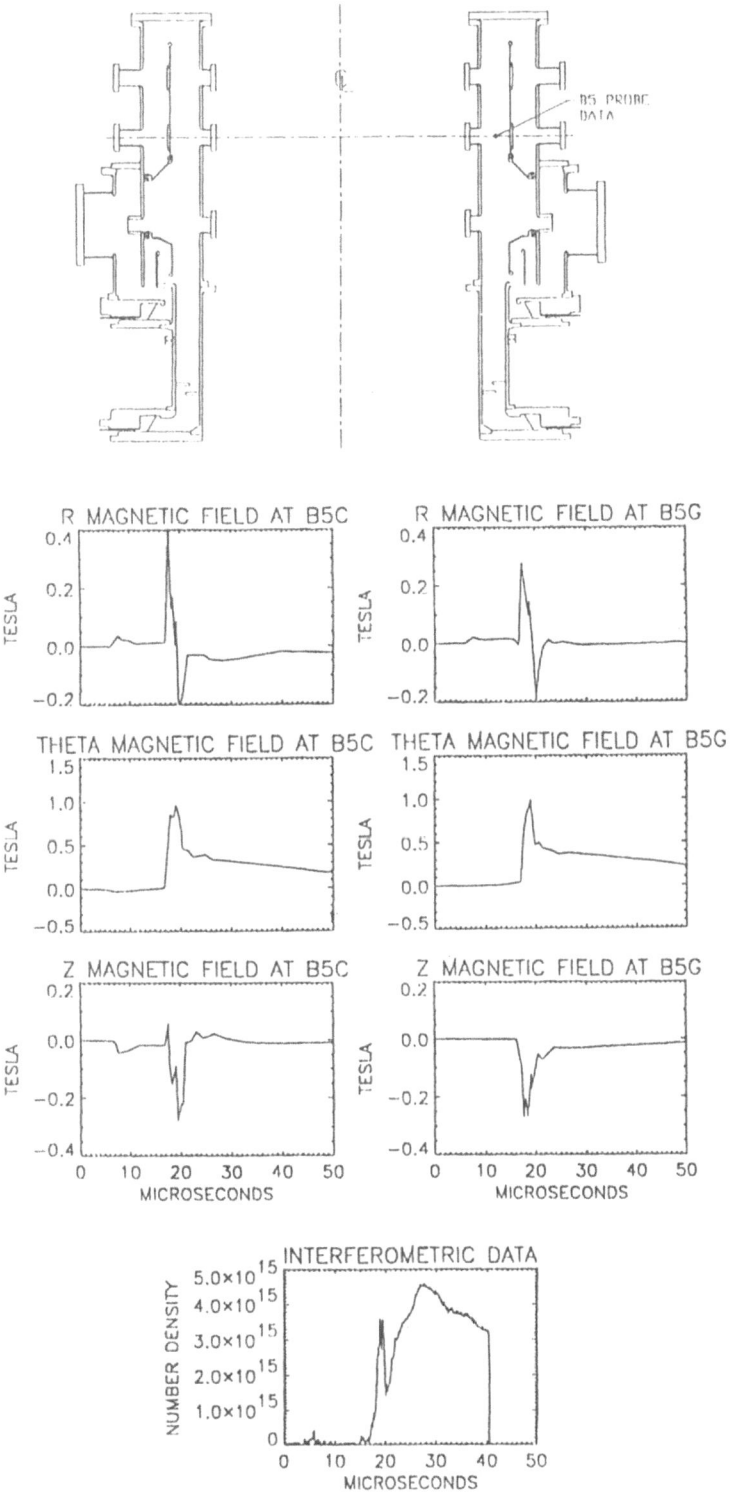

Figure 5. Magnetic probe and interferometer data.

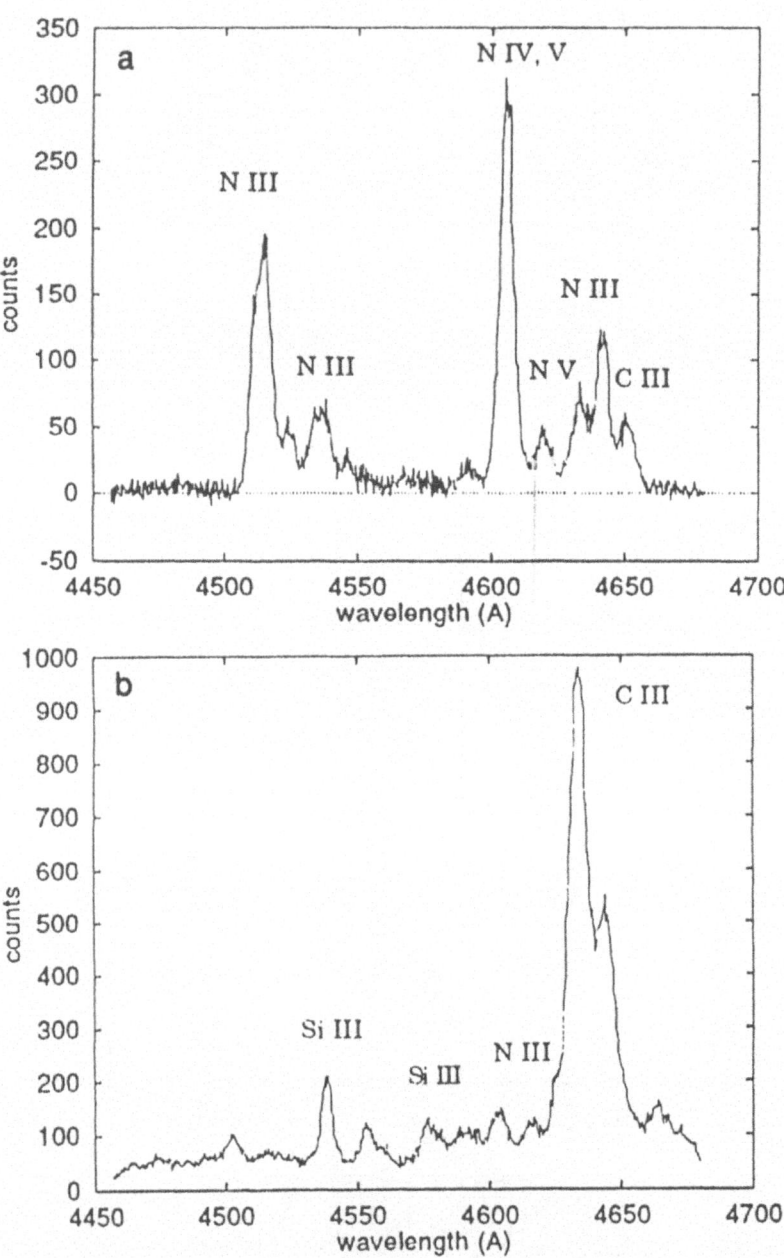

Figure 6. Optical spectroscopy data. a) When the CT is in the field of view, nitrogen (and only a small amount of carbon) is seen between 9–11 µs. b) After the acceleration discharge moves the CT out of the field of view, mostly carbon is seen between 13–14 µs.

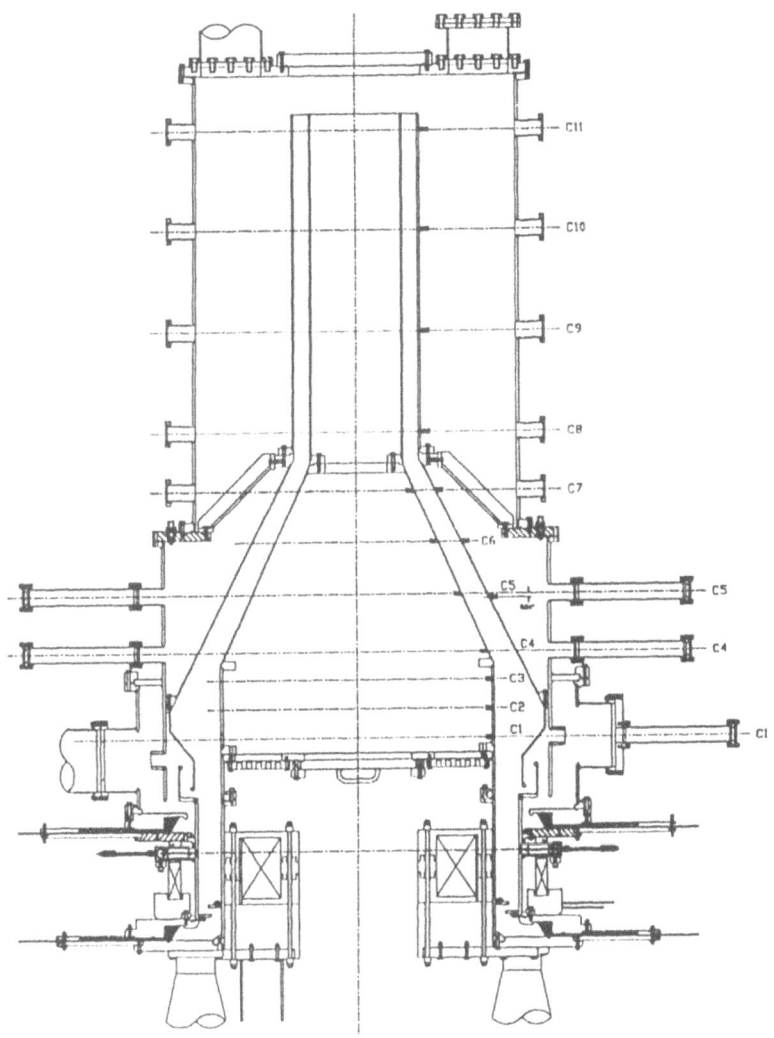

Figure 7. Three times radial compression, non-self similar geometry.

We have experimentally explored conical-coaxial compression/acceleration of compact toroids in non-self similar and self-similar (gap proportional to radius) electrode geometry. An example of the former is shown in Fig. 7, which includes the features of three times radial compression from initially 1 meter major diameter, followed by a straight coaxial electrode section. Poloidal magnetic field probe position vs time-of-arrival trajectories, such as shown in Fig. 8, indicate acceleration with interruption at the abrupt transition in cone angle at the three times radial compression location (or "corner"). The data indicate that the toroid remains intact and resumes acceleration after this interruption, but subsequent designs avoided such abrupt transitions in cone angle. Both inductance growth[6] and probe time-of-arrival data[7] indicate that milligram toroids are accelerated to ~40 cm/µsec for such geometry electrodes with ~1 megajoule acceleration/compression discharge energy. Similar results were obtained for several gas species.

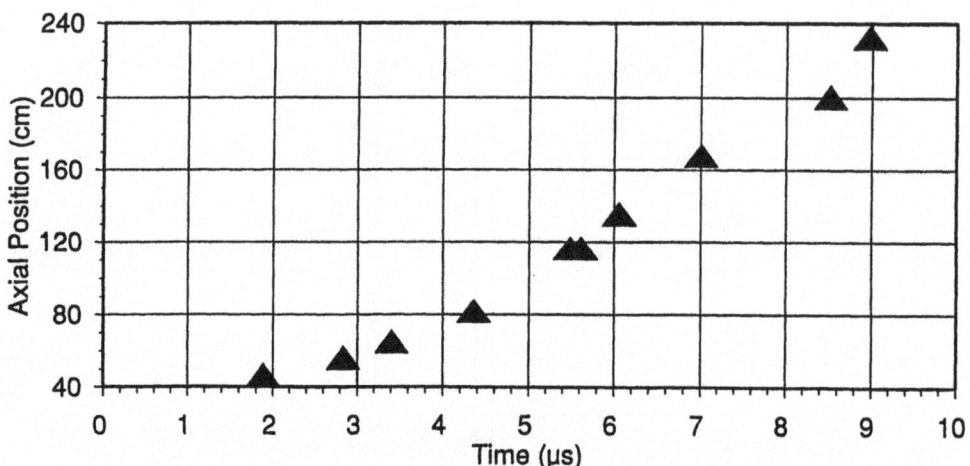

Figure 8. Poloidal magnetic probe position vs time-of-arrival data for geometry in Fig. 7. Time is from start of acceleration discharge (15 μs delay, 70 kV acceleration).

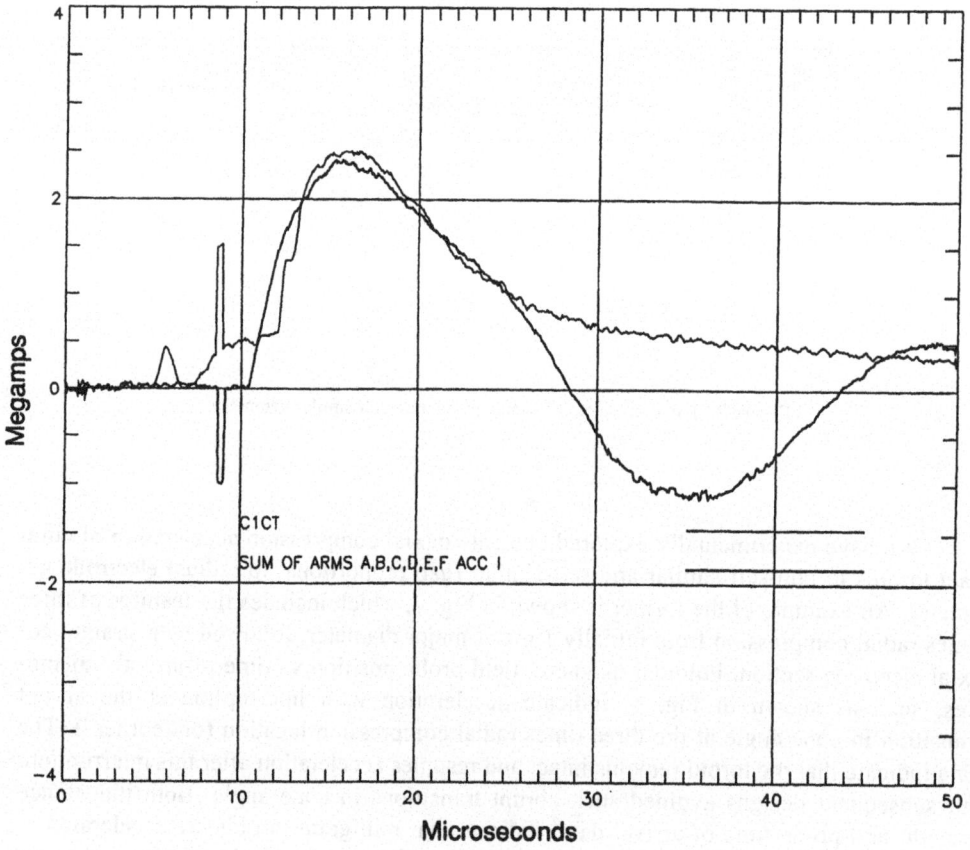

Figure 9. Overlay of total and downstream current. See text.

Evidence for full acceleration current delivery is obtained for both non-self similar and self-similar conical-coaxial compression/acceleration in the ~1 megajoule discharge energy range. An example of such evidence is shown in Fig. 9—an overlay of a Rogowski current probe signal (damped sinusoid) and an azimuthal magnetic probe signal (abrupt rise with precursor). (Both probes were passively integrated with 100 μsec time constants.) The azimuthal magnetic probe precursor is due to the compact toroid self-field toroidal component, which precedes the acceleration discharge "piston" magnetic field. This piston field pressure pushes the toroid. Typically, full acceleration current delivery lasts until the toroid exits the downstream coaxial gun muzzle. The absence of full current delivery in early experiments correlated with current joint problems.

The electrode geometry for self-similar nine times radial compression experiments is shown in Fig. 10. Again, this has an ~1 meter initial major diameter. Data from axial and azimuthal magnetic probes mounted flush with the electrode surface, close to the muzzle, show that the poloidal (axial) field precedes the toroidal (azimuthal field = piston

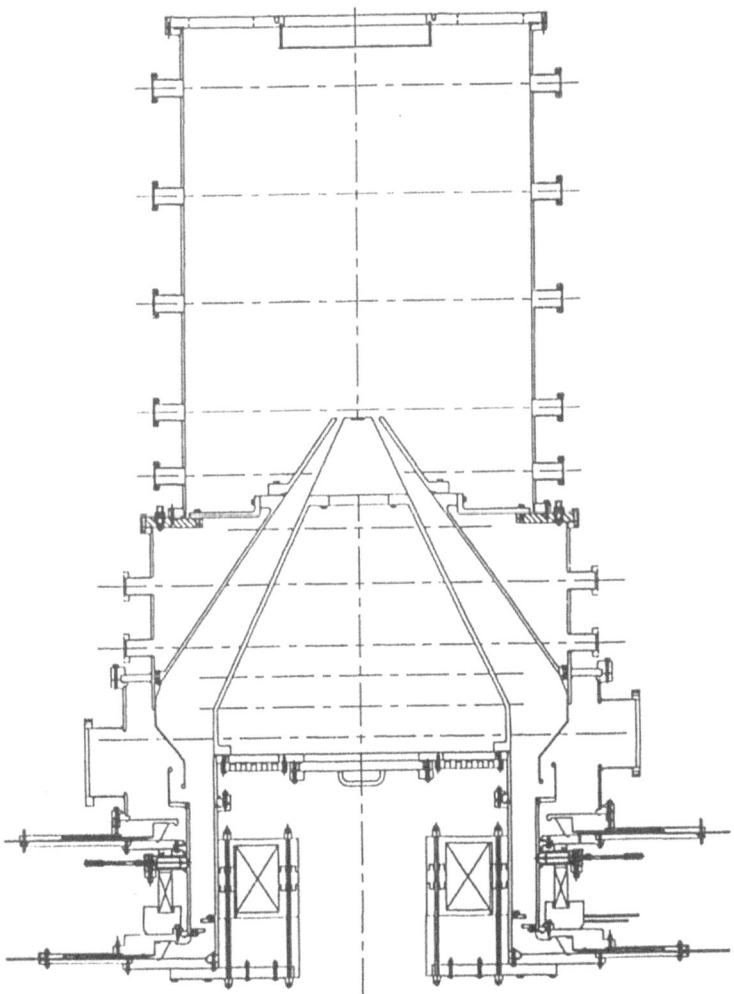

Figure 10. Nine times radial compression, self-similar geometry.

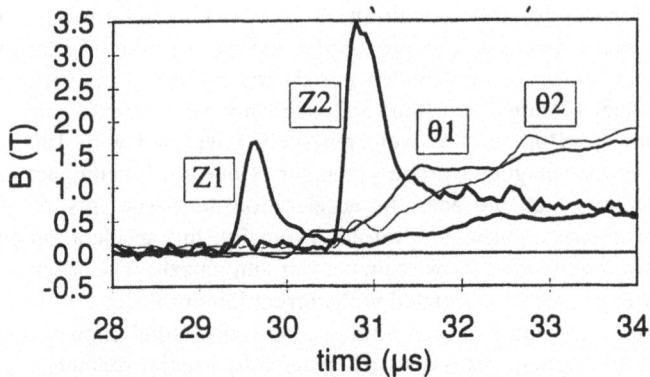

Figure 11. Flush-mounted magnetic probe data near conical-coaxial gun muzzle. Time is from start of formation discharge, with 20 μsec formation to acceleration discharge delay.

field), as it should. That is, the piston field has not "blown-by" the compact toroid field. An approximate requirement to avoid such undesirable "blow-by" is that the compact toroid field exceed the piston field.

The nine times radial compression, self-similar design is intended to achieve higher density with modest velocity. Data indicates density compression ~100 and exit velocity ~10–15 cm/μsec for 0.6 megajoule acceleration discharge operation. This is consistent with 2D-MHD simulations. Experiment-simulation comparisons are shown in Fig. 12. Deceleration is observed and calculated, due partly to compact toroid field compression and partly to electrode effects. Experiments to date have not included measures to reduce electrode plasmas other than a small number of conditioning shots. Performance to date is at the entry level of velocity-density parameter space for observing helium-like neon radiation due to target stagnation. Present efforts include fielding varying cone angle geometries and implementing electrode plasma reduction measures, such as those by Brown and Bellan[15].

The compressions can be self-similar (3-dimensional) in principle, that is, with proper matching of the driving current waveform with compression electrode geometry. The accelerating-compression electrode geometry is conical-coaxial, with either a single or varying cone angle. A single cone angle is appropriate for an exponentially rising current waveform. A varying cone angle is more appropriate for a capacitor discharge (sinusoidal) current waveform. Experimental density compression in a nine times radial compression, self-similar, single cone angle electrode geometry is approximately 100, consistent with 2D-MHD simulation. 2D-MHD simulations for a particular but unoptimized varying cone angle geometry driven by a capacitor discharge predict a density compression of 400 for a radial compression of 10 times. It is quite possible that an optimized varying cone angle geometry for a capacitor-driven current waveform will result in a density compression that is fully three dimensional. Experimental velocities achieved with 0.6 to 1 megajoule acceleration discharge are 40 to 50 cm/μsec with three times radial compression. For nine times radial compression, single cone angle geometry, at these energies, the toroid is decelerated to 15 cm/μsec or less, also consistent with 2D-MHD simulations. This deceleration is due, in part, to toroid magnetic field compression and other internal energy compression. Such simulations predict 35 cm/μsec velocity in combination with 400 times density compression for the previously mentioned varying cone angle electrode geometry, driven with 0.6 megajoule accelerator bank energy.

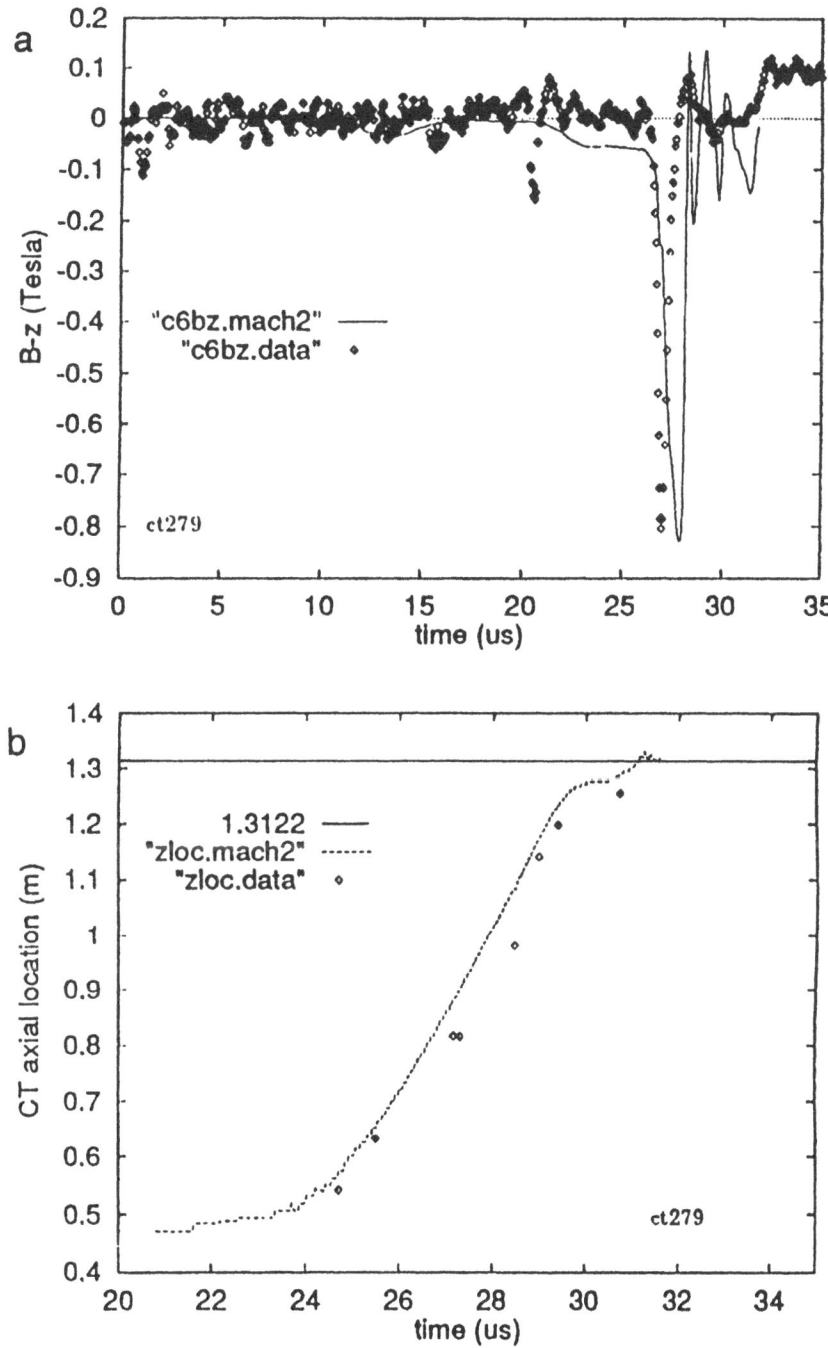

Figure 12. Data - 2D-MHD calculation comparison for poloidal magnetic probe signal (top) and compact toroid axial position vs time (bottom). Time is from start of formation discharge. (a) Magnetic probe data at the C6 location–approximately 2/3 of the way down the 9:1 self-similar compression cones.

The progress in developing compact toroid acceleration-compression at the megajoule energy level (for non-fusion applications) encourages us to consider extrapolation of that technology to the hundred megajoule range for fusion application. The potential for stable acceleration makes velocities high enough for advanced fuel fusion conceivable, if electrode plasma contamination is low enough for direct fusion. Higher levels of electrode plasma contamination of compact toroids are tolerable for the indirect fusion approach. Electrode plasmas are the principal technical issue limiting the energy and performance scaling of compact toroid compression and acceleration. Strategies for managing electrode plasma effects include choosing electrode geometry, driving current waveform, and initial toroid magnetic energy combinations for which substantial acceleration precedes substantial compression, as well as electrode material choice and cleaning measures.

SOLID LINERS/WORKING FLUIDS

We have used 12 to 14 megamp, 8 μsec rise time, 5 megajoule capacitor-driven axial discharges to electromagnetically implode aluminum shells in cylindrical, conical, and spherical geometry[8-11]. These shells typically have initial radius of 4 cm, thickness 1 mm, and implosion time of 13 to 15 μsec. Radiography and inductance growth diagnostics agree well with 2D-MHD simulations. Experimental radiographs and 2D-MHD simulation of synthetic radiographs and inner/outer surface boundaries at experimental radiograph times are shown for a spherical liner implosion in Fig. 13. The spherical liners have tapered thickness, proportional to cosecant squared of the polar angle or co-latitude, so that the ratio of magnetic pressure to liner mass per unit area is independent of polar angle. Such compressible MHD simulations predict inner surface implosion velocities exceeding 2 cm/μsec and stagnation pressure pulses of 10 megabars in cylindrical geometry, 50 to 100 megabars in spherical geometry. The pulsed power to drive such implosions is eco-

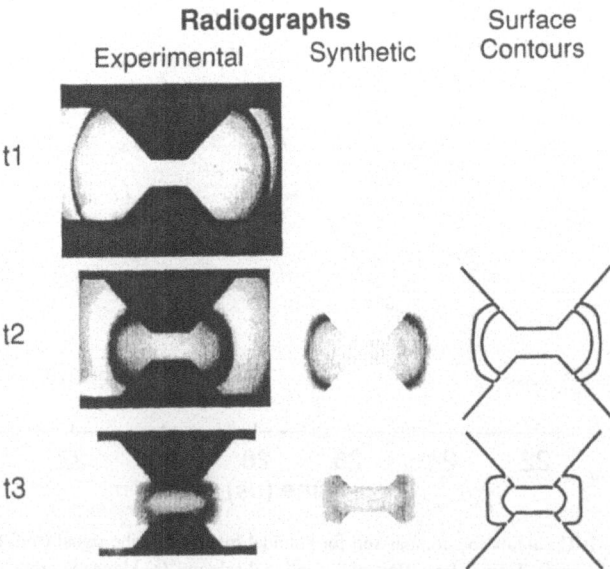

Figure 13. Experimental radiographs and 2D-MHD simulated radiographs, boundary contours of spherical liner implosion at time=0, 12.7, 14 μsec.

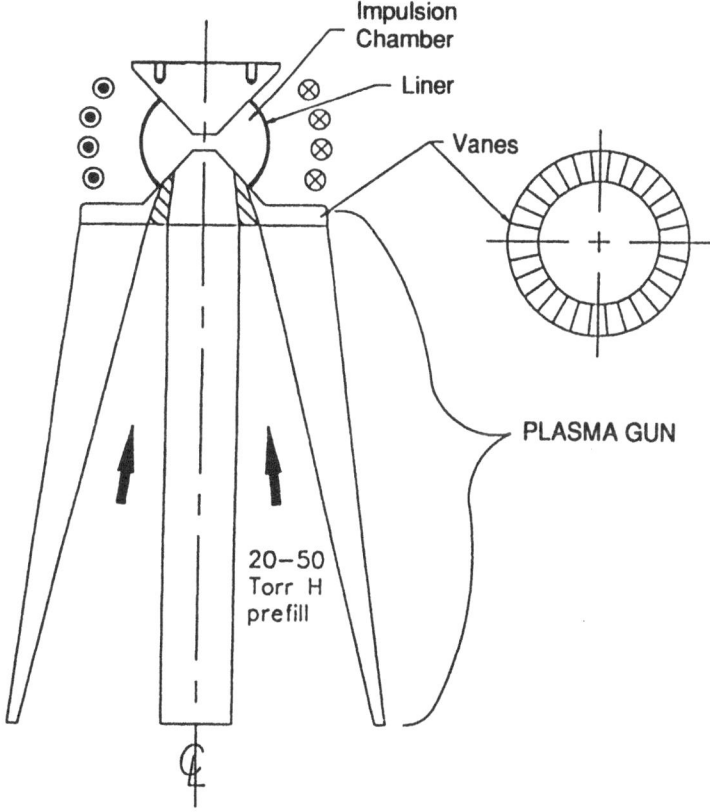

Figure 14. Working fluid formation/injection. The working fluid formation compression requires two pulsed electrical discharges: A~100 kJ pulsed electric plasma gun ionizes a neutral hydrogen gas and pushed it downstream into an implosion chamber; A~5 MJ pulsed discharge implodes an axisymnmetric, hollow metal container (a liner) upon the hot gas (the working fluid) to create high pressure without shocks.

nomically scalable to much higher energy—hundreds of megajoules. At these higher energies, solid or heavy liner implosions can compress magnetized plasmas to MTF conditions.

One application of such solid liner implosions is to compress a plasma working fluid. If the speed of sound in the fluid exceeds the inner surface liner implosion velocity, and if the fluid is *not* magnetized, the compressed fluid will have uniform pressure. This uniform pressure can be used to compress an inner spherical target. We have performed experiments to form hot hydrogen working fluids inside liners, using modest energy coaxial plasma guns operated at high pressure[10,11]. Injection into the liner interior is through a circular array of vanes in the liner implosion electrode. The vanes are to strip away the magnetic field from the coaxial gun discharge. The working fluid formation process is illustrated in Fig. 14. Computer-processed microchannel plate tube framing camera photos of working fluid injection inside a 4 cm radius cylindrical liner are shown in Fig. 15. For these experiments, one implosion electrode was replaced with a transparent lid for axial photographic access. H-beta and H-delta wavelength filtered, 200 nanosecond exposures are shown, taken shortly after first appearance of luminous plasma in the liner interior, and 10 μsec thereafter. In the earlier photos, the spatial modulation due to the array of in-

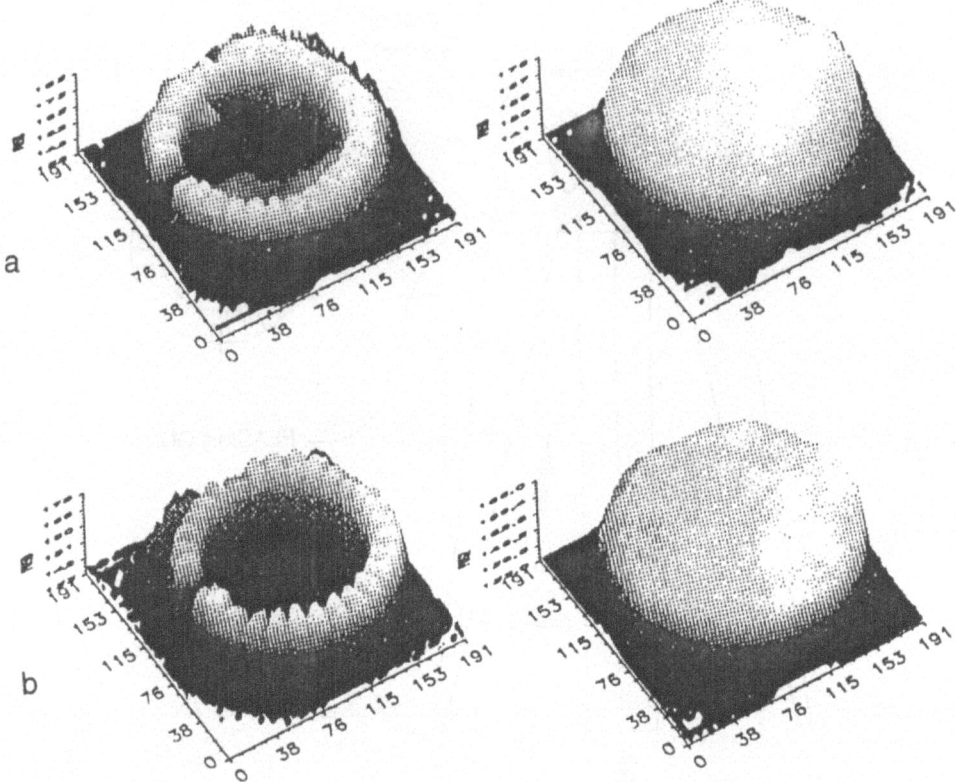

Figure 15. Fast photography of working fluid injection. 0.2 μsec frames; a) H-beta t=10,20 μsec, b) H-delta t=10,20 μsec.

jection vanes is evident, as is the shadow of a re-entrant probe. The later photos suggest a more uniform distribution. Though opacity effects require caution in interpreting this, 2D-MHD simulations also predict substantial uniformity of injected working fluid 10 μsec after first appearance. Piezoelectric probe data (Fig. 16) indicates 50 to 100 bar pressures for the range of experiment operating parameters. This is consistent with 2D-MHD simulations. Such working fluid initial pressure is sufficient to expect compression to the 10 megabar range for inner target compression.[10,11,16]

A concept for using a spherical solid liner implosion with working fluid compression to drive a concentric spherical MTF target is illustrated in Fig. 17. The interior of the target is magnetized by a modest energy coaxial transmission line coupled discharge. Variants of this concept include a straight or twisted cryogenic DT fiber on axis to form, respectively, a simple pinch or mini-compact toroid precompression plasma within the inner spherical target. The mini-compact toroid (or spheromak) approach can only work for an initially large system due to flux decay time scaling with size. Lithium wires in a DT gas fill may also work. A simpler lower energy experiment is to use only the outer spherical liner with magnetized deuterium injection instead of unmagnetized hydrogen working fluid injection. These concepts are somewhat similar to the Mokhov-Chernyshev MAGO concept,[17] which employs a coaxial discharge-driven preheated, magnetized DT plasma subsequently compressed by a very high energy cylindrical or quasi-spherical liner implosion.

Formation, Compression, and Acceleration of Magnetized Plasmas

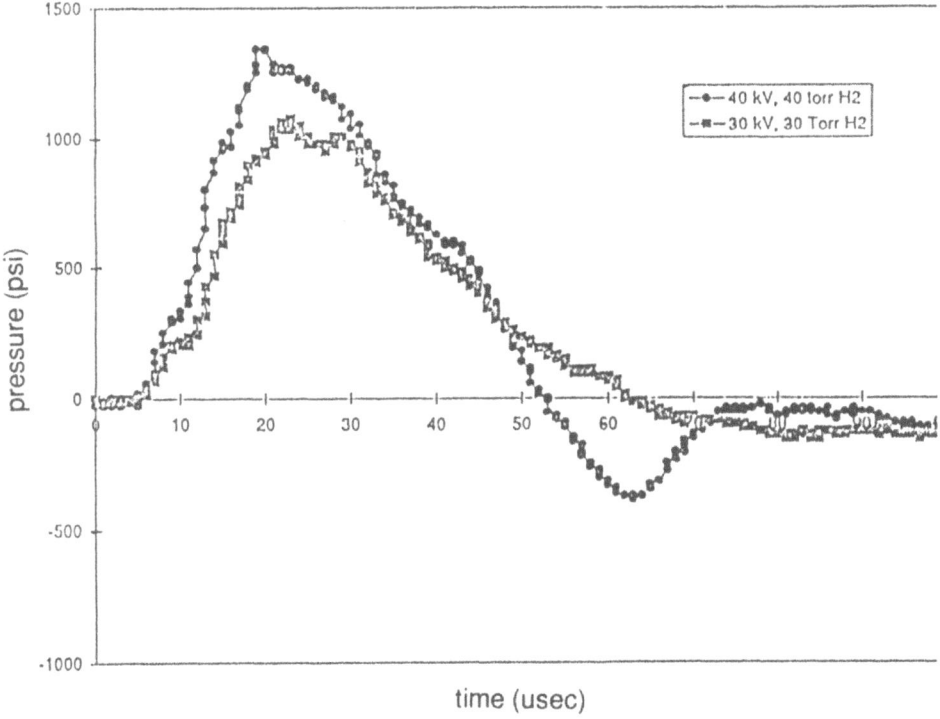

Figure 16. Piezoelectric pressure probe signals from working fluid injection.

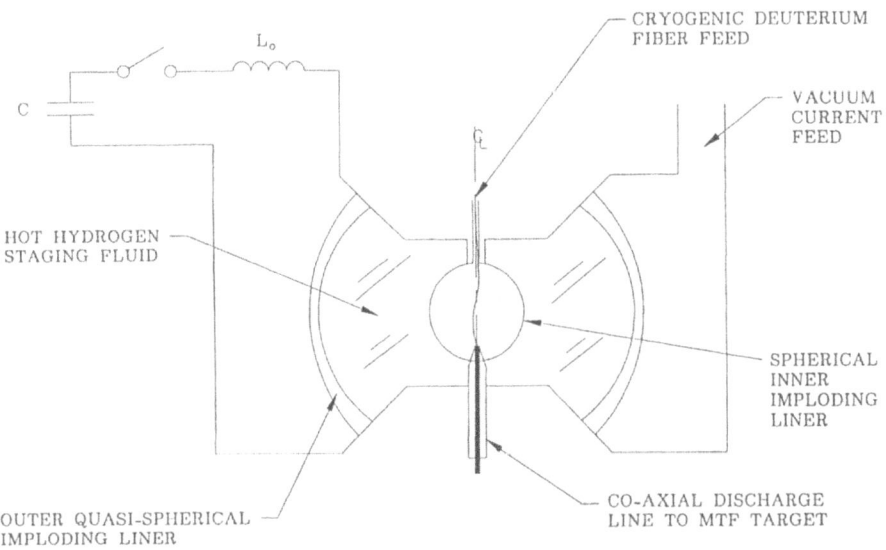

Figure 17. Concept for double spherical liner compression of MTF target.

PULSED POWER DRIVEN FUSION COMMENTS

The Compact Toroid acceleration and Solid Liner compression approaches are two pulsed power-driven approaches to fusion that employ large currents with conduction times ~10 μsec. They rely on low instability growth and magnetized fuel advantages. They have the major advantage of economic scaling to high energy. Their principal technical risk is electrode plasmas. The financial risk for attempting demonstration of ignition is small compared to the established large tokomak and laser-driven approaches.

The other pulsed power approach to fusion is fast Z-pinches. This approach attempts to overcome instability problems by employing short implosion times which means a small number of instability growth times. It is considerably more expensive—though still much less than tokomaks or lasers. Its principal technical risk is the feasibility limits of high current, very fast pulsed power. Compact toroids have a possible alternative role here as long conduction time, very fast transfer, high current plasma flow switches[18,19]. In such a role, compact toroids may reduce the cost and technical risk of driving fast, high energy Z-pinches. In this role, compact toroids may be less susceptible to electrode plasma problems than in a more direct (smaller final radius) role.

ACKNOWLEDGMENTS

This work was supported by the United States Air Force, Defense Nuclear Agency, and the Air Force Office of Scientific Research.

REFERENCES

1. C.W. Hartman and J.H. Hammer, New type of collective acceleration. *Phys.Rev.Lett.*, 48:929–932, April 1982.
2. J.H. Degnan, R.E. Peterkin, Jr., G.P. Baca, J.D. Beason, D.E. Bell, G. Bird, S.K. Coffey, M.E. Dearborn, D. Dietz, M.R. Douglas, S.E. Englert, T.J. Englert, D. Gale, J.D. Graham, K.E. Hackett, J.H. Holmes, T.W. Hussey, G.F. Kiuttu, F.M. Lehr, G.J. Marklin, B.W. Mullins, D.W. Price, N.F. Roderick, E.L. Ruden, M. Scott, S.W. Seiler, W. Sommars, C.R. Sovinec, and P.J. Turchi, Compact toroid formation, compression, and acceleration. *Phys. Fluids B*, 5(8):2938, August 1993.
3. T.W. Hussey, N.F. Roderick, and D.A. Kloc, Scaling of (MHD) instabilities in imploding plasma liners. *J. Appl. Phys.*, 51:1452, 1980.
4. I.R. Lindemuth and R.C. Kirkpatrick, Parameter space for magnetized fuel targets in inertial confinement fusion. *Nuclear Fusion*, 23(3):263, 1983.
5. J.H. Degnan, B.W. Mullins, J.D. Beason, M.E. Dearborn, D. Dietz, K.E. Hackett, J.L. Holmes, E.L. Ruden, D.W. Price, C.R. Sovinec, G. Bird, S.K. Coffey, S.W. Seiler, G.F. Kiuttu, R.E. Peterkin, Jr., N.F. Roderick, and P.J. Turchi, Compact toroid formation experiments at the weapons laboratory. In *ISPP-8 "Piero Caldirola", Physics of Alternative Magnetic Confinement Schemes*, pages 965–973, Bologna, 1991. SIF.
6. J.H. Degnan, G.P. Baca, D.E. Bell, G. Bird, A.L. Chesley, et al, Compression of compact toroids in conical-coaxial geometry. *Fusion Technology* 27 (2), 107 (1995).
7. G.F. Kiuttu, J.H. Degnan, R.E. Peterkin, Jr., E.L. Ruden, F.M. Lehr, et al, Acceleration and compression of compact toroid plasmas. *Invited paper at Beams '94 Conference*, San Diego, 20–24 June, 1994.
8. J.H. Degnan, W.L. Baker, M.L. Alme, C. Boyer, J.S. Buff, et al, Multi-megajoule electromagnetic implosion of shaped solid-density liners. *Fusion Technology* 27 (2), 115 (1995).
9. J.H. Degnan, F.M. Lehr, J.D. Beason, G.P. Baca, D.E. Bell, et al, Electromagnetic implosion of spherical liner. *Physical Review Letters* 74, 98 (1995).
10. F.M. Lehr, A. Alaniz, J.D. Beason, L.C. Carswell, J.H. Degnan, J.F. Crawford, S.E. Englert, T.J. Englert, J.M. Gahl, J.H. Holmes, T.W. Hussey, G.F. Kiuttu, B.W. Mullins, R.E. Peterkin, Jr., N.F. Roderick, and P.J. Turchi, Formation of plasma working fluids for compression by liner implosions. *J.Appl.Phys.* 75 (8), p. 3769 (April 1994).

11. F.M. Lehr, J.H. Degnan, D. Dietz, S.E. Englert, T.J. Englert, J.D. Graham, J.J. Havranek, T.W. Hussey, J.M. Messerschmitt, C.A. Outten, R.E. Peterkin, Jr., N.F. Roderick, U. Shumlak, P.J. Turchi, The formation of a high density working fluid for solid liner implosions. *IEEE Transactions on Plasma Science* (Accepted).
12. J.B. Taylor, Relaxation of toroidal plasma and generation of reverse magnetic fields. *Phys.Rev.Lett.*, 33:1139, 1974.
13. R.E. Reinovsky, W.L. Baker, Y.G. Chen, J. Holmes, and E.A. Lopez, in *Digest of Technical Papers: Fourth IEEE International Pulsed Power Conference*, 6–8 June 1983, Albuquerque, NM, edited by M.F. Rose and T.H. Martin (The Institute of Electrical and Electronics Engineers, New York, 1983), p. 196.
14. J.D. Graham, D. Gale, W. Sommars, M. Scott, and Y.G. Chen, in *Digest of Technical Papers: Eighth IEEE International Pulsed Power Conference*, 17–19 June 1991, San Diego, CA, edited by R. White and K. Prestwich (The Institute of Electrical and Electronics Engineers, New York, 1991), p. 990.
15. M.R. Brown, P.K. Loewenhardt, J. Yee, D.R. Derkits, and P.M. Bellan, High velocity compact torus injector for the TEXT tokamak. *Bull.Am.Phys.Soc 39*, p. 1596 (1994).
16. P.R. Chiang, R.A. Lewis, G.A. Smith, J.M. Dailey, S. Chakrabarti, K.I. Higman, D. Bell, J.H. Degnan, T.W. Hussey, and B.W. Mullins, Target compression by working fluids driven with solid liner implosions. *J.Appl.Phys.* 76 (2), p. 637 (July 1994).
17. V.K. Chernyshev and V.N. Mokhov, On the progress in the creation of powerful magnetic energy sources for thermonuclear target implosion. In R. White and K. Prestwich, editors, *Digest of Technical Papers: Eighth IEEE International Pulsed Power Conference*, pages 395–410, New York, NY, 1991, Institute of Electrical and Electronics Engineers.
18. W.L. Baker, J.H. Degnan, J.D. Beason, G. Bird, C.B. Boyer, et al, Current delivery and radiation yield in plasma flow switch driven implosions. *Fusion Technology* 27 (2), 124 (1995).
19. R.E. Peterkin, Jr., J.H. Degnan, T.W. Hussey, N.F. Roderick, and P.J. Turchi, A long conduction time compact torus plasma opening switch. *IEEE Trans. Plasma Sci.* 21(5), p. 522 (Oct 1993).

14

PROSPECTS OF MAGNETIC ELECTROSTATIC PLASMA CONFINEMENT

Thomas J. Dolan

IAEA, P.O. Box 100
A-1400 Vienna, Austria
Internet: dolan@ripo1.iaea.or.at

ABSTRACT

Magnetic electrostatic plasma confinement may be thought of as electrostatic plugging of a magnetic cusp plasma confinement system, or as magnetic shielding of the grids of an inertial-electrostatic plasma confinement system. A linear set of ring cusps is the preferred magnetic field configuration. High voltage cathodes repel electrons escaping along magnetic field lines from magnetic cusp confinement systems. A negative potential well confines ions electrostatically. Electron losses are by diffusion across a thin boundary magnetic field and by diffusion in velocity space over the cathode barrier potential. Narrow anode gap widths (~ 4 mm) are used to keep the plasma self-shielding voltage drop $\Delta\phi < 100$ kV. The Jupiter-2M experiment at Kharkov demonstrated good confinement with electron transport across the magnetic field at nearly the classical rate. With near-classical transport, reactor studies indicate that a power gain ratio $Q \approx 10$ may be attainable with plasma radius $r_p = 3$ m, ring cusp magnetic induction $B = 6$ T, and applied voltage $\phi_A = 400$ kV. The main issues to be resolved are plasma purity, electron transport, alpha particle energy confinement, and electrode alignment and voltage holding.

INTRODUCTION

Electrostatic plasma confinement systems suffer from high required currents and overheating of the grid wires. Open magnetic confinement systems suffer from rapid plasma loss along magnetic field lines. A combination of these two types of confinement systems, called magnetic electrostatic plasma confinement, combines some good features of each system. The applied electrostatic fields inhibit plasma losses along magnetic field lines, and the strong magnetic field partially shields the grid from particle bombardment. The "grid wires" become magnet coils, surrounded by magnetic cusp fields, as illustrated in Figure 1. This type of plasma confinement has also been called "electrostatically plugged cusps" and "electromagnetic traps."

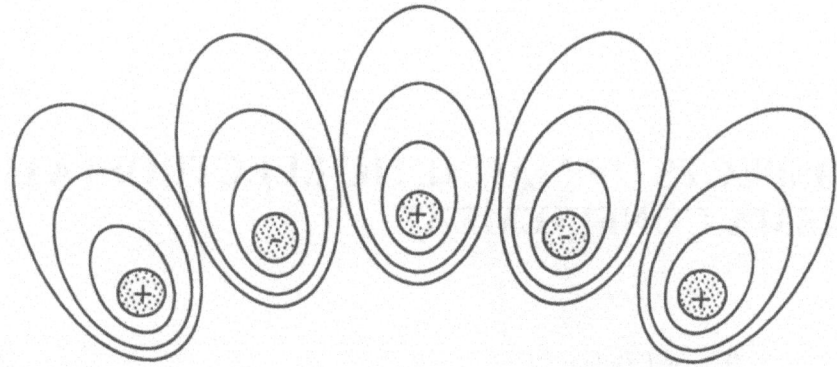

Figure 1. Magnetic protection of a grid. The plus and minus signs denote currents out of and into the plane of the drawing.

Some magnetic cusp configurations suitable for electrostatic plugging are illustrated in Figure 2.

Early experiments were spindle cusps, because they are relatively inexpensive to build, but spindle cusps have small plasma volumes and large loss areas, so they are poor for plasma confinement. Toroidal multipole cusps were studied when it was believed that plasma losses through point cusps might be excessive. The most promising configuration is the linear set of ring cusps. The present article is intended to provide an introduction to the principles of magnetic electrostatic plasma confinement. More details are available in a review paper (Dolan, 1994).

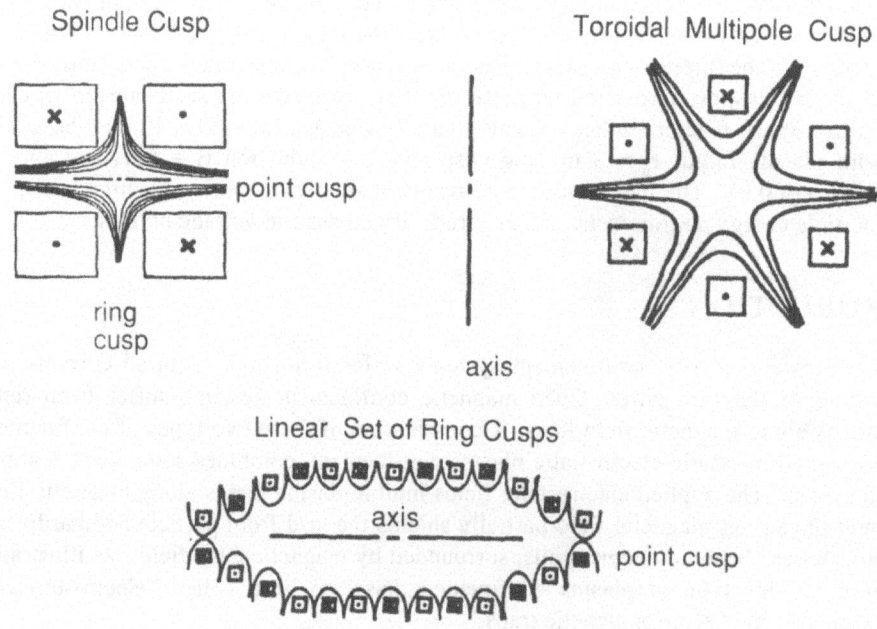

Figure 2. Some magnetic cusp configurations.

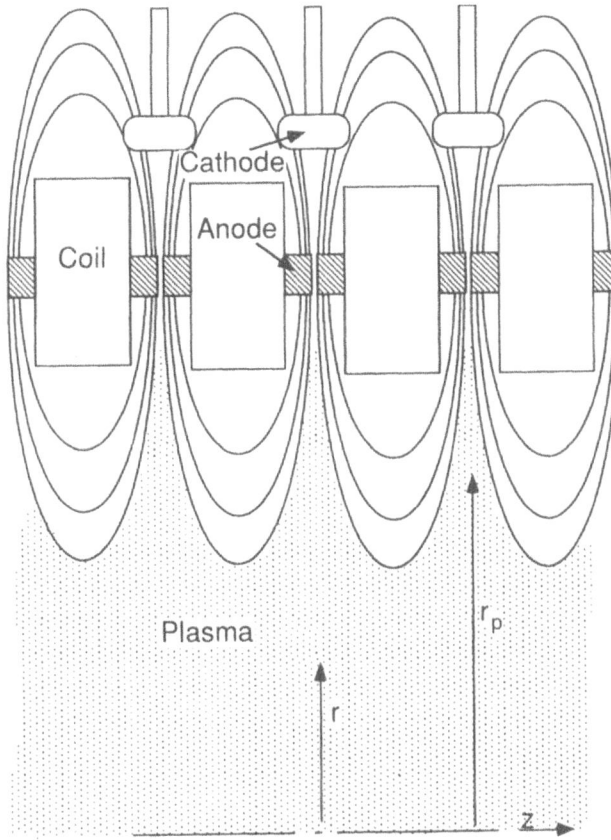

Figure 3. A simplified sketch of electrostatic plugging of ring cusps. The cathodes are at a large negative voltage and the anodes are grounded. The ellipses represent magnetic field lines, and the shaded area represents the confined plasma.

THEORY

Plasma losses through the magnetic cusps are suppressed by high voltage electrodes, illustrated in Figure 3.

Because the plasma is diamagnetic, the magnetic field is excluded from the plasma interior, leaving it relatively field-free and uniform.

The radial variation of electrostatic potential is illustrated in Figure 4.

This potential profile is the same as the axial potential profile in a simple tandem mirror. Magnetic mirrors could also be plugged electrostatically, but the volume of confined plasma would be much smaller than in a magnetic cusp system. Ions in the central plasma are confined by the electrostatic potential well ϕ_i. The ions do not become magnetically trapped in the boundary magnetic field, because its electrostatic potential is above that of the saddle point. When the ions acquire enough energy collisionally to surmount the barrier ϕ_i, then they are lost through the anode gaps to the cathodes.

Large values of $\Delta\phi$ spoil confinement. This voltage drop can be calculated from a solution of the two-dimensional Poisson equation for an assumed electron density distribution. The electron density distribution between the anodes is illustrated in Figure 5.

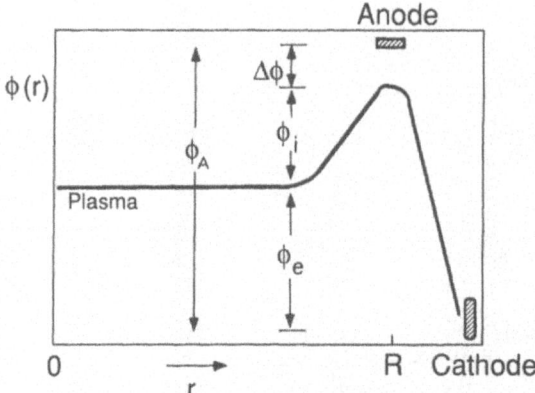

Figure 4. Radial variation of electrostatic potential. Here ϕ_A denotes the applied voltage, ϕ_e is the potential barrier confronting plasma electrons, ϕ_i is the barrier facing ions, and $\Delta\phi$ is the potential drop in the anode gap due to Debye shielding.

The assumed model distribution has a Lorentzian profile that slightly exceeds the electron density profile measured in the Jupiter-1A experiment. The resulting electrostatic potential variation in the anode region is saddle shaped, as illustrated in Figure 6.

If the anode gap is narrow (~ 4 mm), then the magnitude of $\Delta\phi$ can be estimated by solution of a one-dimensional Poisson equation, as a function of magnetic field and anode gap electron density. Such a prediction is shown in Figure 7 for fusion reactor parameters.

Values of B ~ 6 Tesla might be achieved with NbTi superconducting coils. In order to keep the voltage drop $\Delta\phi$ < 100 kV, values of $n_a < 2 \times 10^{19}$ m^{-3} would be required. The central plasma density n can be a factor of 10 higher than n_a. Factors affecting the ratio of n/n_a are listed in Table 1.

The ion density in the anode gap is very low, so their space charge effect is slight. Most plasma electrons are reflected geometrically by the convex magnetic field boundary and do not enter the anode gap region, as illustrated by orbit 1 in Figure 8.

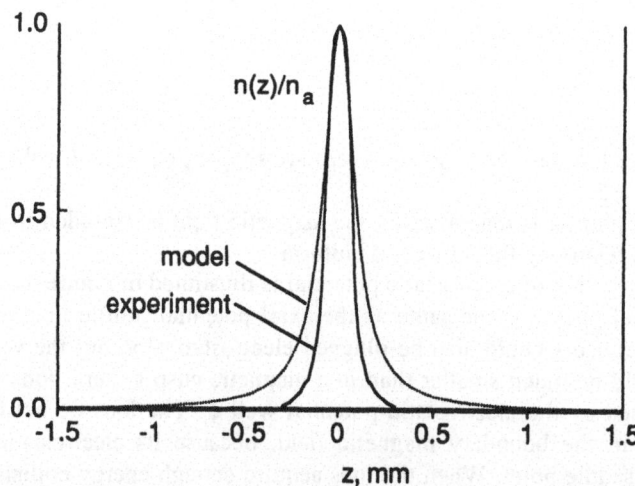

Figure 5. Electron density distribution in the z direction between the anodes. The model distribution is $n(z)/n_a = [1 + (z/\rho_e)^2]^{-1}$, where n_a is the electron density at the center of the anode gap and ρ_e is the electron Larmor radius in the gap. Experimental curve is from the Jupiter-1A experiment (Komarov and Stepanenko 1980), where a diaphragm limiter forms an effective boundary at a = 0.3 mm.

Prospects of Magnetic Electrostatic Plasma Confinement

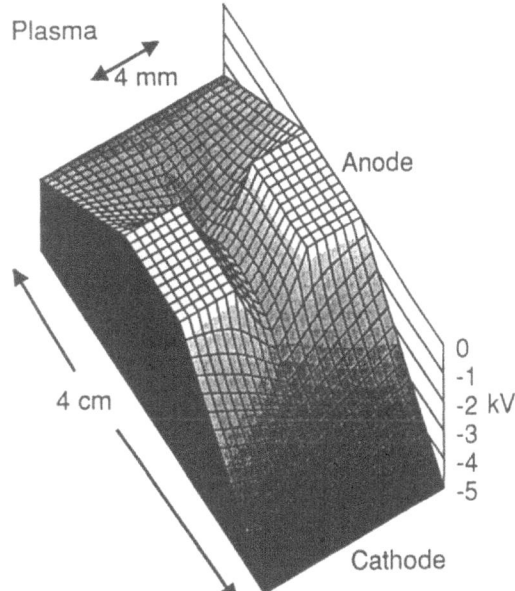

Figure 6. Variation of electrostatic potential in the electrode region. The 4 mm anode gap width has been exaggerated, for clarity. The difference in potential from the anodes to the saddle point is called $\Delta\phi$.

Table 1. Factors affecting the ratio n/n_a (central plasma density/anode electron density)

Ions in the gap	~ 1.02
Magnetic reflection	~ 3
Acceleration	~ 4
Radial focusing	~1.3
Cold trapped electrons	decrease the ratio
Net effect: Expect n/n_a	~ 10

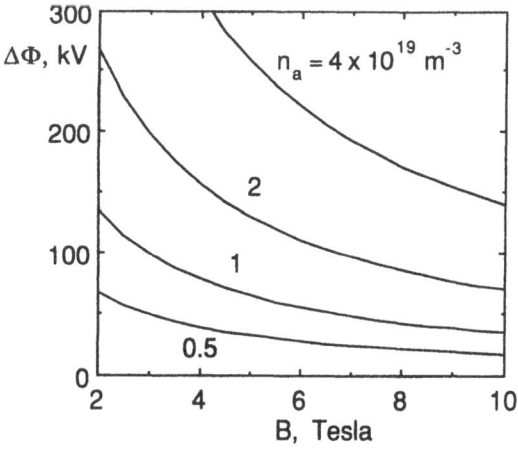

Figure 7. The voltage drop $\Delta\phi$ as a function of magnetic field and electron density in a 4-mm anode gap, assuming the model distribution of Figure 5.

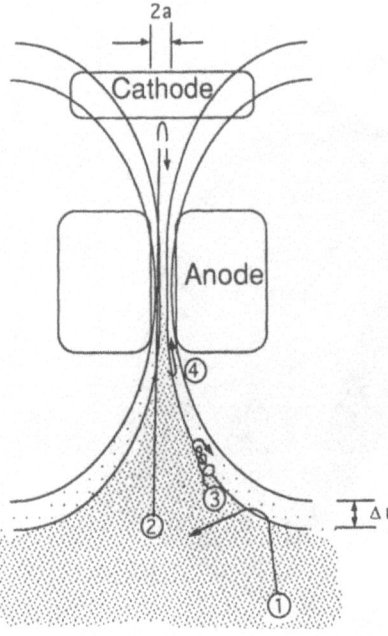

Figure 8. Types of electron orbits. Magnet coils are not shown. Most electrons that do penetrate through the anode gaps are reflected by the cathodes, orbit 2 in Figure 8. Their acceleration as they pass through the anode gaps lowers their density there, as indicated in Table 1. Since the anode radius is typically about 4/3 the nominal plasma radius, the radial electron motion produces a focusing effect of about 1.3 (Table 1). Other electrons become magnetically trapped in the boundary layer (spiral orbit 3, Figure 8), and some cold electrons produced by ionization are electrostatically trapped by the anode potential (orbit 4, Figure 8). These cold trapped electrons tend to decrease the ratio of n/n_a (Table 1). The combination of all these factors leads to an estimated ratio $n/n_a \sim 10$.

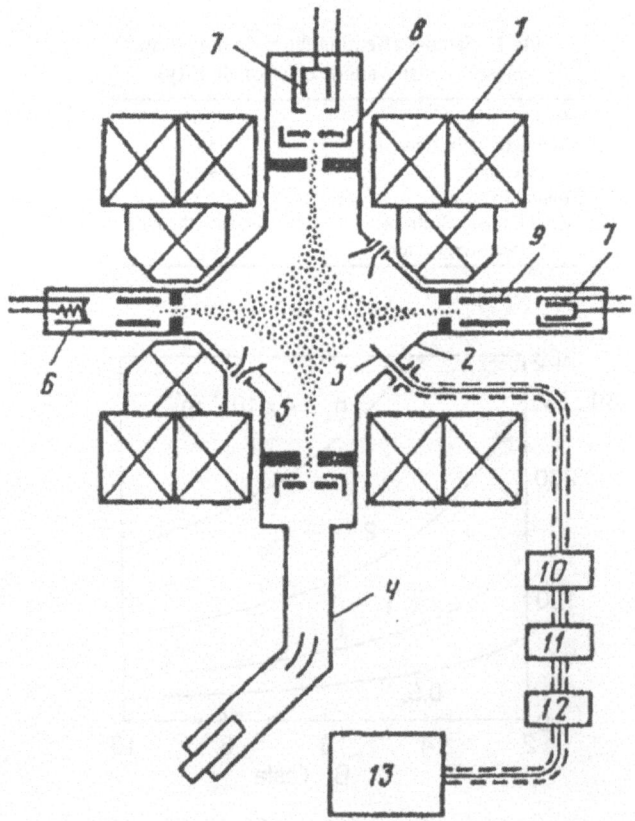

Figure 9. The Jupiter-1A spindle cusp experiment at the Kharkov Institute of Plasma Physics.

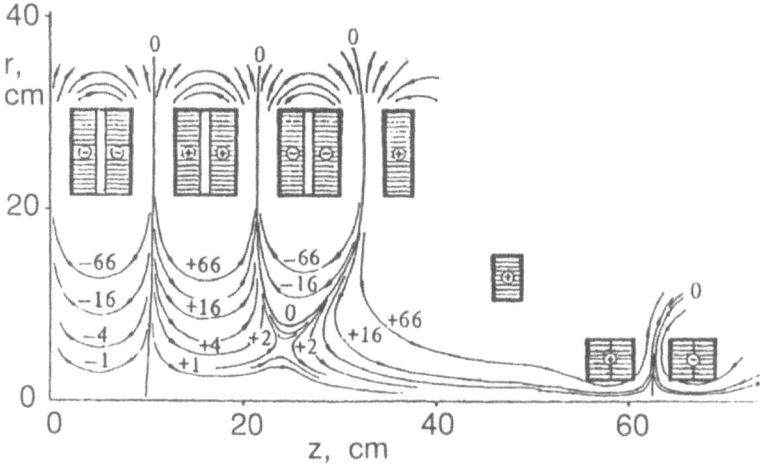

Figure 10. Vacuum magnetic flux surfaces (rA_ϕ, where A_ϕ is the azimuthal component of the magnetic vector potential, arbitrary units) in the Jupiter-2M experiment, Kharkov Insitute of Plasma Physics.

In the Jupiter-1A experiment the measured values were $n_a = 10^{17}$ m^{-3} and $n = 10^{18}$ m^{-3}, which is consistent with this expectation.

EXPERIMENTS

The Jupiter-1A spindle cusp experiment is shown in Figure 9.

In this experiment the plasma was produced by electron beam injection along the axis or by microwave heating. Gridded analyzers in the point cusp and ring cusp measured the electron and ion energy distributions. A microwave interferometer measured the plasma density.

The Jupiter-2M experiment has seven ring cusps with one small ring cusp and a point cusp at each end. The magnetic field configuration for half of this experiment is shown in Figure 10.

Plasma diamagnetism would push these surfaces outward. A cutaway view of the Jupiter-2M experiment is shown in Figure 11.

The magnet coils are placed inside the vacuum chamber. Magnetic forces tend to push the magnet coils apart, and the support plates prevent coil motion. The Jupiter-2M experiment has shown that plasma losses through the point cusps are not excessive (about 6% of the total losses). Conservation of angular momentum prevents most particles from escaping through the point cusps.

The key achievement of the Jupiter-2M experiment is its demonstration of nearly classical electron transport in the boundary magnetic field, as illustrated in Figure 12.

If such good confinement could be obtained at higher magnetic fields and voltages, then it might be feasible to build a fusion reactor based on magnetic electrostatic plasma confinement. The ATOLL toroidal quadrupole experiment had carefully aligned coils and electrodes, but suffered rapid plasma losses due to the lower hybrid drift and ion acoustic instabilities. Why the Jupiter-2M set of ring cusps attains much better confinement is not yet understood, but three ideas have been suggested:

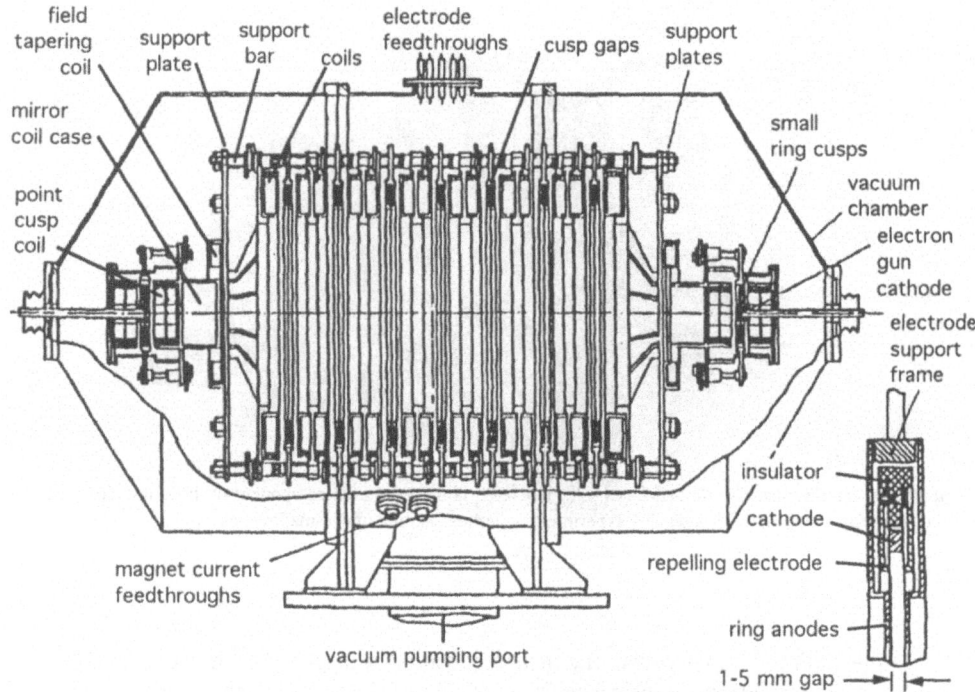

Figure 11. A cutaway view of the Jupiter-2M experiment.

- The Jupiter-2M anodes have only one-dimensional curvature.
- The Jupiter-2M experiment injects electrons only into the central plasma region, and not into the magnetic boundary layer.
- The Jupiter-2M experiment has a large volume of uniform, field-free plasma that may damp instabilities.

Parameters of some magnetic electrostatic plasma confinement experiments are listed in Table 2.

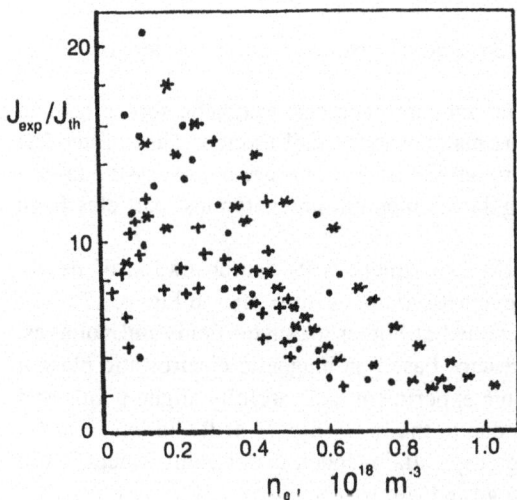

Figure 12. Ratio of experimentally measured loss current to theoretical loss current, based on classical diffusion. This ratio decreases with increasing plasma density, becoming about 2 at the highest density of operation.

Table 2. Parameters of some magnetic electrostatic plasma confinement experiments. Italicized values are estimates

	R cm	a mm	B_c Tesla	ϕ_A kV	n m^{-3}	T_e, keV	T_i, keV	τ s
Spindle Cusps								
S-1, Kharkov	5	1.5	0.4	2.0	3.0E+17	0.27	0.16	4.0E-05
S-3 & S-3M, Kharkov	10	1.5	0.6	5.0	4.0E+17	0.50	0.34	
S-4M, Kharkov	10	*1*	0.25	*1.0*	1.0E+17			
KEMP-II, U. of Quebec	17	1	0.5	6.0	4.0E+17	1.00	*0.30*	5.0E-04
Univ. of Missouri	16	1.5	0.3	5.0	2.0E+16	*0.50*	*0.10*	5.0E-04
Jutphaas	18		0.13			0.20		1.0E-04
Jupiter-1A, Kharkov	23	1.5	0.65	6.0	3.0E+17	1.50	0.40	3.0E-04
Jupiter-1M, Kharkov	10	1.5	1.7	8.0	1.0E+18	0.10	0.05	5.0E-04
Linear Multipoles								
m=4 Multipole, Kharkov	5	1.5	0.15	1.0				
m=8 Multipole, Kharkov	6	1	1.2	1.0	1.0E+17	0.50	0.30	5.0E-05
Toroidal Quadrupole								
ATOLL, Moscow	20	1	1.5	1.7	3.0E+17	0.03	0.07	1.8E-04
Multiple point cusps								
SCIF, San Diego	92		0.2	10				
Linear set of ring cusps								
Jupiter-2M, Kharkov	22	1.6	0.6	2.5	1.0E+18	0.015	0.03	3.0E-03

The linear multipole experiments confirmed theoretical predictions about electron penetration probabilities through the cusp gaps. The Spherically Convergent Ion Focus (SCIF) multiple point cusp experiment lost funding before completing its experimental mission, and it is now at the University of Wisconsin.

All of these experiments, except for the SCIF, have small radii R ~ 0.2 m, magnetic fields B ~ 1 T, and applied voltages ϕ_A ~ 1–10 kV. The plasma densities are typically 10^{16}–10^{18} m^{-3}, and the temperatures are a fraction of the applied voltage. Based upon theoretical predictions and on data from some of these experiments, the approximate scaling relations of Figure 13 have been derived.

To go from n = 10^{18} m^{-3} in Jupiter-2M up to a reactor value of n = 10^{20} m^{-3} would require an order of magnitude increase of B. In those experiments where the magnetic field flat-top is long enough to reach plasma power balance equilibrium, the temperatures should reach about five percent of the applied voltage. Slightly higher temperatures were measured in some experiments, and lower values were found in short-duration experiments.

$$n \propto B^2$$
$$T \approx 0.05\, \Phi_A$$
$$\tau_E \approx 0.02\, \tau_{ei}\, a\, R / \rho_a \rho_b \propto r_p\, \Phi_A / B^{1/2}$$
$$Q \approx 10^{-5}\, B\, \Phi_A^2\, r_p / 3(1 + n_t/n_a)^{1/2}$$

Figure 13. Approximate scaling relations for magnetic electrostatic plasma confinement devices. Here τ_{ei} is the electron-ion momentum-transfer collision time, a is the half-width of the anode gap, R is the anode radius, ρ_a and ρ_b are the electron Larmor radii in the anode gap and plasma boundary, r_p is the nominal plasma radius, ϕ_A is the applied voltage (kV), Q is the ratio of fusion power to input power, and n_t is the cold, trapped electron density in the anode region.

Table 3. Parameters of the proposed Jupiter-3 experiment and of a fusion power reactor, compared with the Jupiter-2M experiment

	Jupiter-2M	Jupiter-3	Reactor
Plasma radius r_p, m	0.07	1	3.5
Plasma length L, m	1.4	8	36
Ring cusp field B, T	0.5	4	6
Applied voltage ϕ_A, kV	2.5	100	400
Plasma density n, $10^{19}/m^3$	0.1	5	10
Ion temperature T, keV	0.03	5	20
Electron lifetime τ, s	0.002	0.3	6

FUSION REACTOR ISSUES

Based on these scalings, the parameters of a proposed Jupiter-3 experiment and a fusion power reactor are shown in Table 3.

A reactor is illustrated in Figure 14. This reactor is comparable in size to the ITER Tokamak, Figure 15. The potential applications of such a reactor are listed in Table 4.

The anode gap voltage drop is fairly well understood (Figure 7). The main issues to be resolved are plasma purity, electron transport, alpha particle confinement, and electrode alignment and voltage holding.

- Plasma purity. Since multiply-charged ions would be well confined in the electrostatic potential well, the gradual accumulation of impurity ions in the central plasma is a serious problem. A high Z_{eff} would reduce the reactor Q below the value of Figure 13. Means to control impurity buildup need to be developed.

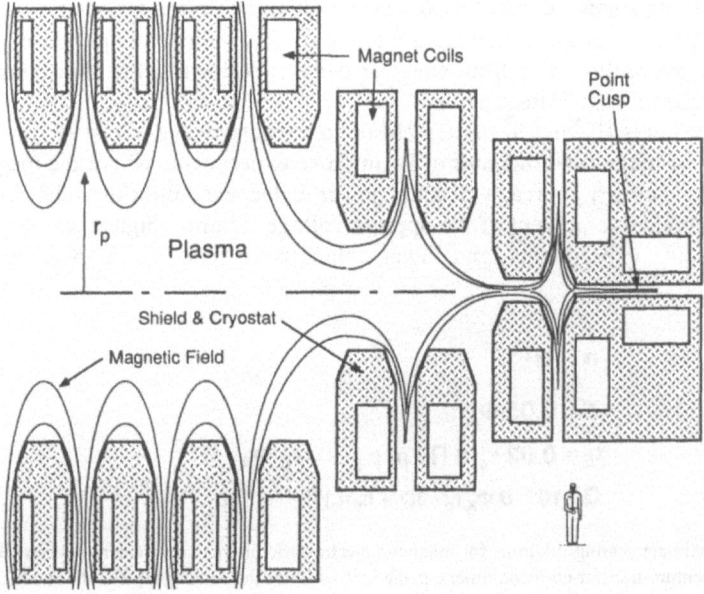

Figure 14. One end of a fusion reactor based on magnetic electrostatic plasma confinement. The electrodes are not shown.

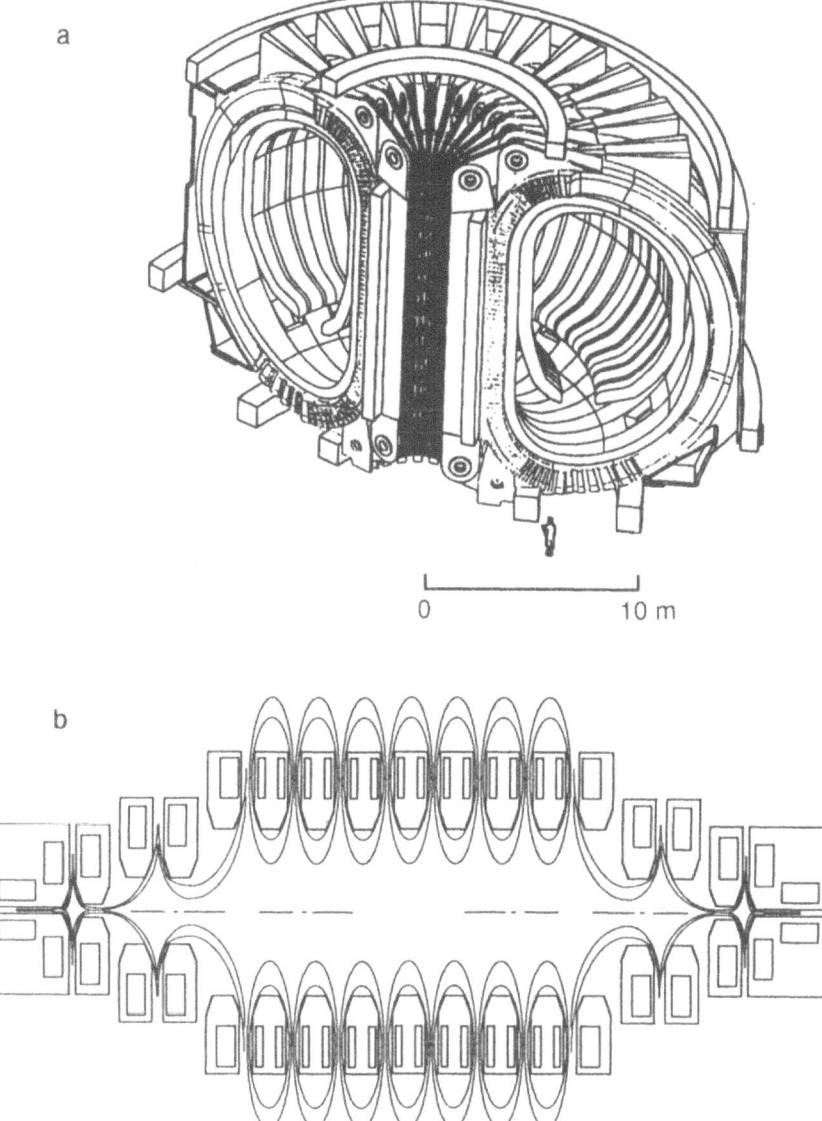

Figure 15. Comparison of the magnetic electrostatic fusion reactor (b) with the ITER Tokamak (a).

- Electron transport. Electron transport rates about twice the classical values were measured in the Jupiter-2M experiment. Whether this good confinement will continue at higher plasma parameters remains to be determined experimentally.
- Alpha particle confinement. Since the alpha particle energy greatly exceeds ϕ_i, the alpha particles are not electrostatically confined. Partial confinement of the alpha particle energy by the cusp magnetic field would enhance the reactor Q value above the value of Figure 13. Studies of alpha particle orbits and slowing down are needed to assess this effect. When an alpha particle hits the wall about 5–10% of its kinetic energy is converted directly into electricity.

Table 4. Potential applications of magnetic
electrostatic plasma confinement

Electrical power generation

Radiolytic hydrogen production

Neutron source
   materials testing
   medical therapy
   isotope production

Space propulsion

Heavy ion source

- Electrodes. Precise electrode alignment was demonstrated in the Jupiter-2M experiment, but it needs to be demonstrated in a larger device. The anodes should be made of a material with low thermal expansion coefficient and actively cooled. Prevention of high-voltage breakdown at an applied voltage $\phi_A \sim 400$ kV will require careful design of the electrode system. A neutron shield for the magnet coils, a radiation collimator, and a voltage divider will be needed (Figure 16).

The PSP-2 rotating plasma experiment at the Budker Institute of Nuclear Physics, Novosibirsk, is not a "magnetic electrostatic plasma confinement" device, but it has demonstrated operation at 400 kV in the presence of hot plasma (Abdrashitov, 1994).

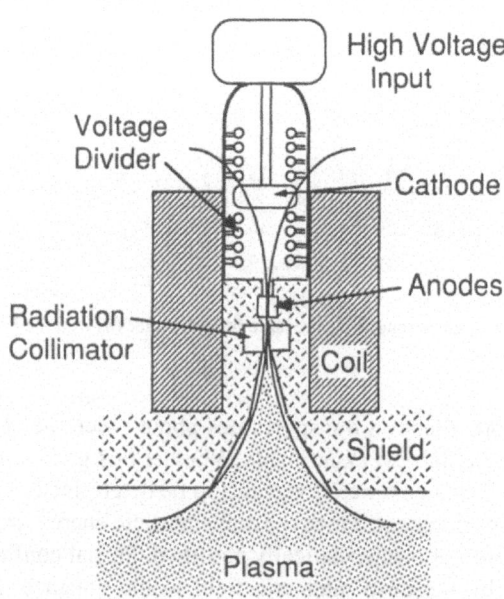

Figure 16. Schematic diagram of the electrode region. (The voltage divider would be more complex than shown in this simplified diagram.)

CONCLUSIONS

Magnetic electrostatic plasma confinement is a combination of an open magnetic confinement and electrostatic plasma confinement that inhibits plasma losses along magnetic field lines and protects the grid from bombardment. The following conclusions may be drawn:

1. A linear set of ring cusps is most promising for a magnetic electrostatic plasma confinement fusion reactor.
2. The central plasma density can be a factor of 10 higher than the anode electron density.
3. If the scaling relations of Figure 13 and Table 3 hold at higher values of magnetic field and applied voltage, then a fusion reactor based on magnetic electrostatic plasma confinement would be possible.
4. The nearly classical electron transport observed in Jupiter-2M is promising, and voltage holding at 400 kV has been demonstrated in a rotating plasma experiment.
5. The most troublesome issue is plasma purity.
6. A steady-state experiment with higher magnetic field and applied voltage is needed to help resolve the issues of plasma density and confinement time scaling, plasma purity, electron transport, and high voltage holding.

ACKNOWLEDGMENTS

This research was supported in part by the U.S. Department of Energy, Office of Fusion Energy, under Contract Number DE-AC07-94ID13223. This field of plasma research was invented by O. A. Lavrent'ev, who derived the theory and conducted successful experiments in spite of meager funding.

REFERENCES

Abdrashitov, G. F., et al., 1994, Review of results and discussions of the hot rotating plasma experiments (experiment PSP-2), *International Conference on Open Plasma Confinement Systems for Fusion*, Novosibirsk, Russia, June 14–18 1993, A. A. Kabantsev, Editor, World Scientific Publishing Co., Singapore.

Dolan, T. J., 1994, Magnetic electrostatic plasma confinement, *Plasma Physics and Controlled Fusion* 36:1539–1594.

Komarov, A. D., and Stepanenko, I. A., 1980, Ukrainian Journal of Physics 25, 1432–1435.

15

ANALYSIS OF THE FUSION BREAKEVEN CONDITIONS FOR D-T PLASMAS OF PRESCRIBED TEMPERATURE EVOLUTION

E. Panarella[*]

Advanced Laser and Fusion Technology, Inc.
189 Deveault Street, No. 7
Hull, P.Q. J8Z 1S7, Canada

ABSTRACT

In the past several years, the Lawson criteria (Proc. Phys. Soc. *B70*, 6, 1957) have provided guidelines to the fusion community on the conditions required for energy breakeven from a hot plasma. These criteria state that, in a pulsed system, breakeven can be achieved if the plasma temperature exceeds a certain critical number (30 million °K for deuterium-tritium reactions), and the reaction is sustained for a sufficient time τ_E such that $n\tau_E > 10^{14}$ cm^{-3} sec, where n is the particle density. In the present study, the complete pulsed system is analyzed from which breakeven is required, i.e., source from which energy flows into a plasma sink, and plasma sink from which energy returns to the source. The case is considered where the α particles escape from the plasma, and the alternative case where they are retained. The analysis departs from the conventional one because it introduces conduction losses from heat transfer theory rather than through the $3nkT/\tau_E$ term, and therefore the containment time τ_E does not play an explicit role in the breakeven conditions. These are now determined by particle density, temperature temporal profile, and dimension of the plasma. The analysis confirms the requirement of a threshold temperature for breakeven. However, it adds another threshold element, a particle density in excess of 10^{16} cm^{-3}, below which breakeven cannot be achieved under realistic reactor conditions. Although a spherical geometry is considered here, such as the spherical pinch plasma configuration (J. Fusion Energy *13*, 45, 1994), the analysis can be extended to the toroidal or other plasma geometries, when the appropriate heat conduction loss term is introduced.

---

[*] Also with the Department of Electrical and Computer Engineering, University of Tennessee, Knoxville, TN, U.S.A., and the National Research Council, Ottawa, Canada.

In light of these results, the analysis of Lawson and successive interpretations have been re-examined. It is found that they have limited validity for a pulsed system, but are better suited for steady-state reactor operation.

1. INTRODUCTION

For 39 years, since J. D. Lawson in 1957 published a paper in the Proceedings of the Physical Society by title "Some Criteria for a Power Producing Thermonuclear Reactor,"[1] the plasma physics community has been guided in its quest for fusion by the directions provided by the 'Lawson criteria' for energy breakeven conditions. These are general criteria, applicable to an idealized situation where a plasma of particle density n is brought instantaneously to a temperature T, which is maintained for a time τ_E after which it is allowed to cool to the original temperature. The criteria state that for breakeven, the temperature has to be above 30 million °K and the product $n\tau_E$ must exceed 10^{14} cm^{-3} sec in deuterium-tritium reactions. The reaction products are not retained in the plasma of the Lawson analysis, and conduction losses are neglected during the plasma burning time. These criteria apply to the case of 'scientific breakeven', in which the energy supplied to heat the plasma and maintain the bremsstrahlung losses is returned to the system, together with the reaction energy, with a recovery efficiency η of 1/3. No allowance is made for the fact that the energy used to heat the plasma and supply the bremsstrahlung losses must come from a conventional source, and that the transfer efficiency is normally much less than 100 percent. Were this considered, the Lawson criteria would become even more difficult to satisfy. Mills[2] has generalized the 'scientific breakeven' conditions by considering several energy recovery efficiencies η, from 2.5 to 60 percent. Some discussion regarding terms in the energy balance equation of Lawson has been carried out in Refs. 3 and 4. Several generalizations of the Lawson criteria are also reported in the standard textbook literature on fusion energy[5–11].

In the present work, the complete system is examined in detail, i.e., source of energy and plasma sink, and various energy transfer efficiencies are considered from the first to the second, as well as from the second to the first. The energy from the source is deposited in a pulse, and the plasma experiences a cycle of temperature values. The analysis departs from the conventional one, in that heat conduction losses are introduced from heat transfer theory, rather than through the $3nkT/\tau_E$ term. In order to ease the inclusion in the energy balance equation of conduction losses, the geometry considered here is the spherical one. More specifically, it is the spherical pinch geometry[12–14]. However, the analysis can be extended to other geometries, including the toroidal, when the appropriate heat conduction loss term is used. We find that the containment time does not play now an explicit role in the breakeven conditions. These are determined by particle number density, temporal profile of the plasma temperature, and dimension of the plasma. At or below ~10^{16} cm^{-3} particle density there can be no breakeven, no matter the plasma temperature, in realistic reactor conditions. Likewise, there can be no breakeven at a temperature below ~3 kV, no matter the particle density. The shape of the plasma temperature excursion has a bearing on the energy breakeven conditions.

Spurred by these results, a thorough re-examination of the Lawson analysis and subsequent interpretations has been undertaken. It is found that the analysis of Lawson suffers difficulties of interpretation when applied to a pulsed system. By contrast, for a steady-state reactor the interpretation is clear and the $n\tau_E$ product acquires a new meaning. In particular, it is found that the term $3nkT/\tau_E$ (3nkT = plasma energy, τ_E = energy confinement time), usually considered as a non-radiative loss term, may not include conduction losses, and should rather be considered as a power input term.

2. ANALYSIS

Fig. 1 illustrates the system under consideration. Energy E from a source is released in a pulse of power W(t). A fraction E_i of the energy (E_i = aE where a is the efficiency of energy transfer from the source to the sink) is deposited in a spherical chamber containing a plasma core and an ionized gas blanket, while the rest (1 - a)E is dissipated in the components of the energy transfer system. The energy

$$\int_0^t W(t)dt$$

that reaches the chamber at time t is used to heat the plasma core, and to provide for the radiation and conduction losses. In the absence of other energy inputs, the energy balance equation is:

$$a \int_0^t W(t)dt = \int_0^t [H(t) + R(t) + C(t)]dt \qquad (1)$$

where H(t) is the rate of heat energy change in the plasma core, and R(t) and C(t) are the rate of energy loss by radiation and heat conduction, respectively. Other losses, such as synchrotron or inverse compton radiation production, are not included here.

A 50 percent mixture of deuterium and tritium is considered here as making up the plasma and surrounding gas. The fusion nuclear reaction taking place in the plasma core is:

$$D + T = He^4 \,(3.5 \text{ MeV}) + n \,(14.1 \text{ MeV}) \qquad (2)$$

where 3.5 MeV is the energy carried by the α particle He^4, and 14.1 MeV is the energy carried by the neutron n.

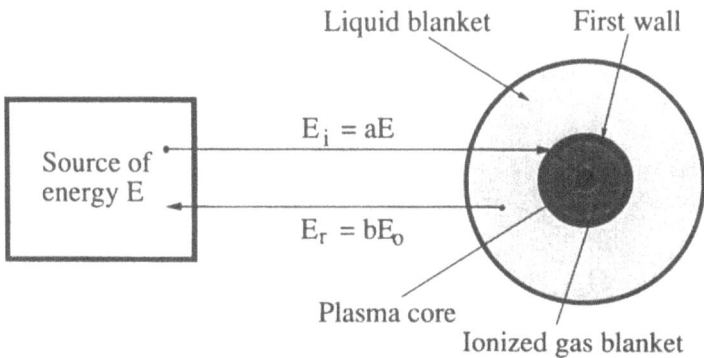

Figure 1. Energy flow system.

At the end of the life t_o of the plasma core, when this has cooled down and returned to the original temperature, an amount of energy E_o is available from the liquid blanket for return to the source. A fraction E_r of this energy ($E_r = bE_o$, where b is the efficiency of energy transfer from the liquid blanket to the source) is the amount that actually reaches the source. Breakeven requires that $E_r = E$. The aim of this study is to find the conditions for breakeven, given the temperature excursion T(t) of a plasma of known particle density n and radius r, and the coefficients a and b of transfer efficiency from the energy source to the sink, and viceversa, respectively, when conduction losses are introduced from heat transfer theory rather than through the $3nkT/\tau_E$ term.

The rate of energy released in the form of α particles and neutrons will be designated as $P_\alpha(t)$ and $P_n(t)$, respectively. Only the α particles may remain in the plasma and further heat it, whereas the neutrons will escape and be absorbed by the liquid blanket. Two cases will be considered:

A. the α particles escape, and
B. the α particles are retained.

A. α Particles Escape

Since both α particles and neutrons escape and are absorbed by the liquid blanket, the energy available at time t for transfer to the source is:

$$E_o(t) = \int_0^t [P_\alpha(t) + P_n(t) + R(t) + C(t)] dt \qquad (3)$$

Breakeven requires that, at the end of the life t_o of the plasma core:

$$\int_0^{t_o} W(t) dt = b \int_0^{t_o} [P_\alpha(t) + P_n(t) + R(t) + C(t)] dt \qquad (4)$$

From eqs. (1) and (4) one gets:

$$\frac{1}{a} \int_0^{t_o} [H(t) + R(t) + C(t)] dt = b \int_0^{t_o} [P_\alpha(t) + P_n(t) + R(t) + C(t)] dt \qquad (5)$$

which can be written as:

$$\int_0^{t_o} H(t) dt = ab \int_0^{t_o} \left\{ P_\alpha(t) + P_n(t) + \left(1 - \frac{1}{ab}\right)[R(t) + C(t)] \right\} dt \qquad (6)$$

At the end of t_o, the plasma core returns to the original temperature. No heat energy remains and therefore:

$$\int_0^{t_0} H(t)dt = 0 \qquad (7)$$

Hence:

$$ab \int_0^{t_0} \left\{ P_\alpha(t) + P_n(t) + \left(1 - \frac{1}{ab}\right)[R(t) + C(t)] \right\} dt = 0 \qquad (8)$$

or

$$\int_0^{t_0} \left\{ P_\alpha(t) + P_n(t) + \left(1 - \frac{1}{ab}\right)[R(t) + C(t)] \right\} dt = 0 \qquad (9)$$

If eq. (9) is satisfied, breakeven is achieved.

B. α Particles Are Retained

Only the neutrons escape now, and the energy available at time t from the liquid blanket for return to the source is:

$$E_0(t) = \int_0^t [P_n(t) + R(t) + C(t)] \, dt \qquad (10)$$

Breakeven now requires that, at the end of the time t_0:

$$\int_0^{t_0} W(t)dt = b \int_0^{t_0} [P_n(t) + R(t) + C(t)] \, dt \qquad (11)$$

Since the α particles remain in the plasma core, one has a new energy source, and the energy balance equation becomes:

$$a \int_0^t W(t)dt + \int_0^t P_\alpha(t)dt = \int_0^t [H(t) + R(t) + C(t)] \, dt \qquad (1A)$$

which can be written as:

$$a \int_0^t W(t)dt = \int_0^t [H(t) + R(t) + C(t) - P_\alpha(t)] \, dt \qquad (1B)$$

From Eqs. (11) and (1B) one gets:

$$\frac{1}{a}\int_0^{t_o}[H(t)+R(t)+C(t)-P_\alpha(t)]dt = b\int_0^{t_o}[P_n(t)+R(t)+C(t)]\,dt \qquad (12)$$

which transforms into:

$$\int_0^{t_o} H(t)dt = ab\int_0^{t_o}\left\{P_n(t)+\frac{1}{ab}P_\alpha(t)+\left(1-\frac{1}{ab}\right)[R(t)+C(t)]\right\}dt \qquad (13)$$

Since $\int_0^{t_o} H(t)dt=0$, breakeven requires the following equation to be satisfied:

$$ab\int_0^{t_o}\left\{P_n(t)+\frac{1}{ab}P_\alpha(t)+\left(1-\frac{1}{ab}\right)[R(t)+C(t)]\right\}dt = 0 \qquad (14)$$

or

$$\int_0^{t_o}\left\{P_n(t)+\frac{1}{ab}P_\alpha(t)+\left(1-\frac{1}{ab}\right)[R(t)+C(t)]\right\}dt = 0 \qquad (15)$$

The first three terms $P_n(t)$, $P_\alpha(t)$ and $R(t)$ in the integrals (9) and (15) are functions of plasma temperature, particle density and plasma volume. Specifically, for a 50% D-T mixture, one has[15]:

$$P_n(n,T,V_p) = 8.31 \times 10^{-24}\,\frac{n^2}{4}\,T^{-\frac{2}{3}}\,e^{-\frac{19.94}{T^{1/3}}}\,V_p \qquad (16)$$

$$P_\alpha(n,T,V_p) = 2.06 \times 10^{-24}\,\frac{n^2}{4}\,T^{-\frac{2}{3}}\,e^{-\frac{19.94}{T^{1/3}}}\,V_p \qquad (17)$$

$$R(n,T,V_p) = 2.14 \times 10^{-30}\,\frac{n^2}{4}\,T^{\frac{1}{2}}\,V_p \qquad (18)$$

where P_n and P_α and R are expressed in watts, T in kV, V_p in cm^3, and n in cm^{-3}.

The rate of heat conduction loss $C(t)$ from the plasma core to the liquid blanket can be derived from heat transfer theory[16] as follows. The rate of heat transfer in a hollow sphere with surface temperature T_1 at r_1, and T_2 at r_2, where r_1, r_2 are the radii of the inner and outer surfaces, respectively, is given by:

$$C_1(r_1,r_2,T_1,T_2) = \frac{4\pi K\, r_1 r_2}{r_2-r_1}(T_1-T_2) \qquad (19)$$

where K is the thermal conductivity of the material that makes up the hollow sphere.

In the present case, heat transfer takes place from the plasma core of radius r_1 and temperature T_1 to the liquid blanket of radius r_2 and temperature T_2. The medium through which heat propagates before reaching the liquid blanket is the ionized gas blanket made up of deuterium and tritium, and the first wall, usually made up of stainless steel[17]. The latter easily transfers heat, so that only the thermal conductivity of the former needs to be considered. Eq. (19) expresses the rate of heat transfer from the entire volume of the plasma core $V_p = 4/3(\pi r_1^3)$. The rate of heat transfer per unit volume is obtained by dividing (19) by V_p. Since the temperature T_1 of the plasma core is much larger than the temperature T_2 of the liquid blanket, one has $T_1 - T_2 \approx T_1$. Likewise, since $r_2 \gg r_1$, one has $r_2 - r_1 \approx r_2$. Eq. (19) thus leads to:

$$c(r_1, T_1) = \frac{4\pi K r_1 r_2 T_1}{(r_2 - r_1)} / \frac{4}{3}\pi r_1^3 \approx \frac{3 K T_1}{r_1^2} \tag{20}$$

and the heat conduction loss rate appearing in (9) and (15) is given explicitly by:

$$C(r_1, T_1, V_p) = \frac{3 K T_1}{r_1^2} V_p \tag{21}$$

where $C(r_1, T_1, V_p)$ is expressed in watts, when T_1 is expressed in kV, r_1 in cm, and K in W/cm kV.

Eq. (21) shows that the rate of heat conduction loss from the plasma core to the liquid blanket is a function of the thermal conductivity K of the ionized gas blanket. Three typical blankets will be considered: a) plasma blanket without magnetic field; b) plasma blanket with magnetic field, and c) non-ionized gas blanket. A comparative analysis will then be made to assess which blanket provides the lowest thermal conductivity, and this will be retained for further analysis.

a. Plasma Blanket without Magnetic Field. In this case, the thermal conductivity is[18]:

$$K = 4.67 \times 10^{-12} \frac{T^{5/2}}{Z \ln\Lambda} \frac{\text{cal}}{\text{sec } ^\circ\text{K cm}} \tag{22}$$

where T is the temperature in °K, and Z is the atomic number. The thermal conductivity therefore increases rapidly with temperature T, and the flow of heat is controlled by the plasma layer with the lowest thermal conductivity. This is the one in contact with the liquid blanket. For a reactor working in a pulsed operation, the temperature of this layer can probably be as high as 10 eV = 1.16×10^5 °K without eroding the wall seriously if the pulse is short. For steady state, the temperature must be much less than 10 eV. For a D-T plasma for which Z = 1 and $\ln\Lambda = 12$ for a particle density of 10^{13} cm^{-3} (although $\ln\Lambda$ does not vary greatly with particle density[19]), one finds:

$$K = 1.78 \frac{\text{cal}}{\text{sec } ^\circ\text{K cm}} = 8.73 \times 10^7 \frac{\text{W}}{\text{cm kV}} \tag{23}$$

This is an enormous thermal conductivity, much higher than the other two cases of the following sections.

b. Plasma Blanket with Magnetic Field. This is the case when the plasma core is magnetically confined. The thermal conductivity is given by[20]:

$$K = 3.54 \times 10^{-25} \frac{A_i^{1/2} Z^2 \; n_i^2 \ln\Lambda}{T^{1/2} B^2} \quad \frac{\text{cal}}{\text{sec °K cm}} \tag{24}$$

where A_i is the atomic weight of the positive ions of density n_i (in cm^{-3}), and B is the magnetic field (in gauss). The thermal conductivity in this case decreases with temperature T and magnetic field B, and increases with particle density n_i. Heat losses are now controlled mainly by the plasma layer with the lowest conductivity, which, being in contact with the hot plasma core, can be identified as having the same parameters as the core itself. Here the temperature can be quite high, say $T = 5$ kV $= 5.80 \times 10^7$ °K. In order to have an estimate of the thermal conductivity in this case, one can consider some favourable but still realistic parameters for a low conductivity, say $n_i = 10^{13}$ cm^{-3} (as in the previous case), $B = 10^5$ gauss, $\ln\Lambda = 19$, $A_i = 2.5$ (averaged for a 50% D-T plasma). One has:

$$K = 1.40 \times 10^{-11} \frac{\text{cal}}{\text{sec °K cm}} = 6.75 \times 10^{-4} \frac{W}{\text{cm kV}} \tag{25}$$

This conductivity is much less than the previous one, and will be retained for further analysis.

c. Non-Ionized Gas Blanket. In this case, kinetic theory gives the thermal conductivity of the gas blanket as[21]:

$$K = \frac{1}{3} m\lambda R \bar{v} c_v \tag{26}$$

where m is the mass of a molecule, λ is the mean free path, R is the number of molecules per unit volume, $\bar{v}$ is their mean velocity, and c_v is the specific heat at constant volume. Since λ is inversely proportional to R, the thermal conductivity is independent of pressure[21]. Although the gas blanket is made up of deuterium and tritium, for simplicity it is considered to be made of hydrogen. At atmospheric pressure, and at a temperature which can be realistically assumed 500 °K near the first wall, the thermal conductivity[22] of hydrogen is:

$$K = 0.25 \frac{W}{\text{m °C}} = 2.90 \times 10^4 \frac{W}{\text{cm kV}} \tag{27}$$

This thermal conductivity is much higher than (25), and therefore should not be retained. However, it will be shown later that, for breakeven purposes, it is the only one that can be retained.

If one now inserts (16), (17), (18), and (21) in (9) and (15), explicit expressions for breakeven are obtained in the two cases where the α particles escape and are retained, respectively:

Fusion Breakeven Conditions for D-T Plasmas

$$V_p \int_0^{t_o} \left\{ 1.04 \times 10^{-23} \frac{n^2}{4} T(t)^{-2/3} e^{-\frac{19.94}{T(t)^{1/3}}} \right.$$

$$\left. + \left(1 - \frac{1}{ab}\right) \left[2.14 \times 10^{-30} \frac{n^2}{4} T(t)^{1/2} + \frac{3KT(t)}{r^2} \right] \right\} dt = 0 \qquad (9A)$$

$$V_p \int_0^{t_o} \left\{ \left(8.31 \times 10^{-24} + \frac{2.06 \times 10^{-24}}{ab} \right) \left[\frac{n^2}{4} T(t)^{-2/3} e^{-\frac{19.94}{T(t)^{1/3}}} \right] \right.$$

$$\left. + \left(1 - \frac{1}{ab}\right) \left[2.14 \times 10^{-30} \frac{n^2}{4} T(t)^{1/2} + \frac{3KT(t)}{r^2} \right] \right\} dt = 0 \qquad (15A)$$

where K is the thermal conductivity given by (25) or (27), and T, r are the plasma core temperature and radius, respectively.

In (9A) and (15A) it has been assumed that, during the short time t_o, only the temperature T of the plasma core changes, whereas n and r remain constant. In other words, it is assumed that the plasma is confined by some means during time t_o, and therefore the plasma radius r and particle number density n do not change. The formulation of the conditions for breakeven, however, as expressed by (9A) and (15A), has general validity and remains the same if n and r also are a function of time. It can have even more validity if n is also a function of space within the plasma core volume. In this case, clearly the integration has to be performed over space and over time.

One can simplify (9A) and (15A) by dividing by V_p to get:

$$\int_0^{t_o} \left\{ 1.04 \times 10^{-23} \frac{n^2}{4} T(t)^{-2/3} e^{-\frac{19.94}{T(t)^{1/3}}} \right.$$

$$\left. + \left(1 - \frac{1}{ab}\right) \left[2.14 \times 10^{-30} \frac{n^2}{4} T(t)^{1/2} + \frac{3KT(t)}{r^2} \right] \right\} dt = 0 \qquad (9A)'$$

$$\int_0^{t_o} \left\{ \left(8.31 \times 10^{-24} + \frac{2.06 \times 10^{-24}}{ab} \right) \left[\frac{n^2}{4} T(t)^{-2/3} e^{-\frac{19.94}{T(t)^{1/3}}} \right] \right.$$

$$\left. + \left(1 - \frac{1}{ab}\right) \left[2.14 \times 10^{-30} \frac{n^2}{4} T(t)^{1/2} + \frac{3KT(t)}{r^2} \right] \right\} dt = 0 \qquad (15A)'$$

3. RESULTS

The three functions that appear in the integrals (9A)' and (15A)' are positive functions, the latter two being subtracted from the first. The integral can therefore have positive or negative values. If the integrand, which is the sum of these three functions, is negative, the integral will also be negative. Likewise, if it is positive, the integral will be positive. One can examine the trend of the integrand by calculating and plotting it as a function of one of its variables T, or n, or ab, or r, assuming prescribed values for the others. In this way, one is able to find for which value of the variable the function (integrand) from negative becomes positive, i.e., assumes a value equal to zero. For this value, or for any positive value of the integrand, breakeven can be achieved. The case where the α particles escape will be considered first [Eq. (9A)'], and then the case where the α particles are retained [Eq. (15A)']. In Sec. 4, the case where T is changing with time will be examined, and various plasma temperature excursions will be considered. The geometry to which these results apply is the spherical geometry. For other geometries, such as the toroidal, the proper expressions for the conduction loss must replace the one used here.

Before proceeding, one needs to make a choice in regard to the thermal conductivity K that one wants to use. From the previous analysis, it would appear that the conductivity given by (24) is the right choice. It will be shown that this conductivity cannot lead to breakeven, under realistic values of the magnetic field B. The reason is the following. Since heat losses are controlled mainly by the plasma layer of the ionized gas blanket with the lowest thermal conductivity, this is the one in contact with the hot plasma core, where the highest temperature and magnetic field can be found. That layer can be identified as having also the same density as the core itself. If one inserts the conductivity given by (24) in (9A)' or (15A)', one finds that n^2 appears now as a common factor in all terms of the integrands, which therefore cancels out, thus making the integrands independent of particle density, and breakeven is not a function of n. Since also the temperature that appears in the denominator of (24) is the same as the one in the conduction loss term of (9A)' or (15A)', then the conduction losses will increase as the square root of the plasma core temperature, thus becoming large at high temperatures. If one now calculates the integrands of (9A)' and (15A)' for T from 1 to 50 kV, r from 1 to 10 cm, a and b from 0.01 to 0.40, and B from 10^3 to 10^5 gauss, one finds that the integrands have always negative values, and breakeven cannot be achieved. Therefore, a magnetized plasma blanket is not suitable for breakeven purposes, and a non-ionized gas blanket, whose thermal conductivity is given by (27), is the only one that can be retained.

Inserting the thermal conductivity K given by (27) in (9A)' and (15A)', the following final expressions are obtained for breakeven in the two cases of α particles escaping or retained, respectively:

$$\int_0^{t_o} \left\{ 1.04 \times 10^{-23} \frac{n^2}{4} T(t)^{-2/3} e^{-\frac{19.94}{T(t)^{1/3}}} \right.$$

$$\left. + \left(1 - \frac{1}{ab}\right)\left[2.14 \times 10^{-30} \frac{n^2}{4} T(t)^{1/2} + 8.70 \times 10^4 \frac{T(t)}{r^2} \right] \right\} dt = 0$$

$$\int_0^{t_o} \left\{ \left(8.31 \times 10^{-24} + \frac{2.06 \times 10^{-24}}{ab} \right) \left[\frac{n^2}{4} T(t)^{-2/3} e^{-\frac{19.94}{T(t)^{1/3}}} \right] \right. \qquad (9A)''$$

$$+ \left(1 - \frac{1}{ab}\right)\left[2.14 \times 10^{-30} \frac{n^2}{4} T(t)^{1/2} + 8.70 \times 10^4 \frac{T(t)}{r^2}\right]\Bigg\} dt = 0 \qquad (15A)''$$

The integrands of (9A)″ and (15A)″ will now be plotted for the cases of α particles escaping and retained, respectively.

A. α Particles Escape [Eq. (9A)″]

A few typical cases will be examined, depending on the variable chosen.

Temperature T as Variable. We begin by choosing T as variable from 0 to 10 kV. In Fig. 2 we have prescribed n = 10^{14} cm^{-3} and r = 0.5 cm. Three curves have been plotted, for a = b = 0.6, a = b = 0.4, and a = b = 0.2, respectively. We find that in all cases the integrand is negative, which means that breakeven cannot be achieved at such low particle density. If one extends the temperature range from 10 up to 500 kV, the integrands are always monotonically decreasing.

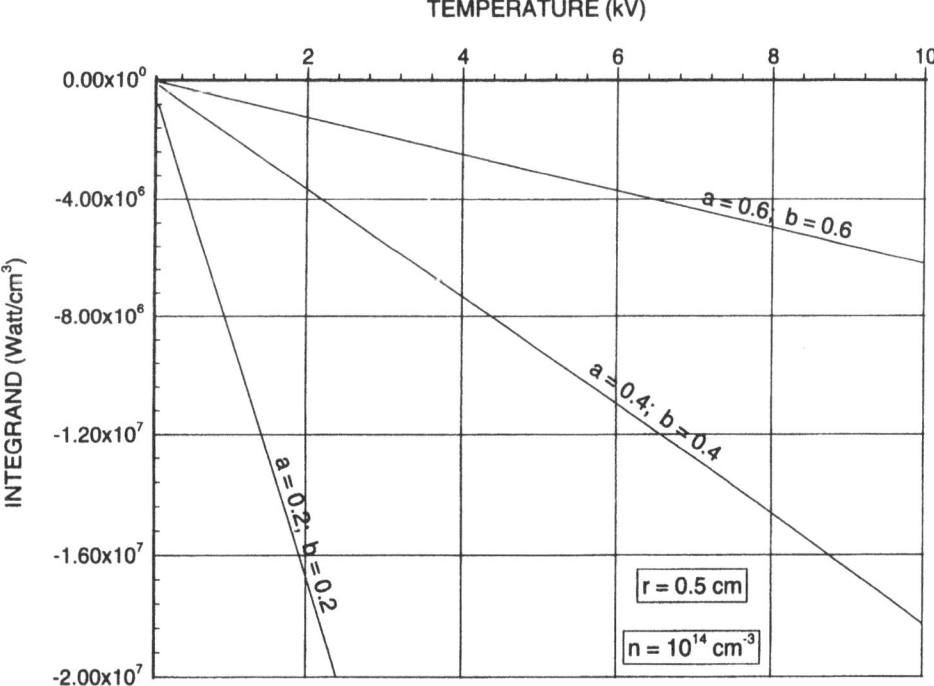

Figure 2. Plot of the integrand in eq. (9A)″ (α particle escape) as a function of temperature T for three values of the energy transfer efficiency (a = b = 0.6; a = b = 0.4; and a = b = 0.2), with particle density n = 10^{14} cm^{-3} and plasma radius r = 0.5 cm. Breakeven cannot be achieved with a particle density of 10^{14} cm^{-3}.

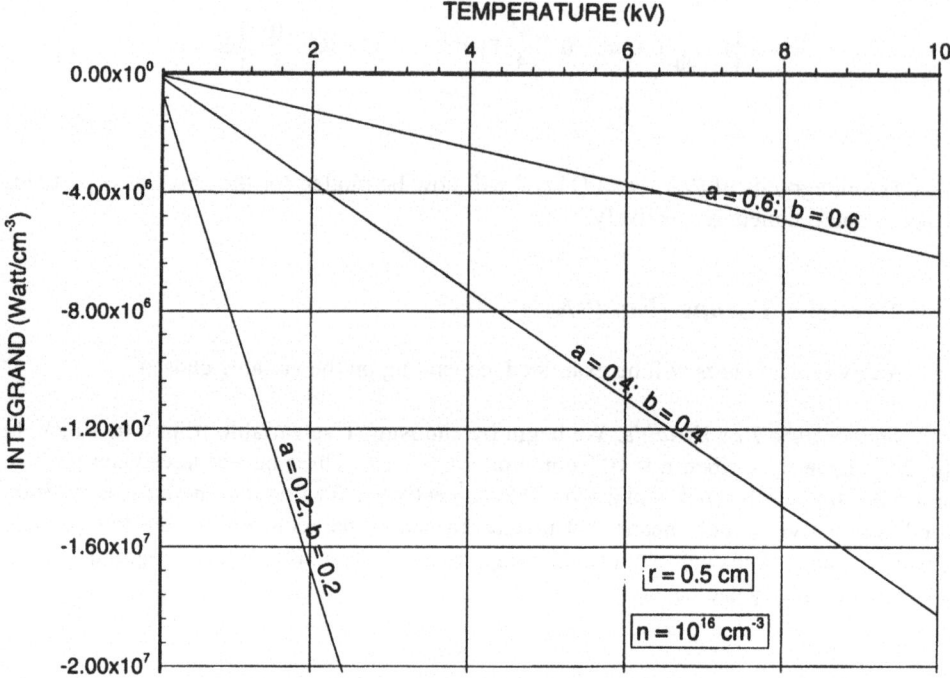

Figure 3. Plot of the integrand in eq. (9A)″ (α particle escape) as a function of temperature T for three values of the energy transfer efficiency (a = b = 0.6; a = b = 0.4; and a = b = 0.2), with particle density n = 10^{16} cm^{-3} and plasma radius r = 0.5 cm. Breakeven cannot be achieved with a particle density of 10^{16} cm^{-3}.

The situation does not improve if one increases the particle density to 10^{16} cm^{-3}, as Fig. 3 shows. Only when one increases the particle density to n = 10^{18} cm^{-3}, then the integrand can become positive (Fig. 4). The temperature for which the integrand begins to be positive is 4.3 kV for a = b = 0.6, and 6.8 kV for a = b = 0.4. For for a = b = 0.2, the curve does not become positive in the temperature range examined.

The integrand becomes positive at lower temperature for n = 10^{20} cm^{-3} (Fig. 5). The temperature is 3.3 kV for a = b = 0.6, 4.7 kV for a = b = 0.4, and 8.7 kV for a = b = 0.2. For n = 10^{22} cm^{-3}, the temperatures do not change: T = 3.3 kV for a = b = 0.6; T = 4.7 kV for a = b = 0.4; and T = 8.7 kV for a = b = 0.2.

Particle Density n as Variable. We can find more precisely the minimum particle density required for breakeven by plotting the integrand of eq. (9A)″ as a function of n (Fig. 6). We find that, for T = 4 kV and r = 0.5 cm, the particle density is n = 1.2 x 10^{18} cm^{-3} for a = b = 0.6. However, for a = b = 0.4 and a = b = 0.2 the function is always negative, which means that breakeven cannot be achieved with these values of transfer efficiency. If one increases the radius of the plasma to 5 cm, leaving the temperature at 4 kV, the minimum particle density for breakeven decreases to n = 1.2 x 10^{17} cm^{-3} for a = b = 0.6 (Fig. 7). Again, for a = b = 0.4, and a = b = 0.2 the function is always negative. If one pushes the plasma radius to r = 50 cm, the results change to n = 1.2 x 10^{16} cm^{-3} for a = b = 0.6. However, we still have negative values of the integrand for a = b = 0.4, and a = b = 0.2 (Fig. 8). When the temperature is increased to 10 kV, and the plasma radius is maintained at 50 cm, the minimum particle density decreases to 3.5 x 10^{15} cm^{-3} for a = b = 0.6,

Fusion Breakeven Conditions for D-T Plasmas

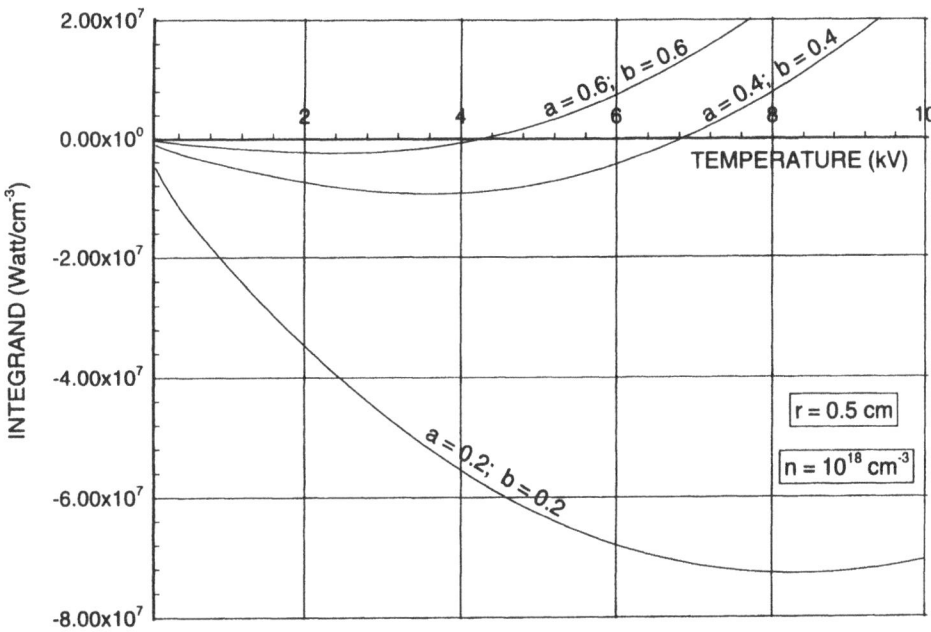

Figure 4. Plot of the integrand in eq. (9A)″ (α particle escape) as a function of temperature T for three values of the energy transfer efficiency (a = b = 0.6; a = b = 0.4; and a = b = 0.2), with particle density n = 10^{18} cm^{-3} and plasma radius r = 0.5 cm. Breakeven can be achieved with a particle density of 10^{18} cm^{-3}, provided the energy transfer efficiency is: a = b = 0.6, or a = b = 0.4.

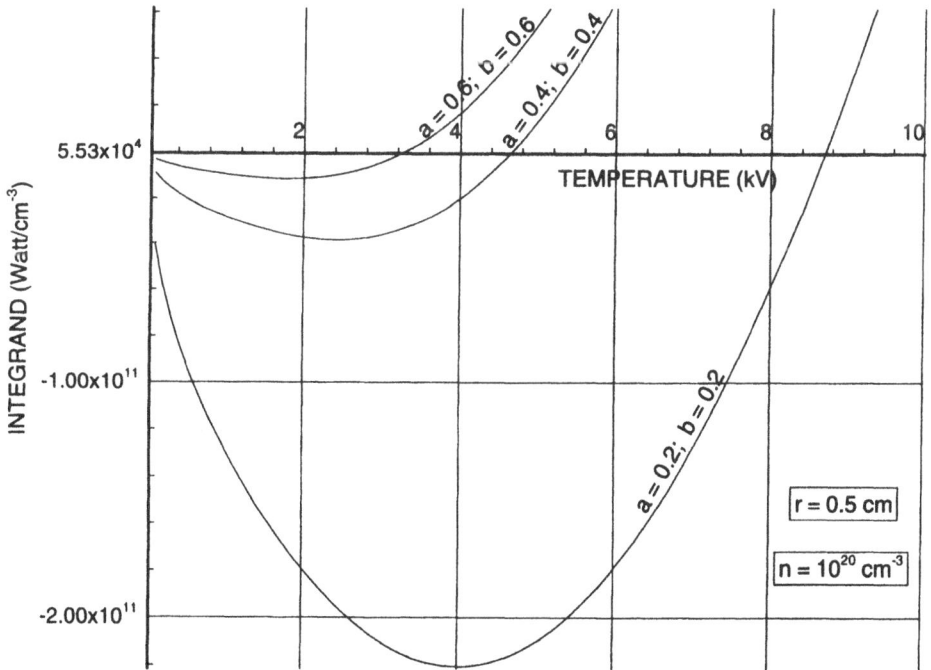

Figure 5. Plot of the integrand in eq. (9A)″ (α particle escape) as a function of temperature T for three values of the energy transfer efficiency (a = b = 0.6; a = b = 0.4; and a = b = 0.2), with particle density n = 10^{20} cm^{-3} and plasma radius r = 0.5 cm. Breakeven can be achieved with a particle density of 10^{20} cm^{-3}, for all three energy transfer efficiencies.

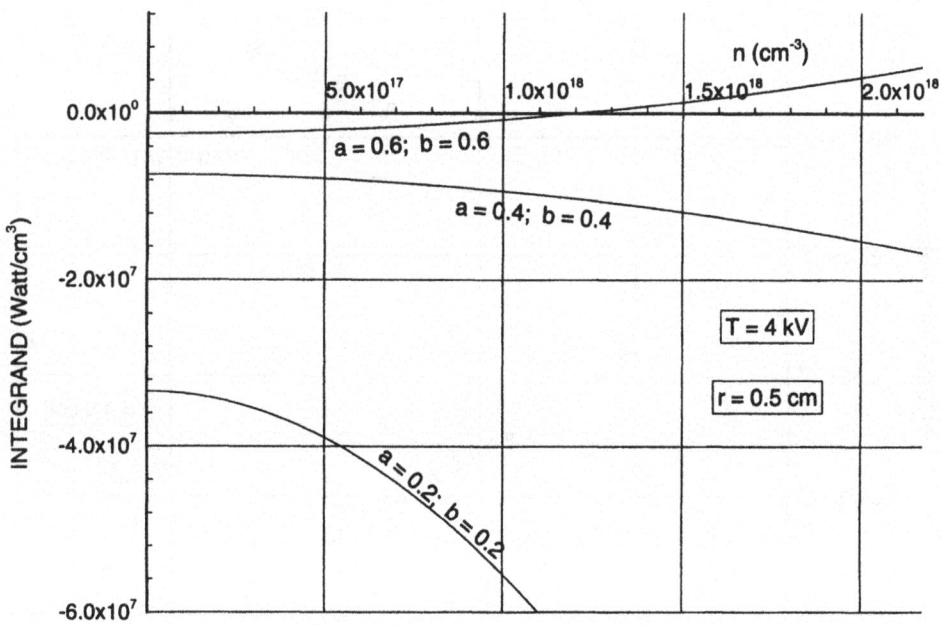

Figure 6. Plot of the integrand in eq. (9A)″ (α particle escape) as a function of particle density n for three values of the energy transfer efficiency (a = b = 0.6; a = b = 0.4; and a = b = 0.2), with temperature T = 4 kV and plasma radius r = 0.5 cm.

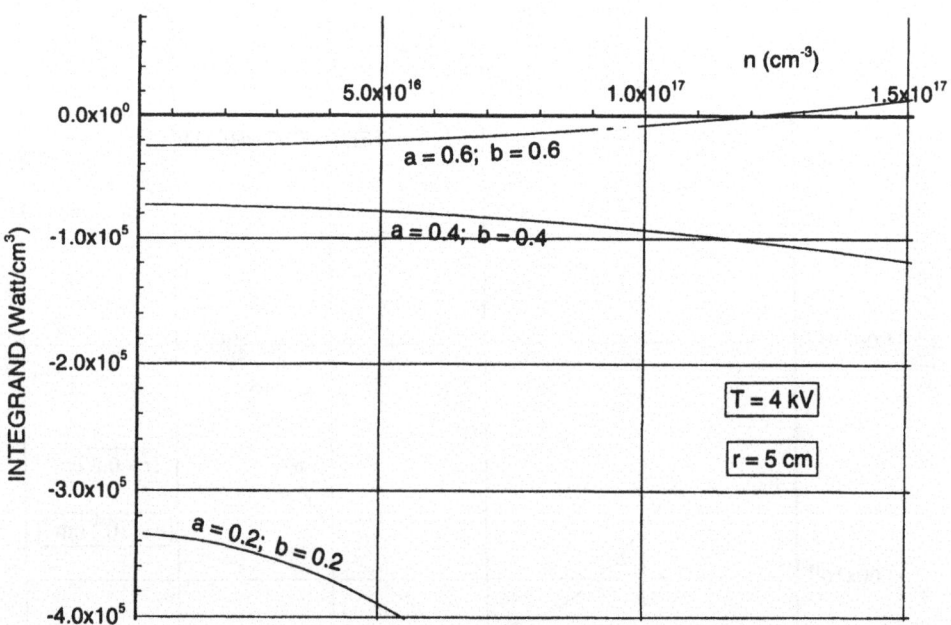

Figure 7. Plot of the integrand in eq. (9A)″ (α particle escape) as a function of particle density n for three values of the energy transfer efficiency (a = b = 0.6; a = b = 0.4; and a = b = 0.2), with temperature T = 4 kV and plasma radius r = 5 cm.

Fusion Breakeven Conditions for D-T Plasmas

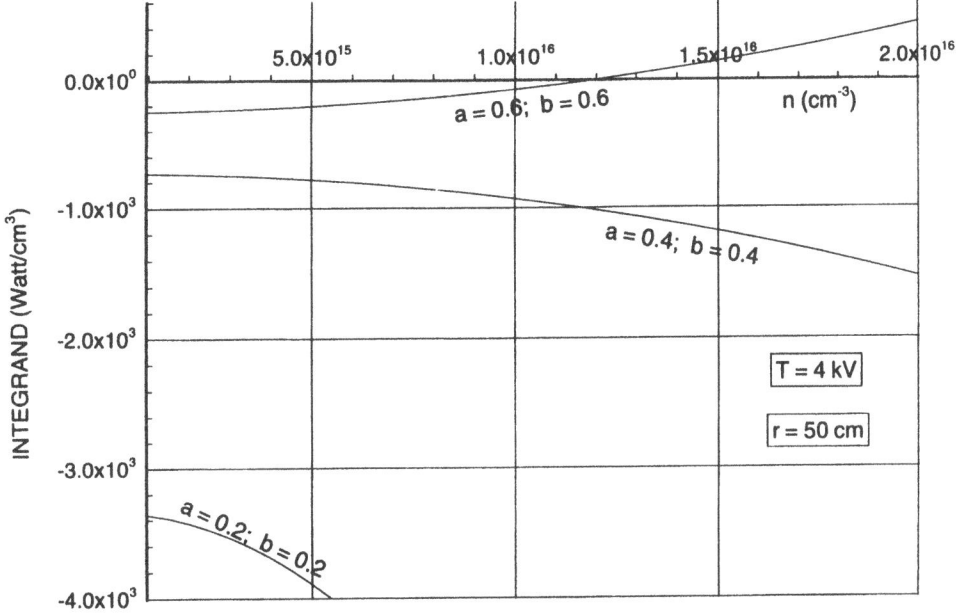

Figure 8. Plot of the integrand in eq. (9A)″ (α particle escape) as a function of particle density n for three values of the energy transfer efficiency (a = b = 0.6; a = b = 0.4; and a = b = 0.2), with temperature T = 4 kV and plasma radius r = 50 cm.

and becomes n = 6.4 × 10^{15} for a = b = 0.4, and 2.5 × 10^{16} for a = b = 0.2 (Fig. 9). If the temperature is pushed to 100 kV and the plasma remains at 50 cm, calculations show that the minimum particle density for breakeven is n = 2.0 × 10^{15} cm^{-3} for a = b = 0.6, n = 3.3 × 10^{15} cm^{-3} for a = b = 0.4, and n = 7.4 × 10^{15} cm^{-3} for a = b = 0.2.

B. α Particles Retained [Eq.(15A)″]

The integrand of eq. (15A)″ will now be used.

Temperature T as Variable. We start by choosing T as variable from 0 to 10 kV. In Fig. 10 we have prescribed n = 10^{14} cm^{-3} and r = 0.5 cm. Three curves have been plotted, for a = b = 0.6, a = b = 0.4, and a = b = 0.2, respectively. We find that, as in the case the α particle escaping, the integrand is negative in all three cases. Only when we increase the particle density to 10^{18} cm^{-3}, the integrands can become positive at T = 3.8 kV for a = b = 0.6, T = 5.0 kV for a = b = 0.4, and T = 6.1 kV for a = b = 0.2 (Fig. 11). If the particle density is increased to 10^{20} cm^{-3}, then T = 2.9 kV for a = b = 0.6, T = 3.6 kV for a = b = 0.4, and T = 4.2 kV for a = b = 0.2 (Fig. 12). If the particle density is increased further to 10^{22} cm^{-3}, the temperature does not change: T = 2.9 kV for a = b = 0.6, T = 3.6 kV for a = b = 0.4, and T = 4.2 kV for a = b = 0.2. The point where the three curves intersect correspond to the plasma temperature for which any change in the overall efficiency ab does not change the value of the integrand. In other words, it is the temperature for which any increase (or decrease) in α particle energy production due to decrease (or increase) of the efficiency ab is exactly compensated by the decrease (or increase) in radiation and conduction losses.

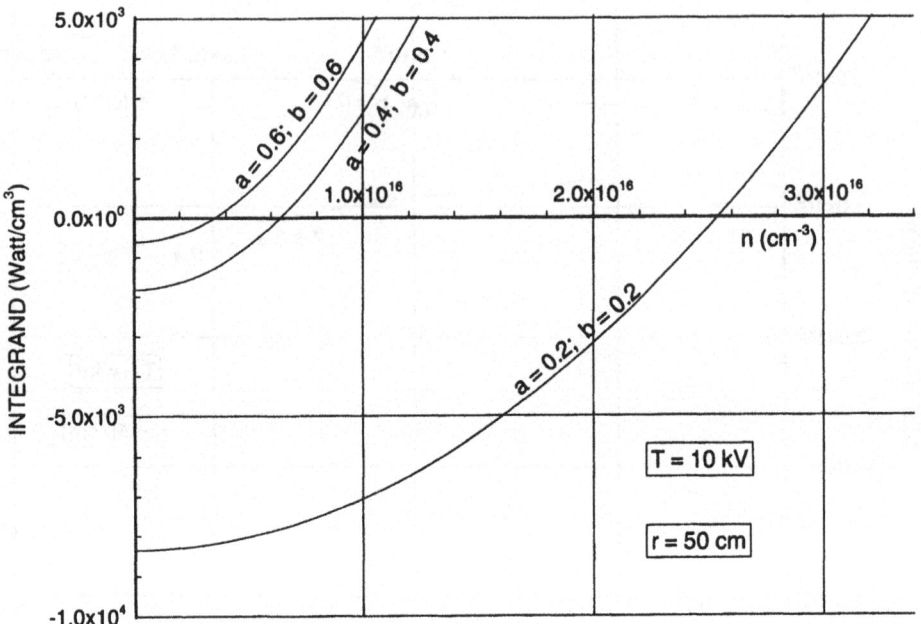

Figure 9. Plot of the integrand in eq. (9A)" (α particle escape) as a function of particle density n for three values of the energy transfer efficiency (a = b = 0.6; a = b = 0.4; and a = b = 0.2), with temperature T = 10 kV and plasma radius r = 50 cm.

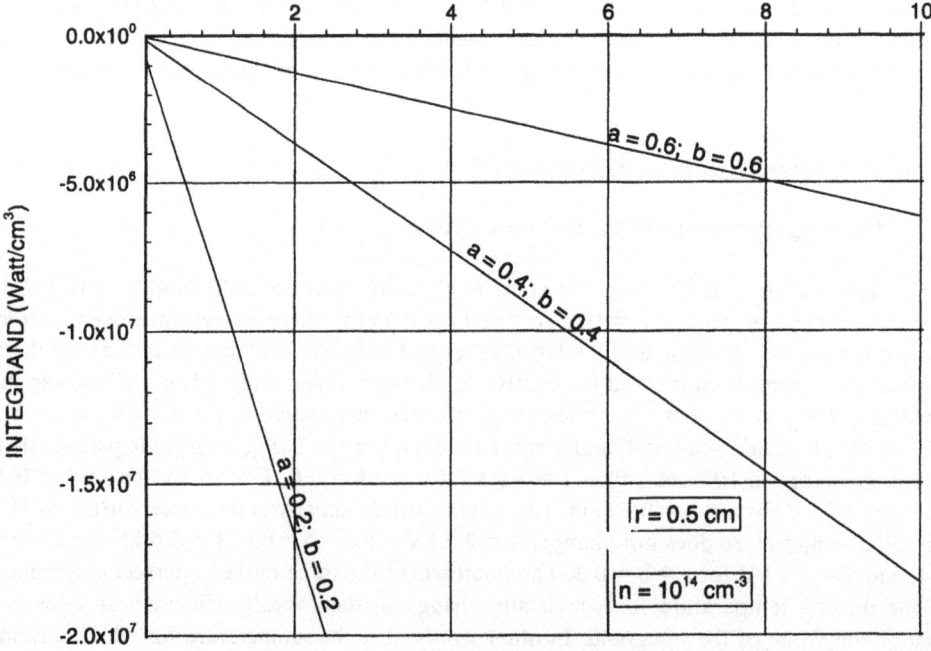

Figure 10. Plot of the integrand in eq. (15A)" (α particle retained) as a function of temperature T for three values of the energy transfer efficiency (a = b = 0.6; a = b = 0.4; and a = b = 0.2), with particle density n = 10^{14} cm^{-3} and plasma radius r = 0.5 cm. Breakeven cannot be achieved with α particle density of 10^{14} cm^{-3}.

Fusion Breakeven Conditions for D-T Plasmas

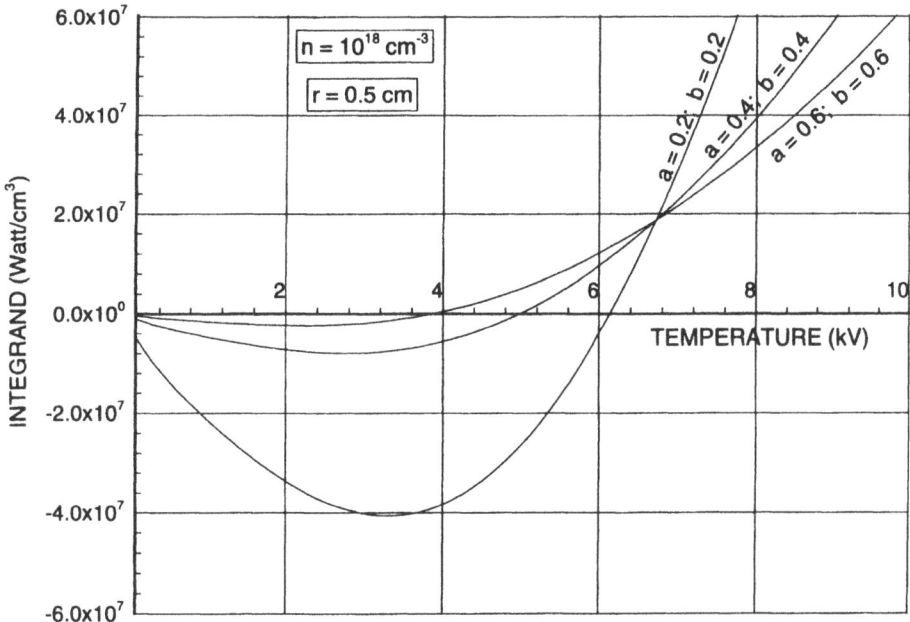

Figure 11. Plot of the integrand in eq. (15A)" (α particle retained) as a function of temperature T for three values of the energy transfer efficiency (a = b = 0.6; a = b = 0.4; and a = b = 0.2), with particle density n = 10^{18} cm^{-3} and plasma radius r = 0.5 cm. Breakeven can be achieved with α particle density of 10^{18} cm^{-3} for all three energy transfer efficiencies.

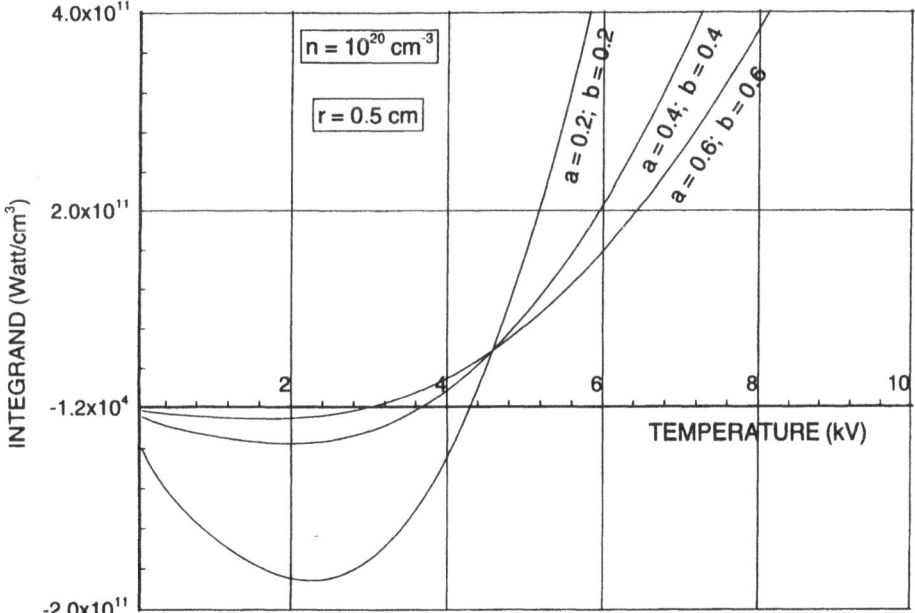

Figure 12. Plot of the integrand in eq. (15A)" (α particle retained) as a function of temperature T for three values of the energy transfer efficiency (a = b = 0.6; a = b = 0.4; and a = b = 0.2), with particle density n = 10^{20} cm^{-3} and plasma radius r = 0.5 cm. Breakeven can be achieved with α particle density of 10^{20} cm^{-3} for all three energy transfer efficiencies.

Particle Density n as Variable. We will now find the particle density for which the integrand becomes zero. For T = 4 kV and r = 0.5 cm, the particle density is 9.1×10^{17} cm^{-3} for a = b = 0.6, and 2.0×10^{18} cm^{-3} for a = b = 0.4. For a = b = 0.2 the integrand is always negative (Fig. 13). When the radius of the plasma is increased to 5 cm, the particle density reduces to 8.8×10^{16} cm^{-3} for a = b = 0.6, and 2.0×10^{17} cm^{-3} for a = b = 0.4. For a = b = 0.2 the integrand is always negative (Fig. 14). If the radius of the plasma is increased further to 50 cm, then the particle density reduces to 8.8×10^{15} cm^{-3} for a = b = 6, and 2.0×10^{16} cm^{-3} for a = b = 0.4. Again, for a = b = 0.2 the integrand is always negative (Fig. 15). If now the temperature is increased to 10 kV, leaving the plasma radius at 50 cm, the integrand becomes positive at $n = 2.9 \times 10^{15}$ cm^{-3} for a = b = 0.6, 4.2×10^{15} cm^{-3} for a = b = 0.4, and 5.6×10^{15} for a = b = 0.2 (Fig. 16).

4. EFFECTS OF THE TEMPORAL EVOLUTION OF THE PLASMA TEMPERATURE ON BREAKEVEN CONDITIONS

The temporal evolution of the plasma has a bearing on the attainment of breakeven conditions. We shall examine various typical plasma temperature excursions in the two cases of α particles that escape and α particles retained.

1. Rectangular Pulse

For a rectangular pulse of duration t_o during which the temperature is constant, breakeven can be attained for any pulse duration t_o, as long as the temperature is equal to

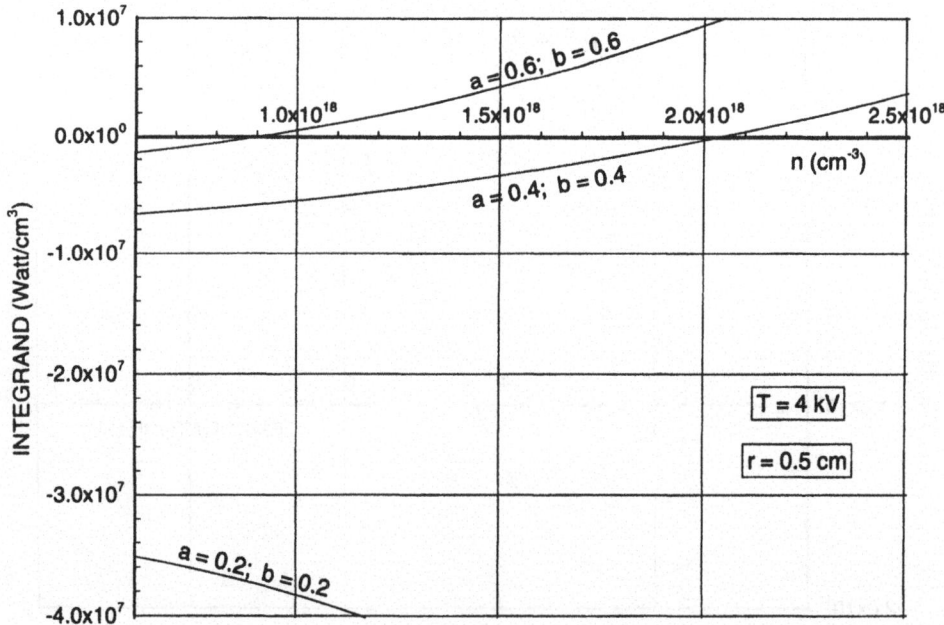

Figure 13. Plot of the integrand in eq. (15A)″ (α particle retained) as a function of particle density n for three values of the energy transfer efficiency (a = b = 0.6; a = b = 0.4; and a = b = 0.2), with temperature T = 4 kV and plasma radius r = 0.5 cm.

Fusion Breakeven Conditions for D-T Plasmas

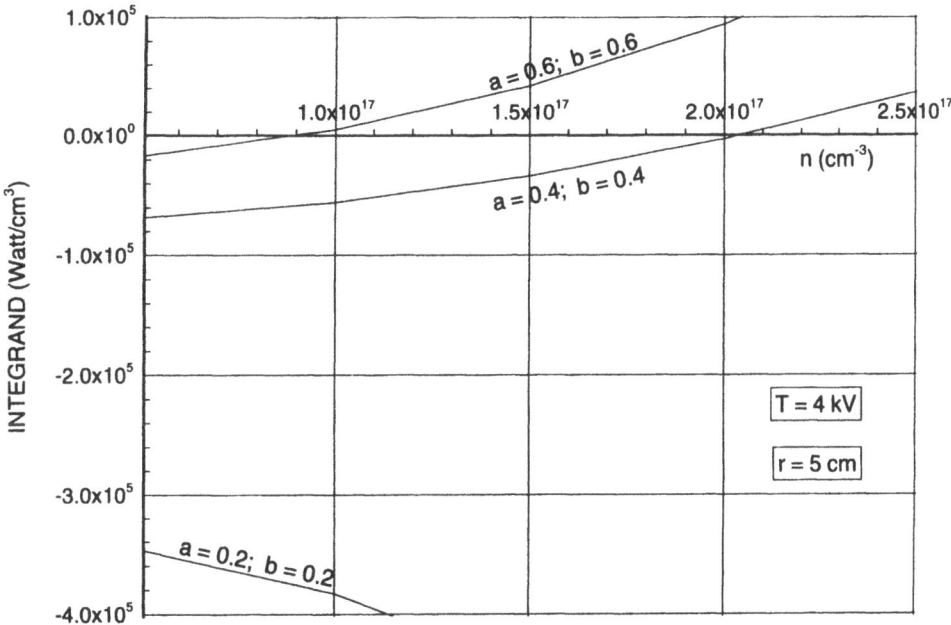

Figure 14. Plot of the integrand in eq. (15A)″ (α particle retained) as a function of particle density n for three values of the energy transfer efficiency (a = b = 0.6; a = b = 0.4; and a = b = 0.2), with temperature T = 4 kV and plasma radius r = 5 cm.

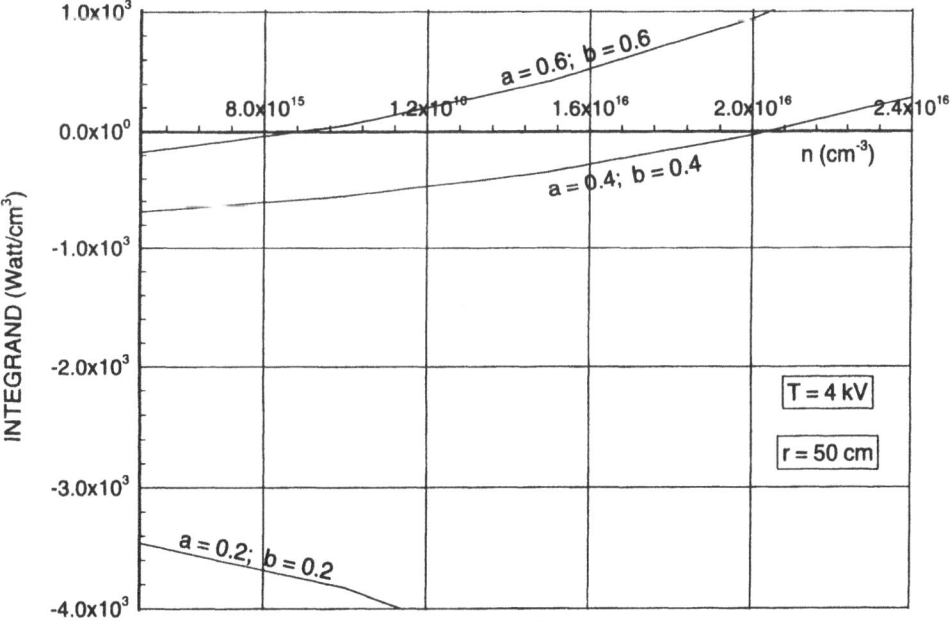

Figure 15. Plot of the integrand in eq. (15A)″ (α particle retained) as a function of particle density n for three values of the energy transfer efficiency (a = b = 0.6; a = b = 0.4; and a = b = 0.2), with temperature T = 4 kV and plasma radius r = 50 cm.

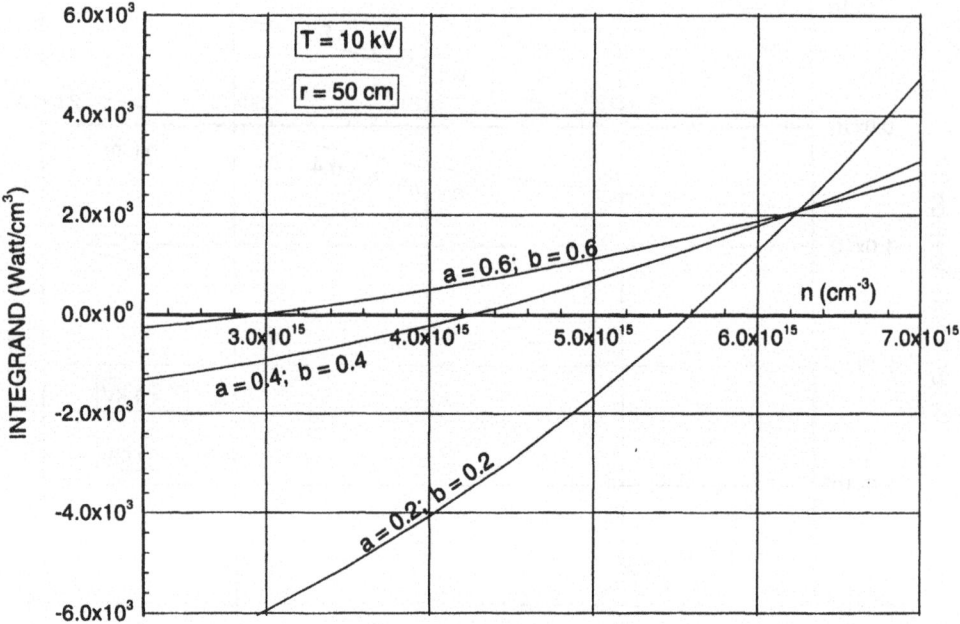

Figure 16. Plot of the integrand in eq. (15A)″ (α particle retained) as a function of particle density n for three values of the energy transfer efficiency (a = b = 0.6; a = b = 0.4; and a = b = 0.2), with temperature T = 10 kV and plasma radius r = 50 cm.

or above those that we have previously found as providing positive integrands in (9A)″ and (15A)″. For instance, if the temperature T is equal to or greater than 3.5 kV for a plasma of radius 0.5 cm, density 10^{18} cm^{-3}, and transfer efficiency a = b = 0.6, the integral (9A)″ is always positive for any t_o of the plasma duration or lifetime. However, if the plasma density is 10^{16} cm^{-3}, breakeven cannot be achieved no matter the plasma temperature or plasma duration or lifetime t_o, because the integrand is always negative under these conditions, unless the plasma radius is very large.

2. Gaussian Pulse

We shall assume a gaussian pulse of this form

$$T(t) = T_p e^{-gt^2}$$

where T_p is the peak temperature and g is the gaussian factor that determines the width of the pulse at half maximum (FWHM) (Fig. 17). In Table 1 we report some results of calculation of the integrals (9A)″ and (15A)″, for various pulse and plasma parameters. We find that, with this type of temperature evolution, breakeven can be achieved for any pulse width, as long as the plasma parameters are favourable. What is important is for the pulse to be of such amplitude as to have a temperature above the minimum temperature for breakeven ($T_{min} \sim 3$ kV) for a sufficient time to compensate for the energy lost during the time when the plasma has temperature below T_{min}.

Table 1. Values of integrals (9A)" and (15A)" for various gaussian pulse and plasma parameters

T_p (kV)	Δt(sec)	n (cm^{-3})	r (cm)	a = b	Integral (9A)"	Integral (15A)"	Remarks on failure to obtain breakeven
3	5.24×10^{-5}	10^{16}	0.5	0.2	-1.40×10^3	-1.60×10^{-1}	Low temperature, low density, and low efficiency
3	5.24×10^{-5}	10^{18}	5.0	0.4	-3.54×10^2	-3.16×10^2	Low temperature
3	5.24×10^{-5}	10^{20}	50.0	0.6	-9.66×10^5	-8.50×10^5	Low temperature
3	5.24×10^{-5}	10^{22}	50.0	0.6	-9.66×10^9	-8.50×10^9	Low temperature
3	5.24×10^{-5}	10^{24}	100.0	0.6	-9.66×10^{13}	-8.50×10^{13}	Low temperature
5	5.24×10^{-5}	10^{16}	0.5	0.2	-2.43×10^3	-2.10×10^{-1}	Low density, low efficiency
5	5.24×10^{-5}	10^{18}	5.0	0.4	-2.72×10^2	-4.48×10^2	Low efficiency
5	5.24×10^{-5}	10^{20}	50.0	0.4	-2.67×10^6	-4.48×10^6	Low efficiency
5	5.24×10^{-5}	10^{20}	50.0	0.6	$+6.16 \times 10^5$	$+1.41 \times 10^6$	Breakeven
5	5.24×10^{-8}	10^{20}	50.0	0.6	$+6.13 \times 10^2$	$+1.41 \times 10^3$	Breakeven
10	5.24×10^{-5}	10^{16}	0.5	0.2	-4.68×10^3	$+7.28 \times 10^{-1}$	Breakeven with α particles
10	5.24×10^{-5}	10^{18}	0.5	0.2	-6.07×10^3	$+7.28 \times 10^3$	Breakeven with α particles
10	5.24×10^{-5}	10^{18}	0.5	0.4	$+9.98 \times 10^1$	$+3.01 \times 10^3$	Breakeven
10	5.24×10^{-8}	10^{18}	0.5	0.4	$+1.00 \times 10^{-1}$	$+3.01 \times 10^1$	Breakeven
10	5.24×10^{-11}	10^{18}	0.5	0.4	$+9.85 \times 10^{-5}$	$+1.99 \times 10^{-3}$	Breakeven
10	5.24×10^{-14}	10^{18}	0.5	0.4	$+2.77 \times 10^{-18}$	$+1.99 \times 10^{-6}$	Breakeven

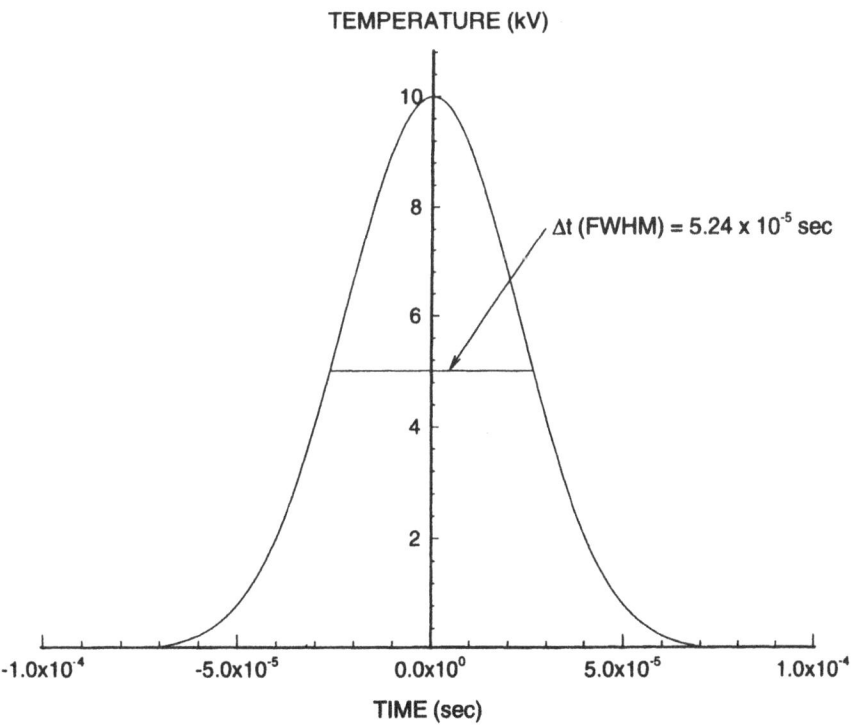

Figure 17. A typical gaussian pulse $T(t) = T_p \exp(-g\, t^2)$, where $T_p = 10$ kV and $g = 10^9$.

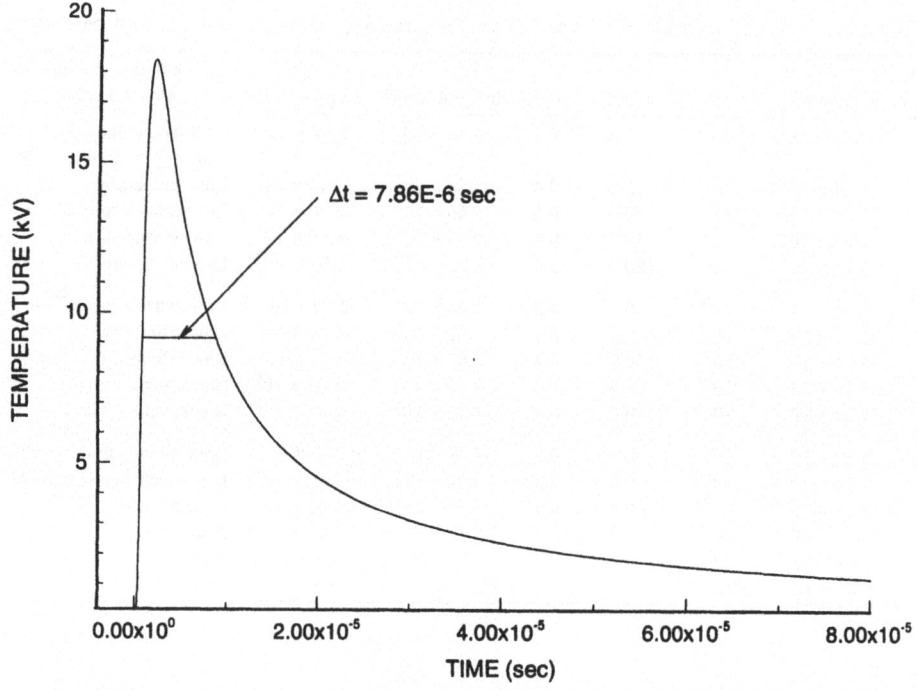

Figure 18. A typical fast risetime-slow decay time pulse of the form T(t) = A (m/t) exp (-n/t), where A = 100, m = 10^{-6}, and n = 2 x 10^{-6}.

3. Fast Risetime-Slow Decay Time Pulse

For a pulse such as the one depicted in Fig. 18, given by the expression

$$T(t) = A\frac{m}{t}e^{-n/t}$$

where the parameters A, m, and n characterize the pulse (A = 100, m = 10^{-6}, n = 2 x 10^{-6} for pulse of Fig. 18), breakeven cannot be achieved, no matter the plasma peak temperature. The reason is that, despite the high peak temperature reached by the plasma, it does not compensate for the energy lost during the long time when the temperature is below the value for breakeven. In the case of the pulse of Fig. 18, the integrals (9A)″ and (15A)″ are always negative irrespective of the plasma parameters.

5. THE ANALYSIS OF LAWSON AND SUBSEQUENT INTERPRETATIONS[1,2, 5–11]

a. The Analysis of Lawson

Lawson considered the case of a pulsed system that followed this idealized cycle:

"…. the gas is heated instantaneously to a temperature T, the temperature is maintained for a time t_o, after which the gas is allowed to cool. Conduction loss is neglected entirely, and it is assumed that the energy used to heat the gas and supply the radiation loss is regained as useful heat".

In the ideal case of Lawson we are then dealing with a plasma of constant temperature T that lasts for a time t_o. The energy 3nkT that the plasma has during its lifetime t_o has been supplied instantaneously at the beginning of the cycle. At the end of the cycle this energy is returned to the system through bremsstrahlung losses, no other loss mechanism being allowed. In particular, conduction losses are excluded.

Lawson finds the conditions for breakeven by considering that the energy from the reactions, together with the bremsstrahlung loss energy and the plasma heat energy, can be returned to the gas, with an efficiency η, thus starting the cycle again. For breakeven, one must have:

$$\eta(E_R + E_i) \geq E_i \qquad (28)$$

where E_R is the reaction energy, and E_i is the input energy. Dividing both sides of this expression by E_i, and defining $R = E_R/E_i$, one gets:

$$\eta(R+1) \geq 1 \qquad (29)$$

For $\eta = 1/3$ (this is the efficiency considered by Lawson), $R \geq 2$. According to Lawson:

$$E_i = E_B + 3nkT = P_B t_o + 3nkT \qquad (30)$$

where P_B is the bremsstrahlung power, and 3nkT is the energy required to heat the gas to a temperature T. During the same time t_o the reaction energy E_R is: $E_R = P_R t_o$, where P_R is the reaction power. The condition for breakeven is therefore:

$$R = \frac{E_R}{E_i} = \frac{P_R t_o}{P_B t_o + 3nkT} = \frac{P_R/3n^2kT}{P_B/3n^2kT + \dfrac{1}{nt_o}} \geq 2 \qquad (31)$$

From this Lawson concludes that, since P_R and P_B are both proportional to n^2, R (a parameter indicative of the attainment of breakeven) is a function of T and nt_o.

In light of the results obtained in our analysis of Sec. 2, we would like to dispute this conclusion. In order to illustrate how the analysis of Lawson can be improved, the cycle to which the gas is subjected should be divided in three phases:

1st phase: Rapid heating of the gas from the original ambient temperature to the fusion temperature T;

2nd phase: Continuous injection of energy in the plasma to maintain the temperature T despite the bremsstrahlung losses;

3rd phase: No more injection of energy and rapid cooling of the plasma back to the original ambient temperature.

The 1st phase lasts for a time t_1, which can be as short as one likes. The 2nd phase lasts for a time t_2. The 3rd phase lasts for a time t_3.

During the 1st phase the input energy e_1 is used to heat the gas to a temperature T and to supply the bremsstrahlung losses e_B. Thus:

$$e_1 = \int_0^{t_1} P_B(t)\,dt + \frac{3nkT}{t_1}t_1 = \int_0^{t_1} P_B(t)\,dt + 3nkT \tag{32}$$

where $(3nkt)/t_1$ is the rate of energy addition during time t_1.

The reaction energy during this time is

$$e_R = \int_0^{t_1} P_R(t)\,dt .$$

During the 2nd phase the input energy is used to keep the plasma at a constant temperature T, despite the bremsstrahlung losses. Hence, the input energy e_2 is equal to the energy leaving the plasma as bremsstrahlung:

$$e_2 = E_B = P_B t_2 \tag{33}$$

and the reaction energy is:

$$e_R = P_R t_2.$$

During the 3rd and last phase, there is no more input energy, the plasma cools down rapidly, and the energy 3nkT leaves the system as bremsstrahlung losses, no other mechanism of energy loss being allowed. Hence;

$$e_3 = 0$$
$$e_B = \frac{3nkT}{t_3}t_3 = 3nkT = \int_0^{t_3} P_B(t)\,dt, \tag{34}$$

where $(3nkT)/t_3$ is the rate of cooling during time t_3. The reaction energy during this last phase is

$$e_R = \int_0^{t_3} P_R(t)\,dt.$$

In light of (34), Eq. (32) can be written as:

$$e_1 = \int_0^{t_1} P_B(t)\,dt + \int_0^{t_3} P_B(t)\,dt, \tag{35}$$

and the condition for breakeven can now be rewritten in this way:

Fusion Breakeven Conditions for D-T Plasmas

$$R = \frac{E_R}{E_i} = \frac{\int_0^{t_1} P_R(t)\,dt + P_R t_2 + \int_0^{t_3} P_R(t)\,dt}{\int_0^{t_1} P_B(t)\,dt + P_B t_2 + \int_0^{t_3} P_B(t)\,dt} \geq 2 \tag{36}$$

This expression can be simplified into the following:

$$R = \frac{\int_0^{t_1+t_2+t_3} P_R(t)\,dt}{\int_0^{t_1+t_2+t_3} P_B(t)\,dt} \geq 2 \tag{37}$$

Since both P_R and P_B are proportional to n^2 (which is considered to be a constant during the plasma lifetime), and the two integrations performed over the same time are not sensitive functions of time, we see that the condition for breakeven, in the case of absence of conduction losses, is independent of particle density n and pulse duration $t_o = t_1 + t_2 + t_3$, but it is only a function of T.

This same result can be retrieved from our analysis. For convenience, we rewrite here Eq. (5):

$$\frac{1}{a}\int_0^{t_o}[H(t)+R(t)+C(t)]\,dt = b\int_0^{t_o}[P_\alpha(t)+P_n(t)+R(t)+C(t)]\,dt \tag{5}$$

where now we have a = 100%, b = 33%, $t_o = t_1 + t_2 + t_3$, C(t) = 0,. Knowing that

$$\int_0^{t_o} H(t)\,dt = 0,$$

we can write Eq. (5) in the following form:

$$\int_0^{t_1+t_2+t_3} P_B(t)\,dt = 0.33 \int_0^{t_1+t_2+t_3} [P_R(t)+P_B(t)]\,dt \tag{38}$$

Dividing both sides by $\int_0^{t_1+t_2+t_3} P_B(t)\,dt$

$$1 = 0.33\,\frac{\int_0^{t_1+t_2+t_3} P_R(t)\,dt}{\int_0^{t_1+t_2+t_3} P_B(t)\,dt} + 0.33 \tag{39}$$

from which:

$$\frac{\int_0^{t_1+t_2+t_3} P_R(t)\,dt}{\int_0^{t_1+t_2+t_3} P_B(t)\,dt} = \frac{1-0.33}{0.33} \approx 2, \tag{40}$$

which is the same expression (37) as before.

The fact that Lawson leaves the plasma heat energy as 3nkT in his expression (31) instead of considering that this energy will leave the plasma as bremsstrahlung losses

$$\int_0^{t_3} P_B(t)\,dt$$

makes us wonder if, perhaps, he was really considering a steady-state rather than a pulsed system, where the initial plasma heat energy is not recycled back into the system through the bremsstrahlung losses and/or other losses. If this conjecture corresponds to reality, then the Lawson criterion may acquire a new meaning: assuming that we have a steady-state reactor in which we have initially invested an energy 3nkT to generate the plasma, energy which will never be returned to us, is it possible to find the rate with which one should feed energy into the plasma for power breakeven condition? The answer is provided by inequality (28), modified as follows:

$$\eta(P_R + P_i) \geq P_i \tag{28'}$$

where the powers P_R and P_i have been obtained by dividing the energies by a certain characteristic time that we shall define shortly. The input power P_i can be generalized as follows:

$$P_i = P_B + \frac{3nkT}{t_o} \tag{41}$$

by including an empirical term equal to the rate of plasma heat addition necessary to compensate for the plasma energy 3nkT that leaks out of the system in a characteristic time t_o by some unspecified mechanism. With this understanding, inequality (28') can be written as follows:

$$\eta\left(P_R + P_B + \frac{3nkT}{t_o}\right) \geq P_B + \frac{3nkT}{t_o} \tag{42}$$

Dividing both sides by $P_B + (3nkT)/t_o$, one gets:

$$\eta\left(\frac{P_R}{P_B + \frac{3nkT}{t_o}} + 1\right) \geq 1 \tag{43}$$

Fusion Breakeven Conditions for D-T Plasmas

For $\eta = 33\%$, we get:

$$R = \frac{P_R}{P_B + \frac{3nkT}{t_o}} = \frac{P_R/3n^2kT}{P_B/3n^2kT + \frac{1}{nt_o}} \geq 2 \tag{44}$$

Inequality (44) says that, in a steady-state reactor, power breakeven is a function of T and nt_o, where t_o is the time to supply an amount of plasma energy equal to 3nkT necessary to compensate for an unspecified equal amount of energy lost in the same time, energy different from bremsstrahlung, that would otherwise cool the plasma. The parameters t_o therefore should not be identified with the plasma lifetime defined in the previous pulsed operation, but it is now a characteristic time in the steady state operation during which an amount of energy 3nkT should be fed into the system in order to maintain the plasma temperature T, which would otherwise drop because of losses different from bremsstrahlung. In short, the term $3nkT/t_o$ is a rate of energy addition, different from the rate P_B of energy addition necessary to compensate for bremsstrahlung losses.

The same conclusion can be retrieved from our analysis. For convenience, we rewrite here Eq. (5) in terms of powers:

$$\frac{1}{a}[H(t) + R(t) + C(t)] = b[P_\alpha(t) + P_n(t) + R(t) + C(t)] \tag{5'}$$

where now we have a = 100%, b = 33%, $R(t) = P_B$ = const., and $P_\alpha(t) + P_n(t) = P_R$ = const. Knowing that, in a steady-state reactor, H(t) = 0, and empirically assuming $C(t) = (3nkT)/t_o$ for an unspecified rate of energy loss different from bremsstrahlung, Eq. (5') becomes:

$$P_B + \frac{3nkT}{t_o} = 0.33 \left[P_R + P_B + \frac{3nkT}{t_o}\right] \tag{45}$$

Dividing both sides by $P_B + (3nkT)/t_o$:

$$1 = 0.33 \left[\frac{P_R}{P_B + \frac{3nkT}{t_o}} + 1\right] \tag{46}$$

which leads to the same expression (44) as before.

b. Subsequent Interpretations

The interpretations of the Lawson criterion in subsequent times have all hinged on the notion that the plasma heat energy 3nkT, even in a pulsed operation, can be returned to the system through another loss term different from bremsstrahlung. This term has been designated by the name of "non-radiative" loss and, in order to have the plasma losing all its energy during its lifetime, the rate of energy loss has been defined as the ratio of the energy 3nkT to the plasma lifetime t_o. With this approach one does not look in detail at the initial plasma formation, maintenance, and final cooling. One considers that the plasma is instantaneously formed at $t = 0$, and instantaneously disappears at $t = t_o$, and that, during the time of its existence, it loses the radiation energy $P_B t_o$ continuously supplied by the source, and the energy $((3nkT)/t_o)t_o$ instantaneously supplied at the beginning of the cycle. The time t_o is now designated by a special name: 'energy confinement time'.

With this notion the Lawson criterion can be rewritten as follows:

$$R = \frac{E_R}{E_i} = \frac{P_R t_o}{P_B t_o + \frac{3nkT}{t_o} t_o} = \frac{P_R}{P_B + \frac{3nkT}{t_o}} = \frac{P_R/3n^2kT}{P_B/3n^2kT + \frac{1}{nt_o}} \geq 2 \quad (47)$$

where now t_o is the energy confinement time. Inequality (47) is the same as (31) and, with this interpretation, Lawson analysis is able to have all losses considered, in a pulsed operation, including non-radiative losses. In particular, one retrieves that the parameter R indicative of the attainment of breakeven is again a function of T and nt_o. It is to be noted that the energy confinement time cannot be different from t_o because, were it smaller, at the end of the plasma lifetime t_o the non-radiative losses would have dissipated more energy than the plasma has and, were it larger, not all plasma energy is dissipated.

We have a number of comments on this interpretation, including the one that the non-radiative rate of energy loss is empirically defined. Since use is made in Lawson analysis of known expressions for the rate of reaction energy P_R and bremsstrahlung loss P_B, it would have been advisable to use a known formula, rather than an empirical one, also for other losses. Moreover, if one admits that there are non-radiative losses, this implies that the source must continuously supply, besides the bremsstrahlung energy $P_B t_o$, also the non-radiative energy at a rate $(3nkT)/t_o$ during the pulse time t_o in order to keep the plasma at the temperature T. When this is done, one finds that, at the end of the pulse, there is still a plasma of energy $3nkT$ to be disposed in some way.

Further analysis clarifies this point and shows that the introduction of an empirically defined rate of non-radiative loss leads to contradiction. Let us divide the pulsed phenomenon of plasma formation, maintenance, and cooling in three phases as before. In order to focus our attention now on the non-radiative losses, we shall assume for the moment that there are no bremsstrahlung losses. During the first phase the input energy e_1 is used to heat the gas to a temperature T and to supply the non-radiative losses. If we indicate the rate of non-radiative losses as $P_{NR}(t)$, at the end of the 1st phase we have:

$$e_1 = \int_0^{t_1} P_{NR}(t)dt + 3nkT. \quad (48)$$

During the second phase the input energy is used to keep the plasma at a constant temperature T, despite the non-radiative losses. Hence:

$$e_2 = P_{NR} t_2 \quad (49)$$

During the third and last phase there is no more input energy, the plasma cools down, and the heat energy $3nkT$ leaves the system as non-radiative losses. Hence:

$$e_3 = 0$$

$$e_{NR} = 3nkT = \int_0^{t_3} P_{NR}(t)dt \tag{50}$$

In light of (50), Eq. (48) can be written as follows:

$$e_1 = \int_0^{t_1} P_{NR}(t)\,dt + \int_0^{t_3} P_{NR}(t)\,dt \tag{51}$$

The total energy E_i supplied to the system during the pulse is therefore:

$$E_i = e_1 + e_2 + e_3 = \int_0^{t_1+t_2+t_3} P_{NR}(t)\,dt \tag{52}$$

This total energy is larger than the non-radiative energy of the third phase, and therefore we can write:

$$\int_0^{t_1+t_2+t_3} P_{NR}(t)\,dt > \int_0^{t_3} P_{NR}(t)\,dt = 3nkT \tag{53}$$

If we empirically define the rate of non-radiative loss as $(3nkT)/(t_1+t_2+t_3)$ we have:

$$\int_0^{t_1+t_2+t_3} P_{NR}(t)\,dt = \frac{3nkT}{t_1+t_2+t_3}(t_1+t_2+t_3) > 3nkT \tag{54}$$

which is contradictory.

There is another argument that indicates that the plasma energy $3nkT$ cannot be dissipated only as non-radiative loss. During the third phase, when the plasma cools down, its energy is really dissipated partly as bremsstrahlung, and partly as non-radiative loss, as follows:

$$3nkT = \int_0^{t_3} P_B(t)\,dt + \int_0^{t_3} P_{NR}(t)\,dt \tag{55}$$

It is difficult to assign a proper role to each loss term, unless their formulation is known, instead of empirically defined.

Because of difficulties of this sort, we believe that they have subsequently led to a different definition of the energy confinement time. It is no more the plasma lifetime $t_o = t_1 + t_2 + t_3$, but it is defined as:

$$\tau_E = \frac{\text{plasma energy}}{\text{non - radiative energy loss rate}} = \frac{3nkT}{P_{NR}}$$

It is therefore a measure of the time it takes for the plasma to lose an energy equal to the plasma heat energy 3nkT. More appropriately, τ_E should be called 'energy replacement time', as apparently originally Lawson called it (W.B. Thompson, private communication). This number is now a measure of the time scale involved in feeding energy continuously into the plasma in order to keep it at a temperature T. The energy confinement (or replacement) time τ_E can therefore be different from the plasma lifetime t_o, and one can have a plasma lifetime longer or shorter than the energy confinement time, depending on the rate with which energy must be fed into the plasma to keep it at a temperature T. For instance, one can think of a wall confined plasma which would rapidly lose energy predominantly by conduction loss at a rate which we assume to be $(3nkT)/\tau_E$. In this case, the plasma lifetime t_o can be longer than τ_E as long as one is able to feed energy at a rate $(3nkT)/\tau_E$ during time t_o.

Whereas the plasma lifetime t_o is a useful parameter to calculate the various energy outputs for breakeven conditions, such as total nuclear energy released during the pulse, total bremsstrahlung energy, etc., the knowledge of the energy confinement time τ_E has relative importance. If $\tau_E > t_o$, i.e., energy confinement time larger than plasma lifetime, it does not mean that we have to keep the plasma at a temperature T for a time $t_o = \tau_E$ for breakeven. It means that we have to supply an energy 3nkT at a rate $(3nkT)/\tau_E$ in order to keep the plasma at a temperature T during time $t_o < \tau_E$. Moreover, assuming that the plasma does not lose non-radiative energy but only bremsstrahlung energy, the energy confinement time is $\tau_E = \infty$. However, the bremsstrahlung losses would force anyway the termination of the life (i.e., of the energy confinement) of the plasma as soon as no more input energy is injected in the plasma. In summary, the empirical definition of $(3nkT)/\tau_E$ as non-radiative energy loss is confusing at best and has limited validity in the assessment of breakeven conditions for a pulsed system.

The validity of this term is however restored when one considers a steady-state, rather than a pulsed reactor, as probably was the case initially considered by Lawson. If one goes back to Eq. (31) and rewrites it for the time τ_E of energy replacement rather than for the plasma lifetime:

$$R = \frac{E_R}{E_i} = \frac{P_R \tau_E}{P_B \tau_E + \frac{3nkT}{\tau_E} \tau_E} = \frac{P_R}{P_B + \frac{3nkT}{\tau_E}} = \frac{P_R/3n^2kT}{P_B/3n^2kT + \frac{1}{n\tau_E}} \geq 2 \qquad (56)$$

one finds that the term $(3nkT)/\tau_E$ is now the rate of heat energy replacement necessary to maintain the plasma at a temperature T. The standard $n\tau_E$ formulation, as derived from (56):

$$n\tau_E \geq \frac{6n^2kT}{P_R - 2P_B} \qquad (57)$$

or

$$\tau_E \geq \frac{6nkT}{P_R - 2P_B}$$

is now a statement that the source must be able to continuously replace an amount of energy equal to 3nkT in a time τ_E in a plasma of density n and temperature T according to inequality

(57) for the reaction to be continuously sustained. Otherwise, the plasma would rapidly cool and the reaction quench. In such a steady-state reactor the initial investment of energy 3nkT into the plasma is never recovered. In a pulsed system this energy is recovered and the conditions for breakeven change, as shown in the analysis of Sec. 2, and in this analysis.

In conclusion, there are two aspects of the analysis of Lawson and subsequent interpretations that emerge from the present study. Firstly, the application of the simple Lawson's analysis to a pulsed system meets with difficulties of interpretation, mainly because the introduction of an empirical term of the form $(3nkT)/\tau_E$ for the non-radiative losses has limited validity in providing a criterion for breakeven. Secondly, when a steady-state system is considered, then the non-radiative loss term $(3nkT)/\tau_E$ empirically defined for a pulsed operation becomes now a rate of energy addition into the plasma, and the standard $n\tau_E$ condition becomes now essentially a statement of the time τ_E required in continuously supplying an amount of energy equal to 3nkT to a plasma of density n and temperature T for the reactor to produce energy. In other words, $(3nkT)/\tau_E$ is a rate of plasma heating requirement.

In Sec. c) below we shall show that the interpretation of $(3nkT)/\tau_E$ as a general non-radiative loss term may meet with other difficulties, because it may not include one of the important loss components, namely conduction.

c. Nature of the Term $(3nkT)/\tau_E$

The term $(3nkT)/\tau_E$, when considered as a loss term, should include all non-radiative losses, i.e., losses different from bremsstrahlung. Therefore, assuming that there are no other non-radiative losses but conduction, the term should be an expression for such losses. In the following, we shall show that this is not the case.

1. Conduction of heat is a boundary phenomenon, and therefore it is expressed as heat flow per unit time per unit area, which we can call C. The total flow of heat out of a volume V having area A is CA. The flow of heat per unit volume is CA/V. Consider two typical cases of plasma geometry: toroidal and spherical. In the toroidal case $A_t = 4\pi^2 rR$, and $V_t = 2\pi^2 r^2 R$, where r and R are the minor and major radius of the torus, respectively. Hence $CA_t/V_t = 2C/r$. In the spherical case $A_s = 4\pi r^2$, and $V_s = (4/3)\pi r^3$, where r is the radius of the sphere. Hence $CA_s/V_s = 3C/r$. In both cases the expression for the conduction losses per unit volume contains a parameter r indicative of plasma volume, which is not present in the term $(3nkT)/\tau_E$.

2. In order for the parameter $(3nkT)/\tau_E$ to represent a conduction loss per unit volume, one must have, in the spherical case, for instance:

$$\frac{3C}{r} = \frac{3nkT}{\tau_E},$$

from which:

$$C = \frac{nrkT}{\tau_E} \tag{58}$$

The total flow of heat out of a sphere of radius r at temperature T in an infinite medium at a temperature much lower than T is given by:[23]

$$Q = 4\pi \, KrT \tag{59}$$

where K is the thermal conductivity of the medium. Hence:

$$C = \frac{Q}{4\pi r^2} = \frac{KT}{r} = \frac{nrkT}{\tau_E} \tag{60}$$

from which:

$$K = \text{thermal conductivity of the medium} = kr^2 \frac{n}{\tau_E}. \tag{61}$$

But K, as expressed by (61), can hardly be a constant of the medium *through* which heat flows. For a plasma near breakeven conditions, Eq. (56) provides:

$$\frac{n}{\tau_E} = \frac{P_R - R P_B}{3RkT} \tag{62}$$

and therefore:

$$K = kr^2 \frac{n}{\tau_E} = r^2 \frac{P_R - R P_B}{3RT} \tag{63}$$

This expression shows that K is not a constant of the medium *through* which heat flows (ionized gas blanket), but rather a function of the radius r, particle density n, and temperature T of the medium *from* which heat flows (plasma core). In other words, expression (63) states that the thermal conductivity of the gas blanket, rather than being determined by its own nature, is determined by the nature of the central plasma from which heat flows. Since one can in principle have a plasma of any temperature T emitting bremsstrahlung at a rate P_B, and reaction energy at a rate P_R with the same blanket surrounding it (one can even imagine a solid wall as a blanket maintained at a constant temperature), expression (63) states that the thermal conductivity of this blanket should change according to the characteristics of the plasma core. This is not possible.

In summary, the above analysis shows that the term $(3nkT)/\tau_E$ may not be considered an expression that includes conduction losses. It seems therefore that the breakeven conditions derived from the original Lawson analysis have implicitly assumed that there are no conduction losses in the system. In other words, despite the attempt at bringing back the conduction losses into Lawson analysis, these have not been really included in a convincing way. If this is so, these breakeven conditions may not have general validity.

6. CONCLUSION

In this paper, the conditions for fusion energy breakeven in a 50% mixture of deuterium and tritium have been derived in a system from which breakeven is required, composed of an energy source and a reacting gas. Several energy transfer efficiencies have been considered from the source to the plasma, and from the plasma to the source. A spherical plasma geometry has been considered, but the analysis can be extended, by inserting the appropriate conduction loss expression, to other geometries, including the toroidal. Two cases have been considered: α particles escaping from the reacting plasma, and α particles retained by the plasma. The major conclusions of this analysis are:

1. Because of the introduction of conduction losses from heat transfer theory, the energy confinement time τ_E does not play anymore an explicit role in the breakeven conditions. The are now determined by particle density, temperature temporal profile, and dimension of the plasma;
2. In a pulsed operation, there is a minimum particle density below which breakeven cannot be obtained, no matter how favourable the other parameters. This particle density is about 10^{16} cm^{-3}. This is because bremsstrahlung and conduction losses are so severe that a sufficient fuel density is required to overcome these losses; and
3. The shape of the temperature temporal evolution of the reacting gas has a bearing on the attainment of breakeven. If the proper temperature is achieved, a rectangular or gaussian pulse can lead to breakeven, whereas a fast risetime - slow decay time pulse may not.

The analysis of Lawson, whose criteria have provided broad guidelines to the fusion community in its quest for fusion in the past several years, has been reviewed. It has been found to be correct if applied to steady-state reactor condditions. In a pulsed system, the criteria meet with difficulties of interpretation.

ACKNOWLEDGMENT

This paper has greatly benefited from useful criticism from H. Chen, G. Enright, J. Lau, P. Savic, and L. Zhang.

REFERENCES

1. J. D. Lawson. Some Criteria for a Power Producing Thermonuclear Reactor. *Proc. Phys. Soc.* **B70,** 6 (1957).
2. R. G. Mills. Lawson Criteria. *IEEE Trans. Nucl. Science* **NS-18,** 205 (1971).
3. B. C. Maglich, and R. A. Miller. Generalized Criterion for Feasibility of Controlled Fusion and Its Application to Nonideal DD Systems. *J. Appl. Phys.* **46,** 2915 (1975).
4. F. F. Chen, J. M. Dawson, B. D. Fried, H. P. Furth, and M. N. Rosenbluth. Comments on 'Generalized Criterion for Feasibility of Controlled Fusion and Its Application to Nonideal DD Systems'. *J. Appl. Phys.* **48,** 415 (1977).
5. T. J. Dolan. *Fusion Research. Principles, Experiments and Technology,* (Pergamon Press, New York, 1982), pp. 73–100.
6. R.A. Gross. *Fusion Energy,* (John Wiley & Sons, New York, 1984), pp. 38–41.
7. T. Kammash. *Fusion Reactor Physics. Principles and Technology,* (Ann Arbor Science Publishing, Ann Arbor, 1975), pp. 21–23.
8. G. H. Miley. *Fusion Energy Conversion,* (American Nuclear Society, 1976), pp. 393–394.
9. J. R. Roth. *Introduction to Fusion Energy,* (Ibis Publishing, Chalottesville, Virginia, 1986), pp. 223–282.
10. J. Raeder, K. Borrass, R. Bünde, W. Dänner, R. Klingelhēfer, L. Lengyel, F. Leuterer, M. Sēll. *Controlled Nuclear Fusion,* (John Wiley & Sons, Chichester, 1986), pp. 210–212.
11. J. Wesson. *Tokamaks,* (Clarendon Press, Oxford, 1987), pp. 8–11.

12. E. Panarella, and P. Savic. Scaling Laws for Spherical Pinch Experiments. *J. Fusion Energy* **3**, 199 (1983).
13. E. Panarella. The Spherical Pinch. *J. Fusion Energy* **13**, 45 (1987).
14. H. Chen, J. Chen, B. Hilko, and E. Panarella. Numerical Comparison Between the ICF and the ICF-Spherical Pinch. *J. Fusion Energy* **13**, 45 (1994).
15. S. Glasstone, and R. H. Lovberg. *Controlled Thermonuclear Reactions,* (D. Van Nostrand, 1960), pp. 20–32.
16. L. C. Thomas. *Fundamentals of Heat Transfer* (Prentice Hall, 1980), p. 9.
17. R.W. Conn. First Wall and Divertor Plate Material Selection in Fusion Reactors. *J. of Nuclear Materials* **76 & 77**, 103 (1978).
18. L. Spitzer, Jr. *Physics of Fully Ionized Gases,* (Interscience Publishers, 1962), p. 144.
19. Reference 18, p. 128.
20. Reference 18, p. 145.
21. J. K. Roberts, and A. R. Miller. *Heat and Thermodynamics,* (Blackie & Son, 1961), p. 289.
22. Reference 16, p. 654.
23. L. R. Ingersoll, O. J. Zobel, and A. C. Ingersoll (1948). *Heat Condution,* (McGraw-Hill, New York), p. 36

16

PROGRESS IN INERTIAL FUSION RESEARCH

Chiyoe Yamanaka

Institute for Laser Technology
2-6 Yamada-oka Suita, Osaka 565, Japan

ABSTRACT

Progress in inertial fusion research in the last 30 years has been briefly described. As Edward Teller said, the ICF research has turned around the third corner to approach the homestretch for the scientific feasibility. Fusion will be the world's ultimate energy source, inexhaustible fuel, worldwide availability, environmental benign and high safety potential, if it can be harnessed economically. Fusion must respond to compete in the marketplace of the 21st century. In this sense, we should have a long prospect of research and development. The concept of ICF has a strong importance for the fusion energy.

The laser fusion has made a great progress in the last 30 years. On the other hand, there has been an active international collaboration to develop the magnetic confinement fusion power research. This kind of action was not observed in the ICF society. Recent inertial fusion experiments on the direct driven fusion at Osaka have successfully got the high fusion neutron yield 10^{13} and the high density compression of 1000 times normal fuel density. The electron degeneracy of core plasma is also observed. The U.S. Halite / Centurion program informed us of indirect driven fusion which will be attainable the high gain for less than the 10MJ of driver. However, the data base is not yet clear to determine the details for high gain. This question can only be solved by using a large laser facility. The U.S. policy on indirect driven fusion program has come to provide the National Ignition Facility with the declassificaton of the experimental data. Experimental and theoretical progress in ICF in the international community has suggested that the time has come to eliminate unnecessary restrictions on information relevant to the energy applications of ICF. Now ICF is in the second stage of the development. The ignition and breakeven are in a scope of the program. The international collaboration will be initiated.

Now, to show the importance of the international collaboration, the world progress of inertial fusion is briefly reviewed setting particular remarks on the Japanese efforts.

INTRODUCTION

Since 1972, our continuous efforts have been performed to organize the international ICF community concerning the IAEA activity and other authorities. In 1988, at

ECLIM in Spain I proposed the Madrid Manifesto to protest the international collaboration in ICF research over the US classification which was supported by many participants of the meeting including the US scientists. In 1990, the US Secretary of Energy James D. Watkins addressed at the thirteenth IAEA Conference on Plasma Physics and Controlled Nuclear Fusion Research according to the reports by Koonin's committee of NAS and Stever's FPAC and announced if inertial fusion has promise as an energy source—and I believe that it does—we should pursue that promise with a sort of cost-effective international collaboration that marks magnetic fusion efforts such as the International Thermonuclear Experimental Reactor (ITER). This address was welcomed by all the participants. A fundamental change was expected. In 1994, DOE finally announced to declassify a mostly part of the ICF progam which was already published by researchers from Japan, Germany and elsewhere.

Now I would like to look back the progress of laser fusion research in a short term.

At the Lebedev Physics Institute in former USSR, subsequent pioneering research on ICF had been performed. It led to the disclosure of laser fusion concepts at the International Quantum Electronics Conference in 1963, followed by a presentation in 1968 of the first detection of fusion neutrons from laser irradiated lithium hydride targets. These progress was due to the development of the lasers, Kalmar and Delfin. At the Lawrence Livermore National Laboratory the pioneering theoretical and experimental works have been performed in the last three decades. Especially a disclosure in 1972 of the implosion physics at the International Quantum Electronics Conference, Montreal by Edward Teller showed that the laser fusion targets could be ignited with much less energy than predicted there-to-fore if the fuel was compressed up to 1000 times of the normal density. And also the understanding of hydrodynamic and instability phenomena associated with the strong compression of ICF targets were prevailed. Series of the Nd glass lasers, Augus, Shiva and Nova were developed to perform the laser fusion experiments. The major glass suppliers for the large Nd glass lasers are in Japan and Germany. LLNL and also LANL have been performed a lot of interesting works in the ICF research which provided an essential guidance for the ICF program in the world. However the US classification policy of inertial fusion especially on the indirect driven fusion was a crucial problem. It hurt the morale of the US scientists who were unable to take credit for their creative work and often must endure the vexation of seeing nearly identical work published in the open literature by workers in Japan, Europe or the Soviet Union. Classification impeded the progress by restricting the flow of information, and did not allow all ICF work to benefit by the open scientific scrutiny. According to the patient efforts of international movement of several countries, the world situation for cooperation is changing to promote.

In France, the CEA laboratory at Lemeil built a Nd glass laser, Phebus, 20kJ which is now open to the public use. Smaller Nd glass laser facilities exist at the University of Rochester, Ecole Polytechnique Palaiseau, the Shanghai Institute for Optics and Fine Mechanics and at several other places around the world. The significant progress has been made over the last five years. Since there is in principle no obstacle to prevent the goal in physics, more intense international efforts should be provided for exploring abundant and affordable energy.

The scientific works at Osaka is internationally recognized. In particular our contributions to ICF theory and experiment culminated in 1987 with remarkable achievement of the record of the compression density approaching $1kg/cm^3$.

In 1972, the Institute of Laser Engineering, Osaka University was established in accordance with the Edward Teller's special lecture on "New Internal Combustion Engine" at Montreal. And also we had timely the first Japan-US Scientific Seminar at Kyoto by the

Japan Society for Promotion of Science and the NSF which was an origin of the international collaboration on the inertial fusion research where 30 scientists from the U.S., Germany, Britain, Soviet Union and Japan gathered together. Our research on the laser plasma initiated in 1963 using ruby lasers and Nd glass lasers. The first issue of research was the laser-plasma coupling. The absorption mechanisms were thoroughly investigated a result of which was to propose the anomalous absorption caused by the plasma parametric instability. Nonlinear plasma instability due to the laser irradiation became a worldwide popular subject. We also investigated the self phase modulation of laser light by plasmas.

In 1975, we invented the so called indirect driven fusion concept "Cannonball Target" at our Daisen Summer Seminar which became later the Institute very popular in the world. At the age of oil crisis, the importance of new energy sources was well understood throughout the country. In a fair wind to fusion research we set LEKKO CO_2 lasers to the Los Alamos group and competed with the Livermore program by GEKKO glass lasers and compared the ideas to the Sandia team by REDEN beam machines.

In 1983 the world largest of glass laser GEKKO XII was completed by the cooperation of the NEC. As for the direct driven ICF, it is potentially more efficient but has significantly more stringent requirements on driver beam uniformity and the control of hydrodynamic instabilities. We had significant progress in this field using a novel type of uniform shell target and a random phasing smooth laser beam.

In 1985, the new idea of LHART (Large High Aspect Ratio Target) was devised by using an implosion simulation code. It could record a super shot of DT fusion neutron yield 10^{13} which was hurriedly after traced by the LLNL group.

In 1987, The green light random phasing 12 beam of GEKKO XII glass laser irradiated a plastic shell target of nearly perfect sphericity to attain the 1000 times normal density. The D-T fuel density reached $200 gr/cm^3$ in absolute. The plasma is some what Fermi degenerated. These details were reported at the IAEA conference in Nice at 1988. Ablative pressure generation and hydrodynamic behavior of compressed fuel were experimentally and theoretically investigated. The implosion performance was optimized by using an appropriate aspect ratio of the target and a suitable laser pulse. The uniformity of laser irradiation as well as the pellet structure was essentially important to avoid the growth of instability.

Since the Edward Teller's lecture at Montreal, it has passed 20 years to attain the high compression densities of fuel predicted. These results give us a high confidence that the ignition and burn of ICF will be attainable with a 300kJ blue laser driver.

Applications of inertial confinement fusion include not only civil energy production but also physics at the laser-atom interaction, nuclear matter under extreme conditions, cosmology, special isotope separation, food preservation, hydrogen production and advanced space propulsion. The pursuit of ICF will contribute substantially to overall scientific strength in several areas.

In the international collaboration, the essential advancement in research and the technology development for fusion shall be carried out in the following items,

1. High-average-power fusion drivers, lasers as well as heavy ion beams
2. ICF target for power-plant and fueling technology including cryogenic methods
3. Material for a reactor chamber and the technology of energy conversion

The development of the ICF reactor is essentially important. The material research including target, tritium and structural materials need an intense technology development to realize a fusion power plant by 2025. The international center for integrating a demonstration power plant of ICF shall be considered. No other alternate energy source holds a

Table 1. Main steps in the history of ICF

Year	Institution	Event
1963	Lebedev Physics Institute (USSR)	Proposal to use lasers for controlled fusion
1968	Lebedev Physics Institute (USSR)	Registration of thermonuclear neutrons in laser-produced plasma
1970	CEA Lemeil (France)	Definite neutron yield observed
1972	Livermore Natl. Lab. (USA) Los Alamos Natl. Lab. (USA) ILE (Japan)	Starting date for the financing for a national ICF program in the USA ICF program in Japan
1974	Lebedev Physics Institute (USSR) (FIAN) Institute of Applied Mathematics	Concept of low entropy compression of shell targets
1975	ILE (Japan)	Indirect drive Cannonball target concept
1977	Livermore Natl. Lab. (USA)	Launching of 10kJ Nd-laser "Shiva"
1978	Los Alamos Natl. Lab. (USA) ILE (Japan)	Launching of 10kJ CO_2-laser "Helios" Launching of 2kJ Nd-laser "GEKKO IV"
1979	Livermore Natl. Lab. (USA)	Density of compressed fuel researches 20g/cm^3
1983	Livermore Natl. Lab. (USA)	Launching of a 20kJ "Novette" 2Nd-laser
1983	ILE (Japan)	Launching of 30kJ Nd-laser " GEKKO XII"
1985	ILE (Japan)	Neutron yield 10^{13} LHART target
1985-1989	Livermore Natl. Lab. (USA)	Launching of 130kJ Nd-laser "Nova" (fuel density, 30g/cm^3; neutron yield, 3×10^{13})
1987-1901	ILE (Japan)	~1000g/cm^3 matter density reached with Gekko XII" laser facility
1994	Livermore Natl. Lab. (USA)	Approval of DOE NIF and declassification

bright promise than fusion and none has ever presented such formidable scientific and engineering challenges.

LASER PLASMA INTERACTION

As the condition for the ICF, we need the well understanding of the two items, the first one is the target physics issues such as laser absorption, energy transport, ablative pressure, stable compression, ignition and high gain of the fusion energy. The second item is the energy driver issues which include the properties of lasers and particle beams. These two items are essentially important for the inertial confinement fusion.

As a goal of the research, the ignition parameters are as follows: the ion temperature is more than 5keV, the compression of fuel density ρ/ρ_s is 500–1000 where ρ_s is the fuel normal density and the aerial density of fuel is larger than 0.3gcm^{-2} which is the α particle range.

Absorption

The electric field of the laser light makes a quivering motion of electrons which is thermalized by the collision with ions. This is the well-known inverse-Bremmstrahlung effect, often called classical absorption.

Increasing the laser intensity, the plasma temperature rises too high to make collisions between particles which leads to a low electric resistivity, that is, the weak absorption of the laser light. Actually the glass laser light of wavelength 1.06µm begins to decrease the absorption beyond the laser intensity 10^{13}W/cm^3 owing to a decrease in collisions. Laser light impinging into a plasma with a density gradient reaches the cut off density region and then reflects back from there.

If the laser intensity increases to much higher than a certain threshold the absorption increases again. This is so-called anomalous absorption which is introduced by nonlinear effects of coupling between the laser light and the plasma waves. This phenomena, which causes the generation of hot electrons, was experimentally found by the author.[1] In the 1970s the anomalous absorption became the main theme of laser plasma research. The parametric instability, Brilliouin scattering and Raman scattering due to the couplings between plasma waves and laser were intensely investigated. Obliquely incident laser light shows a strong resonant absorption when the laser light turns around at the cut-off point where the direction of the light electric field coincides with the direction of the density gradient to produce the plasma oscillation. Fig. 1 indicates general features of the laser light interaction with an inhomogeneous density plasma.

To determine the absorption coefficient, the electron collision frequency must be estimated which depends strongly on the electron temperature T_e. The balance of the laser heating and the thermal conduction loss has a key to determine the electron temperature. Then we can put the thermal flux of the plasma as $fn_{CT}V_eT_e$, where f is so called thermal flux limiting factor. The balance of the energy flow gives the electron temperature,

$$fn_{CT}V_eT_e = I_0\eta_{ab}$$

From this relation we can estimate the dependence of the absorption coefficient upon the incident laser intensity I_0. As an example, Fig. 2 shows the estimated absorption coeffi-

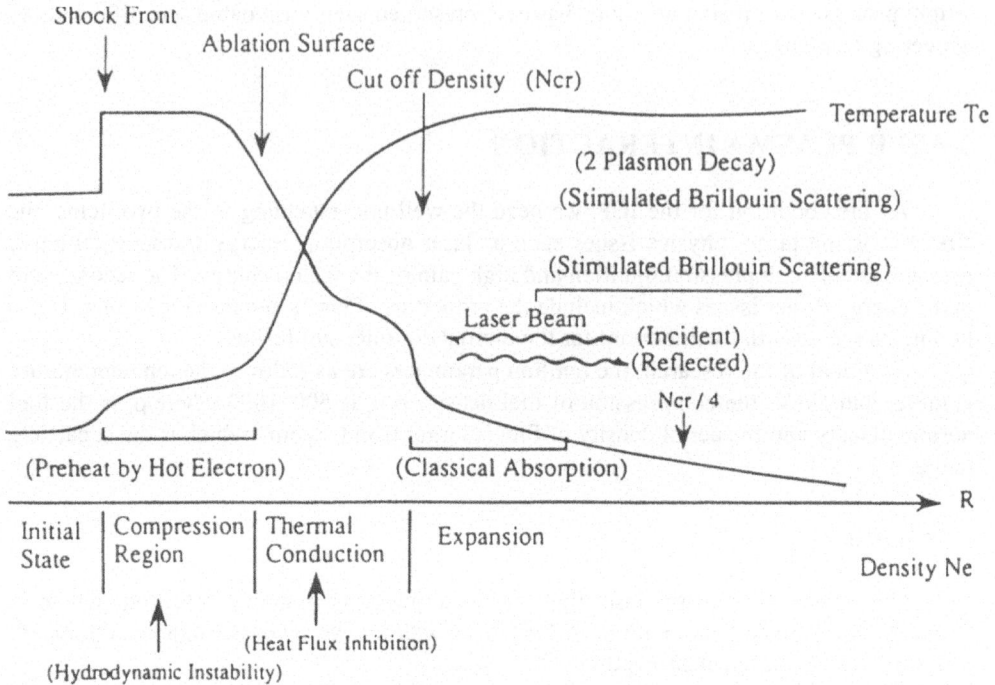

Figure 1. Interaction of laser with plasma.

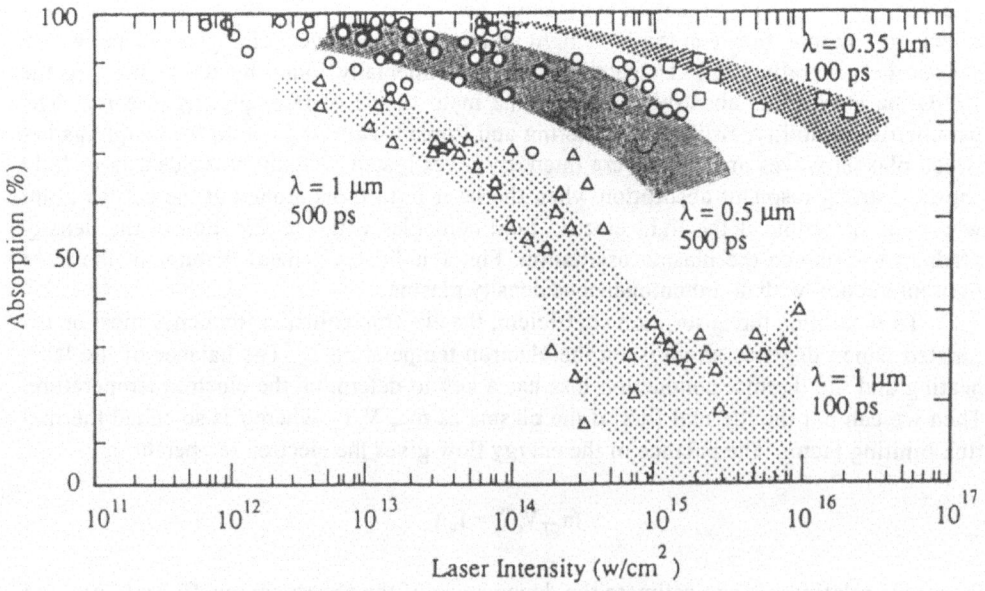

Figure 2. Absorption coefficient and its wavelength dependence.

cient related to the laser intensity with the density scale length L = 100μm and the laser wavelengths 1μm, 0.5μm and 0.35μm which are well accorded with experimental results.

The classical absorption decreases with the increase of the laser intensity as well as the increase of the laser wavelength. In assuming an expanding plasma, a simple model yields the total absorption coefficient η_{ab} in the form[2]

$$\eta_{ab} \propto \frac{f^{0.4}\left(Z^{3/2}\tau_L\right)^{0.6}}{I_L^{0.4}\lambda_L^2}$$

where Z is the ionization level, I_L is the laser intensity, λ_L is the laser wavelength, τ_L is the laser pulse duration and f is the thermal flux limiting factor.

When the classical absorption decreases at the higher plasma temperature produced by the stronger laser intensity, the laser light proceeds to the cut off density region, where the conversion from the laser light to the plasma waves is introduced by the resonance absorption and parametric instability. The plasma waves are then damped by the collective motion of plasma particles. This kind of processes is called Landau damping. The mode conversions from the light wave to the plasma waves are specified into two categories: a linear process and a nonlinear process.

Resonance Absorption

The resonance absorption is introduced by the obliquely incident P-polarized light wave when the density scale length is short in plasmas. In this case a component of the electric vector oscillates in the direction of the density gradient. As $E \cdot \nabla n_e \neq 0$, this oscillation generates fluctuations in plasma which can resonantly induce the electrostatic wave in plasma. This mode conversion from the light wave to the plasma wave finally heats up the plasmas due to the collective interaction between the plasma waves and the electrons. Since the energy is given selectively to suitable parts of the electrons, they obtain much higher temperature than that of the background electrons. They are called hot electrons and this absorption process is not acceptable to ICF, which intends compressing the fuel to an extremely high density by keeping the fuel cold during the laser irradiation.

Actually, the resonance absorption plays an important role in the density profile modification due to the ponderomotive forces of the laser light coupled with the generated plasma wave. The ponderomotive force induces the density scale length near the critical surface to be short which enhances the resonance absorption.[3] Appearance of the hot electrons has been checked by means of particle simulations, and a scaling law of the hot electron temperature T_h has been given as follows:[4]

$$T_h \approx 13 T_e^{1/4} \left(I_L \lambda_L^2\right)^{0.39}, (keV)$$

where T_e in keV and $I_L \lambda_L^2$ in $10^{15} W/cm^2 \mu m^2$. To prevent the generation of hot electrons, the value of $I\lambda_L^2$ must be kept roughly less than $10^{14} W/cm^2 \mu m^2$.

Parametric Instabilities

In laser induced plasmas, various parametric instabilities occur due to the couplings between waves. They are electromagnetic wave, electron plasma wave, and ion acoustic wave. When three of them satisfy the matching condition in energy and momentum, $\omega_0 = \omega_1+\omega_2$ and $k_0 = k_1+k_2$, then the wave energy of mode (ω_0,k_0) is converted to those of mode (ω_1,k_1) and mode (ω_2,k_2). According to the combination of three waves, the parametric instabilities are named as a) parametric decay instability, b) oscillating two stream instability, c) stimulated Brillouin scattering (SBS), d) stimulated Raman scattering (SRS), e) filamentational instability and self-focusing.

The stimulated scattering process plays an important role in plasmas with a longer scale length. The incident laser is reflected back but can not reach the higher density region for deposition energy to produce hydrodynamic pressure.

The two plasmon decay instability and stimulated Raman scattering have also been studied extensively, because they produce super hot electrons. The former is induced by the decay of the incident laser into two plasmawaves at the quarter critical density. In warm plasmas, the plasma waves with relatively long wavelength are predominantly excited. Since the phase velocity of such waves is very large, super hot electrons are efficiently produced.

The analytic treatment of SRS is similar to SBS where instead of the ion wave the electron plasma wave couples with the laser beam. In this case, the scattered electromagnetic wave can propagate in the backward and forward directions. The threshold intensity for SRS in forward is generally much higher than that for SRS in backward. However, SRS in forward produces the plasma wave with higher phase velocity, consequently generating a super hot electron with several MeV energy.

ENERGY TRANSPORT FOR IMPLOSION

The energy transportation from the laser absorbed region to the ablation region is one of the most interesting topics for the implosion process. There are two kinds of transport owing to electrons and radiations.

Electron Energy Transport

Electrons heated by laser light are transported into the region with lower temperature and higher density, while the low temperature electrons come back, forming the heat conduction regions. Basically, the heat flux due to the electrons is described with a diffusion type approximation. However, the plasma is so abruptly heated up by the laser in the gradient length of temperature which is of the order of the electron mean free path. Then we need a new model for the heat transportation instead of the usual diffusion scheme.

As the transport phenomenon belongs to the global hydrodynamics, the flux limited model has been used in the fluid simulations for obtaining a good agreement with experimental results. In this model, the heat flux is;

$$q_e = \frac{q_{SH} \cdot q_l}{|q_{SH}| + q_l}$$

where $q_{SH} = -K_e \nabla T_e$ and q_l is the limited flux defined by $q_l = f n_e v_e T_e$. The factor f is called flux imitation factor and is typically equal to 0.1–0.03.

Heat conduction described by the random walk theory has the diffusivity of a form; $X = \lambda_e v_e/3$, where λ_e is the electron mean free path. The Spritzer's heat flux is roughly given as;[5]

$$q_{SH} \approx \frac{\lambda_e v_e}{2} n_e \nabla T_e$$

in a uniform density plasma. Therefore, defining the temperature gradient scale length $L_T \equiv T_e / |\nabla T_e|$, the heat flux q_{SH} is written as;

$$q_{SH} \approx \frac{\lambda_e}{2L_T} n_e v_e T_e$$

From the above equation, it is easily understood that the heat flux for $L_T/\lambda_e < 1/(2f) \approx 0.1$ can not be described by the Spitzer's formulation.

Physical models of the flux limiter f have been attributed to various anomalous effects; ion wave turbulence[6] which reduces the electron mean free path λ_e, electrostatic field generation due to a hot electron driven return current,[7] magnetic field generation, and so on. Although these models seem interesting, the study of transport has now focused on a more accurate treatment of the Fokker-Planck equation in a steep temperature gradient.

Radiation Transport

In highly ionized plasmas, the energy transport by X-ray radiation appears to be very important compared to the electron energy transport. Radiation processes in medium or high Z plasmas essentially depend upon the atomic state of the plasmas. To study the atomic state of partially ionized plasmas it is required to deal with not only the radiation transport but also fluid dynamics including the equation of state, ionization level, and so on. The radiation transport has been studied extensively in astrophysics.[8] Radiation transport is described by the equation;

$$\frac{1}{c}\frac{dI_\nu}{dt} = \varepsilon_\nu - K_\nu I_\nu$$

where I_ν is the radiation intensity with the energy $h\nu$, ε_ν the emission rate, and K_ν the absorption coefficient. In assuming local thermodynamics equilibrium, the right part is given to be $K_\nu'(B_\nu - I_\nu)$, where $K_\nu' = K_\nu(1-e^{-h\nu/kT})$ is the total absorption coefficient including induced emission and $B_\nu = 2h/c^2 \cdot \nu^3 (e^{h\nu/kT}-1)$ is a Planckian distribution. The absorption coefficient K_ν is due to free-free, bound-free and bound-bound transition processes of the electronic state, and requires the detailed information of the electronic state in the plasmas and atoms.[9]

In solving radiation process in a fluid code, the averaged atom model or opacity tables are commonly used.

IMPLOSION AND IGNITION

A scheme of target implosion driven by the laser irradiation is schematically shown in Fig. 3.

This is for the case of a shell target and the time evolution of the implosion dynamics can be driven into four phases

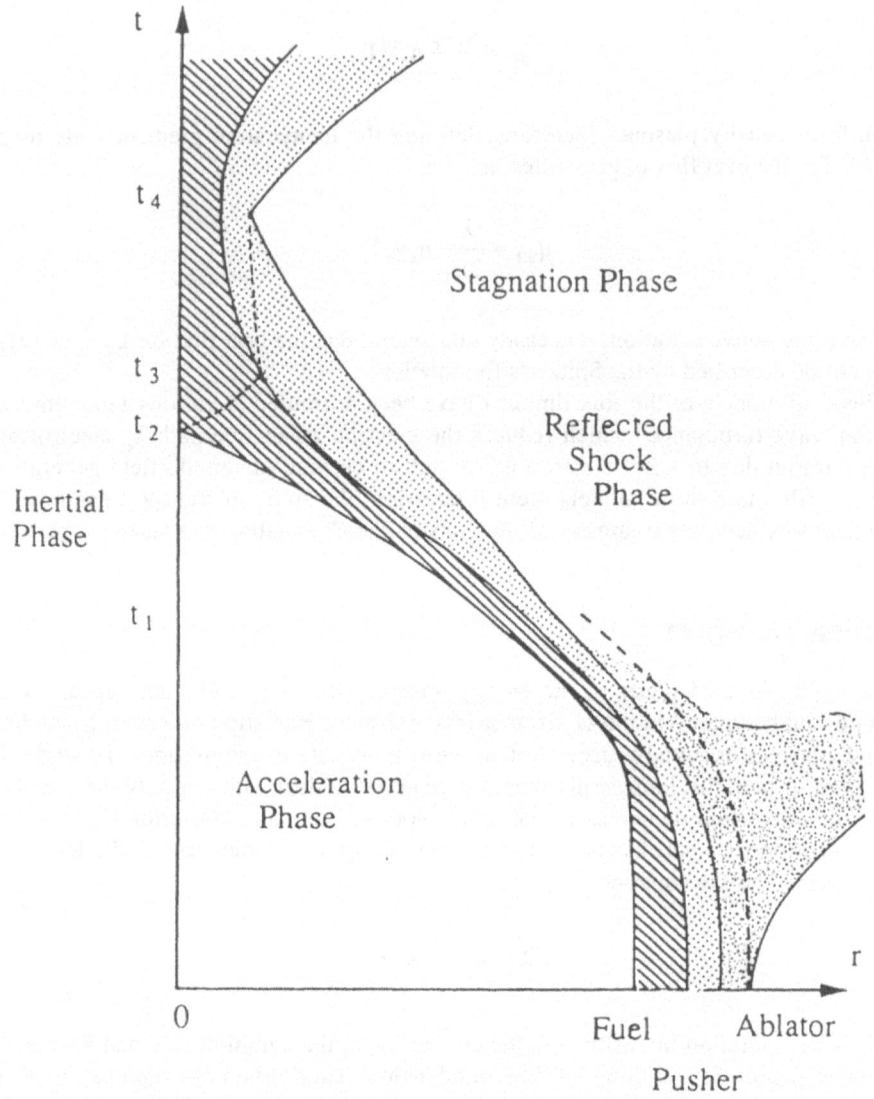

Figure 3. Schematic diagram of ablative driven implosion of a shell fuel target.

1. Acceleration Phase ($0<t<t_1$),
2. Inertial Phase ($t_1<t<t_2$),
3. Reflected Shock Phase ($t_2<t<t_3$),
4. Stagnation Phase ($t_3<t<t_4$).

In the acceleration phase (1), the laser is absorbed to ablate the outer region of the target material and the rocket repulsion drives the pusher and fuel to accelerate inward. In the inertial phase (2), the fuel and pusher freely fall toward the target center at an accelerated velocity V_A, the value of which is required to be $2-3 \times 10^7$ cm/sec.[10] In this phase, the fuel and pusher tend to expand due to their thermal pressure; therefore less preheating of these regions is necessary for the ideal ablative implosion. In the reflected shock phase (3), a void closure of shell triggers a shock wave which propagates outward, heating the fuel by converting directed kinetic energy to thermal energy. At the time when the shock wave arrives at the contact surface, the so-called stagnation phase (4) begins. If at $t = t_3$ the dynamics pressure (ρu^2) of the pusher is larger than the static pressure (P) of the fuel, such discontinuity of the pressure leads to the generation of transmitted and reflected shock waves. Then, the fuel is further compressed and gains energy from the pusher through adiabatic compression in the stagnation phase, if no mixing occurs.

Ablation Process of Pellet

When one considers the ablation of a low-Z material, the electron energy transport plays an important role. Ablated plasmas expand into the vacuum with the velocity of sound $C_s = (P/\rho)^{1/2}$. Then, the so-called deflagration form is introduced in the heated region.[11] We get a scaling law of the ablation pressure;

$$P_A = P_0 f^{1/3} (I_{abs}/\lambda_L)^{2/3}$$

where P_0 is a constant and λ_L is the laser wavelength. In the model of stationary Chapman-Jouguet deflagration wave,[12] the constant is shown to be

$$P_0 f^{1/3} = 12 \text{ Mbar},$$

where I_{abs} is in the order 10^{14}W/cm^2 and λ_L is in μm units. In Fig. 4, the ablation pressure experimentally obtained are plotted, where the solid lines show the scaling above for $\lambda_L = 0.53$ and 1.06μm. The mass ablation rate $\dot{m}$ is also calculated

$$\dot{m} = 1.5 \times 10^5 \, I_{abs}^{1/3} \lambda_L^{-4/3} \text{ (g/cm}^2 \text{ sec)}.$$

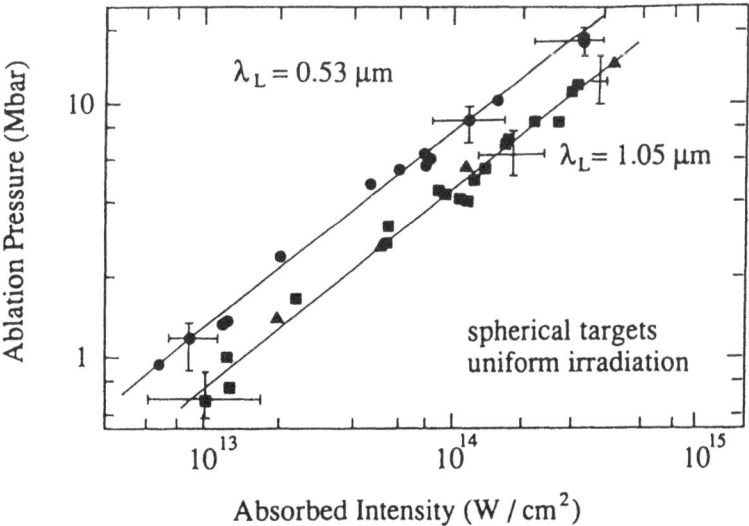

Figure 4. Ablation pressure versus absorbed laser intensity.

Rocket Model for Implosion

When the acceleration of the compressed region has areal mass $M(t)$ and is accelerated to the velocity $V(t)$ at time t, the equation of motion is reduced to the well-known Rocket equation;

$$M(t)\frac{dV(t)}{dt} = P_A$$

In obtaining the above equation, a plane geometry is assumed. Since $M(t) = M_0 - \int_0^t \dot{m}dt$, where M_0 is the initial mass, the equation given above is integrated with the assumption of constant $\dot{m}$ and P_A;

$$V(t) = 2C_s \ln\left(\frac{M_0}{M(t)}\right),$$

where the relation $P_A/\dot{m} = 2C_s$ is used.

Hydrodynamic efficiency is defined by the ratio of the kinetic energy of the accelerated fuel divided by the absorbed laser energy. By relation $I_{abs} = 4\dot{m}C_s^2 \sim 2P_A C_s$, the hydrodynamic efficiency η_H is given by

$$\eta_H = \frac{\frac{1}{2}M(t)V^2(t)}{I_{abs} \cdot t} = \frac{1}{2}\frac{\Phi(\ln\Phi)^2}{1-\Phi}$$

where $\Phi = M(t)/M_0$. It is found the η_H has its maximum at Φ = 20–30%.

Now, we put $\Phi = 1/e$ (~0.37) for obtaining peak value of η_H. Then, we get

$$V = 2C_s, \, d = V^2 M_0/P_A$$

where d is the travel distance. Then we find the following relation between P_A and V;

$$P_A = \rho_0 V^2 \left(\frac{R_0}{d}\right)\left(\frac{\Delta R_0}{R_0}\right)$$

where ρ_0 and ΔR_0 are the initial density and the thickness of a target shell respectively. Regarding R_0 as an initial target radius, $R_0/\Delta R_0$ is the initial aspect ratio of a target. As an example, assuming a target which consists of pure DT shell ($\rho_0 = 0.2 g/cm^2$), the required ablation pressure for obtaining $V = 3 \times 10^7 cm/sec$ is calculated to be 40Mbar, where we set $R_0/d = 2$ and $R_0/\Delta R_0 = 10$. And it is found that the larger aspect ratio ($R_0/\Delta R_0$), the less ablation pressure necessary to accelerate the target. It is noted that from the laser intensity required for P_A = 40Mbar is $I_{abs} = 6 \times 10^{14}\lambda_{\mu m}$ (W/cm^2). Then a condition for the classical absorption $I_L \lambda_{\mu m}^2 < 2 \times 10^{14} W/cm^2$ which prevents the generation of the hot electrons required the wavelength $\lambda_L < 0.33 \mu m$. This is one of the reasons why shorter wavelength lasers have been prepared for implosion experiments.

Instability of Ablation and Stagnation

Instability in the implosion process is a crucial problem for the ICF. We have two cases of instability in the acceleration phase and in the deceleration phase. To prevent the growth of the instability is very important to pursue the implosion.

Ablation Mode. It is important to analyze the Rayleigh-Taylor instability of the dynamics in order to answer whether the implosion scheme is realistic within the symmetry assumption. At the acceleration phase the interface between ablator and pusher is unstable to the Rayleigh-Taylor instability, the growth rate of which is $\gamma = \sqrt{kg}$, where k is the wave number of a perturbation and g is the gravitation which is equal to the acceleration of the ablation front. However, recent studies[13] reveal the fact that the growth rate is reduced due to the ablation and the thermal conduction. An eigenvalue analysis in the self-consistent structure gave the growth rate in a tractable form;

$$\gamma = \alpha\sqrt{kg} - \beta k V_0$$

where $\alpha = 0.9$, $\beta = 3-4$ are constants and V_0 is the flow velocity across the ablation front, which is equal to $V_0 = \dot{m}/\rho_a$, where ρ_a is the maximum density at the ablation front.

Stagnation Mode. In a stagnation dominant implosion mode, the fuel with lower density decelerates the pusher of higher density. The fuel-pusher contact surface tends to be unstable due to the Rayleigh-Taylor instability. This instability has been studied by Hattori et al.,[14] who solves an eigenvalue problem for linear pertubation from a self-similar solution which used the stagnation dynamics in a spherical geometry. They obtained a formula describing the temporal evolution of the perturbation amplitude. An explosive growth of the instability is seen in the stagnation phase, possibly inducing fuel-pusher mixing. In order to avoid the instability accompanied by the stagnation dynamics, it is required to ablate away the pusher until $t = t_3$ for decreasing $\rho_{p,3}$, to generate shock-multiplexing for increasing $\rho_{f,3}$, and to accelerate the shell up to 10^8cm/sec for heating the fuel up to 5–10keV by the reflected shock wave. Such implosion mode is named "Stagnation-free implosion". The world record of neutron yield is attained by this mode using "LHART" target.[15]

Ignition and Burn

A high pellet gain of about $Q = 10^2$ is required for realizing a reactor of ICF. This is because the driver efficiency η_d and coupling efficiency η_c are limited to be $\eta_d\eta_c \ll 1$ and a core gain G = (thermonuclear energy)/(imploded DT fuel energy) in the order of 10^{13} is necessary for a net gain. G of the thermonuclear reaction is calculated to be

$$G \approx 300 f_B,$$

where f_B is the burn fraction. The core gain G cannot get the value 10^3 in a uniform heating. The high gain concept requires an ignitor scheme by which the compressed DT plasma have an inner spark ignition with the temperature 5–10keV and an outer main fuel region with the internal energy as low as possible. Then, the alpha particles produced in the spark region heat the main fuel to trigger the nuclear reaction. The ICF research is requested to demonstrate at first the ignition of the fuel and next to attain the high pellet gain.

Ignition Condition. The ignition condition is given from the energy balance of the DT plasmas which are assumed to be uniformly heated and compressed. A number of simulations of the fusion burn have been preformed by Fraley et al.[16] with uniformly compressed DT fuel. As well-known, a simple estimate of the fusion burn fraction without taking into account the self-heating is[17]

$$f_B = \frac{\rho R}{\rho R + \beta(T_i)},$$

where $\beta(T_i) = 8m_i C_s/\langle\sigma v\rangle_{DT}$ is a function of only T_i. It is noted that for $\rho R \ll \beta(T_i)$, $f_B \approx \rho R/\beta(T_c)$ and the burn fraction increase proportionally to ρR. This ignition point is given by the ρR value 0.4gr/cm^2 at which the f_B increases.

Necessary Energy for Ignition. The DT plasmas with areal density ρR and temperature $T_i(= T_e)$ have internal energy U_{DT};

$$U_{DT} = 10.6 \times 10^9 \frac{1}{(\rho/\rho_s)^2} (\rho R)^3 T_k \quad \text{(Joule)}$$

where ρ_s is the solid density and ρR is in g/cm^2. For a coupling efficiency of 10%, the driver energy of 10kJ is enough for ignition. If a spherically symmetric implosion was done, U_{DT} = 1kJ is an ignition condition at $\rho/\rho_s = 10^3$ for $(\rho R)^3 T_k = 0.1$. However, the implosion symmetry is a key issue and estimation of U_{DT} including the asymmetry effect is essential. As for the direct drive, a substantial ignition may be possible at the nonuniformity of 2–3% in implosion for a few 100kJ driver with coupling efficiency 10%.

In the case of the indirect drive, the coupling efficiency is less than 1%, however, the uniformity of the core is much better than the direct drive case. The required driver energy is expected to be a few MJ. The choice of the direct drive or the indirect drive is a controversial issue for the ICF research.

EXPERIMENTAL STATUS

In the 1970th the laser plasma interaction was one of the most important research items using available lasers of a modest size. These experiments on laser plasmas gave us the basic understanding of laser coupling to plasmas. In the 1980s several huge lasers enabled us to perform the implosion experiments. For the ICF, important issues to investigate accompanied by several items are shown as follows:

1. Absorption mechanism: plasma waves, saturation of instability, electron temperature, ion temperature.
2. Energy transportation: density gradient, electron temperature, X-ray radiation temperature, flux limiting factor.
3. Implosion process: shock velocity, uniformity, mass ablation rate, density, areal density, core size, temperature.

4. Ignition and burn: neutron, α particle, reaction rate, energy yield.
5. Energy extraction: energy balance, efficiency.

To get the information on these items, we have performed several experiments on the fundamental properties of the laser plasma.

Driver Technology

The concepts and system designs of present drivers for inertial confinement fusion experiments are described. An overview of the current status of driver devices in research institutes in the world is given.

Lasers. For glass lasers, Nd doped glass is used as the active medium. Since laser glass can provide a large stored energy as the population inversion, and a large scale laser glass with high homogeneity can be easily obtained, many high power glass laser systems have been constructed in several institutes for ICF experimental research.[18]

The principal high power glass laser systems are summarized in Table 2.

The MOPA (Master Oscillator Power Amplifier) system is generally adopted where several hazards are solved. These are: parasitic oscillation in the amplifiers; the beam degradation due to the diffraction in the long optical path; and the amplification of the retrore-

Table 2. Laser scale glass laser system for ICF in the world

Country	System	Laboratory	Beam number	Output power (TW)	Output energy (kJ)	Pulse width (ns)	Experimental Results
USA	NOVA	LLNL	10	100	120 (ω) 80 (2ω) 70 (3ω)	0.1-3.0	10^{13} neutrons
	OMEGA	LLE U.Rochester	24	15	3 (ω) 2 (ω)	0.1-1.0	10^{10} neutrons, ×200Liq.Dens
	PHAROS-III	NRL	3	1.3	1.4 (ω) 0.8 (2ω)	0.1-1.0	
	CHROM-I	KMS	2	0.6	1	0.1	
Russia	MISHEN	Kurchatov	4	-	1	1.0	
	DELFIN 2 AURORA	Lebedev	216 20	33 -	10 50-500	0.2-3.0 0.03-1.0	planning
Japan	GEKKO-IV GEKKO-XII GEKKO-XII	ILE Osaka	4 2 12	2 7 55	1 2 30 (ω) 20 (2ω) 15 (3ω)	0.1-1.0 0.1-1.0 0.1-1.0	10^8 neutrons 2×10^8 neutrons 10^{13} neutrons, ×1000Liq.Dens
UK	VULCAN	Rutherford Appleton	12	3.6	5(ω) 2(2ω)	0.1-1.0	
France	PHEBUS OCTAL	Limeil	2 8	20 2	20(ω) 1	0.1-3.0 0.1-1.0	
	GRECO	Ecole Poly.Tec.	1	0.25	0.25	0.1-2.5	
China	SHENG GUANG	Shanghai	2	2	2	0.1-1	

Figure 5. Main amplifier chain in the GEKKO XII glass laser system.

flected laser pulse passing through the amplifiers. To avoid these kinds of problems, the laser amplifier chain should be provided an isolation between the amplifiers, the image relay to compensate the diffraction[19] and for rejection of the retro-pulse.[20] According to these requirements, the laser amplifier chain usually consists of several functional components. Here the GEKKO XII glass laser system of Osaka University is shown as an example of a large scale glass laser for fusion applications.[21]

Gekko XII Laser System. The 12 beam GEKKO XII glass laser system is shown in Fig. 5 and the irradiating target system is also given in Fig. 6. This experimental system consists of a laser amplifier chain, 12 beam switching mirror system and two target irradiation chambers. The laser amplifier chain has a large number of components such as the oscillator, the preamplifier chain, the beam dividing optics, the main amplifier chain, the laser alignment system, the laser control system and the laser energy monitoring system. The laser amplifier chain is set for the generation of high power laser pulses with good temporal and spatial profile. The beam switching system can swing the beams from one target chamber to another in a few seconds. The target irradiation system is responsible for focusing on the two target chambers by green or blue laser light. This irradiation system consists of the tuning mirrors, the frequency conversion crystal, the random phase plate, the focusing lens, the beam energy monitor and the irradiation alignment system.

Frequency Conversion. In ICF experiments, shorter laser wavelength light has several advantages in the plasma interactions.[22] Recently, large scale glass laser systems for ICF have frequency conversion systems using nonlinear optical crystals (KDP: Potassium Dihydrogen Phosphate: $KH_2 PO_4$) for the second, third and fourth harmonic conversion.[23] The KDP crystal has several merits in the availability of a large size, high homogeneity, low absorption coefficient and high damage threshold single crystal. In the GEKKO XII laser system, second (2ω) and third (3ω) harmonic generation system are installed to the first and second target chamber, respectively. The conversion efficiency is up to 80% for 2ω and 60% for 3ω.

Progress in Inertial Fusion Research   261

Figure 6. Target irradiation chamber of the GEKKO XII glass laser system.

Uniform Irradiation Schemes. In ILE, the random phase plate is invented to suppress the irradiation nonuniformity of the coherent laser beam.[24] Several alternative methods such as induced spatial incoherence using echelons with a broadband oscillator[25] and multi-lens irradiation concept are investigated.[26] Table 3 shows the development of beam smoothing technologies in various laboratories. The uniform irradiation to prevent the instability during the implosion is a key issue for laser fusion. The effect of the random phase plate and its related technology for beam smoothing is given in Fig. 7. The drive uniform can be also improved by using the large beam numbers and also by optimizing the beam arrangement in geometrical as well as the power balance of the beams. The necessary absorption non uniformity shall be expected in the level of 1%. For the direct drive this level is essentially important. In Fig. 8, these data are indicated.

Table 3. Development of beam smoothing technologies in direct drive implosion

				near future
ILE	Laser+RPP (1983) beam segmentation	ASE+RPP (1989) broadband spatial incoherency	ASE+RPP with angular dispersion (1991)	envelope control polarization control multi aspherical lens array
NRL	ISI (1983) by echelon	echelon-free ISI (1987)	ASE with complete image relay	envelope control
LLE		1D E-O SSD +RPP (1989)	2D E-O SSD +RPP spectral angular dispersion	60 beams zero correlation mask
LLNL			Noisy SSD +RPP	envelope control Kinoform phase plate
Limeil		Optical fiber smoothing temporal and spatial incoherency by optical fiber		

key issue	beam segmentation speckle pattern	→ spatial and temporal incoherency →	beamlet profile control multi beam irradiation

Particle Beams. As discussed in the previous section, the laser drivers have advantages for generating short pulses (order of ns) and high peak power (up to 100TW) with good focusability for small fusion targets (less than 1mm). However, the laser devices have several disadvantages such as high construction cost, low efficiency and difficulties in obtaining the larger output energy. On the other hand, the pulsed power technology can easily generate a large amount of energy with high efficiency and low cost. It may provide a light ion beam for an ICF driver if the ion beam could be bright and transported to the fusion target efficiently.

Since the light ion beam accelerator has the advantage of larger energy deposition rate onto the fusion target as compared with a relativistic electron beam (REB), attempts have been made for developing large scale LIB systems in many countries as shown in Table 4.

For the ion beam diode to produce a bright source is essentially important. Various concepts are proposed.[27] The beam transport to the target is also a crucial issue.

The rapid development of ICF research using lasers and LIB stimulated the application of the heavy ion beam to the ICF driver.[28] This scheme largely depends on the accelerator development by which principally high repetition rate and high efficiency operation are possible. The present output of such a heavy ion beam accelerator is low and the cost is very high. An ambitious program has been performed for constructing the heavy ion RF linac and the storage ring system in GSI of Germany.[29] The induction linac is another promising accelerator of heavy ion because the output power is high and the cost of the hardwear is relatively low. In Lawrence Berkeley Laboratory, the output energy 3.3MJ is designed and the beam transport experiment is going on.[30]

Interaction Experiments

In Fig. 2 the absorption properties of a plasma due to the different wavelength of lasers are shown. The anomalous absorption appears beyond the range of $I\lambda^2 \approx 10^{14} W/cm^2 \cdot \mu m^2$ where

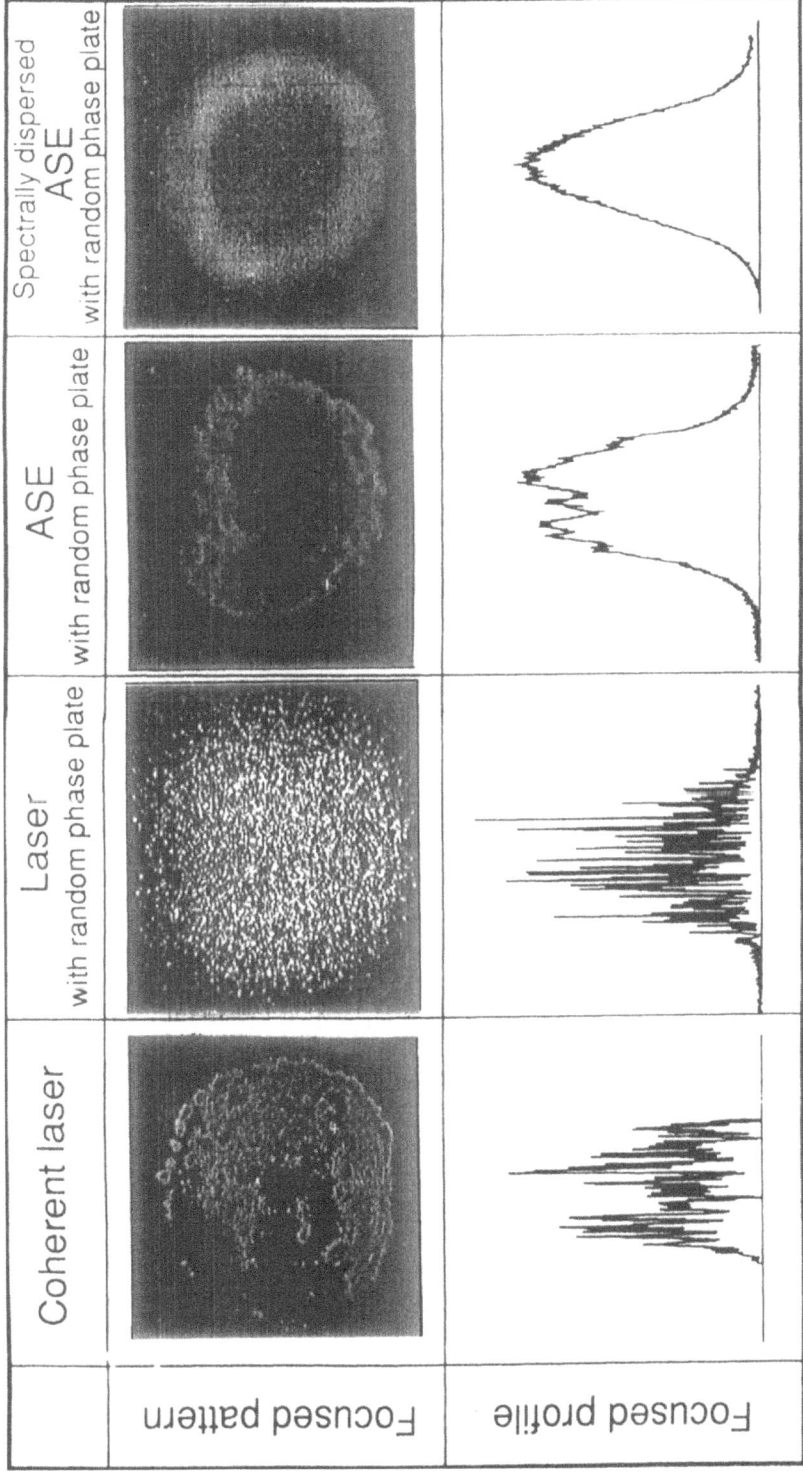

Figure 7. Speckle structure of focused pattern is reduced due to introduced incoherency.

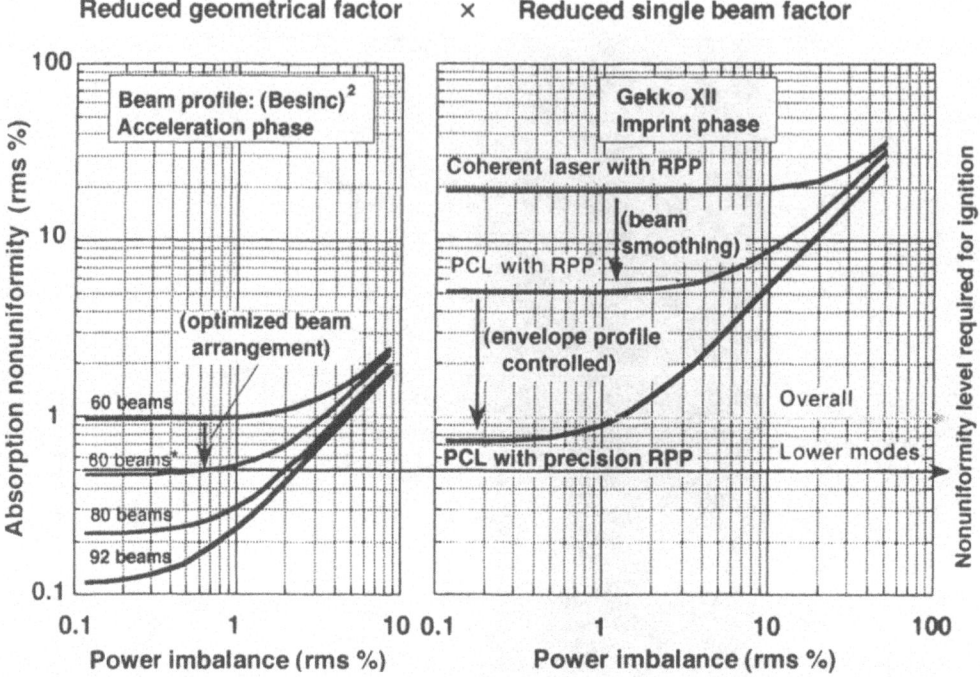

Figure 8. The drive uniform can be improved by the larger beam number, the optimized beam arrangement (geometrical factor), the beam smoothing, the envelope profile control (single beam factor) and the power balance.

I (W/cm^2) is the laser intensity and λ is the laser wavelength. Anomalous absorption has a tendency to produce hot electrons due to the nonlinear process of coupling which is easily introduced by the longer wavelength laser. As well known, the hot electrons induce the preheat of the fuel which prevents the efficient compression of the fuel pellet. This is a reason why the short wavelength lasers become important instead of the highly efficient CO_2 laser. The classical absorption is dominated by the shorter wavelength laser light.

In Fig. 4 and 9, the ablation pressure due to the laser absorption and the compressed fuel density versus the ion temperature are given comparing the effect of the wavelength of lasers respectively. These results also indicate the importance of the shorter wavelength laser.

Direct Driven Experiments

The direct implosion scheme is so efficient at ablating the target that the modest size laser is available for the experiment. The key issue for this scheme is to attain uniform compression which needs uniform irradiation of laser as well as high uniformity of the fuel pellet. To eliminate the nonuniform irradiation due to the diffraction of the coherent laser light, several new methods to suppress the cohere..cy of laser light are introduced such as the random phase plate, the induced spatial incoherence grating with a broadband oscillator and the multi-lens irradiation concept.

As for the pellet uniformity, super high uniform pellets are prepared by glass or plastic microbaloons. The cryogenic DT target is also under investigation for this purpose. Several types of target are shown in Fig. 10.

Table 4. Specifications of large scale LIB systems for ICF application

Country	System	Laboratory	Output parameter	Ion beam output	Diode	Remarks
USA	PROTO I	Sandia (1975)	2.0MV, 0.5MA, 24ns	0.5TW	MID	Foil Cone Target Classical Interaction
	PROTO II	Sandia (1876)	1.5MV, 6MA, 24ns	—	—	—
	PBFA I	Sandia (1980)	2.4MV, 17MA, 24ns	—	—	—
	PBFA II	Sandia (1985)	10MV, 3MA, 40ns	—	—	—
	GAMBLE II A	NRL (1974)	1.0MV, 1.3MA, 80ns	0.4TW	Pinch-Reflex	Disc Target, Classical Interaction and Hydrodynamic Properties
	HERMES III	Sandia (1988)	20MV, 800kA	—	—	—
Russia	ANGARA V-M	Kuruchatov (1979)	2.0MV, 4MA, 80ns	—	—	8 module
	ANGARA V	Kuruchatov	(2.0MV, 25MA, 80ns)	—	—	plan
FRG	KALIF	KfK (1981)	1.9MV, 0.8MA, 50ns	0.5TW	Pinch-Reflex	—
France	SIDNIX	Valduc	1.3MV, 0.7MA, 80ns	—	—	—
Japan	REIDEN IV	Osaka (1979)	1.0MV, 1.0MA, 60ns	0.2TW	Pinch-Reflex	—
	REIDEN SHVS	Osaka (1986)	4MV, 40kA, 100ns	0.1TW	MID	Thin Film Target Classical Hydrodynamics
	ECHIGO	Nagoya (1980)	1.0MV, 0.2MA, 50ns	0.01TW	MID	—

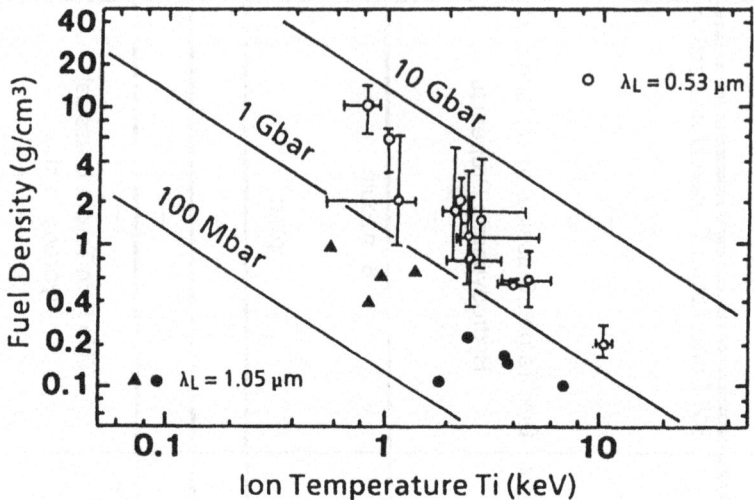

Figure 9. Compressed fuel density versus ion temperature by laser irradiation.

High Neutron Yield Target. Recently we have succeeded in performing a high neutron yield experiment using a novel target LHART: Large High Aspect Ratio Target. This target is a glass microbaloon with an extremely thin shell which contains a fuel gas. When it is irradiated by a long laser pulse, almost all of the pusher layer is exploded until it reaches the maximum compressed condition state. Then, the stagnation free compression is introduced without the pusher-fuel mixing which is introduced by the Rayleigh-Taylor instability.

This scheme was realized by using 'LHART'. In 1986, the D-T neutron yield of 10^{13} was attained by the green beam 13kJ, 1ns, irradiation of GEKKO XII glass laser to the glass microbaloon, diameter 1,235μm, shell thickness 1.31μm, aspect ratio 471 and enclosed DT fuel 6.2 atm. The pellet gain reached 1/500. Fig. 11 shows the neutron yield versus the pellet aspect ratio. From the observation of X-ray pinhole pictures of pellet implosion, using a multi framing camera the image of compressed was clearly recorded for high neutron yield. The largest neutron yield is attained at the aspect ratio 400~500.[31] For the case of the aspect ratio smaller than 400, well known ablative compression is introduced where Rayleigh-Taylor instability becomes dominant to induce the fuel-pusher mixing and to suppress the neutron yield. When the aspect ratio is larger than 500, the shell pellet is thin enough to introduce the burn through of the pusher which prevents the uniform compression. In the case of an aspect ratio of about 450, the stagnation free compression is introduced to get the maximum neutron yield which accords very well with the simulation data given by the ILESTA code. Mixing rate 10% can fit the experimental data.

High Compressed Density Target. A gas-filled target has some limitation to attain the high compressed density. Shell fuel targets such as cryogenic DT shells and plastic CDT shells are very interesting. We have imploded super uniform plastic fuel CDT shells to get the high compressed fuel density. The target has a diameter of 500μm, shell thickness 6~12μm, tritium content >500Ci/cm³ and 5wt % Si added which are assumed constant during the implosion. The green beam 8.5kJ, pulse duration 2ns from GEKKO XII laser with random phase plates is irradiated uniformly. After irradiation we can estimate

Target	η abs	η c
Ablative target 	50 %	1 %
	Higher efficiency by blue	
	High uniformity requirement	
LHART 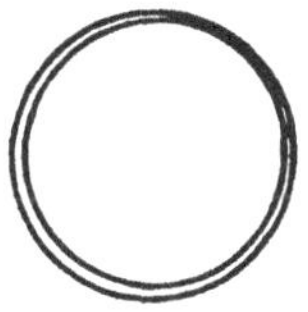	70 %	6 %
	Higher efficiency by blue	
	Stagnation free	
Double shell 	50 %	2 %
	Uniformity improved	
	Velocity multiplexing	
Cannon ball 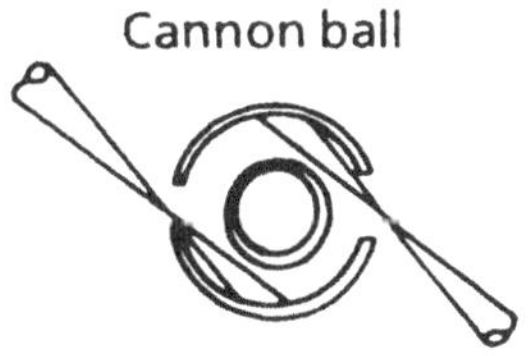	70 %	6 %
	High efficiency	
	Uniformity improved	
Double shell in cannon 	70 %	1 %
	High efficiency	
	Uniformity greatly improved	

Figure 10. Several types of laser fusion target. Laser absorption η_{abs} and core coupling efficiency η_c are given.

the plasma ρR value by the silicon activation method which was reported recently. In this case the activation reaction ^{28}Si $(n,p)^{28}$Al is used.

As well known, the activation is proportional to the neutron yield and plasma ρR. At the same time, multilayer labelled shells with a coating of Mg, Si, Al can show the ablated mass quantity ΔM by the X-ray streak photography. From the experiment we have data on the neutron yield, the Si activation, ρR and mass ablation ΔM for imploding polymer shells which allow us the final fuel density ρ.[32] The measured minimum density ρ_{fuel} is 200gcm^{-3} that is about 1000 times the liquid fuel density. These data are given in Fig. 12 and Fig. 13.

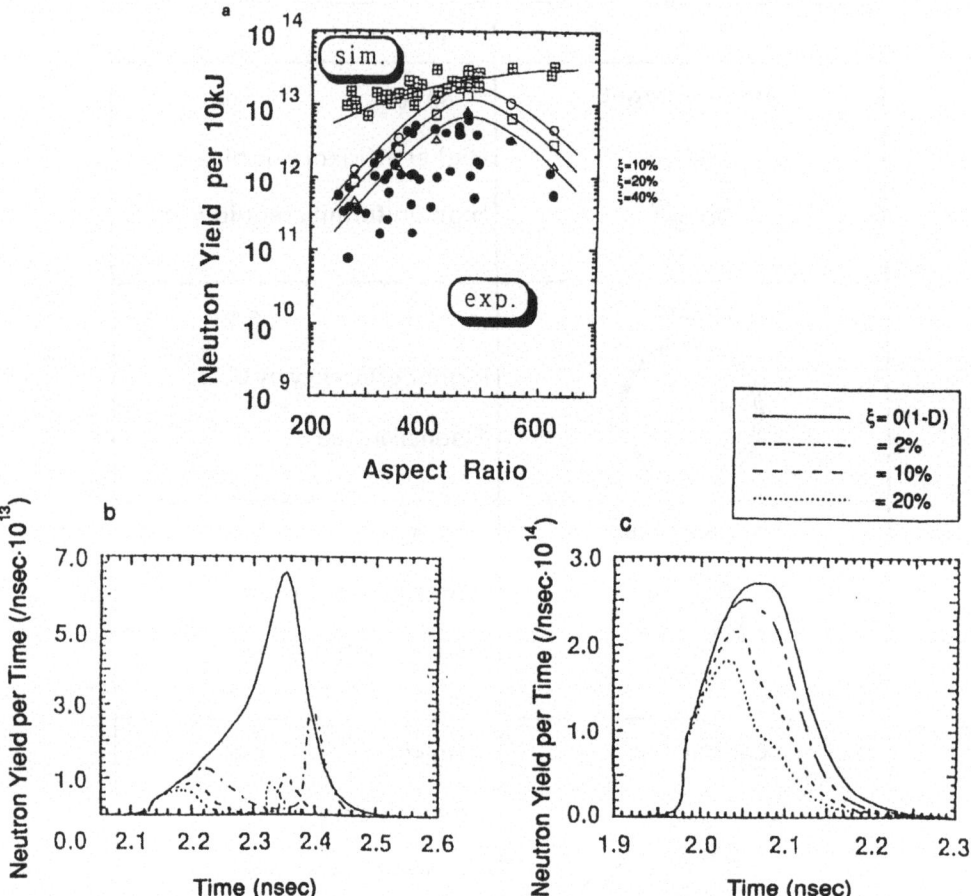

Figure 11. Implosion code with turbulent mixing (ξ = 10–20%) can reproduce neutron yields in LHART-type implosion.

As for the measurement of ablated mass ΔM, one can use the characteristic X-ray emissions from multilayer labelled shell targets.[33] The data on $\langle \rho R \rangle$ and ΔM allow us to determine the final compressed core density ρ_{fuel} by the next relation.

$$\rho_{fuel} = \sqrt{\frac{4\pi/3 \langle \rho_{fuel} R \rangle^3}{(M - \Delta M)_{fuel}}}$$

In Fig. 14 the regions in which knock-on[34] and secondary fusion reaction[35] methods to determine $\langle \rho R \rangle$ can be applied are schematically illustrated. The lower boundaries are set by the sensitivity of measurements. Upper boundaries of the knock-on and the secondary reaction methods are imposed by the stopping of charged particles in the fuel.

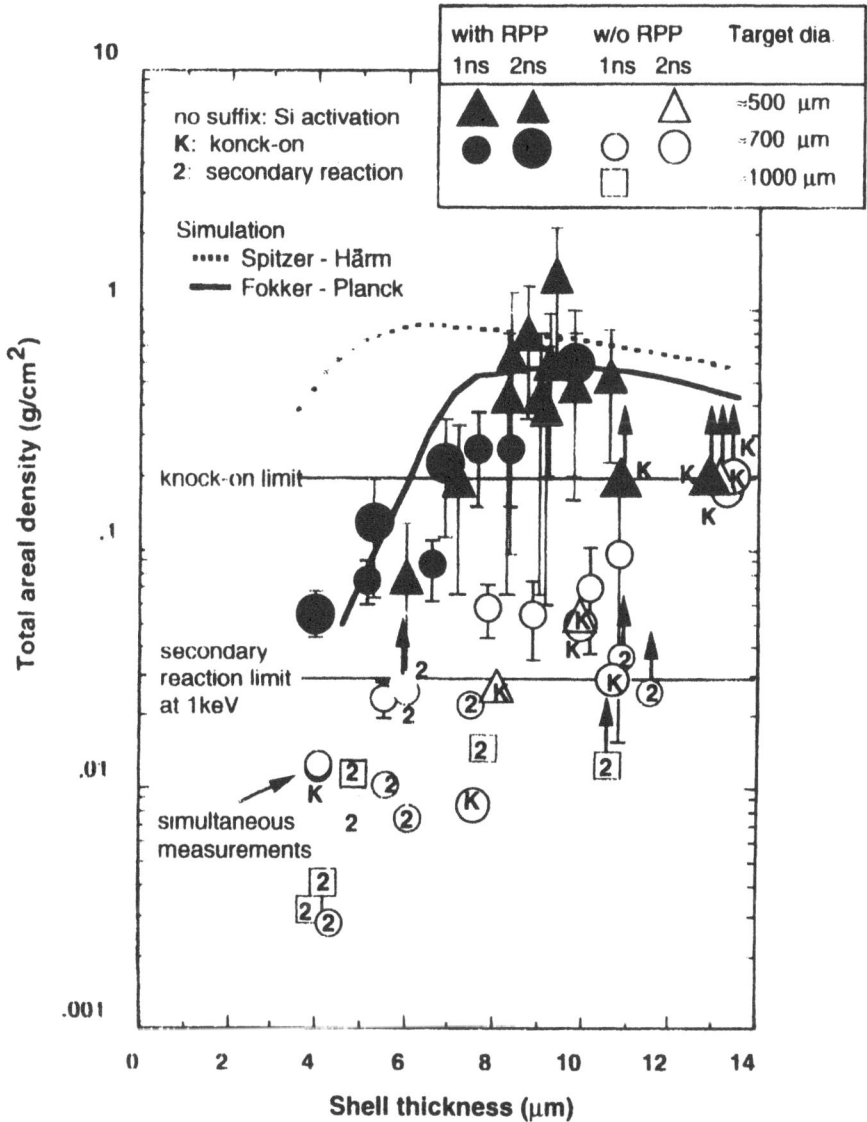

Figure 12. High areal densities were measured with Si activation technique.

Indirect Driven Experiments

Indirect drive is an approach to ICF that relaxes the requirements of implosion uniformity for the direct drive and simplifies the laser configuration of irradiation. These are two categories of indirect driven target: the plasma drive cannonball target and the radiation drive target. The cannonball target is fundamentally a double shell structure, the cavity of which is filled by the plasma or the X-ray which is produced by the impinging laser. These targets are shown in Fig. 10.

The hydrodynamic efficiency of the plasma driven cannonball is proportional to cube of the aspect ratio ρ (ratio of inner and outer radius), so a low aspect target has a poor perform-

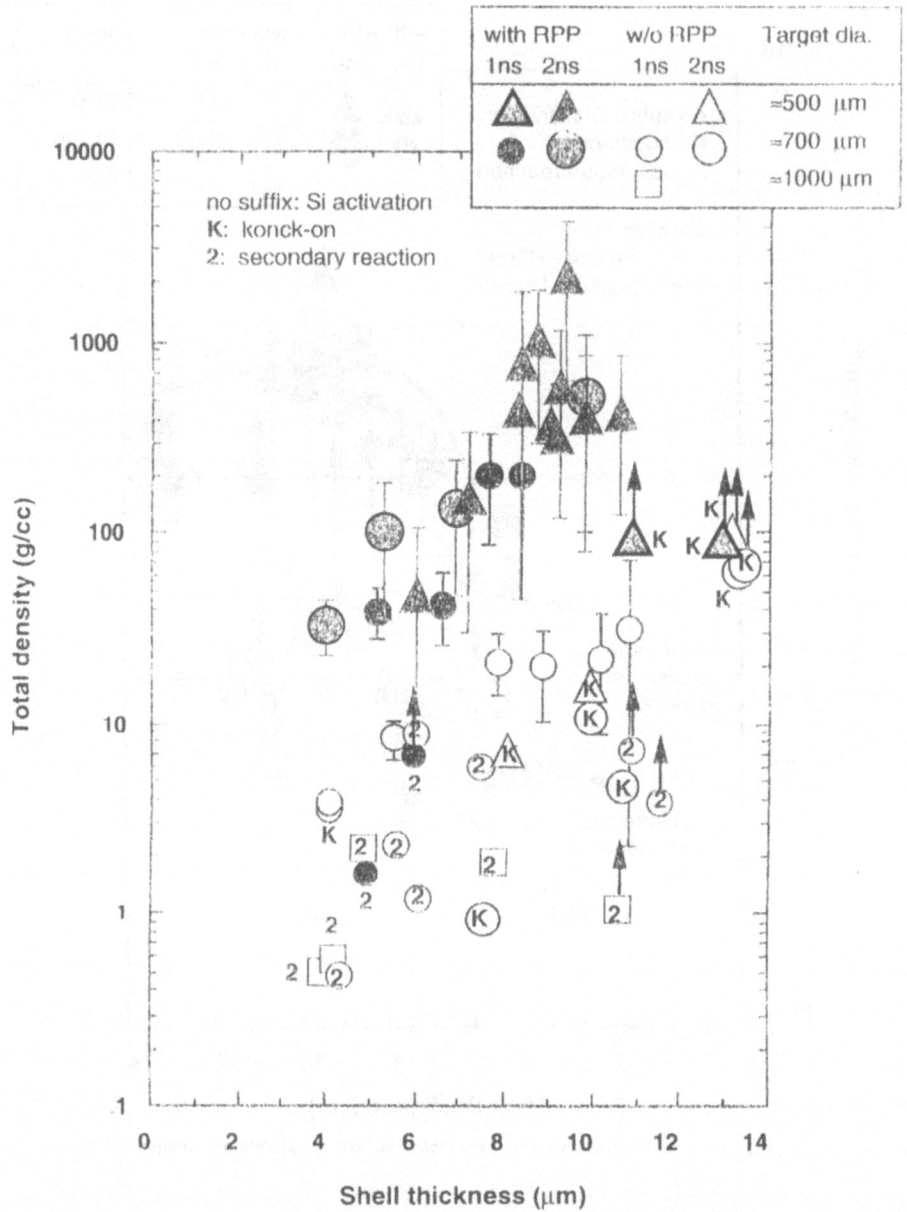

Figure 13. Average densities of 600g/cc almost meet ignition requirement.

ance. The inlet hole of laser beams into the cavity is gradually closed by plasma. In contrast the radiation driven cannonball target has better performance owing to the fact that the absorption of X-rays on the inner fuel pellet is proportional to the square of the aspect ratio ρ. And also the laser light irradiates in two ways: one is through the inlet hole into cavity and the other is to irradiate directly the outer shell where the produced X-ray penetrates into the cavity. The important issues for cannonball targets are the X-ray conversion efficiency from the laser light and the preheat problem of the inner pusher by X-ray.

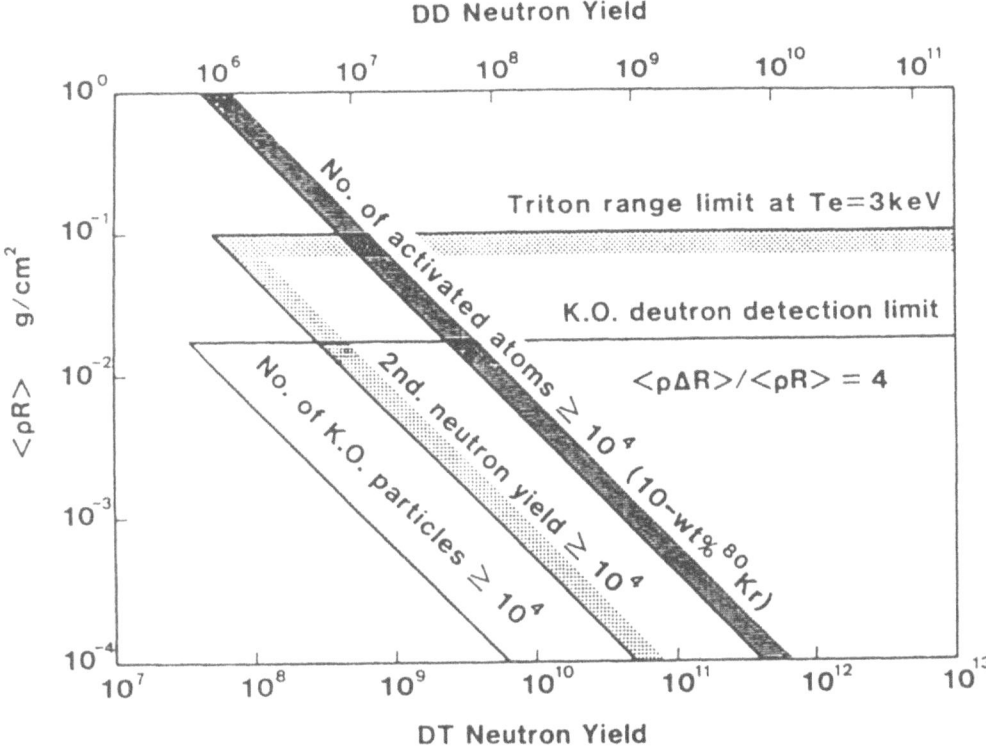

Figure 14. The regions of three methods to determine the ρR value.

The cannonball target has several advantages such as good uniformity, less sensitivity to the laser wavelength and also the laser irradiation configuration. However, the driver power for the cannonball target should be larger that that for the direct drive due to the smaller coupling efficiency. The LLNL is also using a similar concept of the indirect drive named by an old German term "Hohlraum".

X-Ray Conversion. The converted X-ray from the high density and high temperature plasma produced by lasers is very bright in the soft X-ray range of 100~300eV. The experimental and theoretical investigations of the soft X-ray by lasers have been developed.

The ionization equilibrium time τ in the high density plasma is given by

$$\tau \sim \frac{4.5 \times 10^7 Z^3}{n^4 N_e} \left(\frac{kT}{Z^2 E_H} \right)^{1/2} \exp\left(\frac{2Z^2 E_H}{N^3 kT_e} \right)$$

where T_e and N_e are temperature and the density of electrons, n is the principal quantum number of the orbital electron of ions and E_H is 13.6eV. If we put $kT_e \sim 1 keV$, n = 1, Z~10, τ becomes $10^{12}/N_e$. The ionization state of plasma of $N_e \sim 10^{23} cm^{-3}$ will reach the equilibrium in 10ps. The local thermoequilibrium is likely to be established.

However, in a laser produced plasma, the transient properties due to the larger density gradient and also the plasma fluid motion are observed. During the laser heating, the high energy radiation due to the bound to bound level transition is emitted from the low density, high temperature corona. And low energy radiation is dominant from the high density, low temperature region of ablation by the free to bound transition. In a high Z matter, such as gold, multi-ionized states are introduced where the number of states of orbital electrons is so large that the many radiation lines seems to form a quasi-continuum. However they are far from the black-body radiation in the case of a plane target plasma.

In Fig. 15, the converted X-ray radiation spectrum due to the laser irradiation versus atomic number Z is given. The conversion efficiency increases with the increase of atomic number Z and has some undulation. The X-ray radiation spectrum has several humps which corresponds to the electron orbital transitions of the ion such as K, L, M, N and O shells. They show a similar relation of the classical Mosley law which indicates a strong emission at $Z \propto \sqrt{h\nu}$. These relation is indicated in Fig. 16.[36] In a cavity target plasma such as the cannonball target, the laser produced X-ray is absorbed in a cavity and reemitted. The radiation confinement in a cavity can keep a longer time of X-ray emission. Fig. 17 shows the experimental results done by the joint work of ILE Osaka and MPQ Garching which indicates the X-ray intensification in the cavity.[37]

Cannonball Target Experiments. A cannonball target is a double shell target in which laser energy is injected into a cavity formed between the two shells. The outer shell acts as an energy container, whereas the inner shell acts as a pusher for compression of the

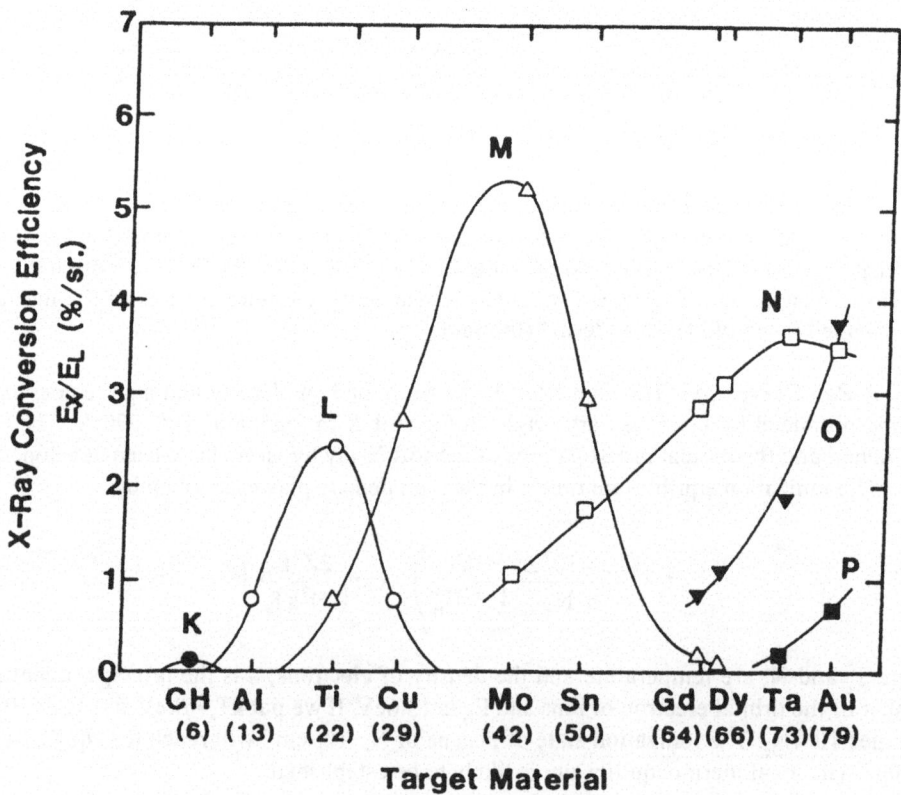

Figure 15. X-ray radiation converted from targets of different atomic number.

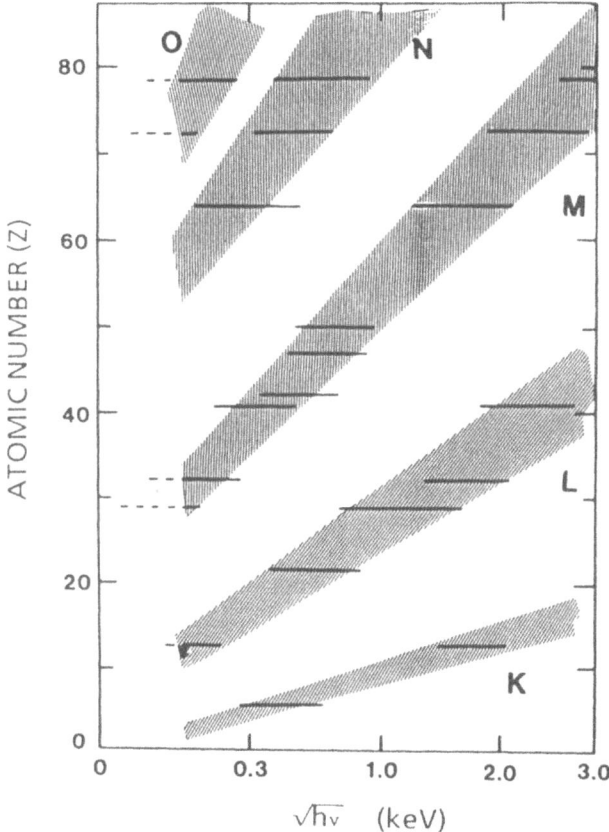

Figure 16. X-ray radiation spectrum corresponding to electron orbital transition in K, L, M, N and O shells of ions.

fuel. The laser energy is introduced into the cavity through the hole in the outer shell or through the outer shell itself. The energy of the laser beams is converted to the plasma kinetic energy and the radiation energy inside the cavity resulting in uniform compression of the fuel. It is the reason why this target is named as a cannonball target.

As mentioned before, there are two modes of compression for the cannonball target: plasma driven cannonball and radiation driven cannonball. When the cavity size is small, it is filled with high temperature plasmas which compress the inner shell by the high plasma pressure. In this mode, the laser pulse has to be short enough to inject the laser energy effectively into the small cavity against the inlet hole closer due to the plasma formation. This scheme needs the high intensity irradiation which introduces the preheating of the fuel by high temperature electrons produced in the cavity.

Nowadays, the radiation cannonball scheme is dominated those cavity size is rather large irradiated by the longer laser pulse. The laser energy can be efficiently converted to the soft X-ray which ablates the inner shell to implode the fuel. The very uniform compression is expected in this scheme by using the shorter wavelength laser light.

Fig. 18 shows the experimental conditions for two types of target geometries for X-ray driven implosion. The closed-cavity type seems to be interesting as well as the cylinder type target. Table 5 indicates the implosion parameters for cannonball target experiments.

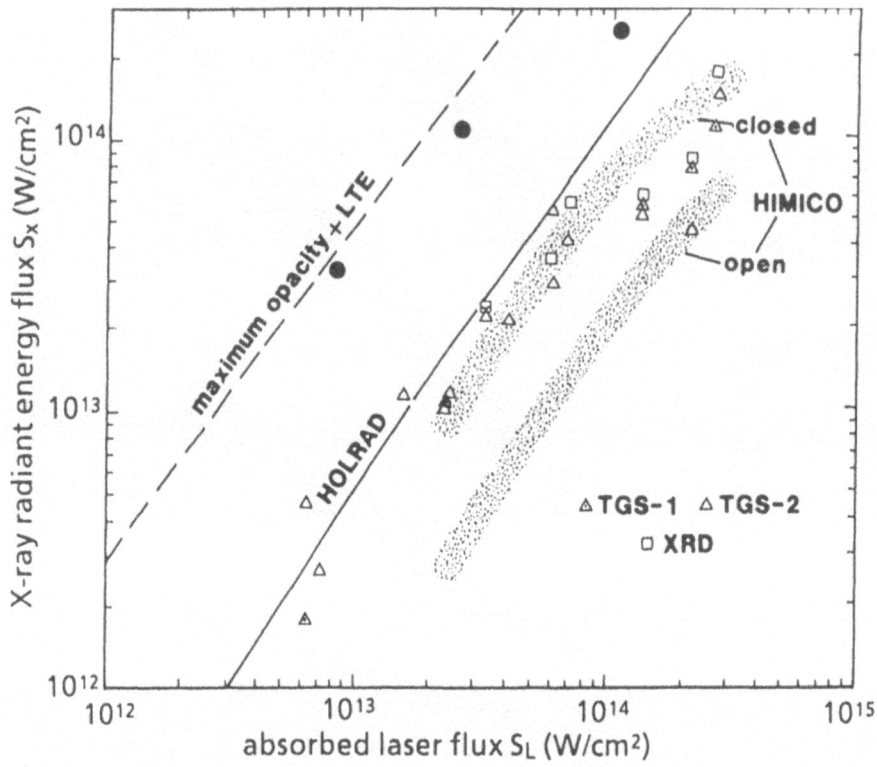

Figure 17. Radiant energy flux S_x through a diagnostic hole of a laser heated cavity versus averaged absorbed laser flux S_L. TGS: transmission grating spectrum, XRD: X-ray diode signal. The open symbols are the data with green laser pulse 0.3ns. The black circles are the recent results by the blue laser of 0.8ns pulse. Several simulation results are also indicated.

FUTURE PROSPECTS

The purpose of inertial confinement fusion research is to established the fusion ignition by irradiating the D-T fuel pellet, to attain a fusion reactor for a heat source and to get an electric power generation plant. Several concepts for inertial fusion reactors are proposed, where the key issue for these is to obtain the high gain of the energy in the DT fuel. The configuration of a fusion reactor by the inertial confinement fusion is estimated to be much easier than that of the magnetic confinement fusion in many aspects. Several scientific results have been attained in the last decades. However there are still technical problems to be solved especially in the development of efficient drivers, economic fuel pellets and materials. Table 6 shows the requirements on implosion fusion plasma parameters.

Since the prophecy of Edward Teller concerning the implosion fusion in 1972, that requested the fuel compression of 1000 times normal density, it has passed 20 years to reach this goal. The ILE Osaka has succeeded this compression experiment of high density 1kg/cm^3. The scientific feasibility of ICF has been accomplished in the way of Teller's issue. However the economic feasibility of fusion energy seems to need a long term of research and development to established the methods and procedures. International collaboration will be very important to proceed the front of research and development.

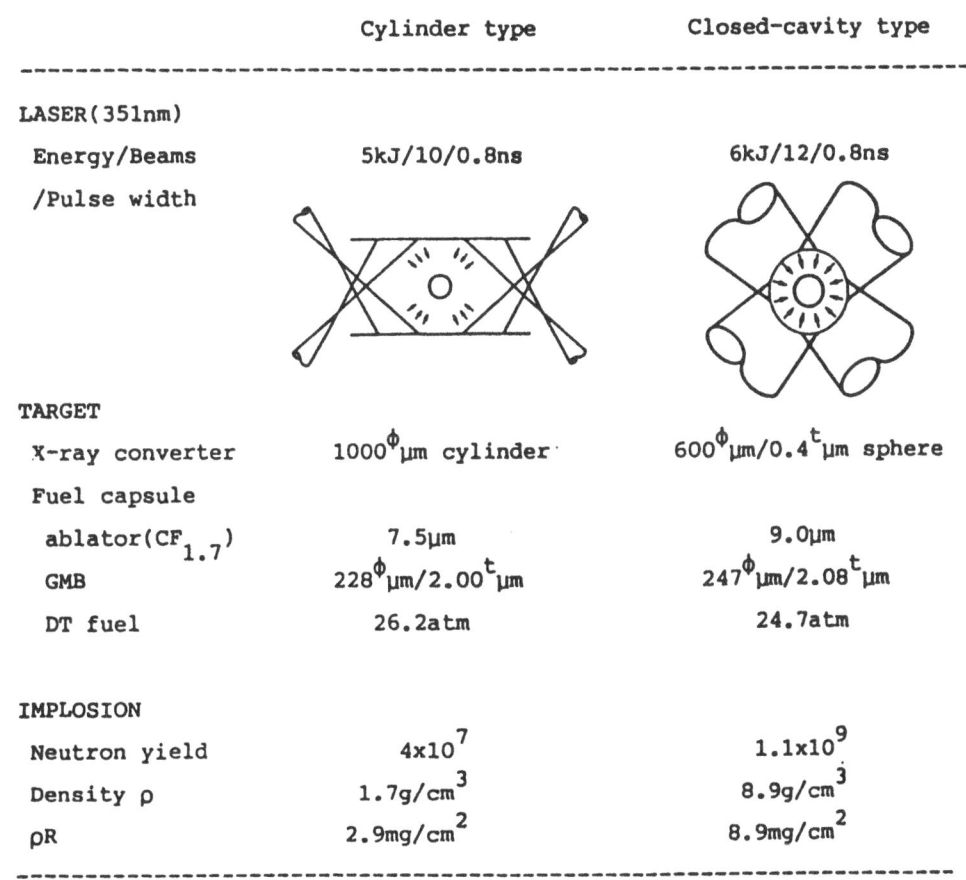

	Cylinder type	Closed-cavity type
LASER(351nm) Energy/Beams /Pulse width	5kJ/10/0.8ns	6kJ/12/0.8ns
TARGET		
X-ray converter	1000^ϕμm cylinder	600^ϕμm/0.4^tμm sphere
Fuel capsule		
ablator($CF_{1.7}$)	7.5μm	9.0μm
GMB	228^ϕμm/2.00^tμm	247^ϕμm/2.08^tμm
DT fuel	26.2atm	24.7atm
IMPLOSION		
Neutron yield	4×10^7	1.1×10^9
Density ρ	$1.7 g/cm^3$	$8.9 g/cm^3$
ρR	$2.9 mg/cm^2$	$8.9 mg/cm^2$

Figure 18. Experimental conditions and results for new types of target geometries for X-ray driven implosion.

Table 5. Summary of implosion parameters

Cavity radiation	$T_R > 200 eV$
Implosion velocity	$v = 2 \times 10^7$ cm/s (1mm diameter cylinder)
Neutron yield	DT 1.6×10^9/shot D_2 5×10^7/shot
Fuel ion temperature	$T_i = 4$ keV (n-TOF)
Convergence ratio	$R_i/R_t = 18$
Fuel areal density(D_2)	$\rho R = 10$ mg/cm^2
Deduced fuel density (D_2)	$\rho = 11 g/cm^3$ (60 × LD)
Implosion uniformity	$\Delta v/v < 5\%$

Table 6. Requirements on implosion fusion plasma parameters

Implosion parameters	Required value
η	$\geqq 5\%$ (coupling)
ε_c	< 10 eV (non preheating) at Liquid Density
R_h / R_0	$> \dfrac{1}{30}$ (compression)
ρ	$\geqq 200 \, g/cm^3$ (fuel density)
T_h	$\cong 10 \, keV$ (igniton)
R_h / R_α	$\geqq 1$ (α range)

REFERENCES

1. Yamanaka, C., Yamanaka, T., Sasaki, T., Yoshida, K and Waki, M. and Kang, H. B., (1972), "Anomalous heating of a plasma by a laser". Phys. Rev., A-6, 2335.[2]
2. Max, C. E., (1982), in Laser Plasma Interaction, Balian, R. and Adam, J. C., pp.388. (Les Houches, Section X X X IV, North Holland, Pub.).[3]
3. Estabrook, K. G., Valeo, E. J. and Kruer, W. L., (1975), "Two-dementional relativistics simulations of resonance absorption", Phys. Fluids, 18, 1151.; Takabe, H. and Mulser, P. (1982), "Self-consistent treatment of resonance absorption in a streaming plasma", Phys. Fuilds, 25, 2304.[4]
4. Forslund, D. W., Kindel, J. M. and Lee, K. (1977), "Theory of hot-electron spectra at high laser intensity", Phys. Rev. Lett., 39, 284.; Estabrook, K. and Kruer, W. L., (1978), "Properties of resonantly heated electron distributions", Phys. Rev. Lett., 40 42.[5]
5. Spritzer, L. and Härm, R. (1953), "Transport phenomena in a completely ionized gas", Phys. Rev., 89, 977.6
6. Manheimer, W. M. (1977), "Energy flux limitation by ion acoustic turbulence in laser fusion schemes", Phys. Fluids, 20, 265.[7]
7. Takabe, H. and Mulser, P. (1982), "Self-consistent treatment of resonance absorption in a streaming plasma", Phys. Fuilds, 25, 2304.[8]
8. Zel'dovich, Ya. B. and Raizer, Yu. P. (1966), "Physics of Shock Waves and High-temperature Hydrodynamic Phenomena" (Academic, New York).; Mihalas, D. and Mihalas, B. W. (1984), "Foundation of Radiation Hydrodynamics", Oxford Univ. Press, Oxford.[9]
9. More, R. M. (1981), "Atomic Physics in Inertial Confinement F'' ", Res. Rep., UCRL-84991, LLNL.[10]
10. Bodner, S. E., (1981), "Critical elements of high gain laser fusion", J. Fusion Energy, 1, 221.[11]
11. Bobin, J. L. (1971) "Flame propagation and overdense heating in a laser created plasma", Phys. Fluids, 14, 2341.; Takabe, H., Nishihara, K. and Taniuti, T. (1978), "Deflagation waves in laser compression I", J. Phys. Jpn., 45, 2001.[12]
12. Takabe, H., Nishihara, K. and Taniuti, T. (1978), "Deflagation waves in laser compression I", J. Phys. Jpn., 45, 2001.;Ahlborn, B., Key, M. H. and Bell, A. R. (1982), "An analytic model for laser-driven ablative implosion of spherical shell targets", Phys. Fluids, 25, 541.[13]
13. Takabe, H., Montierth, L. and Morse, R. L. (1983), "Self-consistent eigenvalue analysis of Rayleigh-Taylor instability in an ablating plasma", Phys. Fuids, 26, 2299..; Takabe, H., Mima, K., Motierth, L. and Morse, R. L. (1985), "Self-consistent growth rate at the Rayleigh-Taylor instability in an ablatively accelerating plasma", Phys. Fuilds, 28, 3676.; Emery, M. H., Gardner, J. H. and Boris, J. P., (1982), "Nonlinear aspects

of hydrodynamic instabilities in laser ablation", Appl. Phys. Lett., 41, 808.; Emery, M. H., Gardner, J. P. and Bodner, S. E., (1986), "Strongly inhibited Rayleigh-Taylor growth with 0.25μm lasers", Phys. Rev. Lett., 57, 703.; McCrory, R. L., Montierth, L., Morse, R. L. and Verdon, C. P. (1981), "Nonlinear evolution of ablation-driven Rayleigh Taylor instability", Phys. Rev. Lett., 46, 336.[14]

14. Hattori, F., Takabe, H. and Mima, K., (1986), "Rayleigh-Taylor instability in a spherically stagnating system", Phys. Fluids, 29, 1719.[15]

15. Yamanaka, C., Nakai, S., Yabe, T., Nishihara, K., Uchida, S., Izawa, Y., Norimatsu, T., Miyanaga, N., Azechi, H., Nakai, M., Takabe, H., Jitsuno, J., Mima, K., Nakatsuka, M., Sasaki, T, Yamanaka, M., Kato, Y, Mochizuki, T., Kitagawa, Y., Yamanaka, T. and Yoshida, K. (1986), "Laser implosion of high-aspect-ratio targets produces thermonuclear neutron yields exceeding 10^{12} by use of shock multiplexing", Phys. Rev. Lett. 56, 1575.; Takabe, H. and Mima, K. (1987), "Numerical study of ignition by stagnation-free implosion", ILE-Report, 8713, ILE, Osaka Univ..[16]

16. Fraley, G. S., Linnerberg, E. J., Mason, R. J. and Morse, R. L. (1974), "Thermonuclear burn characteristics of compressed deutrium-tritium microspheres", Phys. Fluids, 17, 474.[17]

17. Bodner, S. E., (1981), "Critical elements of high gain laser fusion", J. Fusion Energy, 1, 221.[18]

18. Kruple, W. F. (1974), "Induced-emission cross sections in neodymium laser glasses", IEEE J. of Quant. Elect., QE-10, 450.; Stokowski, S. E., Saroyan, R. A. and Weber, M. J. (1978), "Nd-doped laser glass spectroscopic and physical properties", Lawrence Livermore Laboratory, Misc. Report MS-095.; Yamanaka, C., Kato, Y., Izawa, Y., Yoshida, K., Yamanaka, T., Sasaki, T. et al. (1981), "Nd-doped phosphate glass laser systems for laser fusion research", IEEE J. of Quant Elect., QE-17, 1639.; Emmett, J. L., Kruple, W. F. and Davis, J. I. (1984), "Laser R & D at the Lawrence Livermore National Laboratory for fusion and isotope separation applications", IEEE J. of Quant. Elect., QE-20, 591.; Hunt, J. T. and Speck, D. R. (1989), "Present and future performance of the NOVA laser system", Optical Engineering, 28, 461.[19]

19. Hunt, J. T., Glass, J. A., Simmons, W. W. and Renard, P. A. (1978), "Suppression of self-focusing through low-pass spatial filtering and relay imaging", Appl. Opt. 17, 2053.[20]

20. Eidman, K., Sachsenmair, P., Salzman, H. and Sigel, R. (1972), "Optical isolators for high power giant pulse lasers", J. of Phys. E, 5, 56.[21]

21. Yamanaka, C. and Nakai, S. (1986), "Thermonuclear neutron yield of 10^{12} achieved with GEKKO XII green laser", Nature, 319, 757.[22]

22. Yamanaka, C., Yamanaka, T., Sasaki, T., Yoshida, K and Waki, M. and Kang, H. B., (1972), "Anomalous heating of a plasma by a laser", Phys. Rev., A-6, 2335.; Nuckolls, J., Wood, L.,Thiessen A. and Zimmerman, G. (1972), "Laser compression of matter to super-high densities: Thermonuclear applications", Nature, 239. 139.[23]

23. Craxton, R. S. (1981), "High efficiency frequency tripling scheme for high power Nd:glass lasers", IEEE J. of Qunt. Elect., QE-17, 1771.[24]

24. Kato, Y., Mima, K., Miyanaga, N., Arinaga, S., Kitagawa, Y., Nakatsuka, M. and Yamanaka, C. (1984), "Random phasing of high-power lasers for uniform target acceleration and plasma-instability suppression", Phys. Rev. Lett., 53, 1057.[25]

25. Lehmberg, R. H., Schmitt, A. J. and Bodner, S. E. (1987), "Theory of induced spatial incoherence", J. of Appl. Phys. 62, 2680.[26]

26. Den, X., Liang, X., Chen, Z., Yu, W. and Ma, R. (1986), "Uniform illumination of large targets using a lens array", Applied Optics, 25, 377.[27]

27. Goldstein, S. A., Cooperstein, G., Lee, R., Mosher, D. and Stephanakis, S. J. (1987), "Focusing of intense ion beams from pinched-beam diodes", Phys. Rev. Lett., 40, 1504.; Johnson, D. J., Kuswa, G. W., Farnsworth, A. V., Quintenz, J. P., Leeper, R. J., Burns, E. J. T., et al. (1979), "Production of 0.5-TW proton pulses with a spherical focusing magnetically insulted diode", Phys. Rev. Lett., 42, 610.[28]

28. Faltens, A., Hoyer, E., Keefe, D. and Laslett, L. J. (1979), "Design/cost study of an induction linac for heavy ions for pellet-fusion", IEEE Trans. Nuc. Sci., NS-26, 3106.; Fessenden. T. J., Celata, C. M., Faltens, A., Herderson, T., Judd, D. L., Keefe, D. et al. (1987), "Preliminary design of a ~10MV ion accelerator for HIF research", Laser and Particle Beams, 5, 457.[29]

29. Müller, R. W. (1988), "RF linac driver for commercial heavy ion beam fusion", Proc. of 3rd Inertial Confinement Fusion System and Application Colloquiun, Madison.[30]

30. Fessenden, T. J., Celata, C. M., Faltens, A., Herderson, T., Judd, D. L., Keefe, D. et al. (1987), "Preliminary design of a ~10MV ion accelerator for HIF research", Laser and Particle Beams, 5, 457.[31]

31. Yamanaka, C., Nakai, S., Yabe, T., Nishihara, K., Uchida, S., Izawa, Y., Norimatsu, T., Miyanaga, N., Azechi, H., Nakai, M., Takabe, H., Jitsuno, J., Mima, K., Nakatsuka, M., Sasaki, T, Yamanaka, M., Kato, Y, Mochizuki, T., Kitagawa, Y., Yamanaka, T. and Yoshida, K. (1986), "Laser implosion of high-aspect-ratio targets produces thermonuclear neutron yields exceeding 10^{12} by use of shock multiplexing", Phys. Rev. Lett. 56, 1575.[32]

32. Mima, K. (1988), "High density compression of hollow shell target by green and blue GEKKO XII laser", Bull. American Phys. Soc., 33, 1880.[33]
33. Yamanaka, C., Mima, K., Nakai, S., Yamanaka, T., Izawa, Y., Kato, Y. et al. (1987), "Inertial confinement fusion research by GEKKO lasers at ILE Osaka and target design for ignition", Plasma Phys. and Controll, Nucl. Fusion Res. (IAEA, Vienna, 1986), 3, 33.[34]
34. Kacenjan, S., Goldman, L. M., Enterberg, A. and Skupsky, S. (1984), "<ρR> measurements in laser-produced implosions using elastically scattered ions," J. Appl. Phys., 56, 2027.[35]
35. Azechi, H., Miyanaga, N., Stapf, R. O., Itoga, K., Nakaishi, H., Yamanaka, M., Shiraga, H., Tuji, R., Ido, S., Nishihara, K., Izawa, Y., Yamanaka, T. and Yamanaka, C.,(1986), "Experimental determination of fuel density radius product of inertial confinement fusion targets using secondary nuclear fusion reactions", Appl. Phys. Lett. 49, 555.; Bazov, N. G., Vygovskii, O. B., Gus'kov, S. Yu., Il'in, D. V., Lekovskii, A. A., Rozanov, V. B. and Sherman, V. E. (1986) "Diagnostics of laser fusion plasmas on the basis of the products of secondary fusion reactions", Sov. J. Plasma Phys. 12, 526.; Cable, M. D., Lane, S. M., Glendinning, S. G., Lerche, R. A., Singh, M. S., Munro, D. H., Hatchett, S. P., Estabrook, K. G. and Suter, L. J., (1986), "Implosion experiments at NOVA". Bull. Am. Phys. Soc. 31, 1461.; Gamalii, E. G., Gus'kuv, S. Yu., Krokhin, O. N. and Rozanov, V. B. (1975), "Possibility of determining the characteristics of laser plasma by measuring the neutrons of the DT reaction", JETP Lett. 21, 70.[36]
36. Mochizuki, T., Yabe, T., Okada, K., Hamada, M., Ikeda., N. and Yamanaka, C. (1986), "Atomic-number dependence of soft-x-ray emission from various targets irradiated by a 0.53-μm-wavelength laser", Phys. Rev. A, 33, 525.[37]
37. Mochizuki, T., Yabe, T., Tanaka, K. A., Yamanaka, C., Sigel, R., Tsakiris, G. D. et al. (1987), "X-ray confinement in a laser heated cavity", Nuclear Fusion Suppl. 3, 25.

17

HEAVY-ION DRIVEN INERTIAL FUSION ENERGY

R. O. Bangerter and T. J. Fessenden

Lawrence Berkeley National Laboratory
Berkeley, California 94720

ABSTRACT

Heavy ion driven inertial fusion energy is an approach to controlled fusion energy that appears to satisfy the scientific, economic, and environmental criteria for a desirable, long-term solution to mankind's energy needs. Because of projected reliability, efficiency, and cost advantages the U.S. inertial fusion energy program is concentrating on the heavy ion induction accelerator as the inertial fusion driver. This paper describes scaled experiments that have demonstrated many important elements required by induction accelerator/drivers. These include: adequate heavy ion sources; high voltage ion injectors; transport and acceleration of multiple intense ion beams; and beam current amplification with acceleration. Plans for the ILSE accelerator are described. This facility would permit the study of driver beam manipulations at full beam size and intensity but at energies much less than required by a driver.

1. INTRODUCTION

The heavy ion approach to inertial fusion energy (IFE) uses beams of heavy ions to implode and ignite inertial fusion targets. In a power plant this process will be repeated several times per second The resulting fusion energy release will be collected as heat and used to generate electrical energy by conventional methods. Research programs throughout the world, have already put to rest fundamental questions about the basic feasibility of achieving high gain with inertial fusion targets.

Active research toward the development of heavy ion accelerator/drivers began in the late 1970's. Initial experiments at the Lawrence Berkeley National Laboratory established the technical feasibility of key elements of the induction approach to an accelerator/driver. One of the first experiments demonstrated that heavy ion sources of the required quality, intensity and lifetime are available. The Single Beam Transport Experiment (SBTE) [1] established the ability to acceptably transport higher current space-charge-dominated beams than originally though possible with alternating gradient structures. The first multiple ion beam induction linac, MBE-4 [2], demonstrated that current amplification and longitudinal control can be achieved during acceleration. These experiments used beams much smaller than those required by a heavy ion driver.

For the last several years effort has been toward developing the Induction Linac Systems Experiments [3] (ILSE) to test nearly all the beam manipulations required of an induction driver at full scale in beam size and line charge density. A high voltage injector[4] that produces heavy ion beams at driver size and quality has recently been completed. This injector will be used for the Elise program that would build the electric-focused section of ILSE. Elise is the first phase of the ILSE program.

In addition to the linear accelerator we are examining the feasibility of a recirculating induction accelerator[6] that could reduce the cost of a heavy ion driver. Small experiments to test the more complicated accelerator physics of this driver approach are in progress at the Lawrence Livermore National Laboratory. Depending on their outcome, the ILSE accelerator could also be used as an injector for a recirculating induction accelerator expected to accelerate the beams to as much as 100 MeV

These systems are still much too small to be used to ignite a fusion target. The decision to begin construction of a heavy ion accelerator/driver will almost certainly await experimental results from the laser-driven National Ignition Facility NIF [7]. The NIF results should precisely define the parameters required of the heavy ion accelerator/driver but cannot be expected before 2005.

We will give a detailed account of past and current HIF accelerator research in sections 3 and 4. Before that, in section 2, we will consider heavy ion fusion in the context of the key questions that must be answered for the development of any fusion option. Our conclusions are presented in section 5.

2. THE DEVELOPMENT OF INERTIAL FUSION ENERGY

There are three important questions for any fusion option:

1. Does it make scientific sense?
2. Does it make economic sense? This question has two important parts:
    a. What will it ultimately cost to produce energy? and
    b. How much does the research and development cost?
3. Does it make environmental sense?

2.1. Scientific Feasibility

Consider the first question, scientific sense. In addressing this question we must look at the three principal components of a heavy ion fusion power plant, namely the fusion targets, the target chamber (sometimes called the reactor), and the accelerator.

Two experimental lines of research indicate that the target physics does make sense. The first line of research involves laboratory scale experiments using existing laser and light-ion drivers. Progress in this area has been discussed in other papers at the Symposium. The second line of research is the Centurion/Halite Program. This program, using nuclear explosives, demonstrated excellent performance. It put to rest fundamental questions about the feasibility of achieving high gain. In addition to the experimental programs, there has been a very successful target theory program. First-principles calculations (numerical simulations) are in excellent agreement with theory. Finally, the proposed National Ignition Facility [7] is expected to demonstrate laboratory ignition of inertial fusion targets.

Since the National Ignition Facility will be based on a laser, it is natural to ask if the results can be applied to ion beam targets. In the case of radiation (indirect) drive, the answer is yes. In radiation drive, the beam energy from the laser or ion beam is converted to radiation that, in turn, implodes the fuel capsule. To a good approximation, the target implosion process is therefore independent of the type of driver. The one feature that is different is the conversion of beam energy to radiation. Theory, numerical simulation, and experiments with light and heavy ions have established a firm understanding of this aspect of heavy ion fusion.

Consider the target chamber. Many chamber designs [8] for inertial fusion propose to protect the chamber wall with neutronically thick layers of fluids such as lithium, lithium-lead, or lithium-beryllium-fluoride (flibe). For deuterium-tritium fusion there is a profound difference between inertial fusion systems that can use fluid wall protection and fusion systems that cannot. Without fluid walls, radiation damage sets a lower limit on the cost of electricity. It is very difficult to circumvent this limit by increasing power density or changing the confinement scheme. One can lower the cost of the reactor by increasing the power density; but, for a given set of reactor materials, increased power density leads to shorter life.

The driver for an inertial fusion power plant must satisfy several criteria. It must be reliable, durable, efficient, and have a high pulse repetition rate. It must also have acceptable cost and it must deliver good beam quality so the beams can be focused onto the target. Quantitatively, good reliability means that the driver must be available more than 90% of the time—preferably much more than 90% of the time. Good durability means that the driver must last the life of the power plant, typically 30 years or more. Since the repetition rate is several shots per second, a lifetime of 30 years corresponds to 1 to 10 billion pulses. The required efficiency depends on the energy gain of the targets. Simple analyses based on recirculating power fraction indicate that the product of driver efficiency and target gain must exceed about ten for good economics. These analyses also show that the cost of the driver cannot exceed about 1 billion dollars.

The driver is the one area in which heavy ion fusion differs substantially from other inertial fusion schemes such as laser fusion and light-ion fusion. Requirements on beam quality are related to the requirement on the product of driver efficiency and target gain described above. It seems unlikely that the efficiency of an accelerator will exceed approximately 33%. Therefore the target gain must exceed approximately 30. To produce gain greater than 30, the accelerator must put a lot of energy into a small volume in a short time. According to numerical simulations, the required energy lies in the range of 1 to 10 MJ. The required focal spot radius at the target is related to the range (penetration depth) of the ions in the target. Ion ranges of 0.02 to about 0.2 g/cm^2 are appropriate. For small ion ranges, the upper limit on the focal spot radius is approximately 6 mm. For the larger ion ranges the focal spot radius cannot exceed 2 or 3 millimeters. In all cases, most of the energy must be delivered in about 10 ns. Dividing the total energy by this time gives a peak beam power of 100 to 1000 TW. One must choose the beam kinetic energy and the ion current so that their product equals the required power.

Figure 1 shows ion range as a function of kinetic energy for a variety of ions. For heavy ions one can consider kinetic energies as high as 10 GeV (or possibly even higher with advanced target designs). To be specific, assume that the required beam power is 400

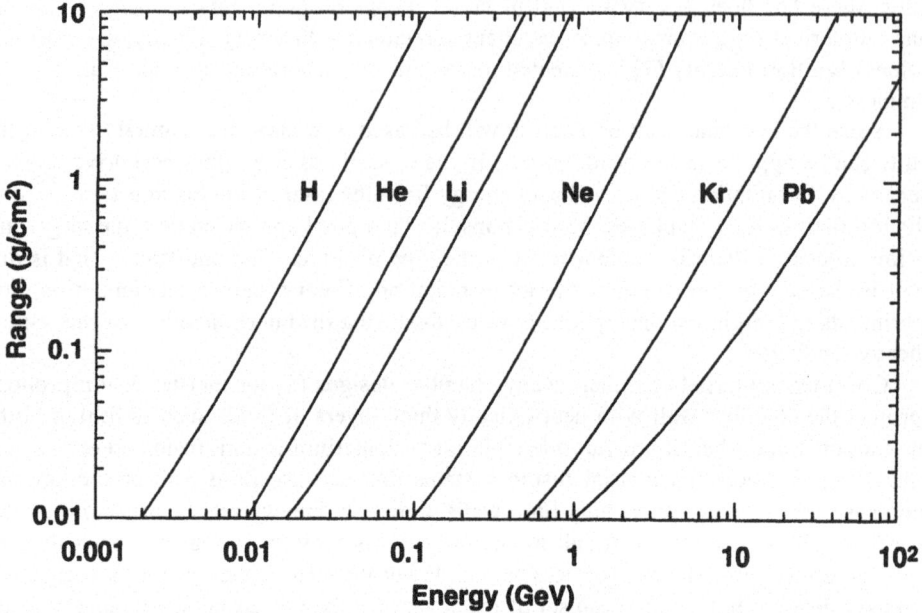

Figure 1. Range-Energy relations for a number of ions. The acceptable range for inertial fusion is approximately 0.02 to 0.2 g/cm².

TW. The required ion current is 400 TW / 10 GeV = 40 kA. For light ions such as lithium, the required beam current is about two orders of magnitude larger. It is difficult to carry even 40 kA in a single beam, so ion driver designs usually have multiple beams. It is not necessary to carry the full ion current in the accelerator itself. Most driver designs employ longitudinal beam compression in the beam lines that lead from the accelerator to the target chamber. In this scheme, the ions at the tail of the beam are accelerated to a slightly higher velocity than the ions at the head of the beam. This velocity differential causes the beam bunches to become shorter as they drift toward the target. A compression factor of approximately 10 is typical. For a ten-fold compression, a 10-GeV accelerator must carry about 4 kA rather than the full 40 kA required by the target.

The driver requirements are summarized in Table I. Conventional high energy accelerators have already demonstrated that they can meet many of these requirements. Theoretical considerations and small experiments indicate that they can also meet the remaining requirements.

High-energy accelerators routinely produce kinetic energies of the order of 10 GeV or even much higher. They readily produce megaJoules of beam energy and they can have high pulse repetition rates, long life, and good reliability. They can produce beams that can be focused to small spots and they can be efficient; however, existing machines have not been designed to deliver tens of kiloamperes of ion current. Delivering high ion current while maintaining the ability to focus the beams is the new feature of accelerator technology required for inertial fusion.

It is noteworthy that electron induction accelerators can deliver about 10 kA of electrons, so the issue is not the ability of the power systems and accelerator structure to carry the current. The issues lie in the beam dynamics of high-current ion beams. The research program to address these issues will be described in sections 3 and 4.

Table I. Typical heavy ion driver requirements

Reliability	≥ 95%
Durability	30 yr. ($10^9 - 10^{10}$ pulses)
Pulse rate	1-10 Hz
Efficiency	Product of driver efficiency and target gain ≥ 10
Acceptable cost	≤ $1B
Beam energy	1-10 MJ depending on target design
Ion range	0.02 to 0.2 g/cm^2*
Focal spot radius	0.2 to 0.6 cm*
Beam power	100-1000 TW
Number of beams	4 to more than 1000

*These are anti correlated

2.2. Economic Feasibility

We now consider the second fundamental fusion question—economics.

Since the inception of the heavy ion fusion program two decades ago, systems studies have consistently shown that heavy ion fusion is competitive with other fusion energy systems. The three most recent heavy ion studies, HYLIFE-II [8], Osiris [9], and Prometheus-H [10], predict lower cost of electricity than recent tokamak studies such as Aries [11]. They also predict lower cost of electricity than similar studies based on krypton fluoride lasers [10, 12].

It is likely that advances in accelerator technology will lead to even lower costs in the future. Superconducting magnets, high voltage insulators, ferromagnetic materials, high power switches, and energy storage are the principal cost centers in fusion accelerators. All have widespread scientific and industrial applications. Continued improvement in these technologies, independent of the fusion application, will certainly occur.

Beam neutralization, improved target concepts, and different accelerator designs may also lead to lower costs. Beam neutralization during the final focusing process would allow the use of lower ion kinetic energy. Of course lower ion kinetic energy requires higher beam currents to meet the target power requirements; but, for linear induction accelerators, current is less expensive than kinetic energy. Directly driven targets have lower beam power and energy requirements than indirectly driven targets. While directly driven ion targets cannot be tested using laser drivers, they may ultimately lead to less stringent accelerator requirements and lower cost. Finally, circular accelerators may be less expensive than the linear accelerators that are normally considered for inertial fusion applications. A recirculating experiment is underway at the Lawrence Livermore National Laboratory as discussed in Section 4.2.

We now consider the second aspect of economics—the cost of research and development. There are at least 8 reasons why the development costs of heavy ion fusion are expected to be low:

1. HIF leverages a large, worldwide investment in accelerator technology.
2. HIF leverages the Defense Inertial Confinement Fusion Program.
3. Accelerators consist of numerous similar or identical components which leads to low development costs
4. Accelerators can be upgraded.
5. Driver, chamber, and targets are largely independent. Modifications are easy. For power production, multiple chambers lead to low cost of electricity.
6. Confinement is not related to chamber size.

7. Each target design is a new confinement system.
8. An expensive neutron test facility of the type required to test and develop materials for magnetic fusion is not needed. Existing alloys are low activation materials when used for inertial fusion.

2.3. Environmental Considerations

Finally, we turn to the environmental aspects of heavy ion fusion. As previously noted, it is possible to protect the first wall of the target chamber with neutronically thick layers of fluid. According to calculations, this type of protection can reduce the wall activation and damage to the point that the chamber can last for the life of the power plant. Moreover, based on existing regulations, the chamber material can qualify for shallow burial.

The accelerator itself is environmentally friendly. Large accelerators such as those at Stanford, Fermilab, Geneva, and Hamburg are near or under farms, villages, or cities.

3. HEAVY ION ACCELERATOR/DRIVERS

Two types of accelerator systems have emerged as potentially attractive for a fusion driver: an r.f. linac feeding a cascaded sequence of storage rings, and a high-current multiple-beam induction accelerator. Studies of the rf linac/storage-rings method are progressing in foreign countries. This approach uses conventional high energy accelerator technology.

Since October 1983, as a result of several studies that suggested technical and economic advantages, attention in the U.S. has focused on the induction approach [13]. This type of accelerator is unique in its ability to continuously amplify both the beam current and energy during the acceleration process. In a conceptual driver, many beams are accelerated in parallel through common induction cores. As the beams gain energy, the beam focusing system can transport higher currents. Therefore, in the linac approach, the multiple beams are combined in groups of four to produce a final total of typically 10 beams. In the recirculating induction approach, several beams are accelerated by a series of accelerator rings to the final energy. During acceleration, the current of each beam increases by a factor of 200 or more, mostly as a result of an increase in ion velocity but also due to the transverse beam combination and a decrease in the length of the beams. A beam power increase by an additional factor of 10 is achieved by drift-compression current amplification between the accelerator and the target as discussed in section 2. Fig. 2 is a sketch of the induction linac driver concept.

3.1. Initial HIF Experiments and Developments

The heavy ion fusion accelerator research strategy in the U. S. is addressing the key issues of induction linacs through a sequence of experiments of increasing scale and sophistication. Beginning in the late 1970's LBNL personnel have conducted a number of experiments to establish the technical feasibility of key components and model novel physics elements of the induction approach to an accelerator/driver.

3.2. Sources

One of the first experiments demonstrated that heavy ion sources of the required quality, intensity and lifetime are available. A multiple-beam induction linac will need 50 or more sources of heavy ions able to emit heavy ions at currents near one Ampere. The quality or nor-

Heavy-Ion Driven Inertial Fusion Energy

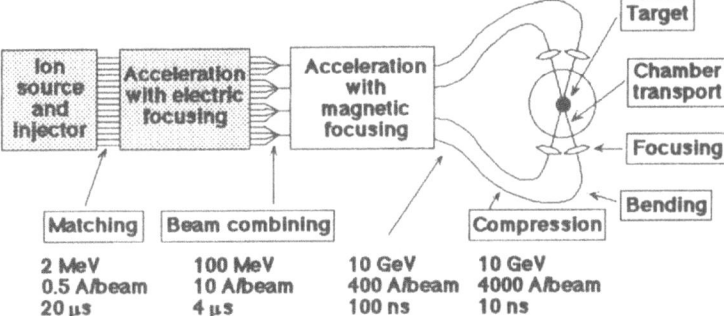

Figure 2. The multiple-beam heavy ion induction linac concept. The shaded areas indicate driver areas that have been investigated by technology developments and/or model physics experiments.

malized emittance of each beam must be much less than the 10 or 20 π mm-mRad limit set by the requirements of focusing the beams onto the inertial fusion target. The source shown in Fig. 3 was developed at LBNL and provided up to 2 Amperes of Cesium^{+1} at a quality approximately 100 times better than required. The Cs$^+$ was obtained by contact ionization from a tungsten plate that was heated to approximately 1000 °C. Cesium beams from this source were subsequently accelerated to 1.5 MeV in a drift tube linac. Experiments since 1980 have obtained beams of Cs^{+1} and Potassium^{+1} at very low emittance from alumino-silicate (zeolite) sources heated to similar temperatures. Fig. 4 shows measurements of Potassium emission as

Figure 3. Picture of the driver size source developed about 1978. This source supplied up to 2 Amperes of Cesium^{+1} at an energy of 500 keV.

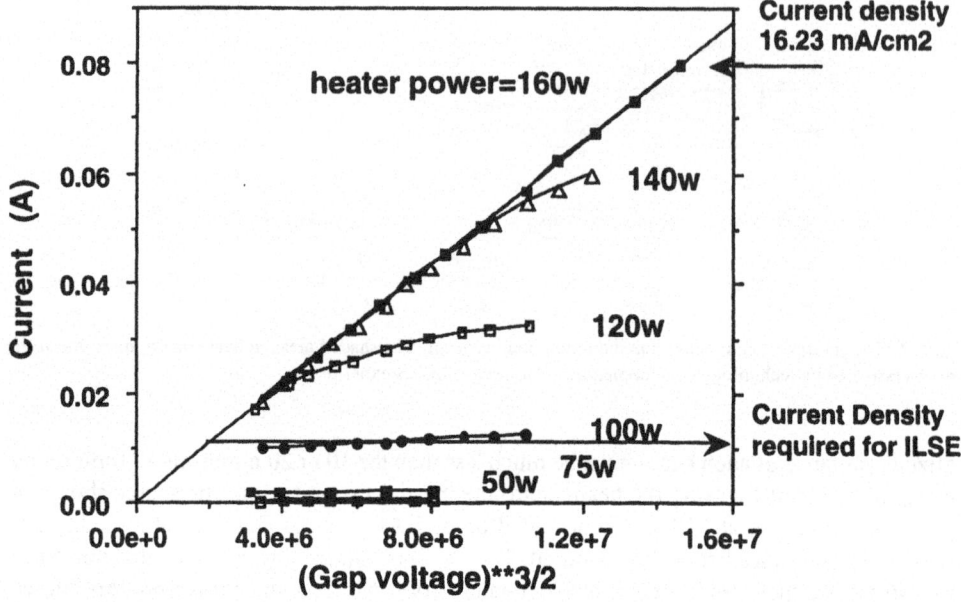

Figure 4. Measurements of Potassium^{+1} emission from a zeolite ion source as a function of applied voltage and heater power.

Figure 5. A picture of the Single Beam Transport Experiment (SBTE) performed at LBNL from 1980–1985. SBTE demonstrated that low emittance ion beams could be transported with no loss in quality by an electric-focus, alternating gradient transport system.

a function of the extraction voltage from a zeolite source developed for use with the ILSE experiments. The experiments showed that this source is able to provide several times the current density required for the ILSE experiments.

3.3. Beam Transport

In 1980 we began fabrication of the Single Beam Transport Experiment. This experiment studied the basic physics of the transport of space-charge-dominated ion beams by electrostatic quadrupole lenses. A picture of the SBTE facility is shown in Fig. 5. A view of the inside showing the electric quadrupoles is presented in Fig. 6. Beam of Cs^{+1} at energies near 125 kV and currents up to 20 mA were transported through 85 electric quadrupoles. Measurements of beam quality versus distance for various adjustments of the focusing voltages showed that much lower emittance ion beams could be transported than initially estimated. These results gave us confidence that high quality ion beams can be transported and accelerated with no loss of quality. This confidence has significantly reduced our cost estimates of multiple-beam induction linac drivers for commercial fusion.

Figure 6. Picture of the inside to the SBTE experiment showing the electrostatic quadrupoles that transported up to 20 mA beams of Cs^{+1}.

Figure 7. Photograph of the MBE-4 current-amplifying ion induction linac. This apparatus accelerated 4 Cs+ beams to nearly one MV while amplifying the total current from 40 to more than 300 mA.

3.4. MBE-4, the First Ion Induction Accelerator

We began fabrication of MBE-4, the first multiple-ion-beam induction linac, early in 1985 and completed fabrication in Sept. 1987. MBE-4 addressed, in a scaled manner, the physics associated with the low energy portion of a much longer induction linac driver. Four space-charge-dominated Cs^+ beams were injected at 0.2 MeV and accelerated to a peak energy of nearly 1 MeV over a distance of approximately 15 m. Figure 7 shows a picture of the completed facility. Experiments were conducted between 1986 and 1992 when the experiment was completed. A diagram of the apparatus is shown in Fig. 8.

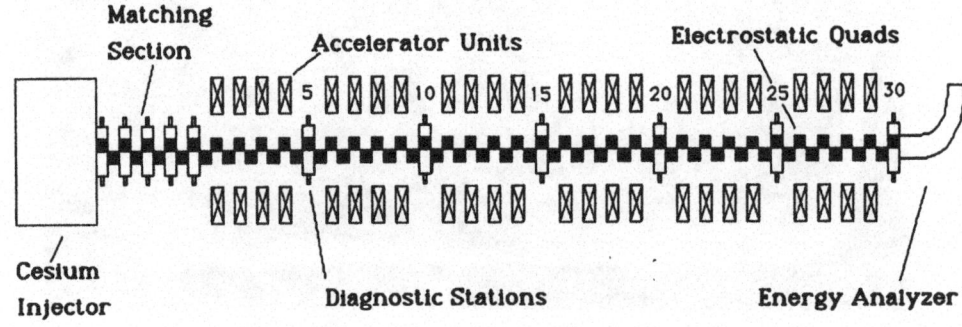

Figure 8. Schematic of the MBE-4 current amplifying induction accelerator.

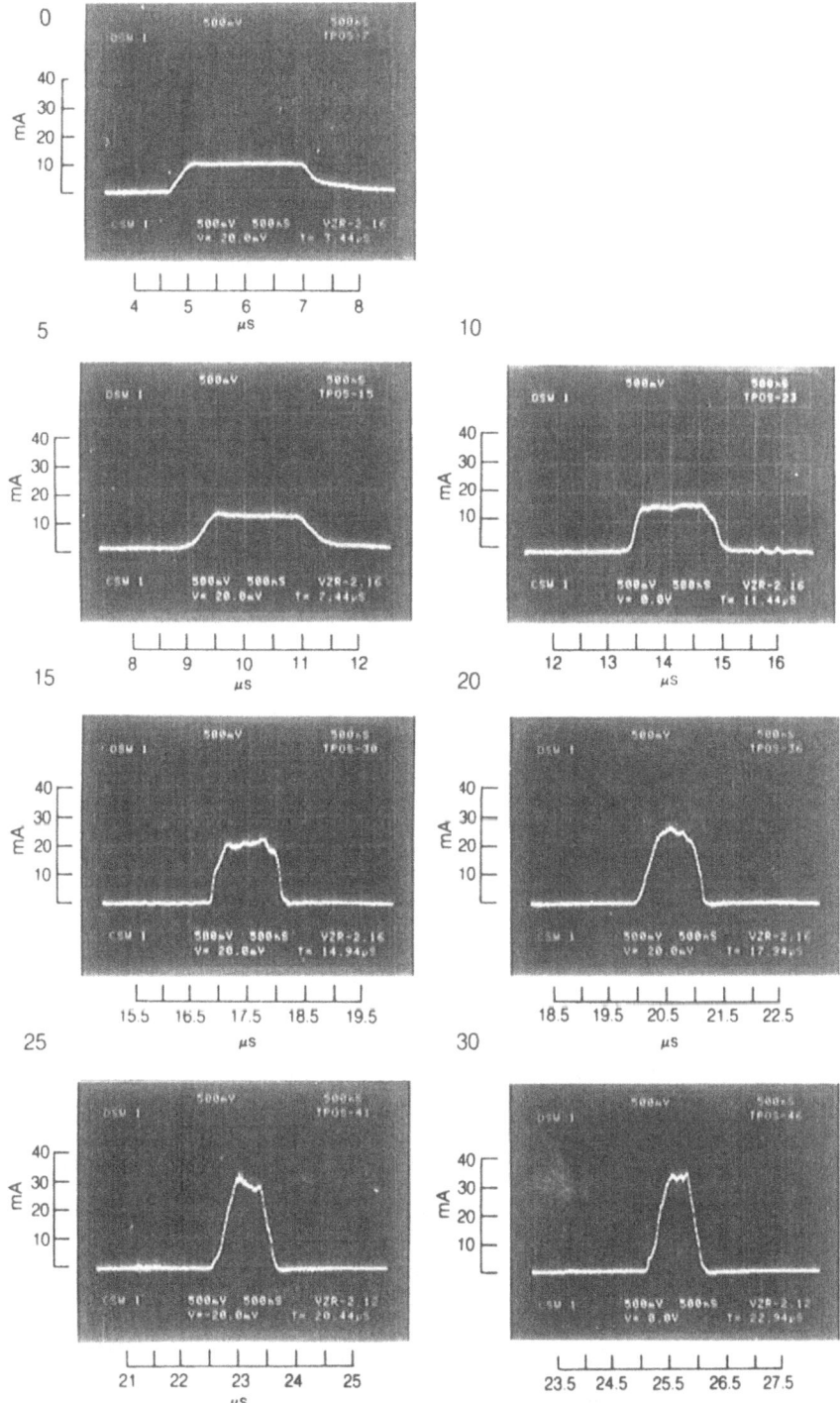

Figure 9. Oscillograms from each of the MBE-4 diagnostic stations showing current amplification along the ion induction linac.

Early MBE-4 experiments with 12 of the 24 accelerator sections showed for the first time that simultaneous acceleration and current amplification of multiple, space-charge-dominated ion beams can be accomplished in an induction linac with adequate beam control and no perceptible degradation of beam quality (increase in normalized emittance). In more recent experiments with the completed facility, current amplification of from 10 to as much as 90 mA/beam has been studied. Fig. 9 shows oscillograms of one of the four Cs^{+1} beams at the diagnostic stations along the accelerator.

One important experiment demonstrated that the ion beams could be accelerated with current amplification while adequately preserving the transverse and longitudinal beam quality. Fig. 10 shows both measurements and calculations of the beam emittance or quality during acceleration through MBE-4. In an ideal accelerator this quantity at best remains constant and does not increase. This result required that the beam be carefully centered and matched in the focusing channel.

These experiments showed that the acceleration waveforms can control beams longitudinally and simultaneously increase the beam energy and current without a loss in beam quality. For these intense beams, space charge reduces but does not eliminate effects due to acceleration errors. However, for the degree of control required in a full-scale driver, better and more agile pulsers than used in MBE-4 will be required.

At present the four beam MBE-4 injector is being used to study the transverse combining of space charge dominated ion beams.

4. CURRENT RESEARCH AND DEVELOPMENT

As beams gain energy in a heavy ion accelerator, the focusing system becomes more efficient and it becomes possible to accelerate increasingly larger currents. By combining

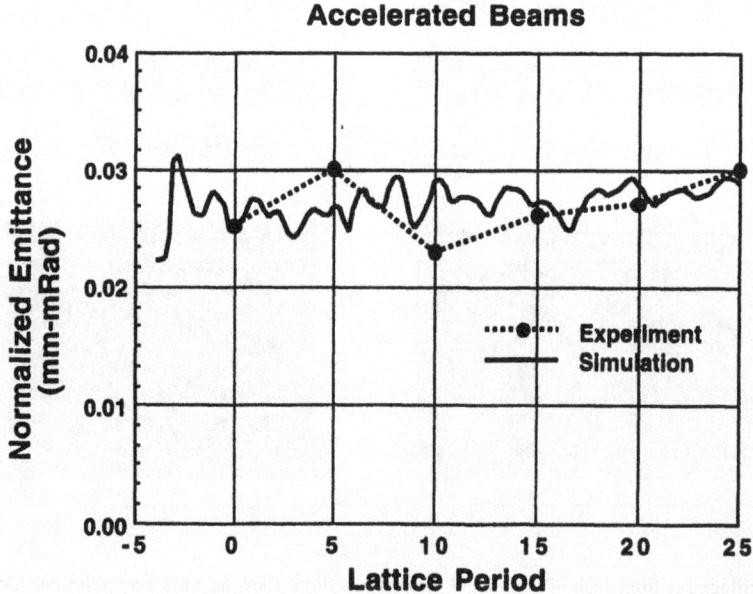

Figure 10. Measurement and simulation of the normalized emittance of a Cs^{+1} beam during acceleration through MBE-4. With care acceleration and current amplification were achieved with no degradation in beam quality.

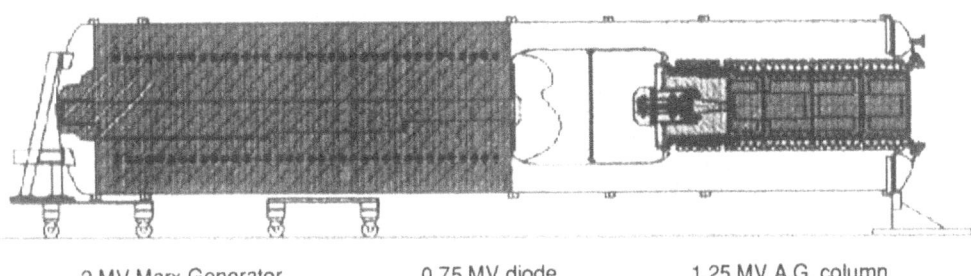

Figure 11. A sketch of the recently completed 2-MV Potassium ion injector. This injector, which will be used initially for the Elise accelerator, produces beams at the energy and intensity needed for a heavy ion driver.

beams once or twice during acceleration, a significant reduction in the fabrication cost of a large heavy ion accelerator can be realized. However, this beam manipulation, if not done carefully, can cause transverse beam heating and a loss of beam quality that could adversely affect our ability to focus the beam onto a fusion target. An experiment to study the physics of transversely combining four beams to one is in progress at the Lawrence Berkeley Laboratory using Potassium beams and the injector used for the MBE-4 experiments. The results of this experiment will be used to assess the limitations of this beam manipulation at driver scale and to contribute to the design of a beam-combining experiment on ILSE.

4.1. Development of Driver Scale Injectors

The development of ion injectors capable of generating beams at the current and energy required by a heavy ion driver was begun at the Los Alamos National Laboratory and transferred to LBNL in the fall of 1987. A 2-MV injector that produces a single potassium^{+1} beam at driver scale current and intensity has recently been developed. This injector, which exceeded its design goals, will be used with the ILSE accelerator. A sketch of the injector is shown in Fig. 11.

4.2. Recirculating Induction Accelerator Development

The development of a heavy ion induction recirculating experiment is underway at the Lawrence Livermore National Laboratory. In this driver concept, four parallel beams of heavy ions are accelerated through approximately 100 turns in a series of three or four circular induction accelerators. By limiting the number of turns and by accelerating very aggressively, the experimenters hope to produce ion beams several orders more intense that can be generated by conventional circular accelerators. Moreover, because the beams effectively "reuse" the accelerator with each turn, this approach could lead to savings in the cost of a heavy ion driver. Because this accelerator concept has not been validated, a small experiment is being fabricated to test the essential physics of this driver approach. A sketch of the apparatus is shown in Fig. 12.

4.3. The Induction Linac Systems Experiments (ILSE)

In 1988, a series of experiments known as the Induction Linac Systems Experiments, or ILSE, were first proposed by the Heavy Ion Fusion Accelerator Research Pro-

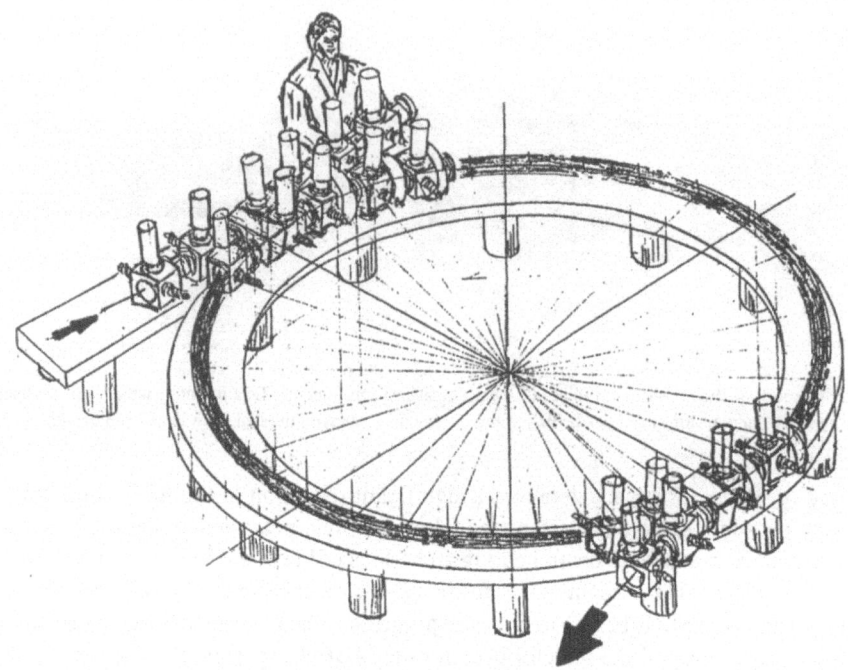

Figure 12. A sketch of the small induction accelerator being fabricated at the Lawrence Livermore National Laboratory. This experiment will be used to examine the physics of this novel accelerator concept.

gram to simultaneously test almost all driver beam manipulations in a single system. The original proposal was to accelerate 16 carbon beams each at approximately 1/4 driver intensity, through an electric-focus accelerator to 5 MeV, transversely combine them to 4 beams, and accelerate these to 10 MV in a magnetic-focus accelerator. In 1992 the ILSE program began to consider accelerating four potassium beams at full driver line-charge density and beam size to 5 MV before combining to a single driver size beam that was accelerated to 10 MV. The experimental facility is designed for assembly in the LBNL building that was constructed for the Bevatron. A sketch of the ILSE accelerator installed in the Bevatron building is shown in Fig. 13.

ILSE is designed to model the elements and beam manipulations of a fusion driver. The ILSE beams will be equal in diameter to those of a driver. To minimize cost, ILSE will be shorter than a driver and have fewer beams. ILSE beams are at full driver scale in size and line charge density ($0.25 \rightarrow 1.0$ μC/m). A lighter ion (K^+) permits magnetic transport at lower beam energy and therefore cost. A short beam pulse length (3.2m) permits longitudinal beam-dynamic studies in a short (≈ 40 m) accelerator. The ILSE induction linac is designed to push accelerator limits to enable cost savings and optimizations of future accelerators. Fig. 14 shows schematically how the ILSE accelerator and experiments will model a heavy ion induction accelerator/driver.

In December 1994, the DOE approved a construction start for Elise [5], the electric focus portion of the ILSE accelerator, to begin in October 1995 with a construction schedule of approximately four years. Elise has an estimated hardware cost of 7.3 M$ (TPC 25 M$). For comparison, the total hardware cost of ILSE is estimated to be approximately 18 M$ (TPC $\approx$ 50 M$) – (94$). Elise will use the 2 MV potassium injector described above.

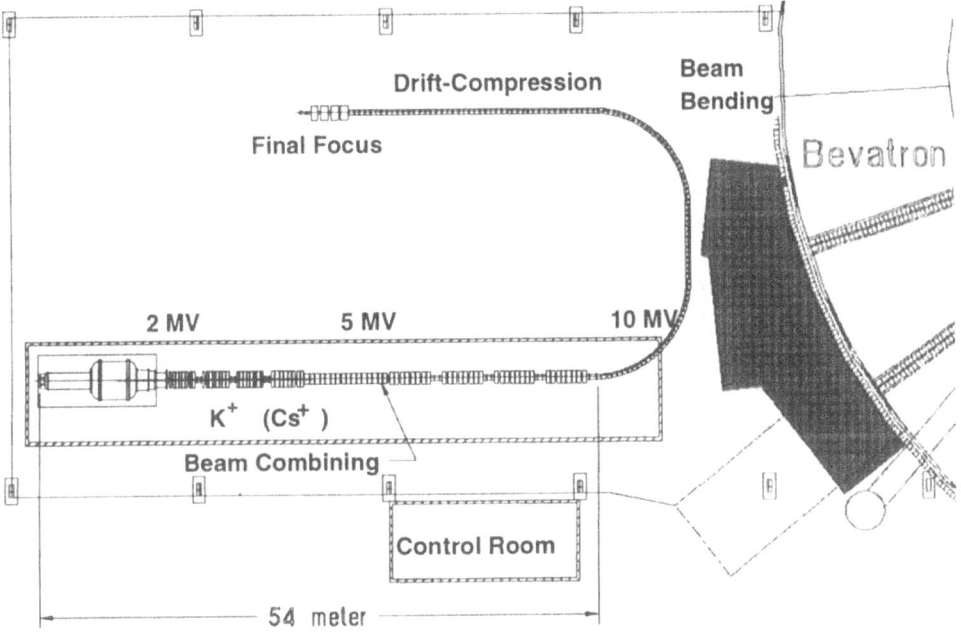

Figure 13. A sketch of the ILSE accelerator and down stream experiments as installed in the Bevatron Building at LBNL.

5. CONCLUSIONS

Heavy-ion-driven inertial fusion is a very attractive path to commercial fusion energy from a scientific, economic, and environmental point of view. Programs at LBNL and LLNL are examining the accelerator physics and engineering developments necessary to develop the induction accelerator as an inertial fusion driver. They have shown that heavy ion sources (single charge state) are available for experiments and at driver scale. Theory and experiment have demonstrated that intense ion beams can be transported by

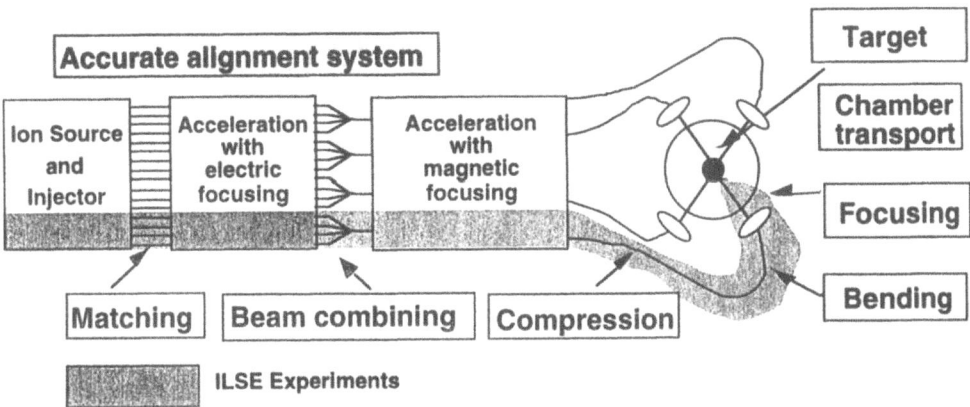

Figure 14. The ILSE experiments will provide valuable information on driver physics and beam manipulations.

electric quadrupoles without loss of quality. A 4-beam ion induction linac has demonstrated current amplification with preservation of longitudinal and transverse beam quality. The program has developed a driver-scale potassium ion injector for use with the Elise accelerator that comprises the electric-focused section of the ILSE accelerator. A model recirculator experiment at LLNL will provide first tests of key recirculator accelerator principles. The ILSE/Elise program will test most remaining HIF accelerator issues—some, particularly the beam line charge and size, at full driver scale. These experiments and developments lead to inertial fusion energy in a way that makes scientific, economic, and environmental sense.

6. ACKNOWLEDGMENT

Work supported by the Director, Office of Energy Research, Office of Fusion Energy, U.S. Department of Energy under Contract No. DE-AC03–76SF00098.

REFERENCES

1. M.G. Tiefenback and D. Keefe, Trans of Nucl. Sci., *NS-32*, 2483, (Oct. 1985).
2. S. Eylon et al., Part. Accel., *37–38*, 235 and 241, (1992).
3. T.J. Fessenden et al., Laser and Part. Beams, *5*, 457, (1987).
4. S. Yu et al., Nuovo Cimento, *106 A*, 1541, (Nov. 1993).
5. J. Kwan et al., "Elise Conceptual Design Report", LBL PUB-5393, (Feb. 1994).
6. J.J. Barnard et al., "Study of Recirculating Induction Accelerators as Drivers for Heavy Ion Fusion", UCRL-LR-108095, (Sept. 1991).
7. The NIF is described in considerable detail on the INTERNET. The net address is <http://www-lasers.llnl.gov/lasers/nif.html>
8. Moir et al., Fusion Tech, *25*, 5, (1994)
9. Meier and Bieri, Fusion Tech, *21* 1547 (1992).
10. L. M. Waganer et al., "IFE Reactor Design Studies, PROMETHEUS-L, H",MDC 92E0008, Vol. I, pES-5 (1992).
11. Krakowski et al., Los Alamos Rpt LA-UR-93–4217Ê(Dec 1993).
12. Meier and von Rosenberg, Fusion Tech, *21* 1552 (1992).
13. D.J. Dudziak et al., Fusion Tech, *13*, (Feb 1988).

18

X-RAY DRIVEN IMPLOSIONS ON THE NOVA LASER

J. D. Kilkenny and the LANL and LLNL ICF team

Lawrence Livermore National Laboratory
P.O. Box 808
Livermore, California 94550
and
Los Alamos National Laboratory
P.O. Box 1663
Los Alamos, New Mexico 87545

The Nova laser at Lawrence Livermore National Laboratory is primarily used for X-ray driven implosions where a shell containing a deuterium-tritium mixture is compressed to a high density and temperature. Experiments on Nova in the last few years have shown a good understanding of the physics of X-ray drive and symmetry in hohlraums, and the physics of hydrodynamic instabilities of implosions. In particular, the difficult parameters of an ignition implosion can all be achieved and seen to be predictable. These parameters include high radiation temperature, high convergence, high aspect ratio and the corresponding high hydrodynamic growth factors and long scale length, 1/10th critical density plasmas.

Ignition with X-ray driven inertial confinement fusion (ICF) requires the simultaneous achievement of several difficult parameters. A high radiation temperature in the range 250 – 300 eV is required to achieve a high ablation pressure and high implosion velocity of the imploding shell. At these radiation temperatures the hohlraums fills with plasma at about 1/10th of the critical density, and the laser beams must propagate through this plasma without too much deleterious scattering. A high radial convergence, in the range of 25 – 35 is required to achieve a sufficiently large areal density of fuel for laboratory ignition, and achieving this degree of convergence requires a level of symmetry of the X-ray drive in the range of 1 - 2% together with keeping the shell on a low adiabat by a shaped drive. Experiments on Nova cannot simultaneously demonstrate all of these parameters, principally because Nova only has ten beams: however separately values of these parameters in the ignition range have been demonstrated.

Nova focuses 20 – 30 KJ of 0.35 μm light into a gold cylinder to produce a hohlraum with an isotropic, quasi-Planckian radiation field. The walls of the hohlraum have a high albedo and efficiently re-radiate, producing radiation temperatures, T_r in the range 200 – 300 eV. Measurements of T_r are made by time resolved X-ray diode arrays, shock break-out from wedged and stepped witness plates and foil acceleration. An array of X-ray diodes makes an absolute spectral measurements using several spectral channels from 100 eV to 2.5 KeV, with a temporal resolution of 150 psec. Shock break-out from an Al witness plate on the wall of the hohlraum measures the drive pressure and therefore the effective drive temperature. There is good agreement between the two measurements. The X-ray drive has a quasi-Planckian spectrum and can be shaped in time to provide a low isentrope for compressing a capsule. With a shaped drive peak temperatures up to 220 eV have been attained. With a 1 nsec flat topped drive, peak radiation temperatures up to 300 eV have been measured. There is a good agreement between the measurements and calculations.

A possible problem with laser driven hohlraums is too much laser scattering from the plasma that fills the hohlraum, leading to energy loss or uncontrollable energy deposition on the hohlraum walls. We have emulated the plasma we expect to form in an ignition hohlraum using gas filled, thin wall plastic balloons and gas filled hohlraums. Plasma have been formed with lengths of several mm at 10% of critical density and electron temperatures of 2–3 keV. A low level of stimulated Brillouin scattering is observed from this plasma. The dependence of the scattering and plasma filamentation on the conditioning of the laser beam is measured.

The X-ray flux, sT_r^4, incident on a capsule in the hohlraum is in the range of 10^{14} – 10^{15} W/cm^2. Compared to direct laser irradiation, X-ray heated matter has a higher mass ablation rate M, producing, by the rocket reaction, a higher "ablation" pressure.

The implosion of a capsule by ablation is Rayleigh-Taylor unstable to small perturbations. Fortunately, the mass ablation reduces the instability growth rate from the classical $g = \sqrt{kg}$ to $g = \sqrt{kgkM/r_c}$, where k is the perturbation's wave number, g is the inward acceleration of the capsule, and r_c is the peak density of the capsule. A detailed set of measurements confirm this reduction of the instability growth rate to an acceptable level.

Implosion experiments have also been used to test predictions of hydrodynamic growth rates. Spectroscopic tracer materials are included in the inner layer of plastic shells filled with gas that are imploded in a hohlraum. When the inner layer of the plastic mixes with the hot compressed gaseous fuel, X-ray emission from the tracer material is used to quantify the level of mix. The level of mix is varied by varying the amplitude and mode structure of the surface finish of the plastic shell by photo-ablation.

To compress a capsule, symmetric irradiation is necessary. Symmetry results from both good beam balance, and the intrinsic smoothing of hohlraums. The Precision Nova project has demonstrate 5 – 8% r.m.s. deviation of power over the ten beams of Nova. Precision Nova and the intrinsic smoothing of a hohlraum, has produced time integrated drive symmetry of 1 – 2%. Symmetry swings occur and techniques for measuring and controlling the time varying swings have been developed.

A high convergence set of implosions has been performed on Nova with radial convergences in the range 20 – 30. This convergence is achieved by using a 8:1 shaped pulse drive, minimizing shock heating and keeping the shell on a low adiabat. Thermonuclear yields, shell and fuel densities and burn durations are measured and closely match the calculated conditions. The yields agree in detail with simulations indicating that the level of mix of the cold shell into the hot compressed gas is in the range 30 – 50%.

19

PRESENT STATUS AND FUTURE PROSPECTS OF LASER FUSION RESEARCH AT OSAKA

Chiyoe Yamanaka

Institute for Laser Technology
2–6 Yamada-oka Suita
Osaka 565, Japan

ABSTRACT

Recent progress of inertial confinement fusion has led us to begin a new research program towards laboratory demonstration of the thermonuclear ignition and burn. We have done an intensive R&D project addressing a number of physics and technical issues. They are (1) to evaluate the required laser energy for ignition and burn as well as the tolerable level of irradiation nonuniformity and target imperfections, (2) to develop uniform irradiation technology with existing twelve beams of GEKKO XII, to provide a stringent test of the theoretical and computational modeling and (3) to design and develop a new laser facility necessary for the ignition and burn.

INTRODUCTION

The implosion experiments by GEKKO XII demonstrated a high temperature compression of DT fuel up to 10 keV, neutron yield 10^{13} and a high density compression of CDT hollow shell pellet to reach 1000 g/cm^3. The records of implosion parameters, together with various fundamental physics data base have been utilized to examine our theoretical and computational modeling and to increase the credibility and reliability of simulation codes.

We have made extensive comparison of the implosion experiments with one-dimensional (1D) simulation including a mixing model, and the model prediction of the mixing width was consistent with the experimental results. We have also made straight forward tests of our two-dimensional (2D) simulation, which inherently includes various hydrodynamic and transport processes, by using planer target experiments.

The key issues toward ignition and burn have been investigated by using 1D and 2D simulation codes which have been reinforced by experimental results. The main concerns are the gain scaling of direct drive targets and its sensitivity to the nonuniformity of irradiation and to the imperfection of pellet. Stability analysis together with the efficiency consideration gives us variety of design of pellets and drivers. The ignition is expected with the drive energy of 0.1~1MJ.

As improvement of uniformity the indirect drive (Cannonball target) experiments are adopted. Cryogenic target for liquid or solid hydrogen fuel and/or cooled target for high density gas fuel have been developed, the qualities of which well meet for the quantitative implosion experiments. New diagnostics with higher time and space resolutions (10ps, 15μm) have been developed to get detail information of implosion.

As the final goal of the Inertial Fusion Energy (IFE), we have completed the conceptual design work of a laser fusion reactor KOYO based on a LD pump solid state laser as a driver. The current conclusion is that the reactor is technically feasible with a pellet gain of 100~150 by a 4 MJ blue laser. In Table 1 the recent progress of ILE fusion research is presented.

DRIVER ISSUES

Laser driver for fusion has several important issues to be cleared. Recently very high power lasers 100kJ~MJ are available to use. The efficiency and also repetition rate in operation are still crucial problems for a reactor driver.

As for the breakeven experiment in direct drive, one of the Key issues is the uniformity of irradiation to targets. At Institute of Laser Engineering a strong campaign of research and development has been performed on the improvement of laser irradiation uniformity.

The laser absorption nonuniformity of multibeam systems can be decomposed into a single beam factor and a geometrical factor for each Legendre mode, when the absorbed pattern of a single beam is axially symmetric on a spherical target.[1]

IMPROVEMENT OF IRRADIATION UNIFORMITY OF DIRECT DRIVE LASER FUSION TARGET

An optimum design of a random phase plate enables us to minimize the lower-mode nonuniformity. The beam smoothing techniques such as the partially coherent light with angular dispersion of spectrum and the two-dimensional SSD can substantially reduce the higher-mode nonuniformity. Thus, it is expected that the absorption nonuniformity of 1–2% will be achieved by the improvement of power balance in beams.

Envelope Profile Control of Focused Beam Pattern

The lower mode nonuniformity can be completely suppressed if the absorbed laser intensity on the target surface obeys $\cos^2\theta$ or $\cos^4\theta$ distribution, where θ is an angle measured from the beam axis.[2]

The envelope profile can be controlled by optimizing the RPP's.[3]

Fig. 1 shows the rms irradiation nonuniformities calculated for the incoherent laser, where d and R are the focal position and the target radius, respectively. We have two choices: one of them is the far-field illumination with an RPP of smaller (1.6-mm) element, and the other is the quasi far-field illumination with an RPP of relatively large (2.4 mm) element. The present choice is 1.6 mm element because the latter case requires a

Table 1. Progress of Laser Fusion with GEKKO XII

FY	Achievement	Driver	Pellet & Diagnostics	Theory & Simulation
Interaction and Ablation				
1983	$Y_n=10^{10}$	Completion of GEKKO XII	Glass microballoon (GMB)	Hot electron
High Temperature Demonstration (Stagnation free compression)				
1984	h_{abs}: Absorption P_{abl}: Ablation	SHG ($\omega \to 2\omega$)	LHART	
1985	$Y_n=10^{12}$			Shock multiplexing
1986	$Y_n=10^{13}$ $T_i=10\,keV$		Fusion yield calibration	
High Density Demonstration (Low isentropic compression)				
1987		THG ($\omega \to 3\omega$)	CD shell	
1988	$\rho=1000\times\rho_s$ (main fuel)	Random phase plate	CDT shell	Nonlocal transport
1989			2nd. reaction p knock-on Si activation	
1990	Fermi degeneracy High density compression			2D, 3D codes
Ignition and Burn Demonstration				
1991	Thermal smoothing	Partially coherent light	Cryogenic target	
1992	RT instability R-M instability			α heating
1993	Implosion stability	GEKKO XII	Ultra-fast frame	
1994		Up grade	Cannonball	

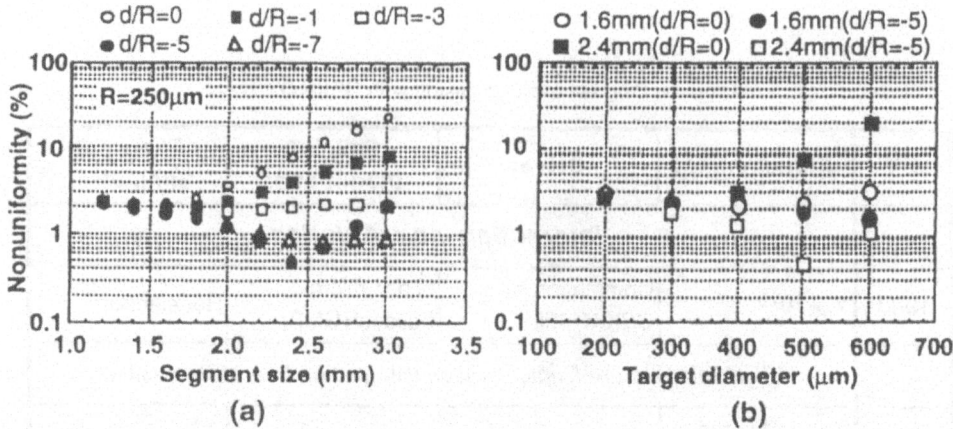

Figure 1. (a): Laser irradiation nonuniformity (σ_{rms}) as a function of hexagonal RPP element size calculated for the incoherent laser with different focusing conditions. The element size is defined by a diameter of a circumscribed circle. (b): Laser irradiation nonuniformity for four cases as a function of the target diameter.

quite uniform near-field pattern. The error of RPP coating thickness was clarified by means of Zernike interference technique to be 1–2% for 0.527-μm laser (2ω).

Furthermore, an aspherical multi-lens array (MLA) of long focal length with specially designed edge shape was demonstrated for the envelope profile control.[4] This MLA is promising for the substantial reduction of the long wavelength intensity modulation that is unavoidable for RPP because of the somewhat correlated interference between neighboring beamlets.

Beam Smoothing and Pulse Shaping

The RPP divides a nonuniform beam into a huge number of small beamlets, then these beamlets randomly interfere with each other on a target surface creating a fully developed, fixed speckle pattern with relatively uniform intensity envelope (see Fig. 2(a)). To suppress the small scale imprint nonuniformity due to this speckle[5], the partially coherent light (PCL, a concept similar to echelon-free ISI)[6] has been developed for the high-power Nd:glass laser as shown in Fig. 2(b).[7,8]

Specifications of PCL are 0.6-nm band width, 32 times diffraction-limited beam divergence (both measured at the fundamental, ω) and the angular dispersion of spectrum of 239 μrad/nm (at the inlet of KDP frequency doubling crystal). The smoothing time in which the single beam nonuniformity reached 10% was measured to be ~40 ps. Therefore, the PCL is suitable for a foot pulse, although it degrades the efficiency of harmonic conversion.

Therefore the smoothing by spectral dispersion (SSD)[9] having 0.6 nm band width (at ω) with 10 GHz frequency modulation was introduced as a main pulse. Moreover a two-dimensional (2-D) SSD has been demonstrated to reduce medium mode nonuniformities caused by a streaked feature along the dispersion direction. Here we adopted the spatial color cycle of which direction was perpendicular to that of angular dispersion of spectrum. Although this modification in SSD scheme is quite simple, the effect of 2-D smoothing is clearly seen in Fig. 2(c).

The irradiation nonuniformities on the target surface were estimated from these patterns. The rms nonuniformity (σ_{rms}) for the case of coherent laser with RPP (Fig. 2(a)) was

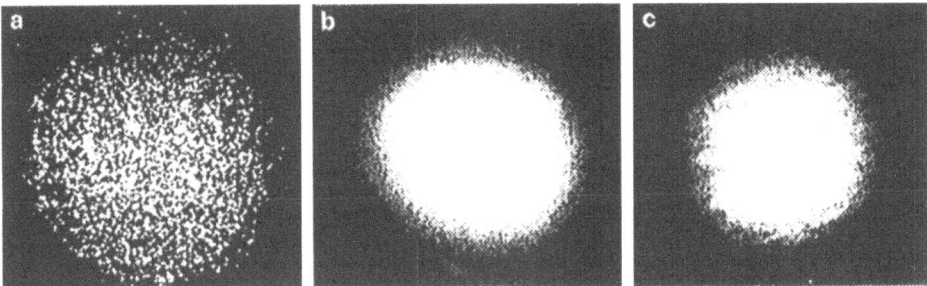

Figure 2. Focused beam pattern measured at a tangential plane of spherical target of 500-μm diameter. (a): coherent laser with an old RPP of 2-mm square elements, (b): angularly dispersed partially coherent light with the old RPP, (c): SSD with a new RPP of 1.6-mm hexagonal elements.

19.7% that was mainly due to the almost white spectrum from medium modes through higher modes (~2% per mode). By the introduction of PCL (Fig. 2(b)) the substantial reduction of intensity perturbation was observed for higher modes (0.05% per mode for $\ell>40$) resulting in a nonuniformity level of 3.8%. For 2-D SSD (Fig. 2(c)) σ_{rms} was estimated to be 4.7%, which seems to be enough small as the thermal smoothing is effective.

In the experimental study of implosion stability, it is required to control the isentrope of imploding shell. If the laser pulse shape can be changed flexibly, one can quantitatively estimate the entropy of the shell compared with other methods such as hot electron preheat or radiation preheat. The system we have developed for the fiber-PCL (Fig. 3(a)) has a capability to deliver an arbitrary pulse shape (such as tailored pulse, square pulse and picket fence) with a fast rise time of order of 100 ps. Before the introduction of spatial incoherence using the step-index multi-mode fiber[3,10], a pulse shaping system consisted of optical fiber couplers was installed. First, a single compressed pulse of ~15 ps was stretched to 50 ps using a grating (G2 in Fig. 3). Next, this square pulse was divided into four pulses with time delay of 50-ps step, and then stacked to a 200-ps flat top pulse. Finally, the 200-ps pulse was divided into 16 channels to be stacked again with different time delays of 200-ps step. Here we changed channel-to-channel intensity ratio using variable attenuators to control the pulse shape. Figure 3(b) shows an example of pulse shape at the optical fiber output that was supplied for the high isentrope compression by a square pulse associated with a separated prepulse. To preserve this pulse shape, the time delay (~300 ps) across the beam diameter due to the grating (G3 in Fig. 3) being used for the angular dispersion of spectrum should be compensated. For this purpose an echelon mirror of 15 steps with 33-ps time separation was set after the pulse shaping optics.

Power Balance Control

Fig. 4 shows the principle of power balance control. The 2ω power balance requires the power density balance of incident ω beam onto the KDP crystal. First, we concentrate

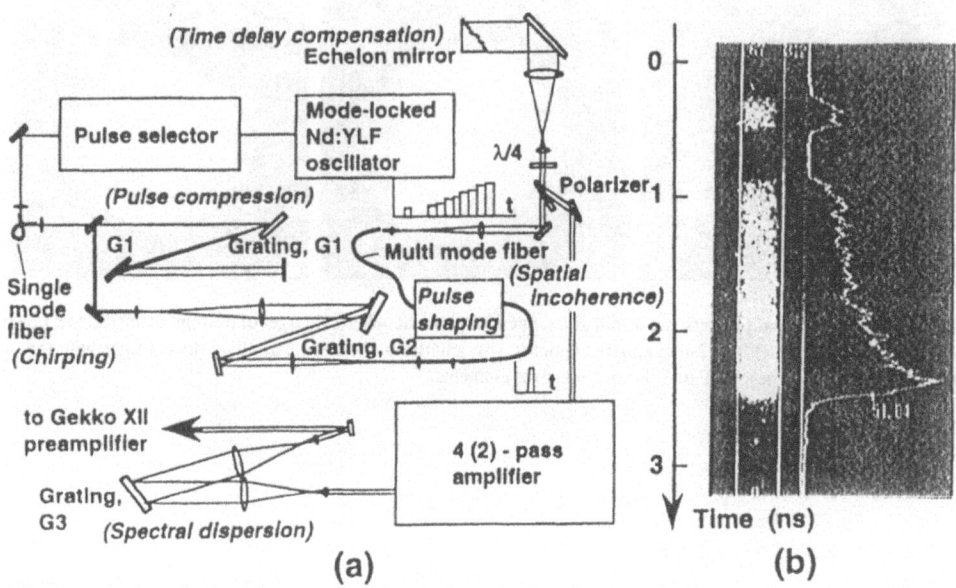

Figure 3. (a): An optical fiber system for supplying PCL with flexible pulse shaping. The system consists of several parts; the frequency chirping (control of temporal coherence), the pulse compression, the pulse shaping, the modal dispersion (induced spatial incoherence) and the compensation for spatial time delay. (b): An example of shaped pulse that was used in high isentrope implosion experiments.

to obtain the similar pulse shapes of ω beams, then ω power density balance can be realized by installing additional variable loss (partial reflection mirror with adjustable tilt angle) at the outlet of amplifier chain. The variable loss compensates not only the output energy imbalance but also the reflectivity imbalance of mirrors being set after the amplifiers. To achieve the similarity of ω pulse, the time dependent power gain is controlled by adjusting the pumping of rod amplifiers at the upper stream and two disk amplifiers at the lower reaches respectively for the small signal region and the saturated amplification region. Thus we can achieve the ω power density imbalance of ~1% level that enables us to obtain the similarity of 2ω pulse with an accuracy better than ~2%. Finally, the differ-

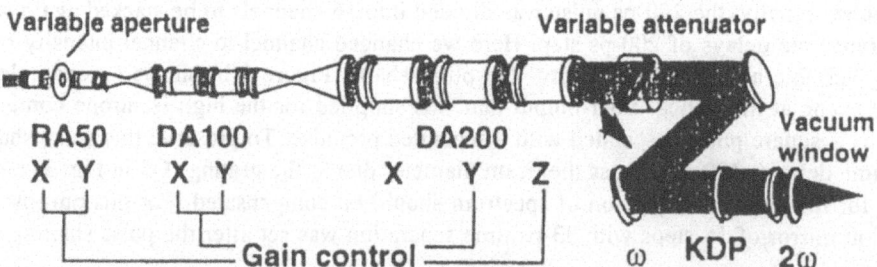

Figure 4. Schematic arrangement of main amplifier chain of GEKKO XII. The power gain is controlled by the rod amplifiers (RA50X and Y simultaneously) and the disk amplifiers (DA200Y and Z independently) for the small signal region through the saturated amplification region to obtain similar ω pulse shapes. By adjusting variable attenuators the ω power density balance is achieved at KDP crystals ensuring the similarity of 2ω pulse shape. The final 2ω power balance on a target is controlled by adjusting diameters of variable apertures.

ences of harmonic conversion efficiency among beams and the imbalance of transmission of focusing optics are compensated by changing the beam diameter using a variable aperture. This procedure of power balance control was confirmed by improving the accuracy (~1%) of energy monitors of ω and 2ω beams and by introducing 4GHz oscilloscopes for the pulse shape measurement.

Fig. 5 shows an initial test result of 2ω power balance control. Preliminary data shows the power imbalance of 4–6% (rms) at around the pulse peak and that of 7–9% (rms) at the foot region. The further improvement is expected by accumulating detailed data.

Overall absorption uniformity

The absorption nonuniformities for different laser conditions were calculated by means of ray tracing.[5] A target simulated was a plastic shell of 600-μm diameter and 6-μm wall thickness irradiated by 6-kJ trapezoidal pulse having a full width at half maximum of 2ns (rise time is 1ns). Here we took into account the fixed power imbalance, and the absorption nonuniformity was evaluated in the imprint phase as plotted in Fig. 6. Expected absorption nonuniformities for PCL foot pulse and SSD main pulse are marked in this figure for two types of RPP designed both for the far-field irradiation and for the profile-controlled irradiation. The achieved overall irradiation (or estimated absorption) nonuniformity was in a few % level, and the further improvement is expected to overcome 1% level. Therefore we have a confidence of the feasibility of ignition demonstration with a 300–500kJ driver.

INDIRECT DRIVE IMPLOSION CANNONBALL EXPERIMENT

Direct drive and indirect drive implosions have very different features, the former has high coupling between laser and target while the uniformity problem for absorption is

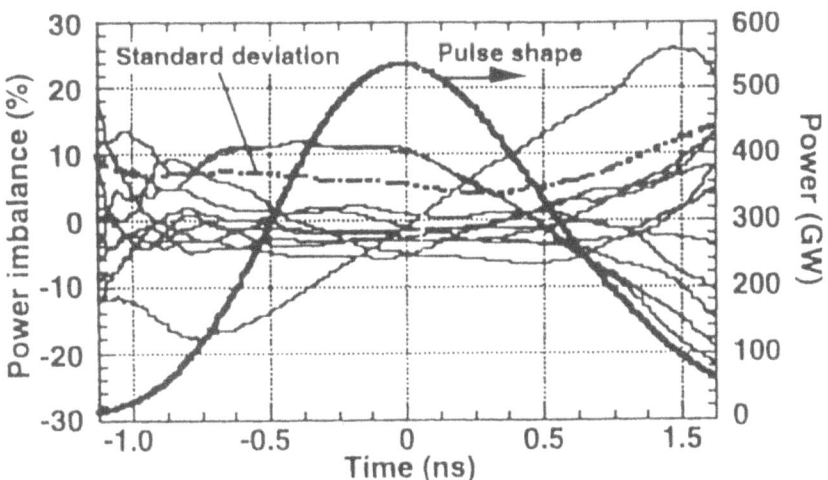

Figure 5. An initial test result of power balance control. Percentage power imbalance of each beam is plotted as a function of time. Here variable attenuators have not been used yet.

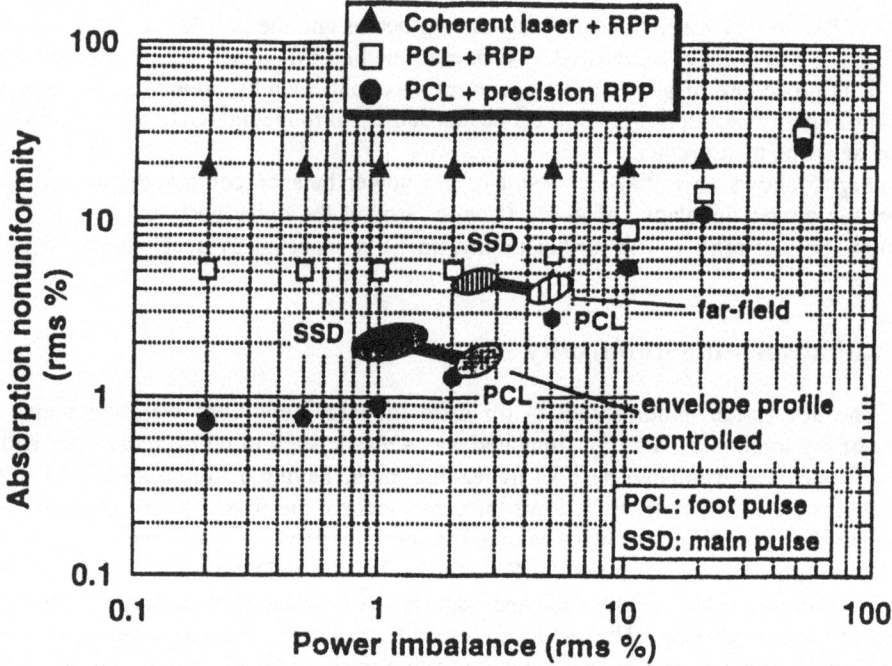

Figure 6. Calculated absorption nonuniformity for typical laser conditions such as the coherent laser with old RPP (shown in Fig. 2(a)), the angularly dispersed partially PCL with old RPP (Fig. 2(b)) and the angularly dispersed PCL with an RPP of 2.4-mm hexagonal elements. Shaded regions show the near-term and future goals of GEKKO XII upgrade.

very crucial which imposes the several precautions to the irradiation system. The latter is low coupling of radiation to target however it has much better uniformity of irradiation. The direct drive fusion is estimated to get an ignition by the 300kJ blue laser if the nonuniformity of irradiation was suppressed to less then 1% RMS. While the indirect drive fusion will need a few MJ driver. The NIF program is expected to demonstrate the fusion ignition by indirect drive. We would like to have a scaling law of indirect drive fusion using the Cannonball target by the GEKKO XII blue laser. The published data on Cannonball experiments of ILE Osaka in these 10 years have a strong influence to the US classification policy of indirect drive fusion which causes the recent disclassification by US DOE Secretary O'Leary.

Experimental Conditions

A typical target consisted of a fuel capulse of approximately 300μm diameter placed inside a cylindrical Au cavity of 800μm diameter 1600μm length. The cavity was irradiated by 10 laser beam of 351nm wavelength with a Gaussian pulse of 0.7–0.9ns width (FWHM). The total laser energy on target was 4kJ with an energy balance of 2% (RMS).

The fuel capulse was either a plastic shell filled with 10atm deuterium with addition of 0.5atm Ar for x-ray diagnostics, or a glass microballoon (GMB) filled with 8atm DT. Both of the targets were coated with teflon whose thickness was varied to change the implosion dynamics. The initial areal density (ρR) of the shell of 1 to 4mg/cm^2 corresponds the capulse radial convergence of 5 to 25, respectively. These shells typically had 1.5%

thickness nonuniformities with low mode numbers of up to 10 and a few-μm scale sueface roughness, due to teflon coating, of 0.1–0.2μm depth.

The irradiation positions of the laser beams on the inside surface of the cavity was chosen to minimize the low-order nonuniformities. The rms nonuniformity of the mode 2, σ(2), estimated from the x-ray transfer calculation[11] is initially 8% and decrease to 2% in 0.5ns as the x-ray is circulated in the cavity. The mode 4 nonuniformity σ(4), which is not sensitive to radiation confinement and laser beam pointing, remains at approximately 2% over the pulse duration.

Observation of Implosion Dynamics at Stagnation

A high gain target design requires isentropic compression of the fuel to a high density and formation of a hot spot at the center of the compressed fuel.[12] One of the major critical issues in this design is the fluid instabilities at the pusher-fuel interface at the deceleration phase which leads to deviation from spherical compression and induced pusher fuel mixing.

We have employed various diagnositc techniques to evaluate the plasma parameters of the compressed core. With increse in the pusher ρR, time integrated x-ray spectra of the Ar-filled targets showed significant broadening of the argon resonance lines accompanied with strong satellites, corresponding to increase in the electron densities from 1×10^{23} to 1×10^{24} g/cm^3.[13] The electron temperature is approximately 1keV with spatial variations over the fuel as observed by a 2-channel monochromatic x-ray imaging camera.[14,15] Using DT filled GMB targets, the emission times of the neutrons and the x-ray were measured simultaneously with their ralative timings calibrated accurately to within 50 ps. The neutrons are emitted almost simultaneously with the x-rays for the pusher ρR of up to 3mg/cm^2, above which the neutron emission terminates earlier than the x-ray emission. This result indicates that the fuel temperature starts to decrease earlier than that of the pusher at the heavily stagnated compression.

We define a "stagnation ratio (SR)" as the ratio of the pusher inner radius at the beginning of deceleration to that at the maximum compression (peak burning time). The stagnation ratio in this experiment was varied from 1 to 3.5. Plot of the observed neutron yield normalized by the 1-D calculated value vs SR (Fig. 7) shows that the normalized neutron yield has close correlation with SR over broad experimental conditions. The normalized neutron yield decreases rapidly with increase in SR and corresponds approximately to the 1-D simulation value at the beginning of stagnation.

The thermal changes of the spatial distributions of the x-ray emissions were observed with a multiple-imaging x-ray streak camera (MIXS) which has a temporal resolution of 10ps and a spatial resolution of 15μm.[16] The x-ray emission framing images of the pusher and the fuel were observed with the DT-filled GMB and the D$_2$/Ar-filled plastic microballoon, respectively. The pusher images show that each segment of the glass shell emits x-rays at different times, varying from -25 to +40ps from the average. The x-ray emission images from the fuel in Fig. 8 also show very rapid changes in the intensity distributions. Nonuniformities with the mode numbers of not only 4 but also of odd numbers such as 1 and 3 are observed in these images.

These results are indicated that the spherical compression of the fuel to a high density at the stagnation phase has not been attained in the present indirect-drive experiments. Analyses on the instability growth show that improvements in the irradiation nonuniformities and in the surface roughness of the target are required to attain spherical compression under strong stagnation.[13] These improvements will be achieved by using smaller diameter, single layer parylene shells which are used for direct drive implosion experiments at ILE.

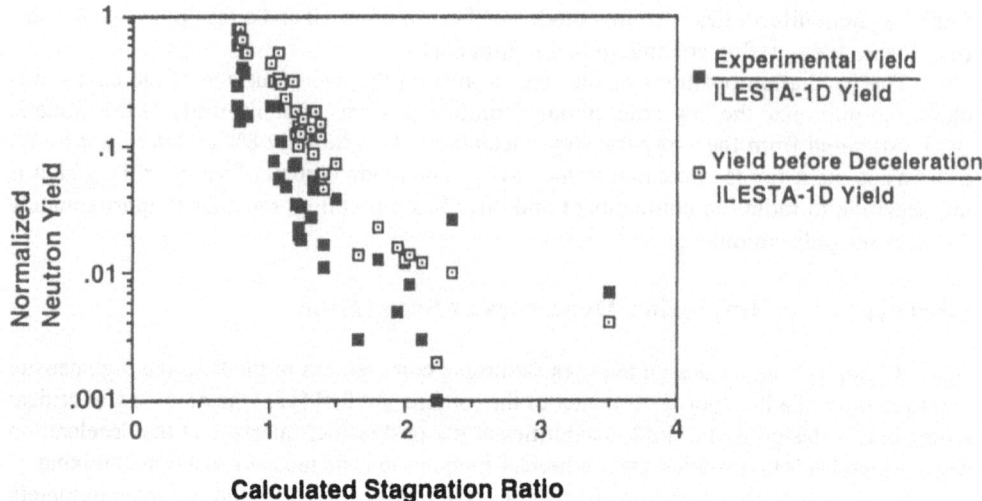

Figure 7. Neutron yield normalized by the value calculated with 1-D ILESTA simulation code, plotted vs. the stagnation ratio. The open squares and closed squares correspond, respectively, to the experimental yield and the calculated neutrons before the start of decelerating.

Irradiation nonuniformities of the low mode numbers which are accompanied with the cylindrical cavities will be reduced by using two bundles of phase laser beams. Higher irradiation symmetry will be attained with a spherical cavity. We note that in one of our early experiments 12 laser beams were introduced into a spherical cavity through 4 entrance holes with tetrahedral symmetry and the fuel capulse in the cavity was successfully compressed.[17] Also in the MPQ/ILE joint experiments, 5 laser beams were introduced

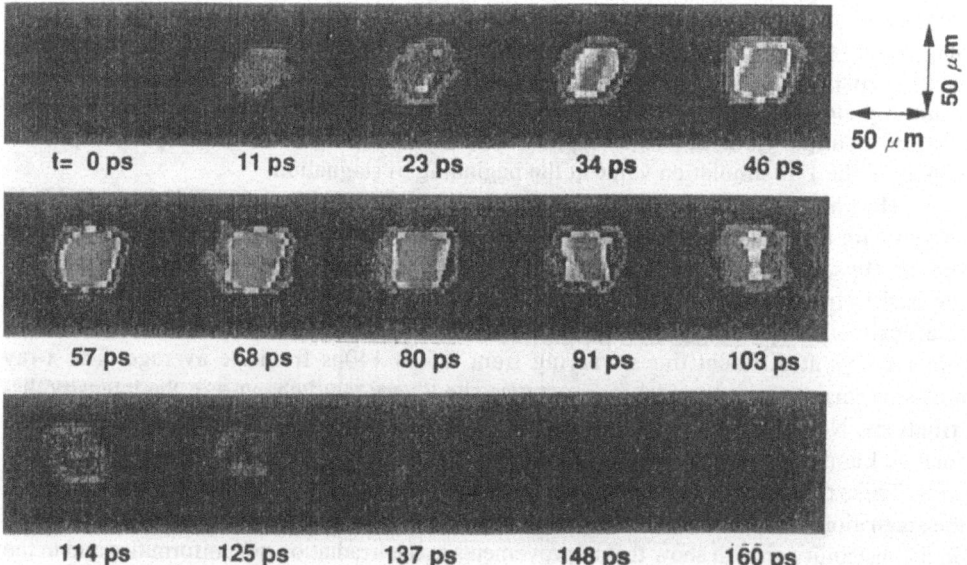

Figure 8. The framing x-ray images, observed with MIXS, of a plastic microballoon filled with D_2 and Ar. The frame interval is 11ps. Imploding core structures can be measured with time-resolution of psec.

through an entrance hole of a spherical cavity and each laser beam was accepted by a trap for conversion to x-ray radiation, resulting in 100eV blackbody radiation.[18] Extremely high irradiation uniformity will be attained by placing a fuel capusle in this type of indirectly driven spherical cavities.

Another possibility is the direct/indirect hybrid target concept. One of the variations in this concept is to use a high-Z doped low density foam in which laser energy is converted to x-ray and transported to a fuel capusle inside the foam layer. Fabrication of the foam shell over a fuel capusle is tested at ILE.

Implosion dynamics at the stagnation phase were studied. A newly developed 10ps x-ray framing camera (MIXS) revealed very rapid changes in the x-ray emission distributions in the pusher and the fuel. These studies indicate that improvements in the irradiation uniformities and the target surface roughness are required to attain the high spherical symmetry under highly stagnated compression.

NUMERICAL STUDIES ON STABILITY AND MIXING IN LASER DRIVEN IMPLOSON

The hydrodynamic instability is the most important issue in laser fusion research. In order to numerically study nonuniform implosion, it is straightforward to carry out direct numerical simulation (DNS) with a high speed computer. However, it is difficult even with near-future super-computers to carry out DNS of the laser driven implosion, because not only an effective Reynolds number of the implosion target plasmas is very high but also sophisticated physical processes such as electron thermal conduction and radiation transport essentially couple with the hydrodynamic phenomena. It is noted that DNS is useful to study the effect of, for example, the energy transport on the turbulence fluid motion by assuming some model situation. It is expected that such energy transport very effective small-scale eddies.

In order to analyze experimental data, we are required to take account of nonuniform implosion as well as the physics essential in laser driven hydrodynamic phenomena, such as laser absorption, electron heat transport, radiation heat transport, the equation of state, nuclear reaction, and so on.[19] From a practical point of view, a large-eddy-simulation (LES) type two dimensional implosion codes should be developed. In LES, relatively large scale nonuniform fluid motion is treated directly, while the eddies smaller than the grid size is treated by a statistical model. Even if this approach is still hard to apply to the laser implosion dynamics because of computer ability, we may be able to employ a two dimensional turbulence model.[20] At the present time, we are developing a two dimensional implosion code and a k-ε type mixing model in parallel. With the two-dimensional code, we study the degradation of fusion gain by nonuniform implosion in a high gain target. The high gain target is a future target to use for fusion reactor and the conceptual study has been carried out with mainly the one-dimensional code. Our idea is to reduce the source of small eddy instability technically by increasing the uniformity of laser intensity profile. Then the nonuniformities which can not be eliminated principally are those with long wavelength. The effect of such nonuniformity on fusion performance may be studied with the two-dimensional implosion code.

However, in order to analyze the experimental results, it is unavoidable to include the effects of mixing due to small scale eddies. This is because the present time the uniformity of laser intensity is poor with many of short scale speckles due to interference and nonlinear amplification. It is rather reasonable to consider that the turbulent mixing is essential to explain the experimental results.[21] Therefore, we have employed a k-ε type turbulence model to couple with the conventional one-dimensional implosion code and studied if the model can explain the fusion yields obtained through a series of experiments carried out at our institute with GEKKO XII laser system.[22]

Turbulent Mixing in Laser Implosion

Since most of experiments carried out in the past were under very nonuniform laser irradiation, we are required to model the effects of turbulent state in laser implosion. The turbulent mixing has been observed in shock tube experiments[23,24] and accelerating tank experiment[25] and has been analyzed with direct numerical simulations[26] or turbulence model.[23,24] It has been reported that the turbulent mixing model based on k-ε type model can well reproduce such experimental results.[23,24] Although we have applied the quasi-linear type diffusion model to study the mixing in laser implosion,[21,27] we try to use the k-ε type model to couple with one dimensional hydrodynamic code ILESTA-1D[19] in order to keep self-consistency of basic equations. We have extended the model in[24] to two temperature case and studied degradiation of implosion performance due to the predominant growth of mixing in stagnation phase.

In our simulation, we have assumed that a turbulence has generated in the acceleration phase with a level of

$$\xi = k / \left(\frac{1}{2} u_0^2\right)$$

where u_0 is the implosion velocity and

$$k = \frac{1}{2}\langle v_t^2 \rangle$$

is the kinetic energy of turbulence, equivalently the total kinetic energy of perturbations. With a constant value of k in space at the beginning of stagnation dynamics, we have solved k-ε type equations in the stagnation phase.

CD-Shell High Density Implosion

It has been reported that almost 1000 times solid density has been achieved by imploding deuterized polyethylene hollow shell targets at ILE Osaka.[28] In the compressed plasma electron is Fermi-degenerated. The dense plasma is so high that the abstrophysical plasma experiment can be expected. Fig. 9 shows the compression state of plasma density.

With this implosion, high density compression has been demonstrated: however, the neutron yield was far below than the 1-D numerical one. It has been phenomenologically explained that this discrepancy is due to no formation of spark-main fuel structure always seen in the conventional simulation of hollow shell implosion.

In order to quantitatively explain this experimental result, we have carried out the same kind of simulation as in the LHART implosion. In Figs. 10 (a) and (b), flow trajectories near maximum compressions are shown for the cases without and with mixing model, where ξ=20% is assumed in starting mixing calculation at t=2.87nsec. By comparing both of them, we can easily understand that with mixing phenomena the central spark, which

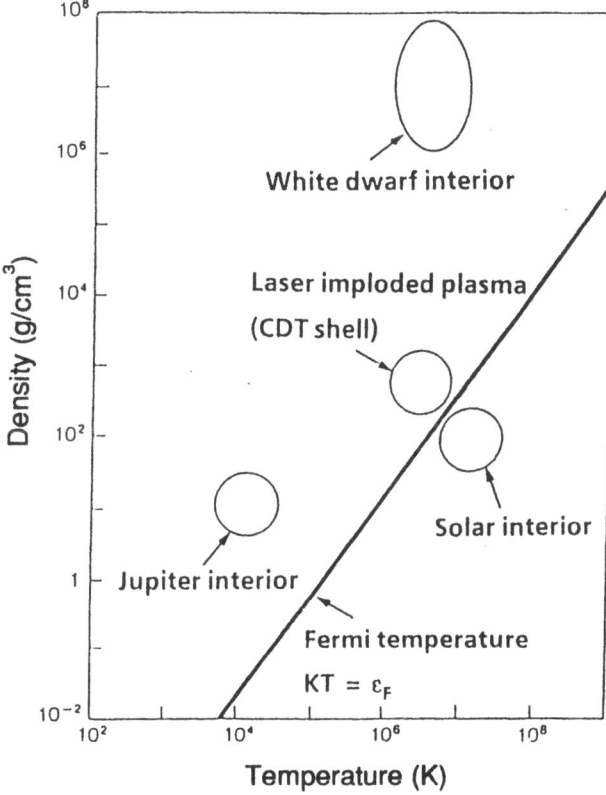

Figure 9. Laser imploded plasma and astrophysical plasma.

plays a role of cushion, is not generated and heavier material directly penetrates into the central region.

On the right hand side of each flow diagram, the density and temperature profiles at maximum compression (t=2.933ns) are shown. With turbulent mixing, the central temperature reduces by a factor of ten. This is the main reason for the discrepancy of neutron yield.

By including the mixing model, it is found that the experimental data of neutron yield (~10^6), spark temperature (~300eV) and ρR (~0.5g/cm^2) are well reproduced with assumption of ξ=20 % at the time of void closure.

Two Dimensional Simulation of High Gain Target

In order to avoid the generation of the small scale turbulence in the acceleration phase, the short-wavelength nonuniformity of laser intensity profile is planed to be eliminated by employing a beam smoothing via spectrally dispersed amplified spontaneous emission. It is numerically confirmed that the rms of the order of 0.1% can be achieved for the short-wavelength modes with the spherical wavenumber ℓ greater than 20.[29] However, it is hard to eliminate the long-wavelength nonuniformity with the wavenumber ℓ less than 10 even with such smoothing technique. In the future high gain demonstration, it is expected that these modes are serious to terminate the fusion burning. We, therefore,

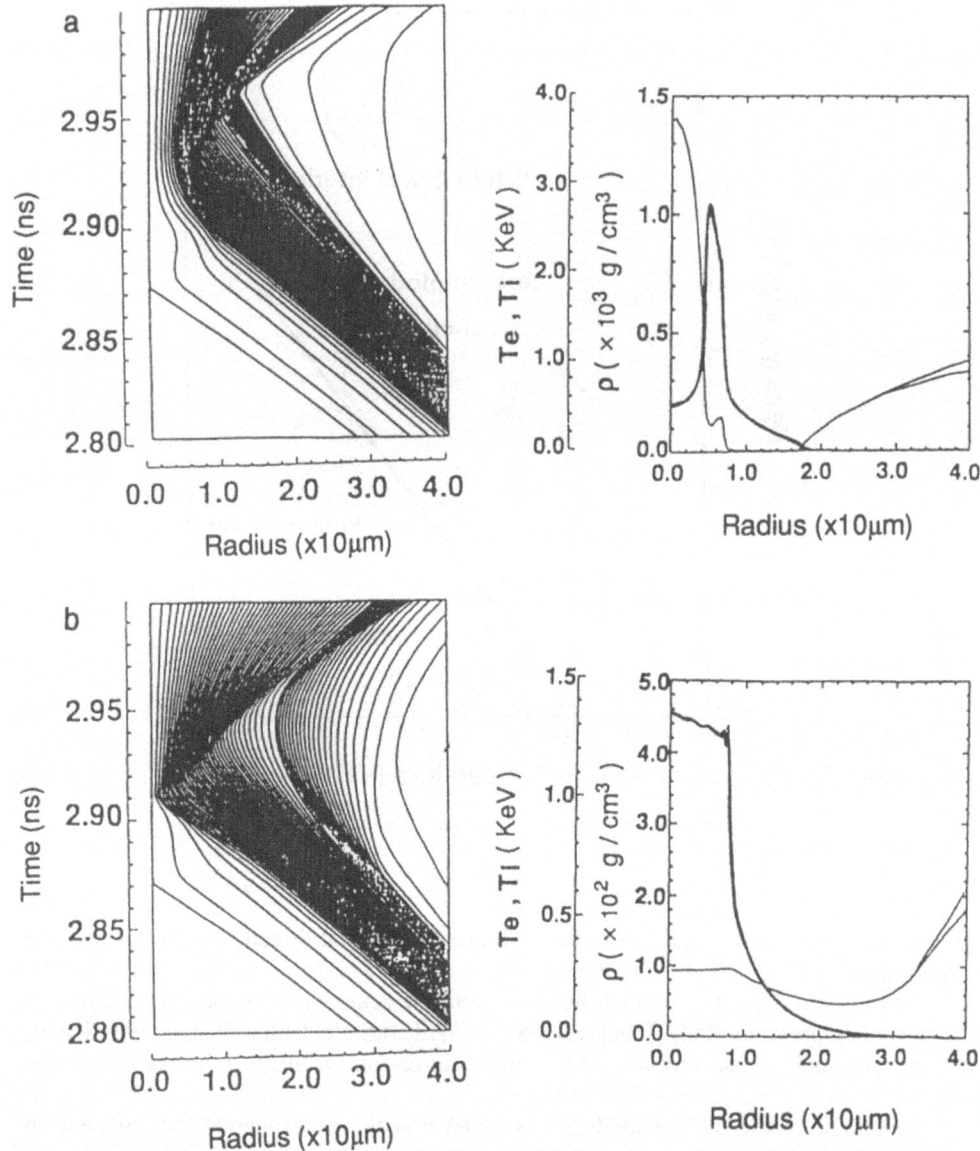

Figure 10. (a) w/o turbulent mixing ; (b) with turbulent mixing.

have studied the effect of long-wavelength nonuniformity in high gain targets. For self-consistent treatment and quantative discussion, it is necessary to cary out multi-dimension simulation with the same physics model as in the one-dimensional implosion code used for high gain target design.[30] In high gain targets, the burning wave is generated before the maximum compression and the alpha-particle heating plays essential role in hydrodynamic evolution of targets. For the case of gain 130 at 4MJ, we have studied a sensitivity of the gain to implosion nonuniformity. In Fig. 11 (a) and (b) , the equi-contours of the density are shown for respective cases without and with alpha-particle heating. By exclud-

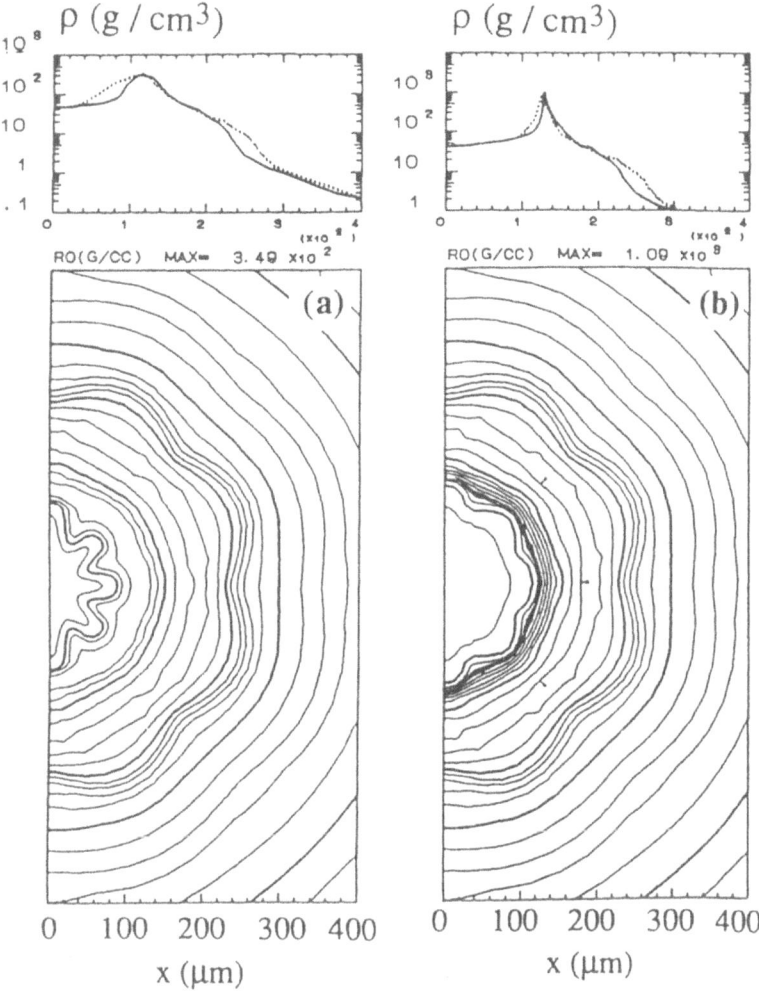

Figure 11. Equi-contours of density near maximum compression of high gain target; (a) without alpha-heating, (b) with alpha-heating. By comparing (a) to (b), the fire-polishing effect due to alpha-heating is seen to make spark structure uniform.

ing the alpha-particle heating, a predominant grow of the instability is seen near the maximum compression at the spark-fuel interface (region of radius about 50μm). By including the alpha-particle heating self-consistently, it is seen that the ignition starts before the maximum compression and the alpha-particle heating near the spark-fuel boundary polishes up non-spherical structure of the boundary as seen in Fig. 11(b).

Therefore, we can expect the thermal smoothing effect to small scale perturbation in the final burning phase. We can control the growth of perturbation small enough in the acceleration phase, we can expect such stabilization effect in the final phase once the fuel ignition takes place.

Effect of nonuniformity is a function of whether the implosion mode is marginal ignition or over ignition mode.[31] We have studied a sensitivity of pellet gain to difference of implosion mode. In Fig. 12, the pellet gains are plotted as a function of implosion ve-

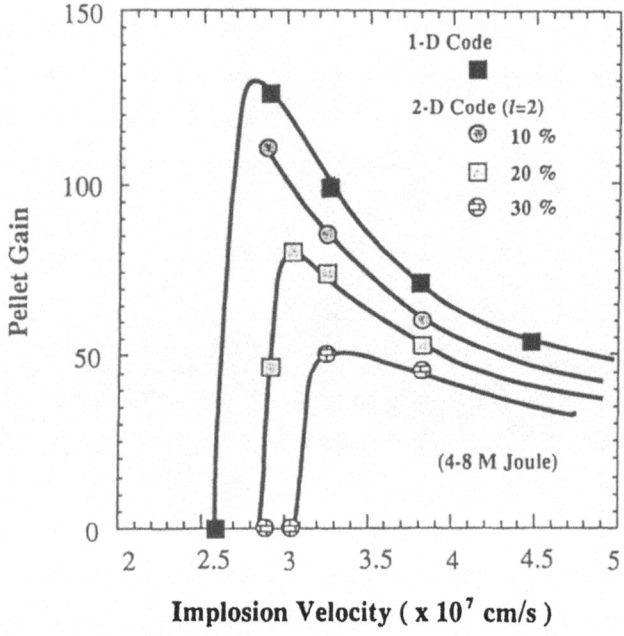

Figure 12. Sensitivity of pellet gain to the implosion velocity and nonuniform implosion dynamics.

locity for the same design as in Fig. 11. In varying the implosion velocity, we have varied the pulse duration of main pulse with the same target.[32] It is typical that the gain increases abruptly when the implosion velocity reached a critical one to ignite the spark. Then, the gain reduces gradually because excess energy is given to the spark region and spark temperature becomes higher than the minimum temperature for ignition.

In Fig. 12, the pellet gain obtained by ILESTA-2D code is plotted for the case of $\ell = 2$ nonuniformity with different value of deformation $\delta R/R_0$ at the void closure time.[30] It can be concluded that the marginal ignition mode yields highest gain within one-dimensional simulation, but very sensitive to nonuniformity. The reduction of the pellet gain becomes less as the implosion mode goes into over ignition mode, while the pellet gain is lower than the marginal ignition even within one-dimensional code.

In high gain target regime, relatively long wavelength nonuniformity becomes important and a two dimensional code is used to study the fire-polishing effect by alpha-particles. It has been reported that once the ignition takes place, a predominant heating by alpha-particles helps to smooth the core structure.

IGNITION FOR DIRECT DRIVE

The 1D simulation predicts the pellet gain of direct drive targets as a function of the laser energy. The key parameters of implosion, such as laser pulse width and shape, the ratio of prepulse to main pulse energy, pellet structure and inflight aspect ratio, have been scanned for optimization in respect to coupling efficiency and stability of implosion. The ignition energy is expected to be as low as 100kJ in an implosion mode of high velocity

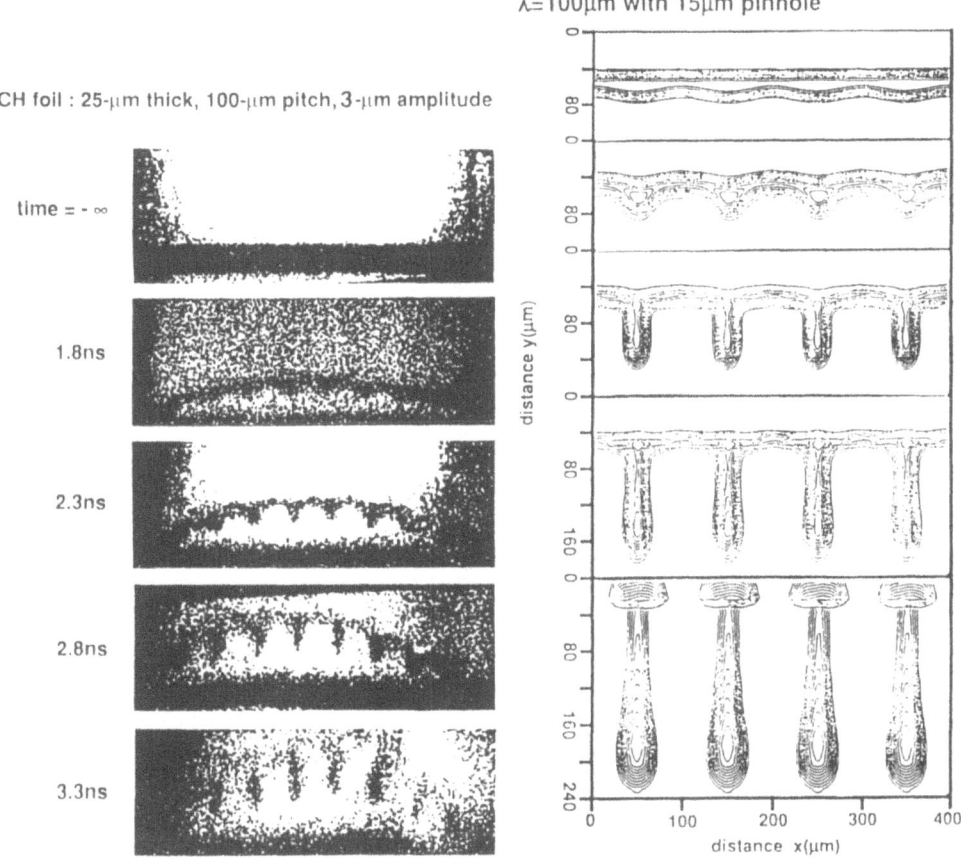

Figure 13. Temporal evolution of Rayleigh-Taylor and simulation results of corrugated target acceleration.

and high inflight aspect ratio. However as actually required energy increases due to various 2D effects. The 2D implosion model that was extensively bench marked with experimental results predict a laser energy of 300–500kJ for ignition and burn. In this case, we assumed a 300Å target surface finish and a 1% (rms) irradiation nonuniformity which are within a scope of the present technology.

Long wavelength instability causes a deformation of implosion sphericity, while short wavelength instability grown as fluid turbulence through nonlinear mode-mode coupling. In order to study the effect of short-wavelength instability, we have developed a turbulent mixing model, which have been compared to the experimental results.

To study the effect of long-wavelength instability, 2D simulation code has been developed, which has been benchmarked with straight tests by using planar target experiments. By using sinusoidally modulated target of varying wave-number and thickness together with varying rise time of laser pulse various types of instability have been investigated. Initial conditions such as imprint pattern shock induced instability and Richtmeir-Meshkof (R-M) instability are followed by Rayleig-Taylor (R-T) instability growth, leading to nonlinear phase which tends to saturation, harmonic generation, bubble and spike formation and ultimately shell breaking.

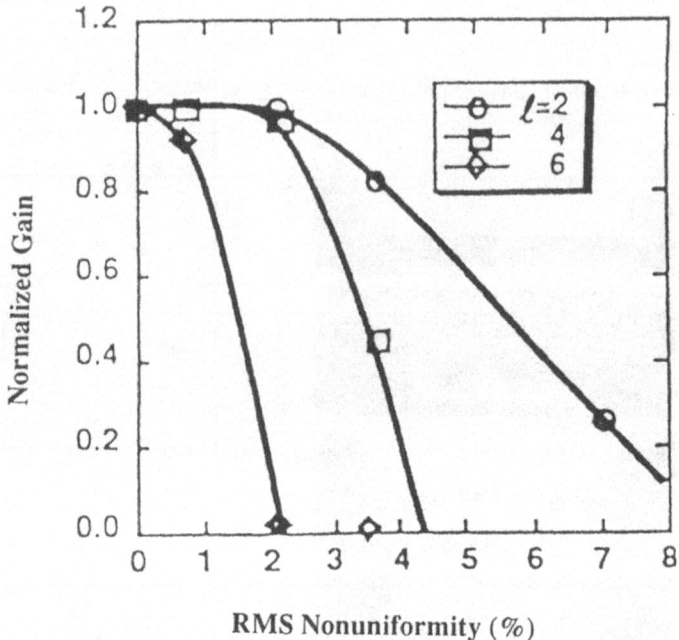

Figure 14. Normalized pellet gain as a function of the laser irradiation nonuniformity with spherical mode l as a parameter at high gain region.

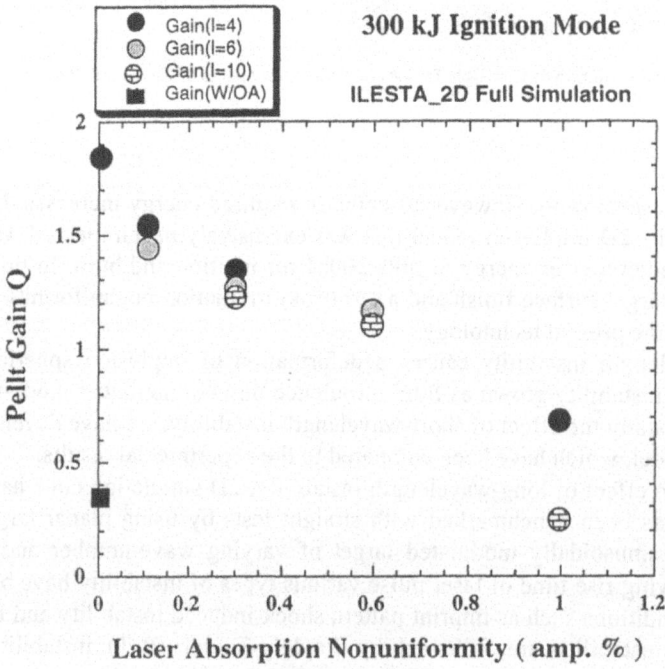

Figure 15. Pellet gain at ignition and no a heating case is shown.

An example of these tests is shown in Fig. 13, where sinusoidally modulated target with single wavenumber is uniformly indicated by partially coherent light. The 2D simulation has well replicated the experimental observation such as bubble-spike structure and perturbation amplitudes.

To evaluate the quantitative effect of low-ℓ mode nonuniformity caused by laser beam imbalance (energy and power imbalance, and pointing and focusing errors), 2D simulation of implosion has been carried out. Fig. 14 shows the gain reduction normalized by the value of the 1D simulation as a function of the laser irradiation nonuniformity caused by laser beam imbalance (energy and power imbalance, and pointing and focusing errors), 2D simulation of implosion has been carried out. It is apparent from the figure that even for the most sensitive $\ell = 6$ mode among the low-ℓ mode, a 1% nonuniformity is tolerable. Fig. 15 also indicates the effect of absorption nonuniformity in various ℓ modes.

The effect of implosion nonuniformity on pellet gain depends on the scale of the implosion, that is total fuel quantity and driver energy. Once ignition occurs at a hot spot, the mixing region of cold fuel can be burned up by a heating. The burning stabilization during the final phase of implosion and expanding phase has been investigated to be effective for operation.

KONGOH-PROJECT

We have done a number of extensive technical R&D for the step of IFE to investigate the ignition and burn.

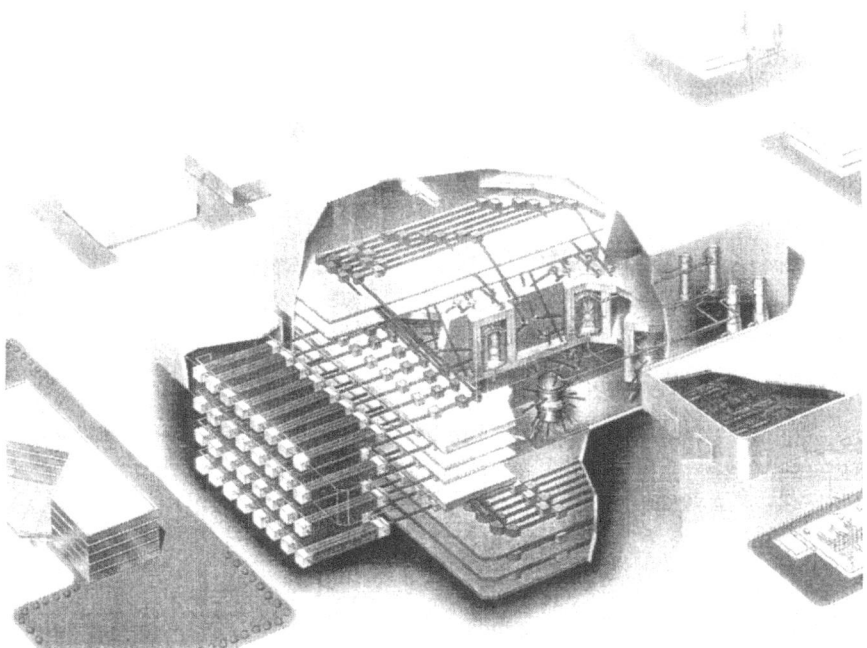

Figure 16. KONGOH project of 300kJ blue GEKKO XII upgrade laser for ignition experiment.

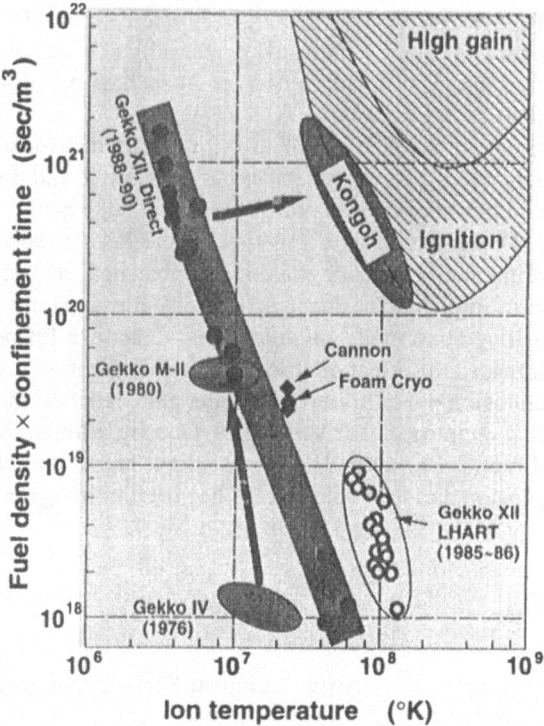

Figure 17. Progress of nτ-T diagram of laser fusion.

Irradiation uniformity of laser on a pellet target is one of the crucial issues in direct drive scheme of laser fusion. Various new technologies for precise operation of GEKKO XII (P-project) have been developed and installed. The project includes (a) beam smoothing techniques by coherent control and spectral dispersion, (b) beam focusing control for an optimum evelope shape of intensity distribution, and (c) the power balance control among twelve beams. The beam smoothing has suppressed the higher mode ($\ell>20$) nonuniformity. The instantaneous power unbalance has been improved to be less than 1% to reduce lower mode nonuniformity. The achieved overall irradiation overcome 1% level that meets the requirement for the ignition demonstration with the planned ignition facility KONGOH.

A higher power glass laser system for the KONGOH project has been designed, and relevant optical technologies have been developed. It is called GEKKO XII Upgrade which can be assembled in the present GEKKO XII building.

The output of the system is designed to be 300kJ of blue light (third harmonics) in 3ns duration of a main pulse. An 8kJ/beam output in a fundamental wavelength is attained with 4 amplifier module (12 sheets of laser glass) in 3 pass amplification of Cassegrainian geometry. The box-type booster amplifier module of 35cm aperture has been designed and tested to obtain the data bases on energy storage, pumping efficiency, small signal gain, high power amplification, extraction efficiency, mechanical and electrical performance.

The technical improvements for better cost/performance of the upgrade system have been studied. They are a new laser glass for 35cm beam aperture, 40cm diameter frequency conversion crystal of KDP or partially deuterated KD*P, 60cm diameter light weight mirror of porous quartz, high laser damage resistant optical components such as Faraday glass, optical glass, new surface finishing and coating techniques, and high energy storage capacitor.

All these technologies have been matured for the construction of the several hundreds kJ system within reasonable cost and period.

A cryogenetically cooled pellet with shell structure fuel is important for the ignition and burn experiment. We have developed a cryogenic target system which can provide a uniform solid fuel layer inside a plastic shell with and without a low density foam layer.

To form a pure fuel layer inside a plastic shell with good uniformity, we have developed new technique to improve the uniformity of frozen fuel layer by plasma formation inside the pellet.

CONCLUSION

The next step to investigate ignition and burn, breakeven and high gain is ready to initiate. The physical data bases and technical developments have been well accumulated for design and fabrication of fuel pellets and drivers for the ignition and burn experiment. The engineering and economical feasibility of laser driven fusion have been evaluated. In Fig. 17, the progress of laser fusion program at ILE is shown for the reference.

REFERENCES

1. S. Skupsky and K. Lee, J. Appl. Phys. 54 (1983) 3662.
2. MURAKAMI, M., et al., Irradiation nonuniformity due to imperfection of laser beams, J. Appl. Phys., 74 (1993) 802–808.
3. MIYANAGA, N., et al., "Partially coherent light for improving irradiation uniformity in directly driven laser fusion experiments", Laser Interaction and Related Plasma Phenomena, Plenum Press, New York, Vol.10 (1992) 251–280.
4. NISHI, N., et al., "Aspherical multi lens array for uniform target irradiation", Laser Coherence Control: Technology and Application, SPIE Vol. 1870 (1993) 105–111.
5. NISHIHARA, K., et al., Recent progress in laser fusion research at Osaka university: Uniformity and stability issues, Phys. Plasmas, 1 (1994) 1653–1661.
6. LEHMBERG, R.H., et al., Theory of induced spatial incoherence, J. Appl. Phys., 62 (1987) 2680–2701.
7. NAKANO, H., et al., Spectrally dispersed amplified spontaneous emission for improving irradiation uniformity into high power Nd:glass laser system, J. Appl. Phys., 73 (1993) 2122–2131.
8. NAKANO. H., et al., Partially coherent light generated by using single and multi mode optical fibers in a high-power Nd:glass laser system, Appl. Phys. Lett., 63 (1993) 580–582.
9. SKUPSKY, S., et al., Improved laser-beam uniformity using the angular dispersion of frequency-modulated light, J. Appl. Phys., 66 (1989) 3456–3462.
10. VERON, D., et al., Optical spatial smoothing of Nd:glass laser beam, Opt. Commun., 65 (1988) 42–46.
11. Nakamura, M et al., "Numerical method for finding uniform irradiation conditions of a fusion capulse driven by x-ray radiation", Laser Particle Beams 10 (1992) 421–433.
12. Lindl, J. McCrory, R. and Cmpbell, E., "Progress toward ignition and burn propagation in inertial confinement fusion", Physics Today 45, September (1992) 32–40.
13. Nishimura, H. et al., "Study of indirectly-driven implosion by x-ray spectroscopic measurements", submitted for pablication.
14. Uschmann, I. et al., "Temperature mapping of compressed fusion pellets obtained by monochromatic imaging", to appear in Rev. Sci. Instrum. Jan. (1995).
15. Nishimura, H. et al., "Symmetry observation and temperature mapping of laser-driven fusion plasma with novel x-ray imaging techniques", IAEA-CN-60/B-P-8, hese Proceedings.
16. Shiraga, H. et al., "Laser-imploded core structure observed by using 2-dimensional x-ray imaging with 10-ps temporal resolution", to appear in Rev. Sci. Instrum. Jan. (1995).

17. Kato, Y., "Implosion of cannonball targets at 1.053μm and 0.526μm with GEKKO XII galss laser system", ILE Quarterly Progress Report, No.14, ILE-QPR-85—14 (1985) 2–25.
18. Nishimura, H. et al. and Sigel, R. et al., "X-ray confinement in a gold cavity heated by 351-nm laser light", Phys. Rev. A44 (1991) 8323–8333.
19. Takabe, H. "Hydrodynamic Simulation Code for Laser Driven Implosion", to be published in Laser and Particle Beams
20. Youngs, D. L. et al., in the proceedings of 4th International Workshop on the Physics of Compressible Turbulence Mixing", ed by P. F. Linden et al. (Cambridge Univ. Press. Cambridge, 1993) pp. 500–507.
21. Takabe, H. et al. in the proceedings of the 14th IAEA Conference, on Plasma Physics and Controlled Nuclear Fusion Research, Vol.3 (IAEA, Vienna, 1993) 143–150.
22. Takabe, H. et al. "Scalings of Implosion Experiment for High Neutron Yield", Phys. Fluids 31 (1988) 2884–2893.
23. Andronov, V. A. et al. "Turbulent Mixing at Contact Surface Accelerated by Shock Waves", Soc. Phys. JETP 44 (1976) 424–427.
24. Gauthier, S. and Bonnet, M. "A k-ε Model for Turbulent Mixing in Shock-tube Flows Induced by Rayleigh-Taylor Instability", Phys. Fluids A2 (1990) 1685–1695.
25. Read, K. I. "Experimental Investigation of Turbulent Mixing by Rayleigh-Taylor Instability", Physica 12D (1984) 45–58.
26. Youngs, D. L. "Numerical Simulation of Turbulent Mixing by Rayleigh-Taylor Instability", Physica 12D (1984) 32–44.
27. Takabe, H. and Yamamoto, A. "Reduction of Turbulent Mixing at the Ablation Front of Fusion Targets", Phys. Rev. A44 (1991) 5142–5149.
28. Azechi, H. et al. "High-Density Compression Experiments at ILE, Osaka", Laser and Particle Beams 9 (1991) 193–207.
29. Nishimura, K. et al. "Recent Progress in Laser Fusion Research at Osaka University : uniformity and stability issues", Phys. Plasmas (1994).
30. Takabe, H. and Ishii, T., "Effect of Nonuniform Implosion on High-Gain Inertial Confinement Fusion Targets", Jpn. J. Appl. Phys. 32 (1993) pp. 5675–5680
31. Atzeni, S., "Implosion Symmetry and Burn Efficiency in ICF", Laser and Particle Beams 9 (1991) pp. 233–245.
32. Takabe, H. et al. "Critical Elements on Laser Driven for Fuel Ignition and High Gain", in the proceedings of the IAEA Tech. Comm. Meet. on Driver for ICF, Osaka, 15–19 April (1991) 160–165.
33. Nakai, S. et al., "Present Status and Future Prospects of Laser Fusion Research at ILE, OSAKA", IEAE-CN-60/B-1-I-1, Oct. (1994).
34. Kato, Y et al., "Indirect Drive Implosion of Cannonball Targets with Blue GEKKO XII Laser", IEAE-CN-60/B-2-II-2, Oct. (1994).
35. Yamanaka, T. et al., "Target Implosion using Deuterium Cryogenic System", IEAE-CN-60/B-2-II-5, Oct. (1994).
36. Nishimura, H. et al., "Symmetry Observation and Temperature Mapping of Laser-driven Fusion Plasma with Novel X-ray Imaging Techniques", IEAE-CN-60/B-P-8, Oct. (1994).
37. Miyanaga, N. et al., "Improvement of Irradiation Uniformity of Direct Drive Laser Fusion Target", IEAE-CN-60/B-P-11, Oct. (1994).
38. Takabe, H. et al., "Numerical Studies on Stability and Mixing in Laser Driven Implosion", IEAE-CN-60/B-P-9, Oct. (1994).
39. Mima, K. et al., "System Analysis of Laser Fusion Reactor KOYO Driven by LD Pumped Solid State Laser", IEAE-CN-60/F-I-5, Oct. (1994).

MAGNETIZED TARGET FUSION

An Overview of the Concept

Ronald C. Kirkpatrick and Irvin R. Lindemuth

Los Alamos National Laboratory
NIS-9, MS B229
Los Alamos, New Mexico 87545

ABSTRACT

Magnetized target fusion (MTF) seeks to take advantage of the reduction of thermal conductivity through the application of a strong magnetic field and thereby ease the requirements for reaching fusion conditions in a thermonuclear (TN) fusion fuel. A potentially important benefit of the strong field is the partial trapping of energetic charged particles to enhance energy deposition by the TN fusion reaction products. The essential physics is described. MTF appears to lead to fusion targets that require orders of magnitude less power and intensity for fusion ignition than currently proposed (unmagnetized) inertial confinement fusion (ICF) targets do, making some very energetic pulsed power drivers attractive for realizing controlled fusion.

INTRODUCTION

The first suggestion for using a strong magnetic field to suppress thermal conduction in a thermonuclear (TN) fusion fuel occurred in 1945 [1], but because a) the goal of fusion explosives was not immediately pursued and b) when TN fusion was finally pursued the availability of fission energy made suppression of thermal conduction unnecessary for achieving fusion, no experimental effort was made to utilize a magnetic field in a fusion fuel until many years later. When controlled TN fusion was first pursued in the late 1950's, the magnetic field was recognized as a means of not just reducing thermal conduction, but totally eliminating conduction to the walls of the plasma container by isolating the plasma physically within the reactor, that is, magnetically confining the fusion fuel. While this approach is very attractive, it has led to limitations on the nature of the fusion fuel and to complex difficulties that even 40 years later are the subject of intense research, without completely satisfactory solutions. The main limitation is the necessity to operate at relatively low pressure, hence low density in the fusion fuel. The simple criterion of Lawson then dictates the necessity of long energy confinement time. This leads to the chief difficulty of magnetic confinement fusion (MCF), that is, that it must be managed for a very long while, thus further aggravate the problem. In an extension of the arguments that led to the Lawson Criterion it has been pointed out that some practical considerations make the requirements on fuel confinement for a useful fusion reactor even more restrictive [2].

Since the time when controlled TN fusion was originally proposed, another controlled fusion concept called inertial confinement fusion (ICF) was invented. It relied on the ability of a laser to deliver a very powerful and intense light pulse to a target and drive it to the extremely high pressure necessary for TN fusion ignition in a very small volume. Because the total energy that would be released is limited by the amount of fuel in the target, and the laser pulse necessary to ignite the fuel is controlled, this explosive form of fusion is considered to be controlled. However, the energy necessary to reach fusion ignition seems just beyond the capabilities of presently available lasers. The light ion beam variant of ICF relies on more energetic technology which presently lacks the intensity necessary for ignition. ICF ignition requires simultaneously sufficient energy, power, and intensity delivered to the target. Attempts to provide the necessary driver for ICF have led to large and expensive programs.

Here, we advocate an approach that is intermediate between MCF and ICF, as currently pursued in the US national fusion programs. It is called magnetized target fusion (MTF) and seeks to capitalize on the advantages of the other two approaches, while avoiding their pitfalls. It is truly a different approach, not just a simple variant of either magnetic confinement or inertial confinement. Both strong magnetic fields and implosion with some form of driver are required. The restrictions and operating ranges thought to be necessary have been mapped out and are discussed below. In addition, previous relevant experiments are cited, briefly described, and discussed. The need for a definitive proof-of-principle experiment and the suitability for various applications is explored, but scarce resources and lack of a working MTF device have limited our ability to do more than speculate about applications. An effort is made to show that some restrictions imposed on the other approaches to controlled fusion and their applications are greatly relieved by MTF. Also, we will explore the possibility of using D-^{3}He as a TN fusion fuel for MTF.

PHYSICS PRINCIPLES FOR AN MTF TARGET

When discussing MTF, the word target is used in its broadest sense. MTF involves implosion of a target plasma, a plasma that must be dynamically formed either inside a region where it is imploded (or otherwise compressed) or outside, followed by injection into that region. For fusion, the target plasma must also be a fusion "fuel", for example, deuterium and tritium (DT) or deuterium and light helium (D-^{3}He). Survey calculations [3] have shown that the target plasma must be created with a temperature T > 50 ev and a magnetic field B > 50 KG, and other calculations have shown that for successful implosion to ignition conditions in the MTF mode, the plasma mass must exceed some minimum mass (~1–10 ug of DT) [4]. This sets lower limits on the energies required for target plasma creation and implosion, but limits that are well within modern pulsed power capabilities.

The physics principles of MTF are:

1. Application of a strong magnetic field to reduce the thermal conduction through the target plasma to the shell (or "pusher") that contains and compresses it.

2. Operation at low density to reduce radiative losses from the target plasma.
3. Sufficiently large size and magnetic field to allow for turning the charged fusion reaction products within the target plasma as ignition conditions are approached.
4. Magnetic flux compression along with the relatively adiabatic compression of the target plasma.
5. Use of an auxiliary energy and current source to heat and magnetize the target plasma.

The last principle (i.e., 5) differs from ICF, where the first strong shock that emerges from the pusher into the DT fuel establishes the temperature and density from which subsequent compression proceeds. This is referred to as setting the adiabat for the compression, since subsequent shocks that follow may be considered to be weak by comparison, and rapid compression by a series of weak shocks approximates an adiabatic compression. Even without a magnetic field, the separate setting of the adiabat for subsequent compression should be very advantageous. This point is discussed in more detail the next section.

REQUIREMENTS FOR FUSION IGNITION

While many fusion concepts don't rely on ignition for economic viability, ignition generally lowers the threshold for economic viability. This point was made in a companion paper in this same proceedings [2]. By examining the energy balance and energy rate equations it is possible to define the physical requirements for fusion ignition. This has been done elsewhere [3–5], so here we will just encapsulate the results of such studies and discuss the implications.

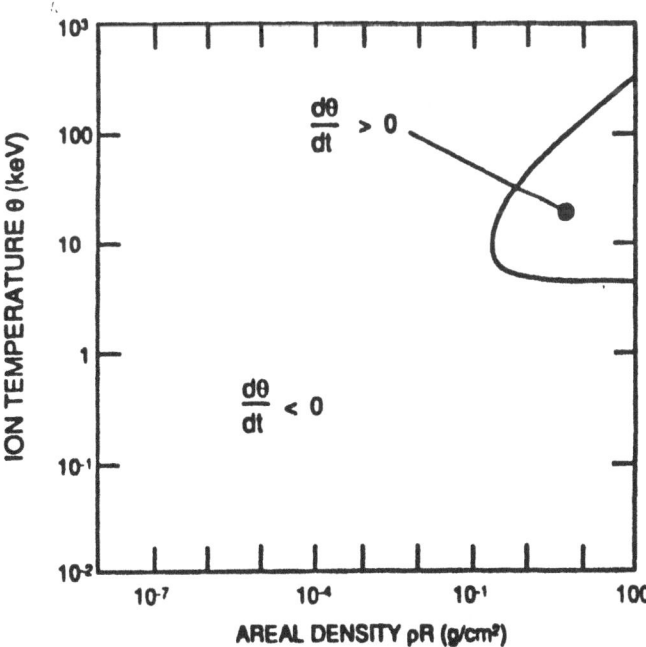

Figure 1. Lindl-Widner diagram for the ICF case (B=0). The fusion region is where fusion energy deposition overcomes energy losses (dO/dt>0) at turnaround (v=0).

The fusion region is insensitive to the DT mass for the target. Figure 1 shows a Lindl-Widner diagram for the ICF extreme of MTF (i.e., B=0). Here, the zero temperature rate contour is shown as a function of plasma areal density ρR and temperature T (where Te = Ti = T for convenience of presentation) for a DT fusion plasma for the case of no external energy being added (-PdV/dt = -3cv v/R = 0). Fusion energy deposition overcomes thermal losses in the forms of bremsstrahlung, heat conduction from the hot fusion fuel to the cold (100 ev) inner wall of the target, and inverse Compton cooling, but only inside the contour in the upper right of the diagram. This contour is remarkably insensitive to other characteristics of the target such as mass and size. It is sensitive to the dominant fusion reaction for the fuel and to the square of the charge carried by the fusion fuel reactants. Almost all ICF studies are done with deuterium and tritium (DT) for the reactants, thereby both maximizing the fusion cross section and minimizing the charge of the reactants. Fusion ignition for ICF relies on compression work being done on the fuel by successive shocks. This moves the fuel from its initial cold state to fusion conditions, similar to the principle of a diesel engine. There are minimum values of T and rhoR for the fusion region (where the fusion energy deposited by the DT alpha reaction products overcomes the losses), and also a minimum value for their product ρR times T which is proportional to the product of plasma pressure times radius PR. For ignition to be achieved, it is necessary to attain or exceed the minimum PR, and to follow a (ρR,T) path for the compression that intersects the fusion region.

The maximum value of PR that can be achieved in an ICF target implosion is necessarily related to the energy in the implosion, and more specifically it depends on the mass of the pusher, the velocity of the pusher, and its material properties. The more energetic the implosion, the higher the the resultant PR. At the end of the compression, ideally from an energy perspective, all the energy of the implosion is shared as internal energy by the pusher and the fusion fuel. A simple analysis of this process shows that under optimum conditions, one fifth of the energy in the implosion can be transferred to the fusion fuel. Four fifths or more must remain in the pusher [6]. For marginal PR, the intersection of the (rhoR,T) path of the compression with the fusion region must be at a single point, dictating that only a very narrow path in rhoR and T plane (i.e., along a single adiabat) can provide ignition. This makes ICF ignition with marginal implosion energy very tricky, because the implosion must provide precisely the right first strong shock, yet must also have sufficient energy to carry the fusion fuel to a sufficient PR. This assumes that the initial density in the ICF target was precisely the design value, which demands rigorous control of the target fabrication and perhaps temperature control in the target chamber. This is why a separate setting of the implosion adiabat for the fusion fuel should be advantageous for ICF, but presently this is only attempted in a limited way through pulse shaping of the laser light.

Exceeding the minimum implosion energy that allows ignition for ICF greatly reduces the difficulty of achieving ignition, because the line of constant PR intersects an increasingly broader part of the fusion region as the value of PR is increased, thus allowing a broader band of adiabats to be followed for ignition. The role of instabilities and potential impurity injection into the fusion fuel adds further complications which have been discussed by Lindl [5] and others. Lindl showed that implosion velocities somewhat in excess of 10 cm/us were required for ICF ignition in the absence of impurities (Fig. 2), and on the order of 20 cm/us was necessary for one level of impurities.

Colgate, et al. [6] have shown that for the case of no impurities the implosion velocity needed for ignition is closely linked to the value of PR needed for ignition and that smaller targets can be ignited if high implosion velocities are used. If fact the minimum

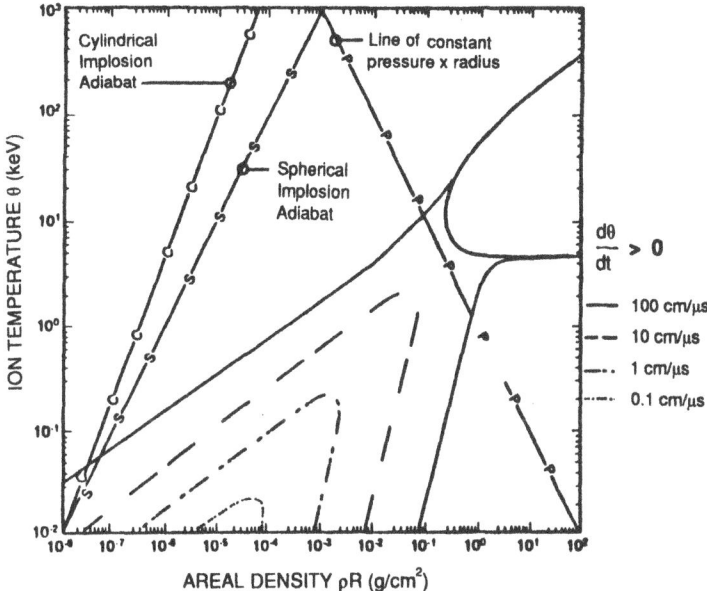

Figure 2. Lindl-Widner diagram with compression heating (v>0). Sufficient implosion velocity allows ignition to be achieved.

mass, hence energy needed, for fusion ignition is very sensitive to the implosion velocity ($E \sim v^{-m}$, where m = 6 to 8). It is always true that at least the thermal energy of the portion of mass of the fusion fuel that actually ignites must be transferred from the pusher for ignition to occur. Because the fuel and the pusher never completely stagnate in ICF targets, more than this irreducible minimum is required.

MAGNETIZED TARGET FUSION IGNITION

When the effects of a magnetic field on the fusion fuel physics is considered, it becomes apparent that the two predominant effects are the reduction of the thermal conductivity and the partial trapping of the fusion charged reaction products. Other less important consequences include the necessity to compress the magnetic field along with the fusion fuel, the additional energy loss due to synchrotron radiation, and the inherent two-dimensional nature of the implosion and transport processes. In fact, the magnetic fields required for MTF are not sufficiently strong to significantly effect the dimensionality of the implosion or to cause significant synchrotron radiation. Compression of the field has great potential benefit in terms of enhancing DT alpha energy deposition (i.e., "self-heating") of the fusion fuel.

Figure 3 shows a Lindl-Widner diagram for the MTF case (B>0). This diagram was constructed in the same way as Figure 1, by plotting the contour of net zero heating plus cooling due to the relevant physical processes, including synchrotron radiation and the thermal conduction as inhibited by the magnetic field. An empirical relation was used to account for the effect of the field on the DT alpha energy deposition. This empirical relation is currently being improved (Fig. 4).

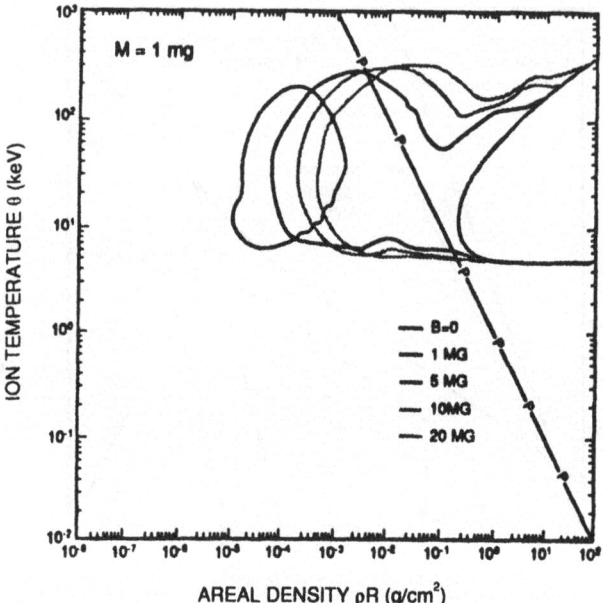

Figure 3. Lindl-Widner diagram for the MTF case (B>0). The size and position of the MTF fusion region depends on both the field strength and the target plasma mass.

For a spherical target with a trapped azimuthal magnetic field, compression causes the field to increase as the inverse square of the radius. The same is true of the areal density ρR. The notable feature of Figure 3 is the appearance of an additional fusion region at much lower ρR than the one for ICF. Also, the size and shape of the ICF fusion region are somewhat modified. In the MTF fusion region, the plasma parameters of importance have very high values: The magnetic field decay time is much longer than an implosion time, the ratio of the thermal energy to the magnetic field energy (beta) is much greater than unity, the product of the electron cyclotron frequency and the collision time is large, and the ratio of the plasma frequency to the cyclotron frequency is large. All these parameters are favorable for classical behavior of the plasma and minimization of plasma instabilities. They are also well within the regime of resistive magnetohydrodynamics (MHD).

One notable aspect of the MTF fusion region is the very low value of PR necessary to access it. This means that the implosion need not be so energetic as is the case for ICF. Still, at least the energy necessary to supply the thermal energy at the time of ignition must be transferred from the pusher to the fusion fuel. Because all the energy loss rates are reduced in the MTF fusion region, the implosion velocity necessary to achieve ignition is drastically reduced. While over 10 cm/us is required for ICF, less than 1 cm/us seems adequate for MTF. This is shown in Figure 5, where the implosion velocity of 0.3 cm/us shows a clear path from an initial 50 ev into the fusion region at turnaround (v=0).

While ignition with such low implosion velocities remains only a theoretical possibility until demonstrated, realization of MTF ignition with these low implosion velocities would have a major impact of the driver requirements for fusion ignition. First, as previously noted, the MTF targets are larger than ICF targets, on the order of a centimeter rather than a millimeter. With very low implosion velocities, the implosion time increases from nanoseconds to microseconds. This means that for about the same mass of fuel, the

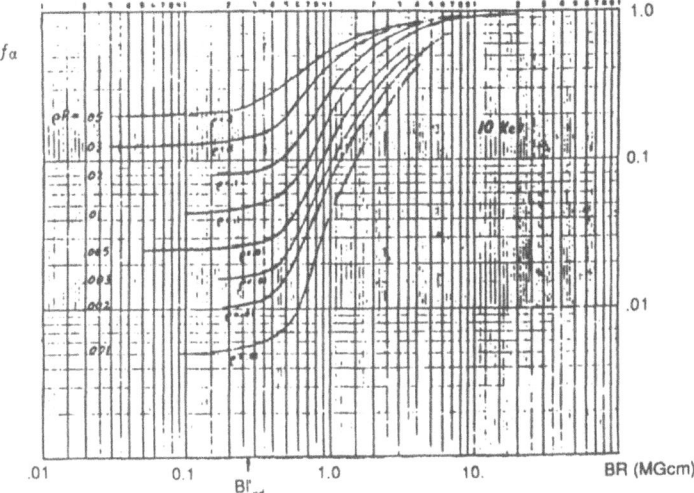

Figure 4. Computed fractional energy deposition for a DT alpha particle in a 10 Kev DT plasma with a uniform magnetic field.

driver power required for achieving ignition is orders of magnitude lower than for ICF. Since the size determines the surface area of the target, the intensity on target can be even more orders of magnitude lower than the intensity needed for ICF. In fact, the minimum hohlraum temperature needed for indirect drive targets in ICF is closely tied to the intensity needed to drive the target to ignition. This means that indirect drive with MTF would require only tens instead of hundreds of ev temperature in the hohlraum. In fact, there is no incentive for indirect drive for an MTF target, because symmetry of implosion is not such an important issue as it is for ICF. Because the initial temperature is determined by the target plasma creation, and it can be made quite high, the convergence necessary for ignition can be quite small ($\sqrt{T_{ign}/T_o}$) ~ 10 to 15) for spherical targets.

PAST EXPERIENCE

The most extensive series of MTF experiments were the Phi-target experiments conducted by Sandia National Laboratory in 1977 [7–9]. These experiments were conducted on a relativistic electron beam machine, which had a non-relativistic prepulse. The previous, simple target experiments conducted by Sandia represented their initial involvement in ICF. Fot their first experiments they had used spherical targets consisting of a thin shell containing deuterium gas mounted on a stalk attached to the anode of the e-beam diode. The relativistic electrons had high enough current to pinch the beam into a pencil sized column that impinged on the target from the cathode. Because the electrons were relativistic, the stopping length was large, and the fields in the diode (as disturbed by the target) caused the electrons to reflex through the target, providing relatively uniform energy deposition. Because the energy deposition was very uniform, the target shell exploded, with part driven radially outward and part inward. This enabled the e-beam machine to implode the targets quite symmetrically. However, for ICF in the usual sense, uniform energy deposition in the shell (and throughout the deuterium gas) limited the degree of con-

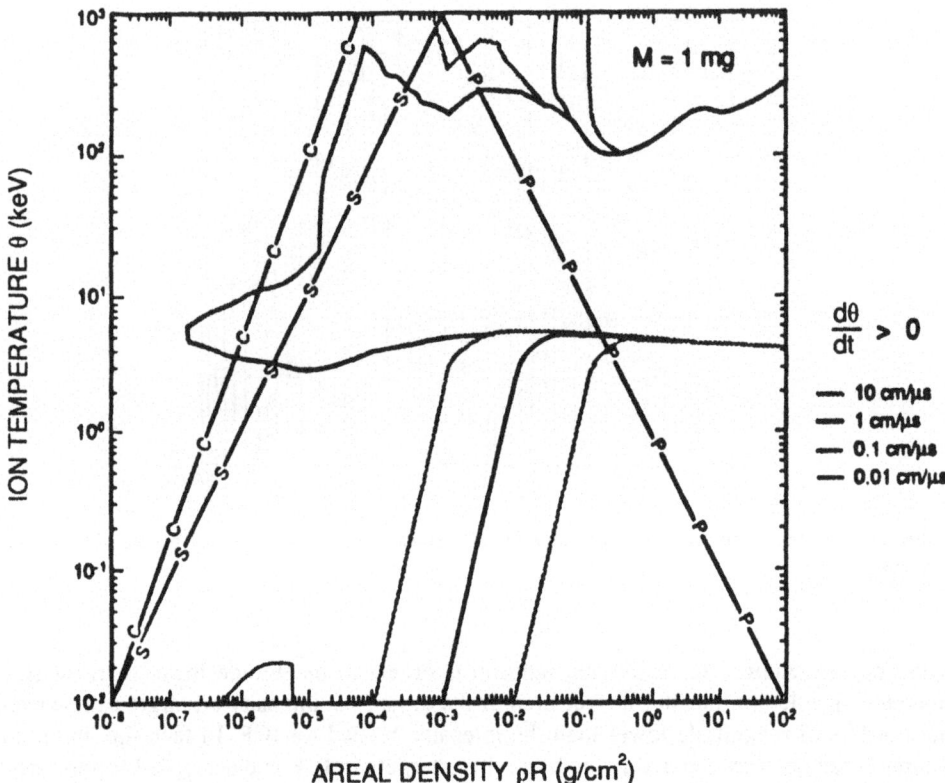

Figure 5. Lindl-Widner diagram with overlayed zero temperature rate contours for various implosion velocities. Also shown on the plot are representative adiabats for cylindrical (C) ahd spherical (S) geometries and a line of constant pressure times radius PR~rhoRxT (denoted by P).

vergence possible. This is because uniform deposition set the shell on a high adiabat, so that the implosion pumped most of the energy into internal energy of the imploding part of the shell, not into the deuterium fuel. Therefore, it was decided to abandon the e-beam machine in favor of a light ion beam machine. The advantage was that the light ions (protons in the first machine) had a very short range (and produced no bremsstrahlung in heavier metal shells), but the disadvantage would be loss of efficiency in producing the light ion beam and the necessity to focus it onto the target, because ions couldn't be pinched into a beam.

While the first proton beam machine was being designed, the Sandia team tried a novel idea on an e-beam machine. An electrode was mounted on the target between it and the cathode, so that it intercepted the non-relativistic pre-pulse from the cathode and discharged a current through the target. In fact, a CD_2 wire was added along the axis inside the target, which exploded due to the pre-pulse current. The collector, CD_2 wire inside, plus the stalk on the anode schematically resembled a Greek Phi, hence the name Phi-target. The discharge created a hot, magnetized plasma. When the pre-pulse ended and the relativistic pulse began, the imploding the shell to compress the magnetized plasma inside up to temperatures sufficient to produce neutrons. These were the first TN neutrons produced by the Sandia team [7].

The experimental campaign involved over two dozen targets, 15 of which were complete. The rest were fielded as null experiments (with bumps, holes, no electrode, etc.) in an effort to determine if some other physics might be responsible for the neutron production besides the compression of the hot, magnetized CD_2 plasma. None of the null targets gave measurable neutrons, but eight of the complete targets did, ranging from 5 to 25 million. These experiments are reported in more detail in an accompanying paper in this symposium [9].

Shortly after the Phi-target experiments some effort was made to model these experiments with one- and two-dimensional (1-D and 2-D) computer codes [10]. One of the few gas filled targets was chosen for modeling, because of computational limitations. First, a 1-D calculation was done to obtain the implosion dynamics of the e-beam driven shell, then a 2-D resistive MHD calculation was done using the interface motion from the 1-D calculation for the history of the outer boundary in the 2-D calculation during the implosion phase. The 2-D resistive MHD calculations showed that during the pre-pulse phase, the plasma had an overturning flow, but during the very short implosion, very little motion occurred. The plasma temperature reached about 400 ev in the calculations. The implosion trapped the field and compressed it up to about 1 MG. A convergence of about 20 was required to match the observed neutron yield. It was concluded that the experimental measurements as analyzed by the 1-D and 2-D codes were consistent with a TN origin for the neutrons.

In addition to the Phi-target experiments, there were liner-on-plasma ideas pursued at Los Alamos and at the Naval Research Laboratory (NRL). The Fast Liner Program in Los Alamos [11] succeeded in producing a plasma at 30 ev with an embedded field of about 10 KG, and in driving a solid liner symmetrically at about 1 cm/us. However, the plasma was thought insufficient for the purpose, and no experiment was done with a plasma inside the liner. The NRL Linus Program proposed to use a liquid liner to compress a fusion plasma, but no integrated experiments were ever done.

Electrically driven shock tube experiments at Columbia University [12] have been interpreted as evidence for classical (Spitzer-like rather than Bohm) reduction of the thermal by a magnetic field in a high (beta) plasma. Particle-in-cell calculations by Dawson, Okuda, and Rosen provide some theoretical understanding for this result [13]. When the plasma frequency is sufficiently high relative to the electron cyclotron frequency, plasma fluctuations don't persist long enough for Bohm diffusion to be an effective transport mechanism.

In Russia, during the days of the Soviet Union, the All-Union Scientific Institute for Experimental Physics (VNIIEF) had embarked on a program of high explosive pulsed power, one of the applications of which was magnetic implosion of a liner, and another of which was the creation of an energetic magnetized plasma. We have come to know the latter as the MAGO experiments [14]. In these experiments they were able to generate neutron producing magnetized plasmas in the range of 200 ev and 50 KG. In recent months, joint experiments with Los Alamos have confirmed these numbers. It would appear that all the requisite factors are in place to allow an integrated MTF experiment with explosive pulsed power, but it should also be possible to do MTF experiments in a beam on target configuration, similar to the Phi-target experiments that Sandia performed. The collaborative experiments are discussed more extensively in another paper in this symposium [15].

DRIVER REQUIREMENTS FOR IGNITION

Ignition of a fusion target that is imploded requires that the driver be able to deliver a pulse of energy that is simultaneously sufficiently intense, powerful, and energetic. Lasers are clearly sufficiently intense and powerful, but currently lack the necessary energy for ICF ignition. Pulsed power drivers such as light ion beams have for a long time had enough energy, possibly had sufficient power, but lacked the focusing, that is, intensity on target. No heavy ion beam driver exists yet, but one of the problems they must face is the necessity to use indirect drive. It appears that heavy ion drivers will be unable to deliver a sufficiently intense pulse to achieve the necessary hohlraum temperature for ICF fusion ignition. So, no driver is available for ICF ignition, except perhaps explosive fission sources. While a fission source could be invaluable for ICF research right now, it is not permitted currently and would not be a politically acceptable choice for a fusion power reactor in the foreseeable future.

Because MTF relaxes the power and intensity requirements needed for ignition in the MTF fusion region, pulsed power drivers become very attractive. The great advantage of MTF in this regard is illustrated by Figure 6, in which the MTF targets are shown to operate at orders of magnitude less power than do ICF targets. It can be argued that MTF targets may have low gain (which should be true only for simple targets), but even if so, this would be more than offset by the fact that pulsed power drivers can have a high efficiently when properly designed to deliver energy to a matched load. In addition, one study of potential MTF targets explores the possibility of obtaining high gain [8].

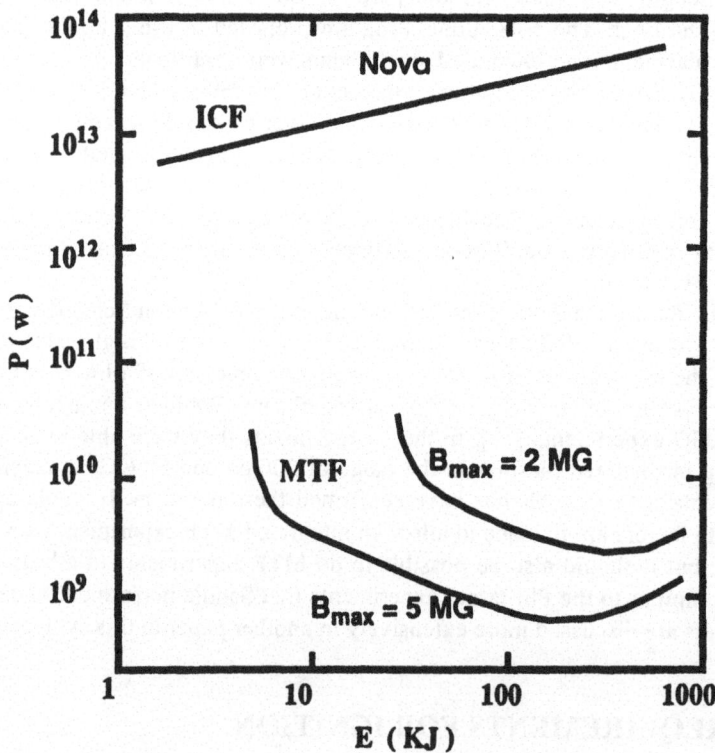

Figure 6. The power and energy necessary for MTF targets. MTF targets should operate at low power (and intensity) on target.

CRITICAL ISSUES FOR MTF

While the previous experiments and calculational analyses are very encouraging, there are a few issues that need to be addressed for MTF.

Plasma instabilities are thought to be suppressed because the plasma is wall supported by the pusher, but hydrodynamic instability of the pusher is a possibility. Because MTF can operate at much lower pressures than ICF, material strength in the pusher may reduce the growth of hydrodynamic instabilities. Also, the pusher will be more nearly incompressible, so that the inner aspect of the pusher may continue to accelerate inward longer than is the case for ICF. While this may be advantageous from the standpoint of hydrodynamic instabilities, if the pusher does not stagnate, then not as much of the the energy in the implosion will be transferred to the target plasma before ignition occurs. This leads to a reduced gain for the target, but should not lead to failure to ignite.

The fact that the target plasma must be magnetized means that the field pressure must augment the particle (or plasma) pressure, so that those regions in the plasma where the field is stronger have some degree of buoyancy, leading to hydrodynamic turnover. This convective behavior could augment the (reduced) conductive heat flow to the wall. The degree to which convection augments conduction depends on the velocity of convection relative to the implosion velocity. A secondary effect of convection is the potential entrainment of wall material in the plasma flowing past the wall. This could have two consequences: the added thermal capacity of the material entrained and enhanced radiative cooling. The physics associated with these processes is understood, but the ability of our codes to model these processes is presently limited. For MTF in its simplest form there is no need to trap the radiation generated in the plasma, so that light elements suffice as ingredients of the wall (i.e., pusher). This means that the maximum value of the impurity nuclear charge Z can be kept to a minimum. The physics of the impurity radiation is well understood, but detailed treatment in a code does involve a significant computational effort. Some simple arguments lead to the conclusion that for light element impurities, the major process that enhances the radiation from the high temperature plasma is bremsstrahlung. The critical question is how much material will be entrained in the plasma. This issue needs to be addressed. If the DT targets have frozen DT layers inside the shell [8], then there would be no impurity, but the problem of added heat capacity would still arise.

Target plasma creation in the Phi-target discussed below was accomplished with a simple electrical discharge through a CD_2 wire or through a deuterium gas. This process was modeled for the case of discharge through the D_2 gas by Lindemuth and Widner [10]. It appears that the resistive nature of the D_2 gas while it was not yet hot allowed the magnetic field to diffuse into the D_2 plasma, thus providing a hot, magnetized plasma. We do not yet know what ranges of temperature, density and field can be produced by this approach. The high density Z-pinch has been modeled by Sheehey [16] and others and is now well understood for a range of conditions, but we still need to explore the range relevant to MTF. In addition, the calculations need to have experimental verification. Finally, the MAGO experiments seem to have produced a magnetized plasma of interest to MTF. These experiments are currently under scrutiny to see if the calculated parameters agree with the experimental measurements. If so, this may be a suitable target plasma for MTF.

THE POSSIBILITY OF AN ANEUTRONIC FUEL FOR MTF

Because both D^3He reaction products (a proton and a helium nucleus) are charged, the magnetic field can help retain up to four times more energy from the D^3He reaction

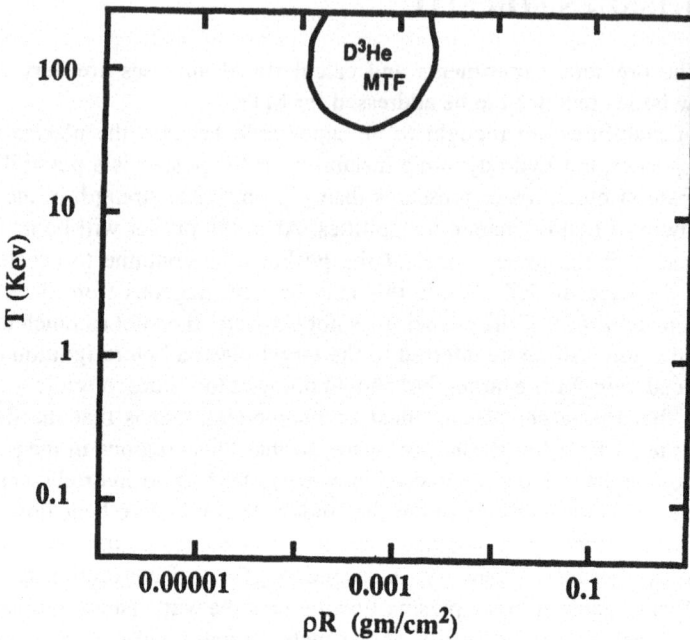

Figure 7. Lindl-Widner diagram for D^{3}He. The conservative assumption that Ti=Te=T is made.

than is possible for the DT reaction. However, this is offset by the higher bremsstrahlung radiation due to the light He nuclei. Marjorie Ward explored the possibility of using D^{3}He as an MTF fuel [17]; her Lindl-Widner diagram for D^{3}He is shown in Figure 7.

Notice that the MTF fusion region still is present for D^{3}He, but it is diminished in size and at higher temperature than for DT. This is partly due to the assumption in the calculations that were used to produce this diagram, that is, that the ion and electron temperatures are equal. This is a very conservative assumption for D^{3}He. Additional calculations are needed to better define the MTF fusion region for D^{3}He. However, there is evidence that D^{3}He might be useful for MTF.

APPLICATIONS

No fusion concept has an application unless it performs adequately for practical use in the proposed application. Currently, there are only a few settings in which fusion performs adequately: the sun and TN weapons are two, but a few very low level, portable neutron sources also exist. Therefore, this discussion of applications for MTF is necessarily speculative, and assumes that MTF will provide a significant (but manageably small) source of TN fusion energy. The characteristics of a working MTF fusion scheme depend on whether a liner on target plasma or a beam on target approach works, as well as the type of driver employed. If the configuration of a working scheme is beam on target, then applications of MTF should be similar to those of ICF, that is, fusion energy production and perhaps space propulsion. If MTF takes the form of a liner on target plasma, then the range of potential applications may be narrow, e.g., some device like that proposed in the NRL Linus Program to provide fusion energy. Edward Teller doubts the energy production application of fusion, but believes fusion is necessary for space propulsion. This is because he believes that fission energy is a reality that is being embraced by the world (outside the US). Until something besides a bomb or the sun is demonstrated at useful power density, then all other applications are speculation. The emphasis should be on the demonstration of a working concept. Once this is done, I predict that many applications will be proposed. One application of fusion that receives little attention is transmutation. Thermalized neutrons can be used for medical isotope production and for converting cheap, abundant elements into rare, valuable ones. We know how well this works for fission reactors. Pulsed neutron sources may be more versitile.

SUMMARY

We have presented an overview of the MTF concept. Two accompanying papers at this symposium have discussed other aspects of MTF: Mary Ann Sweeny provided an account of the 1977 Phi-target experiments at Sandia National Lab in Albuquerque [9] and subsequent calculational studies [8]. Irv Lindemuth discussed the fledgling US/Russia collaboration on the MAGO experiments [15]. Other concepts presented here have overlap with MTF or in some cases may even embody all the basic principles of MTF [18].

The major contribution of MTF is that of matching the target to the presently available drivers. This means that it may be possible to demonstrate fusion ignition in the near term. We currently need a commitment of resources that will allow us to ascertain the truth of MTF. Only a small commitment would go a long way.

The time is ripe for exploring MTF. A relatively small investment of resources is all that is needed. The dividend may be large, and the payoff soon.

ACKNOWLEDGMENTS

The author thanks I.R. Lindemuth, R.E.Reinovsky, and P.T. Sheehey for comments, research support, and computations that contributed to the work presented here. The efforts of many other individuals who are making contributions to the Los Alamos/VNIIEF collaborative MAGO experiments are also appreciated.

REFERENCES

1. Landshoff, R., "Transport Phenomena in Completely Ionized Gas in the Presence of a Magnetic Field", Los Alamos National Lab report LA-466 (1945); Phys. Rev. 79 (1949) 904.
2. Panarella, E., Current Trends in International Fusion Research, Washington, D.C., (these Proceedings, p. 211).
3. Lindemuth, I.R., and Kirkpatrick, R.C., "Parameter Space for Magnetized Fuel Targets in Inertial Confinement Fusion," Nucl. Fusion 23 (1983) 263.
4. Kirkpatrick, R.C., Lindemuth, I.R., and Ward, M.J., "Magnetized Target Fusion, an Overview," to be published in Fusion Technology (1995).
5. Lindl, J.D., "Physics of Ignition for ICF Capsules," Inertial Confinement Fusion Course and Workshop (Varenna, Italy, 1988 A. Caruso and E. Sindoni, ed.).
6. S.A. Colgate, A.G. Petschek, and R.C. Kirkpatrick, "Minimum Energy for Fusion Ignition, A Realistic Goal", Los Alamos National Laboratory report LA-UR-92-2599 (1992).
7. Widner, M.M., "Neutron Production from Relativistic Electron Beam Targets", Bull. Am. Phys. Soc. 22 (1977) 1139.
8. M.A. Sweeny and A.V. Farnsworth, "High-Gain, Low-Intensity ICF Targets for a Charged-Particle Beam Fusion Driver", Nuclear Fusion 21 (1981) 41.

9. Sweeny, M.A., Proceedings, First International Symposium for the Evaluation on Current Trends in Fusion Research, Washington, D.C., November (1994).
10. I.R. Lindemuth and M.M. Widner, "Magnetohydrodynamic Behavior of Thermonuclear Fuel in a Preconditioned Electron Beam Imploded Target," Physics of Fluids 24 (1981) 753.
11. A.R. Sherwood, B.L. Freeman, R.A. Gerwin, T.R. Jarboe, et al., "Fast Liner Proposal," Los Alamos Scientific Laboratory report LA-6707-P (August 1977).
12. B. Feinberg, "An Experimental Study of a Hot Plasma in Contact with a Cold Wall", Plasma Physics 18 (1976) 265.
13. Dawson, J.M., Okuda, H., and Rosen, B., "Collective Transport in Plasmas", in Methods in Computational Physics 16 (1976) 281.
14. A.M. Buiko, V.K. Chernychev, V.A. Demidov, Yu.N. Dolin, S.F. Garanin, V.A. Ivanov, V.P. Korchagin, M.V. Lartsev, V.I. Mamyshev, A.P. Mochalov, V.N. Mokhov, I.V. Morozov, N.N. Moskvichev, E.S. Pavlovsky, S.V. Pak, S.V. Trusillo, G.I. Volkov, V.B. Yakubov, V.V. Zmushko, "Investigations of Thermonuclear Magnetized Plasma Generation in the Magnetic Implosion System MAGO," in "Physics of High Energy Densities," MEGAGAUSS VI, Albuquerque, NM (1992).
15. Lindemuth, I.R., Current Trends in International Fusion Research, (these Proceedings, p. 543).
16. P.T. Sheehey, J.E. Hammel, I.R. Lindemuth, D.W. Scudder, J.S. Schlacter, R.H. Loveberg, and R.A. Riley, Jr., "Two-Dimensional Direct Simulation of Deuterium-Fiber-Initiated Z-Pinches with Detailed Comparison to Experiment," Phys. Fluids 4 (1992) 3698.
17. M.S. Ward, Los Alamos National Laboratory memo X-1-93-.... (1993).
18. Panarella, E., Proceedings, First International Symposium for the Evaluation on Current Trends in Fusion Research, Washington, D.C., November (1994).

21

THERMONUCLEAR FUSION IN A STAGED PINCH

F. J. Wessel,[1] Norman Rostoker,[1] H. U. Rahman,[2] P. Ney,[2] and E. L. Ruden[3]

[1]Department of Physics
University of California
Irvine, California 92697-4575
[2]IGPP
University of California
Riverside, California 92521-0412
[3]Phillips Laboratory
Kirtland AFB, New Mexico 87117-5776

ABSTRACT

Staged pinch implosions provide means to couple energy to a small-diameter fibre on an extremely fast time scale, circumventing the limitations of conventional pinches. In this scheme the generator current initially traverses an intermediate hollow plasma shell which compresses onto the fibre placed coaxially and transfers the current to the fibre with a significantly reduced risetime. The results are impressive as the delivered peak power is increased by several orders of magnitude, the coupling efficiency improves, and the most dangerous plasma instabilities which commonly plague high-density/high-temperature pinches, are eliminated. This technique can be fielded on both fast and slow generators (i.e., 10's of nanoseconds to microseconds) making it feasible to extend the concept to a wide range of presently assembled systems. Staging may therefore present a dramatically new means of pulsed-energy conversion which could find many applications. In addressing the requirements for thermonuclear fusion in a staged Z-pinch, our preliminary calculations based on zero-D models suggest the potential for a significant thermonuclear burn with generator currents of the order of few MA and a risetime of about one μsec. Studies are actively underway at various places around the world (England, France, Germany, Russia) including United States (UCI/UCR) to investigate different aspects of staged pinching and its applications particularly leading to controlled thermonuclear fusion.

1. INTRODUCTION

The staged pinch[1,2,3] represent a generic name for a wide range of pinches (e.g. Z-Z pinch, Z pinch, etc.) that usually follow the similar type of physical mechanism for coupling of energy from one load to the other. The advantage of this method is that one can avoid the early phase of long duration slowly rising current profile and utilize the full thrust of peak current. This results in an unprecedented large value of dI/dt for the current that flows through the final load. For example a current of 2 MA with a risetime of one μsec can be compressed up to 5 MA with risetime of less than one nanosecond by applying only single step staging.[2] The concentration of such a large value of current into a small diameter fibre provides enormous magnetic pressure that compresses and heats the fibre plasma to conditions favourable for thermonuclear fusion. The terminal particle pressure of the fibre plasma ultimately exceeds the magnetic pressure, therefore, the confinement of fibre plasma would be inertial. The most general configuration of a staged Z-pinch is illustrated schematically in Figure 1. It consists of an outer column of gas or plasma and a central-coaxial fibre, straw, or wire, installed between the discharge electrodes of a power generator. An axial magnetic field is also applied in some cases that improves the stability and also provides a mean for energy coupling to the fibre on axis due to the formation of azimuthal current on the surface of the fibre. Since there exists only a modest amount of experimental data with and without an axial magnetic field, it is not yet clear which approach is more effective for the efficient coupling together with a stable implosion.

The staged Z-pinch configuration was first employed for wavelength calibrations of a Kr Z-pinch spectrum obtained from an Al wire plasma on axis.[4] Later on a similar configuration of a staged pinch with an axial magnetic field was studied on the same facility.[5,6] In this configuration when the annular-outer pinch plasma implodes, it traps the axial field compressing it to multi-megagauss levels in a time scale that is much shorter than the characteristic rise time of the axial current delivered by the generator. Because of such a rapidly rising magnetic field, the inductive effects cause the outer surface of fibre

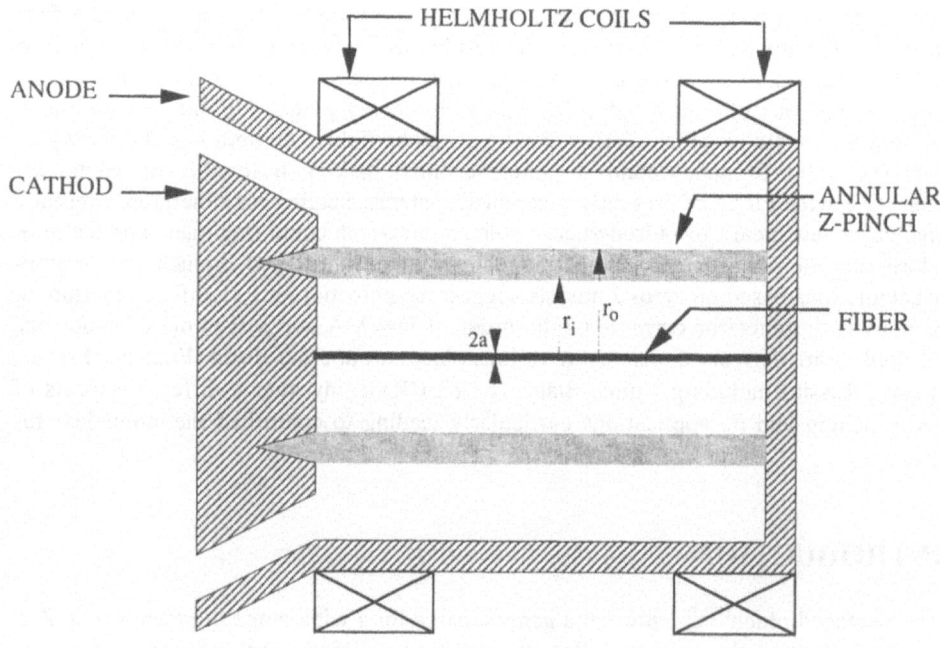

Figure 1.

to ionize, forming a surface θ-current. The risetime of this current which flows around the fibre decreases as the outer pinch stagnates. The concept of axial magnetic flux compression by Z-pinch implosion was also tested on high-power machines at Sandia National Labsoratory[7] and in Russia[8] confirming the theoretical predictions and scaleability to higher values. The maximum reported axial magnetic fields are 1.6 MG (UCI), 2.5 MG (USSR) and 40 MG (Sandia). The UCI studies were the first to observe the formation of a θ-pinch plasma on a coaxial quartz-fibre optic, used to measure the compressed axial magnetic field by Faraday rotation of a laser beam.[9] The fibre plasma formed in this configuration of staged pinch was interestingly free from hot spots and was able to maintain the axial uniformity without going into an unstable regime. Theoretical modelling of this concept exploring the possibility of achieving controlled fusion was carried out by using simple zero-D models.

More recently staged Z-pinch experiments (Z-Z pinch) were conducted at the École Polytechnique, France which were driven by a 0.1 TW, 225 kA, 110 ns pulse-line generator.[10] The outer-plasma column was prepared from an exploded Al foil, collimated into a 3-mm diameter plasma column between the pinch-discharge electrodes and imploded onto 10–20 μm diameter targets of quartz, aluminum, copper, or tungsten wires. Time-resolved X-ray data, discharge current, and time-integrated X-ray pinhole images indicated that current was initially delivered to the aluminum-jet plasma and subsequently transferred to the micron-diameter wire demonstrating excellent energy coupling and heating of the fibre. The resulting pinch was remarkably stable against the sausage instabilities with a 100 : 1 (axial:radial) aspect ratio and was significantly more uniform than an exploding-wire or jet-only pinch. In addition a severalfold reduction in the X-ray pulse risetime was measured with a dramatic shift in the radiated X-ray spectrum to higher energy, from 1.7 KeV (the K-edge characteristic of the outer aluminum liner) to the range, ~ 3 - 5 KeV, demonstrating the existence of a much hotter plasma on axis. Field-stabilized staged Z-pinch experiments were also conducted with a somewhat lower initial axial-magnetic field.[11,12] The data show a severalfold improvement in the X-ray yield for small values of the axial field, approximately 50 G, which is attributed to improved stability.

Recently, a staged Z-pinch was fielded on the Angara-5–1 accelerator in Trinity, Russia at 10 TW, 100 ns risetime, and 2 MA current.[13] The configuration was similar to the French experiments, except the outer shell was a solid-gas jet of argon, xenon, and mixtures with deuterium gas imploded onto $20 - 30$ μm diameter fibres of deuterated polyethylene, aluminum, or copper. The Russian experiments included an extensive array of time and space-resolved diagnostics and confirmed the earlier French results, scaled up a hundredfold in discharge power. The data provided further evidence of densities of the order, $10^{21} - 10^{22}$ cm^{-3}, temperatures of 10 KeV, and perhaps higher. In one experiment using a Xenon gas pinch imploded onto a deuterated-polyethylene fibre the neutron yield was in the range, $10^9 - 10^{10}$ neutrons/shot and was approximately the same value as that observed in a deuterium-gas-mixture pinch (80% deuterium and 20% argon by mass) imploded onto a deuterated-polyethylene target. This data suggest that the neutrons originated inside the target. Another important feature of the Russian data was the fast-time structure of the radiation pulse which displayed a risetime shorter than $1 - 3$ ns with a several ns FWHM; the actual measurement of the pulse time and duration was limited by the bandwidth of the detector-oscilloscope system. This suggests that the energy coupling and thermalization time may be substantially less than the several nanoseconds measured.

The most significant aspect of the above mentioned experiments is that they represent as much as an order of magnitude improvement in the characteristic densities and temperatures commonly attained in simple Z-pinch systems; the performance of which is fundamentally limited by disruptions (instabilities) which occur before peak current. Such disruptions are virtually absent in the staged Z-pinch. Based upon the present experimental results, further studies will inevitably continue in the international community and additional improvements in performance will be obtained. Theoretical understanding of such an efficient energy coupling and stable implosion is somewhat limited and require serious consideration. In the following section we will present some of the preliminary results which are based on simple zero-D models and some qualitative discussion about the stability of the implosion.

2. MODELLING OF THE STAGED Z-PINCH

Although the physics of pinches has been extensively investigated during the last four decades, the concept of pinching in stages is relatively new. Particularly, an efficient way of energy coupling and the stability which are the unique features for this type of pinch configuration has not been fully understood. However, during the past several years some efforts were made to clarify the basic physics issues and performed of the implosion dynamics using simple zero-D numerical codes. In this section some of the key theoretical and numerical results which support the prediction that a staged Z-pinch would produce significant thermonuclear-fusion reactions at the several MA level are presented.

A simple model of staged pinch based on flux compression is proposed to explain the energy coupling from the intermediate pinch to the fibre pinch. This model describes the transfer of current to the central wire plasma at the peak compression, which initially flows through the outer hollow gas puff Z-pinch plasma. The schematic of this model is described in Figure 2a and 2b. Consider a central wire of Cu of radius 100 μm, and length of 4 cm. The initial resistance is 0.021Ω and the inductance is 50 nH. A typical pulse line of 50 kV will have a pre-pulse of about 5% for 1 μsec. This would be sufficient to explode the wire resulting in a plasma with a temperature of about 1 eV expanded to a radius of about 3 mm and density 10^{20} cm^{-3}. The plasma resistance would then be about 0.4 Ω and the current passage through the wire would be about 10 kA assuming a typical pulse line impedance of 2 Ω. We assume these initial conditions for the central wire plasma when the main pulse arrives. It should be noted that with the usual experimental procedure these conditions would not be known precisely. However they could be specified by using an auxiliary pre-pulse circuit for the wire plasma.

The annular Z-pinch plasma is initially a jet of either neutral gas or weakly ionized plasma. We assume it has an initial outer radius $r_0(0) = 2$ cm, inner radius $r_i(0) = 1.8$ cm, a density of 2×10^{17} cm^{-3} and a mass density of 22.6 μgm/cm. As a result of the pre-pulse the initial state of the plasma should be complete ionization and low resistance when the main pulse arrives and makes the plasma implode. Assuming a return conductor radius of 5 cm the initial inductance would be 7.33 nH, much less than that of the wire plasma.

Consider two path integrals illustrated in Figure 2a and 2b. According to Maxwell's equation

$$\oint \mathbf{E} \cdot d\mathbf{S} = -\frac{1}{c}\frac{\partial}{\partial t}\int \mathbf{B} \cdot d\mathbf{A} = 0, \tag{1}$$

and therefore the azimuthal and axial fluxes

Thermonuclear Fusion in a Staged Pinch

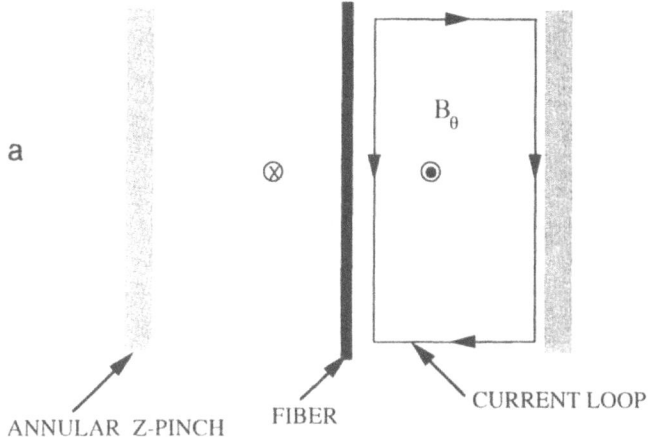

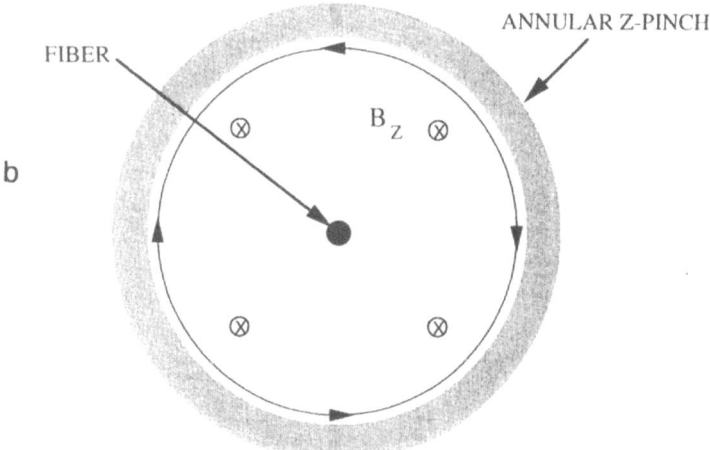

Figure 2. (a) Azimuthal flux compression and (b) axial flux compression.

$$\Phi_\varnothing = \int \mathbf{B} \cdot d\mathbf{A} = \int_{r=a}^{r=r_i} \left(\frac{1I_0}{5r}\right) dr = \frac{1I_0}{5} \ell n \frac{r_i(0)}{a(0)}, \tag{2a}$$

$$\Phi_z = \int \mathbf{B} \cdot d\mathbf{A} = \int_{r=a}^{r=r_i} (2\pi r B_0) dr = \pi B_0 \left(r_i(0)^2 - a(0)^2\right), \tag{2b}$$

are separately conserved. In equation (2) I_0 is the initial current through the plasma in amperes, B_0 is the initial axial magnetic field, $a(0)$ is the initial wire plasma radius, $r_i(0)$ is the initial radius of the outer pinch and l is the length of the pinch. Here, initial means after the pre-pulse and at the beginning of the main pulse. At this time both the wire plasma and Z-pinch plasma have negligible resistivity. As the implosion proceeds both $r_i(t)$ and $a(t)$ decrease. Since flux is conserved, the axial current in the wire plasma and the axial magnetic field must change according to the following relations:

$$I(t) = I_0 \frac{\ln[r_i(0)/a(0)]}{\ln[r_i(t)/a(t)]}, \qquad (3)$$

and

$$B_z(t) = B_0 \frac{[r_i(0)^2 - a(0)^2]}{[r_i(t)^2 - a(t)^2]} \qquad (4)$$

Initially $r_i(0)/a(0) = 18$. Since $r_i(t)/a(t)$ will be close to unity at maximum compression the maximum value of $I(t)$ can be much greater than I_0. Similarly the axial magnetic field $B_z(t)$ will also be much larger than the initial field B_0.

The concept of pressure balance via Bennett equilibrium is quite inapplicable, since the compression of the inner pinch takes place on the Alfven time scale which is much faster than the implosion time of the outer pinch and can be on the order of a several-hundred picoseconds in the final stages of collapse. Therefore, in this case the compressional heating will be more effective than Ohmic heating. Therefore, a dynamic model is required to calculate the dynamics and the heating of the plasma. For simplicity reason, we will present only the results calculated by using a zero-D model. The basic purpose of this exercise is to clarify some of the basic features and the physical process involved in the configuration of the staged Z-pinch. In order to calculate various physical quantities more accurately one needs to use advanced codes like LASNEX which has the capability of using the proper equation of state and radiation hydrodynamics that is treated very poorly in the present model.

The equations that describe the dynamics of the staged Z-pinch, assuming the thin shell approximation (i.e., $R \approx r_i(t) \approx r_0(t)$), are as follows:

$$\frac{d^2R}{dt^2} = -\frac{I_p^2(t)}{100\,R_0^2 M}\frac{1}{R} + \frac{I_0^2}{100\,R_0^2 M}\frac{1}{R}\left[\frac{\ln(R_0/a_0)}{\ln(RR_0/aa_0)}\right]^2 - \frac{B_0^2 R}{4M}\left[1 - \left(\frac{1}{R}\right)^4\right], \qquad (5)$$

$$\frac{d^2 a}{dt^2} = \frac{n_0 T_0}{M_c}\frac{T}{a} - \frac{I_0^2}{100 a_0^2 M_c}\frac{1}{a}\left[\frac{\ln(R_0/a_0)}{\ln(RR_0/aa_0)}\right]^2 - \frac{B_0^2 a}{4 M_c}\left(\frac{R_0^2 - a_0^2}{R^2 R_0^2 - a^2 a_0^2}\right)^2, \qquad (6)$$

$$\frac{dT}{dt} = -2(\gamma-1)\frac{T}{a}\frac{da}{dt} + \frac{(\gamma-1)\times 10^{-20}}{n_0 T_0} a^2\left[\frac{dW_{ohm}}{dt} - \frac{dW_{br}}{dt} - \frac{dW_c}{dt} + \frac{dW_\alpha}{dt}\right], \qquad (7)$$

where $I_p(t) = I_m \sin(\pi t/2t_{1/4})$, I_0 is the initial wire plasma current, I_m is the maximum generator current and $t_{1/4}$ is the quarter period risetime of the current. Equations (5) and (6) are the equations of motion for the outer Z-pinch liner and the inner fibre plasma column respectively. Eqn. (7) is the energy equation for the fibre plasma where most of the relevant terms are included. Heating of the fibre plasma is achieved by Ohmic dissipation, adiabatic compression, and α-particle trapping, since α-particles will easily be trapped due to an extremely strong magnetic fields that exists outside the fibre plasma. Energy losses arise from bremsstrahlung and cyclotron radiation. The Ohmic heating rate can be written as:

$$\frac{dW_{ohm}}{dt} = 2 \times 10^8 \left(\frac{I_0^2}{a_0^2 T_0^{3/2}}\right)\left[\frac{ln(R_0/a_0)}{ln(RR_0/aa_0)}\right]^2 + 1.25 \times 10^{13} \left(\frac{B_0^2}{a_0^2 T_0^{3/2}}\right)\left(\frac{1}{a^2 R^4 T^{3/2}}\right), \quad (8)$$

where the first term is due to the dissipation of axial current and the second term is due to the dissipation of the azimuthal current. The bremsstrahlung and cyclotron radiation losses can be written as:

$$\frac{dW_{br}}{dt} = 2.36 \times 10^{16} \left(n_0^2 \sqrt{T_0}\right)\left(\frac{\sqrt{T}}{a^4}\right), \quad (9)$$

and

$$\frac{dW_c}{dt} = 1.9 \times 10^{14} \, n_0 T_0 \left(B^2 \frac{T}{a^2}\right), \quad (10)$$

respectively. Here,

$$B = \left[B_z^2 + B_\theta^2\right]^{1/2} = \left[B_0^2\left(\frac{R_0^2 - a_0^2}{R^2 R_0^2 - a^2 a_0^2}\right)^2 + \frac{I_0^2}{100 a_0^2}\frac{1}{a^2}\left(\frac{ln(R_0/a_0)}{ln(R_0/a_0 + ln(R/a))}\right)^2\right]^{1/2}. \quad (11)$$

Finally, the α-particle heating can be accounted for by using the following equations provided that all the α's deposit their energy back into the plasma column.

$$\frac{dW_\alpha}{dt} = 3.125 \times 10^{22} \left(\frac{n_0^2}{T_0^{2/3}}\right)\left(\frac{1}{a^2} - 2n_\alpha\right)^2 \exp\left[\left(-\frac{19.94}{T_0^{1/3}}\right)\frac{1}{T^{1/3}}\right], \quad (12)$$

where

$$\frac{dn_\alpha}{dt} = 920 \left(\frac{n_0}{T_0^{2/3}}\right)\left(\frac{1}{a^2} - 2n_\alpha\right)^2 \frac{1}{T^{2/3}} \exp\left[-\left(\frac{19.94}{T_0^{1/3}}\right)\frac{1}{T^{1/3}}\right] - 2\frac{n_\alpha}{a}\frac{da}{dt}. \quad (13)$$

In these equations R is normalized to its initial radius R_0, the radius a and temperature T are normalized to their respective initial values a_0 and T_0, and n_0 is the initial density in units of 10^{20} cm^{-3}. M, M_c are mass per unit length of the Z-pinch and wire plasmas. R_0 and a_0 are initial values of the Z-pinch and wire plasma radii. T is the temperature of wire plasma. The units are as follows: M, M_c are μgm/cm; I, I_0 and MA, B_0 is MG; t, t_0 are nsec; r, a are cm; and T is KeV. The units in Eqns. (8) – (10) and (12) are KeV/ns-cm^3.

This zero-D model is studied numerically for two different cases. In the first case we assumed no axial current through the fibre and only axial magnetic field is allowed to compress. This configuration is called Z-θ pinch. The initial conditions for the outer pinch are I_p = 10 MA, B_0 = 20 kG, t_0 = 50 ns, R_0 = 4 cm, M = 38 μgm, representative of the Sandia National Laboratories Saturn Accelerator. These initial conditions are optimized to attain the highest-compressed magnetic field and the results are displayed in Figure 3. We

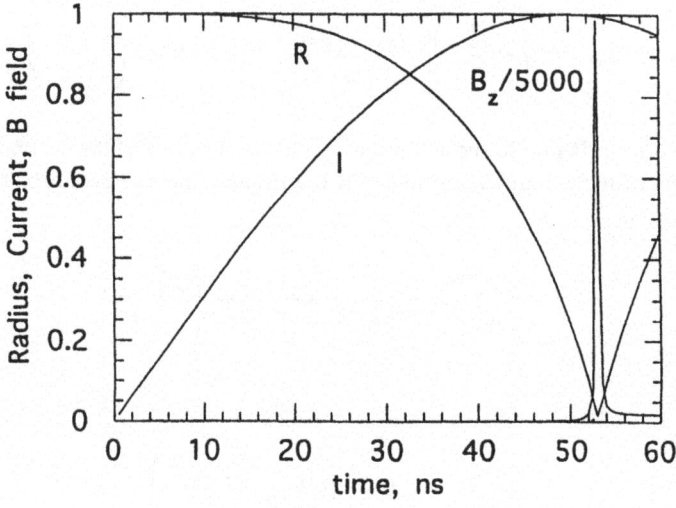

Figure 3.

see that the risetime of $B_z(t)$ is approximately a factor of 100 less than the rise time of the $I_z(t)$ current and the maximum value could be above 100 MG. With such a fast risetime, large-magnetic field and in the presence of current sharing between the parallel discharge loads the fibre will most certainly ionize. Only J_θ induced by such a rapidly changing axial magnetic field B_z heats the plasma initially by Ohmic dissipation and later on by adiabatic compression.

The initial conditions for the inner pinch are assumed as: $a_0 = 200$ μm radius, $n_0 = 10^{22}$ cm^{-3}, and temperature $T_0 = 20$ eV and the results are displayed in Figure 4 and 5. The collapse of the central-fibre plasma (θ/-pinch) occurs on a time scale of $0.1 - 0.2$ ns (den-

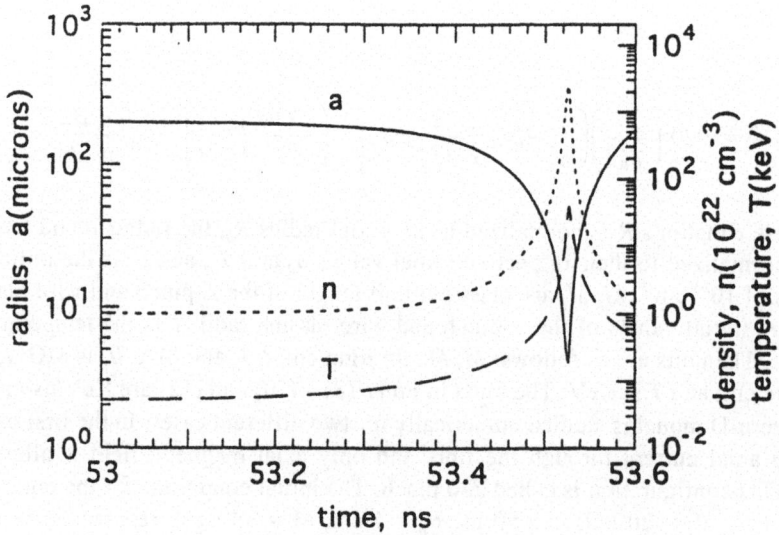

Figure 4.

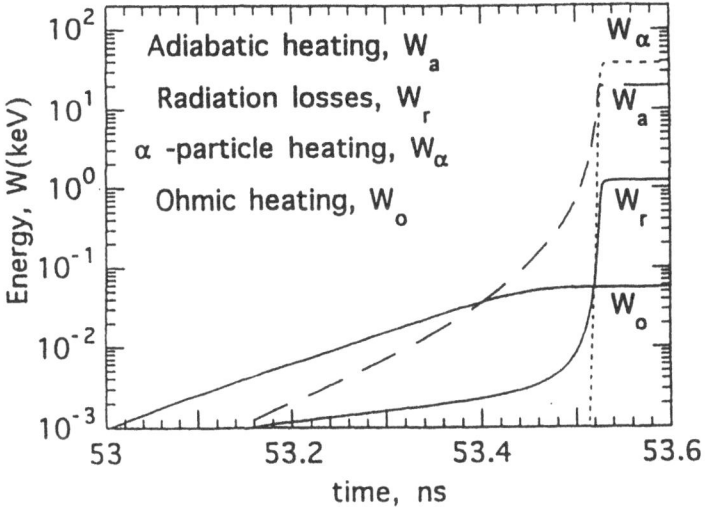

Figure 5.

sity rise time), compared to the field-compression time of few nanoseconds. The maximum parameters of plasma are found to be: $n = 2.5 \times 10^{25}$ cm^{-3}, $T = 40$ KeV, f (burn fraction) $\approx 10\%$, G (gain) > 3 which corresponds to a neutron yield of 10^{18} cm^{-1}. An important thing to note here is that a current of at least 10 MA is required for achieving the break-even in energy[1].

Comparison of different heating and radiation losses show that Ohmic heating starts quite early and the radiation losses are not enough to keep the plasma cold. The main mechanism for heating and compression leading to the fusion conditions is adiabatic compression that appears at later stage of the collapse and is more effective if the initial-plasma temperature (of the fibre) is low. Seeding the plasma with high Z-impurities may increase the radiative loss in this early stage of Ohmic heating and will prevent premature heating of the fibre plasma. Once sufficiently high density and high temperature are reached to allow fusion to take place the α-particle heating will add further energy, as they will be trapped in a highly compressed axial magnetic field. This assumption is valid since the gyroradius for 3.5 MeV α-particles is less than few μms. Even without α-particles trapping one can get a gain of more than 2 with a burn fraction of about 8%.

In the second case a very weak axial magnetic field is applied and an axial current is allowed to flow initially through the fibre. It is assumed that the fibre converts into a plasma and the initial conditions for this plasma are assumed to be: $a_0 = 100$ μm, $n_0 = 3 \times 10^{22}$ cm^{-3}, $I_0 = 200$ kA, and $T_0 = 200$ eV. In this case the compression time scale is so short that both the Ohmic heating and radiative losses are negligibly small. The following parameters are used for the outer pinch: $R_0 = 2$ cm, $t_m = 1$ μsec, and $I_m = 2$ MA. These parameters are relevant to the UCI machine that is presently under construction. The results are shown in Figures 6 and 7. Figure 6 shows the time variation of outer and inner plasma radii, and the density and temperature of the fibre plasma. The fibre is assumed to be made of DT with an equal percentage. At the peak compression the plasma density of 8×10^{24} cm^{-3} and a temperature of 10 KeV can be maintained for at least a period of 1/8 ns. This gives an $n\tau \approx 10^{15}$ which is an order of magnitude higher than the required value for fusion in DT plasma. In Figure 7 the outer current and the inner currents are presented

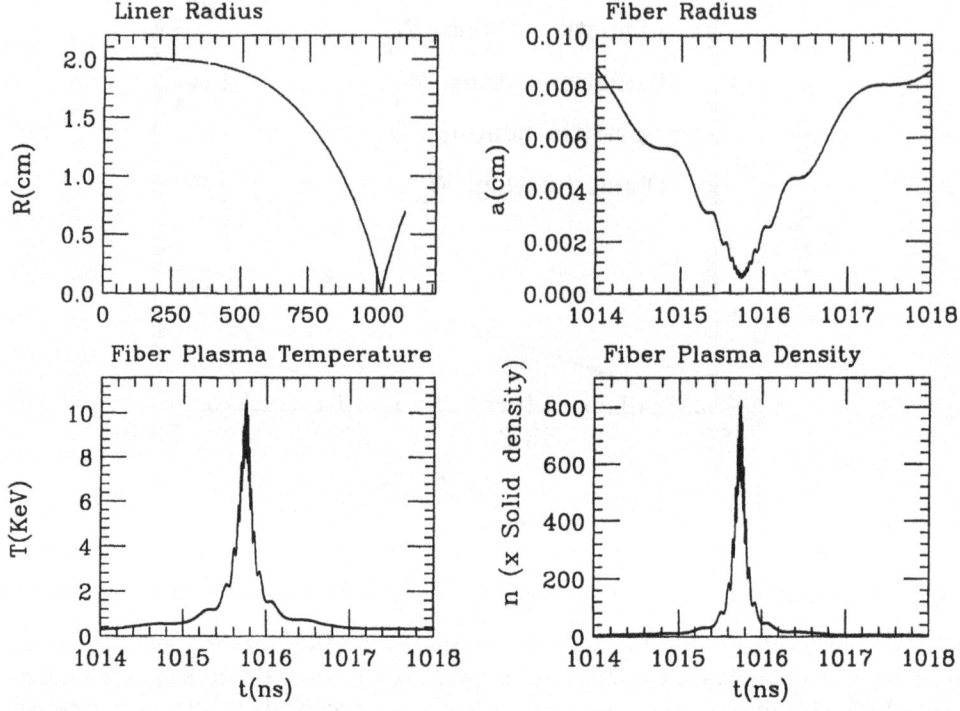

Figure 6.

showing a risetime of the inner current exceeding more than 10^{17} Ampere/sec. Also this current through the fibre seems to oscillate between 2 MA and 5 MA during the period of peak compression. The current in the wire plasma increases because of flux conservation. At maximum compression the peak magnetic field is well above 100 MG in an annular region of thickness of 1.3 μm. The penetration of the magnetic field into the plasma is of order $c/\omega_p \cong 10^{-7}$cm assuming the plasma density is of the order of 10^{25} cm^{-3} at peak compression. Figure 6 also shows the burn fraction and gain in the energy. The result shows that a gain of more than 15 is possible by burning 3% of the fuel.

Further calculations require more elaborate theoretical formulations. The simple calculations presented above suggest that the plasma could be heated to thermonuclear conditions, including only the most basic classical heating mechanisms. These results demonstrate that the characteristic time scales for radiation and thermal losses are long compared to the adiabatic-compression timescale which dominates, leading to a large magnitude increase in the assembled density.

3. STABILITY ISSUES

As discussed in the previous sections, a staged Z-pinch consists of two different types of coaxial plasmas imploding on completely different time scales. Experimental observations have shown that such type of configuration stays quite uniform and stable. However, theoretically it has not yet been shown the reason for the enhanced stability. We now discuss some of the major instabilities that are characteristic of the conventional Z-

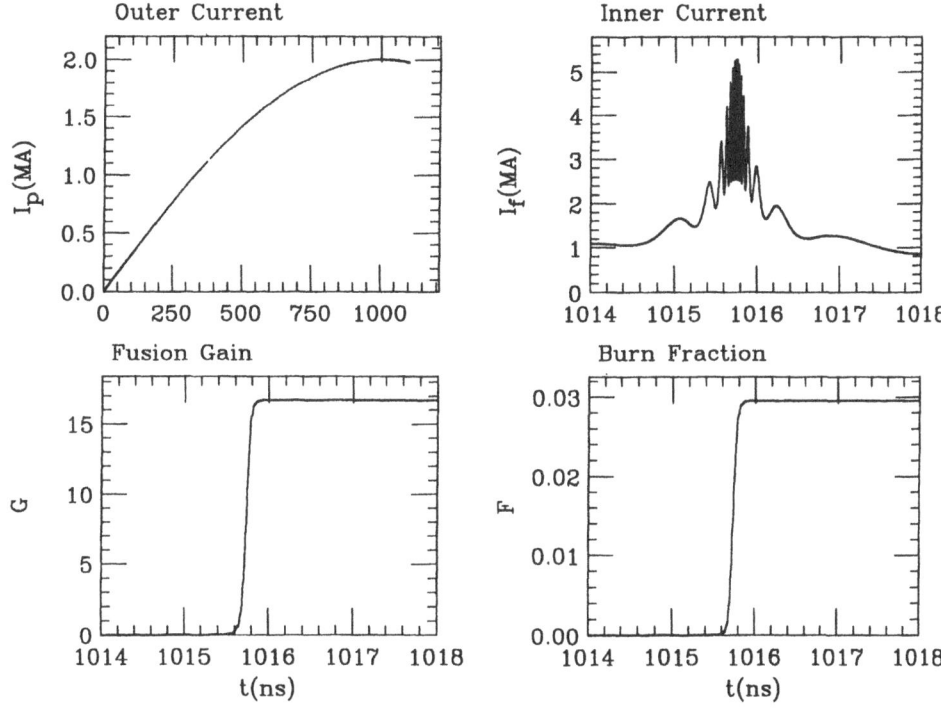

Figure 7.

pinch and the reason for their stabilization in the present configuration. During acceleration of the outer Z-pinch, the Rayleigh-Taylor instability is unavoidable. For a plasma that implodes by magnetic pressure the Rayleigh-Taylor instability grows like,

$$\xi = \xi_o \exp(\gamma t), \tag{14}$$

where ξ_o is some initial perturbation that may appear due to the nonuniform current discharge, $\gamma = \sqrt{gk}$ is the growth rate of the instability with wave number k and $g = B^2 A/8\pi M$ is the acceleration of the imploding plasma. From the equation of motion we can approximate $t = \sqrt{2s/g}$, therefore, Eqn. (14) will reduce to,

$$\xi = \xi_o \exp\sqrt{2sk}. \tag{15}$$

In order to minimize this instability both the initial perturbation ξ_o and the distance s (the initial radius of the plasma column) need to be small. The initial perturbation ξ_o could be minimized by initiating a discharge through a pre-ionized plasma instead of a gas column. On the other hand by choosing the smallest possible radius with sufficient mass density, to give maximum compression at peak current, we can further reduce the growth of the RT-instability. Some of these conditions have been employed in most of the experiments that have shown stable implosions. For example a plasma jet instead of a gas puff was used in the French experiment[10] where the improved stability of the pinch plasma is observed. Pre-ionization of the gas puff before applying the main pulse was also used at UCI experiments and the radius of the outer pinch was optimized at the lowest possible minimum value.[14] This provided a significantly better stability and implosion of the plasma column.

Application of an axial magnetic field will also improve the stability as the B_θ and B_z components combine to provide a resultant magnetic shear. Part of the B_θ magnetic field that accelerates the plasma and of the B_z field produced by the external coil is trapped in the plasma and some in the space interior to the plasma. Although a complete analysis has never been carried out, early UCI experiments have noted, by means of Mach-Zender laser interferometry and Schlieren imaging, that the acceleration phase of Z-pinch is more stable with a trapped B_z-field; similar to experimental observations elsewhere.[4,5,6]

Near the peak compression of the outer pinch the energy is transferred to the coaxial fibre through sheared magnetic flux trapped in between the outer pinch and the inner pinch on an extremely fast time scale. Since the fibre plasma will accelerate and compress over very small radius ($\approx 200 - 100$ μm) the long wavelength modes of R-T instability will not grow at all. However, short wavelength perturbations may be able to grow and need to be investigated because this may cause the diffusion of magnetic flux into the fibre plasma. This will inhibit the compression of the fibre plasma to the desired parameter regime. Magnetohydrodynamic (MHD) instabilities on the other hand may not be able to grow because the implosion time of the fibre plasma may be much faster than the instability growth time. We cannot make any qualitative estimate at this stage because the previous analysis of these instabilities normally assume some sort of static equilibria that may not exist in the present case. The proper analysis of these instabilities should include the dynamic equilibria and multiple time scales that seems to be necessary in dealing with the staged Z-pinch configuration.

4. CONCLUSION

Several laboratories in the U.S., France, and Russia have performed experiments at power levels ranging from P = 0.01 – 10 TW on Z-pinches using the mechanism of staging for transferring the current to the final load. Most of the time these loads are made up of a small diameter ($\sim 10 - 20$ μm) fibre, straw or a column of gas. This results into formation of a pinch plasma column which maintains a remarkably uniform structure that has not previously been observed in any conventional pinch experiment. Also the radiation spectrum from the staged pinch is much harder as compared to the other pinched plasmas suggesting the possibility of a hotter plasma (approximately 20 - 30 KeV).

One of the main advantage of this method is that one can avoid the early phase of long duration slowly rising current profile and utilize the full thrust of peak current for plasma heating and compression. This results in an unprecedented large value of dI/dt for the current that flows through the final load. For example a current of 2 MA with a rise time of one μsec can be compressed up to 5 MA with risetime of less than one nanosecond by applying only single step staging. Usually the load in conventional pinch was either destroyed or used to become unstable before reaching the peak current. The concentration of such a large value of current into a small diameter fibre provides enormous magnetic pressure that compresses and heats the fibre plasma to conditions favourable to thermonuclear fusion. The terminal particle pressure of the fibre plasma ultimately exceeds the magnetic pressure, therefore, the confinement of fibre the plasma would be inertial.

The experiments on staged Z-pinch have also shown the remarkable uniformity of the final plasma column. This may be because the plasma column establishes on a time scale much faster than the growth time of the most disruptive instabilities. The range of parameters that one can achieve by this technique is very interesting for a wide range of applications. For example controlled thermonuclear fusion, intense radiation source and X-ray laser.

ACKNOWLEDGMENT

This research was supported by the Department of Energy, Office of Fusion Energy.

REFERENCES

1. H. U. Rahman, P. Ney, F. J. Wessel, A. Fisher and N. Rostoker, Proc. 2nd Int. Conf. on High Density Pinches, Laguna Beach, April 26–29 (1989), AIP Conf. Proc., p. 195; H. U. Rahman, P. Ney, F. J. Wessel and N. Rostoker, *Comments Plasma Phys. Controlled Fusion 15*, 339 (1994).
2. H. U. Rahman, F. J. Wessel and N. Rostoker, *Phys. Rev. Lett. 74*, 714 (1995).
3. V. P. Smirnov, *Plasma Phys. Cont. Fusion 33*, 1697 (1991); V. A. Gasilov, S. V. Zakarov, and V. P. Smirnov, *JETP Lett. 53*, 85 (1991); V.D. Vikarev, G. S. Volkov, S. V. Zakharov, S. A. Komarov, V. P. Smirnov, I. N. Frolov, and V. Ya. Tsarfin, *Sov. J. Plasma Phys. 16*, 217 (1990); J. G. Linhart, *Nuovo Cimento 17*, 850 (1960), *Nucl. Fusion 10*, 211 (1970), *Nucl. Fusion 13*, 321 (1973), *Nuc. Inst. Methods in Physics Research*, *A278*, 114 (1989).
4. J. Bailey, Y. Ettinger, A. Fisher and N. Rostoker, *Appl. Phys. Lett. 40*, 460, (1982); *J. Appl. Phys. 60*, 1939 (1986).
5. F. J. Wessel, F. S. Felber, N. C. Wild, H. U. Rahman, E. Ruden and A. Fisher, *Appl. Phys. Lett. 48*, 1119 (1986).
6. F. J. Wessel, N. C. Wild, A. Fisher, H. U. Rahman, A. Ron and F. S. Felber, *Rev. Sci. Instrum. 57*, 2247 (1986).
7. F. S. Felber, F. J. Wessel, N. C. Wild, H. U. Rahman, A. Fisher, C. M. Fowler, M. A. Liberman, A. L. Velikovich, *J. Appl. Phys. 64*, 3831 - 3845 (1988).
8. F. S. Felber, M. M. Malley, F. J. Wessel, M. K. Matzen, M. A. Palmer, R. B. Spielman, M. A. Liberman and A. L. Velikovich, *Phys. Fluids 31*, 2053, (1988).
9. N. A. Ratakhin, S. A. Sorokin, S. A. Chaikovsky, *Proc. Seventh Int. Con. on High Power Praticle Beams*, Vol. II, 1204, (1988).
10. F. J. Wessel, B. Etlicher, and P. Choi, "Z-Pinch Implosion of an Aluminum Plasma Jet onto a Coaxial Wire: Enhanced Stability and Energy Transfer," *Phy. Rev. Lett. 69*, No. 21 (1992).
11. F. J. Wessel, B. Etlicher, N. S. Edison, A. S. Chuvatin, L. Veron, C. Rouille, and S. Attelan, "Stabilization of Z-Pinch Plsams using Magnetic Fields," paper presented at the APS Confernce on Plasma Physics, Seattle, WA, Nov. 1992.
12. N. S. Edison, B. Etlicher, A. S. Chuvatin, S. Attelan, and R. Aliga, *Phys. Rev. E 48*, 3893, (1993).
13. V. Smirnov, P. Choi, B. Etlicher, J. Larour, and F. J. Wessel, "Angara 5–1 Experiments," paper presented at the APS Conference on Plasma Physics, Seattle, WA, Nov. 1992.
14. E. Ruden, H. U. Rahman, A. Fisher, and N. Rostoker, "Stability Enhancement of Low Initial Density Hollow Gas Puff Z-pinch by e-Beam Preionization," *J. App. Phys. 61*, 1311, (1986).

NOVEL STAGED Z-PINCH CONCEPT AS SUPER RADIANT X-RAY SOURCE FOR ICF

Vitaly Bystritskii, Frank J. Wessel, Norman Rostoker, and Hafiz Rahman

[1]University of California
Irvine, California, 92697–4575
[2]University of California
Riverside, California, 92521

ABSTRACT

We propose a Super-Radiant X-Ray Source (SRX), which has potential as an ICF target driver. This approach involves an imploding Z-pinch liner, which flux compresses an axial magnetic field. An ion beam with Z >10 and E= 10^5 -10^6 eV is injected through the magnetic cusp, and trapped inside the imploding liner, with a magnetic flux inside the liner and canonical-angular momentum of the ion beam being conserved. As a result of this conservation the ions receive azimuthal energy from the imploding conducting liner, being accelerated to a final energy hundreds times greater, than their initial energy, as $E_{fin} = E_{in} * (r_{in}/r_{fin})^2$, depending on initial r_{in} to final r_{fin} radii ratio. In the final collision with the plasma column, placed co-axial with the liner and rotating ion beam, the former is ionized by the ion beam, and produces radiation with anomalously high input of characteristic M-, L- and K- radiation. It takes place due to the processes of quasi-molecular shell formation in the collisions of multi charged ions, with following promotion of the inner shells electrons to higher energy or the continuum at the expense of the quasi-molecules excess kinetic energy (Garcia, 1973). The approach differs from conventional mechanism of ionization in the Z-pinch via electron-ohmic heating of the plasma, resulting in the radiation of mainly thermal character.

Measurements and calculations give cross-sections for K-, L-, and M- shell vacancy formation with following x-ray production of the order of, $\sigma = 10^{-19} - 10^{-17}$ cm^2, depending on the mass of the colliding heavy particles and their energy (Garcia, 1973).

For the specific example of current discharge amplitude of 60 MA, typical for the JUPITER generator the total radiation yield in the spectral range of hv=10^{-1}–10 keV (M, L, K-shell) is projected to be >10 MJ in a 900 TW pulse, with an over all efficiency of liner energy conversion ≈ 20%.

The SRX configuration should be stable against the most dangerous plasma instabilities typical for most high density/temperature pinches, for short- to long-current risetimes. This stability is a natural consequence of the applied axial-magnetic field and the progressively faster timescales leading to stagnation-energy transfer. The considered intense X-ray radiation could be transported along the axis of imploded Z-pinch to the external target in the hohlraum scheme of implosion

The report analyses in detail various issues of this approach, including injection and trapping of the ion beam, liner implosion with a trapped magnetic field and rotating ion beam, stability of the liner, production of intense bursts of characteristic X-rays and its possible application for driving a hohlraum target.

INTRODUCTION

We wish to produce intense pulses of X-rays in the 0.1–10 keV energy range which has potential as a driver for hohlraum ICF target. High intensity X-ray pulses are presently generated by imploding-plasma liners, foils, and wire-array Z-pinches, driven by high power, short pulse relativistic generators (Pereira, 1988). In the conventional approach, as the plasma liner implodes on-axis and stagnates, the radial ion flow transfers its kinetic energy to the plasma electrons, providing continuous and characteristic radiation (the latter-via inner-shell vacancies generated in the target ions and atoms). Non-thermal processes, due to the presence of instabilities, resulting in generation of transitional high electric field in the Z-pinch could also play an important role in the ionization process.

In the conventional approach to ionize inner-shell transitions efficiently the kinetic energy per ion in the imploding plasma must exceed a threshold value in order to transfer sufficient thermal energy to the plasma electrons (Whitney , 1990). Thus,

$$\eta = E_{ion}/E_{min} > 1 \tag{1}$$

where E_{ion} is the kinetic energy and E_{min} is related to the sum of the ionization energy of the ion plus the electron- and ion- thermal energies.

The strong dependence of η on atomic number places severe constraints on the facility parameters if characteristic radiation from high Z plasmas is desired (for instance from Kr-plasma), since, $E_{min} \propto Z^{3.5}$. Rough scalings for the case of, $h\nu \sim 1-10$ keV with M- K-shells yield of 1 MJ/cm, predict ion's kinetic energies of 1–10 MeV/nucleon range, current risetimes of tens of nanoseconds, and pinch currents, $I_z \sim 100$, MA which are far out of feasible parameters for existing or designed pulsed super power generators (for instance, SATURN, JUPITER).

We attempt to reduce these requirements via magnetic acceleration of a medium atomic weight ion beam and its subsequent non-thermal collisions in a high Z target plasma. The basic configuration is referred to as a Super Radiant Plasma Source (SRX), and it is projected to provide improved X-ray energy-conversion efficiency (over present plasma-radiation source technologies) to satisfy the spectral output, total-energy yield, and radiation-pulse widths to serve for a potential driver of a hohlraum ICF target. Preliminary scalings for the SRX suggest a favorable increase in the ionization efficiency, output yield, and photon energy with driver power and current.

Based upon calculations and preliminary designs presented here for the facility of the JUPITER level, we envision the potential to get hundreds of MeV for magnetically accelerated multi charged ions inside the imploding Z-pinch, interacting with a dense hot target of high-atomic-number material, such as Cu, Xe, W, etc. For the specific case of an Xe ion beam and Xe target over 50 % of the liner energy could be converted into circulating-ion energy (resulting in $E_{Xe} \approx 900$ MeV) and over 20% of the ion energy would be converted into characteristic-target radiation with a total yield of $E_{rad} \approx 9$ MJ per pulse.

The following sections summarizes our key predictions that a SRX Z-pinch would produce significant inner-shell X-ray radiation. The discussion begins with: (1) considering the problem of inner shell vacancy production in ion -atom collisions, and continues with: (2) analysis of dynamics of an imploding-liner Z-pinch, with an axial-magnetic field and trapped-ion beam; (3) analysis of ion-field reversal inside the liner; (4) its stability; (5) discussion of the radiation-conversion process that occurs as the ion beam stagnates onto a dense-target plasma; (6) analysis of the ion beam acceleration design. Some preliminary estimates are made in section (7) on using X-radiation for driving a hohlraum target, placed near the Z-pinch.

1. INNER SHELL VACANCY PRODUCTION IN ION -ATOM COLLISIONS

1.1. Experimental Approach

Figure 1 displays a schematic illustration of the SRX installed between the discharge electrodes of a high-power generator. The scenario of events proceeds as follows: an axial

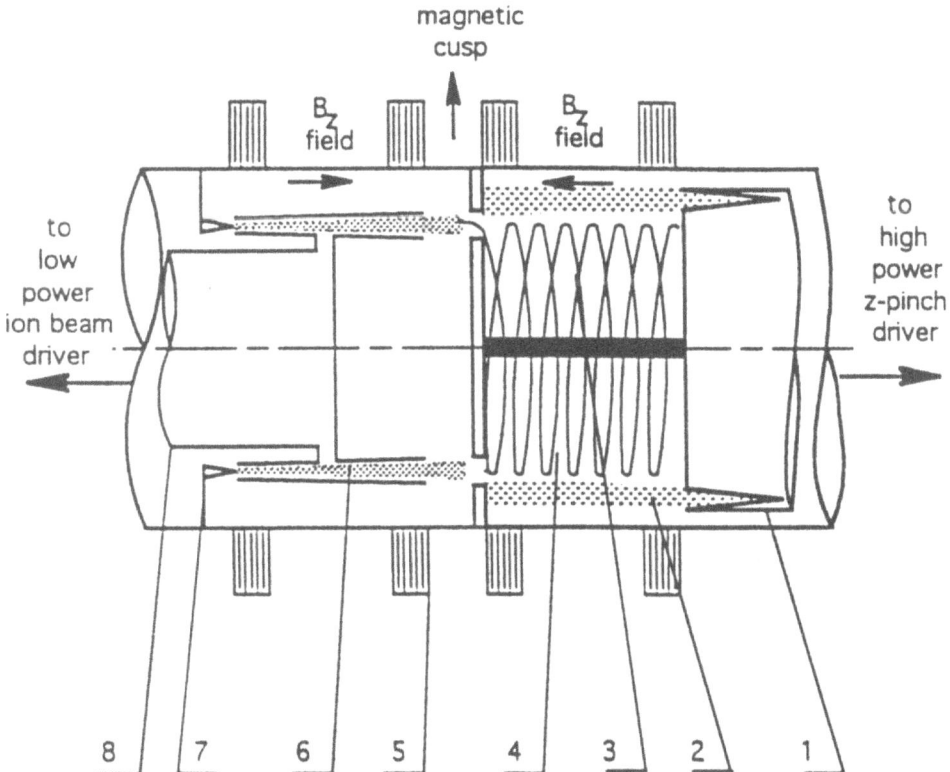

Figure 1. Schematic illustration of the Super Radiatn X-ray Source (SRX): 1) Z-pinch electrode (injector/nozzle/support), 2) imploding liner (gas-puff/plasma jet or foil), 3) co-axial target (high Z wire, foam, or cryogenic fibre), 4) trapped ions, 5) Helmholtz magnetic field coil, 6) ion beam, 7) and 8) cathode and anode of the triode ion beam injector.

magnetic field is injected into the pinch discharge region and shaped near the cathode electrode producing a weak mirror configuration. An oppositely directed magnetic field is applied inside the ion source so that a magnetic-cusp exists between the two pulsed-power systems. After the magnetic fields are established the ion-beam accelerator is energized, injecting a tubular ion beam through the magnetic cusp. The axial velocity of the ion beam is significantly reduced as it passes through the cusp ($v_{fin} \sim 10^{-1} v_{in}$), converting main portion of its axial energy into rotational. After exiting the cusp the ions pass into the axial-magnetic field and are trapped inside the Z-pinch liner. The rotating-ion beam encounters a mirror-magnetic field at the upstream cathode electrode, where it reflects. At the prescribed moment the Z-pinch generator is energized so that the liner begins to accelerate and implode radially inward. In the process of implosion the ion beam azimuthal velocity increases, and the diameter of the ring decreases, due to field compression, reducing almost to zero the escape of ions through the loss-cone. As the liner collapses the magnetic field B_z is rapidly compressed, accelerating the ion beam on a timescale, $\tau_{ac} < \tau_{1/4}$, here $\tau_{1/4}$ is the Z-pinch current rise time, The one particle approximation (valid for the case of ion beam current neutralization) predicts the final energy of rotating ion beam as $E_{fin} \approx E_{in} (R_{in}/R_{fin})^2$. So, for reasonable ratio of radial compression of ~10 -30 we can expect 10^{2-3} enhancement of the ion beam energy. At the instant of maximum compression the high velocity ions stagnate onto a co-axial, high-density target plasma, producing inner-shell radiation. The fastest time scale corresponds to stagnation of the ion beam in a target plasma where it deposits energy, exciting inner-shell transitions, $\tau_{stagn} \ll \tau_{ac}$.

The SRX configuration should be stable against the most dangerous plasma instabilities typical of most high density/temperature pinches, for both short- to long-current risetimes. This stability is a natural consequence of the applied axial-magnetic field and the progressively faster timescales leading to stagnation-energy transfer.

Over twenty years ago the Russian physicist A. Sakharov proposed that such methods be used to accelerate intense ion beams in a magnetic-flux compression liners (Sakharov, 1966). Only few calculations and some experimental attempts on this concept have been performed to date (Bystritskii, 1989; Men'shikov, 1991). Our proposal carries this concept one step further to the final-stagnation, energy-conversion process.

High-current Z-pinches are routinely used in explosive-driven, flux-compression systems at the 100 MA level, with implosion timescales of many microseconds, achieving radial convergence ratios of $R_{in}/R_{fin} \geq 10$ and final magnetic fields in excess of many tens of megagauss (Chernyshev, 1987; Shearer, 1987; Chernyshev, 1991).

At this juncture the fundamental question relates to the energy-conversion process that occurs as the ion beam stagnates in the co-axial target. Although the collision-cross sections of high-energy-ions impacting a dense-target plasma are well established (Lichten, 1967; Dyson, 1990), they are only available for much lower beam intensities than would be attainable in the SRX. Nevertheless, using present best estimates and empirical data on heavy-ion interactions it is clear that such an interaction will radiate profusely and energetically.

1.2. Inner Shell Vacancies Generation

When multi-electron ions experience collisions with other atoms, or ions, the electron shells deeply interpenetrate, developing quasi-molecular orbits (Garcia, 1973; Gardner, 1978), and the inner shell electron individual orbits give rise to two or more molecular orbits each (i.e split, starting from some specific internuclear distance- so called crossing radius R_{cr}), their energy varying as function of inter-nuclear separation, as illustrated in Fig. 2.

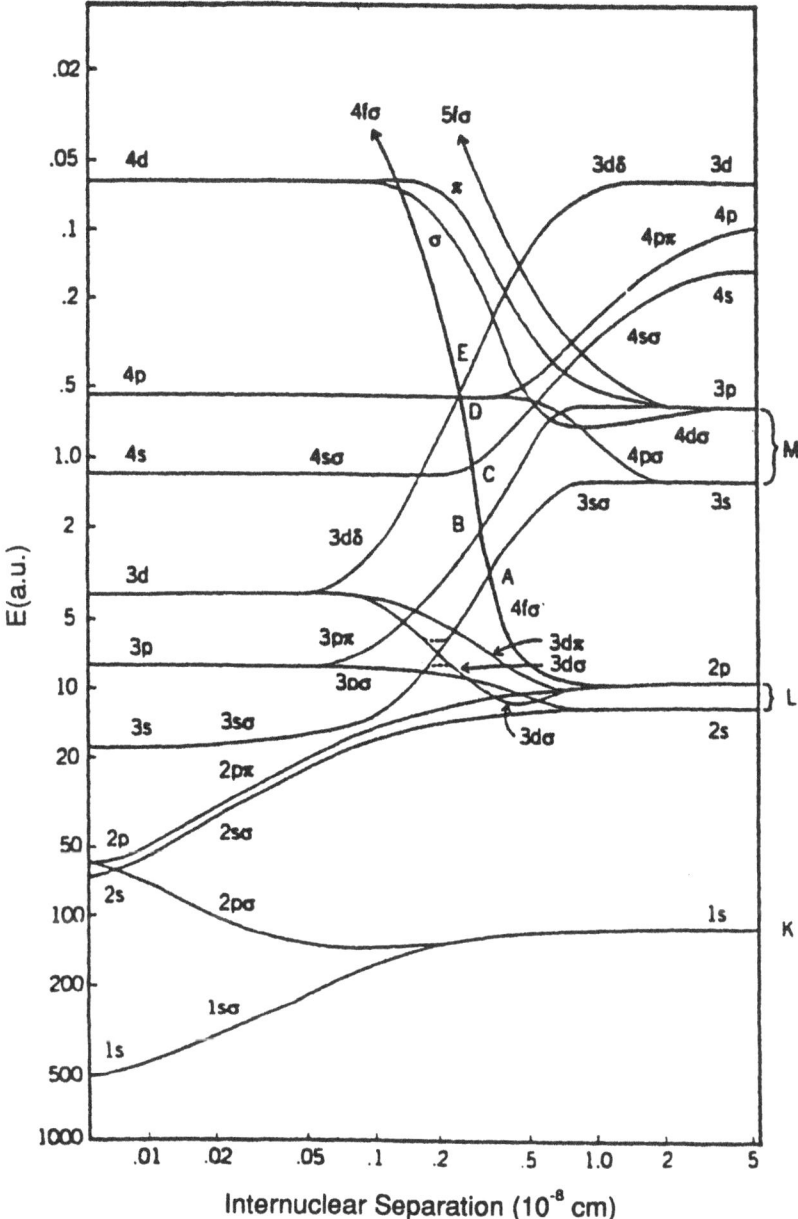

Figure 2. Correlated energy levels for Ar+ → Ar collision. Energy levels at the right are for individual AR atoms and at the left are for the united atom limit of Kr (Ref. 3).

For the case when the $R_{min} < R_{cr}$ (here R_{min} - the nearest nuclei approach distance during collision) there is a finite probability that the electrons occupying the original levels could make adiabatic transitions (i.e promoted at the expense of the quasi-molecule excess kinetic energy) to the higher unfilled levels, or contimuum. This electron promotion will result in the generation of the inner-shell vacancies during the receding of ions at the end of the collision, with respective generation of Auger electrons, and/or X-rays. Measure-

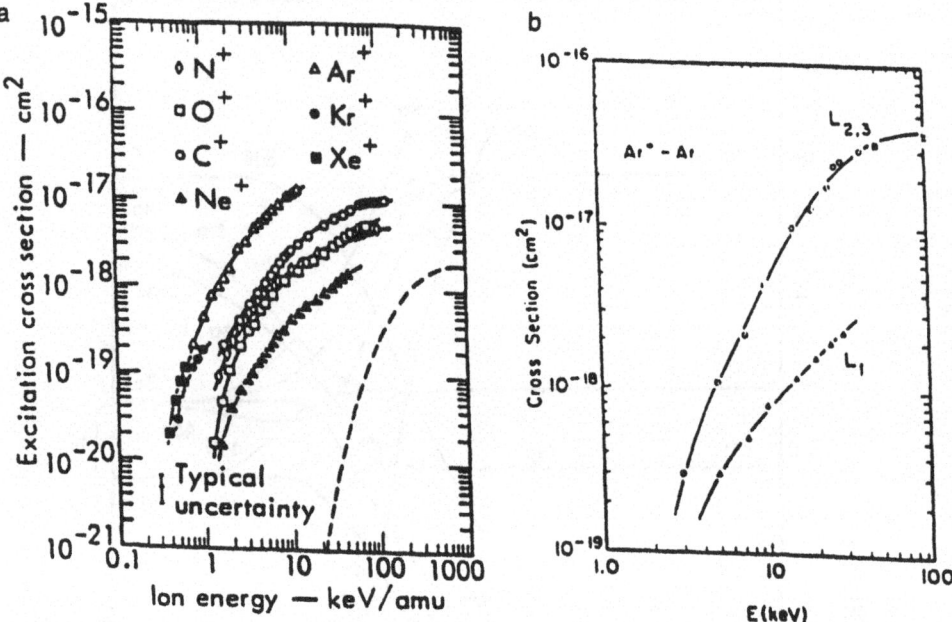

Figure 3. a) Carbon K-shell exitation cross sections as a function of incident ion energy per atomic mass unit for solid targets. The dashed curve is for incident protons. The data for Kr+ and Xe+ may be only the order of magnitude estimates due to the recoil effect in thick, solid target. b) L-shell exitation cross sections for the Ar+-Ar system, (Garcia, 1973).

ments give cross-sections for M-K shell ionization of the order of $\sigma = 10^{-19}–10^{-17}\,cm^2$, depending on the mass of the colliding heavy particles and their energy. It is illustrated in Fig. 3a,b, where experimental data are given for K-, L-shells exitation of different atoms v.s. energy and type of ion-projectiles.

This process is qualitatively different from a binary type of point charge projectile collision with individual bound electron, i.e — from the proton-ion and electron-ion inner shell impact ionization, the latter characterized by cross sections at least one order and more of magnitude smaller, $\sigma = 10^{-22}–10^{-20}\,cm^2$, (Gryzinski, 1964). It is illustrated in Fig. 4.

2. IMPLOSION DYNAMICS

The simplest dynamic model considers a thin-shell implosion of an annular Z-pinch with an entrained-magnetic field and a trapped-ion beam. In this model the ion beam is assumed to be completely current and charge neutralized by the accompanying electrons, so that diamagnetic effects of beam current are absent. In this case an azimuthal momentum balance for the ion shell is given by (Sudan, 1974):

$$r d^2\phi/dt^2 + 2(dr/dt)(d\phi/dt) = Ze[E_\phi - B_z(dr/dt)/c]/m \tag{2}$$

E_ϕ is the azimuthal electric field. B_z is the total magnetic field, $B_z = B_e + B_s$ with B_e is the external field and B_s is the self-field produced by the ion current,

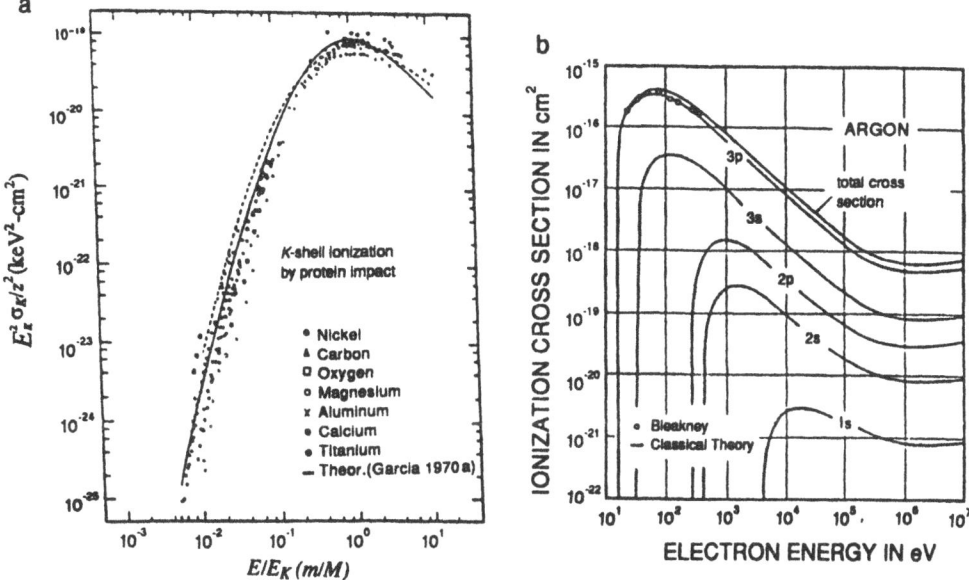

Figure 4. a) The cross section for the ionization of different shells and the total ionization cross section of Ar by electrons up to relativistic energies (Gryzinski, 1964). b) General comparison of Born and binary encounter approximation with experimental K-shell x-ray cross sections by proton impact: dashed curve, Born approximation; solid curve, binary encounter estimate; solid circles, data for elements with $15 \leq Z \leq 70$; inverted triangles, 160 MeV protons; open circles with vertical bar, Ta measurements; open and solid squares, experimental Mg points; open circles, O data (Garcia, 1973).

$$I_i = ZeNU_\phi \tag{3}$$

and the electron current,

$$I_e = -ZcNv_\phi \tag{4}$$

$U_\phi = r(d\phi/dt)$ for ions and v_ϕ is the corresponding quantity for electrons. The electrons have a small gyroradius and drift according to the equation,

$$V_d = c\,[\mathbf{EB}] / B^2 \tag{5}$$

To preserve charge neutrality electrons must drift inward with the ions so that,

$$dr/dt = c\, E\phi/B_z \tag{6}$$

It follows from Eqn. (2), $rd^2\phi/dt^2 + 2(dr/dt)d\phi/dt = 0$, and:

$$U_\phi = rd\phi/dt = R_{in}U_{in}/r,\quad I = I_{in}(R_{in}/r),\ \text{and}\ E_{fin} = 0.5NmU_\phi^2 = E_{in}(R_{in}/r)^2, \tag{7}$$

i.e. the increase of the asymuthal kinetic energy of the ion beam during compression follows the rise of the external B_z-field. It should be noted, basing on Eq. (6), that:

$$E_\phi - (dr/dt) B_z /c = -(1/rc)d(rA_\phi)/dt = -(2\pi rc)^{-1}d[\pi r^2 B_e + cL(I_i + I_e)]/dt = 0, \quad (8)$$

and:

$$\Phi = \pi r^2 B_e + cL(I_e + I_i) = \text{constant}, \quad (9)$$

is the total flux linked by the beam, here $L = 4\pi^2 r^2/c^2$ is the beam inductance per unit length.

The coupled dynamic equations of motion for the outer liner, trapped-ion ring, and ion velocity in this case are written as,

$$d^2R/dt^2 = -I^2/100 m_1 R_1^2 - R_1(B_z^2 - B_{z1}^2)/4m_1. \quad (10)$$

$$d^2R_2/dt^2 = R_2(B_{z1}^2 - B_{z2}^2)/4m_2 + 9.6 Z i V_{\phi o} B_{z1}/A_i + V_{\phi o}^2/R_2. \quad (11)$$

$$dV\phi/dt + V_{\phi o}/R_2(dR_2/dt) = -9.6 Z B_z/A_i(dR_1/dt) + 9.6 Z B_{z1} R_2(R_1 dR_1/dt - R_2 dR_2/dt)/ /A_i(R_1^2 - R_2^2), \quad (12)$$

where : $I = I_z \sin^2(\pi t/2 t_{1/4})$; $B_{z1} = B_z(R_{1i}^2 - R_{2i}^2)/(R_1^2 - R_2^2)$; $B_{z2} = B_z(R_{2i}^2 - a_o^2)/(R_2^2 - a^2)$ t and $t_{1/4}$ are the time and quarter-period risetime in nanoseconds, R_1, R_2, R_{1i} and R_{2i} are the radii and initial radii of the outer-liner and ion-beam annulus respectively in cm, I_z is the maximum-pinch current in MA, m_1 and m_2 are the mass-per-unit-length of the liner and ion beam in μg/cm, B_z is the initial axial magnetic field intensity in MG, $V_{o\phi}$ is the initial-azimuthal velocity of the ion beam in cm/ns, Z is the ion-charge state, and A_i is the ion mass in atomic-mass units.

In the present case the flux-compressed dynamics of the target plasma is not considered, however our previous calculations show that a solid-density plasma with temperatures of 10 keV range can be easily produced in this configuration. The equations (2–4) were numerically solved for the parameters of two point designs characteristic of the UCI Z-pinch and JUPITER facilities. Table 1 summarizes the results.

For the UCI Z-pinch the following initial conditions were used: $m_1 = 60$ μgm/cm, $I_z = 1.5$ MA, $B_z = 2.2$ T, $\tau_{1/4} = 150$ ns, $R_1 = 2$ cm, $L_{pinch} = 4$ cm, initial asimuthal ion energy, $E_{i\phi} = 113$ keV (i.e., 90 % of the injection energy), number of injected ions, $N_i = 10^{15}$ cm^{-1}, using He^{+2}. The Fig. 5a-h display the results for the outer-column radius, ion-shell radius, their respective velocities and the ion energy as a function of time. The outer Z-pinch liner reaches a final radial velocity of, $V_{1f} = 0.042$ cm/ns and radius, $R_{1f} = 1.6$ mm (Figs. 5a,b). The ion shell achieves slightly smaller values of radial velocity and final diameter (Figs. 5c,d). As shown the final azimuthal velocity of the ions approaches, $V_{\phi f} = 2.5$ cm/ns, for an equivalent energy of, $E_{if} = 12.5$ MeV, and $E_{if}/A_i = 3.13$ MeV/nucleon (Figs. 5e,f). The maximum kinetic energy of the liner is, $Wl_f = 21$ kJ and for the ion beam, $W_{if} = 8$ kJ. Thus, 38% of the liner-kinetic energy is converted into ion-azimuthal energy. The remaining energy is stored in the compressed-magnetic field.

For the Jupiter facility the initial conditions are: $m_1 = 22$ mgm/cm, $I_z = 60$ MA, $B_z = 55$ kG, $\tau_{1/4} = 150$ ns, $R_1 = 4$ cm, $L_{pinch} = 4$ cm, $E_{i0} = 810$ keV, $N_i = 2*10^{16}$ cm^{-1}, using Xe^{+8}. These optimized initial conditions insure that the ion trajectories are initially contained within the liner, the implosion occurs at maximum current, and that the final-ion energy was in the range, $E_{if}/A_i \geq 10$ MeV/amu.

Table 1. Point design parameters for the Super Radiant X-Ray Source

	U. C. Irvine	Jupiter (SNL)
liner parameters:		
W_{bank} (MJ)	0.05	100
V_{bank} (kV)	0.06	3
I_z (MA)	1.5	60
$\tau_{1/4}$ (ns)	150	150
R_1 (cm)	2	4
M_1 (mgm/cm)	0.06	22
B_{zo} (T)	2.2	5.5
L_{pinch} (cm)	4	4
injected ion beam parameters:		
A_i (amu)	He^{2+}, 4	Xe^{8+}, 131
τ_i (ns)	110	220
I_i (kA)	12 (Z = +)	931 (Z = +8)
E_{io} (keV)	125	900
V_{zo} (cm/s) x10^{-8}	2.5	1.2
trapped ion beam parameters:		
N_i (10^{16} cm^{-1})	0.1	4
V_{zf} (cm/s) x10^{-7}	7.5	3.7
$V_{\theta o}$ (cm/s) x 10^{-8}	2.3	1.1
W_{io} (kJ)	0.08	23
target:		
M_{target} (amu)	Cu, 63.5	Xe^{8+}, 131
$E_{K-shell}$ (keV)	8.9	34
$E_{L-shell}$ (keV)	1	5.4
final parameters:		
B_{zf} (kT)	0.22	5.5
V_{lf} (cm/s) x 10^{-6}	42	105
W_{lf} (MJ)	0.021	48
$V_{\theta f}$ (cm/s) x 10^{-9}	2.5	3.8
E_{if} (MeV)	12.5	980
W_{if} (MJ)	0.008	24
$\varepsilon_{ions} = W_{ions}/W_{liner}$ (%)	38	50
$W_{k-shell}$ (MJ)	0.003	9.6
$\varepsilon_{radiation} = W_{rad.}/W_{liner}$ (%)	14	20

Fig. 6a-h display the results computed for JUPITER. In this case the outer Z-pinch liner reaches a final-radial velocity of, V_{lf} = 0.105 cm/ns and final radius of, R_{1f} = 1,3 mm (Figs. 6a,b) (compression ratio is 30), and similarly for the ion ring (Figs. 6c,d). As shown in Figs. 6e,f the final-azimuthal velocity of the ions approaches $V_{\phi f}$ = 3.8 cm/ns for an energy of E_{if}/A_i = 7.6 MeV/nucleon (E_{if} = 980 MeV). The final kinetic energy of the liner is, W_{lf} = 48 MJ and in the ion beam, W_{if} = 24 MJ, which is 50 % of the liner-kinetic energy. The final magnetic field $B_{zf} = B_{zo}(r_{in}/r_{fin})^2$ = 55 MG. Magnetic field values of comparable,

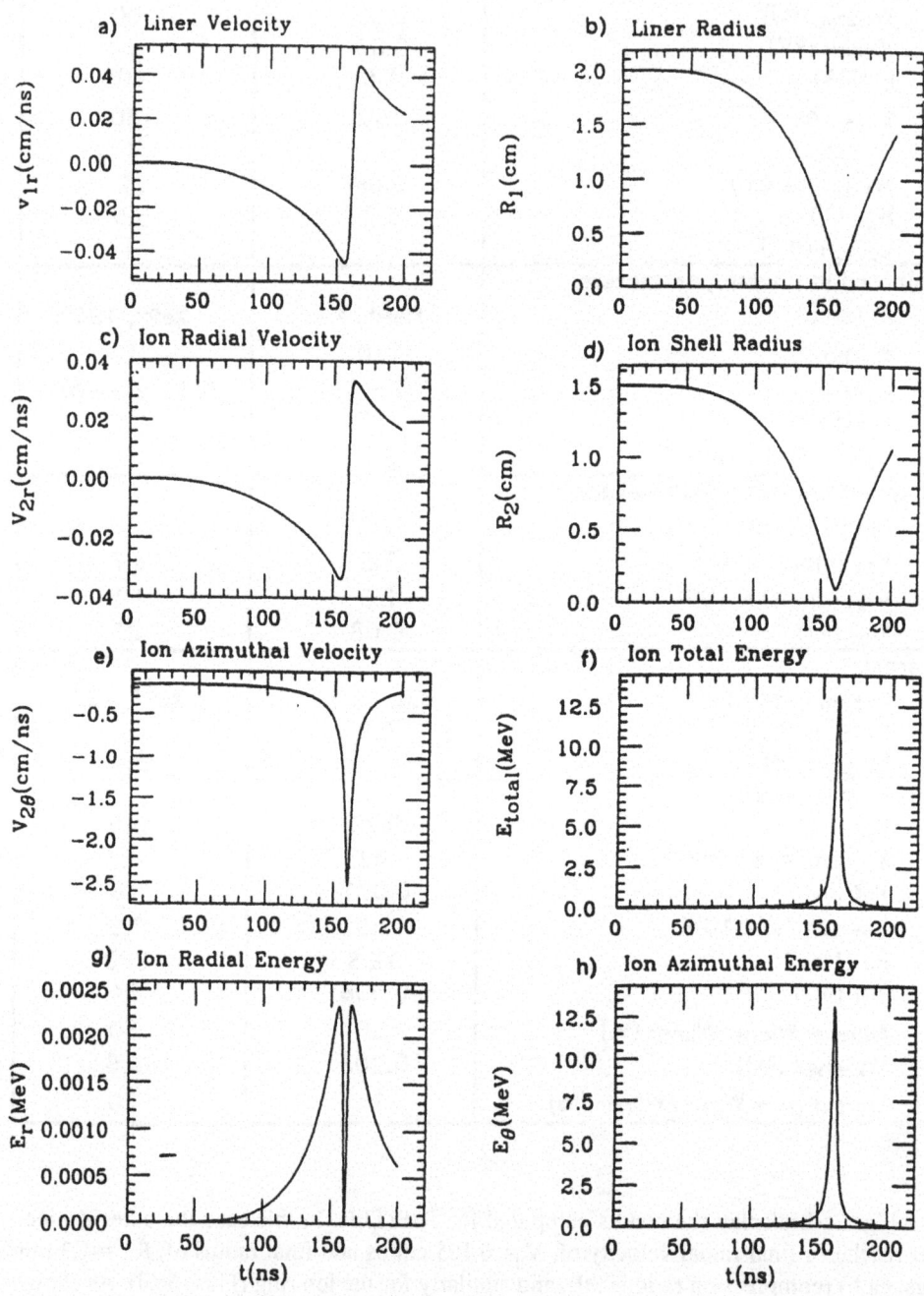

Figure 5. Implosion dynamics for the SRX fielded on the UC Irvine Z-pinch: a) outer liner velocity, b) outer liner radius, c) ion radial velocity, d) ion shell radius, e) ion azimuthal velocity, f) ion total energy, g) ion radial energy, h) $M_l = 6 \times 10^{-5}$ g/cm, $N_i = 10^{15}$ cm^{-1}, $E_{io} = 110$ keV, He^{+2}.

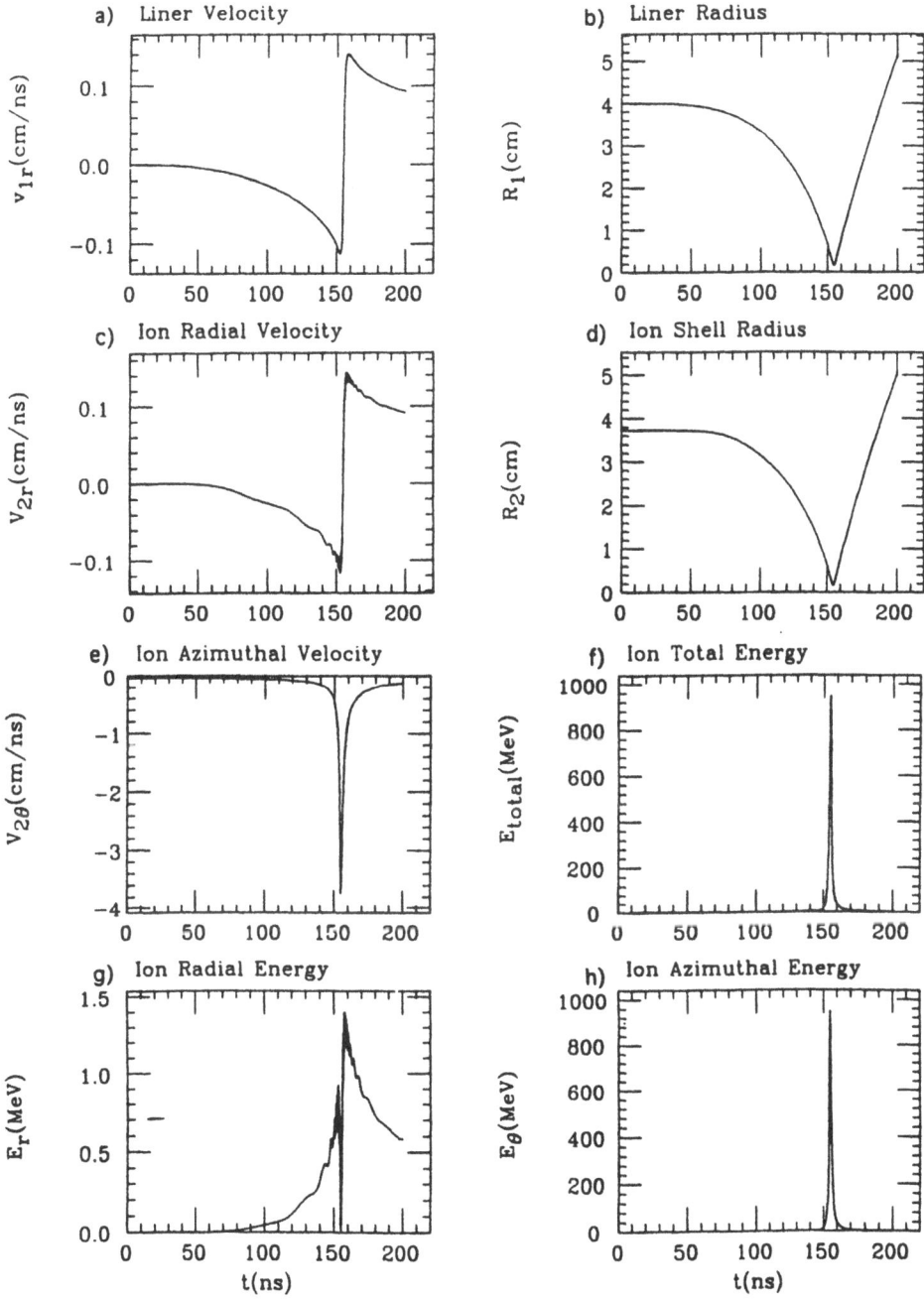

Figure 6. Implosion dynamics for the SRX fielded on the Jupiter Z-pinch: a) outer liner velocity, b) outer liner radius, c) ion radial velocity, d) ion shell radius, e) ion azimuthal velocity, f) ion total energy, g) ion radial energy, h) ion azimuthal energy. Initial conditions: I_z = 60 MA, B_z = 5.5 T, R_1 = 4 cm, R_2 = 3.8 cm, M_1 = 2.2 × 10^{-2} g/cm, N_i = 4 × 10^{16} cm^{-1}, Eio = 880 keV, Xe^{+8}.

but lower magnitude have already been reported in a Z-pinch flux compression experiment on the Proto-II accelerator at Sandia Laboratories, confirming the essential features and scaleability of this means of flux compression (Felber, 1988); values of several MG were also reported at lower power (Wessel, 1986; Rahman 1989; Ratakhin, 1988).

3. FIELD REVERSAL

If a charge neutralized-ion beam is trapped inside the pinch plasma then the dynamics may be different from that discussed above. The increase in ion energy from 0.8 MeV to 0.98 GeV was based upon the assumption of complete current neutralization of the ion beam. Emerging from the ion diode the beam would be completely neutralized, electrically and magnetically by slow electrons. Passing through the cusp would not alter this, provided that the beam thickness, $a > c/\omega$, where $\Omega = \omega_p^2/\Omega_e = 4\pi nec/B_z$. With this condition fulfilled a polarization radial-electric field E_r makes the electrons $E_r \times B_z$- drift at the same speed as the ions, resulting in an electron current equal and opposite to the ion current. and respectively in $B_{self} = 0$. A similar situation arises in the propagation of a neutralized-ion beam across a magnetic field with no deflection of the former, due to electric polarization. (Wessel, 1990). In practice a laboratory vacuum of 10^{-5} torr is sufficient to maintain the polarization for times of the order of a microsecond.

If a background plasma is present, this polarization could be shielded by plasma electrons moving along field lines in a time scale ω_p^{-1}. In this case the resulting self field of the ion current for our parameters could be very large, (i.e., $I_{i0} = ZeNV_\phi \sim 3 \cdot 10^5$ A/cm, and $B_{self} \sim 1.2\, I_{i0} = 0.36$ MG). However, it would not be the case if the background-plasma density in the region of ion beam rotation will be not high enough during the main portion of the liner implosion prior to collision of the beam with the central coaxial target.

During the final phase of compression and heating of the target collisions could drive the electron and ion distributions towards the same average velocity and therefore, currents that cancel. The neutralization condition is that the electrons are collisional, namely,

$$\Omega_{ce} = 2*10^{11} B(T) \leq \nu_{ei} = 2.9*10^{-6} n_e(\text{cm}^{-3})\, T^{-3/2}\, \text{Ln}\Lambda(\text{eV}), \tag{13}$$

where Ω_{ce} and ν_{ei} are the electron cyclotron frequency and electron-ion collision frequency. Substituting $T \sim 10^{2-3}$ eV, $n_e \sim 10^{23}$ cm^{-3}, ln$\Lambda \approx 10$, we obtain, $\nu_{ei} = 10^{15-16}$ s^{-1}. For a final magnetic field value of, $B_f = 5.5*10^3$ T, $\omega_{ce} = 10^{15}$ s^{-1}, thus, $\nu_{ei} \ni \Omega_{ce}$. This question being crucial for scenario of liner implosion dynamics, needs experimental verification.

4. STABILITY ISSUES

During the acceleration of the outer Z-pinch the Rayleigh Taylor instability grows like, $\xi = \xi_o \exp(\gamma t)$, where $\gamma = (kg)^{1/2}$, k is the wave number of the perturbation, $g = B_z^2/8\pi m$, m is the mass/unit area of the Z-pinch. Since $r_a \approx 0.5\, gt^2$ we obtain $\xi \approx \xi_o \exp(2r_a k)^{0.5}$, where r_a is the distance over which the liner accelerates. This instability can be controlled by reducing the initial perturbation ξ_o and the distance r_a, within acceptable experimental limits. The initial perturbation is reduced by initiating a discharge through a preionized plasma instead of a gas column, while the smallest radius, with sufficient mass density, is chosen to give maximum compression at peak current.

In the SRX pinch B_z and B_ϕ components of the magnetic field combine to produce a magnetic shear near the outer surface of the liner Z-pinch. The sheared magnetic field should stabilize the outer surface of the Z-pinch. The inner surface should become unstable when it is decelerated by the trapped B_z field. However, the deceleration takes place over a very short distance of the order of a few millimeters and instability should not be a problem.

There is an additional stabilizing effect due to increasing the mass of the Z-pinch as it "snowplows" during acceleration (Gol'dberg, 1993). Since the wavenumber is related to the radial-mass gradient (i.e., proportional to the inverse of the liner-shell thickness), thicker shells are predicted to be more stable. Moreover, since the plasma becomes highly collisional at final stagnation betatron oscillations of the ring will not occur and the ring should be stable (c.f., Eqn. 13).

5. IONIZATION AND RADIATION

The target plasma could be a gas jet, plasma jet, solid fiber, or straw. During current build-up and implosion of the outer liner the target would be irradiated by ultraviolet light and may support a small quantity of the total Z-current. In any case it would be ionized during liner implosion. Experiments on the Russian Angara 5–1 accelerator at 2 MA and 100 ns risetime, involving an imploding gas shell Z-pinch onto solid wires of copper or aluminum, (Smirnov, 1992) observed the formation of on-axis low-temperature plasma along the central wire during the last 10–20 ns of the implosion; similar observations of surface preionization were evident prior to stagnation at lower power (Wessel, 1992) and in a flux-compression pinch with a central target of quartz(Rahman, 1989). Such data confirm that the target conditions may be adjusted to prepare a high density, relatively cold target plasma prior to stagnation.

For JUPITER we assume that the target plasma is Xe^+ and also the ion beam - and at the moment of stagnation would be compressed by the magnetic field to an ion density of, $n_t = 10^{22-23} cm^{-3}$ with a radius of 1 mm. We see from Figs.6 that the ion velocity at this moment would be, $V_{\phi f} = 3.6 *10^9 cm/s$, the final magnetic field would be, $B_{zf} = 55$ MG, and the ion gyroradius would be, $\rho_i = 1.1$ mm so that the ion beam would circulate about the target. Ion collisions with the lower density plasma surrounding the target would further ionize the ion beam and reduce the ion gyroradius so that the ion beam would penetrate progressively deeper into the plasma of target ions.

As it was discussed earlier, inner-shell ionization by collisions between high Z ions involves in the general case a formation of transitional quasi-molecular orbits and a complicated transfer of energy between colliding particles. No present theoretical models yield accurate inner-shell ionization cross sections for atomic targets over a wide mass and energy range. In case of high- energy-multi electron-ion collisions a reliable conservative estimates of the former still could be made in the binary encounter or Born approximation. In this approach the ion energy required to excite inner-shell transitions is usually of the order of, E_i/A_i ~few MeV/amu for high Z targets with a distance of closest approach in the range, $R_{min} = 0.01$ to 1 a.u. If the orbital period of the K-shell electrons is shorter than the time the colliding nuclei are within a few atomic radii of each other, then a quasi molecule is formed as an intermediate state, and the cross-section for ionization can be comparable to the geometric-cross section. In this case the velocities of the inner-shell electrons, in their respective ions do match, which we speculate is related to the existence of large cross sections, measured to be of order $\sigma \sim 10^{-16} - 10^{-20}$ cm^2. An estimate of the inner shell ionization cross section is given by,

$$\sigma = 2\pi R_{min}^2 [1-(b/Rmin)\exp(-R_{min}/a)] \quad (14)$$

where, $b = 2Z_1 Z_2 a_0 R_{rid}/E_i$, Z_1 and Z_2 are the total number of electrons screening the incident and target particles, $a_0 = 5.3 * 10^{-9}$cm is the Bohr radius, $R_{rid} = 13.6$ eV, the Rydberg constant of energy, E_i is the incident ion energy, and $a = a_0/(Z_1^{2/3} + Z_2^{2/3})^{1/2}$, is the atomic-screening radius. Typical values for the M-K-shell ionization cross section for light onto a heavy ion are in the range, $\sigma \sim 10^{-20}$–10^{-16} cm^2 for, $E_i/A_i \sim$ few MeV/amu (Garcia, 1973; McDaniel, 1991). For instance, L-K -shell exitation in collisons of $S^{+16} + Si$ have a cross section, $\sigma \sim 10^{-18}$cm^{-2} when the incident ion energy is $E_i/A_i = 3$–8 MeV/amu.

For the present problem of, $Xe^+ + Xe$ we assume $\sigma =$ (total for K-M shell ionization) $= 10^{-18}$cm^{-2} and determine how long the peak compression is required to ionize all of the target ions.

$$dn_{K-M}/dt = n_p n_b \sigma_{tot} V = n/\tau \quad (15)$$

where the collision time is, $\tau = 1/n_p \sigma V = 0.278 * 10^{-13}$ sec. The ion beam density is on average, $n_b = N/r_0^2 = 0.52 \times 10^{18}$cm^{-3}. Each beam ion has an energy of, $Eif = 0.88$ GeV and is capable of ionizing the K-M -shells of 10^{4-5} target ions. The time required to ionize the K/M-shells of all the target ions is then, $\delta t = n_p/n_b \tau \sim 1.9 * 10^4 = 0.52$ ns. This is the time that maximum confinement must persist. From the results of Fig. 6 the confinement time would be many times this value and there would be sufficient time to ionize the K-M shells of all the target ions. In this case the total K-M shell radiation would be more than $W = 9$ MJ.

This calculation suggests a way to produce large quantities of energetic radiation, efficiently, without a large inventory of unused magnetic energy or relying on thermal excitation schemes and presents an interesting case for the SRX; even at the level of 0.5 MJ/pulse. Indeed, it may possible to attain even higher yield if a larger number of target ions and target densities were used. Further advances and improvement in these estimates require more elaborate experimentation and theoretical formulations.

6. HARDWARE REQUIREMENT

A unique feature of the SRX is that it is suitable for a wide range of pulsed-driver energy, current, and risetime. For a set of generator current, risetime, and initial-liner radius, the specific requirements are that the ion-injection energy, atomic mass, initial-charge state, and axial-magnetic field must be chosen to match the ion gyroradius with the dimensions of the system.

The hardware required to validate the SRX concept involves a combination of a high-current, Z-pinch facility, preinjected, axial-magnetic field, and a low-energy, ion-beam injector. The largest component of stored energy would be that in the Z-pinch driver, whereas the energy stored in the ion beam accelerator would be several orders-of-magnitude smaller. Table 1 summarizes the point-design parameters for an SRX fielded at 1.5 MA, 150 ns, on the University of California, Irvine, Z-pinch facility and for the future DNA/Sandia JUPITER Z-pinch facility.

6.1. Z-pinch

The Z-pinch accelerator would be a conventiontional design, modified to include a magnetic coil which injects an axial-magnetic field between the discharge electrodes. The outer liner would be annular and could be an injected gas, foil, or a solid liner, as would be the case at high current. The initial liner radius is of the order of several centimeters and the discharge length is similar. The magnet coil would be mounted around the discharge region, inside or outside the vacuum chamber. Normally the field risetime is long compared to that of the pinch, and the coil system is energized prior to the pinch.

Several experiments in the discharge-power range, 0.01–10 TW (nanosecond to microsecond risetime pulse-power drivers), have demonstrated that imploding liner Z-pinches conserve axial-magnetic flux, compressing the field at least hundredfold in magnitude(Felber, 1988; Wessel, 1986; Rahman, 1989; Ratakhin, 1988). Explosive-driven magneto-cumulative generators have achieved even more impressive values of compressed-magnetic field with pinch-current risetimes of many microseconds with current levels approaching 200 MA (Chernyshev, 1987; Shearer, 1987; Sakharov, 1966).

6.2. Ion Beam Accelerator

The ion beam accelerator must inject the ions inside the imploding liner of several centimeters. For constant ion energy the Larmor radius is, $r \sim m_i/Z^2$ which restricts the initial ion charge state and mass. For the UCI 1.5 MA Z-pinch facility we would require 120 keV ions of He^{+2} or 40 keV H^+ for the 0.6 MA facility. In this range we propose to use an intense, ion-beam source, or high velocity plasma gun with 100 ns pulse duration; similar sources exist at UCI although with longer (400 ns) pulse duration, and were studied several years ago(Drum, 1992). At higher pinch current, for example 60 MA, 900 keV Xe^{+8} would be used. Conventional ion diodes generate beams of low-charge state, typically Z = 1–2 due to the short lifetime of the ions in the vicinity of the cathode and the presence of a low density electron cloud. Typically the number of electrons is, $N_e = (j/e)\tau \approx 10^{15}$. For JUPITER we would propose a reflex triode ion source with a razor-edge electron source, as shown in Figure 1. The electron beam would be radially confined by a strong axial magnetic field (magnitude,~few Tesla) and would oscillate between the virtual cathode and cathode, providing multiple ionization of the gas cloud inside the tubular-anode electrode. The requirement for production of $Z_i \geq 10$ for wide range of ion masses is, $(J/e)\tau \geq 10^{17}$ - 10^{18}(Donets, 1992). It corresponds to electron current densities of, $j_e = 10^6 A/cm^2$, with total cathode surface area of 1 $cm^{2,}$ and pulse duration of few hundred nanoseconds or less. A similar ion source/injector assembly was previously analysed (Bystritskii,1985). Unidirectional extraction of the multi-charged ions would result by tapering the inner and outer diameters of the hollow-anode structure in the direction of the virtual cathode (and direction of the Z-pinch) to achieve an axial-voltage gradient,

$$\nabla \phi = \pi j_e \xi \alpha \delta d_o / 2c \tag{16}$$

where j_e is the electron current density, is the average number of electron oscillations inside the anode, d_0 the annular gap between the anode cylinders, δ the thickness of the hollow e-beam, and α the divergence angle of the anode cones. To estimate $\nabla \phi$ we assume: j_e = $10^5 A/cm^2$, ξ= 10, δ= 10^{-2} cm, d_0 = 1 cm, and α = 0.01, which gives, $E_z \approx 4.5$ kV/cm.

Under the action of this electric field the ions will reach the far end of the coaxial gap in a time,

$$t_p = (2l_{an}/a)^{1/2} = (2l_{an}A_i/ZeE)^{1/2} \tag{17}$$

Substituting for t_p and A_i from parameters given in Table 1 the anode length should be in the range, $l_{an} \sim 1-10$ cm for light to heavy mass ions. The ion beam can also be bunched at the exit of the anode by shaping the voltage pulse applied to the diode, following the approximate relation, $V(t) \sim t^{1/2}$. Such wave form shaping of the voltage pulse could be provided by a regular L-C filter placed in parallel with the diode load after the Marx generator (Akerman, 1986).

6.3. Injection and Trapping

The ion beam is created in a strong-axial field and passes through a magnetic cusp prior to entering the pinch. The effect of the cusp is to transform a significant fraction of the axial ion energy into rotational velocity. The requirement on ion beam energy is,

$$E_{beam} \leq (2m_i)^{-1} * (r_L e Z_i B_z/c)^2, \tag{18}$$

where r_L is the radius of the liner, B_z is the magnetic field, is the average charge state of the ions, and m_i is the ion mass. Experiments on ion injection through magnetic field cusps have demonstrated better than 90 % conversion of the axial energy into rotational, corresponding to a ion-axial velocity of, $V_{final} \leq 30\% V_{initial}$. With a helium-ion beam of energy, $E_i = 0.1$ MeV and charge state, $Z_i = 2$, the axial velocity after entering the liner will be $V_{axial} = 7.5*10^7$ cm/s. Decreasing the axial velocity increases the volumetric-ion density approximately three fold.

Proper synchronization of the Z-pinch and ion beam accelerators is required for efficient trapping of the ions in the liner-compressed, rising-magnetic field. The main requirement is to inject the ions just before the outer liner begins to accelerate inward to insure that the ions nearly complete a double transit along the pinch length, as the liner collapses. Thus, the precise requirement is to delay the liner compression by a time:,

$$\Delta t = 2L_{pinch}/V_i \tag{19}$$

With $L_{pinch} = 4$ cm and $v_{axial} \approx 10^8$ cm/s, then $\Delta t \approx 60$ ns. The trapping efficiency would be,

$$\varepsilon = 2L_{pinch}/V_i \tau \tag{20}$$

where τ is the FWHM pulse width of the ion beam pulse. For $\tau = 50$ -100 ns we estimate $\varepsilon = 30$ \%. With a typical time delay of $\delta t = 10$–30ns, timing jitter in the overall system should not be a concern.

7. DRIVING HOHLRAUM ICF TARGET

As it follows from our calculations the characteristic radiation burst takes place in the stagnation phase during a few nanoseconds. In case if the outer liner is opaque to this radiation, the hohlraum scheme of driving an ICF target can be discussed. Let us consider this possibility.

During the implosion of the outer high -Z liner, such as Xe, it remains optically thick (Bailey, 1986), radiating copiously like a black body and therefore stays relatively cold (less than 10 eV). The coaxial target, heated in the final stage of the implosion by the high energy ion beam to temperatures much higher than the outer liner will become optically thin, though it will radiate as a black body, being inclosed in the outer cold liner. For the latter the cold absorbtion coefficients should be a good approximation:

$$\mu = 10^5 \text{ cm}^2/\text{g at 100 eV, and } \mu = 4*10^3 \text{ cm}^2/\text{g, at 180 eV (Dyson, 1990)}.$$

For higher energies the value of μ should decrease slowly and remain above 10^3 cm^2/g in the few keV range Dyson, 1990).

The absorption length $\delta_{abs} = N_{Av}/n_i A\mu$ (N_{Av} -Avogadro number, n_i - liner density) for photons in the relatively cold liner would be: 0.05 μm and 0.11 μm, and lesser than 4.4 μm for photons of 100 eV, 180 eV and few kilo electron volt energy respectively. Here we assumed the terminal density of 10^{22} cm^{-3} for the Xe liner, which is consistent with zero dimensional calculations (Rahman, 1994). It is therefor plausible that the imploding liner of Xe could provide the opaque wall at the hohlraum for radiation temperatures up to few kilo electron volts. Transmission of the radiation would take place through the ends of the pinch. Control over the build up of the black body radiation and its release could be accomplished by the disc placed at the pinch output, the former is initially opaque and then bleaches, becoming transparent on a time scale, defined by its thickness.

The approach proposed herein could provide a directional multi megajoule photon beam on a time scale of the order of nanosecond. It would not be coherent, but it would be of an energy and time scale comparable with the laser proposed for the National Ignition Facility at LLNL.

REFERENCES

Akerman, D.R., V. M. Bystritskii, S. N. Volkov, Ya. E. Krasik, A. M. Tolopa, Sov. Phys. Tech. Phys. **31**, p. 1156(1986).

Bailey, J., A. Fisher, and N. Rostoker, J. Appl. Phys., **60**, 1939, (1986).

Bystritskii, V.M., Sov. Phys. Tech. Phys. **30**, p. 1198(1985).

Bystritskii, V.M., Yu. A. Glusko, G. A. Mesyats, AIP Conf. Proceedings 195 on Dense Z- Pinches, N. R. Pereria, J. Davis, and N. Rostoker, editors, AIP New York, p. 522,(1989).

Chernychev, V.K., N. P. Bidylo, A. I. Lyudaev, A. A. Petrukhin, M. S. Protasov and V. A. Shevtsov, Megagauss Physics and Technology, Edited by C. M. Fowler, R. S. Caird, and D. J. Erickson, p. 707(1987).

Chernyshev, V.K., V.N. Mokhov, 8th IEEE Intl. Conf. Pulsed Power, Editors, R. White and K. Prestwich, p. 395(1991).

Donets, E. D., Sov.J.Part. Nucl., 13(5), pp 387–405, (1982).

Drum, S., W. W. Heidbrink, and F. J. Wessel, Rev. Sci. Inst. **63**, p. 2690(1992).

Dyson, N.A., X-Rays in Atomic and Nuclear Physics, Cambridge University Press, 2nd Edition, New York, 1990.

Felber, F.S., M. M. Malley, F. J. Wessel, M. K. Matzen, M. A. Palmer, R. B. Spielman, M. A. Liberman, and A. L. Velikovich, Phys.Fluids **31**, 2053 (1988).

Fowler, R. S. Caird, and D. J. Erikson, p. 637(1987).

Garcia, J.D., R. J. Fortner, and T. M. Kavanagh, Rev. Mod. Phys. **45**, p.111(1973).

Gardner, R,K., and T. J. Grey, Atomic Data and Nuclear Data Tables **21** p. 515(1978).

Gol'berg, S.M., and A. L. Velikovich, Phys. Fluids **B5**, p. 1164(1993).

Lichten,W., Phys. Rev. **64**, 131 (1967).

McDaniel, E.W., J. B. A. Mitchell, and M. E. Rudd, Atomic Collisions: Heavy Ion Particle Projectiles, p. 313, 1991.

Pereira N.R. and J. Davis, J. Appl Phys. **64**, p.R1(1988).

Rahman, H.U., P. Ney, F. J. Wessel, A. Fisher and N. Rostoker, Proc.2nd Int. Conf. on High Density Pinches, Laguna Beach, April 26–29 (1989), AIP Conf. Proc., p. 195, (1989).

Rahman, H. U., D. Ney, F. J. Wessel, and N. Rostoker, Comments Plasma Phys. Controlled Fusion, **15**, 339, (1994).

Ratakhin, N.A., S.A. Sorokin, and S. A. Chaikovsky, Proc. 7 th Intl. Conf. High Power Particle Beams, Karlsruhe, July, 1204 (1988).

Sakharov, A., Usp. Fiz. Nauk (in Russian), **88**, p. 275(1966).
Shearer, J.W., and D. J. Steinberg, Megagauss Physics and Technology, Edited by C. M. A. Smirnov, V.P., P. Choi, B. Etlicher, J. Larour, and F. J. Wessel, "Angara 5–1 Experiments," paper presented at the APS Conference on Plasma Physics,Seattle, WA, Nov. 1992 and at the 4th Intl. Conf. Dense Z-Pinches, London, England, May 1993.
Sudan, R.N., and E. Ott, Phys. Rev. Lett. **33**, 355(1974).
Wessel, F.J., F. S. Felber, N. C. Wild, and H. U. Rahman, Appl.Phys.Lett. **48**, 1119 (1986).
Wessel, F.J., N. Rostoker, A. Fisher, H. U. Rahman, and J. H. Song, Phys. Fluids **B2**, p. 1467 (1990).
Whitney K.G., Thornhill, K.G., J. P. Apruzese, and J. Davis, J. Appl. Phys.**67** p. 1725 (1990).
Wessel, F.J., B. Etlicher, and P. Choi, "Z-Pinch Implosion of an Aluminum Plasma Jet onto a Coaxial Wire: Enhanced Stability and Energy Transfer," Phy. Rev. Lett. **69**, No.21(1992).

23

FUSION, THE COMPETITION, AND THE PROSPECTS FOR ALTERNATIVE FUSION CONCEPTS

L. John Perkins,[*] James. H. Hammer, and R. Paul Drake

Lawrence Livermore National Laboratory
P.O. Box 808 (L-637)
Livermore, California 94551

ABSTRACT

Although we have achieved great progress in the scientific understanding of fusion, there is some question whether our present, mainline approach to a fusion reactor will lead to a sufficiently attractive commercial product able to compete in the energy market place of the 21st century. This is a result of the projected low power density, high complexity, large unit sizes, and very high development costs. We assess these attributes relative to other future competitive energy sources and offer the thesis that any step change in the reactor product may lie in the exploration of novel, relatively unexplored or revisited *physics* concepts rather than in refined engineering for the present single approach. We then provide an overview of potential alternative fusion concepts under the following classification system: (1) Low density magnetic confinement, (2) Inertial confinement concepts, (3) High density magnetic confinement schemes, (4) Non-thermonuclear concepts, and (5) Coulomb barrier reduction concepts. To be considered a serious contender, any new candidate fusion scheme must point the way to an engineering realization that is a step-change in reactor attractiveness from our present approach. Specifically, this means we must clearly identify the potential for significantly lower capital costs, complexity, and development costs relative to those projected for the conventional tokamak.

> *"There are more things in heaven and earth, Horatio, than are dreamt of in your philosophy"*
> W. Shakespeare, *Hamlet.*

---

[*] (510) 423–6012, Email: perkins3@llnl.gov

THE COMPETITION

Fusion, advanced fission and solar-electric are the only unlimited, non-fossil options for sustainable, baseload electricity generation in the long term. A successful, economic fusion reactor realization would be a profound contributor to the well-being of future humanity, with major global export opportunities in the multi-10-trillion-dollar energy industry of the next century. In the past four decades, we have achieved great progress in the world program in the scientific understanding of fusion. We are now producing considerable scientific and technical information that will be of value in the development of fusion power. However, it has been recognized on a continuing basis that the present single approach—the conventional tokamak—may not lead to a fully viable commercial reactor product able to compete in the energy market place of the next century [see, for example, Refs. 1 – 10]. This approach to a reactor faces the following difficulties:

- Low power density ($\Rightarrow$ High capital cost per kW produced)
- High complexity ($\Rightarrow$ Low perceived reliability of hard-to-maintain, high technology hardware; low perceived availability)
- Pulsed operation for inductively-driven tokamak reactors
- Minimum unit sizes of $\geq 2 GW_e$ to realize even marginal economics
- Not necessarily radioactively benign (tritium plus neutron-activated structural waste)
- Very high development costs (>$10B) for the next stage (the ignited engineering test reactor), notwithstanding that the economics of the ultimate power reactor anyway appears questionable

Advanced fission, the other nuclear option, appears to be *the* competition to fusion mid-way through the next century and beyond. Although, at present, fission is not a growth industry in the U.S., fission plants are due to start up in seven countries in 1995, while active construction is proceeding in 12 countries [11]. The perceived "problems" with fission, namely safety and waste disposal, are actually solved problems from a technical viewpoint and the present political opposition to fission (actually a minority viewpoint) will surely evaporate when and if it is really needed in the next century to sustain a burgeoning world population.

It is important to note that only ~1/3rd of the capital cost of a present-day fission plant is invested in the reactor plant equipment (i.e., the fission power core plus associated equipment within the reactor building and primary loop) with the rest in the balance-of-plant and buildings. Moreover, only ~6% of the capital cost is in the fission power core itself (reactor core plus pressure vessel plus control rods). By contrast, our present paper projections of conventional DT tokamak reactors [12–14] put the reactor plant equipment at ~70% of the total capital costs, with ~50% of the capital cost in the fusion power core itself. Given that the absolute cost of the conventional balance-of-plant will be about the same in both cases per thermal MW_{th} generated, the economic disparity is clear. In addition, a tokamak fusion power plant is significantly more complex than a fission plant of comparable power. Distributing mean-time-to-failure and mean-time-to-repair over the significantly greater number of complex systems implies that target capacity factors of >75% may be rather hard to realize.

To realize clear economic viability, it is not necessary that a fusion reactor be smaller and less complex than a fission core (which would probably be impossible anyway for any rational D-T fusion reactor concept). Nonetheless, it has to move sufficiently far in these two important directions, such that when the potential advantages of safety/environmental features and the fuel cycle are factored in, the overall economic envelope is perceived to be competitive by the potential customer. Otherwise, no one will buy them. A suitably designed fusion reactor constructed of appropriate, low activation advanced materials (e.g vanadium alloy), should have a competitive advantage over fission in the area of safety and environment. However, in the case of the conventional tokamak fusion plant, these potential advantages are by no means enough to offset the capital cost and complexity disadvantages discussed above.

Conceptual fusion reactor designs have been performed on a continuing basis over the past 20 years or so and recent studies indicate cost-of-electricity (COE) projections for tokamak power reactors of about a factor of two or more larger than present fission experience [see, for example, Refs. 12–14]. However, there are three large uncertainties in these COE numbers that fusion reactor studies have not adequately confronted to date:

1. The uncertainty that the physics and technology will work as projected on paper
2. The uncertainty in the capital cost of the extrapolated technology
3. The issue of reliability, maintainability and, therefore, availability of the tokamak power core.

This latter factor, a consequence of complexity, is particularly worrisome. That is, the perception by utilities that a conventional tokamak fusion reactor is not maintainable in any realistic time could make them non-viable on this aspect alone. Thus, we may be lucky to obtain a COE of only twice that projected for advanced fission.

The problem of complexity/reliability also tends to negate the economy-of-scale argument to "build them big". This suggests that tokamak reactors could at least be built in 2–3 GW_e unit sizes to realize a reasonable COE through the non-linear economy-of-scale which applies up to about 3 GW_e [15]. One could certainly argue that the utility structure in the next century may be different from today and that large, multi-GW_e electricity reservations may become the norm, certainly in the world as a whole. However, no utility—private, public or government-owned—will invest in a large, single heat source which is perceived to be vulnerable to frequent outages and significant downtimes for repair. Hydroelectric plants are commonly found in multi-GW_e sizes. However, these have the crucial difference of redundancy, i.e., each plant has the modularity of parallel water feeds to a number of independent turbine-generator sets.

THE NEED FOR INVESTMENT IN VIABLE ALTERNATIVE CONCEPTS

Fusion is the only indigenous energy source that will last as long as the earth lasts. Thus we have the potential for a limitless energy source with low environmental impact, providing a way can be found to harness it economically. It is our contention that any breakthrough leading to a truly viable fusion reactor product will probably lie in the investigation of alternative *physics* concepts rather than in refined engineering for the present approach. Note that this is particularly true if we are ever to realize economic viability with the advanced (non-DT) fusion fuels. The potential advantages of advanced fuels such as D-D, D-3He and especially p-^{11}B, will probably not be realized in a thermonuclear tokamak, and novel physics approaches will almost certainly be required.

Accordingly, we must continue to vigorously pursue the physics of alternative fusion reactor concepts at some viable level within the present world fusion program. And, very importantly, we must earmark R&D funds for conceptual and computational development to the stage where new, definitive experiments can be defined. This requires the promotion of intellectual stimulation in *breadth* to encourage parallel approaches with the acknowledgment that only a very few, if any, may ultimately be successful. It is simply too early and, given the utility of advanced fission, unnecessary, to put all our eggs in one basket at this stage. However, before embarking on a detailed experimental program for a proposed alternative concept, we should, at a minimum, be required to demonstrate *quantitatively* its potential for:

(a) Lower capital costs per kW_e produced through higher power density and a smaller fusion power core;
(b) Higher perceived availability through lower complexity, simpler blanket configurations and minimal inter-linked hardware;
(c) Attractive economy-of-scale for unit sizes <$1000MW_e$;
(d) Modest costs for the experimental development path to a demonstration reactor (DEMO) through small, low cost reactors and short construction times.

In Fig. 1, we give one world view of the operating space for nuclear fusion to illustrate that the potential domain is, in principle, very large. This operating space in Fig. 1 accommodates all forms of fusion that have been observed or predicted. Note, however, that the vast majority of the world fusion research effort has, to date, been concentrated at just one point in this space, namely that at the low density end of the DT thermonuclear line. That is, low density magnetic confinement fusion characterized by n~10^{20} m^{-3}, τ_E~seconds, T~10keV. Some appreciable effort has also been expended at the other end of this line, i.e., inertial confinement fusion characterized by n~10^{30} m^{-3}, τ_E~10^{-10}s, T~10keV. Thus one interesting question to ask is whether there are viable fusion concepts worth pursuing outside of these two mainline directions?

In Table 1 we present a more or less complete list of fusion concepts which are, or have recently been, under study at some level in the world program and classify them according to their main operating principles. (Note that "cold fusion" is not included in either Fig. 1 or Table 1. The prevailing opinion is that, even if there is deemed to be a net energy gain from such systems, it probably does not originate from nuclear fusion, i.e. the release of nuclear binding energy from the rearrangment of nucleons).

WHAT CONSTITUTES A VIABLE ALTERNATIVE CONCEPT?

Of the list of concepts in Table 1, we could ask: which, today, are perceived to be viable alternatives to the conventional tokamak from a reactor viewpoint even though the physics basis is certainly at a much more rudimentary stage? That is, which ones are worth investing research funds in to explore to the next stage? We suggest that a useful way of determining this "worth" is to demonstrate, quantitatively and objectively, that a proposed alternate passes the following initial tests:

- *The Reactor Test* – Assuming the physics "works", it must indicate a significant reactor advantage over the conventional tokamak, specifically demonstrating the potential for a step change in cost and complexity.
- *The Physics Test* – Based on the theoretical and experimental results available to date, the physics must be seen to be plausible.
- *The Development Path Test* – The experimental development path: physics experiments → ignition → engineering test reactor → DEMO must show significant advantages in cost and schedule over that for the tokamak.

Fusion, the Competition, and the Prospects for Alternative Fusion Concepts 369

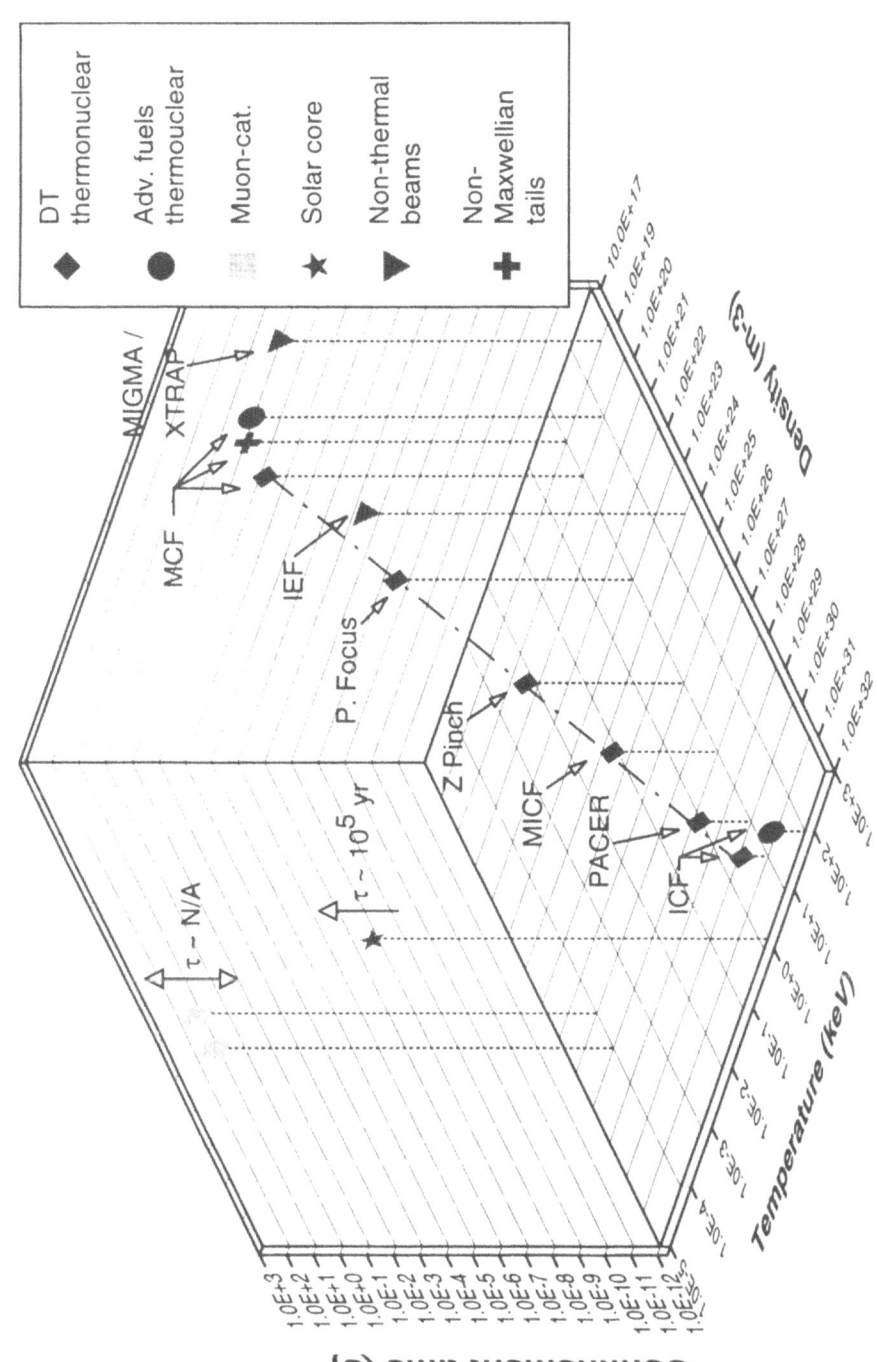

Table 1. Classification of alternative fusion concepts

Low density magnetic confinement:
  Field reversed configuration
  Spheromak
  Spherical tokamak
  Reversed field pinch
  Stellarator
  Mirror
  Tokamak

Inertial confinement fusion:
  Conventional IFE (heavy-ion, laser, ...)
  Advanced, decoupled-ignition, target systems
  Magnetized-target IFE
  High yield pulsed systems (batch burn, propagating burn, various drivers,...
  Fission-driven, high yield (PACER, etc)

High density magnetic confinement:
  Pulsed z-pinches (fiber, assisted, z-θ pinch, staged,...)
  Plasma focus
  Continuous flow pinches
  Wall-confined, magnetically-insulated (various drivers ...)

Non thermonuclear:
  Inertial electrostatic confinement
  Colliding beam systems

Coulomb barrier reduction schemes:
  Muon catalysis
  Others (Shape-enhanced, antiproton catalysis, ...)

Note that an alternative which passes the first of these tests should also have a good probability of passing the third. Such tests could, for example, be applied to discriminate between proposals for different alternatives, especially in the present climate of scarce research funds. In particular, they would suggest that we should not fund alternates which have no better reactor prospects than the tokamak and which will have an equally expensive development path.

Note that the stellarator, for example, may not pass the "Reactor Test" because it is similar in size, cost and complexity to the tokamak. At present, at least, there is barely sufficient funding to pursue one example of this class of fusion machine to the reactor-relevant stage. However, as discussed below, the field-reversed-configuration (FRC), for example, appears to have a good chance of passing the Reactor Test, providing the physics can be shown to "work". We have the means today to quantify the reactor advantages (or otherwise) of many of these alternatives both in terms of capital cost and simplicity relative to the tokamak, providing the physics should prove favorable. So we might inquire why some of the alternatives with a greater *potential* reactor viability are not funded at the rather modest level required to determine whether favorable physics does ensue? If it should, the return on investment would be enormous, and the payoff for humanity profound.

A BRIEF OVERVIEW OF ALTERNATIVE FUSION CONCEPTS

The Low Density Magnetic Confinement concepts in Table 1 are characterized, for D-T fuel, by $n \sim 10^{20}$ m^{-3}, $\tau_E \sim$ seconds, T$\sim$10keV. They are typified by reactors with confining magnetic fields (typically static) at the plasma of a few Teslas and have solid first walls. Concepts at the upper end of this list, i.e. the spherical tokamak, the spheromak and FRC, offer a quantitative potential reactor advantage over those concepts at the lower end with, however, a much less well established physics basis. A cheaper development path is certainly conceivable for ignition-capable devices based on these schemes. In the case of the spheromak [16, 17], the absence of a central solid core circumvents the radial build penalty of linked toroidal coils and provides the potential to ameliorate fusion power flux limitations (edge and/or first wall) through the use of extrapolated divertors and, possibly, liquid wall concepts. The FRC [18] provides, perhaps, the best reactor prospects of all of the low density, magnetically confined schemes. It promises very high beta (β_{eng} = <P>/$(B_{coil})^2$ ~ 1), high power density, a simple, maintainable linear geometry with no linked magnets, and a natural, axial flow divertor for heat exhaust outside the core. Its high beta potential also makes it probably the only magnetic-confinement device with a credible potential for advanced fuel operation [19]. However, near-term basic research is required on its MHD stability characteristics.

The Inertial Confinement Fusion concepts in Table 1 are characterized, for D-T fuel, by n~10^{30} m^{-3}, τ_E~10^{-10}s, T~10keV. The magneto-inertial schemes which employ pre-emplaced magnetic fields before compression for electron and alpha particle insulation, typically operate at intermediate densities and confinement times of ~ 10^{26}–10^{29}m^{-3} and 10^{-6}–10^{-9}s, respectively [20]. In general, ICF schemes side-step the complexity and expense of the plasma/vacuum/solid-wall/magnet interface of conventional, low-density, magnetically-confined fusion [21]. Also, ICF plants could be built in large, multi-GW$_e$ sizes with several, independent target chambers. This realizes the economy-of-scale advantages of a large fusion plant but with the redundancy of multiple, nuclear-grade target chambers (the single driver will, however, remain, a vulnerable item). Conventional heavy-ion-driven ICF schemes operate with target gains of >100 and driver energies of a few MJ [21]. The advanced, decoupled-ignition target systems employ methods of "fast-ignition" so that the input energy required to compress the fusion fuel is decoupled from the ignition requirements of the fuel itself [22]. In principle, this could permit significantly higher gains for the same conventional driver energy, or, equivalently, a significantly lower input driver energy for the same conventional gain.

The High Density Magnetic Confinement schemes shown in Table 1 are typified by the Z-pinch with n~10^{26} m^{-3} and τ_E~10^{-6}s, and the majority are pulsed. The Z-pinch and its close relations were one the earliest fusion concepts to be studied but were plagued by the classic m=0/1 MHD instabilities. Studies are now proceeding on a higher gain, "staged" Z-pinch where an annular plasma is imploded by means of an axial current onto a central axial fiber [23]. Also, interest has been revived [24, 25] in a parallel concept, the continuous flow pinch (CFP). This offers the potential advantages of the linear pinch schemes,—i.e., a compact linear geometry with high power density and the absence of complex magnet systems—but, importantly, with a flowing plasma stream for heat exhaust in a steady-state system. In particular, new theory [25], suggests that the sausage and kink instabilities of the conventional Z-pinch can be stabilized in a CFP via plasma flow without the need for a B_z field (and, therefore, without limitation on the pinch current), providing there is a radial shear in the flow velocity. If a sufficiently stable flow state can be found under fusion conditions, the CFP promises a simple and inexpensive route to controlled thermonuclear fusion.

To introduce the class of schemes under Coulomb Barrier Reduction, we refer above to the significant differences in size and complexity between a conventional thermonuclear tokamak reactor and a fission reactor per unit thermal power generated. The size, cost and complexity of projected thermonuclear fusion reactors are governed by the requirement to sustain a minimum value of the temperature T of the thermalized Maxwellian plasma in the face of significant loss processes. These minimum temperatures are necessary so that energetic ions in the tail of the Maxwellian have sufficient energy to begin to appreciably tunnel through the mutually repulsive Coulomb barrier and, therefore, induce acceptable fusion reaction rates. By contrast, fission is propagated by neutrons which are not subject to an electrostatic Coulomb barrier and will proceed at zero fuel temperature. Even for the fusion reaction with the highest reactivity, i.e., the D-T reaction, minimum thermonuclear temperatures of ~10keV are required. For even marginal performance with the so-called "advanced" fusion fuels, e.g., D-^{3}He, D-D, etc., the minimum fuel temperatures must be raised considerably, ranging from ~50keV to greater than 100keV. This further exacerbates the engineering complexity of the reactor. This suggests that one approach to achieving a step-change in the physics and, perhaps, the economics of fusion power is to attempt to circumvent, at some level, this high temperature threshold set by the barrier penetration requirements of the conventional fusion cross-section. This would be particularly advantageous if we are to realize economically attractive fusion reactors based on the advanced fusion fuels.

One existing method of Coulomb barrier reduction which does work at some level is muon-catalyzed fusion [26] which employs the negatively-charged muon to screen the repulsive Coulomb potential between fuel ions. However, at present this appears to fall short of economic viability due to the sticking of the muon to an outgoing alpha particle in a fraction of the fusion events, thus limiting the maximum number of DT fusions per muon consumed to ~150. We have recently studied other methods of Coulomb barrier amelioration for fusion, including shape-enhanced fusion [27], which has applicability to certain advanced fusion fuels involving deformed nuclei with large quadrupole moments, and antiproton catalysis [28]. Can we conceive of other analogous methods of barrier reduction that could be realized in a practical reactor system? Because of the very large increases in reactivity with modest increases in barrier penetration probability, this is a potential avenue to explore for a true step-change approach to fusion energy. Moreover, it may offer the only glimmer of hope for the truly advanced fuels, such as p-^{11}B, which otherwise have a prohibitively-high barrier for application in a conventional thermonuclear system.

CONCLUSION

The appeal of alternatives to the mainline fusion approach is not new. Over the years, many different ideas have been examined both theoretically and experimentally. We recognize that there is no guarantee of ultimate success. Nonetheless, we believe the payoff of a successful alternative to be so high as to warrant a continued and dedicated effort in the conceptual development of new concepts. In particular, we suggest that a successful alternative, if there is one to be had, will come from new or revisited physics approaches rather than in refined engineering for the present path. However, it is very important that the physics of any proposed alternative concept be coupled with an engineering realization of the reactor embodiment to clearly identify the potential for a step change in capital cost, complexity and development path relative to our present approach. That is, based on best the available models and experimental data, why does it make a better reactor?

ACKNOWLEDGMENTS

The authors are please to acknowledge informative discussions with David E. Baldwin, E. Bickford Hooper, B. Grant Logan, Keith. I. Thomassen, Alan L. Hoffman, Ralph W. Moir, and Charles W. Hartman.

REFERENCES

1. J. P. Holdren, *Science*, **200**, 168 (1978)
2. L. M. Lidsky, "The Trouble With Fusion", *Technology Review*, p33, Massachusetts Institute of Technology (Oct 1983)
3. K. H. Schmitter, "The Tokamak: An Imperfect Frame of Reference", p47 in *Unconventional Approaches to Fusion*, B. Brunelli and G. G. Leotta (Eds), (Plenum Press, NY, 1981)
4. R. Carruthers, "Criteria for the Assessment of Reactor Potential", *Ibid.*, p39
5. R. L. Hirsch, *J. Fusion Energy*, **5**, 101 (1986)
6. D. Pfirsh and K. H. Schmitter, "On the Economic Prospects of Nuclear Fusion with Magnetically Confined Plasmas", Max Planck Institut Fur Plasmaphysik, Garching, IPP 6/271 (1987)
7. C. Sweet, "Criteria for the Assessment of Fusion Power", *Energy Policy*, **17**, 419 (Aug 1989)
8. J. Reece Roth, "A Critical Evaluation of the DT Tokamak as a Power Plant for the Electric Utilities", *Proc 13th Symp on Fusion Engineering, Knoxville, TN, Oct 1989*, IEEE, 89CH2820-9 (1990)
9. W. D. Kay, "The Politics of Fusion Research", *Issues in Science and Technology*, p40, Winter 1991–92
10. C. G. Bathke, R. A. Krakowski and R. L. Miller, "A Need for Non-Tokamak Approaches to Magnetic Fusion Energy", *17th Symp. on Fusion Technology, Rome, Italy*, Sept 1992
11. "World List of Nuclear Power Plants", in *Nuclear News*, **38**, No. 3, 27 (March 1995)
12. F. Najmabadi et al., "The ARIES-I Tokamak Reactor Study", University of California Los Angeles, UCLA-PPG-1323 (1991)
13. R. A. Krakowski, C. G. Bathke, R. L. Miller, K. A. Werley, *Fusion Technology*, **26**, 1111 (1994)
14. J. G. Galambos, L. J. Perkins, S. W. Haney, J. Mandrekas, *Nucl. Fusion*, **35**, 51 (1995).
15. T. J. Dolan, *Fusion Technology*, **24**, 97 (1993)
16. T. R. Jarboe, *Plasma Phys. Control. Fusion*, **36**, 945 (1994)
17. R. Hagensen, R. A. Krakowski, *Fusion Technology*, **8**, 1606 (1985)
18. M. Tuszewski, *Nucl. Fusion*, **28**, 2033 (1988)
19. H. Momota, et al., *Fusion Technology*, **21**, 2307 (1992)
20. See, for, example, R. D. Jones, W. C. Mead, *Nucl. Fusion*, **26**, 127 (1986)
21. R. W. Moir, et al., *Fusion Technology*, **25**, 5 (1994)
22. M. Tabak et al., *Phys. Plasmas*, **1**, 1626 (1994)
23. H. U. Rahman, F. J. Wessel, N. Rostoker, *Phys. Rev. Lett.*, **74**, 714 (1995)
24. L. J. Perkins, C. W. Hartman, "The Application of a High-Power-Density Fusion Neutron Source, Based on the Continuous Flow Pinch to Plutonium Disposition, Fission Waste Transmutation and Driven, Sub-Critical Fission", submitted to *Fusion Technology* (1995)
25. U. Shumlak, C. W. Hartman, "Sheared Flow Stabilization of the m=1 Kink Mode in Z-Pinches", submitted to *Phys. Rev. Lett.* (1995)
26. See, for example, S. E. Jones et al., *Phys. Rev. Lett.* **51**, 1757 (1983)
27. L. J. Perkins, "Shape Enhanced Fusion: Increasing the Reactivity for Some Advanced Fusion Fuels", Lawrence Livermore National Laboratory, UCRL-JC-118675 (1994), submitted to *Phys. Lett.* (1997)
28. D. E. Morgan, L. J. Perkins, S. W. Haney, "Antiproton Catalyzed Fusion", to be published in *Hyperfine Interactions* (1995)

24

IDEAS FOR FUTURE RFP EXPERIMENTS

James A. Phillips, Don A. Baker, and Robin F. Gribble

Los Alamos National Laboratory
Los Alamos, New Mexico 87545

ABSTRACT

This note discusses the future of the Reversed Field Pinch, RFP, in the USA controlled fusion research program. In the late 1980's the RFP was a strong alternate approach to the tokamak. Starting at Los Alamos National Laboratory (LANL) with the small ZT–1 experiment, passing through ZT–S, ZT–40, and ZT–40M, the results led to the approval and initiation of the $70M experiment, ZTH. With favorable reports by review committees and support by the reactor Titan-RFP reactor study, the construction of the experiment was well on its way to completion. "Budget" reasons stopped the RFP program at LANL in late 1990. As the RFP remains one of the more developed alternate approaches to controlled fusion research; it deserves serious attention. We propose improvements in the RFP design based on recent analysis of the ZT–40M data.

INTRODUCTION

The simple Z-pinch has been studied since the early 1940's (1) and 50's (2) as a possible system for the confinement of high temperature plasma for use in controlled thermonuclear fusion. The pinch has been examined in linear and toroidal systems. In pinch experiments, an applied longitudinal electric field starts a current discharge, I_ϕ. With enough I_ϕ, the self magnetic field compresses and confines ionized gas along the axis. Experimenters quickly found, however, that the Z-pinch is violently unstable. The most unstable modes were called the "sausage" and "kink" instabilities. Theorists in 1956 (3–5) suggested that the sausage mode could be stabilized by an included longitudinal magnetic field, B_ϕ, and the kink by a conducting shell immediately outside the discharge. Radial motion of the discharge is inhibited by the compression of the poloidal magnetic field, B_θ, between the discharge and shell. RFP research has been carried out internationally with important contributions from all the programs. For brevity, we shall give a historical overview of the Los Alamos program.

THE LOS ALAMOS PROGRAM

The Los Alamos controlled-fusion experimental program began in 1952 with the construction of the simple toroidal Z-pinch experiment called the Perhapsatron (6) by G. Gamow because "perhaps it will work." These early experiments exhibited the rapidly growing instabilities.

With the ideas of stabilizing the pinch, a series of Perhapsatrons was constructed with Perhapsatron S-4 (7), S for stabilized discharge, shipped to Geneva and run at the Second U.N. International Conference on the Peaceful Uses of Atomic Energy. These experiments showed sustained pinches to > 24 µs, Doppler broadening of He ions ~180 eV, and electron temperatures (line intensity ratios) ~130 eV. The experiments, however, had unacceptable radiation losses (7) due to impurities from the fused quartz discharge tube and the use of mercury diffusion pumps. The magnetic fields diffused at an anomalous rate and high frequency fluctuations were detected on magnetic probe signals. Because of these difficulties the Z–pinch program at LANL was stopped in 1961 in favor of other approaches.

In 1966 the Los Alamos group revived the Z pinch addressing the temperature limit seen in the earlier experiments. Shock heating (8) was used with new switching techniques and the results encouraged the construction of the toroidal Z-pinch experiment, ZT–1 (9).

Advances had also taken place in the understanding of RFP formation, field diffusion, and analytic stability theory (10). The relaxation theory of Taylor (11) gave strong support to the RFP concept. This theory postulated a minimum energy state for a stabilized pinch with a reversed magnetic field between the discharge and the wall. Schematic magnetic-field profiles for a toroidal RFP are shown in Fig. 1. (Note the reversal of B_ϕ at the wall.) The Taylor theory predicts that, given small (unspecified) dissipation, the plasma in a pinch discharge enclosed in a cylindrical toroidal flux conserving shell will relax to a force free configuration of minimum energy, described by the Bessel function model. The distribution is characterized by the pinch parameter $\Theta = B_\theta(\text{wall})/B_\phi(\text{average})$ that determines $F = B_\phi(\text{wall})/B_\phi(\text{average})$, the field reversal ratio. Thus the observed phenomenon of self-reversal of the B_ϕ was predicted theoretically. Fig. 2 illustrates the dependence of F on Θ for the pressureless Bessel function model and for a ZT–40M discharge. (The shift to higher Θ as seen in the ZT–40M discharge departs from the Taylor profile.) The Taylor model assumes passive relaxation. In experiments, the plasma is continuously driven and sustains a reversed-field state with a high (10–25%) poloidal beta, i.e., the ratio of the average plasma pressure to the poloidal field pressure at the wall.

These theoretical advances encouraged us to convert ZT–1 to a slower current-rise RFP experiment. The reversed-field programming increased the discharge lifetime and the reversed-field region remained a thin annulus near the wall.

These results encouraged the construction in 1979–1982 of the experiment ZT–40 with 40-cm minor diameter. The original design was a fast experiment with an insulating vacuum liner. The device was converted to a low voltage experiment using a metal liner and called ZT–40M (12), after favorable operation of the smaller Italian RFP called ETA–BETA II (13) using a metal liner was obtained. The design goals of ZT–40M were a pinch duration of ~150–200 µs and electron temperature, T_e, of 100–200 eV; durations of 35 ms and T_e of 200–400 eV were obtained.

Extensive RFP experimental research was carried out by the ZT–40M team (14). A sample listing of subjects and references are as follows: the startup of RFP's (15), the dynamo effect (16), confinement properties (17), role of impurities (18), carbonization of walls (19), profile and helicity modeling (20), refueling and density control (21), energetic

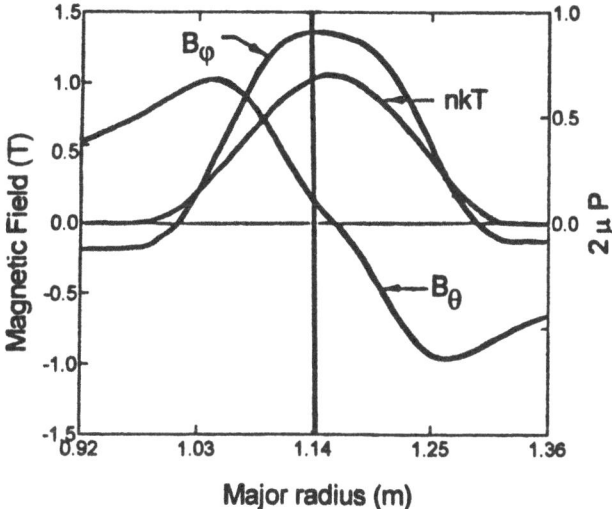

Figure 1. Schematic profiles as a function of major radius for the toroidal-magnetic field B_ϕ, poloidal-magnetic field B_θ, and the plasma pressure for an RFP.

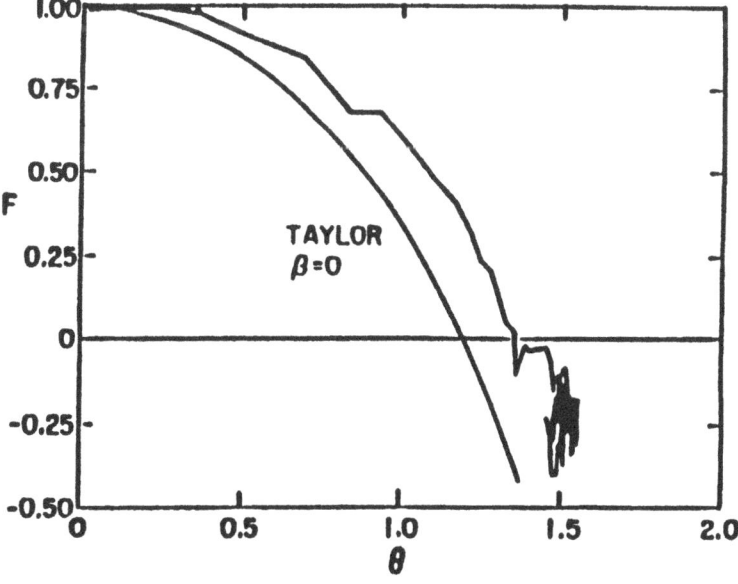

Figure 2. The F–Θ curves for the Bessel function model and a ZT–40M discharge. $F = B_\phi(\text{wall})/B_\phi(\text{average})$ is the field-reversal ratio. $\Theta = B_\theta(\text{wall})/B_\phi(\text{average})$ is the pinch parameter.

Table 1. Titan-II RFP Compact-Reactor design

Major radius	3.9 m
Minor radius	0.6 m
Plasma current	17.8 MA
Poloidal field plasma edge	5.93 T
Toroidal field plasma edge	0.36 T
n τ_E	1.5×10^{20} s/m^3
Poloidal beta	0.2
Electrical power out	900 MW
Cost of electricity	38 mill/kWh
Wall loading[a]	5.5 MW/m^2

[a] This wall loading is high, but the low external magnetic fields allow adequate circulation of the required coolants.

edge electrons (22), beta- limited confinement (23), equilibrium and field errors (24), plasma-wall interactions (25) plasma-circuit interactions (26), and experiments on oscillating current drive (27,28).

Theoretical contributions to the RFP understanding are extensive and can be found in review papers and their references (29,30).

Of primary importance was the demonstration that a high-beta RFP configuration can be sustained by the "dynamo" relaxation (16,31,32) against resistive field diffusion and high temperatures obtained by ohmic heating using slow rising currents and a metal vacuum liner.

RFP REACTOR POTENTIAL

The RFP has the potential of being a useful compact-fusion reactor with high poloidal beta, use of non-superconducting copper coils, ohmic heating to ignition, and the possibility of current drive to sustain the discharge current. Studies (33) of the reactor potential of the approach have found it attractive. Several parameters for an RFP reactor are shown in Table 1.

Experimentally, comparisons of the *published results* of RFP experiments (34) show steady progress toward reactor parameters as the experiments are increased in size and current, Fig. 3. A linear fit to the data on a log-log plot is shown. The Trident Reactor is also shown for comparison. As noted in the figure, the reported RFP energy confinement times, τ_E, are less than a ms and extrapolate to the reactor regime, if the trend shown continues as the current, size, and plasma density increase.

THE NEXT STEP—ZTH

The favorable properties of the RFP encouraged the Los Alamos group to construct and build a still larger and more powerful RFP experiment. The ZTH experiment (35) was proposed, reviewed, and approved with the parameters listed in Table 2.

Construction of ZTH began in 1986. The largest item was the electric generator. One that was available at a canceled nuclear plant in Tennessee was moved to Los Alamos and installed close to the experimental hall. Magnetic field coils for ohmic and toroidal fields were designed and made in Switzerland. The conducting shell, vacuum liner, and

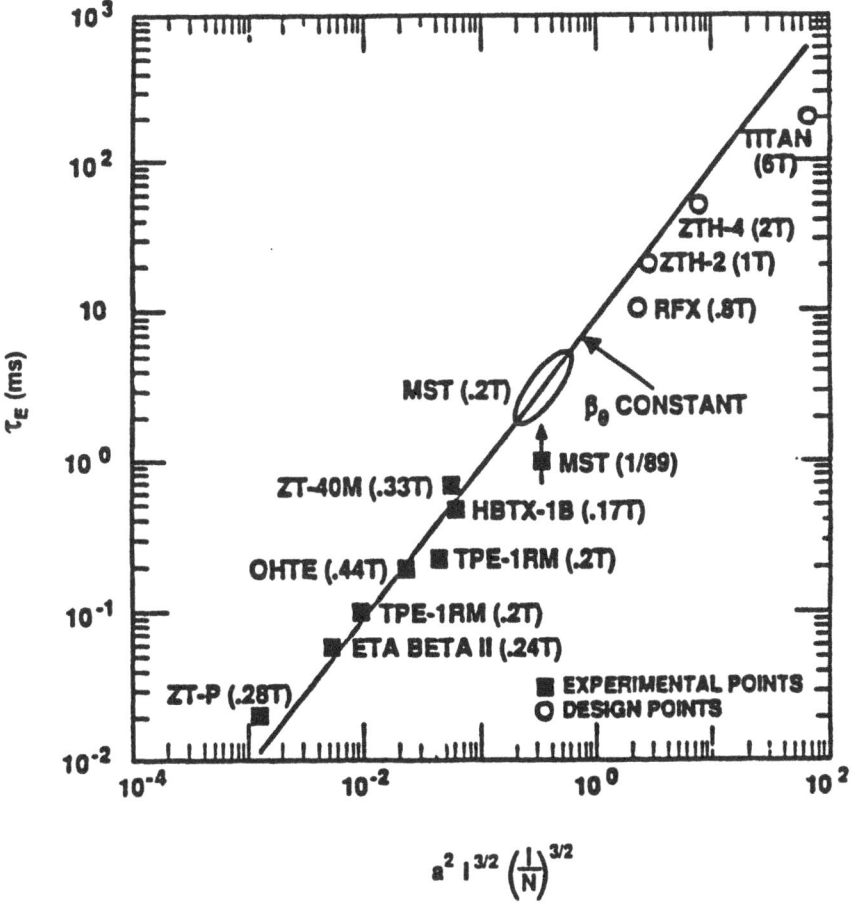

Figure 3. A fit of the energy confinement time, τ_E (ms), for the **published data** of RFP experiments vs $a^2 I_\phi^{3/2}(I_\phi/N)^{3/2}$ in units of m, MA, and 10^{20}m^{-1}. Note the predictions for ZTH and the Titan reactor.

the support structure were being made. Scientific committees designated by Department of Energy (DOE) periodically checked on progress and were strongly supportive.

In 1990, however, the DOE abruptly stopped all magnetic-fusion experiments at Los Alamos including ZTH. The closure was not made for technical reasons but budgetary ones. The DOE decided to fund the tokamak. Mothballing ZTH was not permitted and the staff moved to other positions in the laboratory.

Table 2. ZTH design parameters and predicted operation

Major radius	2.4 m
Minor Radius	0.4 m
Plasma current	4 MA
Generator rating	14,000 MW
Design T_e	4.0 keV

ANALYSIS OF ZT–40M BEHAVIOR

When the experimental data from different RFP experiments are examined a consistent picture is not readily seen (29). For example, wide differences exist in the published dependence of the resistance of flat-top discharges on I_ϕ, varying between $I_\phi^{-3/2}$ and $I_\phi^{-1/2}$. An uncertainty also exists in the scaling of the poloidal beta, β_p. Some experiments report a β_p independent of I_ϕ whereas others show a reduction of β_p with I_ϕ. Much effort has been spent to explain these differences by the role of impurities, the ratio of current to plasma density, low frequency magnetic fluctuations, discharge equilibrium, and magnetic field errors.

Since 1988, the present authors have been concerned with these differences and on analyzing the behavior of ZT–40M are able to give suggestions for these differences. The analysis (30) of the ZT–40M data, accumulated by the RFP team during its operation, shows that for static gas filling, flat-topped discharges, the ZT–40M RFP results are described by two independent parameters, for example, I_ϕ and the pinch parameter Θ, that are under the control of the physicist. Once these parameter are chosen, the other variables are dependent and are determined by the experiment. The magnitudes of the dependent variables including the electron temperature, plasma resistance, toroidal flux, I/N, and their fluctuation amplitudes exhibit minima as a function of Θ for constant I_ϕ. The Θ value at these minima decreases with increasing I_ϕ.

Scaling of the variables with I_ϕ is not unique but depends on the chosen Θ value. Thus, an I_ϕ–Θ trajectory can be chosen to meet an RFP operational constraint. A significant result is that the design of different experiments and the choice of an I_ϕ–Θ trajectory can markedly change the variation of different dependent variables with I_ϕ and make comparisons between experiments difficult or impossible.

As an example (30) of what can be done, we were interested in the I_ϕ–Θ trajectory that gave a constant β_p with I_ϕ, that is, discharges that had β_p between 0.12 and 0.16. Discharges meeting this criterion were selected, the I_ϕ–Θ trajectory determined, and the values of the dependent variables plotted as a function of I_ϕ. In Table 3 the variables of these discharges are listed with their dependence on I_ϕ and amplitudes at I_ϕ = 200 kA.

In contrast, averaging over the wide spread of β_p for the analyzed discharges, β_p varies as I_ϕ to the power -0.70 ± 0.06!

The known correlation of low-frequency fluctuations with the loop voltage of the discharge has focused attention on their source. We find that the fluctuations can be described with a helical traveling wave perturbation with toroidal flux passing in and out of the toroidal and poloidal gaps. With the diverted flux, the hot plasma is lost and bombards the walls releasing impurities. The loss of energy increases the loop voltage and reduces the energy-confinement time.

Table 3. Variation of dependent variables with I_ϕ at nearly constant β_p (power-law fits)

Parameter	Exponent of I_ϕ	Value at 200 kA
β_p	0.01±0.05	0.14
T_e	1.00±0.12	179 eV
n_e	0.99±0.11	3.9×10^{13} cm^{-1}
τ_E	0.81±0.09	0.29 ms
$n_e\tau_E$	1.80±0.14	1.2×10^{10} cm^{-3}s
I_ϕ/N	0.010±0.14	4.2×10^{-14} A/cm
V_ϕ	0.18±0.07	50 V

Before the demise of ZT-40M, an attempt was made to reduce low-frequency fluctuations by connecting the toroidal field coils in parallel rather than in series (36). This connection improved the conservation of toroidal flux inside the shell, as MHD theoretical studies have assumed, and it reduced the low-frequency fluctuations and loop voltage. On examining other experiments, we found strong correlation with the designs of their shells and their discharge parameters. These considerations suggest changes in the design of RFP experiments that will improve operation.

We conclude that the confining shell of RFP experiments has not received enough attention. We admit, however, that the design of a stabilizing shell with minimum field perturbations, especially the poloidal and toroidal gaps, and allowing a slow vertical field for equilibrium to pass through the shell is no easy task. On the other hand, the difficulty has been found and the RFP physics and engineering expertise exists.

Two criticisms against the RFP should be mentioned:

- The "hole" in the doughnut required for the flux change that drives I_ϕ, is objectionable from the engineering viewpoint. The ideal RFP geometry is an infinite straight cylinder, which is obviously impractical, and electrodes might be added, but they would bring in impurities. Bending the discharge into a doughnut requires equilibrium control, but the major radius of the torus is not constrained to be small, as in the tokamak, and can be as large as necessary.
- Some theorists suggest that the large I_ϕ, which provides ohmic heating and plasma confinement, may drive instabilities as I_ϕ is raised to that required in a reactor. An experiment (37) with average current densities exceeding 10 MA/m^2, within a factor 2 to 3 that in the Titan reactor design, to date shows no serious degradation of RFP performance.

RECOMMENDATIONS

To improve the operation of RFP experiments, we recommend analyses by experts on important issues that include:

- The shell and the poloidal and toroidal gaps, as described above.
- Field errors and locking modes of the discharge. These perturb the discharge causing wall bombardment with the release of impurities.
- Startup modes of the RFP require passage through the high-resistance stabilized-pinch regime that increases the V·s consumption. The ramp mode (15) reduces this consumption.
- Diverter designs proposed to exhaust spent fuel and impurities in the RFP need to be engineered and tested.

Finally we recommend the design and construction of a low-cost **engineering** RFP experiment dedicated to understand how an optimized RFP might operate. The experiment should be simple to allow rapid changes with a minimum of diagnostics to reduce field errors. The experiment should be sufficiently large to give significant results (~10-cm minor radius) with peak I_ϕ to 350 kA with constant current for greater than ~150 ms. The experimental program would at first focus on the shell and test the many designs that have been—and will be—suggested. The improved confinement by implementing ideas that result from the above considerations, can reduce extrapolations in plasma parameters needed to reach reactor conditions.

ACKNOWLEDGMENTS

We appreciate the contributions of Ken Klare to the preparation of this paper. Work supported in part by the U.S. Department of Energy.

REFERENCES

1. Bennett, W. H., Phys. Rev. **45** (1934) 890.
2. Cousins, S. W. and Ware, A. A., Proc. Phys. Soc. (London) **64A** (1951) 159.
3. Kruskal, M. and Schwarszchild, M., Proc. Roy. Soc. (London) **A223** (1954) 348.
4. Rosenbluth, M., Los Alamos Scientific Laboratory report, LA-2030 (1956).
5. Shafranov, V. D., J. Nucl. Energy **5** (1957) 86.
6. Burkhardt, L. C., et al., J. Nucl. Energy, **28** 5 (1957) 519.
7. Karr, H. J., et al., Phys. Fluids, **4** 4 (1961) 424.
8. DiMarco, J. N. and Burkhardt, L. C., IV European Conference on Controlled and Plasma Physics, Rome (1979) 53.
9. Baker, D. A., et al., Proc. 4th IAEA Int. Conf. on Plasma Physics and Controlled Nucl. Fusion Research (1970) Madison, IAEA, Vienna, **1** (1971) 203.
10. Robinson, D. C., Plasma Physics **13** (1971) 439.
11. Taylor, J. B., Phys. Rev. Lett. **33** (1974) 1139.
12. Baker, D. A., Buchenauer, C. J., et al., Proc. 10th IAEA Int. Conf. on Plasma Physics and Controlled Nuclear Fusion Research (1984), London, IAEA, Vienna (1985) 439.
13. Buffa, A. A., et al., Proc. 8th IAEA Int. Conf. on Plasma Physics and Controlled Nuclear Fusion Research (1980), Brussels, IAEA, Vienna (1981) 275.
14. ZT–40M experimental team members: D. A. Baker, C. J. Buchenauer, L. C. Burkhardt, J. N. DiMarco, J. N. Downing, R. M. Erickson, R. F. Gribble, A. Haberstich, R. B. Howell, J. C. Ingraham. E. M. Little, R. S. Massey, G. Miller, C. P. Munson, J. A. Phillips, M. M. Pickrell, K. F. Schoenberg, A. E. Schofield, K. S. Thomas, R. G. Watt, P. G. Weber, D. M. Weldon, and G. A. Wurden.
15. Phillips, J. A, Baker, D. A., Gribble, R. F., and Munson, C. P., Nucl. Fusion **28** 7 (1988) 1241.
16. Caramana, E. J. and Baker, D. A., Nucl. Fusion **24** 4 (1984) 423.
17. Weber, P. G., Schoenberg, K. F., et al., Proc. 12th IAEA Int. Conf. on Plasma Physics Controlled Nuclear Fusion Research, Nice, 1988, IAEA, Vienna **2** (1988) 419.
18. Weber, P. G, Phys. Fluids, **28** (1985) 3136.
19. Downing, J. N., Buchenauer, C. J, Ingraham, J. C., et al., J. of Nuclear Materials **128** and **129** (1984) 517.
20. Schoenberg, K., Los Alamos National Laboratory report, LA-UR-87-3454 (1987).
21. Wurden, G. A., Weber, P. G., et al., Int. School of Plasma Physics, Workshop on Physics of Mirrors, Reversed Field Pinches and Compact Tori, Varenna, Italy **1** (1987) 411.
22. Ingraham, J. C., Ellis, R. F., Varenna, School of Plasma Physics, Workshop on Physics of Alternative Magnetic-Confined Schemes, Varenna, Italy (1991) 859.
23. Pickrell, M. M., Phillips, J. A., et al., Proc. 16th European Conf. on Controlled Fusion and Plasma Physics, Venice, Italy **13b** II (1989) 749.
24. Massey, R. S., Buchenauer, C. J., Burkhardt, L. C., et al., Los Alamos National Laboratory report, LA-5967-MS (1983).
25. Mondt, J. P., Plasma Physics and Controlled Fusion, **27** 3 (1985) 339.
26. Phillips, J. A., Baker, D. A., Gribble, R. F., and Munson, C. P., Nucl. Fusion **31** 8 (1991) 1556.
27. Bevir, M. K. and Grey, J. W., "Relaxation, Flux Consumption and Quasi-Steady State Pinches" (Proc. Reversed-Field Pinch Theory Workshop, 1980) Los Alamos National Laboratory report LA-8944-C (1982) 176.
28. Schoenberg, K. F., et al., Phys. Fluids **31** 8 (1988) 2285.
29. Bodin, H. A., Nucl. Fusion **30** (1990) 1717.
30. Phillips, J. A., Baker, D. A, and R. F. Gribble, Nuclear Fusion **35**, 935 (1995).
31. Moffat, H. K., Magnetic Field Generation in Electrically Conducting Fluids, Cambridge University Press (1978).
32. Gimblett, C. G. and Watkins, M. L., Pulsed High Beta Plasma, Pergamin Press, Oxford (1976) 279.
33. The Titan Reversed-Field-Pinch Fusion Reactor Study, UCLA report, UCLA-PPG-1200 (1990).
34. Dimarco, J. N., Los Alamos National Laboratory report, LA-UR-Revised-88–3375 (1988).

35. Dimarco, J. N. and Thomson, D. D., CPRF/ZTH Technical Design Criteria, Los Alamos National Laboratory Report, LA-UR-89–3166 (1990).
36. Phillips, J. A., Baker, D. A., Gribble, R. F., and Munson, C. P., Nucl. Fusion **31** (1991) 1556.
37. Nordland, P. and Mazur, S., Alfvén Laboratory report, TRITA-ALF-95:02, Royal Institute of Technology, Stockholm, Sweden.

25

DENSE Z-PINCHES FOR FUSION

David Scudder and Jack Shlachter

Los Alamos National Laboratory
Los Alamos, New Mexico 87545

Z-pinches have a long history within the experience of fusion research. Their simplicity made them a natural starting point for researchers in the early days of fusion research, and even Tokamaks can trace their ancestry to schemes conceived to stabilize Z-pinches. This paper will concentrate on recent work on linear Z-pinches operating at high density. This is an area which has been pursued at a small level-of-effort by a number of groups over the years. While the state of the art is far from a promising working reactor, innovative solutions to classic Z-pinch problems continue to be developed.

The most commonly pursued means of stabilizing a Z-pinch is to add shear to the magnetic filed. This has the side effect in a linear pinch of causing the magnetic field-lines to be, no longer, closed. To reclose the field lines one then bends the pinch into a torus, and one follows along the familiar path to tokamaks, reversed-field pinches and even field reversed theta pinches. This paper will deal with very different plasma configurations one is led to if one opts to stay with a linear Z-pinch without shear.

High densities are desirable in such a pinch to minimize the number of instability growth times the pinch must hold together to produce a useful fusion yield. As one increases the density, the Lawson time (i.e. the burn time to produce fusion breakeven) drops as $1/n$. The instability growth time for modes that are typically the most troublesome also decreases with increasing density, but as $1/\sqrt{n}$, so the ratio improves at higher density. Advances in the art of high-voltage pulsed power have led to generators with the short risetimes, high voltages and high currents appropriate for atmospheric (or somewhat higher) density Z-pinches.

During the 60s and 70s several groups pursued gas-embedded Z-pinches[1,2,3]. In this approach a discharge channel is initiated in a static fill of gas, typically at atmospheric density. A variety of initiation techniques were pursued including formation of a vortex in the gas and preionizing with electron beams and with lasers. This work was eventually abandoned because the strong electric field along the column needed to drive a fast-rising current also drove an ionizing accretion wave into the embedding gas, thereby ruining the needed inventory control.

Subsequently several groups initiated research into Z-pinches formed from fibres of frozen deuterium. This approach solved the accretion problem by strictly limiting the line density. These initial experiments showed unexpected stability to both $m=0$ and $m=1$

modes[4,5]. This result, so contrary to widely held expectations, has generated a great deal of interest in the theoretical community[6,7,8].

The theoretical effort motivated by these results has produced two candidate stabilizing mechanisms: resistivity and magnetized viscosity. Resistivity is predicted to produce a stabilizing effect when the magnetic Reynolds number $S = \mu_0 \sigma a v_A$ (where σ is the conductivity, a is a scale length, typically the radius, and v_A is the Alfvén velocity), is less than a critical value estimated at about 100. This effect favours small radii and hence high density through its dependence on a. Since S increases with temperature, this effect is more effective in the early or low-current phases of pinch development than in the later phases. Magnetized viscosity, represented by the Alfvén-Reynolds number $L = \rho a v_A / \mu$ (where ρ is the mass density, a the radius, v_A the Alfvén velocity and μ is the viscosity) provides, in some ways, an effect complementary to resistivity. Its stabilizing influence is predicted to become important at the high temperatures characteristic of a pinch at fusion conditions. Calculations by Cochran and Robson[7] have shown that if the product SL, which depends only on N, the line density, is less than about 1000 a pinch can be stabilized throughout its temperature history. The value of N required to meet this condition is $N < 2 \times 10^{16}$ cm^{-1}, less than has been used in experiments to date.

Whether these theoretical predictions correctly describe the stability seen in the first fibre pinch experiments has not been unambiguously determined. Since then experiments have been performed with a more energetic generator to push plasma parameters closer to fusion conditions but using line densities two orders of magnitude higher than reference[7] would suggest, and, indeed, the familiar $m=0$ reappeared[9].

Currently the only fibre pinch experiments being pursued are at Imperial College in London and are just beginning to produce results. A related approach, Puff on Wire, is being pursued by Frank Wessel and Hafiz Rahman at UC Irvine and UC Riverside, and will be described by them at this meeting. The experiments at NRL and LANL, which were supported primarily by internal funds, have both died as programmatic priorities changed.

Significant unanswered questions remain about the nature of the stability seen in initial experiments, the applicability of predictions of stabilizing effects and other possibilities for stabilizing such pinches. The simplicity of a potential Z-pinch reactor, the qualitatively different nature of such a device, and the existence of tantalizing untested physics predictions make this a ripe area for modest fusion research.

REFERENCES

1. E. A. Smars, *Arkiv for Fysik* **29**, 97 (1965).
2. C. W. Hartman et al., *Plasma Physics and Controlled Nuclear Fusion Research 1974* (5th Conference Proceedings, Tokyo, 11 - 15 November 1974) Vienna IAEA 1975, Vol. II, p. 653.
3. J. E. Hammel, D. W. Scudder, J. S. Shlachter, *Nuclear Inst. and Methods* **207**, 161 (1983).
4. J. E. Hammel, *Proceedings of the 2nd International Conf. on Dense Z-Pinches, Laguna Beach, 1989* N. R. Pereira, J. Davis and N. Rostoker eds. 303 (1989).
5. J. D. Sethian, A. E. Robson, K. A. Gerber and A. W. DeSilva, Phys. Rev. Lett. **59**, 892 (1987) and ibid 1790.
6. A. H. Glasser and R. A. Nebel, *Proceedings of the 2nd International Conference on Dense Z-Pinches, Laguna Beach, 1989*, N. R. Pereira, J. Davis and N. Rostoker eds. 226 (1989).
7. F. L. Cochran and A. E. Robson, *Physics of Fluids B* **5**, 2905 (1993).
8. I. D. Culverwell, M. Coppins and M. G. Haines, *Proceedings for the 2nd International Conference On Dense Z-Pinches, Laguna Beach, 1989*, N. R. Pereira, J. Davis and N. Rostoker eds. 246 (1989).
9. D. W. Scudder, J. S. Shlachter, J. E. Hammel, F. Venneri, R. Chrien, R. Lovberg and R. Riley, *Physics of Alternative Magnetic Confinement Schemes*, S. Ortolani and E. Sindoni, eds., Società Italiana di Fisica 519 (1991).

ASSESSMENT OF FIELD-REVERSED-CONFIGURATION STABILITY

Richard E. Siemon

Los Alamos National Laboratory
University of California
Los Alamos, New Mexico 87545

The field line geometry of the Field Reversed Configuration (FRC) can be pictured as follows: Imagine a pure Z pinch with its current surrounded by circular field lines. Next wrap the current around an axis to form a torus, and then stretch it out along the axis to form an oblate torus called the FRC. Given the resemblance to a Z-pinch, it should not be surprising MHD stability analysis predicts unstable modes linked basically to the unfavourable curvature of the confining field. What is surprising are the many experimental studies that do not show the expected modes of instability.

The desire to understand this configuration stems from its unusual prospects for application to controlled fusion. The FRC intrinsically has a very high beta plasma. Plasma pressure is typically 90% of the confining field pressure as opposed to the few percent typically obtained in tokamaks. Thus, for a given plasma density, the magnetic field is smaller than in most magnetic confinement systems, and the troublesome synchrotron radiation that dominates energy balance with advanced fuels can be reduced. The geometry also permits an unusual degree of engineering flexibility. The only required external confining field can be made with a solenoidal magnet (as opposed to toroidal). Once created, the combination of plasma and field that make up the FRC can be translated along the axis of the magnets. Reactor designs become possible where the FRC is formed and heated in one location and then moved to a second location where the wall materials could be selected to accommodate high heat flux and intense radiation from fusion reactions. For example, the materials proposed for lithium wetted walls in inertial fusion systems might be used.

The progress on understanding stability described in this paper is based primarily on the Los Alamos FRC program that was active from the late 1970s through about 1990. The results are reasonably well documented in the literature[1-4]. In particular Tuszewski's review paper[1] gives and excellent summary of work through about 1987. The methods of plasma formation, equilibrium properties, stability and confinement are discussed in detail. The next three papers summarize what was being learned when the program was terminated in 1990. While it must be admitted that the abrupt termination of the research leaves unresolved issues, questions that hopefully will be addressed in the future, it should also be noted that a soft consensus developed among most of the scientists at Los Alamos about the nature of FRC stability.

The following qualitative picture of FRC stability emerged. In the formation methods studied, FRCs were created with enough temperature that ion orbits were comparable to size to the plasma radius. Under that circumstance, kinetic effects act like a strong viscosity in the fluid equation, and the predicted methods of MHD instability have a much reduced growth rate. The effect was demonstrated clearly by Barnes and colleagues[5] using numerical modelling. As initial parameters were varied experimentally to increase the ratio of plasma size to orbit size (which can be shown to be necessary for obtaining the improved energy confinement needed for fusion), plasma confinement inevitably was found to degrade in some manner. However, the fairly extensive diagnostics that were available did not show for many years the theoretically expected tilt mode, characterized by major plasma perturbations at the ends of the FRC where the field lines are most strongly curved. In fact, it seemed that MHD predictions for instability were somehow flawed, and that plasma properties depended in a major way upon details of plasma formation. Kurtmullaev's group in the former Soviet Union advocated this interpretation very strongly[6]. A breakthrough occurred in 1988 when an array of 64 external magnetic probes were installed on the Los Alamos experiment. The probes were positioned in an array near the ends of the FRC and were precisely oriented perpendicular to the main field so that even small signals represented a departure from symmetry. With this method clear evidence was found for the expected tilt mode. Roughly the correct growth rate was observed. Interestingly, the perturbation amplitude (taken as a measure of MHD turbulence) was correlated statistically with the conditions of worst confinement. The experimental interpretation changed from no tilt mode occurs to tilt modes nearly always occur and their amplitude increases when kinetic stabilization becomes weak.

In summary, the FRC has attractive properties in terms of a fusion system, but a challenging issue with respect to MHD stability. Methods have been proposed for resolving this issue including the use of injected high-energy particles to provide the kinetic stabilization. Thus the FRC presents increasing opportunities for future experiments that would build on the base of understanding from previous work. However, it should be realized that the FRC concept does not represent a panacea any more than the other known alternatives, and it will in all likelihood require significant commitment of resources before an improved fusion system could be developed.

REFERENCES

1. M. Tuszewski, *Nuclear Fusion* **28**, 2033 (1988).
2. M. Tuszewski, et al., *Physics of Fluids B* **3**, 2844 (1991).
3. M. Tuszewski, et al., *Physics of Fluids B* **3**, 2856 (1991).
4. D. J. Rej, et al., *Physics of Fluids B* **4**, 1909 (1992).
5. D. C. Barnes, et al., *Physics of Fluids* **29**, 2616 (1986).
6. A. G. Es Kov, et al., *Plasma Physics and Controlled Nuclear Fusion Research* **2**, IAEA, Vienna, p. 187 (1979).

MUON-CATALYZED FUSION IN 1996

Steven E. Jones

Department of Physics and Astronomy
Brigham Young University
Provo, Utah 84602

ABSTRACT

Muon-catalyzed fusion is an alternative approach in which fusion is induced without the need for high temperatures. Indeed, it is the only bona fide approach to cold fusion. Data regarding key parameters are provided, and the possibility of applications is discussed.

1. INTRODUCTION

Muon-catalyzed fusion research is motivated both by a curiosity about nature and by the possibility of applications. After all, exothermic nuclear fusion is readily induced by negative muons. And muon-catalyzed fusion is indeed curious in that it proceeds without the need for plasmas or high temperatures. Research groups throughout the world have pooled efforts on uncover numerous surprises in a tapestry of exotic atomic and molecular processes, unexpected resonances, and extremely rapid nuclear interactions. It is remarkable that a fundamentally nuclear process can be strongly affected by changing the temperature and composition of the environment. This phenomenon demonstrates the subtle interplay of atomic and nuclear physics inherent in muon-catalyzed fusion (μcf).

Theoretical and experimental efforts have also dovetailed to expand our understanding of muon catalysis. The theoretical breakthroughs ago achieved by Leonid Ponomarev and his colleagues fifteen years motivated experiments involving μcf in mixtures of deuterium and tritium. Observed temperature, density and d/t ratio effects in turn led to refinements in the theory. We can say that much progress has been made, but that several questions remained unresolved.

Overall, we can look back and conclude that muon-catalyzed fusion yields have significantly exceeded expectations, leading to renewed speculation regarding applications. To guide our discussion of progress in μcf research, let us consider a straightforward equation:

$$1/Y = \lambda_o/\lambda_c + W \tag{1}$$

where

Y = yield, the number of fusions per muon (average);
λ_o = muon-decay rate (0.455 per microsecond);
λ_c = muon-catalysis cycling rate (1/time between fusion neutrons); and
W = the probability of muon loss per catalysis cycle, for any cause.

It is informative to interpret this governing equation as a sum of probabilities:

1/Yield = Probability of muon decay during + Probability of muon-scavenging   (2)
any stage of the catalysis cycle   due to dead-end processes

Clearly, to increase to fusion yield one must increase the catalysis cycling rate λ_c, and minimize muon losses W. We will here review what we have learned about these important parameters, then examine the current fusion yields vis-a-vis energy applications for μcf.

2. MUON CATALYSIS CYCLING RATE

Figure 1 displays a subset of data obtained at the Los Alamos Meson Physics Facility since 1982 regarding the observed (unnormalized) muon catalysis cycling rate. (See

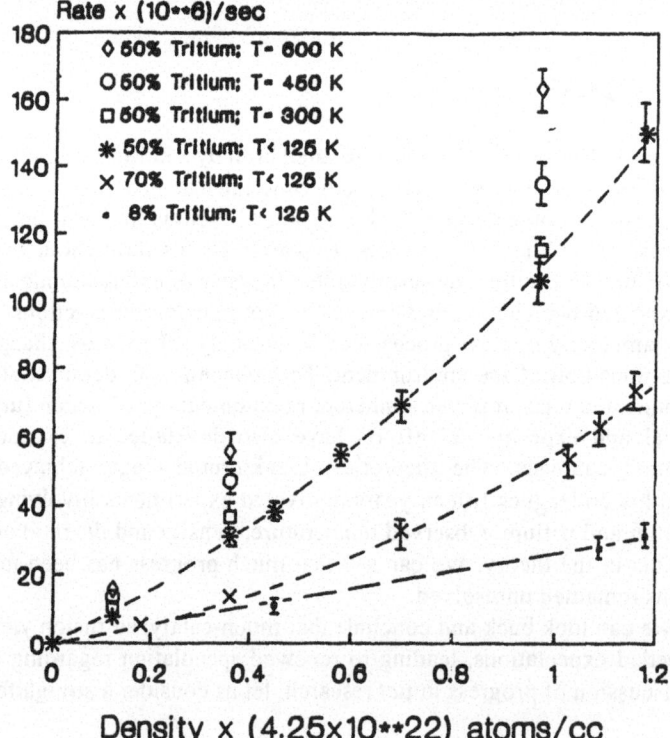

Figure 1. Observed muon catalysis cycling rate as a function of density and temperature of the deuterium-tritium mixture in which the muon is injected.

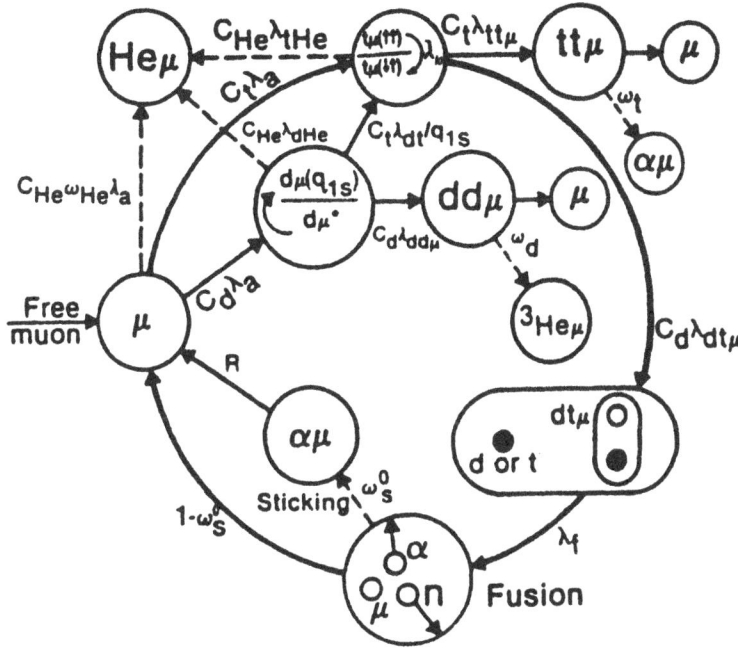

Figure 2. Scheme of the muon-catalyzed fusion cycle, showing reactions which occur when negative muons stop in a mixture of the hydrogen isotopes (d and t) and helium (He) with respective fractions C_d, C_t, and C_{He}. Reaction rates are labeled with λ and muon-loss probabilities are labeled with ω.

Ref. [1].) We see that λ_c depends on density of the deuterium-tritium mixture as well as on its temperature and composition.

Why is this so? Coordinated theoretical and experimental studies have led to a picture of the μcf cycle which is portrayed in somewhat simplified form in Figure 2. Important reaction rates and muon-loss probabilities are labeled on this diagram. Note that C_p, C_d, C_t, and C_{He} represent the atomic fractions of the three isotopes of hydrogen (p,d,t) and helium present in the reaction chamber. The cycling rate can be broken down into component terms according to the prescription:

$$\lambda_c^{-1} \approx \left[\frac{q_{1s}C_d}{\lambda_{dt}C_t} + \frac{0.75}{\lambda_{10}C_t} + \frac{1}{\lambda_{dt\mu}C_d} \right] \phi^{-1} \tag{3}$$

for temperature $\leq 500\ K$, and where ϕ = the density of the target mixture.

Parameter q_{1s} represents the probability that the dμ atom will reach the ground state before the muon is transferred to a triton to form a tμ atom, an energetically favorable reaction. The dμ→tμ transfer reaction is faster for a smaller q_{1s}. But experiments[1,2,3] show that q_{1s} is larger than predicted[4] and decreases more slowly with increasing tritium fraction and density than expected. This transfer reaction is a relatively slow one (requiring typically a few nanoseconds), so it is relevant to understand why q_{1s} is as large as it is seen to be, and how it could be reduced. Understanding the density, temperature, and d/t ratio dependences of q_{1s} remains an important goal. The hyperfine-quenching rate λ_{10} also remains to be fully understood.

Of particular significance is the fact that the rate of formation of dtμ molecule depends strongly on temperature and density, as well on whether a tμ atom collides with a D_2 or a DT molecule (see Figure 2). These effects are reflected in the dependences seen in the muon catalysis cycling rate (Figure 1) and are consistent with the model of resonant dtμ formation developed by Ponomarev and collaborators.[6] Progress in measuring and understanding dtμ-formation has been rapid and gratifying for both theorists and experimentalists. It should be remembered that Ponomarev's predictions of fast, resonant dt formation were largely responsible for the renaissance of μcf research activity during the last decade. Experiments underway at TRIUMF and Rutherford Laboratory, led by Glen Marshall and by Ken Nagamine, should do much to further elucidate these critical parameters in muon catalysis.[7]

The striking density-dependence of $\lambda_{dt\mu-d}$ provides evidence of significant resonant dtμ formation via three-body collisions. Menshikov and Ponomarev[8] have discussed a mechanism for three-body resonant molecular formation, e.g. tμ = D_2 + D_2 + [(dtμ) d2e]* + D_2 + ΔE. The singlet (tμ + D_2) collisions are special in having their strongest resonances just below threshold, where they are not accessible in two-body collisions. By absorbing some kinetic energy, the spectator molecule (D_2, DT, or T_2) moves these strong resonances above threshold, allowing them to contribute to very rapid dt molecular formation.

Consider, then a (D_2 + T_2) mixture at ~30K, compressed to a density of approximately 2.3 LHD (nearly twice the density now achieved at LAMPF, SIN or KEK). The optimum tritium fraction for D_2-DT-T_2 at high temperature equilibrium is C_t ~ 30–40%. Using (D_2 + T_2), we can increase λ_c by moving to higher C_t (~ 50%) so as to increase λ_{dt}/q_{ls}. Based on an extrapolation of our results from LAMPF described above (and assuming no saturation of $\lambda_{dt\mu-d}$ at high density), we would then expect for these conditions:

$$\lambda_c^{obs} \approx 7 \times 10^{-4} s^{-1} \quad \text{(extrapolation)} \tag{4}$$

It is useful to look at the fusion yield which this rate would produce if sticking losses were zero (see equation 1):

$$N_{w=0} = \frac{\lambda_c^{obs}}{\lambda_o} \approx 1500 \text{ fusions}/\mu - \quad \text{(extrapolation)} \tag{5}$$

Thus, it now appears possible to achieve over 250 fusions per muon based on what we have learned about increasing the muon-catalysis reaction rates if sticking can be sufficiently reduced. This result is a clear departure from theoretical predictions of slow cycling rates expected just fifteen years ago, and is an exciting confirmation of the resonance model of dtμ-formation developed by S.S. Gershtein, L.I. Ponomarev and their co-workers.[8,9,10]

We hope to perform a set of experiments with these ultra-high density conditions at Dubna, Russia. Before further discussing the new experiments, however, it is important to turn our attention to the muon-loss factor W which is now expected to limit the achievable number of fusions per muon.

3. MUON-CAPTURE LOSSES (W)

Various ways in which muons may be lost from the catalysis cycle are shown in Figure 2. The muon may be captured and retained by a helium nucleus synthesized during dtμ, ddμ, or ttμ fusion, with sticking probabilities ω_s, ω_d and ω_t, respectively. In addition, small amounts (typically less than 1%) of protium are present, resulting in pdμ and ptμ fusion, with muon sticking probabilities ω_{pd} and ω_{pt}. The muon may also be scavenged by ambient helium in the hydrogen-isotope mixture, as indicated in Figure 2. All of these processes contribute to W, the total muon-loss probability per cycle[1]:

$$W \approx \frac{q_{1s}}{\lambda_{dt}C_t + \lambda_{dd\mu}C_d}\left(0.58\lambda_{dd\mu}C_d\omega_d + \lambda_{pd\mu}C_p\omega_{pd} + \lambda_{dHe}C_{He}\right)$$
$$+ \frac{1}{\lambda_{dt\mu}C_d}(\lambda_{tt\mu}C_t\omega_t + \lambda_{pt\mu}C_p\omega_{pt} + \lambda_{tHe}C_{He}) + C_{He}\omega_{He} + \omega_s^{eff} \qquad (6)$$

Experimentally measured values of W as a function of density are displayed in Figure 3. Results from LAMPF[1] and PSI[11] regarding the "raw sticking" W are in remarkable agreement and point to a rather striking density dependence.

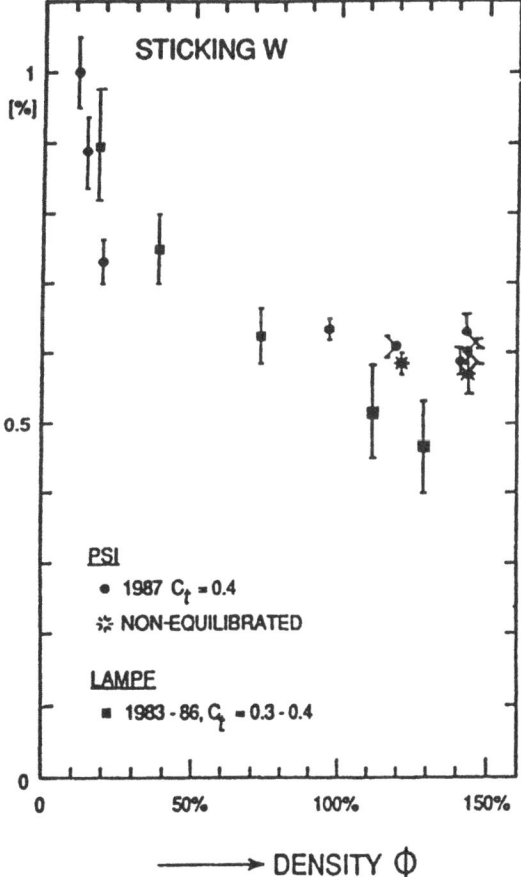

Figure 3. Juxtaposition of data acquired at LAMPF and PSI[11] reveal that the raw sticking loss W decreases with increasing d-t density. Clearly, one or more of the muon-loss processes is density-dependent.

What causes the obvious density-dependence of W? Looking closely at equation (6), we observe that some parameters such as $\lambda_{d t \mu}$ and possibly q_{1s} are significantly density-dependent, and in such a way that W will decrease with increasing target density, as observed. Until all muon-loss terms are fully understood, we cannot be certain whether ω_s, representing alpha-muon sticking following muon-induced d-t fusion and subsequent slowing down of the alpha-muon ion, is density-dependent or not.[12] Indeed, if one assumes that q_{1s} as predicted by Menshikov and Ponomarev[4], then one can account for much of the observed density-dependence of W. However, analysis of the LAMPF data has in fact shown only a weak density-dependence of q_{1s}, leaving a residual density-dependence in ω_s.[1] Thus, until q_{1s} and other interrelated parameters of equations (3) and (4) are sorted out completely, we cannot resolve this question. However, we can agree that W is indeed significantly density-dependent (Figure 3). After all, it is W rather than ω_s alone which influences the fusion yield (see equation 1).

4. DIRECT MEASUREMENT OF ALPHA-MUON STICKING

The data regarding W (Figure 3) were extracted by observing fusion neutrons, which results in a sensitivity to all processes which remove muons from the catalysis cycle. To measure ω_s alone, it is sufficient to count the number N of each of the charged products of the d-t fusion reaction, namely the α^{++} and $(\alpha\mu)^+$ ions:

$$\omega_s = \frac{N(\alpha\mu)}{N(\alpha) + N(\alpha\mu)} \quad (7)$$

Equation (4) expresses the muon loss fraction due only to $\alpha-\mu$ capture and retention following d-t fusion (note that ions are detected in coincidence with 14 MeV neutrons). Muon-stripping processes affect ω_s measured in this way, but complications stemming from competing dd and tt fusion channels, and muon scavenging by helium or other impurities (see Figure 2), can be excluded. Moreover, the ratio of equation (4) does not depend on absolute detector calibrations, a feature which reduces some systematic errors.

The layout for this experiment is portrayed in Figure 4 and is described in detail elsewhere.[13] We recorded both the energy and the arrival time (relative to a fusion neutron) of each ion detected at the surface barrier detector.

In order to gather better statistics, we separately collected alpha-muon ions at 1800 Torr and alpha ions at 490 Torr.[14] The result obtained in this way is:

$$\omega_s^o = (0.80 \pm 0.15 \pm 0.12 \text{ systematic})\% \quad (8)$$

The measurements should be refined in continuation experiments at Rutherford Laboratory led by Professor K. Nagamine.

We can already draw two important conclusions from these new results. First, the measured *initial* alpha-muon sticking probability appears to be in reasonable agreement with published theoretical calculations[15,16]:

$$\omega_s^o = (0.88 \pm 0.05)\% \quad \text{(theoretical)}. \quad (9)$$

Secondly, the directly measured value of sticking evidently rules out the possibility that sticking is very small.

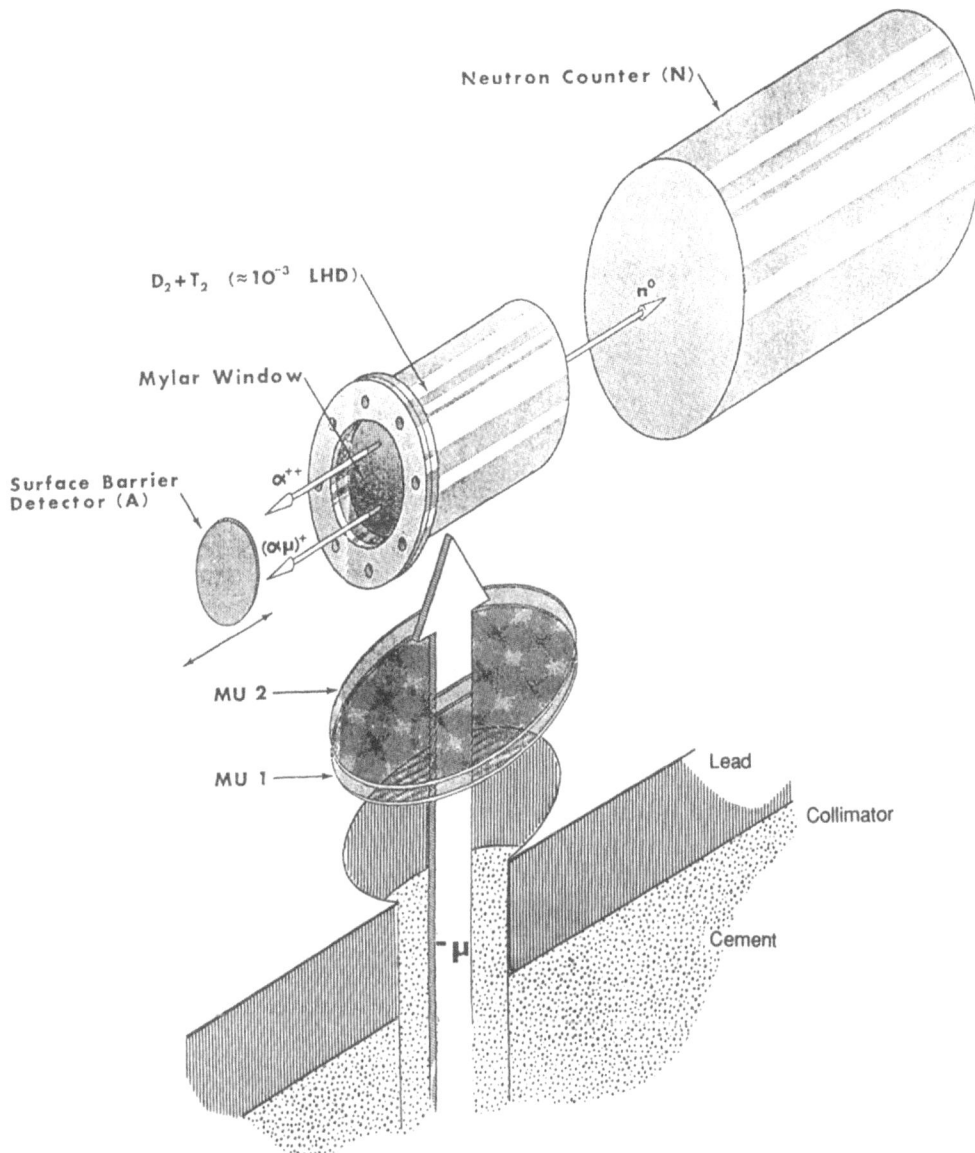

Figure 4. Scheme of the experiment of Jones, Anderson et al. to measure alpha- muon sticking directly by detecting charged and neutral end-products in delayed coincidence.

Before the measurement was made, one could argue that some process not included in equation (6) was making W large, while the sticking term ω_s^o was actually small. It is now clear that the initial sticking is large, around 1%, and that muon stripping as the alpha-muon ion slows down in the gas (density-dependent regeneration R) reduces this value very significantly:

$$\omega_s = \omega_s^o(1 - R\,[\text{density - dependent}]) \tag{10}$$

5. EVALUATION OF STICKING IN HIGH C_T TARGETS

A reliable value of sticking is obtained with the $C_t=0.7$ targets for which the cycling rate is largest and the helium-3 concentration smallest:[17]

$$\omega_s = (0.43 \pm 0.05 (\text{statistical}) \pm 0.06 (\text{systematic}))\% \text{ LAMPF} 1992. \tag{11}$$

Our value may be compared with the latest value from experiments conducted at PSI[18]

$$\omega_s = (0.45 \pm 0.04)\% \text{ PSI} 1992. \tag{12}$$

The results are in very good agreement.

Furthermore, there is good agreement also with our value presented much earlier in 1984 at the Jackson conference:

$$\omega_s = (0.4 \pm 0.1)\% \quad \text{LAMPF} 1984 \tag{13}$$

In 1984, the theoretical expectation was that the sticking probability after muon regeneration (R) due to slowing of the alpha-muonic ion in the deuterium-tritium mixture would have a much larger value:[19]

$$\omega_s = (1-R)\omega_s^o, \omega_s = 0.91 - 0.95\% \text{ Theory} 1984 \tag{14}$$

Owing to the disagreement with theory, our 1984 value was doubted and vigorously challenged as being too small. Indeed, it required nearly a decade for the 1984 LAMPF value to have received full credence.

During this same time interval; the theoretical value for sticking changed appreciably and came down closer to the experimental values (at liquid-hydrogen density):[20]

$$\omega_s = (1-R(\phi))\omega_s^o = 0.58 - 0.65\% \text{ Theory} 1990 \tag{15}$$

Theoretical and various measured $\alpha-\mu$ sticking values versus d-t target density are provided in fig. 5.[21,22] The density (ϕ) dependence of sticking is evidently less than we thought several years ago [1], although it is clearly present due to the density-dependence of the regeneration term ($R(\phi)$ in eq. (14)).

The long-standing disagreement between theoretical and experimental values first enunciated by our group in 1984 means that the muon-catalyzed fusion process is not yet fully understood. We have proposed an experiment to measure ω_s at extremely high densities ($\phi=2.3$ liquid-hydrogen density, LHD) to bring more light on the subject.[23] At such high densities, especially in non-equilibrated mixtures, the cycling rate will be very large and the muon-loss probability will be completely dominated by ω_s as competing channels become insignificant relative to the main d-t cycle (see fig. 2):

$$\omega_s \cong W \text{ (observed sticking) at} \phi = 2.3 \text{ LHD} \tag{16}$$

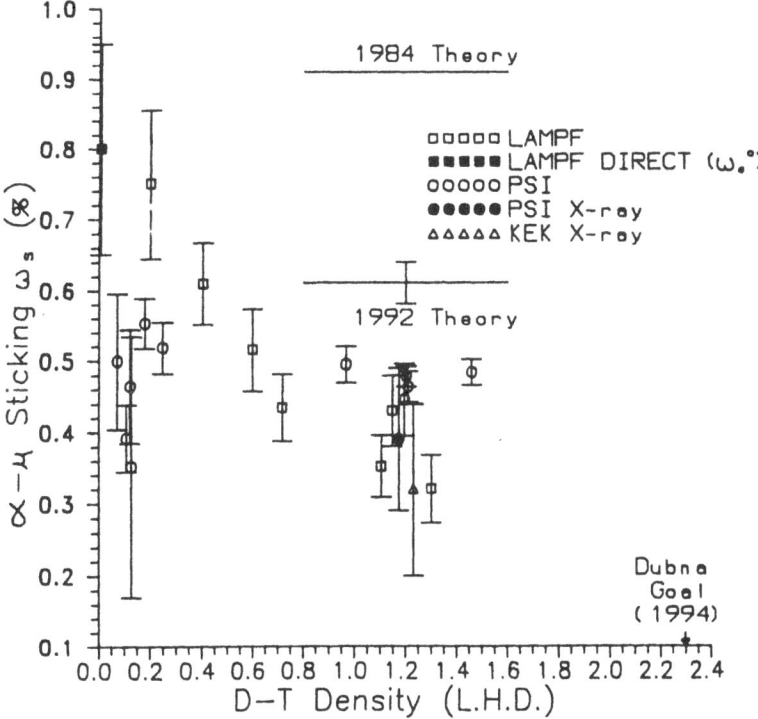

Figure 5. Juxtaposition of theoretical and measured α–μ sticking values versus d-t target density. The extent of the density dependence of ω_s and its value at very high densities can only be resolved by further experiments.

The extent of the density-dependence of ω_s and its value at very high densities can only be resolved by further experiments.

Fig. 6 shows the striking increase of the muon-catalyzed fusion yield with increasing d-t density. We wish to evaluate the fusion yield when the density range of the figure is nearly doubled. The proposed experiment will likely be carried out in Dubna, Russia, as a joint Russian-American-British-Italian-Dutch collaboration.

7. EXPECTED FUSION YIELDS AND CONCLUSIONS

Since the catalysis cycling rate ω_s increases whereas the overall muon loss probability W decreases with increasing d-t density, we expect from equation (1) that the fusion yield will grow rapidly with increasing density. This is indeed the case as demonstrated in Figure 6. In fact, the observed yield exceeds theoretical expectations of a few years ago by a comfortable margin.

But is it enough? Yuri Petrov has shown (24) that a hybrid reactor using cf in conjunction with fission processes could generate power commercially when the cf yield reaches about 150 fusions per muon. Figure 6 shows that this level has been reached in experiments. However, one suspects that fusion-fission hybrid reactors will remain unattractive as long as uranium remains inexpensive, particularly since hybrids partake of many of the problems of conventional fission reactors.

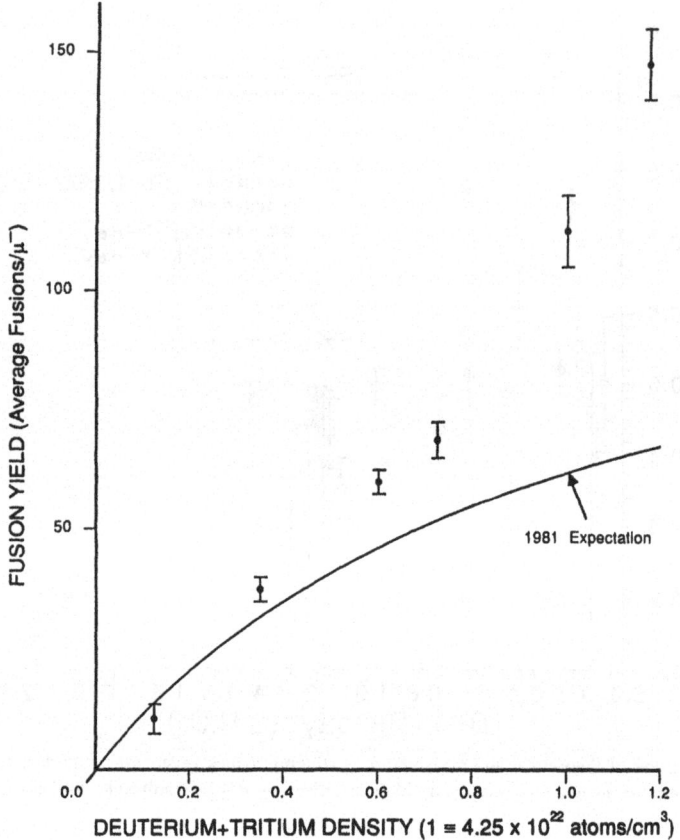

Figure 6. The average number of muon-catalyzed d-t fusion cycles observed by the BYU/Idaho/Los Alamos collaboration as a function of density, for cold (T<100 K), equimolar deuterium-tritium mixtures. Note that the observed yield exceeds theoretical predictions at high densities.

To produce power commercially using μcf alone would probably require an order-of-magnitude increase in the yield per muon, assuming current technology for muon production. Such a jump seems unlikely now because of the barrier imposed by alpha-muon sticking.

However, one way of dramatically decreasing sticking looks quite promising, although very difficult to realize. Menshikov and Ponomarev[25] observed that the stopping power in a plasma medium is much less than in molecular hydrogen isotopes. Hence, alpha-muon ion slowing in plasmas is by nuclear collisions. Consequently, alpha-muon stripping will be greatly enhanced. Eliezer and Henis[26] have recently followed up on this notion and conclude that for a plasma at T ~ 100 eV, the final sticking fraction can be reduced to approximately 10^{-7}. This would allow millions of fusion per muon, except that the dtμ-formation rate becomes small for plasma conditions.

A possible solution to this dilemma lies in the suggestion of a "dusty" tepid plasma.[27] In this scenario, D-T droplets would be injected into a tepid (T ~ 100 eV) plasma to form the dusty plasma. In the regions of cool D-T (T < 4 eV, which would last for perhaps milliseconds), dt -molecular formation would take place very rapidly. Fusion would rapidly ensue (in picoseconds) following dtμ formation. Then muon-alpha ions would often travel into the tepid plasma region where muon-stripping would very likely occur. The freed muon would preferentially stop once more in the cool D-T regions, to continue the muon catalysis cycle. It appears that this process may provide conditions for more than 1000 fusions per muon, which is in the range of interest for serious energy considerations. However, the daunting technical problem of maintaining the cool D-T regions in the tepid plasma environment remains unsolved. Experiments to explore these predictions are under discussion.

ACKNOWLEDGMENTS

Discussions with numerous μcf colleagues have contributed to this summary. I particularly acknowledge valuable input from Ken Nagamine, Alan Anderson, Jim Cohen, Jim Bradbury, Michael Paciotti, Stuart Taylor, Antonio Bertin, Gus Caffrey, John Davies, Mel Leon, Claude Petitjean, Yuri Petrov, Leonid Ponomarev, Johann Rafelski, Antonio Vitale, Alexi Vorobyov, Lali Chatterjee, Piotr Froelich, David Jackson, V. Bystritsky, Gerry Hale, Tom Case, and Glen Marshall.

REFERENCES

1. S.E. Jones, et al., *Phys. Rev. Lett.* **56**, 588 (1986); S.E. Jones, *Nature* **321**, 127 (1986).
2. P. Kammel, *Muon Catalyzed Fusion* **3**, 483 (1988).
3. D. V. Balin, *Muon Catalyzed Fusion* **2**, 163 (1988).
4. L. I. Menshikov and L. I. Ponomarev, *Pis ma Zh. Eksp. Teor. Fiz.* **39**, 542 (1984) [*Sov. Phys. JETP Lett.* **39**, 663 (1984), and **42**, 13 (1985)].
5. W. H. Breunlich et al., *Phys. Rev. Lett.* **58** (1987) 137.
6. S. S. Gerstein and L. I. Ponomarev, *Phys. Lett.* **72B**, 80 (1977); L. I. Menshikov and L. I. Ponomarev, *Phys. Lett.* **16B**, 141 (1986).
7. G. Marshall et al., *Hyperfine Interactions*, **82**, 529 (1993); and K. Nagamine et al., *Hyperfine Interactions*, **82**, 539 (1993).
8. L. I. Menshikov and L. I. Ponomarev, *Phys. Lett.* **167B**, 141 (1986).
9. S. S. Gerstein and L. I. Ponomarev, *Phys. Lett.* **72B**, 80 (1977).
10. S. I. Vinitskii et al., *Sov. Phys. JETP*, **47**, 444 (1978).
11. C. Petitjean et al., *Muon Cat. Fusion* **2**, 37 (1988).
12. L. N. Somov, *Muon Catalyzed Fusion* **3**, 465 (1988).
13. S. E. Jones, et al., *Muon Catalyzed Fusion* **1**, 21 (1987); R. Gajewski and S. E. Jones, *Muon Catalyzed Fusion* **2**, 56 (1988).
14. M. A. Paciotti, et al., *AIP Conf. Proc.*, S. E. Jones, J. Rafelski and H. J. Monkhorst eds., **181**, 38 (1989).
15. J. S. Cohen, *Muon Catalyzed Fusion* **1**, 179 (1987).
16. L. I. Ponomarev, *Muon Catalyzed Fusion* **3**, 629 (1988).
17. S. E. Jones, S. F. Taylor, and A. N. Anderson, *Hyperfine Int.* **82**, 303 (1993).
18. C. Petitjean, et al., *Hyperfine Int.* **82**, 273 (1993).
19. S. S. Gerstein et al., *Zh. Eksp. Teor. Fiz.* **80** (1981) 1960 [*Sov. Phys. JETP* **53**, (1981) 872]; L. Bracci and G. Fiorentini, *Nucl. Phys. A* **364** (1981) 383.
20. J. S. Cohen, *Phys. Rev. Lett.* **58**, 1999 (1987).
21. W. H. Breunlich, P. Kammel, J. Cohen and M. Leon, *Ann. Rev. Nucl. Sci.* **39** (1989).
22. L. I. Ponomarev, *Contemp. Phys.* **31**, 219 (1990).
23. S. E. Jones, Ettore Majorana Series #33, eds. B. Brunelli and G. G. Leotta (Plenum Press, New York, 1987), p. 73.
24. Yu. V. Petrov, *Nature* **285**, 466 (1980); and *Muon Catalyzed Fusion* **3**, 525 (1988).
25. L. I. Menshikov and V. A. Shakirov, Muon Catalysis in Dense Low-Temperature Plasmas, *Zh. Eksp. Teor. Fiz.* **95**, 458 (1989).
26. S. Eliezer and Z. Henis, *Fusion Tech.* **26**, 50 (1994).
27. V. M. Bystritsky and J. Cohen, Private Communications, 1994.

EXPERIMENTAL INVESTIGATION OF MUON-CATALYZED FUSION IN MIXTURES OF HYDROGEN ISOTOPES

V. M. Bystritsky

Joint Institute for Nuclear Research
141980 Dubna, Russia

ABSTRACT

The brief review of experimental investigations devoted to study of mu-atomic and mu-molecular processes in the hydrogen isotope mixture is given. The open question in the muon catalyzed fusion (μCF) field requiring explanations or further investigations are discussed.

One of the possible μCF applications to energy production is considered. A tentative programme of further investigations of mu-atomic and mu-molecular processes in Russian research centers is given.

The history of muon-catalyzed nuclear fusion reactions of hydrogen isotopes (μCF) dates back to 1947, when I. M. Frank put forward a hypothesis[1] of pdμ molecule formation followed by a fusion reaction in it. The hypothesis arose from interpretation of experimental results about the $\pi^- \to \mu^- + \tilde{\nu}\mu$ decay. Undoubtedly, it was tempting to use muon catalysis for creating an alternative power source operating at normal temperatures[2] (as is known, fusion reactions with hydrogen isotopes proceed at a noticeable rate at temperatures T>10 keV, because it is necessary to overcome the Coulomb repulsion barrier between charged particles). In 1948 A. D. Sakharov[3] considered muon catalysis in deuterium and estimated the lifetime of a ddμ molecule. In 1954 L. B. Zeldovich independently inferred that muon could induce nuclear fusion reaction of hydrogen isotopes.[4] It was shown in Ref. 4 that the cross section for an in-flight nuclear reaction (e.g. $t\mu + d \to\ ^4He + n + \mu^-$) is small, and formation of muonic molecules practically always results in a nuclear fusion reaction of hydrogen isotopes. The phenomenon of muon catalysis was first observed experimentally in a H_2+D_2 mixture by L. Alvarez in 1957.[5] This discovery encouraged a great deal of detailed theoretical and experimental investigations embracing all kinds of mu-atomic and mu-molecular processes in mixtures hydrogen isotopes (see reviews Refs. 67).

Interest in mu-atomic and mu-molecular processes in mixtures hydrogen isotopes arises from the following.

First, mu-atomic and mu-molecular processes largely control the efficiency of muon catalysis of nuclear fusion reaction and further steps of negative muon capture by nuclei of hydrogen isotopes. Secondly, analyzing time and energy distributions of products from nuclear fusion reactions in muonic molecules one can gain information both on characteristics of strong interaction between hydrogen isotope nuclei at energies close to zero and on level energies in muonic molecules, which allows the contribution from vacuum polarisation to be found. Thirdly, X-ray radiation from mu-atoms can yield information not only on the nuclear structure but also on the Lamb shift of energy levels in mu-atoms, which allows quantum mechanics methods to be tested. From the theoretical point of view interest in studying all steps of muon catalysis arises from the fact that the quantum mechanics problem of three bodies interacting by the Coulomb law appears in its pure form in collision of hydrogen mu-atoms.

Mu-atomic and mu-molecular processes induced by a negative muon stopped in a hydrogen isotopes mixture occur in the following order. Being small ($a_\mu = \hbar^2/m_\mu e^2 \approx 2.56 \times 10^{-11}$ cm is the Born radius of a hydrogen mu-atom, m_μ is the muon mass) and electrically neutral, mu-atoms of hydrogen isotopes arising from Coulomb capture of muons easily penetrate through electron shells of environment molecules and approach their nuclei as close as about a mesoatomic unit length a_μ. As a result, a series of mu-atomic and mu-molecular processes occur.

1. Elastic scattering of muonic atoms by nuclei of hydrogen isotopes.

2. Transitions between hyperfine structure levels in muonic atoms.

3. Muon transfer from light to heavy hydrogen isotopes (isotope exchange reactions).

4. Muon transfer from nuclei of hydrogen isotopes to nuclei of elements with charge Z>1.

5. Formation of muonic molecules $pp\mu$, $pd\mu$, $dd\mu$, $dt\mu$, $tt\mu$ and $pt\mu$.

Underbarrier nuclear reaction of hydrogen isotope fusion take place in muonic molecules. As a result a muon can get free and induce another muon catalysis chain

$$[H_2 + D_2]\mu^- \longrightarrow \begin{array}{c} \to p\mu \\ \to d\mu \end{array} \lambda_{pd} \downarrow \xrightarrow{\lambda_{pd\mu}} pd\mu \xrightarrow{\lambda_f^{pd}} \begin{array}{c} \xrightarrow{\alpha} {}^3He + \mu^- \\ \xrightarrow{1-\alpha} {}^3He\mu + \gamma \end{array} \text{(5.5 MeV)} \quad (1)$$

$$[D_2]\mu^- \longrightarrow \begin{array}{c} \to (d\mu)_{3/2} \\ \to (d\mu)_{1/2} \end{array} \lambda d \downarrow \xrightarrow{\lambda_{dd\mu}} dd\mu \xrightarrow{\lambda_f^{dd}} \begin{array}{c} \xrightarrow{\beta} \begin{array}{c} \xrightarrow{1-\omega_d^0} {}^3He + n + \mu^- \quad (3.3 \text{ MeV}) \\ \xrightarrow{\omega_d^0} {}^3He\mu + n \end{array} \\ \xrightarrow{1-\beta} \begin{array}{c} \xrightarrow{1-\tilde{\omega}_d} t + p + \mu^- \quad (4.03 \text{ MeV}) \\ \xrightarrow{\tilde{\omega}_d} t_\mu + p \end{array} \end{array} \quad (2)$$

$$[D_2 + T_2]\mu^- \longrightarrow \begin{array}{c} \to d\mu \\ \to t\mu \end{array} \lambda_{dt} \downarrow \xrightarrow{\lambda_{dt\mu}} dt\mu \xrightarrow{\lambda_f^{dt}} \begin{array}{c} \xrightarrow{1-\omega_s^0} {}^4He + n + \mu^- \quad (17.6 \text{ MeV}) \\ \xrightarrow{\omega_s^0} {}^4He\mu + n \quad (14.1 \text{ MeV}) \end{array} \quad (3)$$

$$[T_2]\mu^- \longrightarrow \begin{array}{c} \to (t\mu)_{F=1} \\ \to (t\mu)_{F=0} \end{array} \lambda t \downarrow \xrightarrow{\lambda_{tt\mu}} tt\mu \xrightarrow{\lambda_f^{tt}} \begin{array}{c} \xrightarrow{1-\omega_t} {}^4He + 2n + \mu^- \quad (11.33 \text{ MeV}) \\ \xrightarrow{\omega_t} {}^4He\mu + 2n \end{array} \quad (4)$$

$$[H_2 + T_2]\mu^- \longrightarrow \begin{matrix} \to (p\mu) \\ \to (t\mu) \end{matrix} \lambda_{pt} \downarrow \xrightarrow{\lambda_{pt\mu}} pt\mu \xrightarrow{\lambda_f^{pt}} \begin{matrix} \to ^4\text{He} + \gamma & (19.8 \text{ MeV}) \\ \to ^4\text{He} + e^+ + e^- & (18.8 \text{ MeV}) \\ \to ^4\text{He} + \mu^- & (19.8 \text{ MeV}) \\ \to ^4\text{He} + \mu^- + \gamma & (19.8 \text{ MeV}) \end{matrix} \quad (5)$$

- $\lambda_p d$ is the rate of muon transfer from $p\mu$ atoms to deuterons; $\lambda_d t$ is the rate of muon transfer from $d\mu$ atoms to tritons;

- $\lambda_f^{pd}, \lambda_f^{dd}, \lambda_f^{dt}, \lambda_f^{tt}, \lambda_f^{pt}$ are the nuclear fusion rates in $pd\mu$, $dd\mu$, $dt\mu$, $tt\mu$ and $pt\mu$ molecules respectively;

- $\lambda_{pd\mu}, \lambda_{dd\mu}, \lambda_{dt\mu}, \lambda_{tt\mu}, \lambda_{pt\mu}$ are formation rates of $pd\mu$, $dd\mu$, $dt\mu$, $tt\mu$ and $pt\mu$ molecules respectively;

- λ_d, λ_t are rates of the transition between hyperfine structure levels in $d\mu$ atoms ($F = 3/2 \to F = 1/2$, F is the total spin of the $d\mu$ atoms) and in $t\mu$ atoms ($F = 1 \to F = 0$);

- $\omega_d^0, \omega_s^0, \omega_t^0$ are the coefficients of muon sticking to He nuclei in reactions (2)–(4);

- β is the relative probability for a neutron-yielding fusion reaction in a $dd\mu$ molecule;

- α is the relative probability for a fusion reaction in a $dd\mu$ molecule with the formation of a 5.5 MeV γ-quantum;

- Formation of $pp\mu$ molecules does not lead to the nuclear fusion reaction $p + p \to d + e^+ + \nu_e$ because the muon lifetime is very short ($\tau_\mu = 2.2 \times 10^{-6}$ s and the pp fusion probability is determined by weak interaction. The results of first experiments with a H_2+D_2 mixture and pure deuterium indicate that it is impossible to create a power source based on muon catalysis because mu-molecular formation rates are comparable with the free muon decay rate ($\lambda_0 = 0.455 \times 10^6$ s^{-1}) and sometimes even smaller.[8] This conclusion moderated enthusiasm for muon catalysis investigations.

Yet, one question remained open.

The $dd\mu$ molecular formation rates measured with bubble chambers[9,10] were almost an order of magnitude smaller than those measured with a diffusion chamber.[11]

To explain this sharp discrepancy of experimental values, it was assumed[12] that there is a resonant $dd\mu$ molecular formation mechanism. Essentially, it means that $dd\mu$ molecules are formed in excited vibrational ($\nu = 1$) and rotational (L = 1) states and the binding energy of $dd\mu$ molecule is spent to excite vibrational levels of a peculiar kind of a molecule with the $dd\mu$ molecule as its one center and a deuteron as the other. This mechanism is only possible if in the $dd\mu$ system there exists as level with binding energy below the dissociation energy of the D_2 molecule (about 4.6 eV) (see Fig. 1).

In 1978 a group of theorists from Dubna quite accurately calculated the dependence of the $dd\mu$ molecular formation rate on the deuterium temperature.[13] To check the shape of the dependence $\lambda_{dd\mu}(T)$ experimentally was of interest not only for clarifying $dd\mu$ molecular formation mechanisms but also for checking the precise calculations of the three-body system energy. In 1979 an experiment was carried out in Dubna.[14] The goal was to investigate the dependence $\lambda_{dd\mu}(T)$ in the temperature range 120–400 K. The result (Fig. 2) was in good agreement with the theoretical prediction based on the resonant $dd\mu$ molecular formation

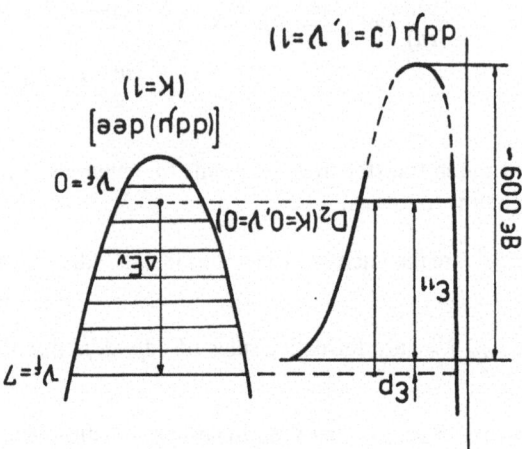

Figure 1. Levels scheme of resonant ddμ formation.

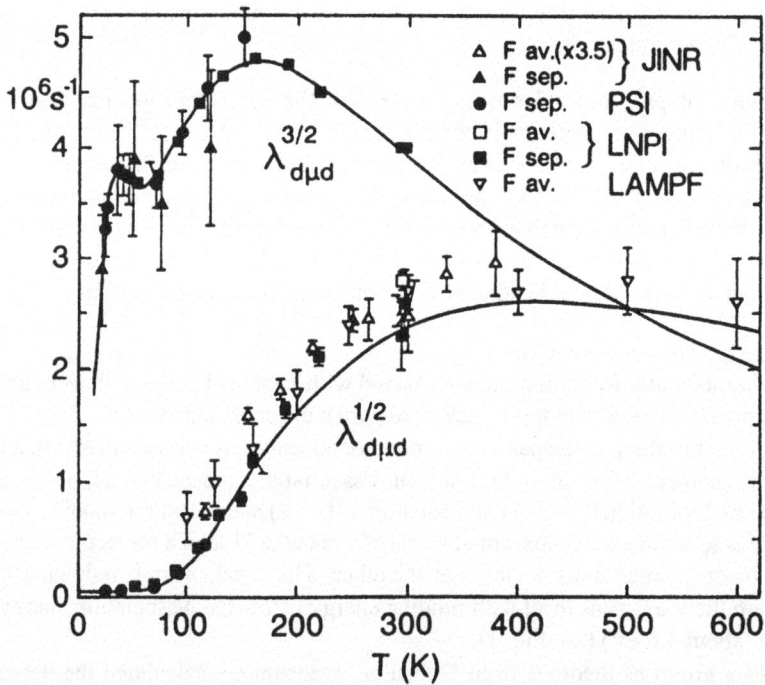

Figure 2. Measured ddμ mesic molecule formation rates for the two dm hfs spin stated vs. D_2 temperature T. The original Dubna data[11,14] were incorrectly calibrated and are therefore renormalized by a factor 3.5. The data sets[14,25,26] (open points) represent hfs averaged γddμ values, while the data[16,27-29] (full dots) and the theory curve[30] are given separately for each hfs state. (Taken from Ref. 63).

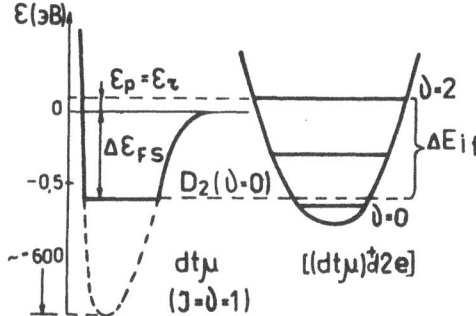

Figure 3. Levels scheme of resonant dtµ formation.

mechanism. It showed that a new phenomenon of resonant formation of deuterium mu-molecules was experimentally discovered.

The discovery became a stimulus for thorough µCF investigations in many laboratories of the world. At the meson factory PSI[15–17] they found hyperfine structure of ddµ molecular energy levels and confirmed resonant dependence of the ddµ molecular formation rate on the deuterium temperature, which was earlier found in Dubna. Besides, they measured rates of ddµ molecular formation from dµ atomic states with $F = 3/2$ and $F = 1/2$ in the temperature range 25–150 K and rates λ_d of transition between hyperfine structure levels in dµ atoms.

All currently available data on ddµ molecular formation rates are given in Table 1 and Fig. 2. Table 1 shows the dynamics of µCF investigations in mixtures of hydrogen isotopes over 10 years (from 1982 till 1992). As evident from Fig. 2, there is good agreement between the experimental results and current theory. A high-pressure ionization chamber[26] was used to measure the neutron-to-proton production branching ratio of the dd fusion reaction ($R = 1.39 \pm 0.4$) and the coefficient of muon sticking to ^{3}He nuclei produced in the reaction ($\omega_d = 0.126 \pm 0.004$). At that time discovery of the resonant ddµ molecular formation mechanism could be regarded as a serious argument for existence of a theoretically predicted dtµ molecular formation mechanism of the same type (see Fig. 3):

$$t\mu + D_2 \to [(dt\mu)dee]^* \quad \text{(process rate } \lambda_{dt\mu-d}) \quad (6)$$

$$t\mu + DT \to [(dt\mu)tee]^* \quad \text{(process rate } \lambda_{dt\mu-t}) \quad (7)$$

The calculation[13] based on the dtµ molecular formation mechanism demonstrated that the dtµ molecular formation rate may be as large as $\lambda_{dt\mu} \sim 10^8$ s^{-1} (for the liquid hydrogen density), i.e. 2 orders of magnitude higher than the muon decay rate. It means that during its lifetime in the D/T mixture one muon can successively induce over 100 nuclear fusion reactions (3) in a dtµ molecule. To check it experimentally was of great interest at that time as it would reveal if mu-catalysis of the dt fusion reaction could be used for power industry purposes.

In 1979 an experiment[31] with a D/T mixture was for the first time carried out in Dubna. Its goals were

1. to find muon catalysis of deuterium and tritium nuclear fusion reactions;

2. to measure the yield of reaction (3), the dtµ molecular formation rate and the rate of muon transfer from dµ atoms to tritium nuclei.

Table 1. The Development of the Knowledge of Crucial μCF Quantities from $\sim$ 1982 to 1992. All rates λ_i are normalized to $\phi = 1$ and are given in units (s^{-1}), the sticking values in (%) and X per muon[63]

Quantity	1982[20]		1992[11–15]	
	Theory	Experiment	Theory	Experiment
λ_a	$\sim 10^{12}$	—	4×10^{12}	$3.93(3) \times 10^{12}$
λ_d	6×10^6	7×10^6	55×10^6	$36(1) \times 10^6$
λ_t	9×10^8	—	13×10^8	$10(2) \times 10^8$
λ_{pd}	1.7×10^{10}	$1.4(2) \times 10^{10}$	1.4×10^{10}	$1.4(2) \times 10^{10}$
λ_{pt}	7.5×10^9	—	5.8×10^9	—
λ_{dt}	1.9×10^8	$2.9(4) \times 10^8$	2.7×10^8 (see Fig. 9)	$2.8(2) \times 10^8$
$\lambda_{d\mu d}^{F=0}$				
25 K	0.04×10^6	0.08×10^6	0.04×10^6	$0.045(3) \times 10^6$
300 K	1×10^6	$0.76(11) \times 10^6$	2.4×10^6	$2.6(1) \times 10^6$
$\lambda_{d\mu d}^{F=1}$	—	—	$3\text{–}4.8 \times 10^6$ (see Fig. 3)	$3\text{–}4.8 \times 10^6$
$\lambda_{p\mu d}$	5.9×10^6	$5.8(3) \times 10^6$	5.63×10^6	$5.6(2) \times 10^6$
$\lambda_{p\mu t}$	6.5×10^6	—	6.38×10^6	$6.0(4) \times 10^6$
$\lambda_{t\mu t}$	3×10^6	—	2.64×10^6	$1.8(6) \times 10^6$
$\lambda_{d\mu t-p}$	—	—	$0\text{–}200 \times 10^8$	$> 3\text{–}15 \times 10^8$
$\lambda_{d\mu t-d}$	$\sim 10^8$	$> 10^8$	$0\text{–}70 \times 10^8$	24 K: $3.3(4) \times 10^8$
$\lambda_{d\mu t-t}$	—	—	$0\text{–}100 \times 10^8$ (see Fig. 7)	$0.1\text{–}5 \times 10^8$
λ_c	—	—	(see Fig. 4)	$0\text{–}2 \times 10^8$
$\lambda_f^{d\mu t}$	$\sim 10^{12}$	$> 10^8$	1.1×10^{12}	$> 10^9$
$\lambda_f^{p\mu d}$	$0.7(2) \times 10^6$	$0.305(10) \times 10^6$	$0.43/0.107 \times 10^6$	$0.41(2)/0.11(1) \times 10^6$
$\lambda_f^{d\mu d}$	$\sim 10^{11}$	—	$0.46(7) \times 10^9$	$0.31(4) \times 10^9$
$\lambda_f^{p\mu t}$	—	—	$\sim 10^4$	$0.065(7) \times 10^6$
$\lambda_f^{t\mu t}$	—	—	10^7	$1.5(2) \times 10^7$
$\omega_s (\%)$	~ 1	—	$0.060\text{–}0.066$ (ϕ dependence, see Fig. 5)	$0.59(7)$
$\omega_d (\%)$	14	> 14	12.3	12.2(3)
$\omega_t (\%)$	5–18	—	5–18	14(3)
X_c^{max}	~ 100	—	160	170(2)

The schemes of processes involving the muons in the D/T mixture and of experimental setup are shown in Figs. 4, 5. after negative muon stop there is shown in Fig. 3. The measurements were carried out at different temperatures of the gaseous D/T mixture (93–613 K), tritium and deuterium concentrations. The tritium concentration varied in the range from 0.81×10^{-2} to 3×10^{-2} and the D/T mixture pressure from 6.6 atm to 66.2 atm. The values of $\lambda_{dt\mu}$ and λ_{dt} derived from analysis of the experimental data were:

$\lambda_{dt\mu} > 10^8$ s^{-1} (at the 90% confidence level);

$\lambda_{dt} = (2.9 \pm 0.4) \times 10^8$ s^{-1}.

They agreed with the theory well.[13]

The first results[31] of μCF investigations in the D/T mixture showed that two most important conditions for effective muon catalysis of deuterium and tritium nuclear fusion are satisfied: the values of $\lambda_{dt\mu}$ and λ_{dt} are large enough.

These conclusions follow from the expressions for the muon catalyzed dt fusion cycle

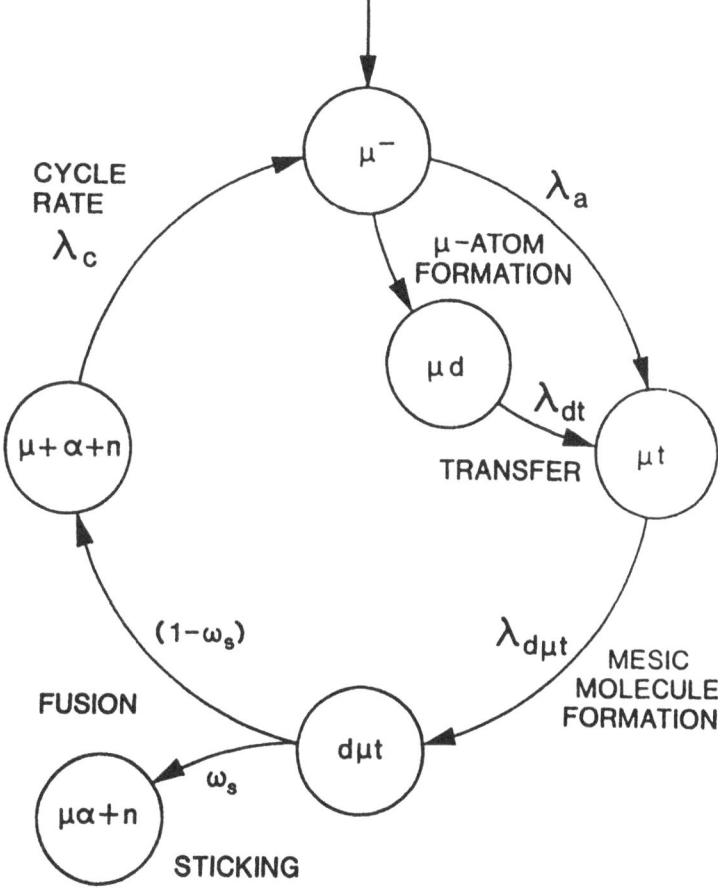

Figure 4. Simplified scheme of the deuterium–tritium fusion cycle.

rate and the reaction yield (X_c) per muon[18,25,32-34]

$$\frac{1}{\lambda_c} \approx q_{1s}\frac{C_d}{\lambda_{dt}C_t} + \frac{1}{\lambda_{dt\mu}C_d} \tag{8}$$

$$\lambda_{dt\mu} = C_d\lambda_{dt\mu-d} + C_t\lambda_{dt\mu-t},$$

$$X_c = \frac{\varphi\lambda_c}{\lambda_0 + \omega\varphi\lambda_c}, \tag{9}$$

where q_{1s} is the probability for transition of dμ atoms from the excited to the ground state; C_d, C_t are the relative deuterium and tritium concentrations ($C_d + C_t = 1$); λ_{dt} is the rate of muon transfer from dμ atoms to tritons reduced to the liquid hydrogen density; $\lambda_{dt\mu}$ is the rate of dtμ molecular formation in the D/T mixture[32] reduced to the liquid hydrogen density ($n_H = 4.25 \times 10^{22}$ cm^{-3}); φ is the D/T mixture density reduced to liquid hydrogen density ($\varphi = 1$ corresponds to $n_H = 4.25 \times 10^{22}$ cm^{-3}); $\omega = (\omega_s +$ other muon loss factors) is the fraction of muons lost in each μCF cycle due to sticking to charge products of fusion reaction

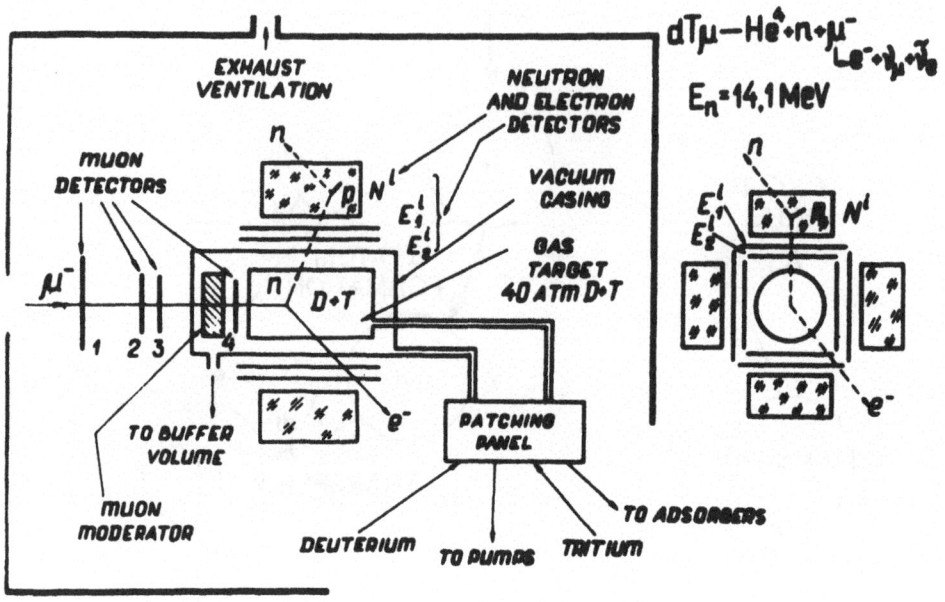

Figure 5. The scheme of the experimental set up.

in ddμ, dtμ, ttμ molecules or transfer from muonic transfer from muonic atoms of hydrogen isotopes to impurity nuclei.

The results obtained in Dubna[3] renewed the interest in more thorough μCF studies in DT mixtures in many laboratories. The goal was to find out if the phenomenon could be of any practical use. Obviously, it made one carry out comprehensive studies of μCF in a deuterium-tritium mixture, namely 1) measure the probability ω_s for muon sticking to ^{4}He nuclei produced in reaction (3) at different DT mixture densities and tritium concentrations, 2) determine optimum experimental conditions to allow large multiplicity of neutrons from the mu-catalyzed dt fusion reaction, 3) investigate the relation between the dtμ molecular formation rate and temperature, D/T mixture density, tritium concentration, 4) investigate the slowing-down kinetics of dμ, tμ, Heμ atoms in the D/T mixture, 5) measure rates of competing ddμ, ttμ molecular formation processes and sticking coefficients ω_d and ω_t under the chosen optimum conditions allowing large multiplicity of neutrons from reaction (3), 6) find rates of muon transfer from dμ and tμ atoms to He nuclei, 7) measure rates of transition between hyperfine structure states in dμ and tμ atoms.

In 1982 extensive studies of μCF in a D/T mixture began at meson factories LAMPF[25,33-36] and PSI.[37-40] Characteristics of μCF in a DT mixture were measured in a wide range of temperatures (15-800 K), densities ($\varphi = 0.12-1.45$) and tritium concentrations. In 1986 the like studies began at the meson factory KEK[41,42] and the PINP phasotron.[26,43] In 1988 the meson factory RAL joined the activities.[44-46]

In Figs. 6-8, there are the results of measuring the cycle rate of muon catalyzed dt fusion reaction from Dubna,[31] PSI[37,47,48,39,40,38] and LAMPF.[25,33-36] It is seen that a) at low D/T mixture temperatures there occurs mainly resonant dtμ molecular formation via process (6) at a rate of about 4×10^8 s^{-1} at $\varphi = 1.2$ and quite strong dependence of the dtμ molecular formation rate on the mixture density; it can be due to three-body collisions (tμ+D$_2$+D$_2$) during formation of dtμ molecules; b) at T>300 K $\lambda_{dt\mu-d}$ is seen to increase with growing temperature, and the contribution to formation of dtμ molecules via process (7) also increases;

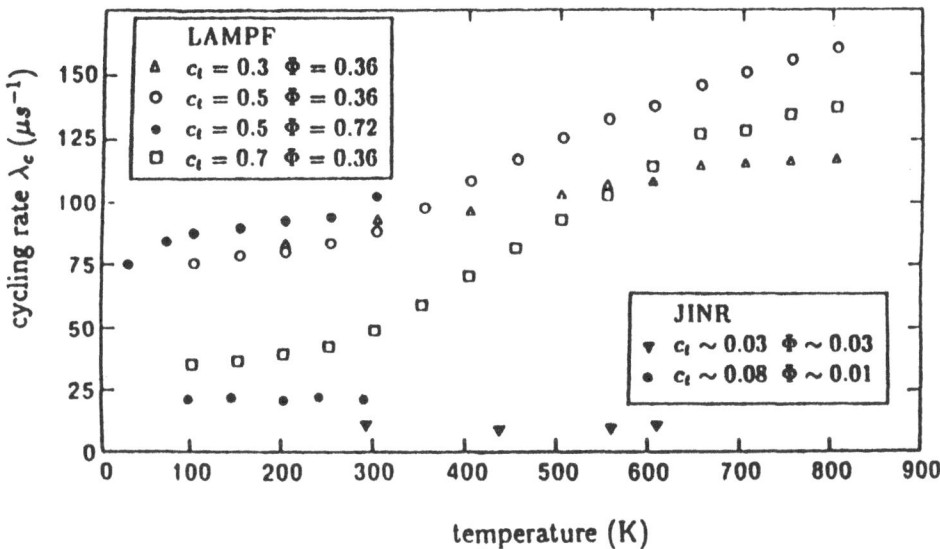

Figure 6. The dependence of normalized dt cycling rate γ_c vs. D/T mixture temperature at different tritium concentrations C_t and density of D/T mixture (The Dubna experimental data were deduced from the original publication [31]).

c) too weak dependence of $\lambda_{dt\mu-d}$ on the mixture density in an unequilibrated mixture at low temperatures remains unclear; d) at tritium concentrations $C_t < 0.2$ the rate of muon transfer from $d\mu$ atoms to tritons is virtually independent of the mixture density and of its molecular composition.

The yield X_c of the dt fusion reaction per muon was found to be

$$X_c = \begin{matrix} 150 \pm 20 \text{ (Ref. 25)} & (\varphi = 1, 2, \text{in liquid } D_2 + DT + T_2) \\ 113 \pm 10 \text{ (Ref. 37)} & (\varphi = 1, 2, \text{in liquid } D_2 + DT + T_2) \\ 124 \pm 10 \text{ (Ref. 48)} & (\varphi = 1, 45, \text{solid } D_2 + T_2) \end{matrix}$$

Figure 9 shows the measured values of the effective muon-to-helium sticking coefficient ω_s ($\omega_s = \omega_s^0(1-R)$, where $\omega_s^0 \approx (0.926 \pm 0.005)\%$[49,50] is the probability for initial sticking of muons to ^{4}He nuclei in reaction (3b), $R \approx 0.3$[51–53] is the portion of muons returned to μCF by being stripped off from Heμ atoms as they slow down: ^{4}Heμ + N $\to$ ^{4}He + μ + N' (N,N' $\equiv p,d,t$), the initial energy of Heμ atoms is 3.5 MeV).

As to measurement of ω_s^0, there are only tentative results from LAMPF[45,46] and RAL[44,57–59]

$$\omega_s^0 = (1.2 \pm 0.2(\text{stat.}) \pm 0.1(\text{syst.}))\% \quad (P = 640 \text{ Torr})$$

$$\omega_s^0 = (0.8 \pm 0.15(\text{stat.}) \pm 0.12(\text{syst.}))\% \quad (P = 1800/940 \text{ Torr})$$

The values found to the effective sticking coefficient ω_s indicate that the largest number of muon catalyzed dt fusion cycles induced by one muon does not exceed $X_c \approx \frac{1}{\omega_s} \approx 200$.

There are different ways proposed to reduce the sticking coefficient ω_s. They involve acceleration of ^{4}Heμ atoms in strong electric and radio frequency fields[60]; muon catalysis in a dense low-temperature plasma under certain conditions, which may allow an increase in

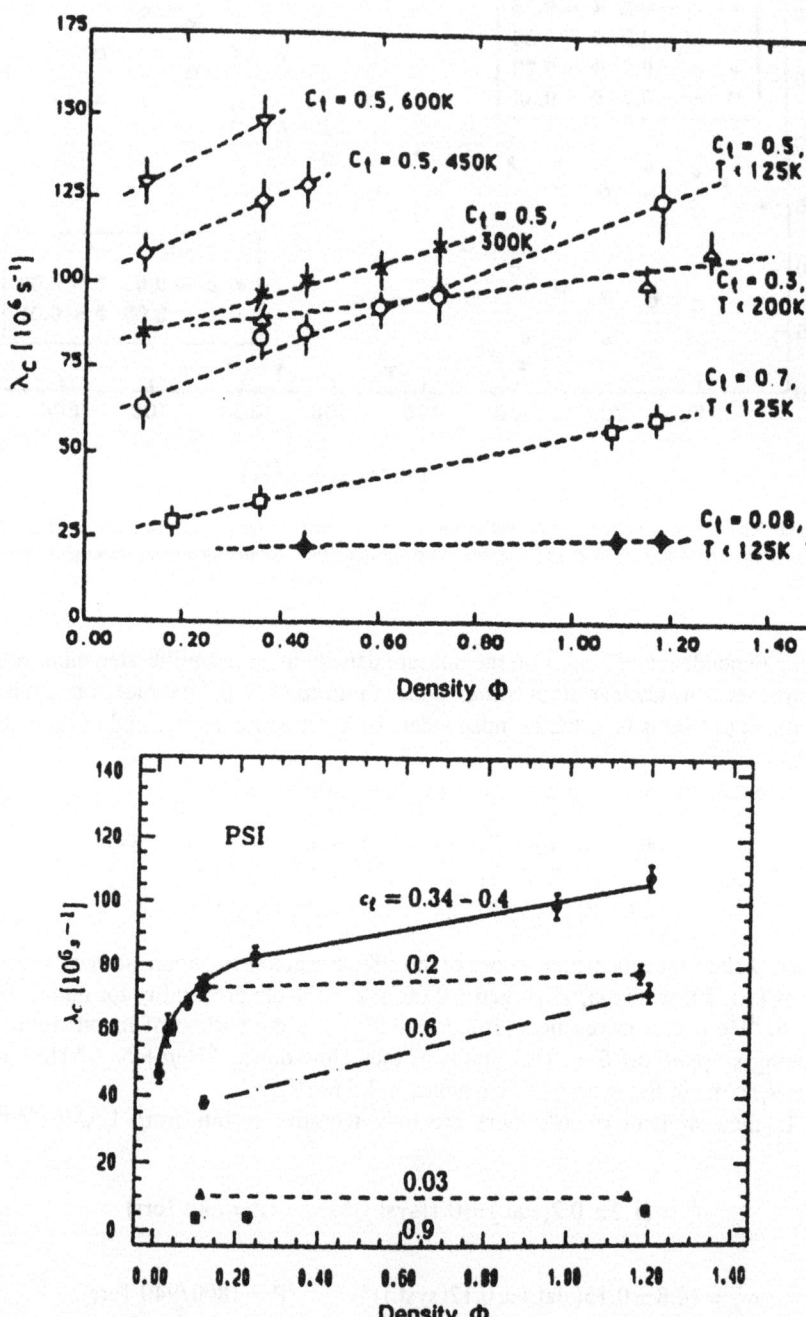

Figure 7. The dependence of normalized dt cycling rate γc vs. D/T mixture density at different tritium concentrations C_t and temperatures. top) LAMPF's data; bottom) PSI's data. Target temperatures are between 20 K and 45 K.

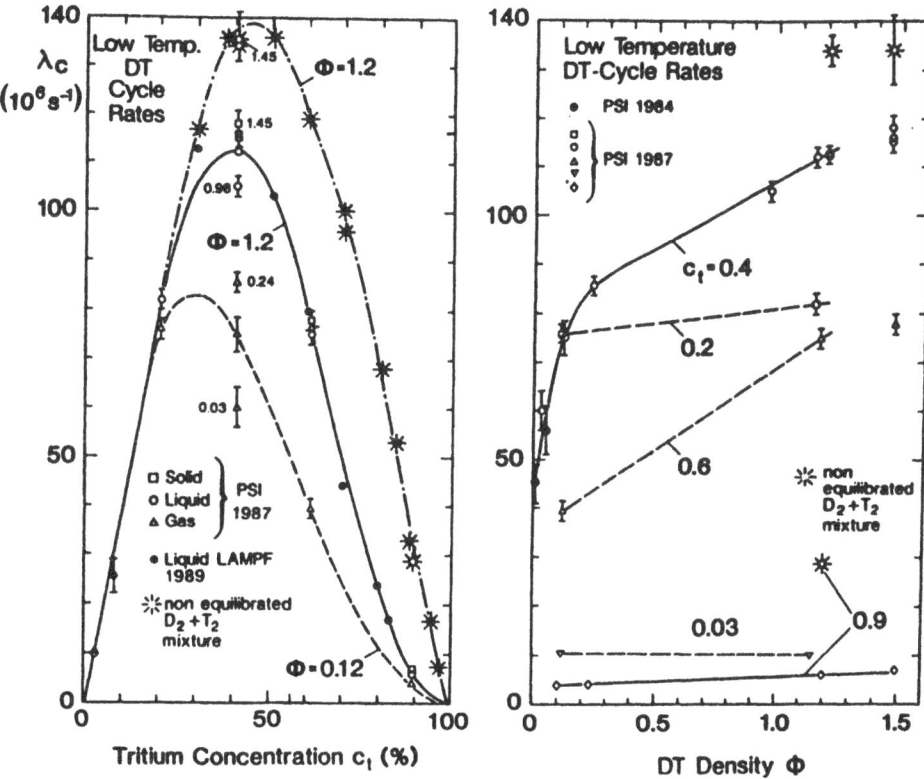

Figure 8. Overview of normalized dt cycle rates measured at temperatures 12 K–35 K. (a) Plotted versus C_t; the curves follow measurements of the same density in liquid mixture $D_2 + T_2$ (dash–dotted) curve, in liquid equilibrated mixtures $D_2 + DT + T_2$ (full curve), and in equilibrated gas (dashed curve); (b) same data plotted versus dt density ϕ for various tritium concentrations C_t (taken from Ref. 63).

the number of dt cycles of muon catalyzed fusion to 10^3.[61] Yet, any of them can hardly be implemented in the near future.

Muon processes in a triple mix of H_2, D_2 and T_2 are of great interest from the point of view of increasing the efficiency of muon catalysis the dt fusion reaction. Recent theoretical calculations[62] indicate that at $t\mu$ atom energies 0.1–2 eV formation of $dt\mu$ molecules in collisions with D_2, DT and HD molecules is of resonant character and occurs at rates well above the rates of these processes at $t\mu$ atom energies below 0.1 eV. The rate of the process

$$t\mu + \mathrm{HD} \longrightarrow [(dt\mu)pee]^* \qquad \text{(process rate } \lambda_{dt\mu-p}) \tag{10}$$

is much higher than the rates of processes (6) and (7) (see Fig. 10). Thus, there arises an opportunity to increase the efficiency of muon catalysis the dt fusion reaction without heating the D/T mixture to a high temperature (about 2×10^4 K).[23]

In this case the tritium concentration is substantially lower than in the binary DT mixture, which also increases safety of the apparatus. Yet, the final conclusion about using the triple mix can only be drawn after correct calculations and experimental studies of all competing processes (e.g. formation of $dd\mu$, $pt\mu$ and $pd\mu$ molecules).

During all the above-mentioned μCF studies a lot of experiments on investigation of mu-atomic processes were carried out. Their goals were a) to measure cross sections for $p\mu$

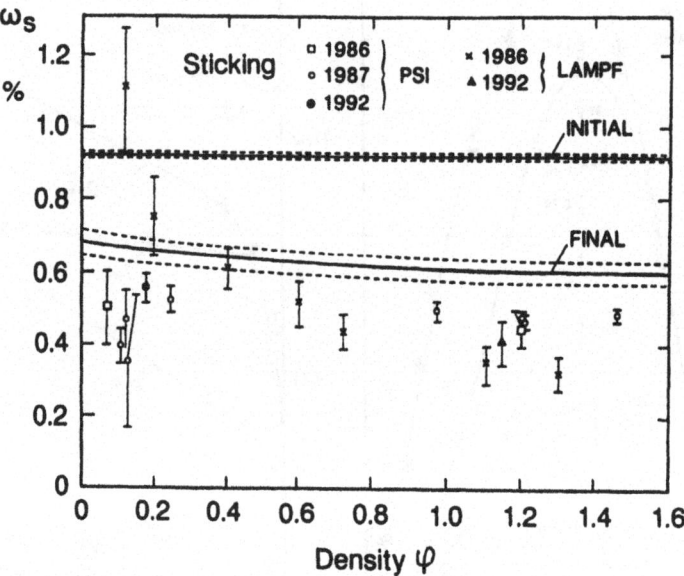

Figure 9. The dependence of effective muon-to-helium sticking coefficient ω_s in (%) vs. D/T mixture density reduced to liquid hydrogen density. PSI data) Refs. 373948555663; LAMPF) Refs. 353654. Solid lines) result of calculations; dashed lines) estimated uncertainties of theory.

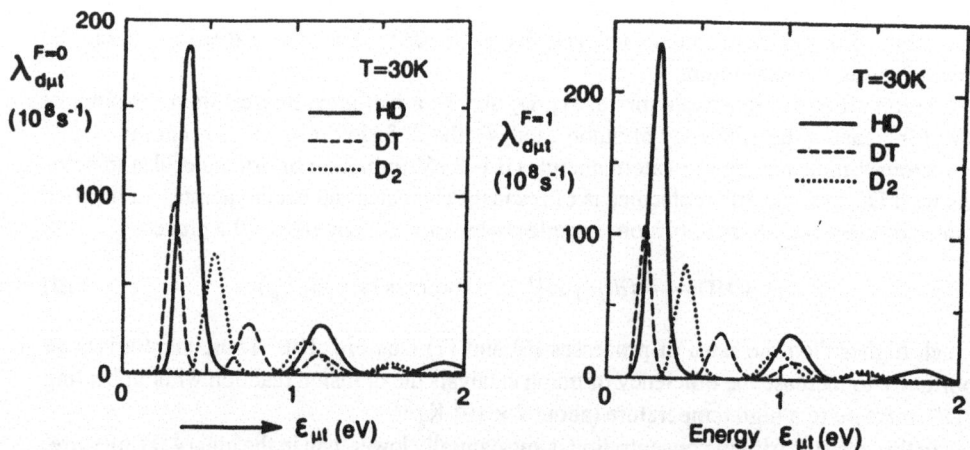

Figure 10. The dependences of resonant dtm formation rates in the collisions of $t\mu$ atoms [in both (hyperfine structure) spin states of $t\mu$ atom, $F = 0$ and $F = 1$] with HD, DT, and D_2 molecules vs. $t\mu$ atom energy.

and dμ atoms in hydrogen and deuterium; b) to measure rates of transition between hyperfine structure levels in pμ, dμ and tμ atoms; c) to study schemes of cascade transitions in muonic atoms of hydrogen isotopes; d) to measure rates of muon capture by protons and deuterons; e) to measure rates of muon transfer from mu-atoms of hydrogen isotopes to nuclei of elements with $Z > 1$; f) to measure rates of isotope exchange p$\mu \to$ dμ, d$\mu \to$ tμ, p$\mu \to$ tμ; g) to study atomic capture of muons in hydrogen mixture with other elements. Information on the results of these investigations can be found in reviews Refs. 21226364.

Another important process competing with dtμ molecular formation deserves attention. It is muon transfer from mu-atoms of hydrogen isotopes to He nuclei. The results of earlier experiments[65–67] and calculations[68] indicate that the muon transfer rate is small or at least comparable with the free muon decay rate.

In 1982 a theoretically predicted[71] phenomenon of molecular charge exchange of ground-state hydrogen isotope mu-atoms on He nuclei was experimentally observed in Dubna.[69,70] This experimental result showed that the transfer rate was more than 2 orders of magnitude higher than previously thought. Earlier in 1977 the muon transfer process from excited hydrogen isotopes to He nuclei was experimentally observed in Dubna.[70] Thus interest in these processes greatly increased because

1. knowing parameters of muon transfer from hydrogen isotope mu-atoms to He nuclei one can correctly interpret the results of experiments on measurement of μCF characteristics in a D/T mixture and estimate the frequency of He removal from the D/T mixture and the necessary degree of its purity during exposures to muon beams (muons captured by He nuclei are lost for μCF);

2. one needs correct description of cascade transition kinetics for hydrogen mu-atoms in mixtures of hydrogen and He.

Up to date this process has been dealt with in quite a lot of investigations. Their results are given in the review Ref. 72. As to theoretical investigations of μCF, new mathematical methods have been developed to solve the problem of three bodies interacting by the Coulomb law (adiabatic representations,[73] variational calculations of sets of basis functions,[74,75] the nonadiabatic coupled-rearrangement-channels approach,[76] the adiabatic hyperspherical basis approach[77]). Practically all characteristics of mu-atomic and mu-molecular processes were calculated by these methods with a high accuracy. It not only provides correct comparison with the experiment and sometimes explanation of some nontrivial experimental results but also allows predicting new specific features of the μCF process (see reviews Refs. 21226364).

Now let us briefly consider the ideas of using μCF for power production. The first practical scheme of μCF-based power generation was proposed by Petrov.[78] According to the estimations,[79–82] the power to be expended for production of one muon stopped in a D/T mixture is about 8 GeV/μ^- nucleon (see Fig. 11). Obviously, the energy released after 130–150 dt cycles of muon-catalyzed fusion in the D/T mixture (150 × 17.6 MeV = 2.6 GeV) cannot compensate for this power expenditure and the energy balance is negative. However, using a ^{238}U blanket around a D/T target[78,81] one can have a positive energy balance (it will be shown below). At present there are different μCF application projects.[81,83–86]

A mesocatalytic hybrid reactor[78,81,84] (MCHR) consists of 1) a pion-producing target (Li, Be, C), 2) a converter for pion to muon conversion ($\pi^- \to \mu^- + \tilde{\nu}_\mu$), 3) a synthesizer (a target filled with a D/T mixture), 4) a ^{238}U blanket, 5) solenoid's magnets 6) magnetic mirror (see Fig. 12). A scheme of mesocatalytic power generation is shown in Fig. 13. It is supposed to produce π^--mesons in an accelerator of heavy hydrogen isotopes (d, t) with energy E = 1–2

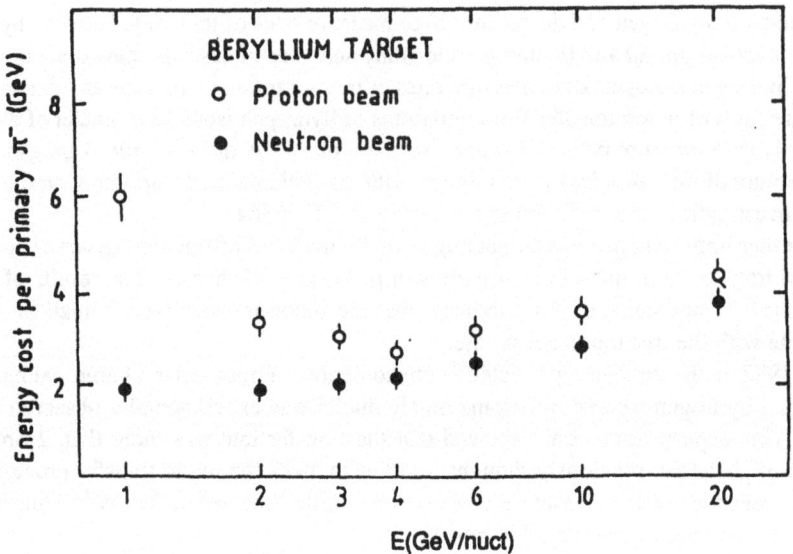

Figure 11. Energy needed to produce a negative pion within an infinite beryllium target as a function of the projectile particle kinetic energy (proton and neutron).

GeV/nucleon. The yield of π^--mesons for the above-mentioned targets is $Y_{\pi^-} \approx 0.17\pi^-$/GeV nucleon. Muons, arising with a probability of 75% from the in-flight decay of π^--mesons in the converter, arrive at the synthesizer ($Y_\mu \approx 0.13\mu^-$/GeV nucleon). In the synthesizer each muon induces about 100 dt fusion cycles, while each 14.1 MeV neutron induces 0.86 fission of ^{238}U nuclei in the ^{238}U–Li blanket (fission energy of ^{238}U and ^{239}Pu nuclei is about 200 MeV) and produces 2.3 ^{239}Pu nuclei and 0.7 tritium nucleus. Then ^{238}Pu produced in this way is burnt in a thermal neutron reactor (the fuel breeding coefficient for this type of reactor is $\psi = 1.7$ fiss./Pu). Then the total amount of electric energy released via the mesocatalytic channel (MC) is

$$E_{MC} = (13 \cdot 0.86 \cdot 0.4 + 13 \cdot 2.3 \cdot 1.7 \cdot 0.34) \cdot 200 \text{ MeV} = 44.3 \text{GeV/GeV nucleon}$$

(the efficiency of thermal-to-electric energy conversion is supposed to be 0.4 for fast reactors and 0.34 for thermal ones).

Since only 30% of the initial deuteron beam energy are expended for production of π^--mesons, the rest of the beam energy is spent for fission of ^{238}U nuclei and production of ^{239}Pu nuclei (each nucleon of the 1 GeV deuteron beam induces 13 fissions and produces 45 ^{239}Pu nuclei in a uranium target, it is normal electronuclear breeding). In this case the total amount of electric energy released via the electronuclear channel (EN) is

$$E_{EN} = 0.7(13 \cdot 0.4 + 45 \cdot 1.7 \cdot 0.34) \cdot 200 \text{ MeV} = 4.4 \text{ GeV/GeV nucleon}$$

The total energy released via both channels is 8.7 GeV/GeV nucleon. It is seen that the MCHR almost doubles the energy generated via the EN channel with the deuteron beam of the same energy. The efficiency of high-current accelerators taken to be 0.6,[81] it is seen that power supply of the accelerator and its auxiliary equipment is about 20% of total released energy. The electric energy generated by burning plutonium at thermal nuclear power plants (3.6 + 3.5 = 7.1 GeV) is about 4 times as large as the energy spent for its production. Hence it

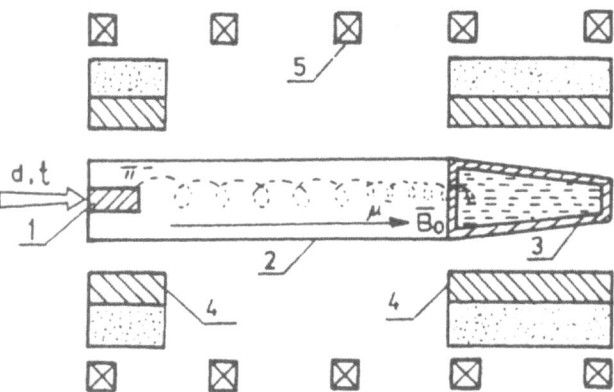

Figure 12. Conceptual version of a hybrid mesocatalytic reactor. 1) pion-producing target; 2) convertor; 3) synthesizer; 4) blanket; 5) superconducting solenoid.

follows that the MCHR can produce nuclear fuel for 4 thermal nuclear reactors of the same power as the MCHR. This multiplier is 5 times as large as that for modern fast breeders of the same power. A 1 GW MCHR requires an accelerator with deuteron current about 50 mA and deuteron beam power about 200 MW ($E_d = 4$ GeV).

Thus, the mesocatalytic method for energy generation and plutonium production is comparable with other current methods (fast breeders, thermonuclear breeders, electronuclear breeders). Yet, apart from cost and technological problems there are many problems related to reprocessing of tritium, radiation damage of the deuterium-tritium synthesizer shielding, efficiency of the whole system with respect to other nuclear breeding methods.

At present there are also projects of using μCF in a simplified MCHR in order to create powerful sources of 14.1 MeV neutrons for transmutation of nuclear wastes and for examination of radiation resistance of various materials.[87,88]

In summary, it is seen that a great deal of μCF investigations have been carried out. They allowed much better comprehension of all processes proceeding after stops of negative muons in pure hydrogen isotopes and their mixtures. Yet, there remain problems which call for explanation or investigation.

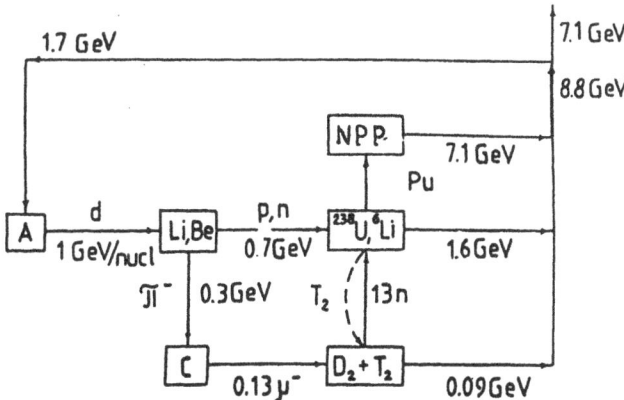

Figure 13. Scheme of the mesocatalytic power generation. A) accelerator; Li, Be) pion-producing target; C) convertor ^{238}U, ^{6}Li) blanket; $D_2 + T_2$) synthesizer; NPP) nuclear power plant.

1. It is necessary to discriminate between contributions from two-body and three-body collisions to dtμ molecular formation in a wide temperature range.

2. Nonlinear dependence of the dt cycle rate of muon-catalyzed fusion on the D/T mixture density at low temperatures needs further investigations.

3. It is surprising that at low tritium concentrations ($C_t < 0.2$) in the D/T mixture dependence of the rate for muon transfer from excited dμ atoms to tritium nuclei on the density and molecular composition of the mixture does not manifest itself.

4. There is disagreement between new experimental values of the effective muon-to-helium sticking coefficient ω_s and theoretical results.

5. There is disagreement both between rates for ddμ molecular formation in collisions of dμ atoms and HD molecules measured in different experiments and between these experimental results and theoretical predictions.

6. Difference is observed between experimental and theoretical rates for transition between hyperfine structure states in dμ atoms.

7. Precise measurement of the initial sticking coefficient ω_s^0 must be carried out.

8. It is of great importance to study kinetics of mu-atomic and mu-molecular processes in a triple mix of H_2, D_2 and T_2 and to determine multiplicity of μCF cycles.

9. It is necessary to investigate μCF in pure tritium in order to gain highly accurate information on μCF characteristics.

10. The dtμ molecular formation rate and the effective sticking coefficient ω_s must be measured at high DT mixture densities ($\varphi \sim 2.5$).

11. Direct measurement of cross sections for scattering of dμ and tμ atoms by H_2 molecules must be carried out (a minimum in the scattering cross sections ($\sigma \sim 10^{-21}$ cm^2) at collision energies 3–5 eV—the so-called Ramsauer–Taundsend effect—is theoretically predicted).

12. Kinetics of excited mesohydrogen in a hydrogen-helium mixture remains almost unstudied so far.

13. There is disagreement both between cross sections for scattering of pμ atoms in hydrogen and of dμ atoms in deuterium measured in different experiments and between these experimental results and theoretical predictions.

14. Experimental neutron-to-proton production branching ratios of the dd fusion reaction lack any satisfactory explanation.

This list is far from being complete, of course. Many problems are omitted since they require detailed explanation of all aspects related to kinematics of mu-atomic and mu-molecular processes, which is beyond the scope of the present paper.

So far we have been dealing with muon catalysis of nuclear fusion reactions in charge-symmetrical muonic molecules of hydrogen isotopes. It is also of interest to study nuclear fusion reactions in charge-asymmetrical muonic molecules like ^{3,4}Hedμ, ^{3,4}Hetμ, ^{6,7}Lidμ, ^{6,7}Litμ, formed by muon transfer from dμ, tμ atoms to He and Li nuclei.[89]

Finally, we point out the milestones in the programme — further studies of mu-atomic and mu-molecular processes in Russian research centers (JINR, SPINP, KIAE).

1. Direct measurement of the muon-to-helium sticking coefficient ω_s at high densities of a D/T mixture (liquid, T=21 K) and high tritium concentrations ($C_t = 25 - 75\%$).

2. Direct measurement of the neutron yield in triple mix of H_2, D_2 and T_2 (to check theoretically supposed existence of 'hot' atoms under conditions corresponding to effective occurence of the dt fusion reaction).

3. Investigation of spin dependence of the ddμ molecular formation rate in solid deuterium at temperatures from 5 to 20 K.

4. Investigation of temperature dependence of the pdμ molecular formation rate in a gaseous H/D mixture and measurement of the rate for fusion from different spin states of the pdμ molecule in a wide range of deuterium concentrations and temperatures.

5. Investigation of temperature dependence of the ddμ molecular formation rate in collisions of dμ atoms with H/D molecules in a wide range of temperatures and deuterium concentrations.

6. Measurement of the rate for formation of ttμ molecules, the rate for the fusion reaction in them, the coefficient ω_t for muon sticking to ^{4}He nuclei resulting from the tt reaction.

7. Investigation of muon transfer from excited mu-atoms of hydrogen isotopes to He nuclei.

8. Measurement of cross sections for scattering of dμ and tμ atoms by H_2 molecules, measurement of the pdμ and ptμ molecular formation rates in the collision energy range 0.1–45 eV using pure "beams" of dμ and tμ atoms emitted to vacuum from a layer of a solid H_2+0.1% D_2 (T_2) mixture (this measurement is planned to be carried out at the accelerator TRIUMF by the collaboration JINR (Russia)—Institute of Physics of Fribourg University (Switzerland)—TRIUMF (Canada)—IPNI (Poland)—University of California Berkeley—PSI-University of Victoria.

9. Investigation of nuclear fusion reactions in charge-asymmetrical muonic molecules like ^{3,4}Hedμ, ^{3,4}Hetμ, ^{6,7}Lidμ, ^{6,7}Litμ. Scientific justification of the project is submitted by the collaboration JINR (Russia)—INFN (Italy)—IPNI (Poland)—SPINP (Russia).[89] Experiments are planned to be performed at the meson factory of RAL.

10. Development of new methods for solving the three-body problem, precise calculation of energy levels in muonic molecules of hydrogen isotopes, characteristics of mu-atomic, mu-molecular and μCF processes. Creation of theoretical models for description of μCF kinetics in systems of different hydrogen isotope composition.

References

[1] F. C. Frank, Nature 160 (1947) 525.
[2] G. M. G. Lattes et al., Nature 160 (1947) 453.
[3] A. D. Sakharov, Lebedev Rep. (1948).
[4] Ya. B. Zeldovich, Doklady AN SSSR, 95 (1954) 493.
[5] L. W. Alvarez et al., Phys. Rev. 105 (1957) 1127.
[6] S. S. Gershtein and Ya. B. Zeldovich, Usp. Fiz. Nauk 71 (1960) 581.
[7] S. S. Gershtein and L. I. Ponomarev, Muon Physics, VIII. Eds. Hughes V. W., Wu C. S. - N. Y., London: Academic Press, 1975, p. 141.
[8] J. D. Jackson, Phys. Rev. 106 (1957) 330.

[9] J. H. Doede, Phys. Rev. 132 (1963) 1782.
[10] J. G. Fetkovich et al., Phys. Rev. Lett. 4 (1960) 570.
[11] V. P. Dzhelepov et al., Zh. Eksp. Teor. Fiz. 50(1966) 1235.
[12] E. A. Vesman, Pisma Zh. Eksp. Teor. Fiz. 5 (1967) 113.
[13] S. I. Vinitsky et al., Zh. Eksp. Teor. Fiz. 74 (1978) 849.
[14] V. M. Bystritsky et al., Zh. Eksp. Teor. Fiz. 76 (1979) 460.
[15] W. H. Breunlich, Nucl. Phys. A353 (1981) 201.
[16] P. J. Kammel et al., Phys. Lett. 112B (1982) 319;
     P. J. Kammel et al., Phys. Rev. A28 (1983) 2611.
[17] P. J. Kammel et al., Atomkernenergie/Kerntechnik 43 (1983) 195.
[18] L. I. Ponomarev, Atomkenenergie/Kerntechnik 43 (1983) 175.
[19] C. Petitjean, Fusion Engineering and Design 11 (1989) 255.
[20] W. H. Breunlich et al., Ann. Rev. Nucl. Part. Sci. 39 (1989) 311.
[21] L. I. Ponomarev, Contemporary Physics 31 (1990) 219.
[22] S. S. Gershtein et al., Sov. Phys. Usp. 33 (1990) 591.
[23] L. I. Ponomarev and C. Petitjean, Fusion Technology 20 (1991) 1022.
[24] V. P. Dzhelepov et al., Sov. Phys. JETP 19 (1964) 1376 and 23 (1966) 820.
[25] S. E. Jones et al., Phys. Rev. Lett. 56 (1986) 588.
[26] A. A. Vorobyov, MCF 2 (1988) 17;
     D. V. Balin et al., MCF 2 (1988) 163;
     D. V. Balin et al., MCF 7 (1992) 1.
[27] V. V. Filchenkov and L. Marczis, MCF 5/6 (1990/91) 499.
[28] H. Zmeskal et al., MCF 1 (1987) 109; Phys. Rev. A42 (1990) 1165.
[29] D. V. Balin et al., MCF 2 (1988) 241.
[30] M. Faifman, MCF 2 (1988) 247.
[31] V. M. Bystritsky et al., Sov. Phys. JETP 53 (1981) 877.
[32] S. S. Gershtein et al., Sov. Phys. JETP 51 (1980) 1053.
[33] S. E. Jones et al., Phys. Rev. Lett. 51 (1983) 1757.
[34] S. E. Jones et al., Atomkernenergie/Kerntechnik 43 (1983) 174.
[35] S. E. Jones et al., MCF 1 (1987) 21.
[36] A. J. Caffrey et al., MCF 1 (1987) 53.
[37] W. H. Breunlich et al., Phys. Rev. Lett. 58 (1987) 329; MCF 1 (1987) 67;
     J. Werner et al., MCF 5/6 (1990/91) 209.
[38] H. Bossy et al., MCF 1 (1987) 115.
[39] C. Petitjean et al., MCF 1 (1987) 89.
[40] W. H. Breunlich et al., Preprint Lawrence Berkeley Laboratory, University of California 21366, 1986.
[41] K. Nagamine et al., MCF 1 (1987) 137.
[42] T. Matsuzaki et al., MCF 2 (1988) 217.
[43] D. V. Balin et al., MCF 1 (1987) 127.
[44] F. D. Brooks et al., MCF 2 (1988) 85.
[45] S. E. Jones et al., AIP 181 (1989) 2.
[46] M. A. Paciotti et al., AIP 181 (1989) 38.
[47] W. H. Breunlich et al., Phys. Rev. Lett. 53 (1984) 1137.
[48] C. Petitjean et al., MCF 2 (1987) 37.
[49] M. Kamimura, AIP 181 (1989) 330.
[50] L. N. Bogdanova et al., Sov. J. Nucl. Phys. 50 (1989) 848.
[51] C. D. Stodden et al., Phys. Rev. A41 (1990) 1281.
[52] J. S. Cohen, MCF 3 (1988) 421.
[53] H. Takahashi, MCF 3 (1988) 453.
[54] S. E. Jones, Hyp. Int. 82 (1993) 303.
[55] T. Case et al., Hyp. Int. 82 (1993) 295.
[56] K. Lou et al., Hyp. Int. 82 (1993) 313.
[57] F. D. Brooks et al., Hyp. Int. 82 (1993) 327.
[58] B. Alper et al., Proc. μCF'89, Oxford Rep. RAL-90-022, p. 35.
[59] J. D. Davies, MCF 7 (1992) 195.
[60] H. Daniel, MCF 5/6 (1990/91) 335.
[61] L. I. Men'shikov and V. A. Shakirov, Zh. Eksp. Teor. Fiz. 95 (1989) 458.
[62] M. P. Faifman and L. I. Ponomarev, Phys. Lett. B265 (1991) 201;
     M. P. Faifman et al., MCF 4 (1989) 1.
[63] C. Petitjean, Nucl. Phys. A453 (1992) 79C.
[64] L. I. Ponomarev, MCF 3 (1988) 629.

[65] M. Shiff, Nuovo Cim., 22 (1961) 66.
[66] O. A. Zaimidoroga et al., Zh. Eksp. Teor. Fiz. 44 (19663) 1852.
[67] A. Bertin et al., Lett. Nuovo Cim., 18 (1977) 381.
[68] A. V. Matveenko and L. I. Ponomarev, Zh. Eksp. Teor. Fiz. 63 (1972) 48.
[69] V. M. Bystritsky et al., Zh. Eksp. Teor. Fiz. 84 (1983) 1257.
[70] V. M. Bystritsky et al., Mesons in matters. Proc. of the Int. Symp. on Meson Chemistry and Mesomolecular Processes in Matter. D1,2,14-10908, Dubna, p. 220.
[71] Yu. A. Aristov et al., Yad. Fiz. 33 (1981) 1066.
[72] V. M. Bystritsky. Preprint JINR, E1-93-451; to be published in Yad. Fiz.
[73] L. N. Bogdanova et al., Nucl. Phys. A454 (1986) 653.
[74] Chi-Yu Hu, Phys. Rev. A36 (1987) 4135.
[75] S. E. Haywood et al., Phys. Rev. A37 (!989) 3393.
[76] M. Kamimura, AIP 181 (1989) 330.
[77] V. V. Gusev et al., MCF 4 (1989) 155.
[78] Yu. V. Petrov. In: Proc. of 14th Winter School LNPI, February 27–March 6, Leningrad (1979), p. 130.
[79] Yu. V. Petrov and Yu. M. Shabelski, Yad. Fiz. 30 (1979) 129.
[80] A. Bertin et al., Europhys. Lett. 4 (19787) 875; MCF 5/6 (1990/91) 349.
[81] Yu. V. Petrov, MCF 1 (1987) 351.
[82] V. V. Kuzminov et al., Hyp. Int. 82 (1993) 423.
[83] E. A. Garusov et al., MCF 3 (1988) 583.
[84] Yu. V. Petrov and E. G. Sakhnovsky, MCF 3 (1988) 571.
[85] H. Takahashi, MCF 1 (1987) 375.
[86] S. Elieser et al., MCF 5/6 (1990/91) 357.
[87] S. Monti, "A proposal for 14-MeV Neutron Source Based on Muon-Catalyzed Fusion", ENEA NUC-RIN Bologna (1991).
[88] T. Kase et al., MCF 5/6 (1990/91) 521.
[89] V. B. Belyaev et al., Preprint JINR, D15-92-323, Dubna, 1992.

MAGNETOELECTRIC TOROIDAL CONFINEMENT

J. Reece Roth

UTK Plasma Science Laboratory
Department of Electrical Engineering
409 Ferris Hall
University of Tennessee
Knoxville, Tennessee 37996-2100

ABSTRACT

In toroidal magnetic fusion experiments, including the tokamak, gross confinement is provided by static magnetic fields, which cannot do work on the plasma. The addition of electric fields along the minor radius of a toroidal plasma, to provide magnetoelectric toroidal confinement, has been reported in an Electric Field Bumpy Torus (EFBT), and in recent papers at a meeting of the IAEA Technical Committee on Tokamak Plasma Biasing. Such radial electric fields introduce the possibility of doing work on a toroidal plasma. This work, which can be provided by relatively inexpensive DC power supplies, can manifest itself in charged-particle transport radially inward against the density gradient, with concurrent improvements in density or containment time; this work can also manifest itself in heating the ions and electrons to high energies by E/B drift.

INTRODUCTION

In the vast majority of toroidal magnetic fusion experiments, including the tokamak, gross confinement is provided by static magnetic fields which cannot do work on the confined plasma. For this reason, an attempt at confinement by magnetic fields alone is ultimately doomed to failure, since the plasma will be transported toward the walls by the operation of the second law of thermodynamics. The marginal performance of current radial transport mechanisms in tokamaks (whatever they are) suggests strongly that more active intervention in the transport process is needed, by mechanisms which can do work on the plasma. The addition of electric fields along the minor radius of a toroidal plasma, to provide magnetoelectric toroidal confinement, introduces a mechanism by which relatively inexpensive DC power supplies can do work on the confined plasma.

Magnetoelectric toroidal confinement is most commonly effected by biasing the entire plasma to a high DC potential. In the various biasing schemes, one electrode is usually a limiter or the collector plates of a divertor, and the other electrode is the grounded walls of the vacuum vessel in which the plasma is confined. These biasing arrangements characteristically produce a toroidal plasma with an electric field along the minor radius that varies from tens of volts to several kilovolts per centimeter. The use of magnetoelectrically contained plasmas has a long history in open-ended devices such as the Penning discharge and tandem mirrors.

The characteristics and performance of the NASA-Lewis Electric Field Bumpy Torus (EFBT) experiment will be reviewed an example of magnetoelectric toroidal confinement. This experiment was operated at the NASA Lewis Research Center from 1972 to 1978 (Refs. 1,2) and was the first magneto-electric toroidal experiment to be extensively diagnosed. In the EFBT, gross confinement was provided by a bumpy torus magnetic field, while plasma heating, macroscopic stabilization, and enhanced confinement were provided by strong radial electric fields originating from a high DC bias potential maintained by external DC power supplies. In this plasma, the radial electric field exceeded 1 kV/cm at the plasma boundary and penetrated inward to at least one-half the plasma radius. These high DC radial electric fields drove strong low frequency (less than 500 kHz) fluctuations of the density and potential. Under the proper conditions, these fluctuations could produce radially inward fluctuation-induced transport against the radial plasma density gradient. The ions and electrons also received energy from external power supplies through the imposed radial electric field, and the E/B drift velocities thermalized to kinetic temperatures which could reach several kilovolts. Magnetoelectric toroidal containment was subsequently applied to tokamak experiments, which were summarized at a meeting of the IAEA Technical Committee on Tokamak Plasma Biasing in 1992 (Ref. 3).

CONSEQUENCES OF MAGNETOELECTRIC CONFINEMENT OF TOROIDAL PLASMAS

The full benefits of magnetoelectric toroidal confinement are available only if the toroidal plasma is biased by external power supplies to electrostatic potentials which exceed the "natural" plasma potential that would exist in the absence of external bias. Well confined toroidal plasmas typically will assume a positive plasma potential equal to a small multiple of the electron kinetic temperature of the plasma in electron volts, as a result of the radial transport of more mobile electrons across the confining magnetic field, leaving an excess of less mobile ions behind. This electron loss continues until a positive plasma potential builds up, and an ambipolar steady state exists in which the rate of ion and electron loss is equal. In order to significantly modify or improve the confinement times and number densities of a toroidal plasma in such a steady state, one must bias the toroidal plasma to potentials that differ significantly from this natural plasma potential. Magnetoelectric toroidal plasma confinement therefore requires that the entire toroidal plasma be biased by DC power supplies to form a toroidal electrostatic potential well, the maximum potential of which is significantly different from the natural plasma potential.

Experience with the NASA-Lewis Electric Field Bumpy Torus plasma (EFBT) (ref. 1) has shown that the number density and particle confinement time of this magnetoelectrically confined plasma were greatly enhanced if the plasma was negatively biased, to provide a negative, toroidal electrostatic potential well that preferentially traps the ion population. In such a potential well, the radial electric field points inward along the minor radius of the torus, in a direction to push ions inward against the radial density gradient.

The electrical circuit of a DC-biased toroidal plasma is completed by a current that flows between the biasing electrode and the grounded external walls of the vacuum tank. The biasing electrode functions as a Langmuir probe in either electron or ion saturation, depending on whether the electrode is positively or negatively biased, respectively. If other ion losses from the toroidal plasma are minimized, an electrode that biases the plasma negatively will collect all the ions lost along the minor radius of the plasma, and the electrical circuit to the DC power supply will be completed by an equal current of electrons that find their way from the surface of the plasma (distant from the biasing electrode) to the grounded walls. With opposite polarity, when the toroidal plasma is positively biased, the electrode will collect all the electrons which escape along a minor radius of the plasma, and the electrical circuit to the power supply will be completed by ions transported across the confining magnetic field to the grounded walls of the vacuum system. Once the toroidal plasma is biased to high negative potentials (high = a biasing potential equal to several times the numerical value of the electron kinetic temperature in electron volts), one is in a position to do work on the toroidal plasma through the electric fields, maintained by external power supplies, along the minor radius of the plasma. This work done on the plasma can have a number of important consequences, some of the most significant of which are discussed below.

Ion Heating

The electric field along the minor radius of the toroidal plasma is at right angles to the confining toroidal magnetic field, thus leading to E/B drift, according to the equation

$$V_d = \frac{E \times B}{B^2} = \frac{E_r}{B} \text{ meters / sec.} \quad (1)$$

This azimuthal drift velocity has the interesting property that it is independent of the charge, mass, and energy of the charged particles undergoing the drift. Unless finite gyroradius effects are important, the ions and electrons will drift in the same direction with the same velocity in such crossed electric and magnetic fields. The kinetic energy associated with this drift velocity is given by

$$E = \frac{1}{2} m V_d^2 = \frac{1}{2} m \frac{E_r^2}{B^2} \text{ Joules} \quad (2)$$

Equation 2 makes the very interesting prediction that the energy imparted to each species is directly proportional to its mass. The ions therefore receive nearly all of the energy imparted to the plasma by the DC radial electric fields. If the electric fields are a few kilovolts per centimeter, then deuterium ions can acquire energies on the order of keV, while electron kinetic temperatures will remain at low levels determined by turbulent energy coupling between the ion and electron populations.

In the EFBT plasma discussed in ref. 1, it was found that deuterium ion energies characteristically ranged from a few hundred electron volts to a few keV, under conditions for which the electron kinetic temperature was in the range from 2eV to 15eV. In addition to the fact that magnetoelectric toroidal heating deposits most of the work done on the plasma in the ion population, the cost of the DC electrical power needed to bias the plasma is less than any other form of energy input used to heat plasmas. Typically DC electrical power is available at the multi-kilowatt to megawatt level at equipment costs that range from as little as ten cents a watt to no more than fifty cents per watt.

Radially Inward Transport

Once a toroidal plasma has been electrostatically biased to a high enough potential to affect the radial transport across the magnetic field, the ions will experience a radial electric field which tends to push them inward if the plasma is negatively biased, and to push them outward if the plasma is positively biased. Probably because of their much larger mass, it is more effective to toroidally confine the ions than the electron population, and this is best done by biasing the toroidal plasma negatively with respect to the surrounding grounded walls of the vacuum system. It was shown that in the magnetoelectrically confined NASA Lewis Electric Field Bumpy Torus plasma, fluctuation induced transport dominated the radial transport process (refs. 1–3). This transport mechanism has also been identified as dominant in the outer edge of some tokamak plasmas (ref. 3).

Fluctuation induced transport can carry ions radially inward against the density gradient in the presence of very strong, inward-pointing radial electric fields. In the EFBT plasma, all the ions were lost from the plasma across the sheath between the plasma and the negative biasing electrode. The capability to effect radially inward transport is potentially important, because it allows one to keep the ions trapped in a restricted toroidal volume, rather than allowing the plasma to diffuse radially outward across the magnetic field to the walls, which is the ultimate fate of all toroidal plasmas trapped only in static magnetic fields.

Favorable Scaling of Particle Containment Time

In a properly negatively biased toroidal plasma, the only loss of positive ions will be across the sheath between the plasma and the negative biasing electrode, on which the ions are collected. The ion containment time in the plasma is therefore inversely proportional to the area of the biasing electrode exposed to the plasma, and directly proportional to the plasma volume. For example, if one were to double the volume of a toroidal plasma in contact with a fixed negatively biased electrode, it would take the average ion twice as long to find the electrode and be lost, and the particle confinement time would therefore be doubled. Conversely, if the electrode surface area in contact with the plasma were to be cut in half in a plasma of fixed toroidal volume, then the particle containment time would be doubled, since it would take the average ion twice as long as previously to reach the negatively biased electrode and be lost. This scaling of the ion containment time in the plasma is very different from Alcator, Kaye-Goldston, or other purely magnetic fusion containment approaches. The containment time in magnetoelectric toroidal confinement does not depend on the toroidal magnetic field, provided that it is strong enough to suppress finite gyroradius effects, nor is the containment time in magnetoelectric containment proportional to the electron kinetic temperature.

OPTIONS FOR BIASING A TOROIDAL PLASMA

The biasing of a toroidal plasma possessing plasma parameters of fusion interest is not a trivial matter. The power flux on the surface of any solid electrode immersed in a fusion grade plasma is far higher than any steady state heat transfer mechanism can cope with. A major problem in magnetoelectric toroidal confinement is therefore the development of means by which a fusion grade plasma can be electrostatically biased by electrodes that will not be damaged by the plasma. This problem has been dealt with by accepting very high heat transfer rates for experimental, non-fusion-grade plasmas; by biasing toroidal plasmas at their edges; and by using indirect biasing mechanisms. Methods which have already been reduced to practice include biasing toroidal plasmas with a water-cooled electrode rod inserted across the minor diameter of the plasma; by using an electrode ring around the minor circumference of the plasma, shaped to conform to the local particle drift surfaces; by having a single biased limiter at the plasma boundary, an approach which has been widely used in tokamak research; and to electrostatically bias a Tokamak plasma by biased magnetic divertor plates. Other toroidal plasma biasing schemes which have been suggested include injecting energetic ions or electrons into the plasma from a biased ion or electron source, an approach which was briefly attempted with electrons from a heated filament on the Macrotor tokamak (ref. 4), and to scrape off large gyroradius ions, to leave a toroidal plasma with an excess of electrons and therefore negatively biased.

At this stage of magnetoelectric toroidal plasma research, the most widely used method of biasing the plasma has been the direct use of electrodes or limiters. This approach has the advantages that it is the simplest way to bias a toroidal plasma for study; such electrodes allow the plasma to be biased either positively or negatively for exploratory research; and one can monitor the total ion or electron loss from the plasma by measuring the DC current flowing to the electrode. The use of electrodes or limiters has a number of disadvantages however, including a relatively high power flux on their surface at high density plasma conditions. It is likely that the effectiveness of the electrode or limiter as a biasing electrode would be greater, the closer it is to the dense interior of the plasma, but this produces a tradeoff with the higher power fluxes to be expected as one penetrates inward to higher densities. The bombardment of negative electrodes by plasma ions will lead to electrode erosion, which can shorten the lifetime of the electrodes by thinning their walls, and the sputtered atoms from the electrode can introduce impurities into the plasma and contaminate it with electrode material. Most of these problems or disadvantages, however, can be addressed by using biased impingement plates in magnetic divertors, or using biased ion or electron sources located either in a magnetic divertor or at the edge of the plasma.

TOKAMAK BIASING

Within the past fifteen years, numerous attempts have been made to electrostatically bias tokamak plasmas (Ref. 3). The usual objective of such biasing experiments is to affect the L to H mode transition, or to otherwise improve the confinement properties of the tokamak plasma.

Plasma Bias Experiments on the Macrotor Tokamak

The first attempt to electrostatically bias a tokamak plasma was reported by Oren et al. in 1979 (ref. 4). A schematic drawing the Macrotor Tokamak and the biasing circuit used in this experiment is shown in Figure 1. The Macrotor Tokamak was a small exploratory research device operated at the University of California at Los Angeles. In these experiments, the entire plasma was biased negatively to a potential up to -300 volts by a single electron-

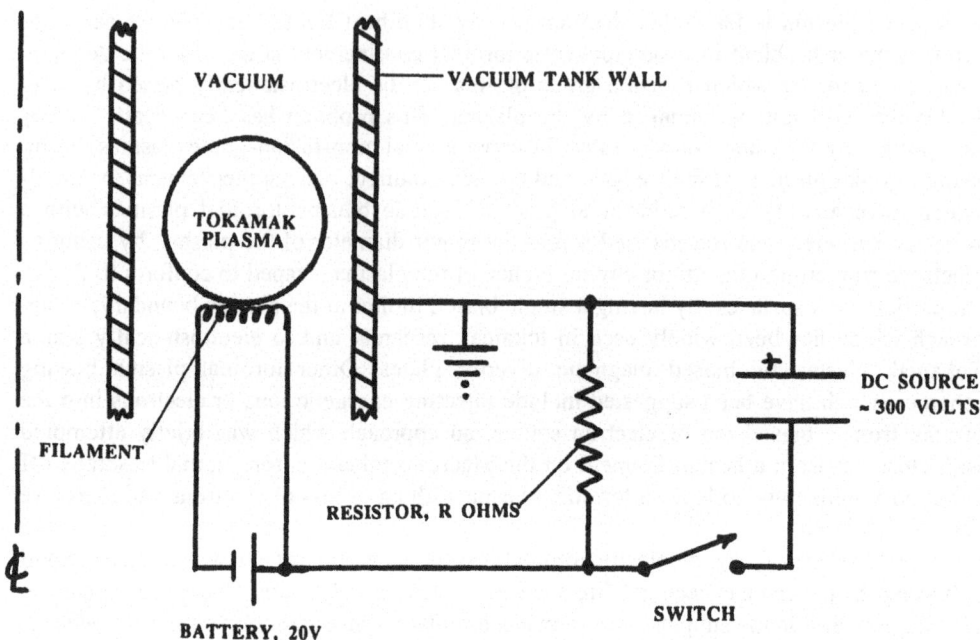

Figure 1. Arrangement used to impose a negative bias on the UCLA Macrotor Tokamak Plasma (Ref. 4).

emitting filament electrode. It was found that there was essentially no voltage drop along the magnetic field lines in the tokamak plasma, and that as a result of the bias, a radially inward pointing electric field was observed, with a magnitude of approximately 50 volts per centimeter. The effects of this bias on the plasma characteristics were remarkable. The Macrotor group reported that the average electron number density was increased a factor of 10 by the bias, that the particle containment time in the tokamak plasma improved by a factor of 5 to 10 times, and that the ion kinetic temperature in the plasma increased by about 20 electron volts. This group also observed that there was a poloidal drift velocity of 2×10 centimeters per second. The only detrimental effect of bias in these experiments was an increased impurity level when the bias voltage was applied.

In this biasing configuration, an electron-emitting filament was in contact with the edge of the plasma, serving as an electron emitting limiter to the plasma edge. The heated filament was powered by a 20 volt power supply. This filament was connected to ground through a resistor, across which a negative potential of up to three hundred volts could be imposed from a DC power supply.

It is perhaps surprising that the results reported by the Macrotor group (ref. 4) did not immediately initiate a vigorous research effort on electrostatically biased tokamak plasmas to replicate the greatly improved confinement which they reported. However, this line of research did not flourish, even at UCLA, and it was not until the late 1980's that other research groups finally turned their attention to the electrostatic biasing of tokamak plasmas.

IAEA Technical Committee Meeting on Tokamak Plasma Biasing

After a hiatus of approximately 10 years following the report of tokamak plasma biasing by Oren et al. from UCLA, the tokamak community slowly became engaged in the subject of tokamak plasma biasing after 1989. The usual motivation for electrostatic biasing of a tokamak plasma was to affect the L to H mode transition, hopefully in a direction which would improve plasma containment by broadening the range of parameters under which the plasma would operate in the H mode. The results of many of these experiments were reported at an IAEA Technical Committee Meeting on Tokamak Plasma Biasing which was held in Montreal, Canada on September 8 to 10, 1992 (Ref. 5). By that time, a significant worldwide effort on tokamak plasma biasing was underway. The IAEA meeting had forty-three registrants from 11 countries representing 28 research laboratories. There were a total of 34 papers on various aspects of tokamak and toroidal plasma biasing, of which eight were theoretical or computational papers, and 26 were experimental papers, most of them relating to tokamak plasmas. A detailed summary of this conference is available in reference 3.

The tokamak plasma biasing reported at the Montreal meeting was accomplished either by biased limiters in contact with the edge of the plasma, or by biased impingement plates located in magnetic divertors. Among these experiments, bias voltages up to ± 1000 volts were applied to the biasing electrodes, and bias currents of up to 14 kiloamperes were measured. Among the favorable results reported were improvements in the containment time and plasma number density of up to a factor of 10. In the most ambitious of these tokamak plasma biasing experiments, performed on the Doublet III-D tokamak plasma, bulk heating of the electrons was observed in the outer regions of the plasma at a level of several megawatts. This heating resulted from the application of up to 700 volts, and bias currents of up to 14 kiloamperes (Ref. 6).

The papers presented at the Montreal meeting collectively made a strong case for the beneficial effects of electrostatic biasing of toroidal plasmas, especially tokamaks. Magnetoelectric containment has been shown to improve particle confinement time in toroidal devices, to increase the electron number density in the core plasma, to reduce impurity levels, to heat a surface layer of shield electrons in the outer regions of the toroidal plasma, and even to raise ions to kilovolt kinetic energies. Unfortunately, the electrostatic biasing experiments at nearly all tokamak laboratories were a peripheral rather than a central element of their research program. The resources devoted to magnetoelectric toroidal containment of tokamak plasmas then and since do not appear to the present author to be commensurate with the promise of this approach, as revealed by the papers at the Montreal meeting.

THE NASA-LEWIS ELECTRIC FIELD BUMPY TORUS AS A MAGNETOELECTRICALLY CONFINED TOROIDAL PLASMA

The more recent magnetoelectrically confined toroidal plasma experiments on tokamaks and other plasmas described above, in reference 3, and in the proceedings of the Montreal conference (ref. 5) were preceded by the NASA Lewis Electric Field Bumpy Torus (EFBT) experiment, which was initiated in 1967, and operated at the NASA Lewis Research Center until 1978. This experiment was conceived solely as a magnetoelectrically confined toroidal plasma and produced a wide variety of physics results, many of which may be relevant to electrostatically biased tokamak and other toroidal plasmas (see Ref. 7).

The EFBT plasma had a volume of 82 liters, a major diameter of 1.5 meters, and a maximum minor diameter of about 16 cm. Gross confinement was provided by a bumpy toroidal magnetic field with twelve sectors, a mirror ratio of 2.5:1 in each sector, and a maximum induction on the magnetic axis of 3.0 Tesla. The EFBT plasma could be biased with one or more electrode rings/limiters to potentials of up to ± 50 kilovolts by power supplies capable of delivering up to 20 amperes. The radial electric field could exceed several kilovolts per centimeter, and its sign and magnitude were both independent variables. This plasma was highly turbulent as the result of strong E/B drift, and this turbulence gave rise to fluctuation-induced transport, both radially inward and radially outward. It was found that in this plasma, fluctuation-induced transport was the dominant form of transport, accounting within a factor of two for all the current collected by the biased electrode rings.

The magnetoelectric containment of plasmas has a long history which has been reviewed in references 1 and 2. The work reported below is the result of a 15 year research program on magnetoelectrically contained plasmas of fusion interest which was performed at the NASA-Lewis Research Center between 1963 and 1978. From 1963 until 1972, experimental research was focused on the modified Penning discharge, a magnetoelectric magnetic mirror concept described in references 8 thru 11. In 1967 the author proposed to the NASA-Lewis management the construction of a toroidal version of the modified Penning discharge, in which a plasma would be confined by a bumpy toroidal magnetic field configuration, and biased to high potentials, on the order of kilovolts. This proposal was approved, and the NASA-Lewis Electric Field Bumpy Torus (EFBT) project produced its first plasma in 1972. Between that year and 1978, a series of investigations on this magnetoelectrically confined toroidal plasma were published in references 12 thru 32.

Since the plasma was confined by a bumpy toroidal configuration, no large plasma currents were required, as in the tokamak, to produce a rotational transform or closed drift surfaces, thus allowing the experiment to be operated in the steady-state. The biasing of the plasma was accomplished with one or more electrode rings or rods maintained at positive or negative potentials of up to 50 kilovolts. Thus, both the magnitude and the polarity of the radial electric fields were independent variables. The magnitude of the potentials and of the radial electric fields was considerably larger than those characteristic of present tokamak biasing experiments (Ref. 5). An important feature of the Electric Field Bumpy Torus research at NASA-Lewis was the implementation of a fluctuation induced transport diagnostic originally proposed in references 16 and 17. The NASA-Lewis work included a collaboration with Prof. Edward J. Powers, Jr. and his students at the University of Texas at Austin who, under NASA contract, produced software which analyzed the fluctuation-induced transport data from the EFBT plasma. This software has been used in subsequent studies of fluctuation-induced transport in tokamaks.

The essence of the EFBT concept is illustrated in Figure 2. Figure 2a shows the basic bumpy torus magnetic containment geometry which, in the NASA-Lewis experiment, consisted of 12 superconducting coils in a toroidal array. The lack of axial symmetry (or bumpyness) of the magnetic field along the magnetic axis causes individual charged particles (but not magnetic field lines) which circulate around the major circumference to drift around the magnetic axis, thus providing an effective rotational transform and closed drift surfaces. The plasma confined in the bumpy torus magnetic field geometry was biased with electrode rings/ limiters which were maintained at positive or negative potentials of kilovolts by external DC power supplies. Figure 2b shows schematically the magnetoelectrically confined EFBT plasma with positive bias. Extensive data are available from this experiment for both biasing polarities, for a wide variety of geometric shapes of the individual electrode rings, and for the use of from 1 to 12 electrode rings to bias the plasma.

Magnetoelectric Toroidal Confinement

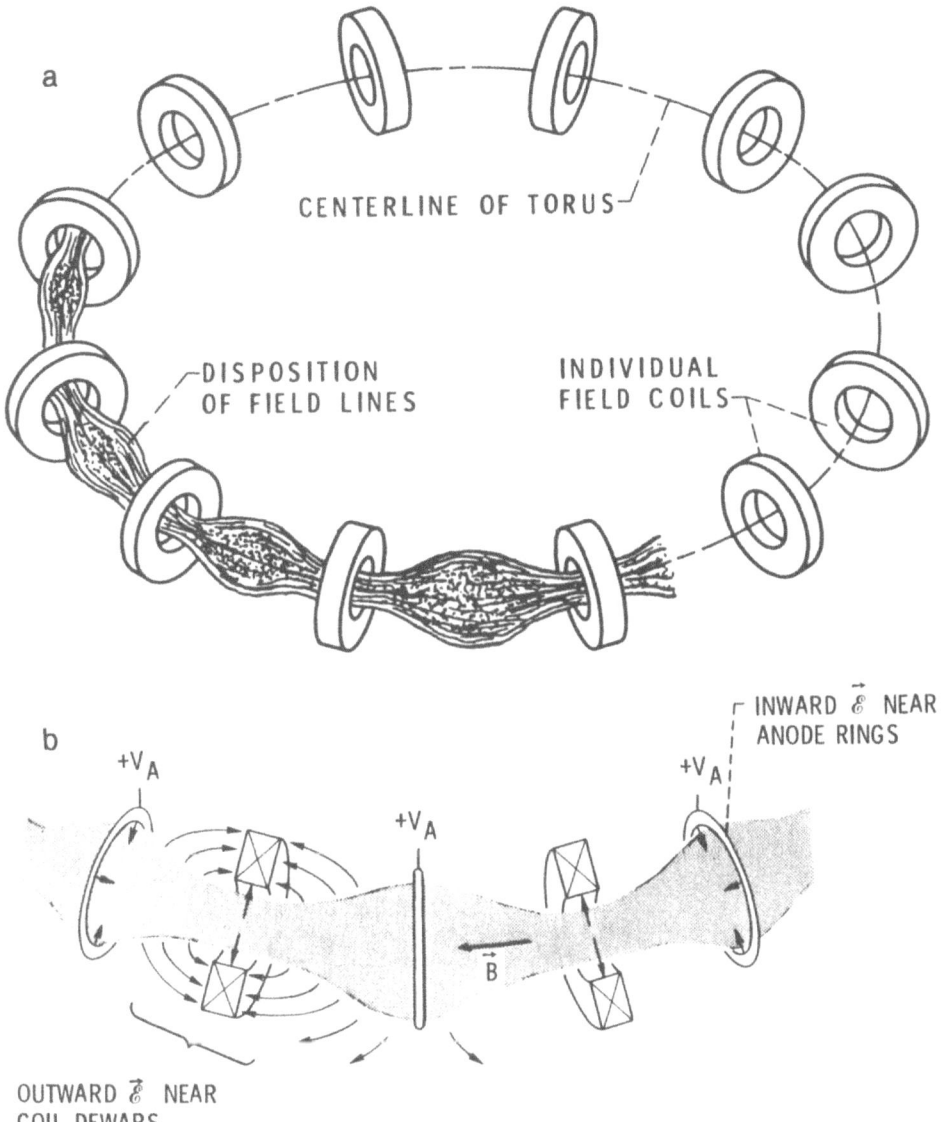

Figure 2. Schematic of the Electric Field Bumpy Torus (EFBT) concept. a) Gross confinement provided by a bumpy toroidal magnetic field configuration. b) Electrostatic biasing of the magnetoelectrically confined and heated EFBT plasma.

The NASA-Lewis EFBT experiment utilized the first superconducting magnet facility to generate a toroidal magnetic field for a fusion related experiment. This magnet facility was built at NASA-Lewis and is described in references 21 and 22. A cutaway drawing of this facility is shown in Figure 3, and its characteristics and performance are listed in Table 1. A photograph of the EFBT plasma is shown in Figure 4. This photograph was taken through one of the viewpoints in the equatorial plane, under conditions for which an electrode ring was located in each of the 12 sectors. The plasma was normally operated with deuterium gas and in the steady-state. Some of the characteristics of the EFBT

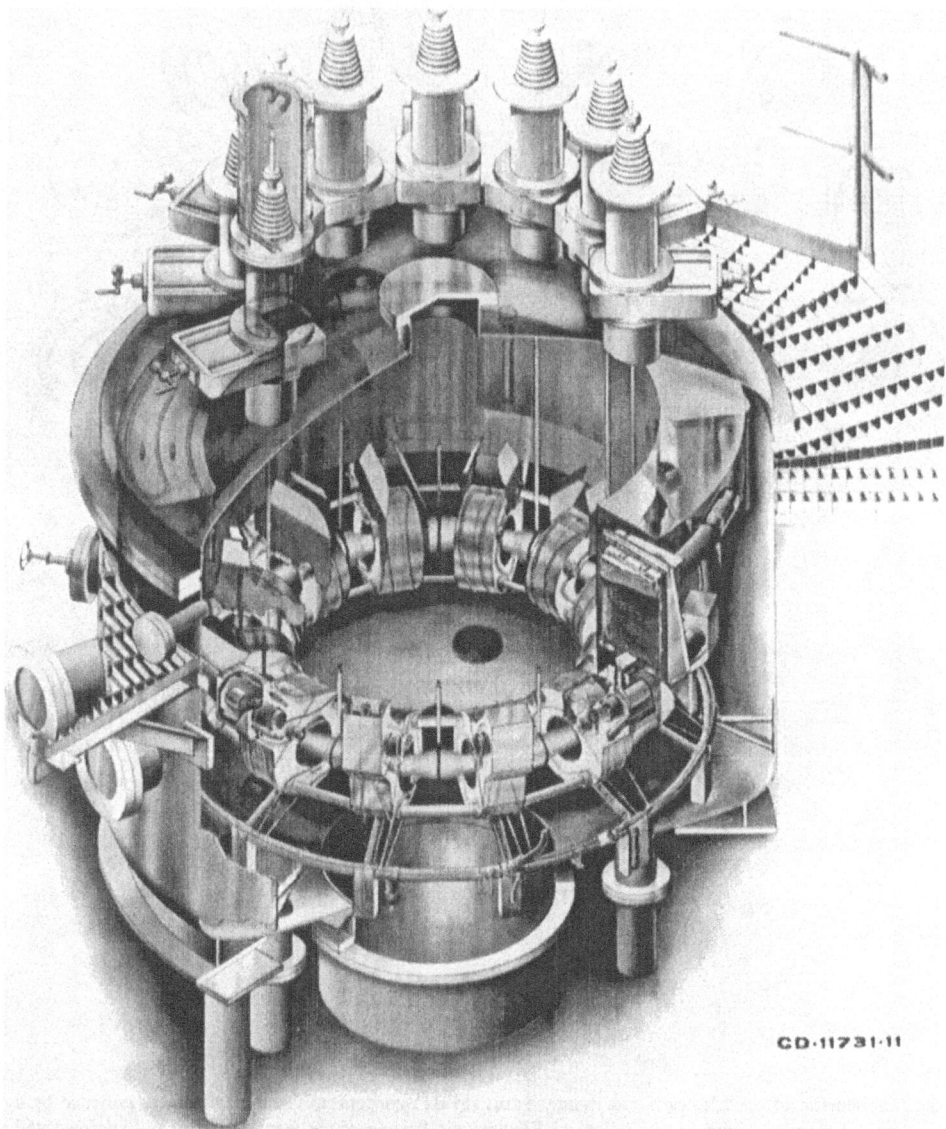

Figure 3. The Superconducting Bumpy Torus Magnet Facility, operated at the NASA Lewis Research Center from 1972 to 1978. The twelve superconducting coils each were capable of 3.0 Tesla at the midplane. The major diameter of the toroidal array was 1.5 meters.

plasma are listed in Table 2. The *particle* containment times were several milliseconds, approximately a factor of 10 higher than the Bohm containment times for EFBT plasma conditions. Another significant characteristic of the EFBT plasma was its large ratio of ion to electron kinetic temperature. This was a consequence of the E/B drift heating mechanism; in crossed electric and magnetic fields, particles acquire drift energies proportional to their masses.

Table 1. Characteristics and performance of the NASA-Lewis superconducting bumpy torus magnet facility

Characteristics	
Major diameter of torus	1.52 m
Inside diameter of coil spoolpiece'	21 cm
Inside diameter of coil dewar	19 cm
Inside diameter of coil winding	21.8 cm
Outside diameter of coil winding	30.5 cm
Axial width of coil winding	12 cm
Maximum magnetic field on magnetic axis	3.0 T
Mirror ratio on magnetic axis	2.48 : 1
Plasma confinement volume	≈82 liters
Performance Data	
First coil operation	April 24, 1972
First plasma	December 5, 1972
Final shutdown	March 31, 1978
Days of operation with coils charged	436
Working days since first plasma	1337
Utilization fraction	33%
Total hours experimental operation	2620 hours
Number of coil normalcies	189

Plasma Biasing Methods and Results

A wide variety of electrode/limiter geometries were used during the course of the NASA-Lewis EFBT experiments. In Figure 5 are shown 3 electrode configurations used for most of the data taken from this experiment. Figure 5a shows a D-shaped electrode which approximately conforms to the midplane cross-sectional profile of the plasma. The flat part of the D faces inward along a major radius, and the copper tubing was water-cooled. Figure 5b shows a water-cooled stainless steel electrode which was inserted across the minor diameter of the EFBT plasma. Another similar geometry had two adjacent stainless steel tubes, also inserted across the plasma minor diameter. Since the particle containment time is theoretically predicted to be inversely proportional to the electrode contact area with the plasma, some measurements were made in which only a single biasing electrode was used, consisting of an

Table 2. Characteristics of the EFBT plasma

Highest plasma densities	$\bar{n}_e = 3.1 \times 10^{12} / cm^3$ $n_{e_{max}} \times 6.2 \times 10^{12} / cm^3$ $\}$ $\tau_p = 2.52$ msec
Highest particle containment time	$\tau_p = 6.0$ msec at $n_{emax} = 1.0 \times 10^{12}/cm^3$
Highest simultaneous $n_{emax} \tau_p$	$n_{emax} \tau_p = 1.6 \times 10^{10}$ sec/cm^3
Ion kinetic temperatures for Deuterium	
For above conditions	$360 \leq T_i \leq 520$ eV
Highest ever	$T_i = 2500$ eV
Electron kinetic temperatures	
For above conditions	$2 \leq T_e \leq 15$ eV
Highest ever observed	$T_e \approx 150$ eV
Plasma volume	
Typically	82 liters

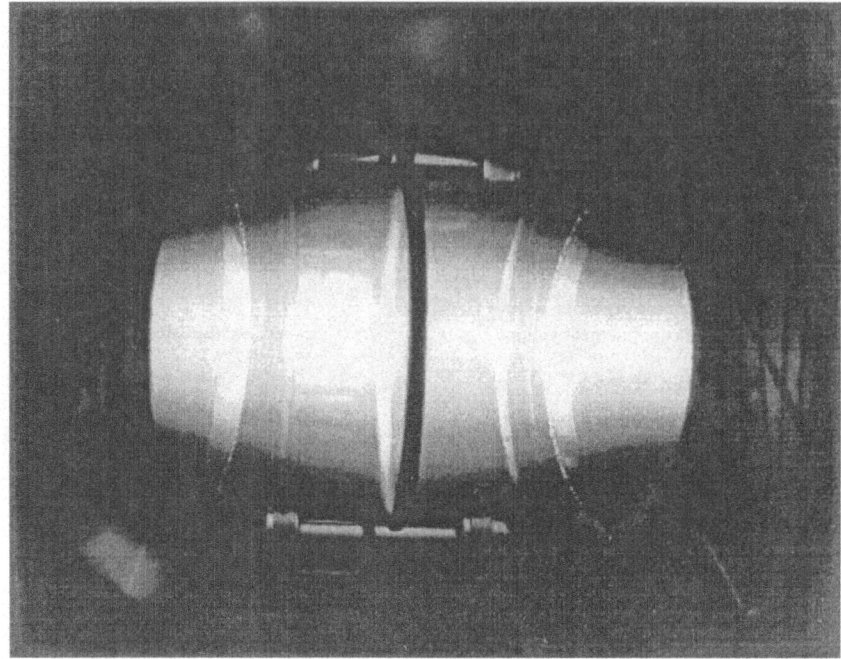

Figure 4. A photograph of the EFBT plasma showing a biasing electrode in the foreground, and the sector at the opposite end of a major diameter in the background.

uncooled tungsten wire across the plasma minor diameter, as shown in Figure 5c. In normal operation of the plasma shown in Figure 4, limiters or electrodes in the throats of the magnetic field coils were not necessary or desirable. The DC electrical circuit to the high voltage power supply was completed by radially diffusing ions or electrons which were transported rapidly across the magnetic field lines to the grounded inner surface of the superconducting magnet dewars at the magnetic mirror throats.

A major issue in the EFBT experiment was the radial profile of electrostatic potential in the plasma, and the magnitude of the radial electric field at the plasma boundary, which was chiefly responsible for confining and heating the plasma. The radial potential profile was sampled with a hydraulically actuated floating Langmuir probe, drawn to scale in a plan view in Figure 6. Since the potentials being measured were on the order of kilovolts, and the electron kinetic temperature of the plasma was perhaps 10eV, the difference between the floating and plasma potential in these measurements should not be large.

Some results of these probing experiments are shown in Figure 7. This figure shows the radial potential profile with the plasma biased by a single electrode ring with a positive potential of 10 kilovolts, located 2 sectors (60°) and 4 sectors (120°) away from the sector in which the probe was located. There is some evidence here of a relatively small axial electric field from sector to sector within the plasma, of magnitude 15 to 20 volts/centimeter. Within a few centimeters of the visible boundary of the plasma, also approximately the anode ring radius, the radial electric field is on the order of kilovolts per centimeter, and extends at least 4 centimeters radially into the plasma interior. The slight fall-off of potential near the axis indicated on the upper profile may have resulted from the perturbing effect of the actuating probe deeply penetrating the plasma.

Magnetoelectric Toroidal Confinement

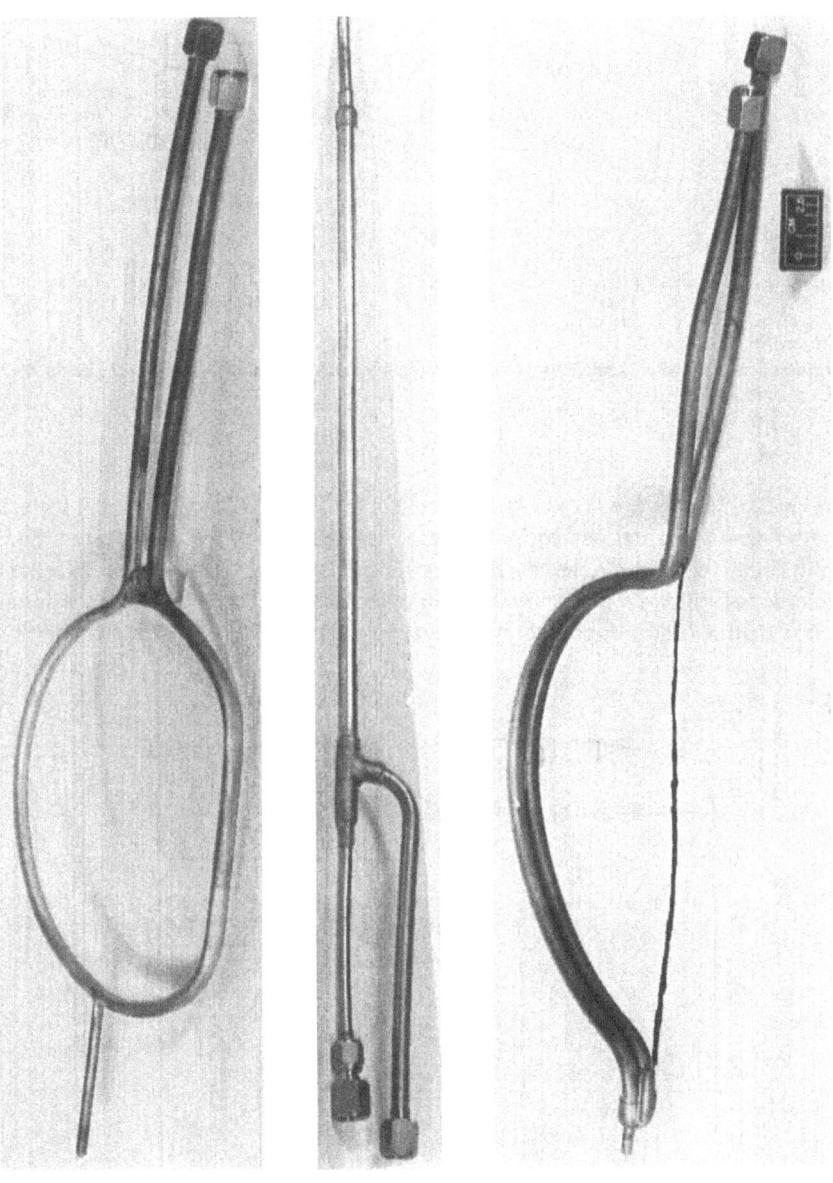

Figure 5. An illustration of three major types of biasing electrodes used on the EFBT plasma. a) D-shaped ring electrode; b) a tubular electrode across the minor diameter of the plasma; and c) an uncooled wire across the minor diameter of the plasma. All electrodes but the wire were water cooled.

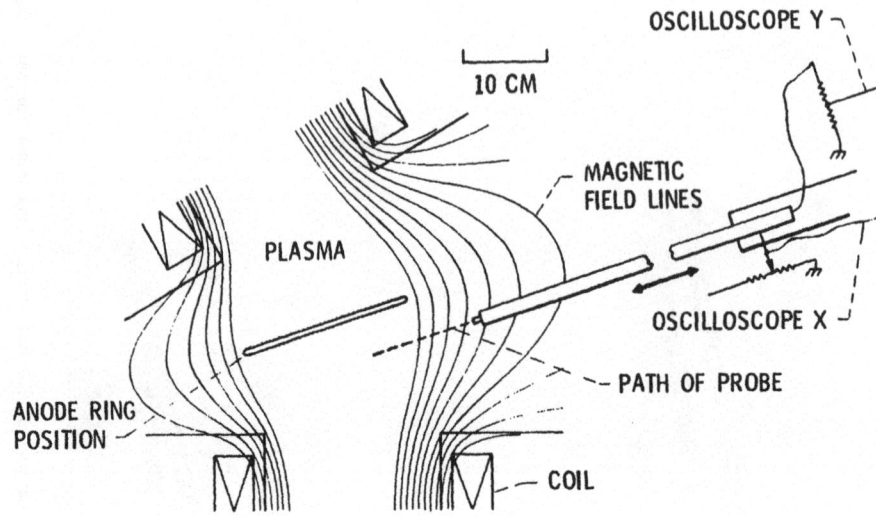

Figure 6. Hydraulically actuated Langmuir probe for measurement of floating potential profiles in the EFBT plasma.

Similar profile data are shown in Figure 8 for ± 10 kilovolts applied to 11 electrode rings, with no electrode ring located in the sector in which the probe was located. For both positive and negative biasing voltages, the electric fields in the vicinity of the plasma edge are on the order of kilovolts per centimeter. In both cases, the drop-off in potential at minor radii less than 5 centimeters may be an artifact of the perturbing effect of the probe.

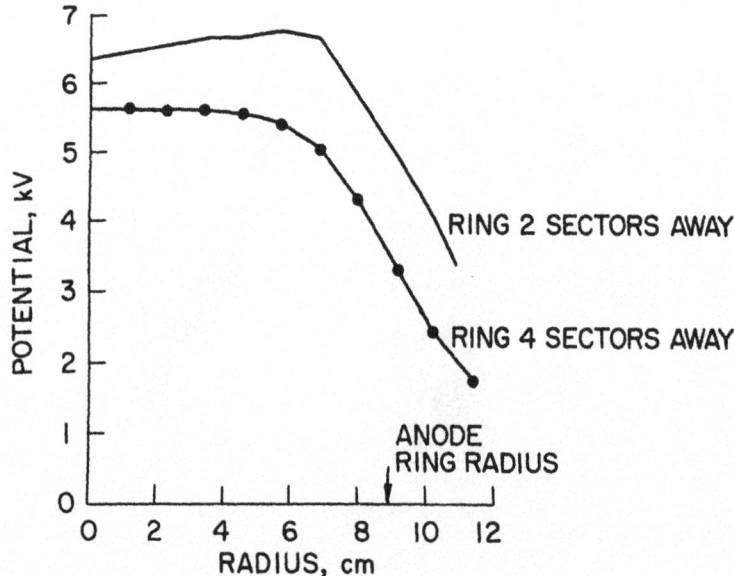

Figure 7. The floating potential profile of a positively biased plasma with one anode ring. Profiles are shown for the single anode ring two and four sectors (60° and 120°) away from the sector containing the probe. The anode and visible plasma radius were approximately 9 cm.

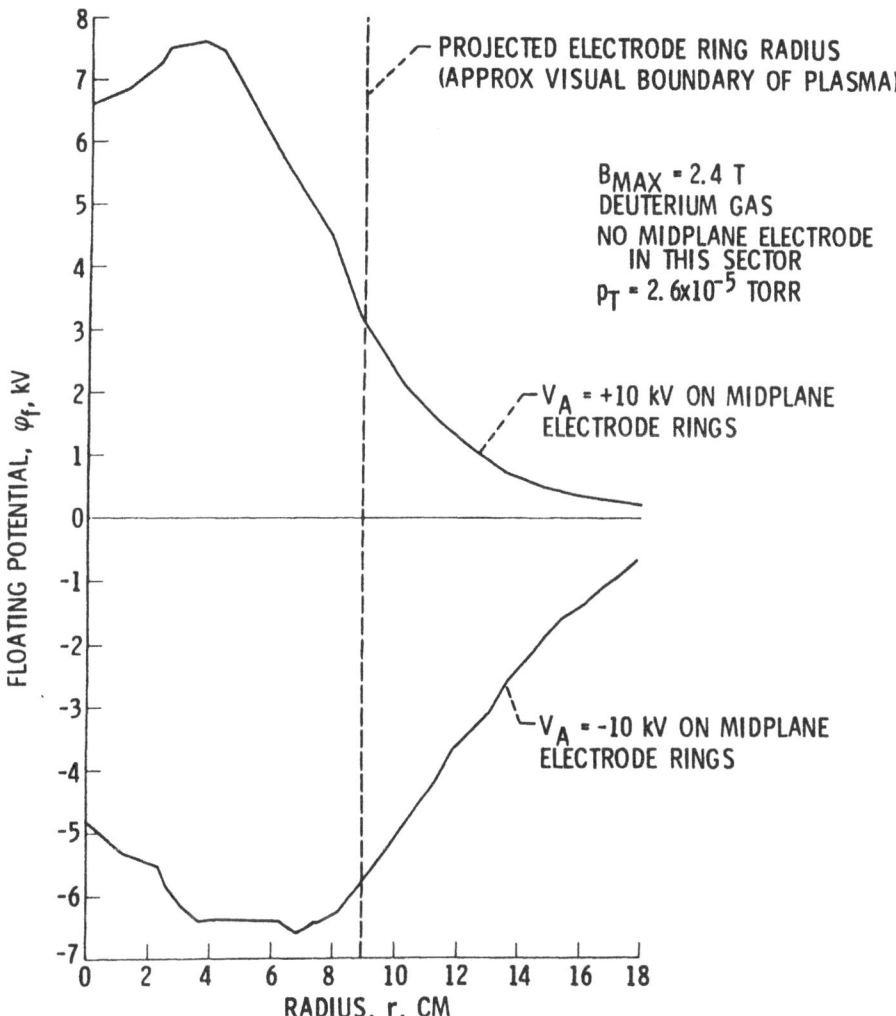

Figure 8. Radial profile of floating potential, with positive and negative electrode rings in the eleven sectors not occupied by the radial probe. The eleven electrode rings were biased to potentials of plus or minus 10 kV. Data nearer than 4 cm to the plasma minor axis may have been affected by the probe pertubation.

These data illustrate quite clearly that electrostatic biasing of toroidal plasmas to kilovolt potentials can result in very strong electric fields which are capable of penetrating at least half way from the edge to the plasma interior.

A later series of experiments was made with a hydraulically actuated Langmuir probe system used for our fluctuation induced transport measurements. The zero reference position for the edge measurements made with a floating and ion saturated Langmuir probe are shown in Figure 9. Data taken with the hydraulically actuated probe shown in Figure 9 are shown on Figure 10 and 11. This figure illustrates two characteristic experimental runs with positive electrode polarity, and two with negative electrode polarity. The ion saturation current is shown as a function of the minor plasma radius, with the zero position approximately at the edge of the plasma. It is interesting that in most cases, the rela-

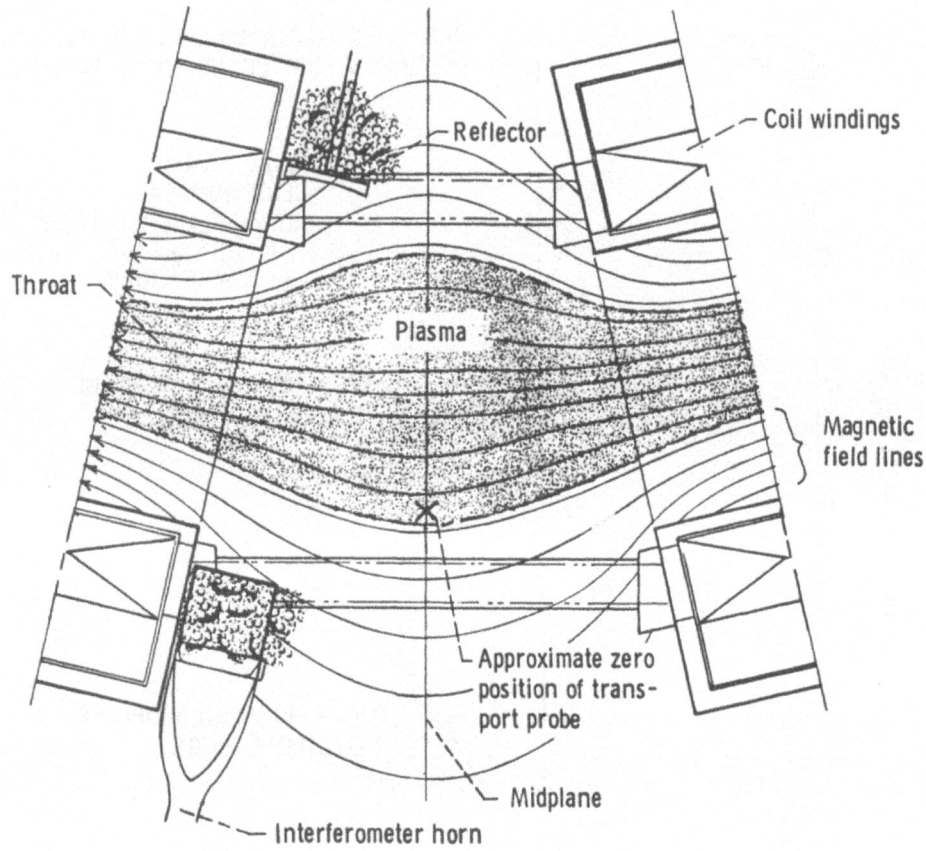

Figure 9. Experimental arrangement used to measure the chordwise plasma number density, radial floating potential, and relative electron number density at the midplane edge of the plasma.

tive electron number density drops off exponentially, with e-folding distances which range from 2.6 to 10 centimeters for the data shown in this figure. The floating potential measured with the Langmuir probe system shown in Figure 9 is shown on Figure 11 for the same experimental runs shown in Figure 10. Consistent with Poisson's equation, the drop-off of electrostatic potential is also exponential for many of these experimental runs.

Current-Voltage Characteristics of the EFBT Plasma

The current-voltage curves of the EFBT plasma share many characteristics with the classical low pressure DC normal glow electrical discharge, which has been studied for over 100 years. On Figure 12 is shown the current-voltage curve of the EFBT plasma for negative and positive polarities and for 2 electrode configurations, shown in Figure 5a and b. These low voltage current-voltage curves obey a power law of the form

$$V \sim I^k, \quad 2 \leq k \leq 12. \tag{3}$$

This power-law relation, with values of k from 2 to 12, yields a practically flat (or vertical) voltage-current relationship, a defining characteristic of the classical normal

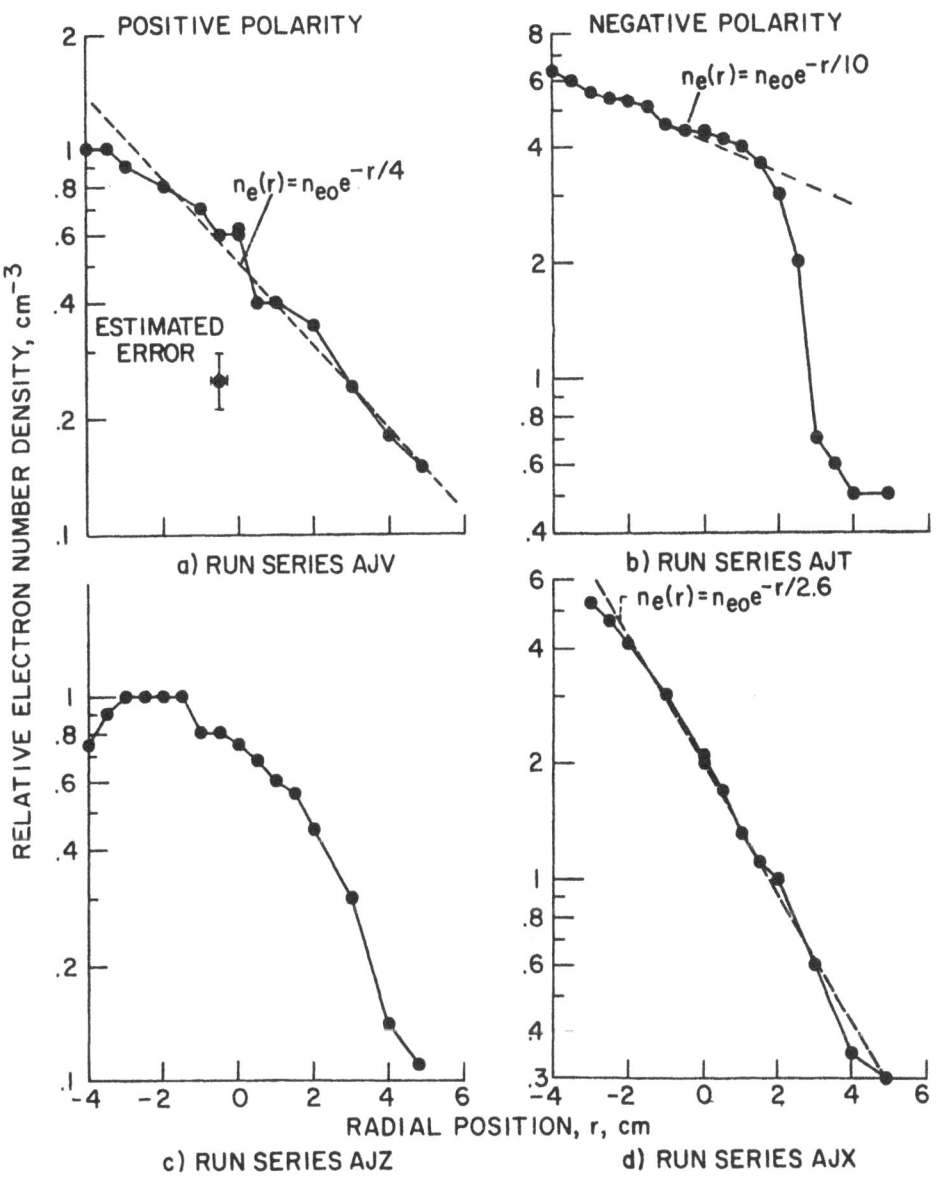

Figure 10. Ion saturation curves taken from the plasma edge indicated in Figure 9. Runs a and c taken with positive polarity; runs b and d taken with negative electrode polarity.

glow DC electrical discharge. Curves similar to those shown in Figure 12 have been seen in many DC normal glow discharges and high intensity electrical arcs used in industrial applications (Refs. 33, 34).

On Figure 13 are shown current-voltage curves over a wider range of voltages and currents, for positive and negative electrode polarities, and for several different background pressures of deuterium gas. These curves illustrate the characteristic power law relationship between voltage and current, as well as a discontinuous mode transition from

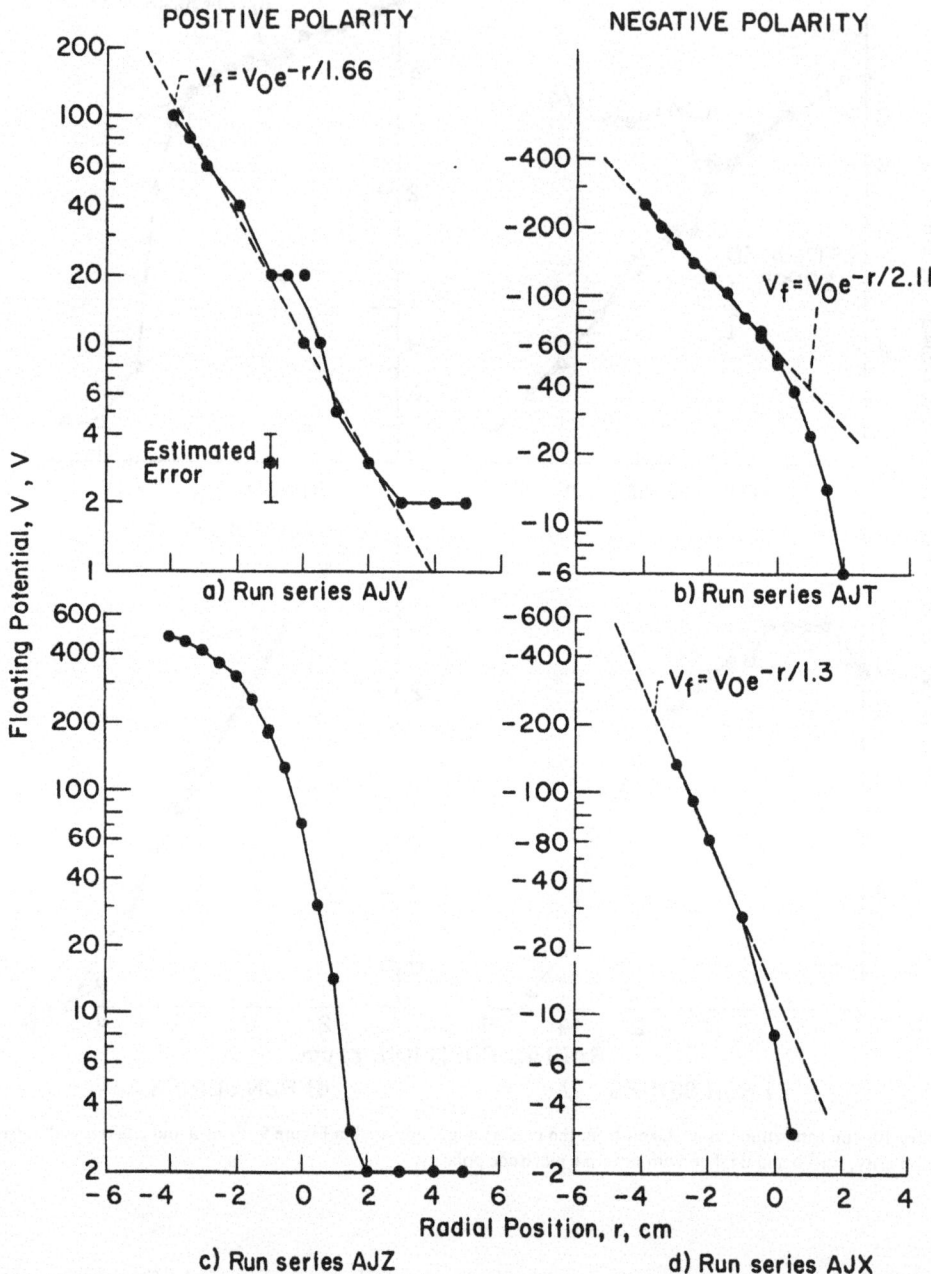

Figure 11. Floating potential measurements made with a Langmuir probe as in Figure 10; runs a and c for positive polarity; runs b and d for negative polarity.

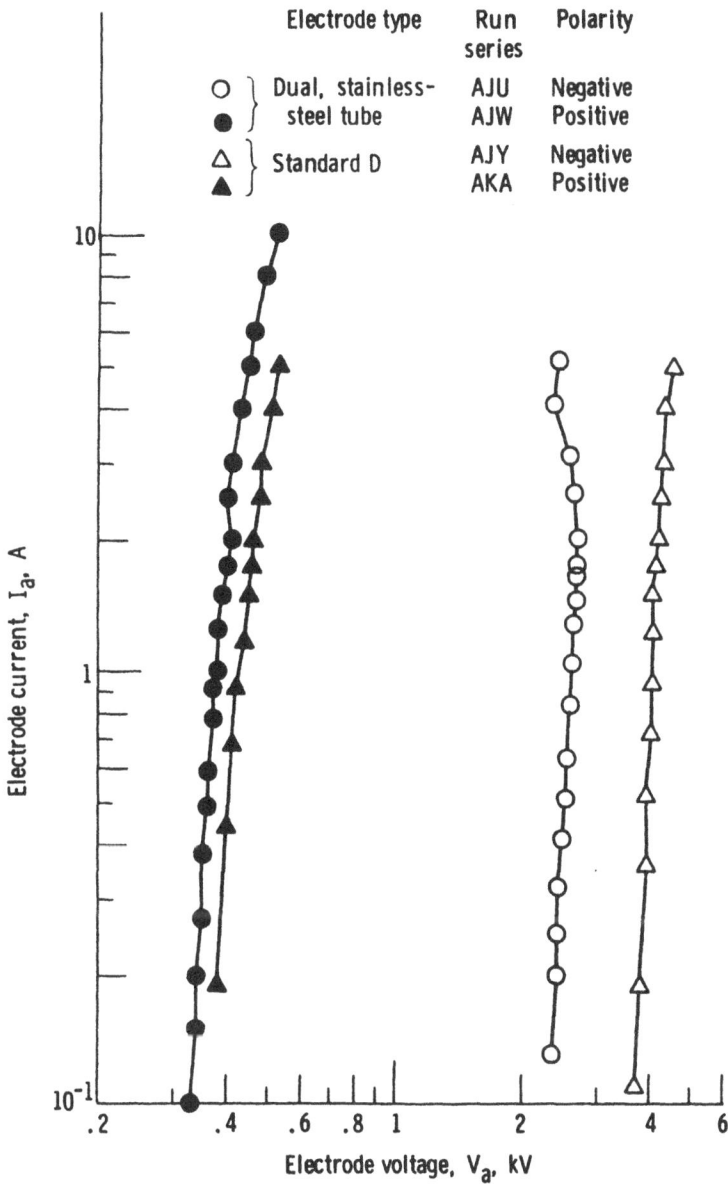

Figure 12. Low voltage current-voltage curve for a single electrode with positive or negative polarity, and two different electrode configurations.

what was at NASA-Lewis referred to as the "high pressure" to the "low pressure" modes. The physical process responsible for this mode transition was never identified, although it is known that the high pressure mode, to the upper left of these curves, is the higher density, more interesting mode of operation. Similar mode transitions are also seen in industrial glow discharge plasmas.

One issue in biasing a toroidal plasma is whether, if the plasma is biased with more than one electrode ring, the current would be equally divided among electrode rings, or all

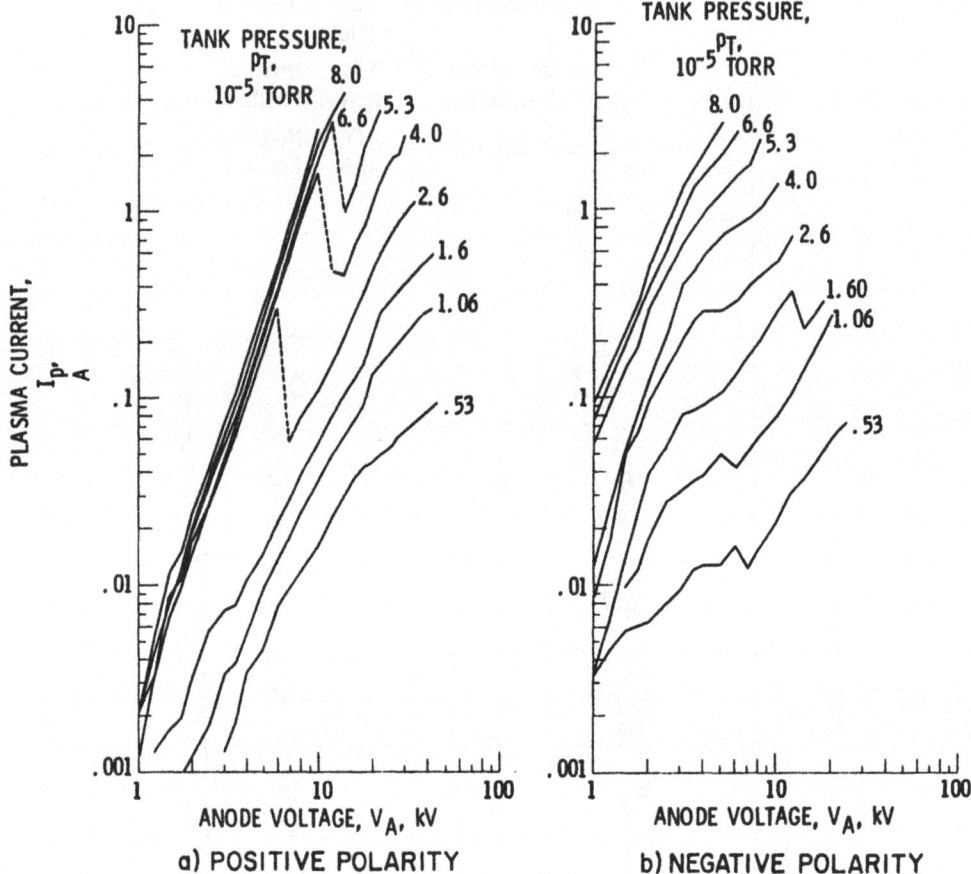

Figure 13. Current-voltage curves of the EFBT plasma, illustrating mode transitions (dotted lines) as a function of the background pressure of deuterium gas, with twelve electrode rings, and a maximum magnetic field of 2.4 Tesla. a) Positive polarity, b) Negative polarity.

flow to a single preferred electrode. To answer this question, the EFBT plasma was operated with all 12 electrode rings in place, and the current flowing to each electrode ring individually metered. The currents flowing to the 12 individual electrodes are shown in Figure 14 for negative polarity, a maximum magnetic field of 2.4 Tesla, a background gas pressure of 5×10^{-5} Torr of deterium gas, and 3 electrode voltages. It is clear that these electrodes draw current equally around the major circumference of the the EFBT plasma, with no one electrode dominating the current collection.

It was possible to bias the EFBT plasma with as few as one electrode at positive or negative polarity. When this was done, it became an interesting question whether the plasma was of equal density around the major circumference of the torus, or whether the electrostatic biasing effect is confined near the sector where the biasing electrode was located. To resolve this question, the plasma was biased with a single electrode. That electrode was moved from sector to sector around the major circumference of the torus, with respect to the sector in which the microwave interferometer (which measured the plasma density) was located. The results showed that biasing at one sector was sufficient to produce an essentially uniform plasma density around the entire circumference of the torus.

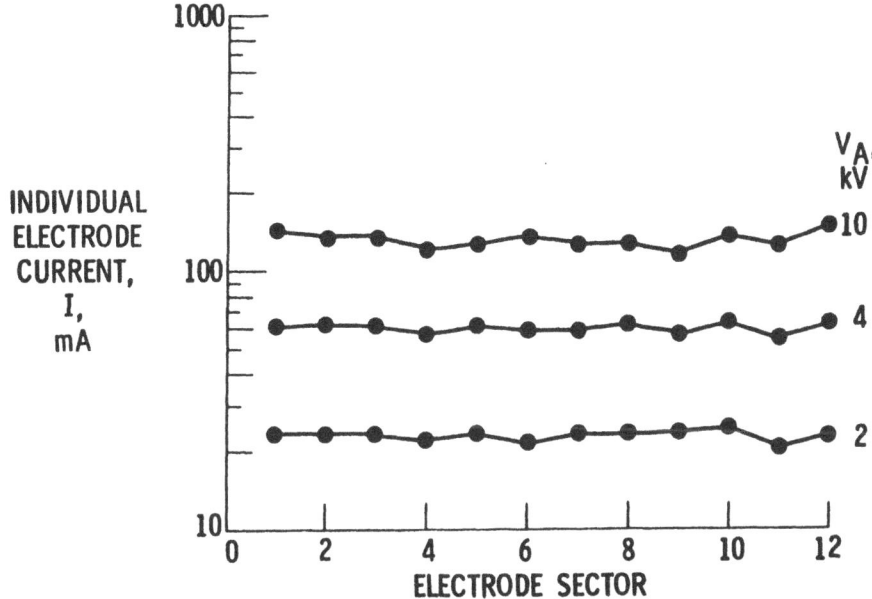

Figure 14. The current flowing to each of twelve individual electrodes for three negative biasing voltages under operating conditions with a maximum magnetic field of $B_{max} = 2.4$ Tesla, and 5×10^{-5} torr of deuterium gas.

The total electrode/limiter area in contact with the plasma is a major parameter determining the particle containment time and number density. It can be argued that if one doubles the electrode/limiter contact area, then the average particle being collected by the electrode will take only half as long to find the "hole" in the plasma, thus cutting the particle containment time in half. This general trend is illustrated in Figure 15, which shows the particle containment time as a function of average electron number density for two electrodes, one the fine tungsten wire shown in Figure 5c, and the second electrode with a much larger surface area consisting of two stainless steel tubes, like the single tube shown in Figure 5b, adjacent to each other. Figure 15 clearly shows the improvement of density and containment time possible by decreasing the electrode surface area in contact with the plasma. This improvement was not in direct ratio to the electrode surface area, probably because the sheath area (of undetermined size) determined the effective area ratio, and not the surface area of the electrodes themselves.

The Lewis EFBT experiment allowed the polarity to be changed as an independent variable, and thus to assess the effectiveness of radially inward or outward electric fields in confining the plasma. A characteristic example of data showing the effect of electrode polarity on particle containment time is shown in Figure 16. The data with negative electrode polarity, which created radially inward electric fields that pushed ions into the plasma, has a far greater density and confinement time than the positive polarity, for which the electric field pointed radially outward and pulled ions out of the plasma. This large factor of improvement, a factor of 10 or more in particle containment time, was accompanied by a similar increase in number density. Such results were a consistent feature of the paired comparison of polarities from the EFBT plasma. It became clear during this course of experiments that the plasma was far better contained by strong radially inward electric fields than electric fields that pulled ions out of the plasma.

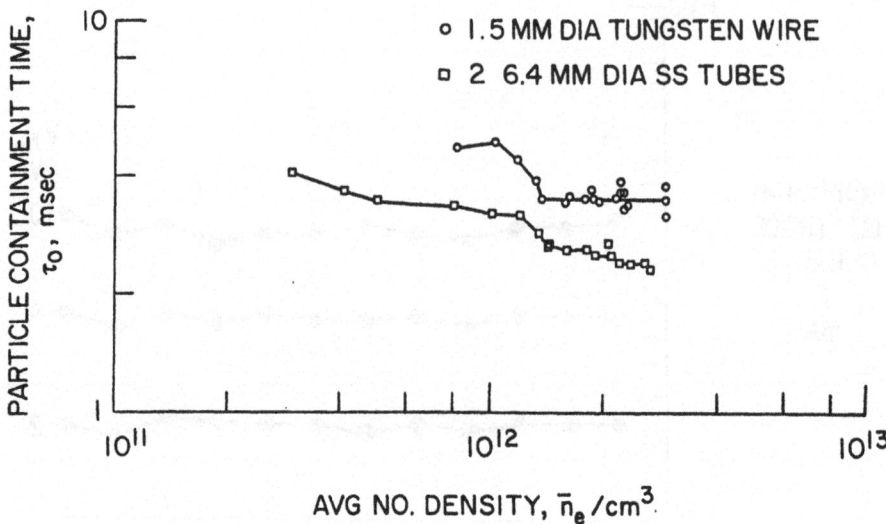

Figure 15. The particle containment time as a function of the average electron number density for biasing by a single electrode consisting of a 1.5 millimeter diameter tungsten wire like that shown in Figure 5c, and two 6.4 millimeter diameter stainless steel tubes like the one shown in Figure 5b.

Scaling Law for Particle Containment Time

Measurement of the particle containment time in the EFBT plasma is particularly simple, because the negatively biased electrodes/limiters intercept essentially the entire ion current from the plasma. A series of investigations (Refs. 1,2) indicated that there were no significant parasitic currents flowing outside the plasma volume which were not intercepted by the electrodes, and the very high potentials used assured that no ambipolar currents flowed to either electrode. Under these conditions, the total DC current flowing to the electrodes/limiters is given by

$$I_A = \frac{\bar{n}_e e V_p}{\tau_p} \text{ amps,} \quad (4)$$

where $\bar{n}_e$ is the average electron number density in the plasma, V_p is the plasma volume, approximately 82 liters in this case, and τ_p is the particle containment time. When the electrodes are negatively biased, the electrode current consists of ions leaving the plasma and being collected on the electrodes/limiters.

Equation 4 can be manipulated to give the particle containment time in terms of experimental variables,

$$\tau_p = \frac{\bar{n}_e e V_p}{I_A} \text{ sec.} \quad (5)$$

Thus, the particle containment time in the EFBT plasma can be determined by a measurement of the plasma volume (which is virtually constant at 82 liters independent of operating conditions); a measurement of the average electron number density with the

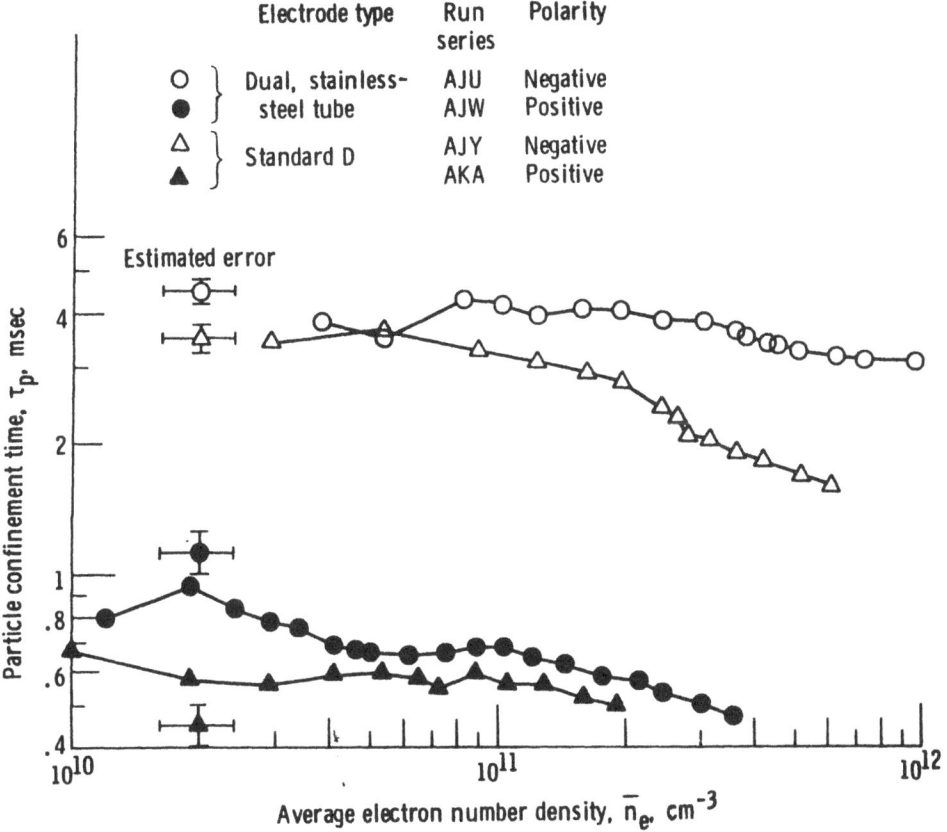

Figure 16. The average particle containment time as a function of the average electron number density for positive and negative polarities of two different electrode ring configurations.

microwave interferometer shown on Figure 9; and a measurement of the total current flowing to the electrode rings/limiters. The particle containment times so determined range from a few tens of microseconds for positively biased plasmas which pushed ions out of the containment volume, up to 6 milliseconds, for negatively biased plasmas in which radially inward fluctuation-induced ion transport occurred.

The total ion current flowing to the electrodes/limiters may be written as

$$I_A = \frac{eNA\bar{v}_i\tilde{n}_i}{4} = \frac{eNA\tilde{n}_i}{4}\sqrt{\frac{8eT_i'}{\pi M}} \text{ amps}, \qquad (6)$$

where N is the total number of electrodes/limiters, A is the contact area of each electrode/limiter with the plasma, v_i is the ion thermal velocity, and $\tilde{n}_i$ is the ion number density at the sheath boundary surrounding the biasing electrode/limiter. The ion thermal velocity v_i may be written in terms of the ion kinetic temperature T_i' and the ion mass M as shown in Equation 6. Equation 6 indicates quite clearly that the total current drawn by the power supply is proportional to the contact area of the plasma with the electrodes/limiters, and to the total number of electrodes used to bias the plasma.

By substituting the right-hand member of Equation 6 into Equation 5, one can obtain a scaling law for the particle containment time as a function of the plasma parameters as follows,

$$\tau_p = \frac{4\bar{n}_e V L_p}{NA\tilde{n}_i} \sqrt{\frac{\pi m}{8eT_i}} \cdot \sec.$$

(7)

This scaling law is independent of the magnetic induction, in agreement with experimental measurements on the EFBT plasma. If the plasma is quasineutral, and singly ionized (as the deuterium plasmas of the EFBT experiment were), the ratio of the average electron number density, $\bar{n}_e/\tilde{n}_i$, to the ion number density at the edge of the electrode sheath, $\tilde{n}_i$, will probably be a constant somewhat greater than unity. Equation 7 therefore suggests also that the particle containment time will be nearly independent of electron number density, since $\bar{n}_e$ and $\tilde{n}_i$ should be proportional to each other as the plasma conditions are changed. The scaling law of Equation 7 indicates quite clearly that the particle containment time is inversely proportional to the number of electrodes used to bias the plasma and to the contact area A of an individual electrode/limiter with the plasma. The particle containment time also is directly proportional to the plasma volume. These latter three dependences, on plasma volume, number of electrons, and electrode contact area, are the expected result of the ion containment time depending on the size of the electrode loss area available to it, just as the time required to empty a tank of water is dependent on the number and size of the holes in the tank.

If we define τ_o as the particle containment time resulting from the contact of a single biasing electrode with the plasma, the containment time resulting from the loss to N electrodes will be given by

$$\tau_e = \frac{\tau_o}{N}.$$

(8)

If there is an intrinsic loss process in addition to loss of ions on the electrodes, and if that intrinsic ion containment time is τ_{in}, then the total particle containment time τ_p which results from loss on the electrodes and the intrinsic loss process is given by

$$\frac{1}{\tau_p} = \frac{1}{\tau_e} + \frac{1}{\tau_{in}}.$$

(9)

Rearranging, the particle containment time in Equation 9 is given by

$$\tau_p = \frac{\tau_o \tau_{in}}{\tau_o + N\tau_{in}}.$$

(10)

From Equation 4, one can write the average electron number density in the plasma in the form

$$\bar{n}_e = \frac{I_A \tau_p}{eV_p} = \frac{I_A \tau_o \tau_{in}}{eV_p(\tau_o + N\tau_{in})},$$

(11)

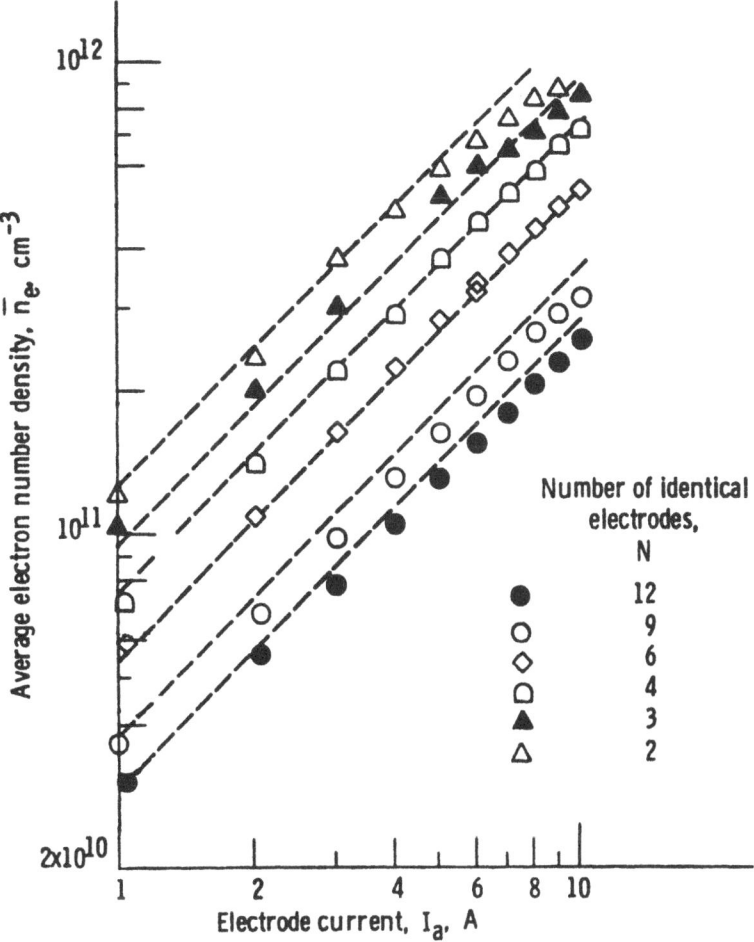

Figure 17. The average electron number density as a function of electrode current and the number of identical electrodes used to bias the plasma.

thus indicating that the average electron number density should be directly proportional to the current flowing to the electrodes, and inversely proportional to the total plasma volume, relationships which have been experimentally confirmed.

On Figures 17 thru 19 are some confirmatory data of the relationships discussed above. Figure 17 shows the average electron number density as a function of the electrode current for 6 different numbers of identical electrode/limiters, N. These data illustrate the direct proportionality of electron number density to current, confirming the relationship of Equation 11, and these data also show the inverse relationship between the electron number density and number of electrode rings in contact with the plasma, implied by Equation 7.

On Figure 18 is a plot of the particle containment time as a function of the electrode current for various numbers of identical electrode rings, N. These data show that the particle containment time is independent of electrode current, implied by Equation 7 if the ratio $\bar{n}_e/\bar{n}_i$ is approximately constant. The data in Figure 18 also exhibit the inverse dependence of the particle containment time on the number of electrode rings, N. or, alter-

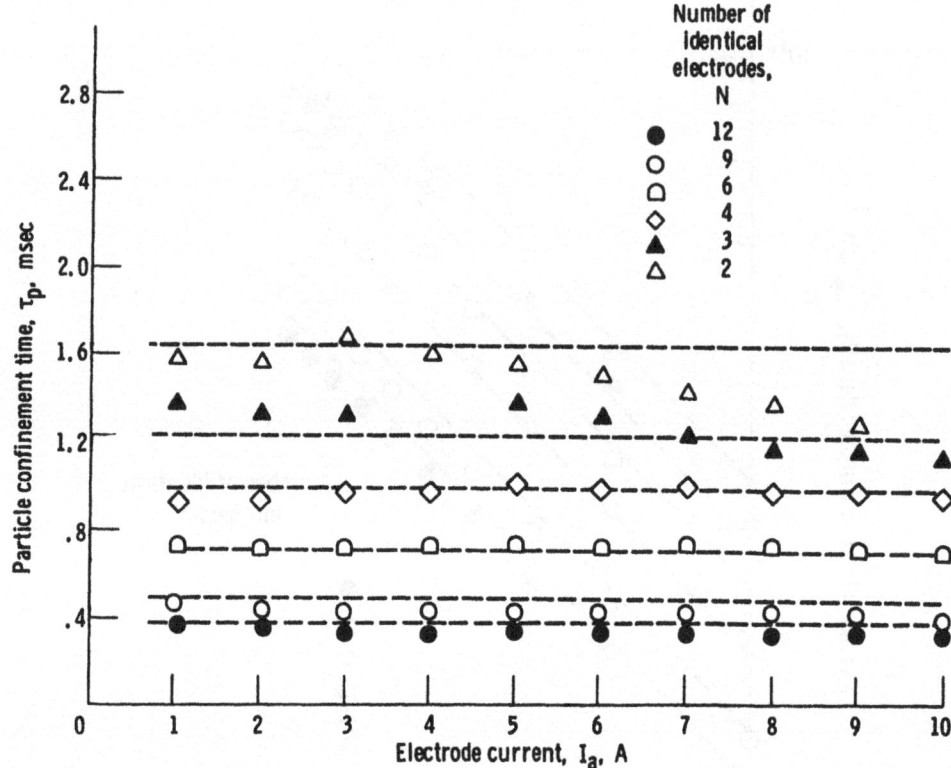

Figure 18. The particle containment time as a function of the electrode current and the number of identical electrodes used to bias the plasma.

natively, the inverse relationship of the particle containment time on the total area of electrode/limiters in contact with the plasma, NA. Figure 19 confirms the relationship predicted by Equation 11, in which the particle containment time is directly proportional to the specific plasma volume, the cubic meters of plasma per biasing electrode. Two sets of experimental data are shown, for which the intrinsic particle containment times, τ_{in}, are respectively 5.0 and 12.4 milliseconds. If the data fell along the line for which $\tau_{in} = \infty$, that would imply that there were no losses in addition to those of ions on the electrodes.

Magnetoelectric Heating in the EFBT Plasma

The strong radial electric fields apparent in Figures 7 and 8 result in a high E/B azimuthal drift velocity of the plasma. The radial electric fields, on the order of kilovolts per centimeter, are sufficiently high that deuterium ions can acquire energies on the order of kilovolts. It is this E/B heating mechanism, discussed above in connection with Eqs. 1 and 2, that is responsible for the observed fact that the ion kinetic temperatures in the EFBT plasma are at least two orders of magnitude greater than the electron kinetic temperatures, as indicated in Table 2.

A series of experiments were conducted on the EFBT plasma with both positive and negative electrode polarities to determine the nature of the rotating spokes responsible for ion heating in this plasma. The placement of capacitive probe arrays at several locations

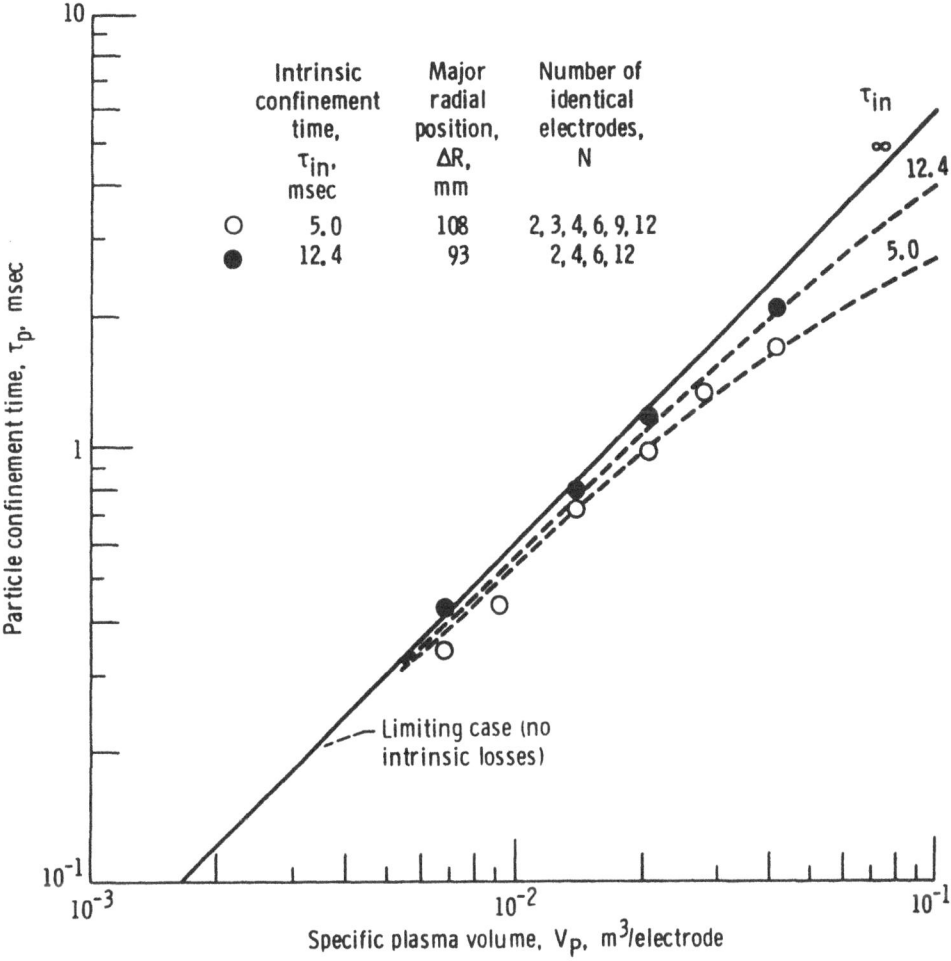

Figure 19. The particle containment time as a function of the specific plasma volume, in cubic meters per electrode, illustrating the proportionality of particle containment time to the plasma volume.

around the major circumference of the torus, and at several locations within each sector, showed that the rotating spokes in the EFBT plasma usually (but not always) were coherent around the major circumference of the torus, in the manner illustrated on Figure 20 for negative midplane electrodes. When the plasma was positively biased, the direction of rotation of these spokes was opposite to that shown in Figure 20. Coherency of the spokes around the major circumference was the usual way in which the plasma operated, even when the plasma was biased by a single electrode in one sector.

As implied by Equation 2, the velocity of the E/B rotating spokes should be proportional to the square root of the ion energy. Some data illustrating this for the high pressure mode are shown in Figure 21, in which the ion spoke rotation frequency, measured by a capacitive probe, is plotted as a function of the ion kinetic temperature, measured by a charge exchange neutral energy anlayzer, for deuterium gas. The data for a deuterium plasma with B_{max} = 2.4 Tesla indicate a proportionality of the spoke rotation frequency to the square root of the ion kinetic temperature with no adjustable parameters.

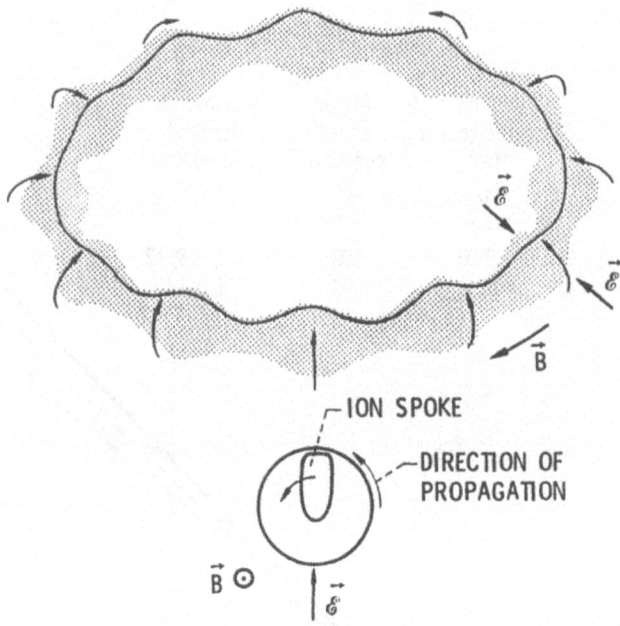

Figure 20. Experimentally observed ion spoke rotation with negative midplane electrodes and a negative bias on the plasma. The rotating spokes were in nearly all cases coherent around the major circumference of the torus. The spokes rotated in the opposite sense for positive plasma bias.

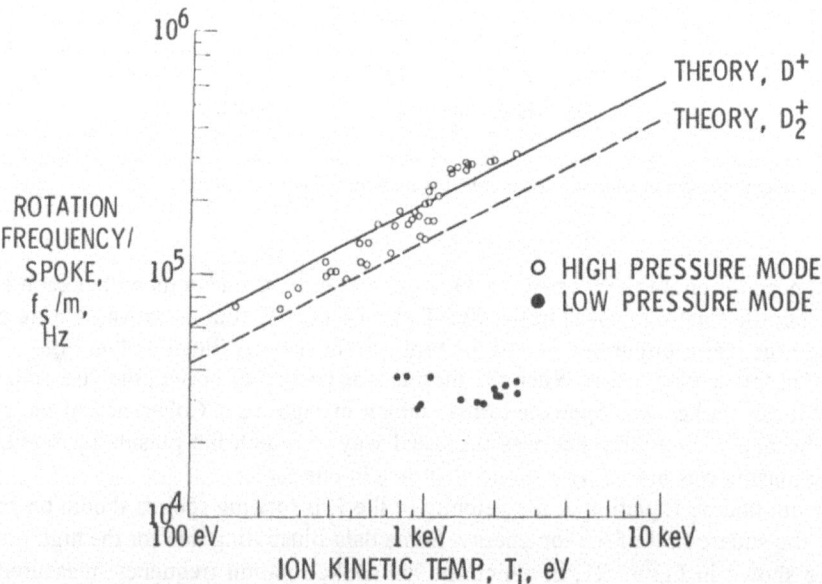

Figure 21. The relation between ion spoke rotation frequency and ion kinetic temperature, exhibiting the square root dependence expected for magnetoelectric E/B heating. Data shown are for a single positive midplane electrode with $B_{max} = 2.4$ Tesla in deuterium gas.

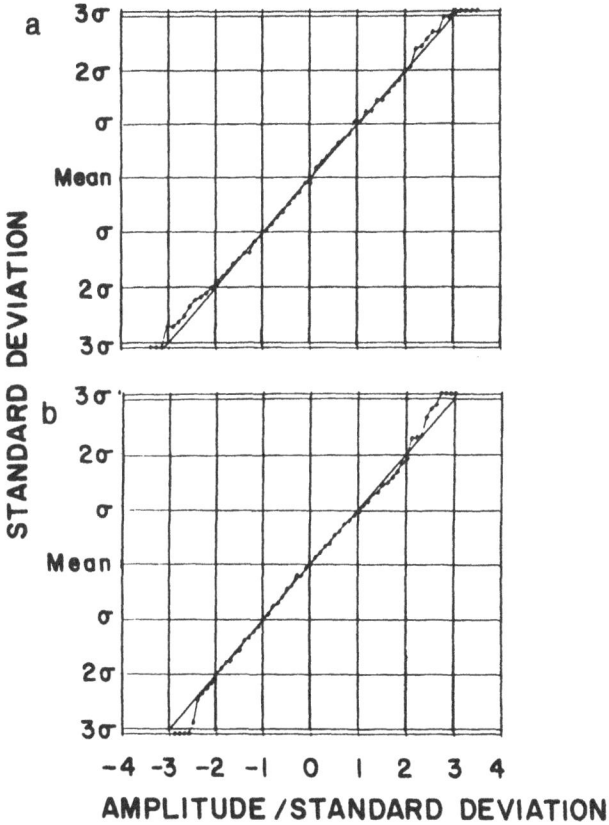

Figure 22. Fluctuations of a) potential and b) electron number density at the edge of the plasma, plotted on probability paper. The straight lines indicate thermalized, Gaussian distributions of fluctuations of these two quantities.

Analysis of the time series containing the electrostatic potential and electron number density fluctuations showed that under most circumstances of operation, the fluctuations were random Gaussian noise. An illustration of this is shown in Figure 22, in which the cumulative probability distribution function is plotted as a function of the amplitude, normalized with respect to the standard deviation, to yield a probability plot. The fact that this cumulative distribution is a straight line on the probability plot over 3 sigma indicates Gaussian random noise, which was characteristic of this plasma over much of its operating range.

An autopower spectrum of the turbulent fluctuations of electrostatic potential is shown in Figure 23 over a frequency range from 100 kilohertz to 1 megahertz, under conditions for which the ion spoke rotation frequency was approximately 300 kilohertz. The background turbulent spectrum has a characteristic power law relationship given by

$$\phi = \phi_o v^{-m}, \tag{12}$$

where the exponent m depends on the plasma operating conditions. Above the spoke or energy input frequency at 300 kHz, the turbulent spectrum is enhanced by approximately a factor of 3 as the result of an energy cascade from the spoke input frequency up to higher

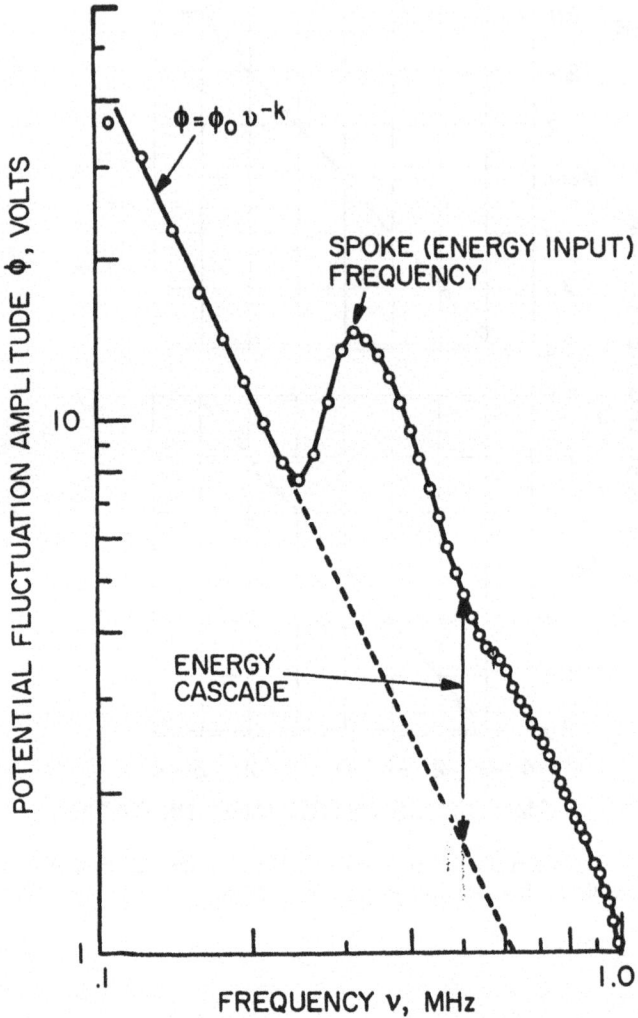

Figure 23. Autopower spectrum of electrostatic potential fluctuations in a magnetoelectrically heated plasma. The spoke (energy input) frequency produces an energy cascade responsible for maxwellianizing the E/B drift energy of the ions.

frequencies. These data are analogous to what one would observe in the autopower spectrum of pressure fluctuations downstream of a rotating fan in ordinary fluid dynamics. A pressure sensitive detector would see fluctuations at the fan rotational velocity, and an enhanced spectrum at frequencies above the fan rotational velocity, as the turbulence decayed into smaller and smaller eddies with higher and higher characteristic frequencies. Ultimately, the energy cascades to such a small scale in both the fluid and plasma cases that it is indistinguishable from thermal motions, and the macroscopic rotational energy of the spoke is therefore converted into random thermal motions of the ions.

Some examples of ion energy distribution functions from the NASA-Lewis magnetoelectrically confined plasmas are shown in Figure 24. Figure 24a is an example in helium gas, in which charge-exchanged helium ions were detected out to energies of 22 keV,

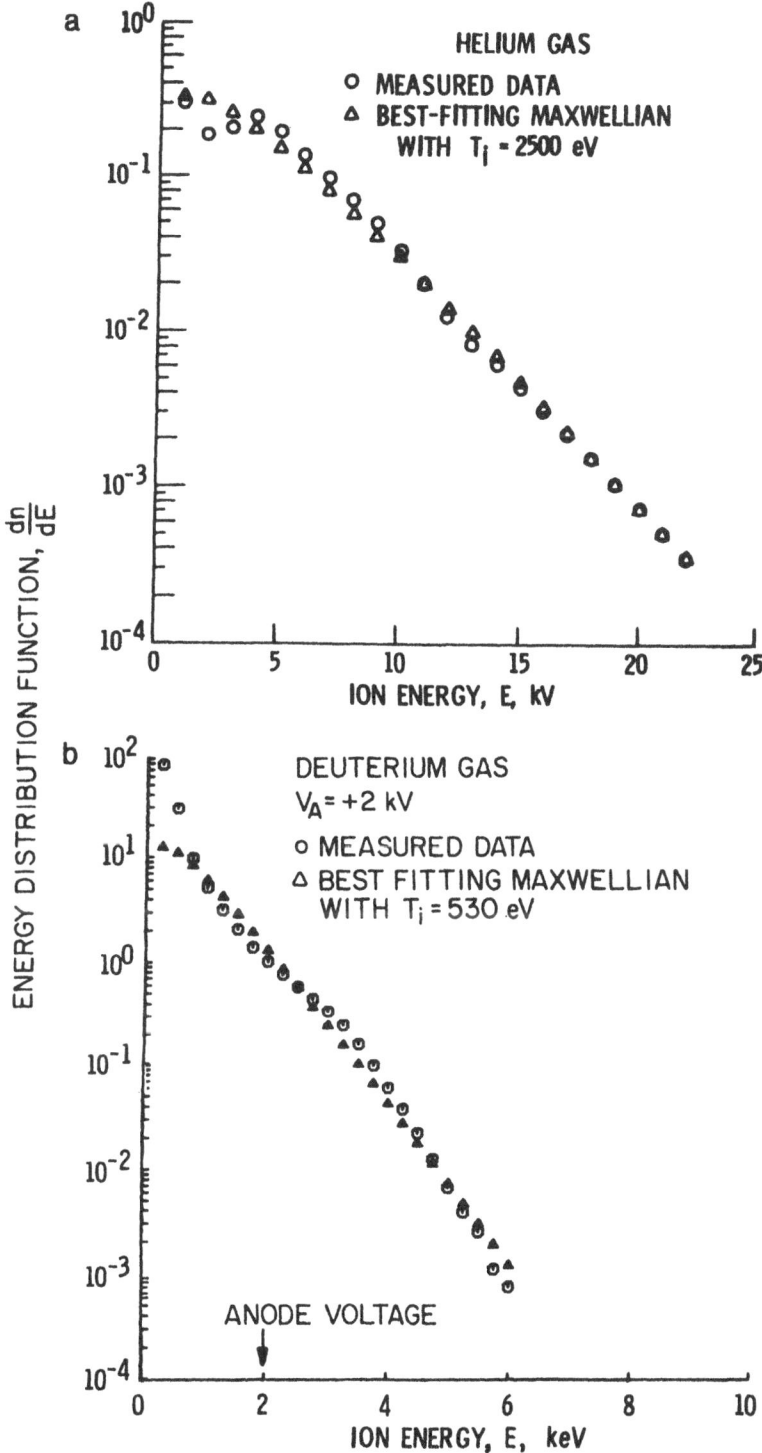

Figure 24. Characteristic examples of the ion energy distribution function in a magnetoelectrically heated E/B plasma. Open circles are measured data, open triangles are the best-fitting maxwellian. a) Data taken in helium gas with T_i = 2500 eV and ions detected out to 22 keV. b) deuterium gas with ion kinetic temperature T_i = 530 eV, an anode voltage of 2 kV, and ion energies observed out to 6 keV.

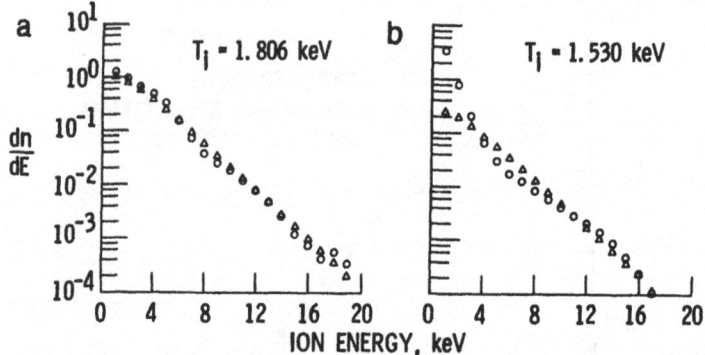

Figure 25. Demonstration that magnetoelectric heating takes place around the entire torus. a) the electrode ring and neutral particle analyzer in the same sector, yielding an ion kinetic temperature of 1.8 keV; b). The anode ring and neutral particle analyzer in opposite sectors of the bumpy torus, yielding an ion kinetic temperature of 1.53 keV.

and the best-fitting maxwellian to the experimental data yielded a helium ion kinetic temperature of 2500 eV. Figure 24b is an ion energy distribution function taken in deuterium gas for which the best-fitting maxwellian had a kinetic temperature of 530 eV. These data are particularly significant, since the anode voltage which produced the data in Figure 24b was only 2 kilovolts, at which there was no anomaly in the ion energy distribution function, indicating that the ion energies were truly thermalized.

An important concern about magnetoelectric heating of toroidal plamas is whether the radial electric fields and the heating mechanism would be sustained with equal effect all around the toroidal circumference, or whether they would be confined to a small region in the vicinity of the biasing electrode. Figure 25 addresses this question, and shows the effect of moving the single biasing electrode ring with respect to the neutral particle energy analyzer from the same sector (the data shown on the left), to the opposite sector of the bumpy torus (shown on the right). This removal of the ion energy measurement point halfway around the torus resulted in only a relatively small decrease in the ion kinetic temperature, amounting to only 15% in this case.

The experimental program at NASA-Lewis yielded many more interesting data related to ion and electron heating in the EFBT, too numerous to discuss here, but which is summarized in the references cited at the end of the paper (Refs. 1,2, and 12–15). Experimentally, electric fields above 100 volts per centimeter are found to penetrate the EFBT plasma to radial distances of at least 7 centimeters, nearly to the plasma minor axis. The e-folding distance of these electric fields ranges from 1.5 to 10 centimeters, at least 10 to 100 times the local Debye length. It was usually found that the ion energy distribution functions are maxwellian. The maxwellian distributions are the result of very strong, Gaussian electrostatic turbulence in the plasma. In deuterium gas, the EFBT plasma has yielded ion kinetic temperatures, measured with a charge exchange neutral energy analyzer, which have ranged from 200 eV to 2500 eV.

Fluctuation-Induced Transport in the EFBT Plasma

One of the major accomplishments of the NASA-Lewis Electric Field Bumpy Torus research program was the identification and experimental investigation of radially inward ion transport in the EFBT plasma, due to fluctuation-induced transport in the strong, radially inward electric fields. This research program was accomplished with the able collaboration of Prof. Edward J. Powers, Jr. and his students at the University of Texas at Austin, during the years 1974 thru 1978. This group developed sophisticated computer software which allowed time series data from 2 capacitive and 1 Langmuir probe to be analyzed by fast Fourier transform techniques. These computer programs produced a wide range of information on the properties of the plasma fluctuations, including the transport of ions across the magnetic field lines. The software developed by the University of Texas group under NASA contract formed the basis for subsequent software used to study fluctuation-induced transport in tokamak plasmas. This work within the Lewis program is discussed in references 1 and 2, 16–19, 26–28, and 30–32.

An example of this collaborative research on fluctuation-induced transport in the EFBT plasma is shown in Figure 26. On the left are the transport and cumulative transport spectrum for radially inward ion transport in the EFBT plasma. This transport was associated with a radially inward electric field produced by negative bias of the plasma, and by proper adjustment of a weak vertical magnetic field applied to the bumpy torus plasma as a whole. In this case, the inward transport was driven by 3 oscillatory modes, with frequencies below 100 kilohertz. On the right side of Figure 26 are the transport and cumulative transport spectrum for positive bias of the EFBT plasma, in which the electric field pulled ions radially outward. The sign of the transport and cumulative transport has reversed with respect to Figure 26a, as a result of the reversal of direction of the radial electric field. In this case the transport is due to a broad spectrum of fluctuations, out to approximately 300 kilohertz. Figure 27 shows how the average plasma density, particle containment time, and cumulative transport rate depend on the polarity of the biasing potential and the electrode current. These data indicate an inward or outward transport that is proportional to electrode current. They also indicate that the number density and containment time are far greater for negative polarity, as a result of inward transport of ions at the surface of the plasma.

One of the unexplained observations made in the EFBT experiment was that the average electron number density of the plasma, as well as the cumulative particle transport, were both strong functions of a weak vertical magnetic field applied to the plasma. Characteristic data relating to this effect are shown in Figure 28. The EFBT vacuum tank shown in Figure 3 was wrapped with 2 coils above and below the equatorial plane of the torus, on the outside wall of the vacuum vessel. These coils allowed a weak vertical magnetic field up to ± 10 millitesla (± 100 Gauss) to be imposed on the plasma. No theoretical consideration led us to expect a strong effect of this vertical field on the plasma confinement characteristics, but in the event, the number density, containment time, and the sign and magnitude of the radial transport all turned out to be a very strong function of this weak vertical magnetic field. Detailed examination of the fluctuation-induced transport data indicated rather clearly that the phase angle between density and potential fluctuations was closely linked to the magnitude and sign of this weak vertical magnetic field. However, the physical process responsible for this linkage was not identified before the experiment was terminated.

DISCUSSION

The ability of the EFBT plasma to operate in the steady-state, and the reliability of the facility, made it possible to take data for a wide range of experimental conditions re-

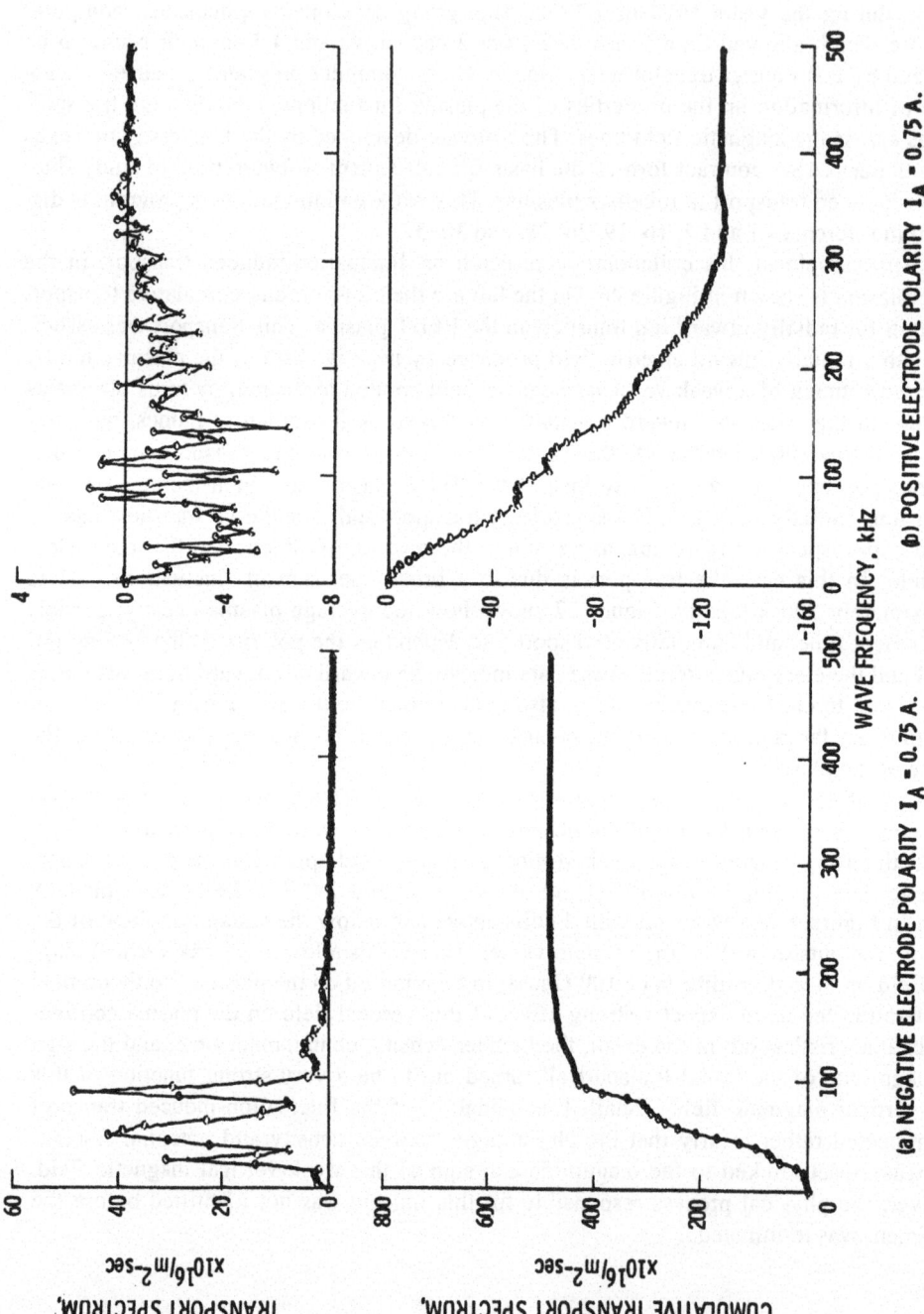

Figure 26. Transport spectra and cumulative transport spectra for a) negative and b) positive electrode polarity, illustrating reversal of the direction of radial ion transport.

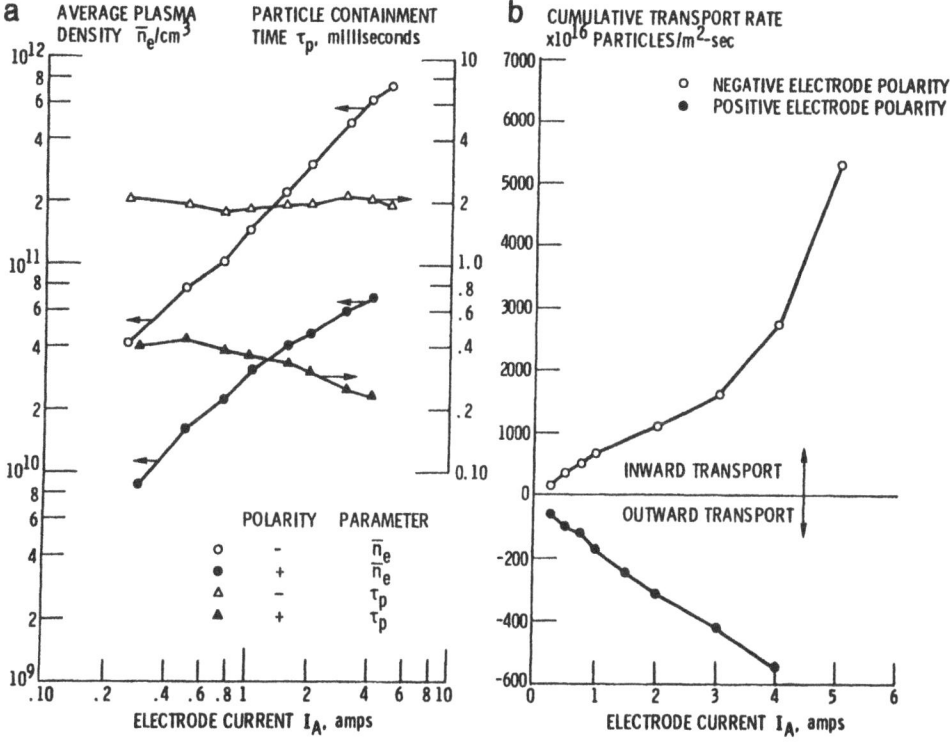

Figure 27. Data illustrating the improved electron number density and confinement time possible when the negatively biased toroidal plasma produces inward ion transport.

lated to electrostatic biasing of this toroidal plasma. It is probably significant that our investigations of the rotating spokes, the heating effect of a single electrode ring, and other parameters were not strong functions of the distance away from a single biasing electrode along the major circumference. The EFBT plasma tended to operate more or less like a good toroidal electrical conductor, with very little asymmetry around the major circumference, in spite of the fact that the plasma was often biased by a single electrode in one sector, and had an axial variation in magnetic induction of 2.5:1 within each sector. This observation is an encouraging indication that asymmetries around the major circumference of axisymmetric devices like tokamaks should not be significant to the outcome of magnetoelectric toroidal biasing. If small or few asymmetries are evident in an EFBT plasma, they should be even less obvious and less significant in axisymmetric plasmas.

The Electric Field Bumpy Torus is also one of the very few fusion grade magnetic containment concepts, the confinement time scaling law for which is known. The data presented in support of the scaling law of Equations 7 and 11 give a relatively simple picture of the confinement in the EFBT plasma. The major ion loss from the plasma is by collection on the surface of a negatively-biased electrode in contact with the plasma. If one cuts this contact area in half, the average particle containment time will double, since the average ion will then take twice as long to find the hole in the plasma through which it can escape to the biasing electrode. It is a major finding of the EFBT experiment that the electrostatic containment of a toroidal plasma is primarily determined by the confinement of ions. If the ions are well confined, then the electrons seem to take care of themselves. It

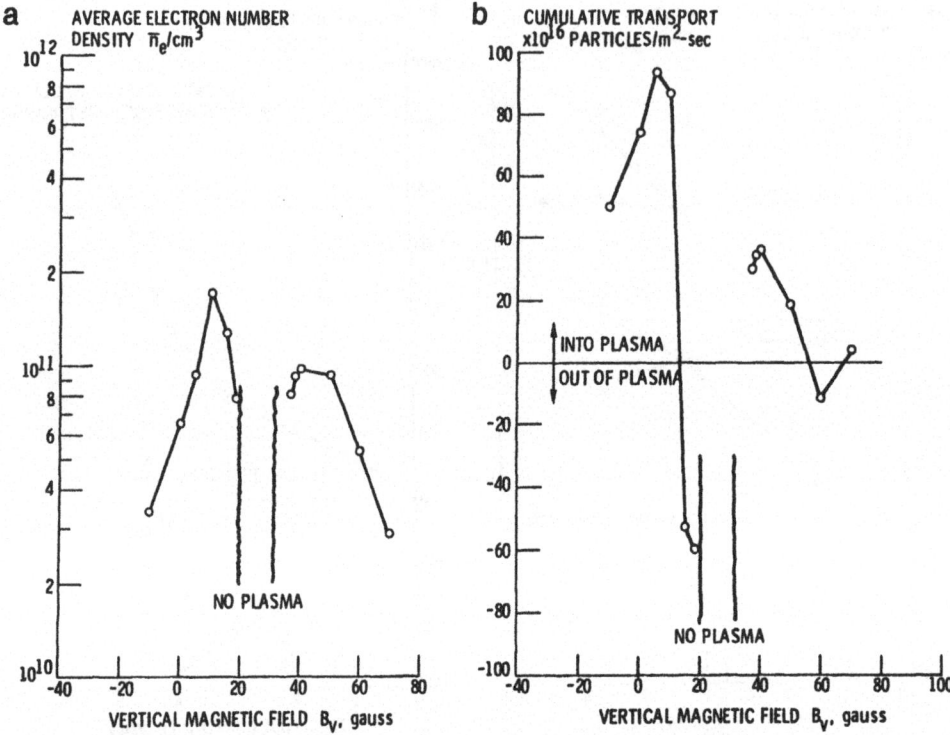

Figure 28. Data which illustrate the effect of a weak vertical magnetic field on a) the average electron neutral number density and b) the magnitude and direction of radial ion transport in the plasma.

appears probable that these results from the EFBT experiment carry over to all magnetoelectrically confined toroidal plasmas.

It is interesting to speculate what might happen if some of the larger machines currently in operation, such as TFTR and JET, were operated as magnetoelectrically confined toroidal plasmas rather than tokamaks. In examining the scaling law for particle containment time of Equation 7, we assume that the ratio of average electron number density to the ion number density at the electrode sheath, $\bar{n}_e/\bar{n}_i$, is the same in JET or TFTR as it is in the EFBT plasma. We further assume that the ion kinetic temperature, ion mass, and total electrode surface area in contact with the plasma, NA, are also the same in the larger experiments. That being the case, the particle containment time in the JET and TFTR, apparatus would be larger by the ratio of the volume of these machines to that of the EFBT.

Table 2 indicates that our best containment times were 6 milliseconds in the EFBT, (already a decade higher than Bohm diffusion) with a volume of 82 liters. The volume of the JET device is approximately 153 cubic meters, a volume ratio about 1,870 times that of the EFBT plasma. If the particle containment time scaled as the volume, under the above assumptions, the particle containment time in JET would be 11.2 seconds! Similarly, if the same assumptions were to hold for TFTR, and the volume of the TFTR plasma is approximately 35 cubic meters, yielding a volume ratio about 430 times that of the EFBT. The volume scaling of Equation 7 would then yield a 2.6 second particle containment time for TFTR if it were operated as a magnetoelectrically confined toroidal plasma, with the same electrode contact area and other conditions observed in the NASA-Lewis EFBT.

The above speculative calculation indicates some of the potential payoffs of the electrostatic biasing of tokamak plasmas. Not only does electrostatic biasing allow one to do work on the plasma, thus forcing ions inward against the density gradient, but also, this inward transport greatly improves the containment time and number density otherwise possible, while at the same time heating the plasma ions preferentially to kilovolt kinetic temperatures. These are all powerful reasons for further research on electrostatically biased toroidal plasmas.

REFERENCES

1. Roth, J. R.: "Ion Heating and Containment in an Electric Field Bumpy Torus (EFBT) Plasma". *Nuclear Instruments and Methods*, vol. 207, Nos. 1, 2 (1983) pp. 271–299.
2. Roth, J. R.; Krawczonek, W. M.; Powers, E. J.; Kim, Y. C.; and Hong, J. Y.: "Ion Confinement and Transport in a Toroidal Plasma with Externally Imposed Radial Electric Fields". NASA TP 1411, (1979).
3. J. R. Roth: "Summary of the IAEA Technical Committee Meeting on Tokamak Plasma Biasing, Montreal, Canada, Sept. 8–10, 1992". *Fusion Technology*. vol. 23 (1993) pp 246–252.
4. L. Oren, S. Talmadge, R. J. Taylor, D. Whelan, and S. J. Zweben: "Electrostatic Confinement in Macrotor", *APS Bulletin*, vol. 24, No. 8(1979) p. 1109.
5. Vladimir Demchenko, Editor: "Tokamak Plasma Biasing", Proceedings of the IAEA Technical Committee Meeting at Centre Canadien de Fusion Magnetique, Montreal, Canada, Sept. 8–10, 1992. Available from: Dr. Vladimir Demchenko, IAEA, Wagramerstrasse 5, P. O. Box 100, A-1400 Vienna, Austria.
6. M. J. Schaffer, et al.: "Edge Effects During DIII-D Divertor Biasing." IBID. Ref. 5 above, pp 340–374.
7. J. R. Roth, "Electrostatic Biasing and Radially-Inward Transport in a Magnetoelectrically Confined Toroidal Plasma". IBID Ref. 5 above, pp 131–187.
8. Roth, J. R.: "Origin of Hot Ions Observed in a Modified Penning Discharge". *Phys. Fluids*, vol. 16, no. 2 (1973) pp. 231–236.
9. Roth, J. R.: "Hot Ion Production in a Modified Penning Discharge." *IEEE Trans. Plasma Sci.*, vol. PS-1, no. 1 (1973) pp. 34–45.
10. Roth, J. R.: "Energy Distribution Functions of Kilovolt Ions in a Modified Penning Discharge". *Plasma Phys.*, vol. 15, no. 10 (1973) pp. 995–1005.
11. Roth, J. R.: "Experimental Study of Spectral Index, Mode Coupling, and Energy Cascading in a Turbulent, Hot-Ion Plasma". *Phys. Fluids*, vol. 14, no. 10 (1971) pp. 2193–2202.
12. Roth, J. R.; Gerdin, G. A.; and Richardson, R. W.: "Characteristics of the NASA Lewis Bumpy Torus Plasma Generated with Positive Applied Potentials". *IEEE Trans. Plasma Sci.*, vol. PS-4, no. 3 (1976) pp. 166–176. (Also NASA TN D-8114, 1976.)
13. Roth, J. R.; and Gerdin, G. A.: "Characteristics of the NASA Lewis Bumpy Torus Plasma Generated with High Positive or Negative Applied Potentials." *Plasma Phys.*, vol. 19, no. 5, (1977) pp. 423–446. (Also NASA TN D-8211, 1976.)
14. Roth, J. R.: "Factors Affecting Ion Kinetic Temperature, Number Density, and Containment Time in the NASA Lewis Bumpy-Torus Plasma." NASA TN D8466, (1977).
15. Roth, J. R.: "Effects of Applied DC Radial Electric Fields on Particle Transport in a Bumpy Torus Plasma." *IEEE Trans. Plasma Sci.*, vol. PS-6, no. 2 (1978) pp. 158–165.
16. Powers, E. J.: "Spectral Techniques for Experimental Investigation of Plasma Diffusion due to Polychromatic Fluctuations." *Nucl. Fusion*, vol. 14, no. 5 (1974) pp. 749–752.
17. Smith, D. E.; Powers, E. J.; and Caldwell, G. S.: "Fast-Fourier-Transform Spectral Analysis Techniques as a Plasma Fluctuation Diagnostic Tool." *IEEE Trans. Plasma Sci*, vol. PS-2, no. 4 (1974) pp. 261–272.
18. Singh, C. M.; et al.: "Fluctuation Spectra in the NASA Lewis Bumpy Torus Plasma." NASA TP-1257, (1978).
19. Singh, C. M.; et al.: "Low-Frequency Fluctuation Spectra and Associated Particle Transport in the NASA Lewis Bumpy Torus Plasma." NASA TP-1258 (1978).
20. Roth, J. R.; et al.: "Inward Transport of a Toroidally Confined Plasma Subject to Strong Radial Electric Fields." *Phys. Rev. Lett.*, vol. 40, (1977) pp. 1450-1453. (Also NASA TM-73800, 1977.)
21. Roth, J. R.; et al.: "A 12-Coil Superconducting Bumpy Torus Magnet Facility for Plasma Research." Fifth Applied Superconductivity Conference, IEEE, Inc., (1972) pp. 361–366.
22. Roth, J. R.; et al.: "Characteristics and Performance of a Superconducting Bumpy Torus Magnet Facility for Plasma Research." NASA TN D-7353 (1973).

23. Roth, J. R.: "A Model for Particle Confinement in a Toroidal Plasma Subject to Strong Radial Electric Fields." NASA TM X-73814 (1977).
24. Krawczonek, W. M.; et al." "A Data Acquisition and Handling System for the Measurement of Radial Plasma Transport Rates." NASA TM-78849 (1977).
25. Roth, J. R.; and Krawczonek, W. M.: "Paired Comparison Tests of the Relative Signals Detected by Capacitive and Floating Langmuir Probes in Turbulent Plasma from 0.2 to 10 MHz". *Rev. Sci. Instrum.*, vol. 42, no. 5 (1971) pp. 589–594.
26. Smith, D. E.; and Powers, E. J.: "Experimental Determination of the Spectral Index of a Turbulent Plasma from Digitally Computed Power Spectra." *Phys. Fluids*, vol. 16, no. 8 (1973) pp. 1373–1374.
27. Kim, Y. C.; and Powers, E. J.: "Effects of Frequency Averaging on Estimates of Plasma Wave Coherence Spectra." *IEEE Trans. Plasma Sci.*, vol. PS-5 (1977) pp. 31–40.
28. Kim, Y. C.; Wong, W. F.; Powers, E. J.; and Roth, J. R.: "Extension of the Coherence Function to Quadratic Models." *Proceedings of the IEEE*. Vol. 67, No. 3 (1979) pp 428–429.
29. Roth, J. R.; Alexeff, I.; and Mallavarpu, R.: "New Mechanism for Electromagnetic Emission near the Geometric Mean Plasma Frequency." *Phys. Rev. Lett.*, Vol. 43, No. 6 (1979) pp 445–449.
30. Kim, Y. C.; Powers, E. J.; Hong, J. Y.: Roth, J. R.; and Krawczonek, W. M.: "The Effect of a Weak Vertical Magnetic Field on Fluctuation-Induced Transport in a Bumpy Torus Plasma." *Nuclear Fusion*, Vol. 20, No. 2, (1980) pp 171–176.
31. Roth, J. R.; Krawczonek, W. M.; Powers, E. J.; Hong, J. Y.; and Kim, Y. C.: "The Role of Fluctuation-Induced Transport in a Toroidal Plasma with Strong Radial Electric Fields." *Plasma Physics*, Vol. 23, No. 6, (1981) pp 509–514.
32. Roth, J. R.; Krawczonek, W. M.; Powers, E. J.; Kim, Y. C.; and Hong, J. Y.: "Fluctuations and Turbulence in an Electric Field Bumpy Torus Plasma." *Journal of Applied Physics*, Vol. 52, No. 4, (1981) pp 2705–2713.
33. J. L. Vossen and W. Kern, Eds.: *Thin Film Processes*, Academic Press, NY (1978) ISBN 0–12–728250–5.
34. R. C. Eberhart and R. A. Seban: "The Energy Balance for a High Current Argon Arc", *Int. J. Heat and Mass Transfer*, Vol. 9 (1966) pp 939–949.

30

BALL LIGHTNING

What Nature Is Trying to Tell the Fusion Community

J. Reece Roth

UTK Plasma Science Laboratory
Department of Electrical Engineering
University of Tennessee
Knoxville, Tennessee 37996–2100

ABSTRACT

Ball lightning has been extensively reported, usually in association with thunderstorms, by chance observers who constitute perhaps five percent of the adult U.S. population. These sightings have been documented by polling such observers, and what is known about the characteristics of ball lightning is the result of information which emerged from these polls. This data base is useful because it reveals a phenomenon with very interesting implications for fusion energy, and it provides important constraints on theoretical models of ball lightning. Unfortunately ball lightning itself has not been accessible to scientific analysis, because it has not been possible to reproduce it in the laboratory under controlled conditions. Natural ball lightning has been observed to last longer than 90 seconds, and to have diameters from one centimeter to more than one meter. The energy density of a few lightning balls has been observed to be higher than 20,000 joules per cubic centimeter, well above the limit of chemical energy storage of, for example, TNT at 2000 joules per cubic centimeter. Such observations suggest a plasma-related phenomenon with significant magnetic energy storage and excellent confinement. Ball lightning should provide a novel paradigm in fusion research, as well as being of great theoretical and practical interest to the plasma research community.

INTRODUCTION

Many approaches to achieving controlled fusion are described in other papers presented at this Symposium. In this paper, I would like to discuss the phenomenon of ball lightning, which is independent of human agency, and which I hope to show has characteristics of interest for the problem of controlled fusion energy. I am not the first to notice the possible relevance of ball lightning to the problem of fusion energy. Early Russian fusion pioneers, including Peter Kapitza (Ref. 1) also took an interest in the subject of ball lightning for this reason. In the United States during the 1960's, J. Rand McNally, Jr. of the Oak Ridge National Laboratory, and Warren D. Rayle of the NASA Lewis Research Center conducted polls of the staff at their respective laboratories to find out more about the properties of ball lightning (Refs. 2, 3) because of its possible fusion relevance. Somewhat later, James Tuck of the Los Alamos National Laboratory (Ref. 4) conducted an exploratory research program which included a one-time successful attempt to generate ball lightning in the laboratory. Unfortunately these pioneering efforts did not flourish, and I therefore take the opportunity of this Symposium to call the attention of the fusion community once again to the subject of ball lightning.

Ball lightning is defined as a glowing, self-luminous sphere moving slowly or floating in the atmosphere, and is usually associated with lightning strokes or thunderstorm activity. Ball lightning is much more rare than a lightning stroke, and its occurrence only at random times and places has made its scientific study very difficult. In the United States, ball lightning has been observed by approximately 5% of the adult U.S. population at some time in their lives. What is now known about the characteristics of ball lightning is the product of a large body of chance observations by untrained observers, many of whom were participants in a frightening experience. This data base is useful, however, because it provides important constraints on theoretical models of ball lightning.

Perhaps because the phenomenon of ball lightning remains a scientific mystery, the literature of ball lightning is voluminous. Ball lightning has been placed in the context of ordinary atmospheric electrical and lightning discharges by Uman (ref. 5) and by Golde (ref. 6). Two additional books, entirely on the subject of ball lightning, have been written by Singer (ref. 7) and by Barry (ref. 8). These latter two books contain some of the better documented sightings of ball lightning, summarize information on its characteristics, discuss inconclusively various theoretical models for ball lightning, and report attempts to reproduce ball lightning artificially under controlled conditions.

The characteristic observation of ball lightning occurs approximately 80% of the time after a lightning stroke or during a thunderstorm. The lightning ball can be from one centimeter to 1.5 meters in diameter; almost all colors of the visible spectrum have been reported; their durations have been from a few seconds to longer than 150 seconds; they appear to have neutral buoyancy, and either sink very slowly to the ground, or float above the ground at a constant height; they rarely make noise until they suddenly disappear, with a noise that ranges from a barely audible popping sound, to a major explosion. An ozone-like or sulfurous odor is sometimes associated with ball lightning, and when they terminate, their behavior ranges from simply winking out of existence, to a major explosion which can do more damage than a hand grenade. In those cases in which lightning balls have come in contact with human observers, the results range from a barely perceptible tingling sensation, to death.

Attempts have been made to obtain quantitative information about ball lightning by detailed polling of adult members of the public who have seen ball lightning at some time in their lives. Typically, such a poll will consist of a first round in which the staff of a major national laboratory or other institution will be asked whether they have ever seen ball lightning. Those who reply affirmatively are asked to fill out a more detailed questionnaire, designed to elicit every type of qualitative or quantitative information which might be available from an untrained observer without scientific instruments. The results of several of these polls are cited in refs. 7 and 8. One such poll, already mentioned, was that of J. R. McNally, Jr., who circulated questionnaires to 15,923 staff members of the Oak Ridge National Laboratory, which elicited 498 reports of ball lightning observations (ref. 2). In a second poll, W. D. Rayle surveyed 4,400 staff members of the NASA Lewis Research Center in Cleveland, Ohio, and obtained 180 reports of ball lightning observations (ref. 3). These, and other similar such polls summarized in references 7 and 8, provide all the available information on ball lightning characteristics. This information is useful for constraining theoretical models of ball lightning, but is not to be compared with the much better knowledge of plasma parameters which would result from diagnostic studies of ball lightning which was reproducibly generated in the laboratory.

BALL LIGHTNING CHARACTERISTICS

The literature of ball lightning, a summary of attempts to artificially generate ball lightning in the laboratory, a survey of the observed features of ball lightning, and a description of early ball lightning models have been presented elsewhere by the present author (refs. 9, 10). The observed features of ball lightning are summarized in Table 1, taken from reference 9. The low and high values reported in Table I are near the extreme wings of the observed distribution of properties; the characteristic values are the most probable or median values reported in the literature, such as references 2, 3, 7, and 8. From the perspective of fusion energy, the long duration of ball lightning of up to 90 seconds or more, is very interesting, since such durations are well beyond the confinement time required for both DT and advanced fuel fusion. Other particularly fusion-relevant ball lightning characteristics include its energy content, its energy density, and its kinetic pressure.

The database of ball lightning observations has been carefully worked over in an attempt to determine the energy content of ball lightning. In some cases trained observers, including military munitions experts, reported information on effects or damage resulting from the final disappearance or explosion of ball lightning, and used such information to estimate its energy content. These investigations have shown that ball lightning may have a total energy content ranging from 0.01 joule, to values greater than 10 megajoules. The energy content of a hand grenade is approximately 1 megajoule, for comparison. A characteristic or median value for the energy content of ball lightning is somewhere in the range of 10 to 100 joules.

In about 13 cases, discussed by Barry (ref. 8), enough information was available to estimate not only the energy content of the ball lightning, but also its energy density. A log-normal cumulative probability distribution for these data are plotted in Figure 1, from Barry (ref. 8). These data have a median energy density of approximately 5 joules per cubic centimeter, and extreme values ranging from less than 10^{-3} joules per cubic centimeter, to values higher than 10^5 joules per cubic centimeter. The very high value of the energy density at the upper end of the distribution is significant, both in its implications for fusion applications, and as a constraint on possible models for ball lightning. Energy densities in

Table 1. Properties of ball lightning

Property	Low value	Characteristic value	High value
Diameter, cm	1 cm.	10-15 cm.	150 cm
Duration, sec.	1 sec.	3-6 sec.	100 sec.
Energy content, J	0.01	10-100	10^7
Energy density, J/cm^3	10^{-3}	5	10^5
Mass density, grams/cm^3	1.0×10^{-3}	1.3×10^{-3}	1.5×10^{-3}
Kinetic pressure, n/m^2	1×10^5	1.016×10^5	1.1×10^5

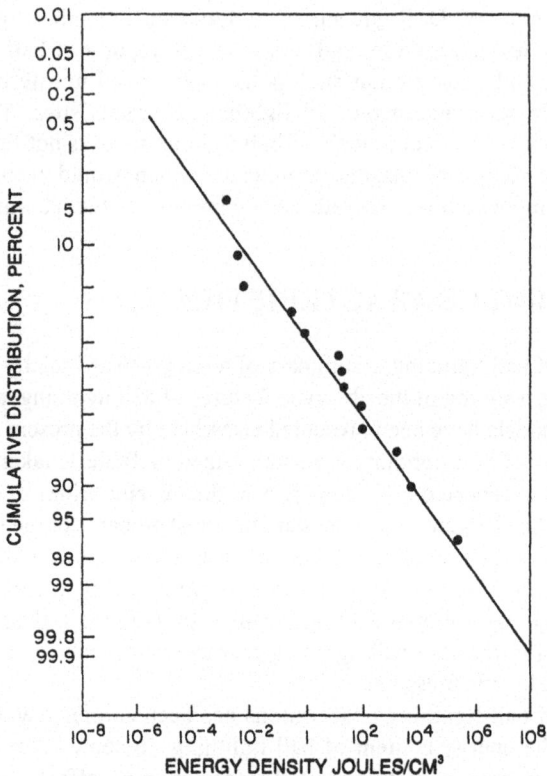

Figure 1. Log-normal cumulative distribution of the energy density of ball lightning, in joules per cubic centimeter (Barry (ref. 8)).

excess of 10^4 joules per cubic centimeter appear to rule out a chemical origin for the more energetic forms of ball lightning. The energy density of TNT, which has one of the highest energy contents of any solid explosive, is about 2000 joules per cubic centimeter, far below the maximum observed for ball lightning. Almost the only energy storage mechanism for energy densities above 1000 joules per cubic centimeter is stored magnetic energy, probably an important clue to the physics of ball lightning.

Ball lightning is observed to maintain an approximately constant diameter during its entire lifetime, thus suggesting a balance between the external atmospheric pressure, and the net kinetic pressure of the plasma/magnetic field configuration inside the ball lightning. Since the external pressure on ball lightning is 1 atmosphere, this places an upper bound on the net kinetic pressure inside the ball lightning of no more than 10^5 newtons per square meter.

IMPLICATIONS OF BALL LIGHTNING OBSERVATIONS

The large body of ball lightning observations strongly indicates that it is a real phenomenon, the physics of which should be explainable in terms of known physical processes. If we are willing to grant the reality of ball lightning, then what are the implications of interest to the fusion research community? One feature of ball lightning which motivated several of the surveys cited previously (refs. 2 and 3) is the long duration of ball lightning, ranging from about one second, up to more than 130 seconds. These times are much longer than the particle containment time required for a net power producing fusion plasma, and so the existence of ball lightning suggests a plasma confinement method which may be of fusion relevance. The physical processes responsible for the long lifetime of ball lightning may be of interest in fusion research for the additional reason that it might tell us with what gases we might surround the fusion plasma in order to minimize charge exchange and electron-neutral scattering losses.

Another feature of ball lightning observations is the large energy density and energy content associated with some ball lightning observations. Energy densities greater than 10^4 joules per cubic centimeter indicate a magnetic energy storage mechanism in ball lightning, and total energies up to 10 megajoules indicate plasmas energetic enough to be of fusion relevance. The magnetic containment method responsible for ball lightning would be of interest to fusion research because of the MHD stability of ball lightning over many of tens of seconds.

FUSION-RELATED BALL LIGHTNING MODELS

The above observations have led to several models for ball lightning, including psychogenic models, chemical models, RF energy models, and the vortex plasmoid, all discussed elsewhere (ref. 9). Probably the most fusion-relevant and most widely accepted in the older ball lightning literature is the vortex plasmoid, a model based on a self-contained magnetically confined plasma, which has been discussed by Singer (ref. 7) and is illustrated in Figure 2, taken from Reference 7. This model sees ball lightning as a luminous, self-contained toroidal plasma configuration having a large toroidal current ring, the magnetic field of which interacts with the earth's magnetic field. One difficulty with this model is that the observed

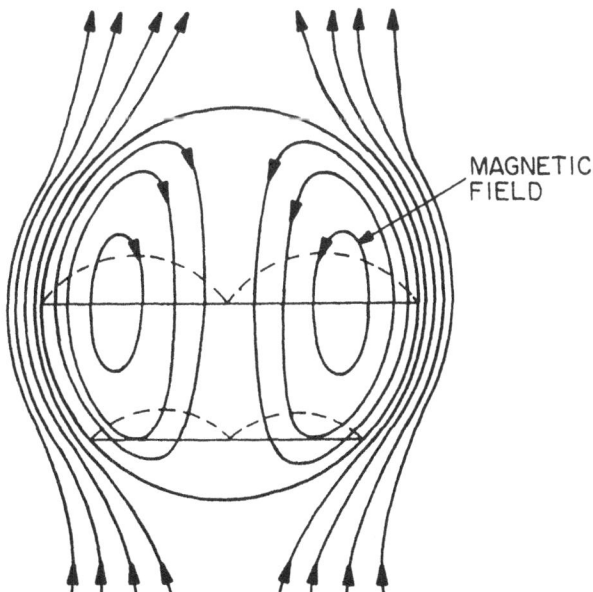

Figure 2. The vortex plasmoid model for ball lightning, from Singer (ref. 7).

motion of ball lightning appears to be independent of the local direction of the earth's magnetic field at the point of observation. This model has inspired more recent attempts to understand the physics of ball lightning, including those based on fusion confinement configurations which resemble the vortex plasmoid, and the Koloc model.

Ball lightning differs from the plasmas currently used in magnetic fusion research in several significant ways. First, there is no analog in fusion research to the existence of a confined plasma at one atmosphere. Fusion experiments operate in a moderately good vacuum. In fusion experiments, variation of the surrounding neutral background gas pressure has little effect on the equilibrium or stability of the plasma, beyond the possible introduction of impurities. Also in fusion research, there is no analog to the free-floating plasma bubble which is the characteristic feature of ball lightning. All magnetic fusion experiments at present are tied down to external confining magnetic field coils, even though in some fusion confinement configurations, such external magnetic fields play a relatively minor role in confining the plasma.

Several dozen alternate magnetic containment configurations have been discussed in reference 11, and a few of these have characteristics which tempt one to draw an analogy to ball lightning, even though the most similar of the known magnetic containment configurations differ from ball lightning in the ways described in the previous paragraph. First is the Field Reversed Mirror, in which a torus of plasma containing a large axisymmetric circulating current is confined in the midplane of a magnetic mirror geometry. The plasma current flows in a sense opposite to that in the coils which generate the mirror magnetic field, thereby generating a diamagnetic field, reducing the magnetic field on the axis. If one eliminated the external field coils, and allowed the torus of plasma to form an insulating blanket against the surrounding neutral gas, the result would look very similar to ball lightning.

A similar fusion-related magnetic confinement geometry which resembles ball lightning is generated by the Field-Reversed Theta Pinch (ref. 11). In this configuration, an elongated torus of plasma with large diamagnetic currents is confined by a magnetic mirror field generated by a theta pinch. The plasma configuration and magnetic field geometry look very similar to that of the Field Reversed Mirror, although the Field Reversed Theta Pinch relies on transient means of plasma production.

Finally, the Spheromak configuration (ref. 11) probably resembles ball lightning more closely than any other fusion-related magnetic containment configuration, partly because of its efficient confinement by the self-generated magnetic field produced by currents inside the toroidal plasma, and partly because of the relatively weak external magnetic field necessary to stably restrain the Spheromak plasma.

All three of the fusion-related magnetic containment geometries discussed above are similar, consisting of a toroidal plasma with a large toroidal current, strong enough to generate a poloidal magnetic field which keeps the plasma contained. They all possess a separatrix, and require external magnetic field coils, the strength of which differs from one configuration to the next, to provide plasma equilibrium. In ball lightning, the pressure of the surrounding atmosphere may take the place of the magnetic pressure provided by the external magnetic field required to stabilize these geometries. The relatively highly-developed theory of these three magnetic containment configurations will probably form a useful basis for understanding the physical processes active in ball lightning.

THE KOLOC MODEL FOR BALL LIGHTNING FORMATION

A recent model for ball lightning formation is due to P. M. Koloc (refs. 12, 13). Koloc's model is considerably more detailed than the older models, is better grounded in

Table 2. The Koloc model for ball lightning formation

1.	The Z-Pinch
2.	Bennett Pinch Equilibrium
3.	The Kink Instability
4.	Helix Formation
5.	Torus Formation
6.	Mantle Formation
7.	The Pinch Instability – Torus Isolation

advances recently made in fusion research, and describes in detail the formation process as well as the ultimate ball lightning product. This model has resulted in a United States patent (ref. 14).

According to the Koloc model, the physical processes which produce ball lightning begin with a lightning stroke, and end with ball lightning after proceeding through a series of sequential steps listed in Table 2. In this model, the first step in the formation of ball lightning is the creation of a linear Z-pinch by the ionized channel associated with an ordinary lightning stroke, illustrated in Figure 3. The large currents in the plasma channel squeeze the plasma, due to the azimuthal field generated by the high axial current flowing in the lightning stroke. The plasma forms a Bennett-pinch like equilibrium (ref. 11), between the outward kinetic pressure of the plasma, and the inward force of the poloidal field. This equilibrium is unstable, and will lead to the growth of kink and sausage instabilities. In the kink instability, illustrated in Figure 4, the initially axisymmetric cylindrical channel will tend to bend into a helical spiral, which looks much like a single strand of a multi-strand rope. This unstable equilibrium worsens as a result of the poloidal magnetic field being stronger on the inside of the curved plasma than it is on the outside.

The Koloc model develops as the kink instability grows in a nonlinear manner, resulting in an extreme distortion of the cylindrical channel into a series of helical turns, looking like a helical spring, illustrated on Figure 5. The current in adjacent turns of the coil illustrated in Figure 5 flows in the same direction, and the turns tend to attract each other, leading to collapse and the formation of a single torus. The result is a torus of plasma with a circulating toroidal current greater than that in the original lightning chan-

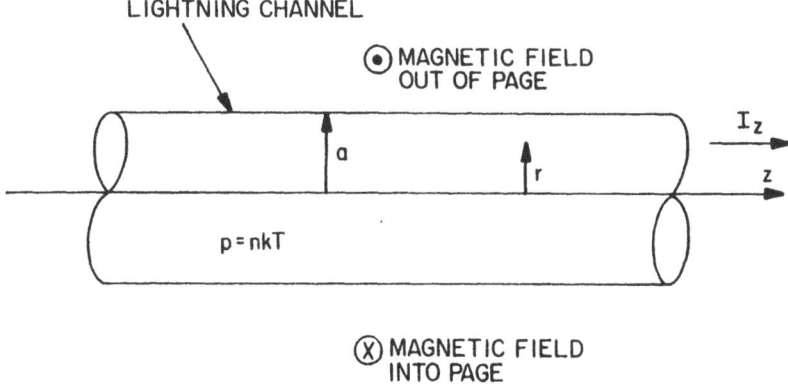

Figure 3. A cylindrical current carrying plasma corresponding to a lightning channel in Bennett pinch equilibrium. From Roth (ref. 11).

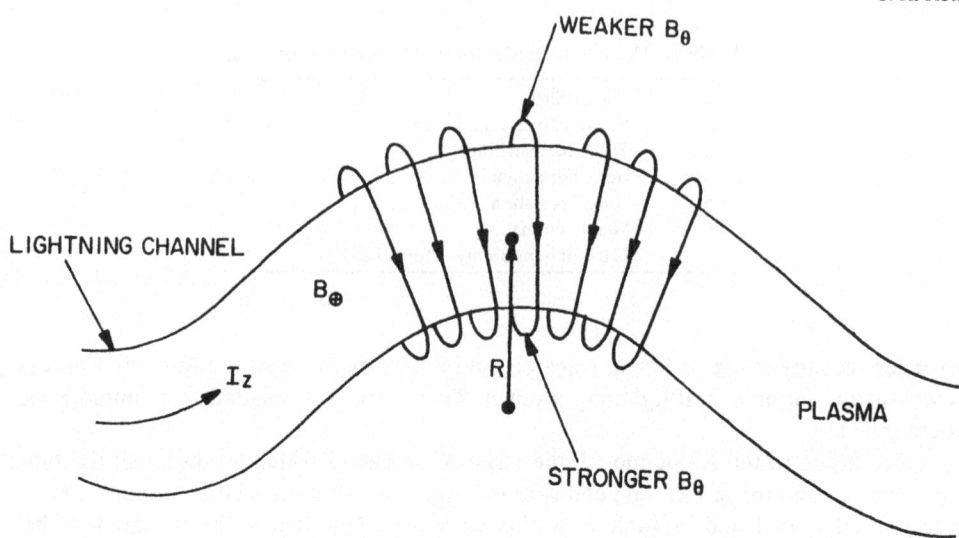

Figure 4. Development of the kink instability in the lightning channel of Figure 3. The helically twisted plasma looks like a single strand of multistrand rope, and the instability is driven by the stronger magnetic field on the inside radius of curvature, from Roth (ref. 11).

nel by a factor equal to the number of turns in the helix that formed it. After the torus forms, the main lightning channel can be pinched off by the "sausage" or pinch instability (ref. 11), illustrated in Figure 7. In the sausage instability, a small perturbation in the diameter of the plasma leads to a change in the current density, and a corresponding change in the azimuthal magnetic field. A small constriction tends to grow as the current density and azimuthal magnetic field increase, and this instability can lead to pinching off the plasma channel above and below the torus. The result of this process is the formation of a free-floating torus of plasma, as shown in Figure 8.

Thus far, the Koloc model depends on well understood plasma phenomena which have been repeatedly observed in the laboratory. However, if a bare torus of highly ionized plasma were to exist in the atmosphere without some form of shielding from neutral atmospheric gases, the electron-neutral scattering and energy dissipation would quench the plasma on a time scale measured in milliseconds. To deal with this objection, the Koloc model assumes that the torus of Figure 8 forms a thin shell of relativistic electrons, as indicated schematically in Figure 9, which insulates the plasma torus from the surrounding neutral gas. This thin layer is called the "mantle" in the Koloc model. These hot electrons are maintained by currents flowing along lines of longitude from one pole of the dipole to the other, which heat the electrons in the thin shell to relativistic energies, high enough to have scattering cross sections consistent with the long lifetimes observed in ball lightning. The energy required to maintain this thin shell of relativistic electrons is supplied by the slow collapse of the dipolar magnetic field and by the stored magnetic energy in the circulating current of the torus.

The outcome of the processes described above is a torus of plasma, containing a large circulating current, which generates a dipolar magnetic field confined within a shell of hot electrons illustrated schematically in Figure 9. Immediately outside the hot electron boundary is a partially ionized gas which fades off into the surrounding neutral atmosphere. In the Koloc model, ball lightning is thought of as a plasma bubble, with number densities lower than that of the atmosphere surrounding it, but with kinetic temperatures

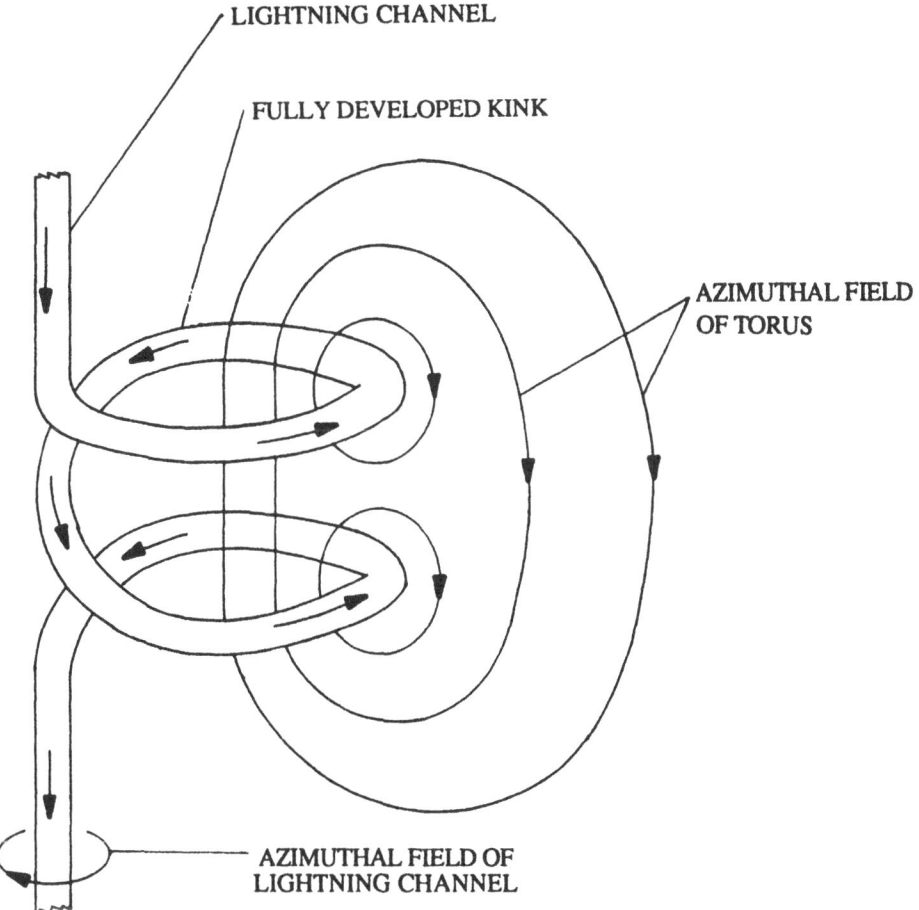

Figure 5. Nonlinear evolution of the kink instability into a helix prior to torus formation. Illustration from Koloc (ref. 14).

and magnetic pressures which provide an expansionary force to keep the plasma bubble inflated against the pressure of the surrounding atmosphere. Detailed models have been developed by Koloc for the hot electron boundary, and for the physical processes in the mantle which allow ball lightning to maintain itself in the atmosphere for long periods of time (refs. 12, 13).

The magnetic containment configuration described by the Koloc model is very similar to that of the Spheromak, the Field Reversed Mirror, or the Reversed Field Pinch, except that it does not require an external magnetic field for confinement. The ultimate confinement mechanism in the case of ball lightning appears to be the surrounding neutral atmosphere. If the Koloc model is correct, the ball lightning plasma could be compressed or rarefied by increasing or decreasing the pressure of the surrounding neutral gas.

The requirement for a relativistic shell of electrons surrounding the plasma torus is probably the least satisfactory aspect of the Koloc model, and one that thus far has no precedent in laboratory plasma physics. The presence of significant numbers of relativistic electrons as an essential feature of a ball lightning model would seem to be inconsistent

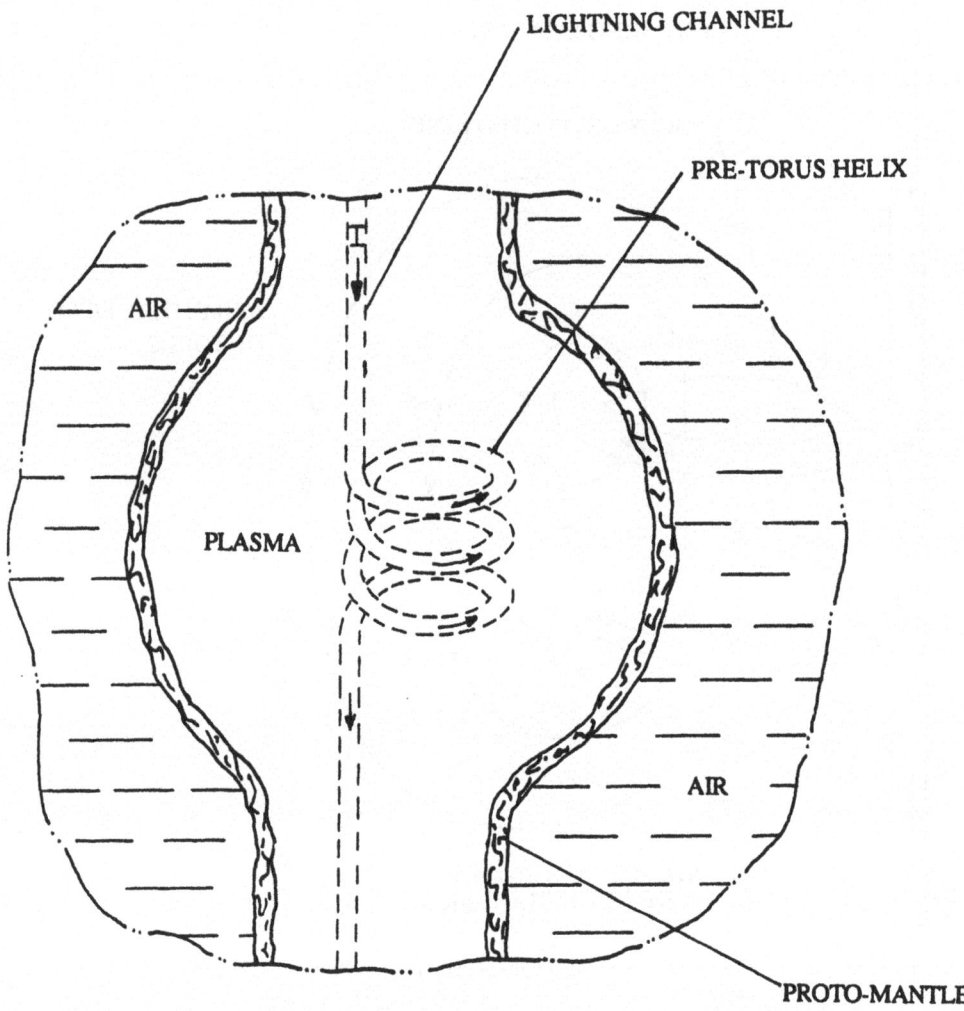

Figure 6. Helix evolution in the Koloc model, mantle formation from Koloc (ref. 14).

with the very low energy content implied by some ball lightning observations. This aspect of the Koloc model was formulated to deal with the very long lifetime which ball lightning is observed to have in atmospheric air.

Another mechanism than relativistic energies by which the electrons could have a very long confinement time and low scattering cross section is if they were scattering in a Ramsauer gas. The Ramsauer gases are those gases which, due to quantum mechanical effects, have very low electron-neutral scattering cross sections at low electron energies. Until recently, there was no reason to think that ball lightning could reasonably contain a Ramsauer gas, and thus have a long electron-neutral scattering time. It has become known however (ref. 15) that some nitrogen oxides are Ramsauer gases. It is possible that during the intense plasma chemistry associated with a lightning stroke, the oxygen and nitrogen of the atmosphere may combine to form nitrogen oxides. These Ramsauer gases may then permit the electron population to remain energetic far longer than would be the case if the electrons were to scatter against ordinary oxygen or nitrogen molecules.

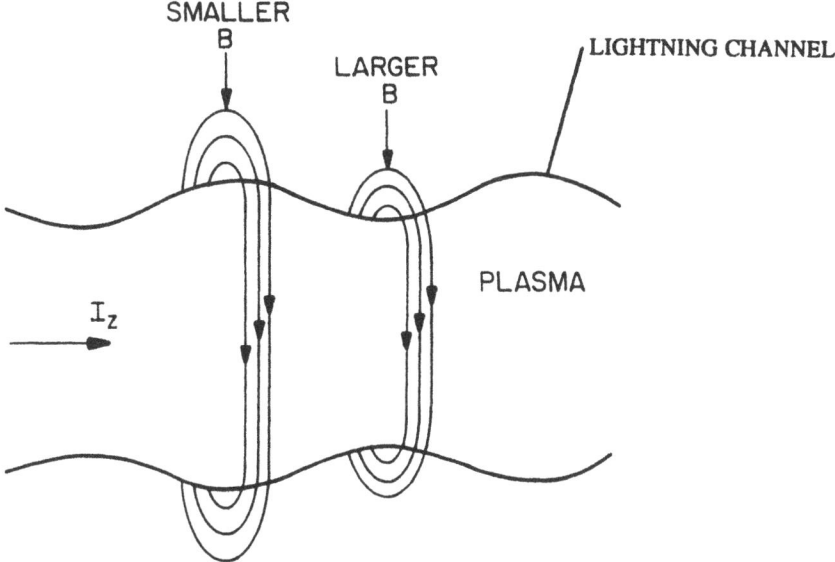

Figure 7. Development of the sausage or pinch instability due to radial perturbations in the plasma column diameter. Such perturbations can cut the plasma torus off from the main lightning channel after the torus formation shown on Figure 5, from Roth (ref. 11).

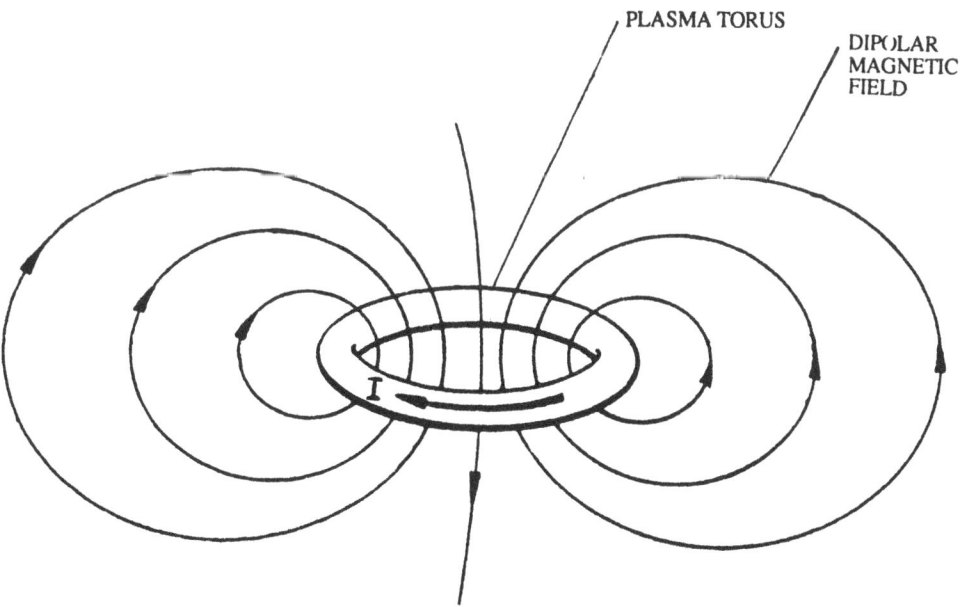

Figure 8. Residual plasma torus after collapse of helix and pinching off of the lightning channel (from Koloc, Ref. 14).

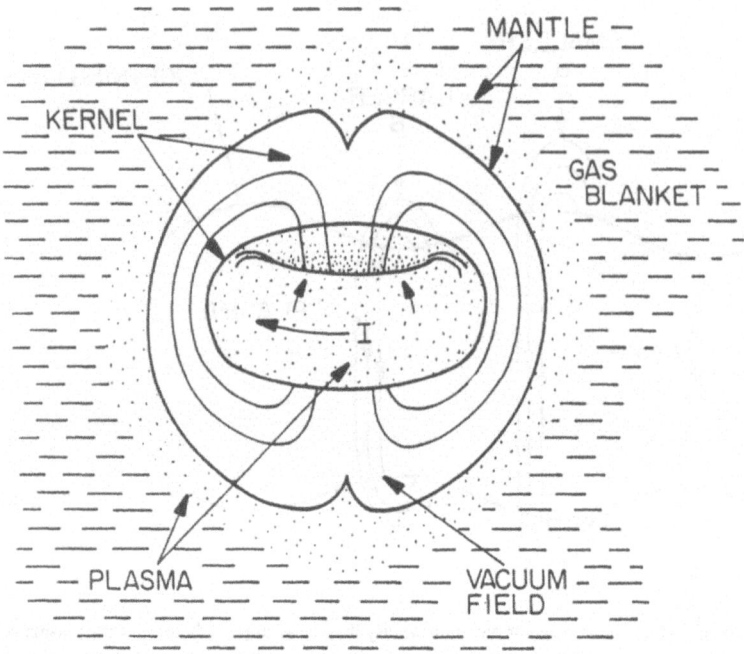

Figure 9. The end point of the Koloc model, showing a toroidal plasma insulated from the surrounding neutral gas by a high energy electron shell. The "kernel" forms a plasma-like bubble, which is maintained by a balance of outward magnetic and plasma pressure against the inward pressure of the surrounding atmosphere.

APPLICATION OF BALL LIGHTNING TO FUSION REACTORS

Suppose that it became possible to generate ball lightning in the laboratory and that the plasma parameters of ball lightning could be made to include the densities, kinetic temperatures, and containment times required to meet the Lawson criterion. As a stimulus to ball lightning research for fusion applications, it is useful to consider the fusion-relevant characteristics of ball lightning, and the potential advantages of ball lightning fusion reactors.

Fusion-Relevant Characteristics of Ball Lightning

Some of the fusion-relevant characteristics of ball lightning are listed in Table 3. In the second column is the range of parameters from the available ball lightning observations for the characteristics listed in the first column, and in the third column are shown the approximate requirements for a net power producing, self-sustaining DT fusion reactor. It is evident that the duration of ball lightning is longer than the particle containment time needed for a net power producing fusion reactor.

The second row of Table 3 indicates that the energy density of ball lightning can be equal to or far higher than the one joule per cubic centimeter characteristic of a self-sustaining DT fusion reaction. The third line of Table 3 indicates that ball lightning can be of a size large enough to provide significant amounts of fusion power, if fusion energy densities of 1 to 10 megawatts per cubic meter could be maintained in a ball lightning fusion reactor. The fourth row indicates that the kinetic pressure of ball lightning, approximately 1

Table 3. Fusion-relevant characteristics of ball lightning, compared with self-sustaining DT fusion requirements

Characteristics	Ball Lightning Observations	Self-Sustaining DT Fusion Requirements
Duration	1 sec to > 100 sec	1 to 10 sec.
Energy Density	10^{-3} to 10^5 J/cm^3	1 J/cm^3
Diameter	~ 1 cm to 2 meters	> 1 meter
Kinetic Pressure	~ 1 atm.	1.6 to 10 atm.
Number Density	Poorly known	~ 10^{14}/cm^3
Kinetic Temperature	Poorly known, > 0.025 eV	10^{14} eV

atmosphere, is comparable to the kinetic pressure, $p = nkT$, of a magnetic fusion reactor, which would range from 1.6 to 10 atmospheres for DT and advanced fuels, respectively.

The ion and electron number density in ball lightning is very poorly known and would have to be somewhere below 2.7×10^{19} particles per cubic centimeter, the particle number density of the neutral atmosphere. In a net power producing DT fusion reactor, an ion number density of about 10^{14} per cubic centimeter would be required, and advanced fuel reactors would require densities of perhaps two or three times this value (ref. 11). Finally, in the last line of Table 3 the kinetic temperature of ball lightning is very poorly known but must be more than the temperature of atmospheric air. For a self sustaining DT fusion reactor, the kinetic temperature of the ions must be about 10 keV, with values of 3 to 5 times this value required to burn advanced fuels.

Potential Advantages of Ball Lightning Reactors

The first potential advantage of ball lightning reactors might be small size and power output. Existing studies of DT Tokamak fusion power plants (ref. 11) suggest that such reactors must produce several gigawatts of thermal power and be so costly and large in size that they cannot be used for anything other than central station electric utility powerplants. The observed size of ball lightning is an encouraging indication that a fusion reactor based on ball lightning might be small enough to be used in mobile applications and of small enough power output that applications of considerably smaller scale than electric utility powerplants could be considered.

The ability of ball lightning to exist in stable equilibrium at one atmosphere suggests that, if desired, a ball lightning fusion reactor could be operated at one atmosphere, like ordinary fossil fuel boilers, or at least without the necessity of an expensive vacuum system. This also implies that a knowledge of the physical processes which allow ball lightning to exist at one atmosphere might have implications for the selection of the gas which should surround a fusion plasma in order to minimize charge exchange or scattering losses. On a more speculative note, the capability of ball lightning to operate at one atmosphere also suggests that more exotic applications of ball lightning fusion powerplants, such as rocket engines for earth takeoff, or fusion powered turbojets, may be possible.

The fact that nature produces ball lightning without costly or complicated equipment is an encouraging indication that, once we understand how ball lightning is formed, the equipment needed to produce a ball lightning fusion plasma will itself be simple and require only relatively simple containment equipment, by current standards of magnetic fusion research.

Another potential advantage of ball lightning fusion reactors is that the balance between the external atmospheric pressure, and the kinetic pressure of the plasma should allow the plasma number density, temperature, and fusion power output to be controlled by compression or expansion of the neutral gas surrounding the ball lightning plasma. If one compresses the surrounding neutral gas, it is reasonable to expect that an adiabatic compression of the confining magnetic field and its associated plasma will occur, leading to a hotter, denser plasma and greater power output. Thus, while the ball of plasma is undergoing fusion reactions, the power output should be capable of being adjusted on demand. This type of reaction control was contemplated by Koloc in his U.S. patent (ref. 14).

Finally, the ball lightning literature contains reports of ball lightning which lasted for durations of well over 100 seconds. There apparently is a tendency for small ball lightning to last for somewhat shorter times than large diameter ball lightning, but on the basis of existing observations, it seems reasonable to expect that ball lightning reactor plasmas, once formed, should last at least 10 to 100 Lawson containment times, and allow a virtually complete fusion burn without refueling. It is not possible to say at this time whether a refueling mechanism could be found which would allow the magnetic field in a ball lightning reactor to regenerate itself and maintain the plasma in the steady state; most likely, a ball lightning fusion reactor will be a repetitively pulsed device, similar to that suggested by Koloc (ref. 14).

Ball Lightning Fusion Reactor Concepts

The potential small size and ability of a ball lightning fusion reactor to operate in the atmosphere open up a large variety of potential applications which are not possible to most other magnetic fusion containment concepts, because of their unavoidably large size and requirement for vacuum operation (ref. 11). Such potential applications include space power and propulsion systems, where power outputs ranging from 100 kilowatts to 200 megawatts are required, a range not easily supplied by the DT Tokamak or other mainline magnetic fusion concepts. The possibility of a ball lightning fusion reactor operating in the atmosphere may open up, for the first time, a realistic possibility of a fusion rocket capable of operating from the surface of the earth to low earth orbit. Prior to the emergence of this concept, fusion power and propulsion systems for space applications were restricted to the vacuum of outer space.

The potential small size and simplicity of ball lightning generation and confinement equipment would open up a wide range of mobile applications of fusion reactors for land, sea, and air transportation. It is intriguing that ball lightning as small as 1 or 2 centimeters in diameter has been observed. If something this small were capable of fusion reactions, fusion engines for automobiles, trains and trucks would not be out of the question. Even if a ball lightning fusion powerplant required plasmas with dimensions of several meters, applications to submarines and surface ships might be possible. The capability of ball lightning to operate in the atmosphere raises the possibility of a fusion energy source for aircraft, with a turbojet engine or fanjet engine using fusion rather than chemical fuel in the combustor. Another application which will follow from the potential small size and simplicity of a ball lightning fusion reactor is that distributed, household primary energy sources might be possible, in which every individual household and apartment building could have its own fusion reactor to provide heating, air conditioning, water purification, and electrical power.

SUMMARY

If the potential of ball lightning plasmas for fusion reactors is to be realized, a beginning needs to be made on the basic research and development work needed to provide the necessary information about ball lightning physics. One of the principal reasons that ball lightning is not better understood is that, as a random natural event which could not be duplicated in the laboratory, it is not accessible to scientific investigation, and it has not been possible, except in rare chance encounters, to apply scientific instruments to make measurements of its characteristics. Clearly, a major task is to reliably and reproducibly generate ball lightning in the laboratory. When this is done, ball lightning should be accessible to scientific investigation, and to the application of modern plasma diagnostic methods. These methods should provide information which will either confirm, modify, or replace the Koloc and other models discussed above. Such plasma diagnostic investigations of ball lightning should also shed light on the physical process by which the core plasma insulates itself from the surrounding atmosphere for long periods of time.

Once information becomes available from experimental investigation of laboratory created ball lightning, then theoretical modeling of ball lightning can proceed. This modeling will probably make rapid progress, in view of the large body of knowledge already available in the fusion community relating to the Spheromak and other similar plasma containment configurations. This theoretical understanding should also allow one to generate ball lightning in the steady state, and to determine scaling laws that relate ball lightning number densities, kinetic temperatures, and containment times so that some indication will become available concerning the region of plasma parameter space that ball lightning is capable of spanning.

REFERENCES

1. Peter Kapitza: *The Nature of Ball Lightning*, Dokl. Akad. Nauk SSSR Vol. 101 (1955) pp. 245–248.
2. J. Rand McNally, Jr.: *Preliminary Report on Ball Lightning*, ORNL Report #3938, May, 1966, STAR Accession #N66-32479.
3. Warren D. Rayle, *Ball Lightning Characteristics*, NASA Technical Note TND-3188, January, 1966, STAR Accession #N66-15316.
4. James L. Tuck: "Ball Lightning: Status Summary to November, 1971", Los Alamos Reprot LA-4847-MS, December, 1971, Star Accession #N72-23428.
5. Martin A. Uman: *Lightning*, McGraw Hill Book Co. New York, N.Y. (1969), pp. 243–248.
6. R. H. Golde, Editor: *Lightning: Vol. 1, Physics of Lightning*, Chapter 12, S. Singer, "Ball Lightning" Academic Press, New York, N.Y. (1977), pp 409–436.
7. Stanley Singer: *The Nature of Ball Lightning*, Plenum, N.Y. (1971), ISBN 306-30494-5.
8. James D. Barry: *Ball Lightning and Bead Lightning*, Plenum Press, N.Y. (1980), ISBN 0-306-40272-6.
9. J. Reece Roth: Ball Lightning: "What Nature is Trying to Tell the Plasma Research Community." *Fusion Technology* Vol. 27, No. 3, May, 1995 pp. 255–270.
10. J. Reece Roth: "*Ball Lightning as a Route to Fusion Energy*", Proc. 13th Symp. on Fusion Energy, Vol. II, IEEE Cat. No 89CH2820–9 (1989), pp. 1407–11.
11. J. Reece Roth: *Introduction to Fusion Energy*, Lincoln Rembrandt Publishing, Charlottesville, VA (1986), ISBN 0-935005-07-2.
12. Paul M. Koloc: "The Plasmak Configuration and Ball Lightning" Proc. International Symposium on Ball Lightning, Tokyo, Japan, July, 1988
13. Paul M. Koloc: "Plasmak Star Power for Energy Intensive Space Applications" *Fusion Technology* Vol. 15, March 1989, pp. 1136–41.
14. Paul M. Koloc: *Method and Apparatus for Generating and Utilizing a Compound Plasma Configuration*, United States Patent 4,023,065, May 10, 1977.
15. E. Witalis: "Ball Lightning as a Magnetized Air Plasma Whirl Structure" *J. Meteorology*, Vol. 15 (1990) pp. 121–128.

31

FUSION IMPLICATIONS OF FREE-FLOATING PLASMAK™ MAGNETOPLASMOIDS

Paul M. Koloc

Prometheus II, Ltd.
Box 1037
College Park, Maryland 20741-1037

INTRODUCTION

Plasma confinement schemes may be viewed as being related through evolutionary trends. The mirror machine represents an open magnetic geometry which was superseded by the closed magnetized toroidal geometries. Additionally, these toroidal geometries evolved by transference of coil current (Stellarators) to plasma currents (tokamaks). The Stellarator has essentially no plasma current, while the tokamak (as represented by ITER in Figure 1) has a major toroidal current. The Spheromak goes further and successfully incorporates a significant fraction of the poloidal current of the tokamak's toroidal field coils as a plasma current.

Each advancing step has improved plasma parameters and stability. The Spheromak has nearly ideal MHD stability if it is nested within a conducting shell of image currents; otherwise, if it relies on tokamak-like vertical field coils, it is subject to combination displacement or tilting instabilities. Notice the huge size of the ITER by comparison to the Spheromak, which is a compact device.

The last evolutionary step in coil current replacement is found in the exchange of the conducting shell (vertical field coil) of toroidal devices by the plasma mantle which distinguishes ball lightning and the PLASMAK™* configuration. The internal dipolar magnetic field and toroidal plasma with internal fields and currents collectively comprise the Kernel, as illustrated in Figure 2. The additional ingredient that makes the magnetoplas-

---

* "PLASMAK™" is derived from "Plasma Mantle and Kernel" and is a trademark of Prometheus II, Ltd. It is used as a reserved adjective to reference technology, features, and other associated matters related to the subject innovation as described in patents and other publications or held in trade secrets.

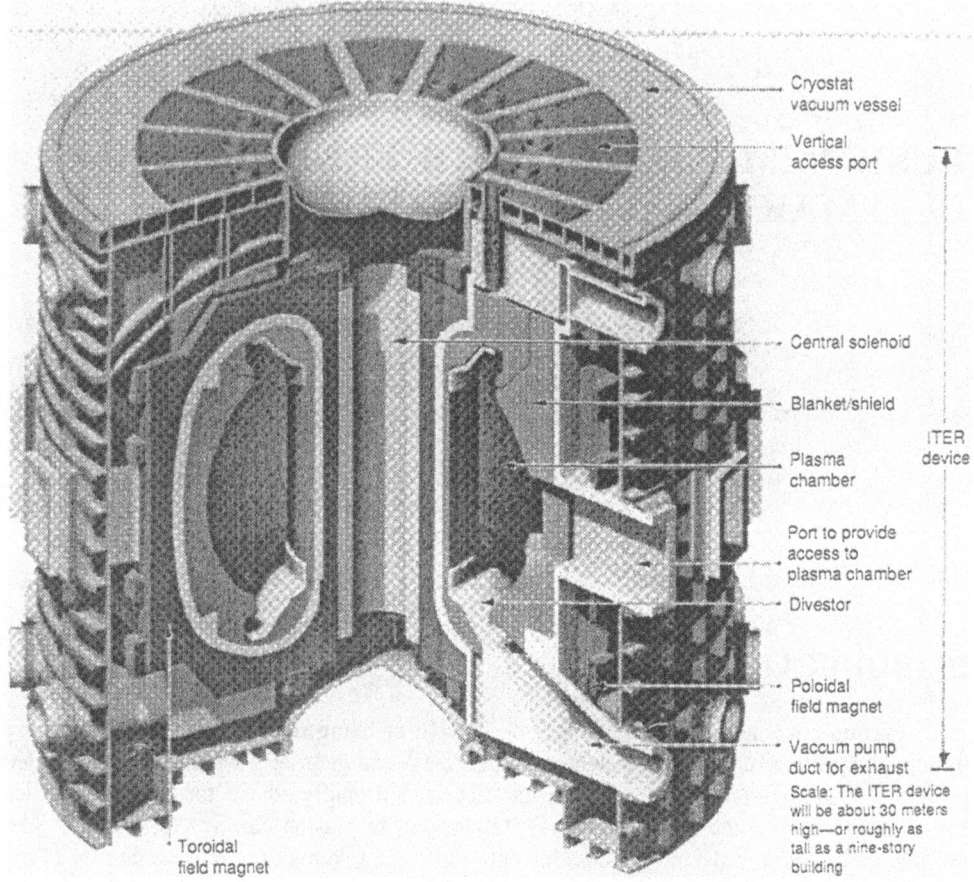

Figure 1. ITER.

moid viable is the inclusion of high conductivity, for without it, the trapped poloidal flux would diffuse away within a couple of tens of microseconds for a plasmoid with a 10 cm radius. The generation and maintenance of energetic currents entirely within the magneto-plasmoid is difficult and took a number of years to implement.

REPRODUCIBLE GENERATION OF PMKS/BALL LIGHTNING

It has been pointed out by Roth (1994) that the understanding of ball lightning has been held back by the inability to reproducibly generate it in the laboratory for scientific study. In this paper, I would like to report the reproducible generation of free-floating

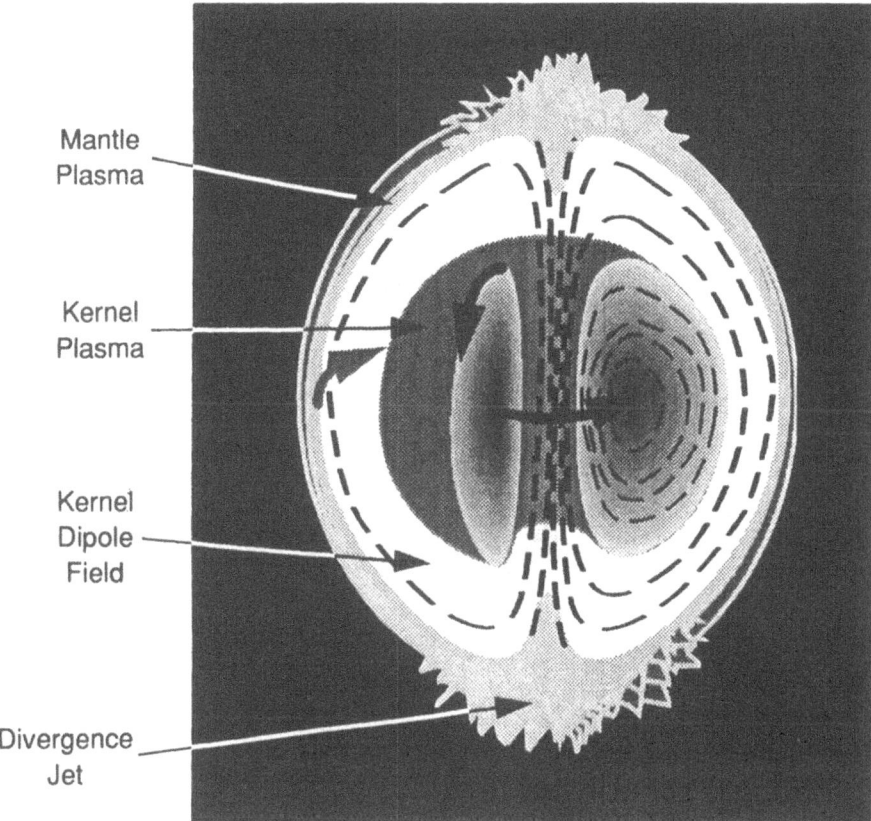

Figure 2. PLASMAK™ topology.

plasma structures in the laboratory, which I call PLASMAK™ magnetoplasmoids, or PMKs†, and which have many of the properties of ball lightning. In addition, these PMKs appear to be formed by the model which I have described previously (1989;1990), and which is reviewed by Roth (1994) in his paper at this symposium.

This model pictures a lightning bolt/Z-pinch as a Bennett pinch which undergoes the kink instability to form a helical plasma column. This helix collapses along its axis to form a torus of plasma which is pinched off and interacts with the surrounding atmosphere to form a free-floating plasma bubble, the PMK/Ball Lightning.

PMKs have been reproducibly generated by entering the formation process at the stage of the plasma helix. A helix of slightly ionized and excited air is formed by creating a real image of a helical flashtube, shown in Figure 3. Ionization density may increase with a short fast rise pulse across the helical image path. The helical plasma produced is connected across a capacitor bank by means of a closing ignitron switch. The developing current and fast rising field within the helical path interact through mutual coupling. These

---

† PMK is a contraction of "**Plasma Mantle and Kernel**" which is a descriptive title of the basic configuration and refers specifically to the physical magnetoplasmoid entity. "PMK", a noun, designates the physical magnetoplasmoid.

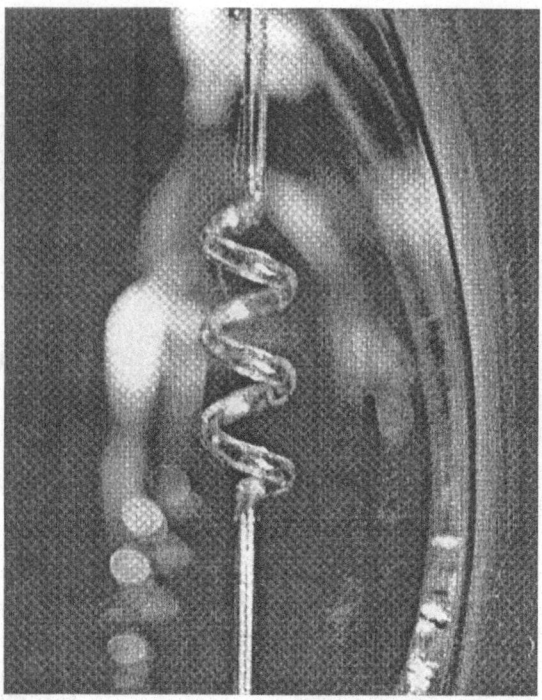

Figure 3. Flashlamp and mirrors.

forces pull the adjacent helical coils together, resulting in the collapse of the helix into the plasma torus. Surrounding ionization in the vicinity also interacts to form a cloaking mantle of plasma which completes the formation process. This results in the free-floating PMKs discussed below.

There is a strong resemblance between external PLASMAK™ topology and that of natural ball lightning (see Figures 2 and 4). The external appearance of ball lightning and

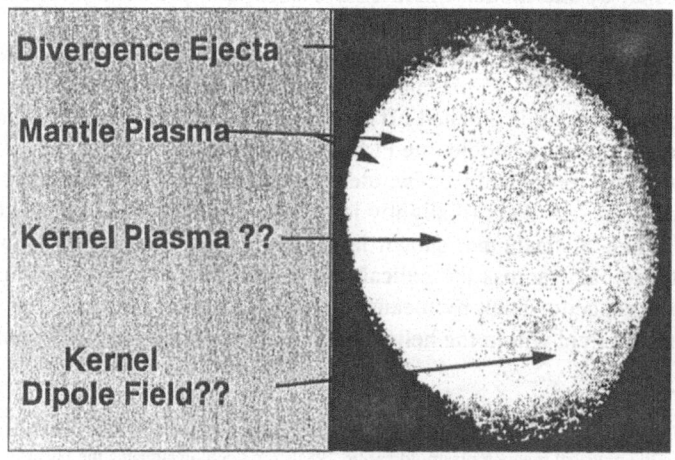

Figure 4. Ball Lightning (Schneidermann Photo, 1933)

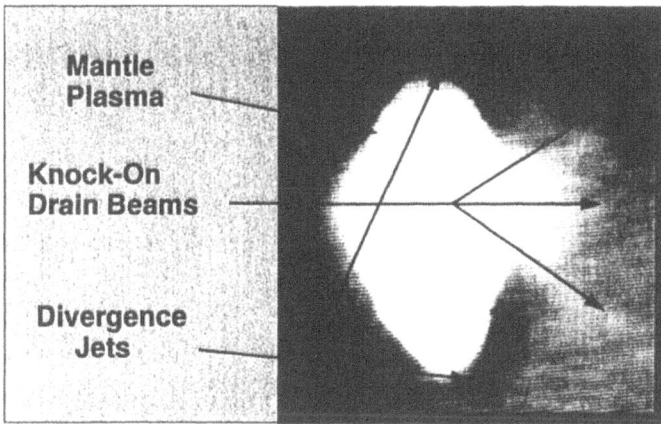

Figure 5. PMK external features.

the free floating PMKs produced in the laboratory are shown in Figures 4 and 5 respectively. Additionally, the stability of the PMK produced in ordinary air is demonstrated by the effect of a magnetic impulse of a few tens of microseconds on a PMK formed just above threshold (Figure 6). The impulse moved the plasmoid out of the formation blowoff cloud, and its image was captured by the second VCR frame (approximately 30 milliseconds). Input energy was two kilojoules, so that the blowoff plasma cloud has half the input to the shots shown in Figures 9 and 10. The magnetic impulse produced a noticeable deformation on the lower end of the PMK in Figure 6.

The generation of the plasma in the blowoff cloud is of interest. In the initial instant of formation, the substantial power input and relatively small volume of plasma produces intense radiation into the surrounding gas blanket. This radiation-generated plasma is a sort of husk which subsequently expands, and later cools and contracts. However, during formation the fully energized husk acts as an opaque barrier and normally prevents viewing of the forming PMK within until adiabatic expansion cooling sets in and the husk be-

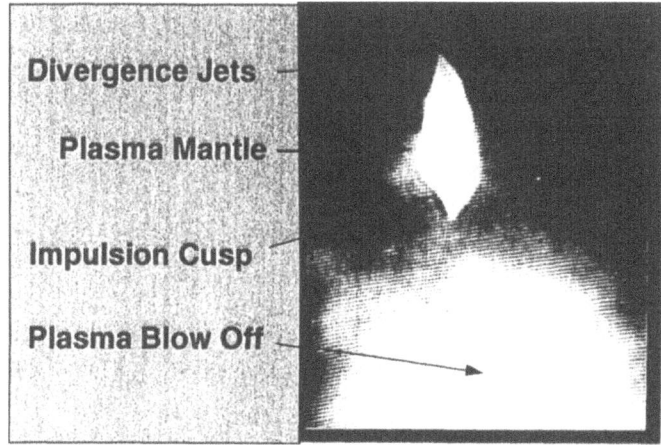

Figure 6. Threshold PMK.

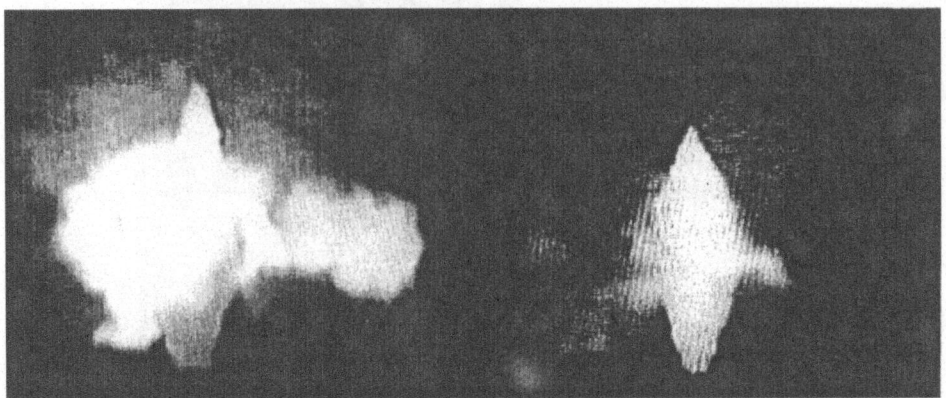

Figure 7. (a and b) Shorted formation.

comes transparent. The blowoff plasma later contracts to essentially its original volume and ceases to emit line radiation, except for long lived persistent fluorescence.

To observe an initial formation state of the PMK, the formation was interrupted and shorted so as to clamp further energy input, and thus produce a reduced amount of external plasma around the forming PMK. Two such photographs are shown in Figures 7a and b, where time between this one frame is 30 milliseconds.

Here in Figure 7b the reduced energy allows viewing. It is noticeable that the Mantle is well formed and still coming to equilibrium, considerably beyond the time for ordinary plasma "sparks" to exist. Except for the noticeable decay of excess plasma, this low energy PMK seems to be almost static. The crowbarred PMKs are about three quarters of the height of the fully formed, six centimeter major axis length, four kilojoule input PMKs (shown below in Figure 9). Due to non-linearity of the plasma parameters during crowbarring, energy available to the PMK cannot be estimated except by comparison of resulting volumes.

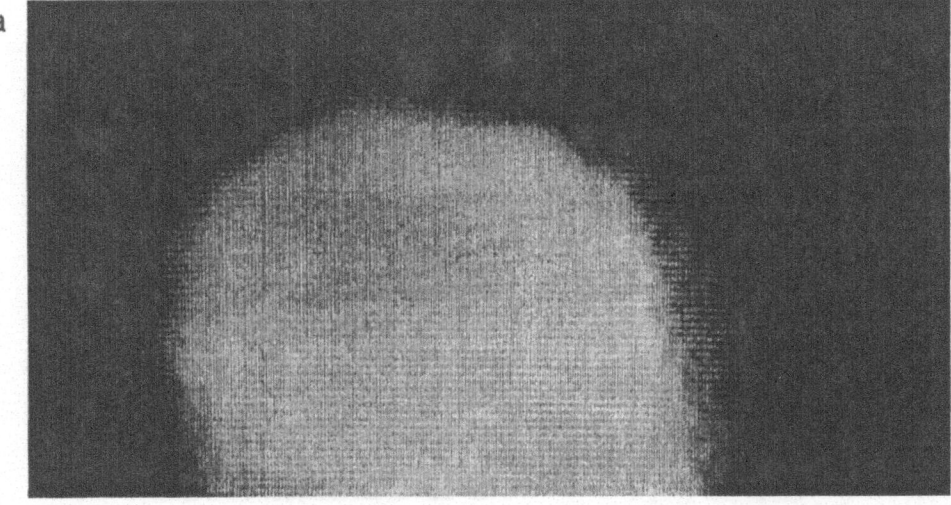

Figure 8a. Early formation.

Fusion Implications of Free-Floating PLASMAK™ Magnetoplasmoids

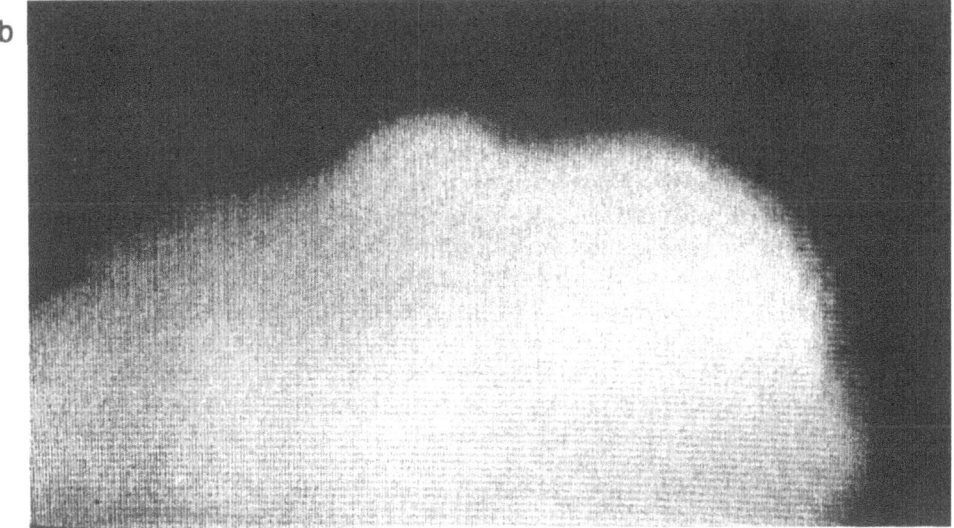

Figure 8b. Early formation.

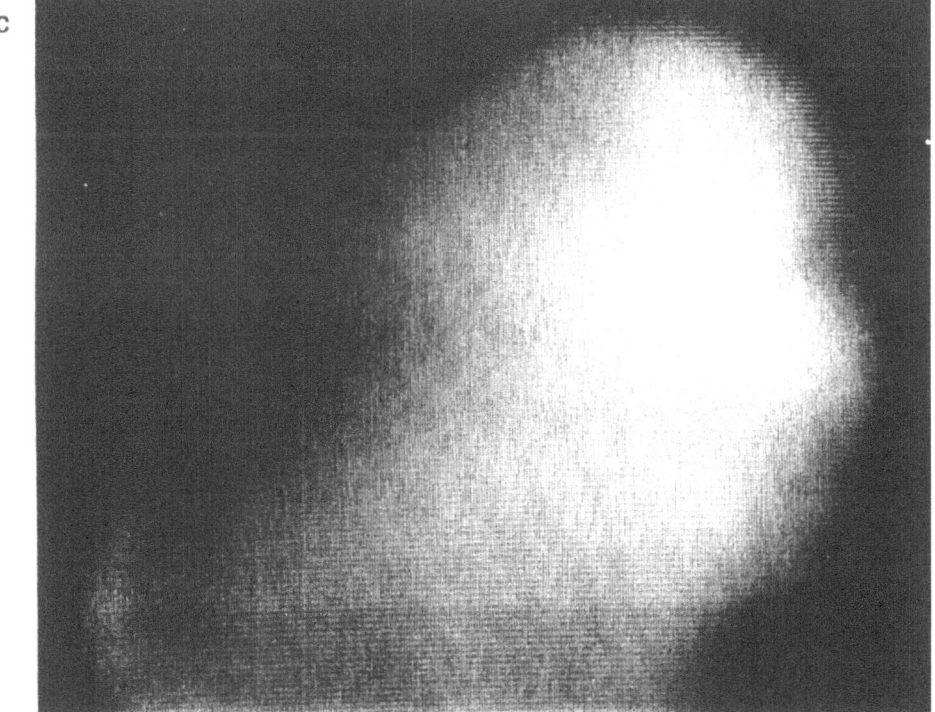

Figure 8c. Early formation.

To illustrate the generation of long lived small PMKs in ordinary atmospheric air, three selected "first frames" of three different shots from a VCR camera are shown in Figure 8(a-c). The smallest blowoff plasma of the three depicted (8a) is the first frame preceding the threshold PMK (Figure 6) discussed above. The others (8b and 8c) are initial frames typical of the four kilojoule input plasmoids. In Figure 8b, the camera view is aimed high so the lower part of the plasma husk is not shown as in 8c.

The PMKs were photographically captured for 100 to 125 milliseconds, and radio-frequency (plasma) noise persisted for 240 milliseconds. Because the camera was fully zoomed for close up detail, further tracking of PMKs in the series was not done. After leaving the view of the camera, the PMKs continued their direct course into the wall of a 45 cm diameter cylinder, which was erected to surround the formation region in order to reduce local air turbulence from circulating fans. A couple of months later, crazing due to energetic radiation (EUV or UV) became visible at the strongly localized interception locations of the PMKs with the wall. Apparently such solarization processes produce strain within the plastic matrix.

Care was taken to eliminate the possibility of image reflection of the PMK or other plasma by the cylinder or any other surface, by maintaining camera operation in the fully zoomed position and by maintaining an effective viewing angle with respect to the axis of the cylinder. Only matte surfaces of radiation blocks were in the view.

A full view of a four kilojoule formation energy PMK is shown in Figure 9 where the plasmoid has moved forward, coming out of the expanded blowoff plasma. In perspective the motion is toward the viewer and upward, as if continued flight would carry it over the observer's left shoulder. The fully expanded blowoff cloud and its upper boundary are shown in the background, and trails of excited neutral copper atoms are spewing from

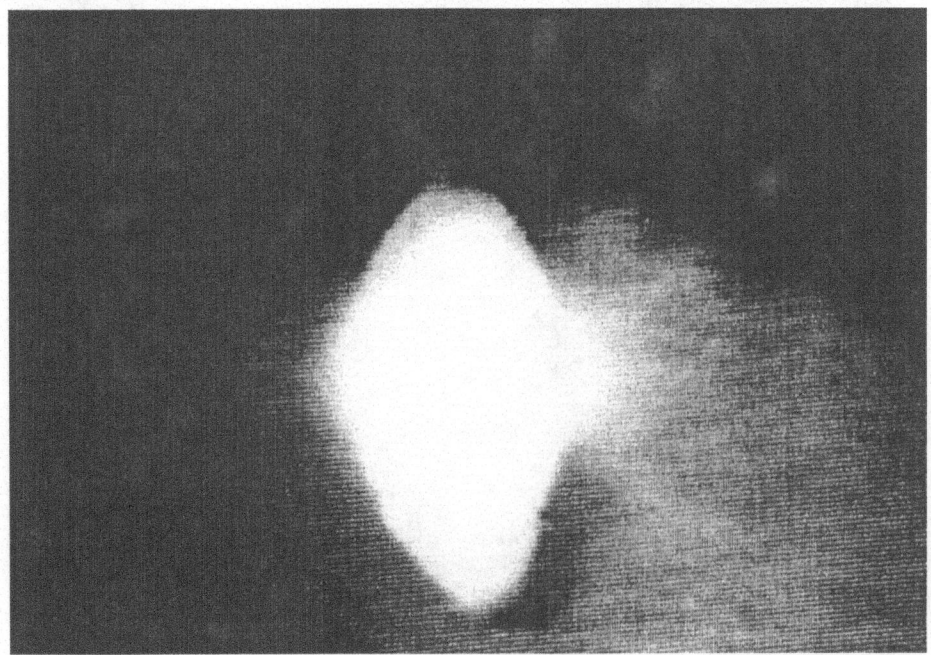

Figure 9. Free floating PMK.

Figure 10. PMK partially off frame.

both poles. Unfortunately, the crisscross of expelled electron knockons has been bleached by the lack of color and the copying process, and their otherwise more definitive tracks are nearly lost.

In the next frame, 30 milliseconds later (shown in Figure 10), the PMK is only about one-third exposed since it has partially risen above the camera view. Here the glowing copper jet trail is still visible, as well as the clearly outlined shape of the lower portion of the PMK. The PMK has risen above the perhaps slightly contracted and cooling blowoff formation husk plasma.

Finally, with the neutral density 2.5 filter removed from the camera, fluorescence on the plastic shield between the formation region and the camera is visible in Figure 11. In this frame, the much brighter PMK has completely disappeared above the edge of the viewing window. The more sensitive optical system has picked up the lower edge of a one to two centimeter thick azure blue region (at the top of the image) caused by nitrogen fluorescence surrounding the PMK (at top of image), that has risen above the frame. Judging by glow thickness, the fluorescing nitrogen is likely excited by electron recombination of O^{++} to O^+ and N^{++} to N^+. Further, only the yellowish fluorescence of the plasma blowoff cloud is seen in its cooled and contracted state, which is essentially identical to its initial formation shape (Figure 8a). Incidentally, the first frames of this formation were bleached by over exposure. With the attenuation filter in place, the other formations do not show this low-light level phenomenon.

An improved apparatus for making PLASMAK™ magnetoplasmoids on a routine basis is essentially complete. First efforts will be made to verify the model and measure basic parameters. Additional diagnostics will include calorimeters, stacked film plates, stereoscopic X-ray pinhole spectroscopy, laser imaging, and high speed video.

Reliability of formation, small size and long life of even airborne PMKs will be verified and will greatly aid in this work. For example, the time of flight into the calorimeter can be varied. Thus, the total power loss rates can be resolved using statistical methods

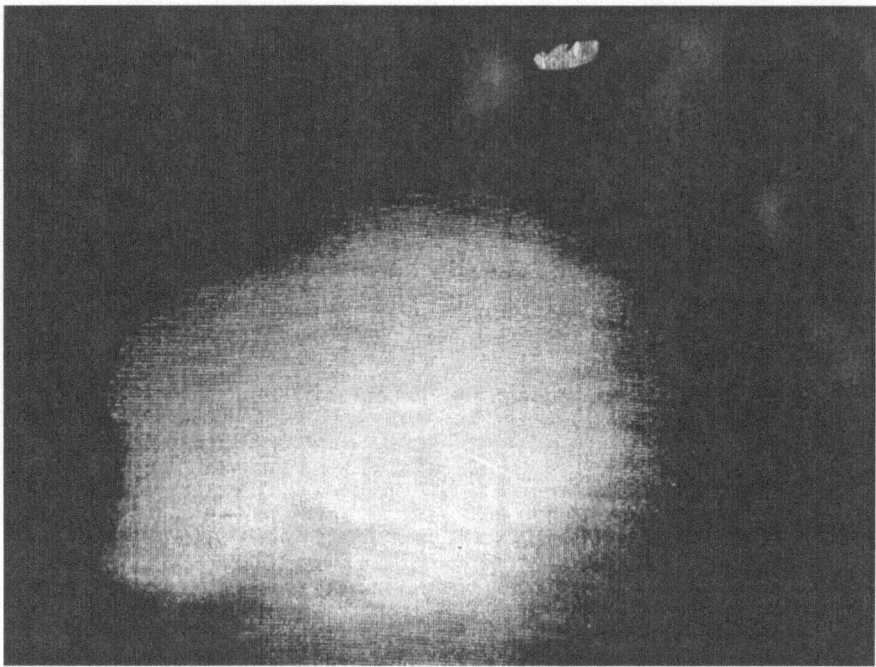

Figure 11. PMK related fluorescence.

on data collected from PMKs with identical initial formation parameters but differing flight times. Loss rate is the calorimetric energy differences over the flight time differences for a set of plasmoid measurements.

Some of the diagnostic work requires more sophistication than is presently available and therefore will await resources. For example, spectroscopy is quite complex and must be planned carefully, since the Mantle represents a gradient of plasma regimes from its inner surface to outer surface. Determining from where, which plasma segment, and percentage is added to the total measured radiation/resorption line contributions will not be trivial. On the other hand, illumination with a nitrogen laser and viewing by narrow band near-UV filter may make the internal plasma Kernel visible. Use of a shuttered stereoscopic X-ray pin hole camera will likely disclose a 3-D image of energetic currents and their densities, sizes and shapes. Attenuator plate/film stacks will yield first crude indications of the X-ray energy spectrum.

COMPRESSION OF PMKS FOR A FUSION BURN

The PLASMAK™ concept was conceived by anticipating its topology as the evolutionary end-point of developments in the fusion program, as mentioned above. An important consideration for engineering physicists is efficient application of external pressure, whether from magnetic field coils in conventional magnetic confinement, or external gas pressure in the case of a PMK, to produce compression which results in a fusioning plasma. One also wishes to leverage this external pressure to compress, ignite and burn a fusion plasma.

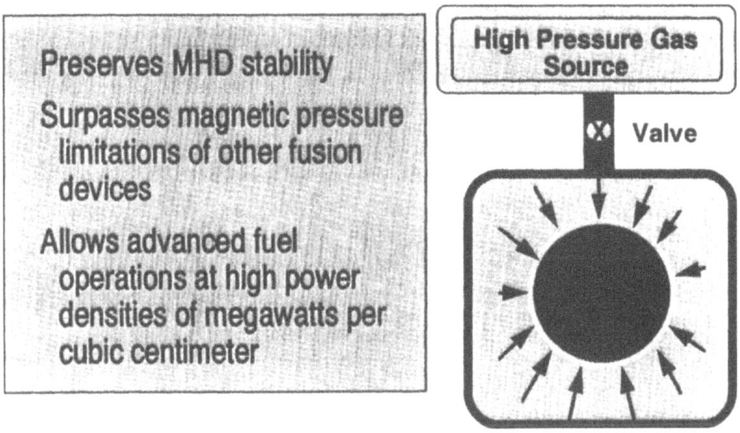

Figure 12. Compression heating.

There has been much presentation of various record values for each of three simultaneously required criteria for a fusion burn: kinetic temperature, fuel density and confinement time. This list can be reduced to two, namely fuel pressure and confinement time, since the required kinetic temperature and ion number density are established from the Lawson criterion. It is important not to trade pressure OR time to gain improvement in the other, since they are both needed simultaneously. Fortunately for ICF approaches, burn time depends on pressure squared; the time compression is more than made up for by pressure increase and corresponding reduction in burn time. The ability for an advanced concept to leverage or amplify fuel pressure can be a most valuable asset, since it yields not only shorter burn times and higher burn densities, but very much reduced burn chamber sizes. The increase in electrical conductivity that comes with adiabatic compression will tend to compensate for the loss of size. Consequently, lifetime may remain relatively flat over the compression and burn.

The PMK can be precompressed adiabatically utilizing simple techniques, as shown schematically in Figure 12. Of the available external pressure brought to bear through increased external gas pressure, the engineering beta (Bussac, 1978) describes the ratio of plasma pressure to the external magnetic/gas pressure that confines the plasma in equilibrium. In Spheromaks, and in the PLASMAK™ topology, this internal to external engineering beta exceeds one (actually over an order of magnitude). This leveraged compression process is confined to the region within the Mantle (depicted by lightest ring in Figure 12).

Another interesting aspect to a compressed PMK relates to energy balance. The external gas-driven adiabatic compression will heat the PMK to fusion ignition. During burn, the fusion driven bremsstrahlung from the Kernel plasma will penetrate into the dense compression blanket, storing its delivered heat. This heat will then generate higher compression and temperatures. Note the arrangement of the dense gas blanket surrounding the highly compressed PMK in Figure 12; this may allow more efficient burning of aneutronic fuels, including $p^{11}B$. Cyclotron radiation through its tenth harmonic should be reflected at the innermost Mantle wall due to the compression enhanced electron densities (Koloc 1989). A few burn parameters are given for D-^{3}He and $p^{11}B$ in Tables 1 and 2, respectively (Book, 1987; Koloc, 1990).

Table 1

Parameters	Initial	D-^{3}He (40 keV)	D-^{3}He (70 keV)
Compression ratio	1	7	10
Kernel volume (ltr)	274	.8	.27
Blanket pressure (ATM)	2	4.6k	20k
Plasma density (w/cc)	1E15	3.4E17	1E18
Beta	.05	.25	.36
Radiation density (w/cc)	.5	.51E6	1.6E6
Burn density (w/cc)	0	.714E6	11.6E6
Energy loss time (sec)	1	5.3	6.4
Total Peak Power(w)	0	.570E9	3.2E9
Total Energy/Pulse(j)	0	160E6	160E6

A PLASMAK™ burner will operate at orders of magnitude higher pressures than conventional concepts such as represented by ITER. The highly fusion heated compressed blanket around a spent PMK can be allowed to expand to inductively drive electromagnetic power conversion at efficiencies exceeding 90% (Stekly, 1964). By controlling currents in magnetic coils, barrier fields can be generated which should prevent significant plasma interaction with expulsion aperture walls. Consequently, the systems can be relatively quickly commercialized.

SUMMARY

PLASMAK™ advantages as a fusion burner are substantial. PLASMAK™ generators will likely be capable of burning aneutronic fuels with high energy conversion efficiency using direct conversion. They will have high power density (megawatts per cc), while their low specific mass (kG/kW) will be ideal for propulsion applications. They will be capable of fifty megawatt to one hundred gigawatt power output, and can operate in thermal propulsion or electric propulsion modes.

The PMK plasmoid itself has substantial engineering advantages. It is ideally MHD stable with no tilt or slip, is self contained and self generated, and is adiabatically compressible. These advantages will result in lower capital and development costs, and will also allow for fast turnaround development and prototype cycles.

Table 2

Parameters	Initial p-11B	p-11B
Compression ratio	1	10
Kernel volume (ltr)	274	.274
Blanket pressure (ATM)	2	20k
Plasma density (w/cc)	1E15	1E18
Beta	.10	.25
Radiation density (w/cc)	.62	8.0E6
Burn density (w/cc)	9E-6	9.15E6
Energy loss time (sec)	3.73	8.6
Total Peak Power(w)	2.5	2.51E9
Total Energy/Pulse(j)	.5	47.8E6

Table 3

Program Phases	Description	Start	End
PoP	PMK Air formation with diagnostics	(18 mo.)	y–0
Phase I	PMK in low Z gases/compression	(12-18 mo.)	y–1.5
Phase II	Strong PMK compression	(12-18 mo.)	y–3
Phase III	Engineering breakeven D-3H	(6 mo.)	y–4.5

PLASMAK™ aneutronic engineering burns may occur within five years, given the proper professional interest and corporate or institute support. A fast-track development plan might appear as in Table 3.

In summary, small but highly stable PMKs *have been formed in STP atmospheric air* with exceptionally long lifetime and stability (adding credibility to their identification with natural ball lightning). PLASMAK™ technology has potential advantages in pressure capability, simplicity and electric conversion efficiency. PLASMAK™ aneutronic engineering breakeven burns could occur in five years with adequate funding.

REFERENCES

J. D. Barry, "Ball Lightning and Bead lightning," Plenum, NY, (1980) ISBN 0-306-40272-6. (contains an 86 page bibliography).

D. L. Book, *NRL Plasma Formulary*, NRL 0084-4040, 1987.

M. N. Bussac, H. P. Furth, et al., "Low-Aspect Ratio Limit of the Toroidal Reactor: The Spheromak," IAEA CN-37, Innsbruck, 1978.

P. M. Koloc, "PLASMAK™ Star Power for Energy Intensive Space Applications" *Fusion Technology*, Vol 15 (1989), pp. 1136–1141

P. M. Koloc, "The PLASMAK™ Configuration and Ball Lightning," *Proc. First International Symposium on Ball Lightning*, Tokyo, Japan, 1988. Published in Y. H. Ohtsuki (ed.), *Science of Ball Lightning (Fire Ball)*. World, Teaneck, NJ, 1989.

P. M. Koloc, U.S. Patent No. 5,015,432, issued May, 1991

P. M. Koloc, U.S. Patent No. 4,891,180, issued January, 1990

J. R. Roth, "Ball Lightning: What Nature is Trying to Tell the Plasma Research Community", First International Symposium on Evaluation of Current Trends in Fusion Research, Washington, D.C., 1994.

J. R. Roth, "Ball Lightning as a Route to Fusion Energy" *Proc. 13th Symp. on Fusion Energy*, Vol. II, IEEE Cat. No. 89CH2820-9 (1989), pp. 1407–11.

J. R. Roth, *Introduction to Fusion Energy*, Lincoln Rembrandt Publishing, Charlottesville, VA (1986), ISBN 0-935005-07-2.

S. Singer, *The Nature of Ball Lightning*, Plenum, NY (1971), ISBN 306-30494-5, [Schneidermann Photograph, 1933]

Z. Stekly, *Plasma Jet Driven Inductive Efficiencies with Superconducting Coils*, AVCO Everett Internal Report, 1964.

32

ALTERNATE FUSION CONCEPTS

Norman Rostoker

University of California
Irvine, California 92717

INTRODUCTION

The most important long term environmental problem is how to supply the energy needs of expanding populations without the destructive side effects of global warming, water and air pollution and radioactivity. Controlled Thermonuclear Fusion involving D-T reactions in a Tokamak is the only long term program currently supported by US-DOE and by the international community. The US effort is very focused and is not expected to lead to a reactor in less than 30 years. Many problems remain involving stability, development of low activation materials, radio-activity, heat pollution, etc. If all of these problems were solved the reactor would not find any customers in today's marketplace because of the size, complexity and cost of the system.[1] Alternate concepts means the investigation of new ideas that might lead to a shorter development time and a reactor that would be competitive and environmentally acceptable now and 30 years hence.

IMPORTANCE OF FUSION RESEARCH

The very fact that one can look back at 40 years of fusion research implies that it has not been considered to have a high priority. The chief reason why it has been supported is because of the belief that the supply of fossil fuels is limited. When I started to work on reactors in 1956 at General Atomic, it was considered that in about 20 years oil would become too expensive to maintain this source of energy and the standard of living in USA. For this reason both fission and fusion reactor programs were well supported. Oil exploration was also well supported and extremely successful. After the discovery of vast mideast oil resources, there were few jobs for oil geologists and it became difficult to keep up the price of oil. In the seventies the oil embargo gave a substantial boost to the support for fusion research and the job market for oil geologists. The price of oil rose rapidly but soon stabilized when the embargo ceased, leading to a steady decline in fusion research support. If fusion research support remains hostage to oil economics and politics, it will not in the foreseeable future have a high enough priority to result in a useful reactor.

It has become clear in recent years that there are more urgent reasons for developing fusion reactors than the possible limitations to the supply of fossil fuels. Since 1956 there has been an important new discovery—the environment and what our industrialized societies can do to it. We have also noticed that the world population is about 6 billion and doubles in about 40 years. Only about 20% have a high standard of living, but the rest have aspirations. The problems of air pollution and global warming associated with fossil fuels will become acute long before they become scarce. Such fuels for basic energy production will have to be abandoned in a few generations or a serious decline in the quality of life will be experienced by rich and poor nations alike. (Consider for example what it would be like if we had stubbornly refused to abandon horses as the basic means of transportation.) Considering the possibility of the per capita consumption of energy of the USA becoming typical for 10 billion people, the only serious candidates to replace fossil fuels are nuclear. Although the technology of fission energy is much more advanced than fusion there are fundamental problems of reactor safety, fission product radioactivity, and public acceptance. Fusion does not have these problems but we do not yet know what works or what the final form and cost of a fusion reactor will be.

The present implementation of energy policy is not commensurate with the importance of the problem of replacing fossil fuel. A well-balanced program is necessary with the kind of serious support previously only directed to defense because like defense, it has to do with survival. At present, the industrialized nations are mainly concerned with short-term problems. However, when a problem is recognized and articulated we do respond. After all, our freeways are not filled with horses.

THE FLAGSHIP FUSION PROGRAM

The Tokamak magnetic confinement system is the only program that receives significant (civilian) support from the Department of Energy (DOE) and there is no investment from industry. Many companies provide technical and engineering support as subcontractors for building large machines. The Electric Power Research Institute (EPRI) supports some research on Cold Fusion. Alternate concepts receive about 1.5 M$ out of about 373 M$ for fiscal 1995. At present the major facilities are D-III-D at General Atomic and TFTR at Princeton Plasma Physics Laboratory. Both programs have been very productive and TFTR has demonstrated Deuterium-Tritium (D-T) burning. The future facilities include TPX at Princeton and the International Tokamak Experimental Reactor (ITER) for which one of the design centers is in San Diego. I refer to these facilities as the flagship program. They are much larger and more expensive than the previous generation of Tokamaks. For example, ITER is supposed to be about 30 meters high, 30 meters in diameter and cost about 20,000 M$, to be shared by USA, Japan, Europe and Russia.

Considering the lack of growth of the US budget, the decision to devote almost all of the budget to the fusion concept that has made the most progress seems to be quite natural. There are still many problems to solve before a Tokamak reactor is feasible: control of kink instabilities and Alfven wave instabilities driven by energetic particles, ignition, removal of fusion products (ash), development of low activation materials with adequate structural properties, etc. It will be at least 30 years before these problems can be resolved and the development of a practical power reactor can be initiated. If all of these problems were solved the reactor would not find any customers in today's marketplace because of the size, complexity and cost of the system.[1] It is, for this reason, that many members of the fusion community believe that alternate concepts should receive an investment from DOE of the order of 10% of the budget. Alternate concepts means the investigation of new ideas that might lead in less time to a fusion reactor that would be competitive and environmentally acceptable now and 30 years hence.

The almost complete focusing of resources on the flagship program has some serious long-term consequences. Basic research is disappearing. University support and student interest are declining. Even innovative ideas on Tokamaks cannot find support. The present and foreseeable future budget is too small for TPX and ITER, let alone for the other needs cited.

HISTORY OF FUSION RESEARCH

The present level of support for Fusion Research is difficult to understand considering the enormous importance of a new energy source for the expanding world population, and the very considerable progress with Tokamaks in recent years. Some insight can be gained by considering the history of Fusion Research.

Research on Controlled Thermonuclear Fusion has been supported by the USA for about 40 years. There have been many reports of progress in the public and scientific communications media. There have been almost no reports in the media of termination of projects due to lack of progress. In fact, most of the fusion projects have been terminated. Typically an idea started with a modest experiment that didn't work. It was always attributed to the fact that the experiment was too small. Larger experiments were then constructed one after another and eventually the absence of progress with a large and expensive machine terminated the project. Examples are the toroidal Z-pinches at LANL and General Atomic, the DCX at ORNL, the magnetic mirror program at LLNL, the toroidal Theta-pinch at LANL and the stellarator at PPPL. More recently projects were terminated before construction was completed; for example, the reverse field pinch at LANL, the MFTF-B at LLNL. Until 1968 fusion research involved many individual investigators/entrepeneurs with many ideas that were not systematically investigated.

The perceived pressures of competition favored escalation rather than investigation. By 1968 the entire fusion program was close to termination. Then L. Artsimovich, the Soviet developer of the Tokamak toured the USA and convinced most laboratories to build Tokamak experiments. Fundamental changes involved the replacement of competition with collaboration nationally and internationally. With the focusing of the program and the consequential implication of progress came large increases in the budget. Almost all other projects, except Tokamak, were eventually terminated and the fusion community was perceived to speak with one voice. The size of the experiments escalated smoothly culminating in the TFTR at PPPL. There was a brief interruption when Robert Hunter, during his tenure as Director of Research at DOE, insisted that the physical basis for anomalous transport should be understood before proceeding to still larger machines. The TFTR has come close to some of the conditions for fusion but it is about ten years and a billion dollars late. JET has succeeded in[2] producing 1 Megawatt of power for 1 sec by using 89% Deuterium and 11% Tritium. The discharge was terminated by carbon blooming indicating that materials problems dominate as soon as any significant energy is involved. More recently[3] TFTR has produced 6.2 MW for about 1 sec with 50% D, 50% T.

If the US program on fusion continues on its present course of supporting only the flagship Tokamak programs, the eventual energy source will be too expensive for widespread use. A similar premature focus of all previous efforts on Fission Reactors has severely limited their use. This decision led to a reactor that lacks public acceptance and an industry that is in decline because the reactor is too expensive. There are some similarities with the Fusion program. The Fission program for power reactors focused quite early (about 1960) on pressurized light water reactors. This was because of the considerable investment that had already been made in this reactor for submarines. In order to save money this seemed to be a natural decision. It turned out to be a very expensive decision. A comparative study of the leading Alternate Fusion Concepts might avoid repetition of this unfortunate history in the development of Fission power. That is indeed the purpose of this conference.

HOW CAN ALTERNATE CONCEPTS BE FUNDED

There is a fundamental perception problem that makes it very difficult to obtain investment in alternate fusion concepts by DOE or any other source. The perception that Fusion power will take at least another 30 years to be developed has been widely disseminated. Only the US government can consider such a long-term investment and it does so with considerable hesitation.

The available literature on the plan for ITER certainly supports such a time estimate and more than 30 years can be expected if there are interruptions in funding. Concerning alternate concepts
that might significantly reduce the development time, much of the fusion community says that based on 40 years of experience, there aren't any. Furthermore, if there were any, they would be a threat to the flagship program.

Since 1958, and until recently, there have been many sources of support outside of DOE for fusion research. For example:

Laboratory	Source of support	Time
General Atomic	Texas Atomic Energy Research Foundation	1958–1967
General Electric	General Electric	1959–1966
Cornell University	Philadelphia Electric Co. and ESADA	1968–1975
INESCO	Penthouse Magazine	1982–1986
Directed Technologies, Inc.	DARPA	1988–1992
Fusion Energy Corp.	High Voltage Engineering Corp.	1972
Advanced Physics Corp.	Swiss Aluminum Company	1974–1977
	Al Faisal (Saudi Arabia)	1975–1978
	15 American Individual Investors	1978–1984
	USAF	1985–1988

This list is not comprehensive, but it is sufficient to illustrate the fact that there were many sources of support in addition to DOE. The reason why there were other sources of support was because of the enormous importance of the problem. It was also important that the perception of how long it would take to achieve fusion was 10–20 years rather than 30–40 years.

A significant fraction of the plasma physics and fusion community does believe that there are alternate concepts that should be supported because they could reduce the time to develop fusion and/or produce a better reactor. These views have been detailed in formal letters as follows:

1. Letter to George Brown, Chairman of the House Committee on Science, Technology and Space from the Coalition for Alternate Fusion, signed by 20 well-known scientists in the fusion community. September (1993).
2. Letter to Bennett Johnson from the University Fusion Association, signed by Stewart Prager, president. April 21 (1993).

Discussions and activity in the US Congress have considered the future of controlled fusion research and the question of how focused the program should be has received considerable attention. For example:

- (1993) Senator Bennett Johnson introduced a bill according to which the DOE fusion program should concentrate on ITER and contain only elements that support ITER such as TPX. If the results are not satisfactory, or one of the major international partners should drop out, the entire DOE effort should be terminated and replaced by a $50 million dollar/year research program. (This bill did not become law).
- (1994) Congressman Swett introduced a bill according to which DOE should terminate TPX and participation in ITER and it should begin a major effort on alternative concepts that have the potential of becoming commercially feasible. (This bill did not become law).
- (1994) Congressman and Chairman George Brown introduced a bill that provides for the support of ITER and TPX. In addition, it provides for an Alternative Fusion Research Program in the amount of 26 M$ for the first year (fiscal 1995) and 31 M$/year for each of the next two years. In addition, it suggests that Alternative Fusion Research be administered by an Assistant Director for Alternative Fusion Research. (This bill passed the house but has been deferred by the Senate to the next session).

It seems likely that the case for alternate concepts is sufficiently strong that there will be at least modest support from DOE. There is, within the Office of Fusion Energy, the sum of approximately 1.5 M$ devoted to alternate concepts which started in 1993 in response to widespread criticism of a previous decision to eliminate everything by Tokamak research.

Even if the Brown bill becomes law, the problem remains that investment in Fusion research is still quite inadequate for ITER, TPX, and alternate concepts. It is also quite inconsistent with the importance of the problem. Progress in each component will be restricted unless additional investment can be found. Moreover, until there is non-government participation in the investment, there will always be a lingering doubt as to whether the objective is really a practical reactor and not just sandboxes for scientists.

With the focusing of the US program on TPX and ITER there is a need for an independent budget and management to support Alternate Fusion Research and related Plasma Physics.

Such an independent administration for Alternate Fusion concepts would be in the national interest; it would permit the fusion effort to continue and would ensure that the considerable funding and intellectual investment the US has made in this program over many years, not be lost in the event that ITER does not proceed as planned.

There are also other themes of particular importance which should be emphasized:

a. Advanced fuels that would significantly reduce the radioactivity of the fuel and the reaction products.
b. Reactor systems suitable for applications such as propulsion and power systems for space.

c. The development of reactor concepts acceptable to industry—at present the Fusion program does not enjoy any investment support from the private sector. A new structure to support alternate fusion concepts should stimulate the interest of the private sector and encourage joint ventures. It is difficult to imagine how Fusion Power could ever be realized without the participation of the private sector.

Considering the importance of joint ventures, the possibility of administering new initiatives in ARPA, NSF and/or the Department of Commerce should be considered. There is an additional resource available for joint ventures with the passage of the National Technology Competitiveness Technology Transfer Act of (1989) - Public Law 101–189. For example, in FY 93, about 141 M$ was available for DOE Technology Transfer Initiatives. At present[4] "60% of the federal R and D budget is devoted to defense programs and 40% to non-defense programs. At the very least, in the next 3 years the federal government should shift the balance between defense and non-defense programs back to a 50–50 balance which would free-up over 7B$ for non-defense R and D....."

WHICH ALTERNATE CONCEPTS SHOULD BE FUNDED

Considering the fact that Tokamak research involves thousands of scientists with adequate financial support for about 25 years, how does one compare an Alternate Fusion Concept that has only experienced modest support for a few years, or a really new idea that has received almost no support?

The present environment is rather hostile to new ideas for fusion. In attempting to find support inventors have access to a limited set of computational tools or experiments. Most of the effort is focused on seeking support which depends on DOE procedures and evaluations. If the ideas are evaluated at all it is usually by committee reviews in which busy individuals give a quick reaction to ideas that they have little time to examine closely. Such committee review processes are unlikely to promote ideas that stray from established paths.

Recognizing this problem a proposal for Advanced Fusion Assessment[5] has been submitted by LLNL that involves the Plasma Physics Research Institute (PPRI), participants from other LLNL groups, and collaborators from other fusion laboratories and Universities. The objective is to seek, generate and objectively examine ideas for fusion systems that offer the potential for a step change in cost and complexity over the present main line programs in both magnetic confinement and inertial confinement fusion.

Conventional alternate fusion concepts such as stellarator, reversed field pinch, mirrors, etc., have received support for many years that has recently been greatly reduced or terminated. There is no evidence that these ideas would ameliorate the problems of size, cost and complexity in any appreciable way. If there is a route to dramatically more attractive fusion systems, it will be in the investigation of new or relatively unexplained physics rather than in engineering refinements of present or recently terminated programs.

How to evaluate alternate concepts for fusion is not a trivial problem. That is why this conference was organized.

REFERENCES

1 L. Lidsky, MIT Technology Review, Oct. 1984, p. 32; D. Pfirsch and K. H. Schmitter, Economic Prospects of Nuclear Fusion with Tokamaks, Max Planck Institute, GARSCHWING, IPP/271 (Dec. 1988).

2. JET Team, Nucl. Fusion 32, 187 (1992).
3. J. D. Strachan et al., Phys. Rev. Lett. 72, 3526 (1994).
4. Clinton/Gore, "Technology: The Engine of Economic Growth. A National Technology Policy for America," National Campaign Headquarters, Little Rock, Ark. (Sept. 18, 1992).
5. L. J. Perkins, R. P. Drake and J. H. Hammer, Proposal on Advanced Fusion Assessment Program, Lawrence Livermore National Laboratory, (Aug. 19, 1994).

33

INERTIAL FUSION DRIVEN BY INTENSE CLUSTER ION BEAMS

C. Deutsch,[1] A. Bret,[1] S. Eliezer,*[2] J. M. Martinez-Val,[2] and N. A. Tahir[3]

[1]Laboratoire de Physique des Gaz et Plasmas†
and GDR 918, Bât. 212
Université Paris XI
91405 Orsay, France

[2]Institute of Nuclear Fusion
Universidad Politecnica de Madrid
José Gutierrez Abascal, 2
28006 Madrid, Spain

[3]Gesellschaft für Schwerionenforschung
Postfach 110552
64220 Darmstadt, Germany

ABSTRACT

The present state of the art concerning the use of intense cluster ion beams for driving an inertial fusion pellet containing a thermonuclear fuel is reviewed. Emphasis is placed on the fragmentation and stopping of correlated ion fragments in dense target material. The direct drive approach is given a hydrodynamic as well as a full one-dimensional simulation treatment. Indirect drive looks highly promising.

I. INTRODUCTION

In the field of particle-driven inertial confinement fusion (ICF), one is now witnessing a persistent interest in very heavy drivers with the smallest possible charge-to-mass ratio [1],[2]. Up to now, the corresponding mass range has extended over a large scale: from heavy atomic ions up to macroparticles (containing 10^{22} atoms) driven to hypervelocities ($\sim$ 50–1000 km·s^{-1}).

---

* Also at Soreq, Yavné, Israel.
† Associé au C.N.R.S.

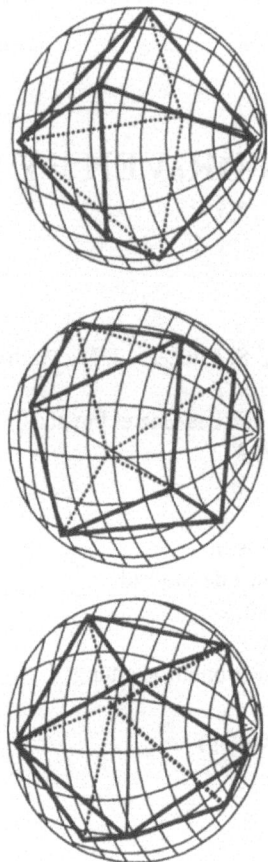

Figure 1. Depiction of calculated charge distribution for an idealized spherically symmetric C_{60}^{7+}, C_{60}^{7+} and C_{60}^{9+} from left to right.

The purpose of this paper is to speculate about driving potentialities afforded by intense cluster ion beams (hereafter referred to as CIB). Today it is well known [3]-[5] that cluster ions containing nearly any number of constituent atomic ions may be easily prepared, identified, selected, and even transported (Fig. 1).

As far as direct drive compression is concerned, a straightforward manipulation of scaling laws based upon a Bethe-like stopping formula displays almost at once the obvious payoffs of playing with a heavy projectile.

The straightforward argument runs as follows [6]. A spherical shell, of radius r, thickness Δr, and density ρ, is assumed to stop an ion beam of ion mass M_i, stripped charge state Z_i, and energy E_i. Since the ions lose their energy predominantly by Coulomb interactions with the electrons in the region Δr, we can write for the ion stopping the following expression:

$$\frac{dE}{dx} \approx -\frac{4\pi N Z_i^2 e^4 M_i}{2mE_i} \ln\Omega, \tag{1}$$

where m is the electron mass and N is the electron number density.

Integrating Eq. (1), by assuming approximate constancy of the Coulomb logarithm $\ln \Omega$, gives the approximate ion range λ as:

$$\lambda = mE_i^2 / 4\pi N Z_i^2 e^4 M_i \ell n\Omega = \Delta r. \qquad (2)$$

The specific energy deposition ε_d, obtained by assuming that most of the target mass is in the shell Δr, is:

$$\varepsilon_d = nN_i E_i / 4\pi r^2 \lambda \rho \approx \text{const}, \qquad (3)$$

where N_i is the total number of ions of energy E_i in n beams entering and stopping in the spherical annulus in a pulse length Δt. For a current I in each beam, $N_i = I\Delta t / Z'_i e$, where Z'_i is the ion charge state in the beam. Putting $\rho \propto N$ and using Eq. (2) in Eq. (3) gives the following approximate scaling law:

$$nIM_i Z \Delta r / r^2 E_i Z'_i \approx \text{const.}, \qquad (4)$$

An enhanced M_i, for instance, could allow for a smaller beam intensity I or a larger neck radius r.

Among a number of additional and intriguing possibilities, CIB would permit direct drive through momentum directly imparted to a pure DT fuel hollow target.

This highlights the momentum rich beam (MRB) [2] concepts. One can thus expect a smoother compression with a lower energy threshold, of the order of 0.2–0.5 MJ requested for ignition. Maschke speculated [2] on an ~ 10 μsec pulse length, for a 100 kA/cm^2 Cs$^+$ beam, accelerated to few hundred keV. Obviously, CIB offer attractive and more flexible alternatives for achieving similar goals.

This paper is mostly dedicated to a theoretical and numerical estimate of cluster ion stopping in cold and hot matter, as well as the elaboration of those results to drive directly a hollow spherical pellet containing a thermonuclear fuel. The classical concepts of particle-driven inertial confinement fusion (ICF) are therefore adopted to the present situation featuring intense cluster ion beams (CIB) [7]. The corresponding faisceaulogy issues are not addressed here.

Sec. II is dedicated to a presentation of the cluster ion stability, and also to a brief discussion of multifragmentation in target. Inflight stability in a dense plasma of the resulting cloud of ion fragments is given some attention.

Correlating ion fragments stopping is worked out à la Bethe for cold matter (Sec. III) [8] and dense plasmas (Sec. IV) [9]. General ideas about the hydrodynamics of a pellet driven directly by intense CIB are outlined in Sec. V.

The various possible options: hot rocket model, rocket model with momemtum transfer and hammerlike compression are then briefly investigated in Secs. VI, VII and XIII. These macroscopic considerations are then used in Sec. IX to work out a specific case. An extensive and one dimension simulation worked out within a Lagrangian code is given in Sec. X. It develops the microscopic approach corresponding to the above. It documents how efficient a CIB compression may look like when compared to its standard heavy ion homolog.

The highly promising prospects of indirect drive are finally given a short glimpse in Sec. XI.

II. CLUSTER ION FRAGMENTATION

A. Stability

Presently, a lot of cluster ions with charges of either sign are routinely produced and then smoothly linearly accelerated [10],[11]. Several studies are now currently underway to investigating the inflight charge stripping and fragmentation of cluster ions through collisions with a residual gas pressure. Amongst many avaible molecular architectures, the carbonlike ones are often given a particular emphasis. For instance, many stable charge states C_{60}^n have already been identified for the ionized stages of fullerene with $-2 \leq n \leq 7$ (Fig. 1). Similar results have also been obtained for C_{70}^n. The highest ionisation stages have a lifetime ≥ 30 µsec larger than a travel time. The linear acceleration of the given ions has been recently probed at Orsay, on a Tandem Van de Graaf machine [11]. First, C_{60}^- has been produced through a 20 keV Cs^+ beam, with a 30% yield, and accelerated up to the terminal where collisions with a few torr of nitrogen produce a double charge exchange into C_{60}^n, with a much smaller yield ~ a few per thousand which features the high stability of weakly ionized fullerene.

Then the second half of the available voltage may be used, so that the fiducial Tandem energy ~ 30 MeV may be transfered to the cluster ion at accelerator exit. It thus appears feasible to plan cluster ion interaction experiments with cold matter or dense plasma target. Similar results are also knowledgable for Au_n^p cluster ions.

B. Multifragmentation

In most cases of interest for ICF applications, i.e., when the projectile kinetic energy per amu is larger than 10 keV, the stopping processes of cluster ions matter are likely to be preceded by a fragmentating event. Such an occurrence is highly dependent on the beam-target pair interaction. However, when the number of constitutive atoms is equal to or larger than three, the combinatorics complexity of the resulting debris increases rapidly with their number. This also allows for a static and quasigeometric analysis emphasizing universal properties.

Elaborating a little bit further, it has to be appreciated that in the given energy range the cluster projectile is highly likely to experience a partial coulomb explosion. This means that the resulting debris will be at least once ionized. Moreover, their relative velocity is expected to be small compared to the projectile one over most of its quasi-linear range within the target. Therefore, these debris are expected to fly in a highly correlated motion with relative distances of order of a Bohr radius a_o.

Experimental evidences that support the argument that a Coulomb explosion actually takes place during the penetration of a cluster into a solid target are presently well documented for swift rigid molecules like ionized methane or ammonium. Vager and his colleagues [12] have recently observed the nearly complete electron stripping of CH_4^+ accelerated at a velocity $V \sim 0.02c$ in a 30 Å formvar foil. The Coulomb explosion thus yields $C^{4+} + 4H^+$. In the center of mass of the moving molecules, the final velocities of the fragment ions after the Coulomb explosion are typically $u \sim 0.0004c$. The molecular fragments are then restricted to a cone half-angle $\theta =\sim u/V \sim 0.02$ radian.

A crucial simplification concerning the subsequent ranges calculations is afforded by the velocities ratio of the ion debris to the initial CIB one.

Two such debris are expected to experience, at most, when located within a a.u. distance, a Coulomb repulsion ~1 Rydberg (13.6 eV). So, an initial CIB with a kinetic energy

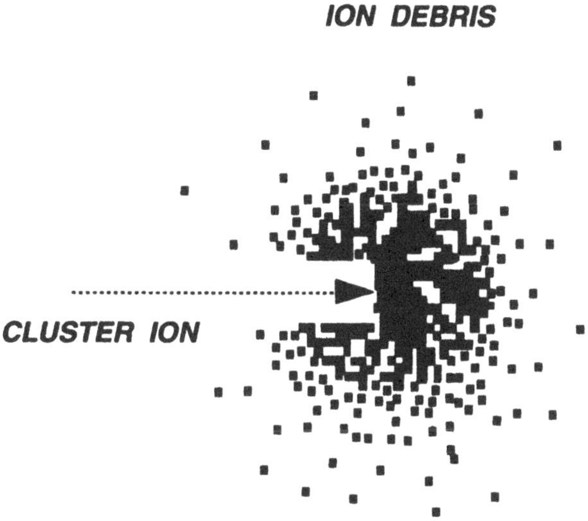

ION DEBRIS

CLUSTER ION

Figure 2. Cluster ion multifragmentation in target.

~ a few tens of keV/a.m.u. will impart a nearly unchanged velocity to the resulting charged debris.

The repulsion velocity of the latter will be nearly two orders of magnitude smaller than the initial CIB.

The given Coulomb explosion takes place on a femtosecond scale length, which supports maximum entropy production within a so-called maximum entropy production (MEP) model.

Other fragmentation scenarios based essentially on combinatorial arguments might also be considered. However, they do retain as an exact asymptotic limit, the MEP scenario.

This corresponds to a sudden projectile-target interaction with a maximum produced disorder compatible with initial conditions fixing energy and momentum conservation (Fig. 2).

Such an approach might be implemented through a kind of Saha-like approach, provided the fragmentation times remain much shorter than that pertaining to the thermal expansion of the N-cluster-target interaction volume V. Then, a temperature concept makes sense for describing the cluster evolution.

Of special relevance to the present work is the thermodynamical realization of the dynamical parameter [13], [14]

$$x = \frac{V}{\lambda_i^3} \times e^{-a_B/k_B T}, \quad k_B T = \text{particle kinetic energy} \tag{5}$$

in terms of the cluster binding energy (if any) a_B and the thermal wavelength λ_i of the monomer subunit given by $\lambda_i = \hbar / (2\pi m_i k_B T)^{1/2}$. Here, V is the cluster-target interaction volume at equilibrium.

In practical conditions of interest for ICF, the cluster kinetic energy should be larger than 10 keV/a.m.u. in order to treat the projectile-target interaction within a stopping formalism. Eq. (5) is derived through a Saha-like approximation for the distribution of cluster ions debris.

Then $a_B \sim$ eV/atom is much smaller than $k_B T$. So,

$$x \sim \frac{V}{\lambda_i^3}.$$

For $x = 0$, one just gets a fused and unbroken cluster, for larger values one witnesses fragmentation in a few relatively large subunits. Many other regimes are also likely to occur when x increases further. For $x \gg A$ (A denotes the number of basic atomic units in cluster projectile) one reaches a particularly interesting situation called multifragmentation.

Then the cluster A explodes into A elementary basic units. Such a behaviour can be associated to the Coulomb explosion alluded to previously provided the resulting ions are not too heavily charged. Such a pattern appears highly desirable for optimizing the correlated stopping picture detailed below.

C. Inflight Coulomb Cluster Stability [15]

In order to assess quantitatively the stability of a given arrangement of ion fragments resulting from a Coulomb explosion in dense matter, we consider the simplified but suggestive situation of a corona made up of N identical and singly charged ions regularly spaced while moving toward a dense and classical plasma target.

Each ion is then submitted to two kinds of fields: the inhomogenous interfacial field $\bar{E}$, and the Coulomb explosion field $\bar{C}$, which increases the cluster size. Clearly, it is the combination of those two fields which changes the shape of the corona. A single Coulomb explosion would just make it grow while the interfacial field alone would just temporarily deform the corona as it passes through the interface. We are now studying separately those two effects.

C.1. Coulomb Explosion. Considering the cluster pictured on Figs. 3 and 4, we perform the calculation of the force $\bar{C}$ acting on the ion 0. This ion experiences from its kth colleague the force

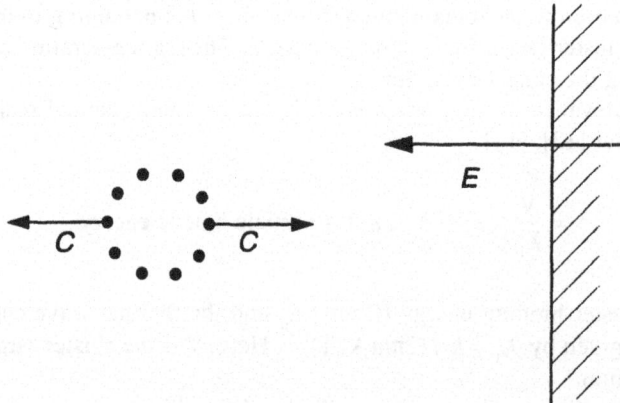

Figure 3. Corona of N identical ions moving towards the plasma. Both the Coulomb repulsion $\bar{C}$ and the inhomogenous interfacial field $\bar{E}$ produce the deformation of the corona. The plasma is represented by the hatched region on the right.

Inertial Fusion Driven by Intense Cluster Ion Beams

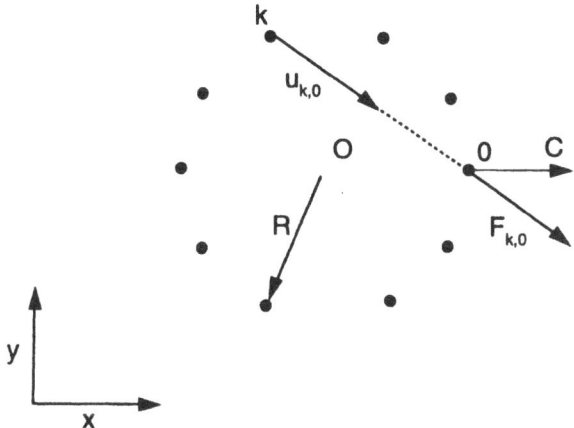

Figure 4. Model of the corona. Coulomb cluster.

$$\vec{F}_{k,0} = \frac{(Ze)^2}{r_{k,0}^2} \vec{u}_{k,0} .$$
(6)

The force acting on 0 is then,

$$\vec{C} = (Ze)^2 \sum_{k=1}^{N-1} \frac{\vec{u}_{k,0} \cdot \vec{x}}{r_{k,0}^2}$$
(7)

Because of the corona symmetry $\vec{C}$ is aligned along the x axis so that,

$$\vec{C} \cdot \vec{x} = C = (Ze)^2 \sum_{k=1}^{N-1} \frac{\vec{u}_{k,0} \cdot \vec{x}}{r_{k,0}^2} .$$
(8)

Let (x_k, y_k) be the coordinates of the kth ion

$$x_k = R\cos\left(\frac{2k\pi}{N}\right),$$

$$y_k = R\sin\left(\frac{2k\pi}{N}\right),$$
(9)

k varying from 0 to N - 1. One then gets,

$$C = (Ze)^2 \sum_{k=1}^{N-1} \frac{x_0 - x_k}{\left[(x_0 - x_k)^2 + (y_0 - y_k)^2\right]^{3/2}} ,$$
(10)

which can be written as,

$$C = \frac{(Ze)^2}{R^2} G(N),\qquad(11)$$

with

$$G(N) = \frac{1}{4}\sum_{k=1}^{N-1}\frac{1}{\sin\left(\frac{k\pi}{N}\right)} \underset{N\to\infty}{\approx}\left(\frac{N}{2\pi}\right)\ell n\frac{2N}{\pi}\qquad(12)$$

The force simply relies on the diameter of the cluster and a form factor depending on the number of ions. Each ion is submitted to the same force $\bar{C}$. The radius thus obeys

$$M\ddot{R} = \frac{(Ze)^2}{R^2} G(N)\qquad(13)$$

with $R(t=0) = R_0$ and $\dot{R}(t=0) = 0$. Replacing the pure Coulomb repulsion R^{-1} by a statically screened expression

$$\frac{1}{R}e^{-R/\lambda_s}$$

makes to appear a more complete expression

$$M\ddot{R} = \frac{(Ze)^2}{R^2} G(N)\cdot\left(1+\frac{R}{\lambda_s}\right)e^{-R/\lambda_s}\qquad(14)$$

if one restricts to classical Debye screening.

We assume the initial radius to be very small compared to the interface size so that the ions get ionized in the same way at the same time. This assumption explains why we are led to neglect in a first approximation plasma screening in the ion-ion potential when introducing the Coulomb force in Eq. (6). A Debyelike screening of the potential in exp(-r/λ_s) is considered as negligible when all relative distances remain much smaller than λ_s. The simultaneous and instantaneous ionization is taken for the origin of times. Integrating once the equation and setting $\psi = R/R_0$, we get

$$\dot{\psi}^2\frac{\psi}{\psi-1} = \frac{2(Ze)^2 G(N)}{MR_0^3},\ \dot{\psi}\equiv\frac{d\psi(t)}{dt},\psi(0)=1,\qquad(15)$$

With screening included Eq. (13) is straightforwardly extended into

$$\dot{\psi}^2 = -\frac{2(Ze)^2 G(N)}{MR_0^3}\cdot\left(e^{-R_0/\lambda_s} - \frac{e^{-\psi R_0/\lambda_s}}{\psi}\right),\qquad(16)$$

We now look for the radius of the cluster once it has entered the plasma. If its velocity is given by V_0, we must look at the radius at time $t \approx 2\lambda_s/V_0$. Eq. (14) can be solved exactly and we just have to consider two limiting cases:

- $\psi(t = 2\lambda_s/V_0) \approx 1$, so that the cluster size remains almost the same. One then has

$$\psi\left(t = \frac{2\lambda_s}{V_0}\right) \approx 1 + \frac{E_p}{E_c}\left(\frac{\lambda_s}{R_0}\right)^2, \tag{17}$$

where $E_p = G(N)(Z_e)^2/R_0$ is the initial Coulomb energy of an ion and E_c its kinetic energy: The Debye corrected counterpart also reads as

$$\psi\left(t = \frac{2\lambda_s}{V_0}\right) \approx 1 + \frac{E_p}{E_c} \cdot \left(\frac{\lambda_s}{R_0}\right)^2 e^{-R_0/\lambda_s}, \tag{18}$$

- $\psi(t = 2\lambda_s/V_0) \gg 1$, so that the cluster size has now increased a lot. We have in this case

$$\psi\left(t = \frac{2\lambda_s}{V_0}\right) \approx 2\sqrt{\frac{E_p}{E_c}} \cdot \left(\frac{\lambda_s}{R_0}\right), \tag{19}$$

for a pure Coulomb interaction and

$$\psi\left(t = \frac{2\lambda_s}{V_0}\right) \cong 2\sqrt{\frac{E_p}{E_c}} \cdot \frac{\lambda_s}{R_0} \cdot e^{-R_0/2\lambda_s}, \tag{20}$$

for a Debye one.

The relevant parameter controling the growth thus appears to be

$$\frac{E_p}{E_c}\left(\frac{\lambda_s}{R_0}\right)^2 \cdot e^{-R_0/\lambda_s} = \Gamma\delta^2 e^{-1/\delta}, \tag{21}$$

with

$$\Gamma = \frac{E_p}{E_c} \text{ and } \delta = \frac{\lambda_s}{R_0}$$

For instance, for $\Gamma = 0.1$ the first inequality is valid when $\delta > 3.6$. For $\Gamma > 0.01$ one needs $\delta > 10.48$.

The growth is important when $\Gamma\delta^2 e^{-1/\delta} \gg 1$ and negligible for $\Gamma\delta^2 e^{-1/\delta} \ll 1$. Since we assume $\delta \gg 1$ (so that all ions get ionized at the same time), the ratio between Coulomb and kinetic energy of the cluster at the beginning of its expansion determines the growth. A high velocity cluster of weakly ionized ion fragments is thus expected to display a small deformation only.

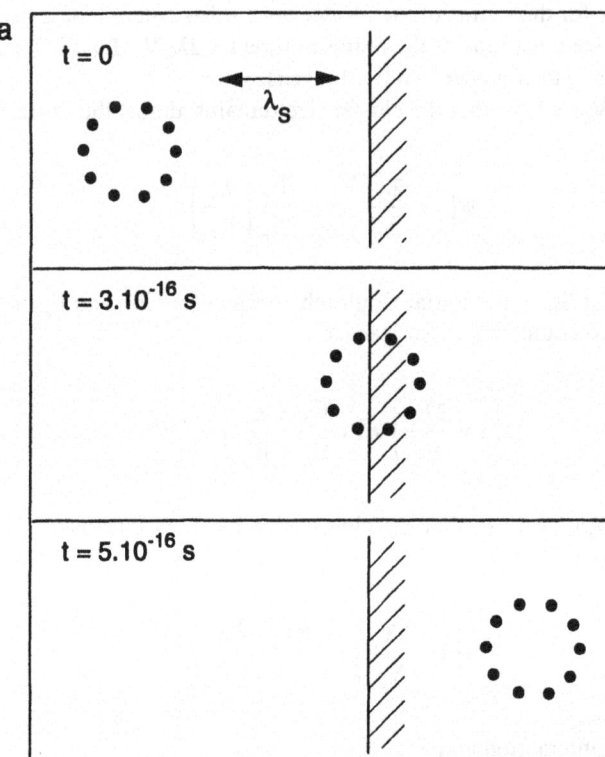
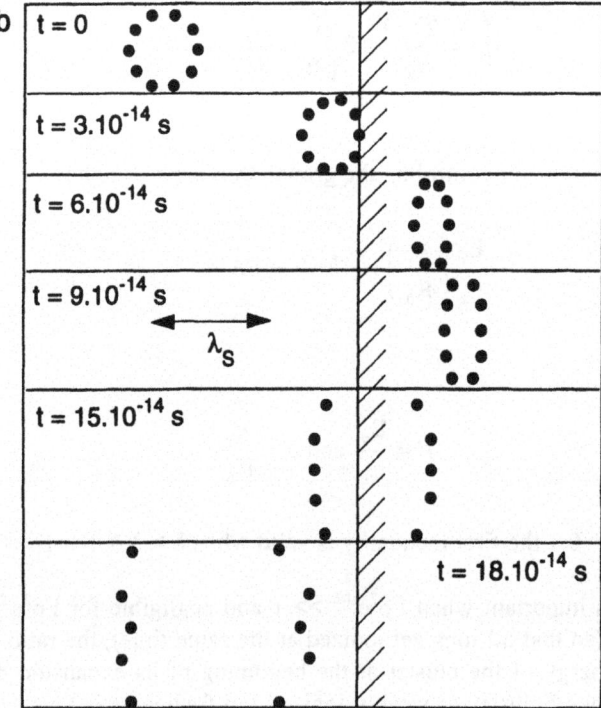

C.2. Computer Simulation. We now turn to computer simulation in order to confirm previous results and investigate others situations. We consider a cluster containing 10 ions of charge Z and mass M with initial velocity V_0. Trajectories are computed through a first order Runge-Kutta algorithm. One therefore has for the ith ion,

$$\vec{X}_i(t+\Delta t) = \vec{X}_i(t) + \vec{V}_i(t)\Delta t$$

$$\vec{V}_i(t+\Delta t) = \vec{V}_i(t) + \vec{A}_i(t)\Delta t,$$

$$\vec{A}_i(t) = \frac{\vec{F}_i(t)}{M}. \tag{22}$$

$\vec{A}_i, \vec{V}_i, \vec{X}_i$ represent respectively the acceleration, velocity and position vectors for ion i. $\vec{F}_i$ is the force acting on ion i, resulting from both the interfacial field and the internal Coulomb force. The time step Δt is constant. We performed three different simulations:

- In the first one (see Fig. 5a), kinetic energy is much more important than the potential barrier ($\Re$ = potential barrier height/ion kinetic energy = 5.10^{-4}) since $V_0 = V_{th}$, the electron thermal electron velocity. The cluster shape does not change.

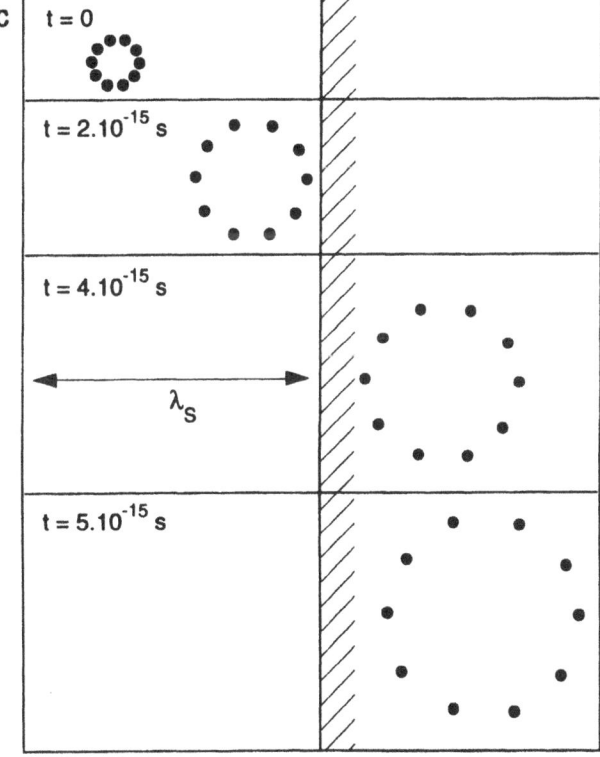

Figure 5. Shape evolution of the corona passing through the plasma-vacuum interface. (a) Z = 1, M = 10 m_p (proton mass), R_0 = 5 a.u., $V_0 = V_{th}$; N_e= 10^{24} cm^{-3}, T = 10 keV so that $\Re=5.10^{-4}$, $\Gamma\delta^2=9.4\ 10^{-7}$ (b) Same parameters as in (a) except $V_0=V_{th}/150$ so that $\Re=12$. (c) Same parameters as in (a) except Z=10, R_0=1 a.u., $V_0=V_{th}/27$ so that δ=14, $\Re$=0.4 and $\Gamma\delta^2\ e^{-1.8}$=7.30 with Γ=0.04.

- In figure 5b the kinetic energy is smaller than the potential barrier. So the cluster cannot enter the plasma and bounces on the interface.
- In figure 5c we consider a ratio R = 0.4 and $\Gamma\delta^2 e^{-1/\delta}$ = 7.30. With Γ = 0.04 the cluster size increases while its shape remains circular, according to our previous theoretical expectations.

III. ION FRAGMENTS STOPPING IN COLD MATTER

The main goal of the present section is to lay the ground for a basic understanding of the strong interaction of the cluster ion debris resulting from an initial cluster ion impacting and getting fragmentated in the target outer layers. The fragmentation issues have been eschewed above. It is therefore taken for granted that the cluster ions get suddenly disrupted under impact, on a femtosecond time scale, through a kind of Coulomblike explosion [7], [8], [14].

Moreover, we focus interest on situations where the initial impacting velocities are nearly fully transmitted to the ion debris. This hypothesis implies that the energy toll given to the Coulomb repulsion between ion debris remains negligible with respect to initial cluster kinetic energy.

Such simplifying assumptions have been demonstrated to be compatible with a bona fide particle-driven compression [7] of a pellet containing a thermonuclear fuel.

As far as the driver-pellet interaction is concerned it should be recalled that it essentially features a standard electromagnetic coupling between an incoming and swift ion projectile with the degenerate target electrons. The ion debris are partially ionized so their individual energy loss appears proportional to the square of the effective charge Z_{eff} (V), a velocity dependent quantity. At this point, the situation is completely analogous to that already well documented for light ion and heavy ion beam drivers making use of uncorrelated atomic projectiles.

The conspicuous novel feature afforded by the multifragmentation of cluster ions in target is the additional correlated stopping due to ion debris flying at same initial velocity and in close vicinity within a cloud extending over a few atomic units only. It has to be pointed out that the relative distributions of ion debris, taken as pointlike for the sake of simplification, build up a manifold of Coulomb clusters with different topologies which can be partly neutralised by the background of target electrons. Such an effect is likely to delay the dispersion of the given cloud of ion debris arising through the Coulomb repulsion.

At this point, our main concern is about the highly rewarding correlated contribution which usually adds up to the individual energy loss arising from nearby flowing ion debris. Such a specific effect has already been given a lot of attention in a cold matter context [16].

Recent numerical simulations have already clearly demonstrated [4], [7] and [8], that an overwhehming majority of ion debris topologies provide a positive and enhanced correlated stopping (ECS in the sequel). Such a behavior is already clearly displayed by the simplest dicluster arrangement of two charges orientated arbitrarily with respect to their overall drift velocity $\vec{V}$. Even in this case, the given ECS gets as large as the individual and uncorrelated contribution. A noticeable but actually harmless counterexample is however afforded by regularly distributed point charges flying rigorously parallel to $\vec{V}$. Then, much of the two-body electrostatic interferences build up destructively. Fortunately, such a topological arrangement is expected to reconformate itself into a zigzag-like shape through the corresponding straggling effect.

Our technical starting point is then a Coulomb cluster of nearby pointlike charges featuring ion debris resulting from an energetic fragmentation of an initial weakly ionized cluster ion accelerated toward a fully degenerate electron target. Each ion charge experiences a stopping force through Coulomb interaction with the jellium fluid. The corresponding physics is considered as pretty well understood. For the last twenty years or so, a rather intense activity has also been given to formulating in various ways the stopping mechanisms of two charges flowing close the one to the other in a jellium or a harmoniclike collection of neutral atoms [16]. In this context, it appears convenient to use atomic units (a.u.) for the target parameters.

In this work, we emphasize out a situation for energetic ion debris with a kinetic energy/nucleon $E \geq 10$ keV/a.m.u., so the overall drift velocity after impact in target should remain at least two orders of magnitude above a so-called transverse velocity due to ion-ion Coulomb repulsion (intercharge distance ≥ 1 atomic unit). In such a case, ion stopping may well be addressed within a kind of Bohr-Bethe-Bohr (3B) formulation with an inverse square projectile velocity dependence. So, anticipating a bit the results detailed below, and taking as a standard stopping material, solid Li (r_s = interelectron distance/Bohr radius = 3.27) in the given velocity range, the stopping time will be $\sim 10^{-12} - 10^{-13}$ sec with a much slower ion-ion relative motion. Such a picture enables us to consider a frozen configuration of ion debris building up a kind of Coulomb cluster interacting mostly with degenerate target electrons. In order to avoid discussing unessential details, we shall restrict in the sequel to regular topologies with same ion-ion nearest neighbour distances.

Then, we consider a given cloud of cluster ion debris in target, as a 2-body superposition. The stopping analysis then puts emphasis on the individual in contribution, and on the correlated one, as well.

The latter is treated here as a superposition of dicluster contributions [4],[8], according to a model due to Basbas and Ritchie [16].

We start our analysis by considering a frozen (polarized) configuration of ion debris flying in target, closely to the initial CIB trajectory. This picture stems from the fragmentation scenario outlined above.

According to the 2-body superposition principle advocated above, the stopping power of a N-cluster is straightforwardly given by

$$S_c = \left(\sum_i Z_i^2\right) S_p + 2 \sum_{1 \leq i < j \leq N} Z_i Z_j S_v(B_{ij}, D_{ij})$$

$$\equiv \text{Point} + \text{Corr} \qquad (23)$$

in terms of the pointlike contribution

$$S_p = \frac{e^2 \omega_p^2}{v^2} \ln\left[\frac{2mv^2}{\hbar \omega_p}\right], \qquad (24)$$

with

$$\omega_p = \frac{3^{1/2}}{r_s^{3/2}} \text{a.u.},$$

and v in atomic unit, and the also correlated contribution $S_v(B,D)$ worked previously for cylindrical geometry [16].

The present jellium modelling is particularly well adapted to the stopping behaviour of a Li target. This latter has a density d = 0.534 g/cm^3, so r_s = 3.27. Moreover, Li exhibits the smallest first energy of ionization (~ 3 eV), and it is very easily turned into a jellium of degenerate free electrons under heavy ion impact. The Bethelike formalism developed above is valid provided the ion debris velocity, nearly collinear to the initial CIB one, remains larger than $(\omega_p/2)^{1/2}$. For the Li shell target, at hand, this implies v ≥ 0.39 a.u.

As a significant result, we evaluate on Fig. 6, the stopping in a Li target of N correlated charges disposed on vertices, centre and face centres, of a cubic box. The latter is taken polarized along the initial impact velocity v. The stopping per charges is shown to be an increasing function of N.

In agreement with the quadratic form (23), the ratio Rap=corr/S_c, is the largest for equal-charge case.

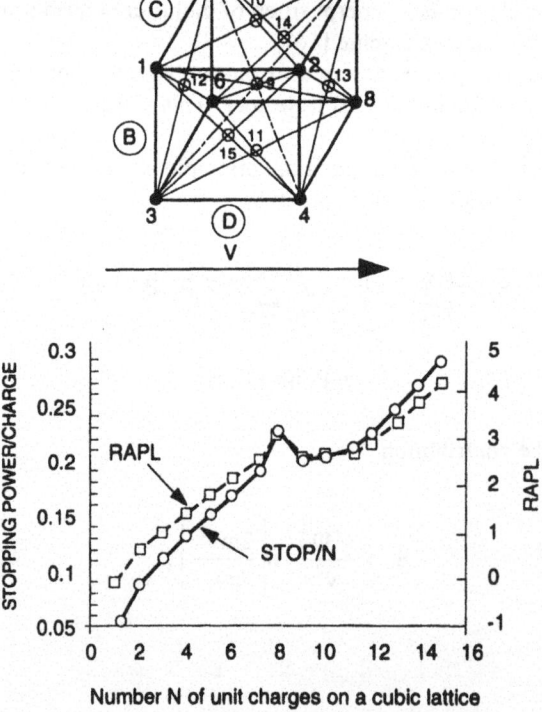

Figure 6. Stopping per charges of 1≤ N ≤ 15 unit charges located at vertices, center and faces center of a cubic box B*C*D with B=C=D=2 a.u. The overall drift velocity is V = 2.4 a.u. Rapl denotes the ratio Corr/Point (see Eq. (23)).

The first and obvious manifestation of the enhanced correlated stopping concerns the N-dependence, for a given spatial charge configuration. In order to highlight the basic physics features of the present model, we first assume that every correlated charge flows with same velocity.

Then if one displays step-by-step 15 charges on a cubic box with B = C = D = 2 a.u., as on Fig. 6, one witnesses a rapid increase of the stopping per charge and of the correlation contribution featured by the ratio Rapl.

This mundane example demonstrates how efficient the charge correlation may be, even with a relatively reduced number of ion debris. The jellium density ($r_s = 3.27$) corresponds to a lithium blanket of a thermonuclear fusion pellet [7].

The nest step in investigating the ECS is afforded by different charge configurations with same N and charge connectivity (number of bonds at a given particle location).

On Fig. 7 a boxlike arrangement is compared to lines of regularly spaced pointlike charges, flowing either along or transverse to an overall drift velocity. As intuitively anticipated, the compact boxlike topology displays a faster increase in stopping with N. Moreover, linear charge arrays transverse to velocity are more efficiently stopped than longitudinal ones.

This example makes it clear that everything else being equal, the repartition of charges with respect to velocity is a significant parameter qualifying stopping performances.

For instance, if one is looking for an efficient hammer like pressure in a given target, one is led to prefer compact Coulomb clusters providing the shortest ranges, with the highest resulting rate of shock waves production.

It is also very important to investigate, for a given cluster geometry, the size dependence of its stopping capabilities in a given target. For that purpose, Fig. 8 features the cubic box depicted in Fig. 6 with a charges nearest neighbour distance ranging from zero (coagulation) to 2.5 (in a.u.). As expected, stopping/charge steadily decays as size grows up, while the correlation ratio R_{ap} = Corr/S_c spans nearly all of its allowed range including the upper bound N-1 at full coagulation. It also decreases rapidly and even goes down to zero at very small target electron correlation length $\sim 1.1\ r_s^{1/2}$.

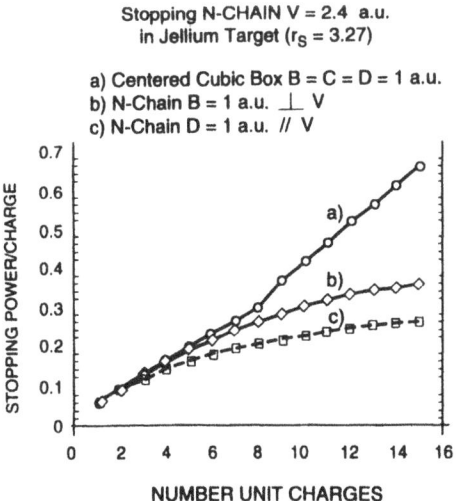

Figure 7. Stopping per charges of $1 \leq N \leq 15$ unit charges, distributed on three different arrangements.

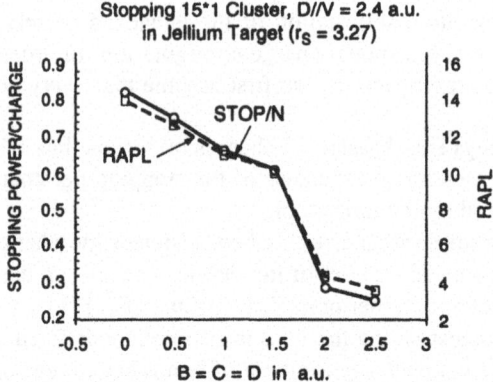

Figure 8. Stopping per charges and correlation ratio Rapl of 15 unit charges located at vertices center and faces on cubic boxes of various sizes. $r_s = 3.27$.

The ECS amply documented in most of the previously considered stopping situations, should lead to much shorter stopping ranges in target. With a view toward further compression of dense matter through intense beams of cluster ions, we picture in Fig. 9 and in physical units of practical interest, range-energy relationships for dense jellium targets pertaining to the same cubical arrangement of eight unit charges and to materials of potential interest as convertor of beam kinetic energy into hard photons (X rays). Here, we make contact with the so-called indirect drive scenario for particle-driven fusion. Qualitatively, results depicted on Fig. 9 look similar to those pertaining to atomic beams. However, as expected, the cluster ranges are shorter by order of magnitude at same projectile energy/nucleon. Fig. 9 features a gold foam made of wires, which when wrapped in a thin

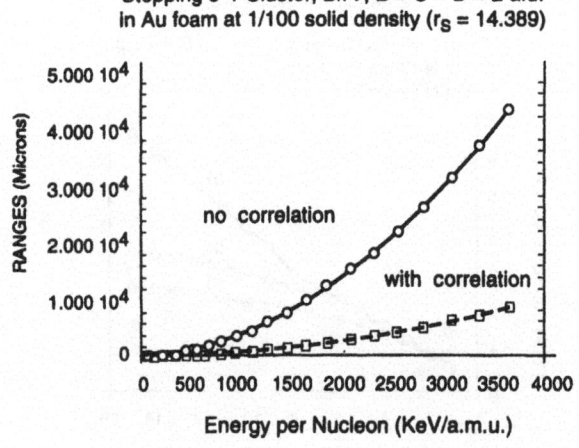

Figure 9. Range energy relationships for cubic boxes with B=C=D=2 a.u. accommodating 8 unit charges at vertices.

solid Au envelope provides an efficient particle-photon conversion in a two-sided illumination scenario.

We have seen how rich and diverse is the physics of the stopping of Coulomb clusters in a jellium target. In most situations of potential experimental interest, correlation amongst ion debris is expected to bring in a very strong stopping enhancement. The latter essentially depends on size, topological arrangement of particle locations, and of charge distributions, as well. The most symmetrical ones are the most effective.

In the direct drive situation, the driver kinetic energy gets converted through the stopping process into hydrodynamic shock waves compressing and finally igniting the inner fuel.

In terms of the projectile kinetic energy E, the driving pressure thus reads as (Energy/volume)

$$P(100 \text{M}bar) = \frac{\left(N_c I(MA) \Delta t(nsec) / Z(Cb)\right)^{1/3} \times 0.0056 M_i' \times E(keV / a.m.u.)}{R(\mu m)}, \quad (25)$$

where N_c is the number of building units (atom or molecule) with atomic mass M_i contained in a given accelerated cluster ion. Beam intensity I is in megaamps and beam pulse duration Δt in nsec. Beam energy is in keV/a.m.u. Stopping range R is in μm. Z(Cb) denotes the static charge, in Coulomb, of the accelerated cluster.

IV. ION FRAGMENTS STOPPING IN DENSE PLASMA [9]

In the foregoing section, we paid a rather thorough attention to the correlated stopping of fragmentated ion debris in a degenerate electron target, essentially featuring a cold Li corona coating a directly driven pellet.

The analysis of the driver-pellet coupling was then grounded on a dielectric function formulation routinely used in the physics of cold condensed matter.

Now, we turn to the opposite case of a high temperature target featuring the compressed payload material during the stagnation and burn phases. A simple and efficient modelling is then achieved by considering a dense but classical and nondegenerate electron fluid again described within a dielectric framework. As above, we neglect any electron-electron correlation in target. So, conceptually speaking, the low and high temperature cases are indeed very similar. Indeed, the two of them are thus based on the same random phase approach underlying the dielectric formulation.

Here, we restrict to a nondegenerate electron stopping medium which is also of obvious relevance for further applications in magnetically confined thermonuclear fusion. Nonetheless, it is now worthwhile to mention that dicluster stopping in an arbitrarily degenerate jellium has been recently given a detailed attention [17].

Dicluster stopping in a classical electron plasma, according to the recent treatment by d'Avanzo et al. [18] is first outlined, as a preliminary step toward N-cluster stopping explained as a linear superposition of dicluster contributions. The enhanced correlation stopping (ECS) is then thoroughly documented. Its specific dependence on N topological ion debris configurations, velocity and target electron coupling are thus stressed out. The cubic box arrangement with eight unit charges often appears as a convenient working example.

Here we briefly review a basic formulation for dicluster stopping in a dense and hot classical plasma [18] which parallels closely a similar derivation for a T = 0 electron jel-

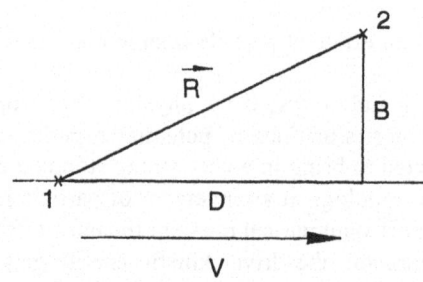

Figure 10. Cylindrical coordinates along projectile velocity $\vec{V}$ for dicluster stopping with intercharges distance $\vec{R}$. B and D denote $\vec{R}$ projections perpendicular and parallel to $\vec{V}$ respectively.

lium detailed above (Sec. III). The correlation contribution to stopping is conveniently highlighted by taking the ion pair as previously in a cylindrical geometry (Fig. 10) defined by the common and overall drift velocity $\vec{V}$ and interparticle distance $\vec{R}$.

In this work, we neglect as before any velocity mismatch between the two charges Z_1 and Z_2. Preliminary considerations [17] show that if one writes the stopping power as $S = -\Delta E/\Delta x$, magnitude of the mean energy-loss per unit pathlength, one can define Ω^2 as the square of the standard deviation of the energy-loss distribution per unit pathlength.

For a plasma target with $k_B T \gg E_F$ (Fermi energy), one thus gets within a good approximation $\Omega^2 \sim 2 k_B TS$. So, computing the stopping power already yields a reasonable estimate for the straggling. Moreover, with a view toward an experimental program developed on the Orsay tandem linear accelerator [11], we think it useful to restrict to classical plasma targets of fully ionized deuterium at a rather moderate temperature $T \cong$ a few eV, while the corresponding free electrons density ranges from 2×10^{17} cm^{-3} up to 10^{20} cm^{-3}. At such modest temperatures, the straggling may thus be neglected in a first approximation.

Dicluster stopping calculation starts with the electrostatic potential expression in a referential system where the two ions are at rest [16,18], and the leading ion is located at the origin. It reads as (see Fig. 10)

$$\phi_1(\vec{r}) = \frac{(4\pi)}{(2\pi)^3}\int d^3k \frac{e^{i\vec{k}\cdot\vec{r}}}{k^2 \varepsilon(k,\vec{k}\cdot\vec{V})}\left(Z_1 + Z_2 e^{-i\vec{k}\cdot\vec{R}}\right), \quad (26)$$

where

$$\varepsilon(k,\omega) = 1 + w\left(\frac{\omega}{k}\right)\frac{1}{k^2}, w(\xi) \equiv X(\xi) + iY(\xi), \quad (27)$$

denotes the longitudinal dielectric function for a maxwellian and classical electron plasma [19].

According to Fig. 10, the dicluster stopping is defined by (N_D = number of electrons within Debye sphere)

$$-\frac{dE}{dx} = Z_1 N_D \frac{\partial \phi_{ind}}{\partial \vec{r}}\bigg|_{\vec{r}=\vec{r}_1(t)} + Z_2 N_D \frac{\partial \phi_{ind}}{\partial \vec{r}}\bigg|_{\vec{r}=\vec{r}_2(t)}, \quad (28)$$

where $\vec{r}_1(t) = \vec{V}t$ and $\vec{r}_2(t) = \vec{V}t + R$ respectively. ϕ_{ind} denotes the electrostatic potential of one unit charge acting on the other. So, from expression (1) and for $Z_1 = Z_2 = Z$ one gets

$$-\frac{dE}{dx} = 4\frac{Z^2 N_D}{\pi} \int_0^{k_{max}} dk\, k^3 \int_0^1 \frac{d\mu\, Y(\mu V)\mu}{(k^2 + X(\mu V))^2 + Y^2(\mu V)} \times$$

$$\times \left[1 + \cos(k\mu D) J_0(k\sqrt{1-\mu^2}\, B)\right],$$

$$\equiv 2Z^2 S(1) + 2Z^2 S_V(B,D) = S_c$$
$$= \text{Point} + \text{Corr} \tag{29}$$

in terms of the ordinary Bessel function $J_0(x)$ and $k_{max} = 4\pi(V^2 + 2)N_D/Z$.

Also, we have split the correlation (Corr) contribution to stopping from the uncorrelated pointlike (Point) one. B and D are respectively the transverse and longitudinal projections of $\vec{R}$ on $\vec{V}$ (Fig. 10). At this juncture, it is worth while to notice that the presently considered frozen geometry (Fig. 10) is well justified because we restrict to a high velocity range for the projectile. For instance, if one considers a dicluster with a 10 keV/A.m.u. kinetic energy, and if we admit that all the Coulomb repulsion potential within ion pair contributes to a velocity transverse to $\vec{V}$, one has approximatively

$$\frac{V_\perp}{V} \cong 0.79\, Z\, A^{-1/2},$$

where A is the cluster ion atomic mass. So, the present polarized approximation appears well justified for heavy ion fragments with $A \gg 1$, and not for hydrogen ones [20].

Now, we consider the simultaneous stopping of N ion debris flowing with same projectile velocity V in the close vicinity of each other, i.e., within a few atomic units of relative distance. A first and obvious extension of the above and polarized 2-body treatment is to superimpose the N(N-1)/2 dicluster energy losses available in a Coulomb N-cluster.

Such a simple generalization has already produced a huge variety of charge configurations leading to the ECS, in cold target stopping [7–8] provided N > 2. Rewriting the second right hand side of Eq. (29) for two distinct charges, the corresponding N-cluster expression is then obtained as (cf Eq. (23) for a cold jellium target)

$$-\frac{dE}{dx} = \sum_{i=1}^N Z_i^2 S(1) + 2 \sum_{1 \le i < j \le N} Z_i Z_j S_V(B_{ij}, D_{ij}) = S_c$$

$$\equiv \text{Point} + \text{Corr},$$
$$= \text{Stop} \tag{30}$$

As shown by previous studies of the ECS in a degenerate jellium target in Sec. III, it will prove also useful to quantify the expected ECS through a few obvious ratios. The first one reads as

$$R_{ap} = \frac{Corr}{S_c} \leq 1 - \frac{\Sigma_{i=1}^{N} Z_i^2}{(\Sigma_i Z_i)^2}, \qquad (31)$$

in terms of the total stopping S_c.

An obviously related quantity of great interest is

$$R_{apl} = \frac{Corr}{Point} = \frac{R_{ap}}{1 - R_{ap}} \leq \frac{(\sum_i Z_i)^2}{\sum_i Z_i^2} - 1, \qquad (32)$$

denoting the ratio of correlated to uncorrelated stopping.

These ratios are independent of target coupling. They also apply to a T = 0 jellium [8] and to a partially degenerate electron fluid as well. They produce a priori guidelines of great interest for the ECS numerical estimation.

For instance, symmetrical charge configurations with $Z_1 = Z_2 = ... = Z_N \equiv Z$, fulfill $R_{ap} \leq 1 - N^{-1}$ and $R_{apl} \leq N-1$. In this case, one has $0.666 \leq \sup(R_{ap}) \leq 1$ with $3 \leq N \leq \infty$. Selecting out a highly dissymmetric configuration with $Z_1 = Z_2 = ... Z_{N-1} = 1$ and $Z_{N=1}$, one witnesses $0.59 \leq \text{Sup}(R_{ap}) \leq 0.75$ with $3 \leq N \leq \infty$. As a consequence, one is led to estimate symmetric configurations as those providing the highest correlated stopping.

Before embarking into an extensive study of N-cluster stopping, it proves already highly instructive to investigate through Eqs. (28),(29) the dicluster ECS, in terms of overall velocity V and relative distance $\bar{R}$. Fig. 11 features a target plasma at T = 2 eV and with $n_e = 10^{17}$ cm^{-3}.

The ratio R_{apl} (Eq. (32)) is then depicted in terms of B/λ_D and V/V_{th} at a fixed intercharge distance $R = \lambda_D$. One already witnesses an ECS of the order of 30 per cent on most of the parameters range. ECS increases steadily with V up to a nearly constant plateau value. Also, for small V (≤ 5) one observes a larger stopping for an oblique dicluster than

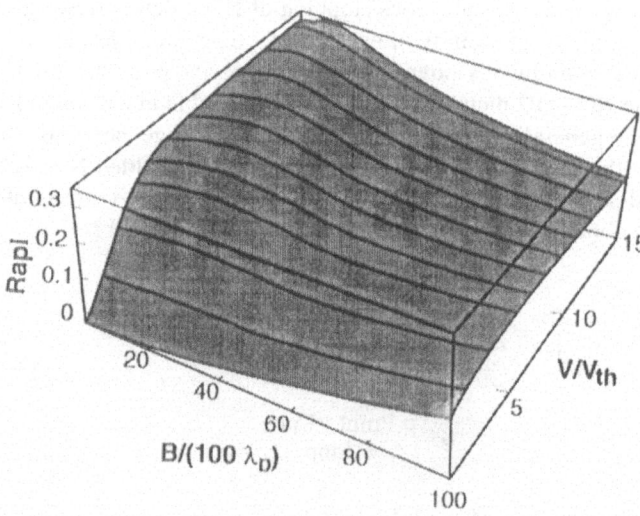

Figure 11. Ratio Rapl (Eq. (32)) for a dicluster with $Z_1=Z_2=Z$ and $R=\lambda_D$ in terms of B/100 in λ_D and V in V_{th}. T=2 eV and $n_e=2\times10^{17}$ cm^{-3}. $R=\lambda_D$.

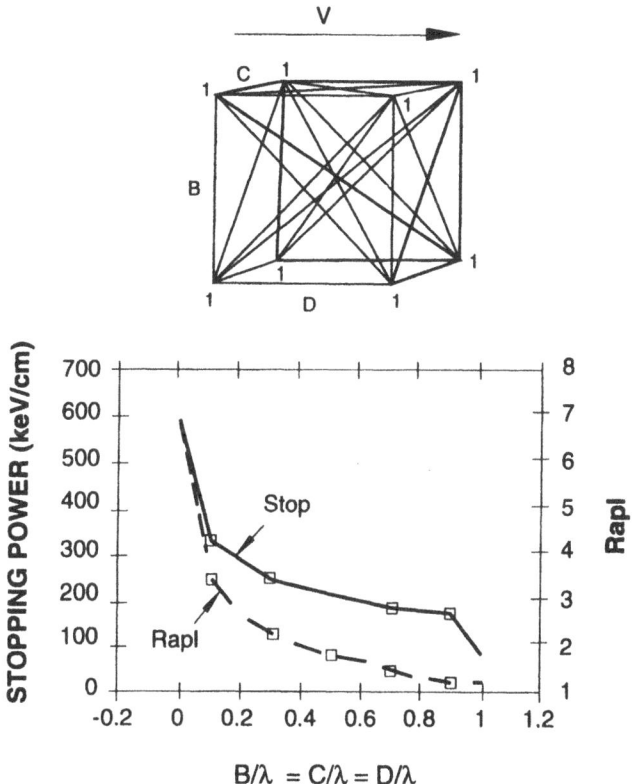

Figure 12. Stopping of eight unit charges and ratio Rapl (Eq. (32)) arranged on a cubic box with side ranging from 0 to λ_D. $V=3\ V_{th}$, T=2 eV. $\vec{D}//\vec{V}$ and 2×10^{17} cm^{-3}.

for a parallel one, with respect to $\vec{V}$. It is useful to recall that a plasma with T = 2 eV and $n_e = 2 \times 10^{17}$ cm^{-3} has V_{th} = 0.25 a.u. and λ_D = 444 a.u.

From the above, it is already clear that a mere two-body vicinage effect already brings in a substantial ECS. So, now we implement the N-cluster stopping through Eq. (30). In order to emulate a typical arrangement of closely packed ion debris, we found it very instructive to look after a cubical display of eight unit point like charges at vertices, as shown on Fig. 12.

The given cubic box has a side B = C = D running from 0 to λ_D. Projectile velocity is $V = 3\ V_{th}$. As previously $\vec{D}/\vec{V}$ while B and C denote the two transverse dimensions.

At complete coagulation (B = C = D =0) the ECS fulfills the upper bounds (31) and (32) for a symmetric charge distribution with R_{apl} = 7 at any target density. When the intercharges distances increase, total stopping and ECS decay as well. Nonetheless, even at the largest size (B = C = D = λ_D), the ECS remains at work with $R_{apl} \geq 1$. It should be kept in mind that for the given plasma targets, the ion debris relative distances are pretty large on a atomic scale (of the order of several tens of a.u.).

We now contrast the cubic box with regular arrays of pointlike charges propagating either parallel or transverse to the overall velocity $\vec{V}$.

Fig. 13 displays a target electron density available on linear plasma column and Z-pinch devices installed on ion accelerator beam lines. A first overall and striking observa-

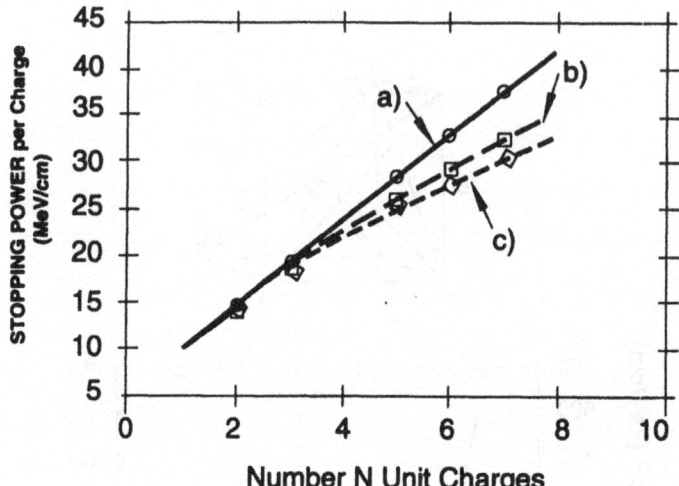

Figure 13. Stopping of N-clusters in terms of N with $B=C=D=0.1\ \lambda_D$ and $V=3\ V_{th}$. $T=2$ eV, $D//V$. $n_e=2\times10^{17}$ cm^{-3}.

tion is afforded by the rather scale invariant increase with N of the stopping/charge although its value differs nearly by an order of magnitude. Despites the three considered topologies share a same connectivity, the cubic box is the best performer because its charge-charge distances enjoy a more compact distribution than in linear arrays. In the last two, the transverse N-chain exhibits the second largest stopping.

Also, it proves rather illuminating to explore the opposite situation (Fig. 14) with a large size ($B=C=D=\lambda_D$). Now, one witnesses behaviour at variance with previous ones, so topology manifests itself globally. The overall N-cluster extends now over several screening lengths.

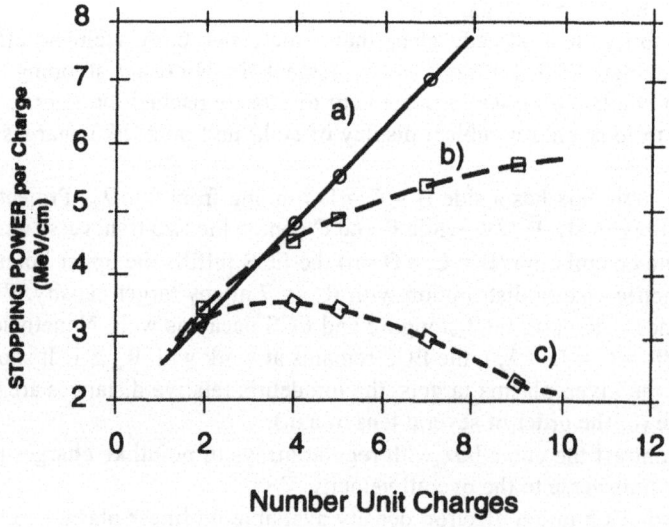

Figure 14. Same caption as in Fig. 4 with $B=C=D=\lambda_D$, $n_e=10^{20}$ cm^{-3}.

The order of relative stopping efficiency remains unchanged.

Curves b and c in Fig. 14 anticipate specific N-chains trends detailed below. Those results demonstrate that the ECS remains nonnegligible at large particle interdistances.

We made use of a complete random phase approximation for the linear response of a dense classical electron target, to estimate the interaction with incoming Coulomb N-clusters of ion debris. The given stopping is taken in a high velocity formalism based on a linear superposition of every correlated dicluster contribution within N-cluster. We restricted to even distributions of unit charges in order to stress out the basic features and properties of the enhanced correlation stopping (ECS) such as ion debris topology and target plasma coupling.

The nearly ubiquitous ECS is thus expected to be of considerable and potential significance in providing new range-energy relationships useful in designing novel scenarios of interest for the cluster ion beam-dense matter interaction in particle-driven inertial confinement fusion. The presently considered model bears a direct relevance to the plasma phases of the compressed pellet containing the thermonuclear fuel.

V. HYDRODYNAMIC REGIMES OF CLUSTER-DRIVEN TARGETS (DIRECT DRIVE) [21]

Beam-target interaction phenomena are very dependent on the type of beam. In the laser-driven case, high temperatures appear in the expanding corona. The target outer material is ablated because of thermal conduction between the critical surface and the ablation surface, and the pressure level in this zone becomes very large. Such ablation pressure drives the compression of the internal part of the target (where the fuel is located). The main effect of the ablation is the acceleration of the afore-mentioned internal part. A similar situation may exist in cluster beam-target interaction. Such hydrodynamic reaction follows the law of the hot rocket model, which is summarized in the following Sec. VI.

Heavy ion beams direct drive is also dominated by energy deposition, but the interaction region is much wider. The external coating of the target for high-gain reactor-size microcapsules is about 100 mg (while the fuel is only 1 mg). The dose rate in the interaction zone is very moderate (as compared to the laser case, where a few milligrams of coating are enough) and the temperature of the ablated material is at least one order of magnitude lower than that of the laser-target corona. The fuel is accelerated inwards by the ablation pressure (as in the laser case) but the thermodynamic gradients are lower, which means that strong shock waves can be avoided [22]. This is why the performance of the target is not very sensitive to pulse shaping in the heavy ion beams case, unlike in the laser-driven case, where a tuned shock multiplexing must be generated by a suitable time pulse shaping in order to get high compression [23]. The hot rocket model is also useful for the theoretical analysis of the heavy ion beams directly-driven targets.

Cluster-driven ICF will exhibit a somewhat hybrid behaviour in the interaction process. Clusters have a much shorter range than ions and thus they will require a thin coating (as lasers) but there will not be in this case a critical surface where most of the energy is deposited.

There is, however, a main dissimilarity between cluster driving and the other driving types. In these ones, momentum transfer effects are negligible as compared to energy deposition effects. This is not necessarily true in cluster driving, where momentum transfer can become comparable with the ablation pressure effects. In Sec. VII some analytical calculations will be presented to characterize the performance of the targets in this case. The hammer model (Sec. VIII) is dominated by the beam momentum and has some analogies with impact fusion.

The performance of cluster-driven ICF depends on the type of regime dominating the interaction phase between the cluster beams and the target. The global analysis of an inelastic scattering points out the existence of two limits:

- When the beam total mass (M_b) is greater than the target mass (M_t), most of the beam energy is transformed into kinetic energy of the compound body formed by the embeddment of the beams particles into the target. It is worth pointing out that those particles must not penetrate too deep into the target as to heat the fuel during this phase. The internal structure of the target is thus particularly important, because steep gradients of pressure will appear, which will launch inward shock waves accelerating the fuel. These phenomena can be studied with the hot hammer model, presented in Sec. VIII.
- When the beam total mass (M_b) is much lower than the target mass (M_t), most of the beam energy is transformed into internal energy of the target coating, where pressure and temperature reach very high levels. A strong ablation process thus takes place, the external part of the target expanding outwards and the internal part of the target being accelerated inwards. This can be studied with the hot rocket model, which is presented in Sec. VI. In Sec. VII, the intermediate case is analyzed for the case of a rocket model plus a nonnegligible momentum transfer.

VI. THE HOT ROCKET MODEL

This model can be described by the conservation equations of mass, momentum and energy, plus the equation of state and some simplifying hypothesis on the expanding corona regime. Fig. 15 depicts the main characteristics of the interaction zone and will be used to calculate the hydrodynamic efficiency of the implosion. It is assumed that the beam clusters penetrate up to a depth x_d, which will be located very close to the surface because the range of the clusters is very short. From the energy deposition region inwards, a heat wave penetrates in the target and enhances the ablation process.

The conservation equations can be written as follows in the frame of the energy deposition region x_d (presuming one-dimensional, planar geometry; see Fig. 15)

$$\rho_{p\ell} v_{p\ell} = \rho v = \dot{m} = \text{constant} \tag{33}$$

$$\rho_{p\ell} v_{p\ell}^2 + \rho_{p\ell} c_{p\ell}^2 = \rho v^2 + \rho c^2 \tag{34}$$

$$\dot{m}\left[\frac{v_{p\ell}^2}{2} + \frac{\gamma}{\gamma-1} c_{p\ell}^2\right] - \dot{m}\left[\frac{v^2}{2} + \frac{\gamma}{\gamma-1} c^2\right] + q_{in} - q_{out} = I \tag{35}$$

where c is the speed of sound (c = $(P/\rho)^{1/2}$) and the subscript pℓ refers to the expanding plasma outwards of the point x_d. Thermal fluxes flowing outwards and inwards are represented by q_{out} and q_{in}. I stands for the beam energy flux (erg.cm^{-2}s^{-1}). $\dot{m}$ denotes the ablation of target material.

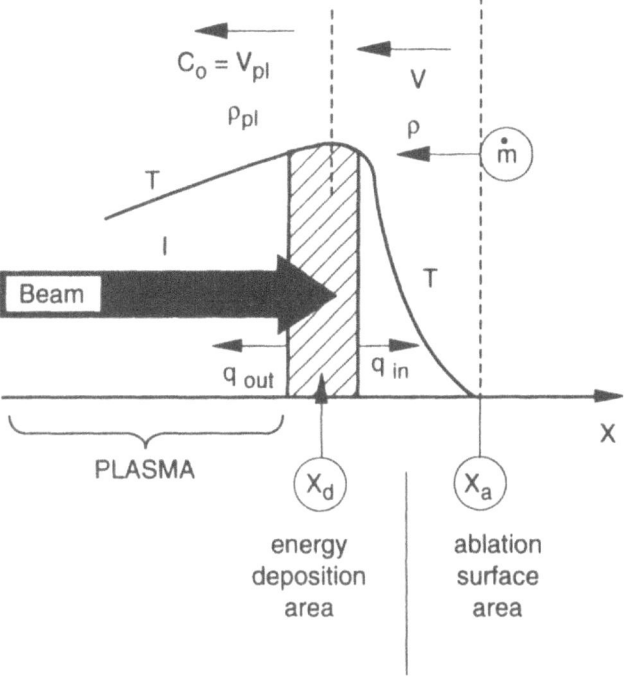

Figure 15. Schematics of cluster ion beam interacting with the outer zone of a thermonuclear pellet.

The conservation equations at the point x_a, the ablation front, can be written as

$$\rho v = \dot{m} \tag{36}$$

$$P_a = \rho v^2 + \rho c^2 \tag{37}$$

$$q_{in} = \dot{m}\left(\frac{v^2}{2} + \frac{\gamma}{\gamma-1}c^2\right) \tag{38}$$

where P_a is the ablation pressure. Assuming that the plasma expands sonically, that is,

$$v_{pl} = c_{pl} = C_o, \tag{39}$$

from Eqs. (33) and (34), one gets

$$\dot{m}(2C_o) = \dot{m}\left(v + \frac{c^2}{v}\right) \tag{40}$$

which can be used to determine the ablation pressure (Eq. (37))

$$P_a = \dot{m}\left(v + \frac{c^2}{v}\right) = 2C_o \dot{m} \tag{41}$$

The energy conservation equations (35) and (38) can be solved with the hypothesis given in Eq. (39) giving

$$I = \dot{m}\left[\frac{C_o^2}{2} + \frac{\gamma C_o^2}{\gamma - 1}\right] - q_{out} \tag{42}$$

By definition, for a free streaming plasma

$$q_{out} = -\dot{m}v_{pl}^2 = -\dot{m}C_o^2 \tag{43}$$

where the negative value of q_{out} comes from the way in which the energy balance (Fig. 15) has been written. From Eqs. (42) and (43) one gets

$$I = \dot{m}\left[\frac{3}{2}C_o^2 + \frac{\gamma}{\gamma - 1}C_o^2\right] \tag{44}$$

which relates the beam energy flux, I, to the ablation mass rate

$$\dot{m}C_o^2 = \left[\frac{2(\gamma - 1)}{5\gamma - 3}\right]I = \frac{1}{4}f_\gamma I, \; f_\gamma \equiv \left[\frac{2(\gamma - 1)}{5\gamma - 3}\right]. \tag{45}$$

which depends on $\gamma(\gamma = C_p/C_v)$. For $\gamma = 5/3$ we get $f_\gamma = 1$ and

$$\dot{m}C_o^2 = \frac{I}{4} \tag{46}$$

while, for example, for $\gamma = 7/5$ one has $f_\gamma = 0.8$ and

$$\dot{m}C_o^2 = \frac{I}{5} \tag{47}$$

The former equations can be used to determine the hydrodynamic efficiency of the hot rocket model. The momentum conservation equation for the target can be written as

$$M_t \frac{dv_t}{dt} = P_a \tag{48}$$

The mass at a time t is given by

$$M_t = M_o - \dot{m}t \tag{49}$$

where we have made use of the hypothesis $\dot{m}$ = constant, and M_o is the initial target mass. Eqs. (41) and (48) can be written as

$$(M_o - \dot{m}t)\frac{dv_t}{dt} = P_a = 2C_o\dot{m} \tag{50}$$

and integrating over time

$$v_t = -\int_0^t dt \; \frac{2C_0}{(M_0 - \dot{m}t)} \frac{d}{dt}(M_0 - \dot{m}t) \tag{51}$$

$$v_t = 2C_0 \ln\frac{M_0}{M_t} \tag{52}$$

At the end of the ablation phase, the target will be imploding at the highest speed, and the remaining mass will be the payload mass, M_p, which will be slightly larger than the fuel mass. The hydrodynamic efficiency of the implosion will be

$$\eta_H = \frac{M_p v_p^2 / 2}{It_p} \tag{53}$$

where t_p is the beam pulse duration and v_p is the payload speed. Eq. (46) relates I to the physical variables of the ablation process and can be used with Eq. (52) and Eq. (53) to yield

$$\eta_H = \frac{1}{2}\frac{M_p}{(M_o - M_p)} \ln^2\left(\frac{M_o}{M_p}\right) = \frac{x\ln^2 x}{2(1-x)} \tag{54}$$

where $x = M_p/M_o$. It is worth quoting that the efficiency given in Eq. (54) is equal to 1/2 that of the cold rocket model, i.e., the pure hydrodynamic model which does not take into account the thermal energy required to feed the ablation process.

The hydrodynamic efficiency η_H, Eq. (54), has a maximum of about 1/3 for a value of the payload mass $M_p = 0.2\ M_o$. It is even more important to notice that such a maximum is very broad. For $M_p = 0.1\ M_o$, the efficiency is still 30%, and drops to 23.5% for $M_p = 0.05\ M_o$. It is only 11% for $M_p = 0.01\ M_o$ which is a value representative of heavy ion beams driving. For higher values of M_p, which are more interesting for the cluster-driven scheme, the efficiency goes down slowly. It is still 30% for $M_p = 0.3\ M_o$, 24% for $M_p = 0.5\ M_o$ and 20% for $M_p = 0.6\ M_o$, but it goes down to 13% for $M_p = 0.7\ M_o$.

VII. THE ROCKET MODEL WITH BEAM MOMENTUM TRANSFER

The direct pressure (Π) produced by the momentum transfer of the beam may be comparable to the ablation pressure. In this case the rocket equation can be rewritten as

$$M_t \frac{dv_t}{dt} = P_a + \Pi \tag{55}$$

where Π is related to the energy flux I as follows

$$\frac{\Pi}{I} = \frac{\sqrt{2m\varepsilon_o}}{\varepsilon_o} = \sqrt{\frac{2m}{\varepsilon_o}} = \alpha = \frac{2}{v_o} \tag{56}$$

where m and ε_o are the mass and the energy of an individual cluster of the driving beam. In this context we can see clearly the difference between heavy-ion-driven and cluster-driven ICF. Ions will have a mass of 200 (atomic units) and an energy of several GeV. Clusters will have a much larger mass (may be 2000 atomic units or more) and much smaller energy (of some MeV, and even less). Therefore, α is almost zero for heavy ions and will be much higher for clusters.

We can use $\dot{m}C_o^2 = f_\gamma I/4$ to relate Π to C_o through I with

$$\Pi = \alpha I = \frac{\alpha 4 \dot{m} C_o^2}{f_\gamma} \tag{57}$$

where $f_\gamma = 4\left[\frac{2(\gamma-1)}{5\gamma-3}\right]$ and $\gamma = C_p/C_v$. Eq. (57) can be rewritten as

$$(M_o - \dot{m}t)\frac{dv_t}{dt} = 2\dot{m}C_o + \frac{4\alpha \dot{m} C_o^2}{f_\gamma} = 2\dot{m}C_o\left(1 + \frac{4C_o}{f_\gamma v_o}\right) \tag{58}$$

where $4C_o/f_\gamma$ must be smaller than v_o (the cluster speed) for our hypothesis to be met.

The hydrodynamic efficiency can then be expressed as

$$\eta_H = \frac{M_p v_p^2/2}{It_p} = \left(1 + \frac{4C_o}{f_\gamma v_o}\right)^2 \frac{x \ln^2 x}{2(1-x)} \tag{59}$$

where $x = M_p/M_o$.

The factor multiplying the bracket is the same as in the hot rocket model without momentum transfer, Eq. (54). For $x = 0.25$, Eq. (54) gives an efficiency of 1/3. Hence, should the bracket be close to 3, the hydrodynamic efficiency would be approaching 1. However, our hypothesis $4C_o/f_\gamma < v_o$ sets an upper bound for the bracket of Eq. (59) which must not be higher than ~1.1. Nevertheless, as it is squared, that would mean an efficency 20% higher than that of the hot rocket model. As an example, for $C_o = 0.10\, v_o$ and $\gamma = 5/3$ the hypothesis previously used would be fulfilled. With a sound speed $C_o \sim 10^7$ cm/s, $v_o \sim 10^8$ cm/s and assuming a cluster mass $m = 2000$ atomic units, the cluster energy would be 10 MeV, which is a reasonable choice for this purpose. For cluster speeds much higher than 10^8 cm/s, the momentum transfer term would become negligible. For speeds much lower than that, this term would be dominant and a hammer model would be appropriate, which is analyzed in the following Sec. VIII. However, before discussing the hammer model, it is shown that under the assumption of this section, a «detonation» profile wave can be formed.

Since the absorption layer is very small [7], i.e., the cluster stopping range R_a is of the order $R_a \sim$ a few µm, the momentum and the energy of the cluster are absorbed in the same region. In this case the cluster can induce singularities (e.g. shock waves or detonation waves) in the pressure and the density profile. The «jump» conditions, in a frame of reference moving with the absorption layer, are

$$\rho_o v_o = \rho_1 v_1 \tag{60}$$

$$P_o + \rho_o v_0^2 = P_1 + \rho_1 v_1^2 + \Pi \tag{61}$$

where 0 and 1 denote the undisturbed and the disturbed regions respectively.

In Eqs. (60) and (61) a steady state solution was assumed. This is justified since the time scale for a disturbance to cross the absorption layer is about R_a/c (~ 10 ps) which is much smaller than the beam pulse duration (~ few nsec).

The Π term in Eq. (61), as well as in Eq. (55), is important if it is of the order of the ablation pressure P_1 (in Eq. (61) or P_a in Eq. (55). This implies

$$\frac{\Pi}{P_1} = \frac{4C_0}{f_\gamma v_0} \sim 1 \tag{62}$$

or equivalently

$$\Pi \sim P_1 \sim \sqrt{\frac{2m_p A}{\varepsilon_o}} I \tag{63}$$

where m_p is the proton mass, A is the atomic number of the cluster, ε_o is the cluster energy and I is the irradiance cluster in erg/(s.cm²). On the other hand

$$P_1 \sim \rho_0 c_0^3 \tag{64}$$

where ρ_o is the target density. For $\rho_o \sim$ 1g/cm³ and $c_o \sim 10^7$ cm/sec, Eqs. (63) and (64) require a beam irradiance of about 10^{15} W/cm² for A = 1000 and ε_o = 10 MeV. Under these circumstances, the energy «invested» in the absorption layer is so high that an isothermal «jump» (in density and pressure) will occur.

VIII. THE HAMMER MODEL

A schematic comparison between the rocket model and the hammer model is given in Fig. 16. The pellet consists of a deuterium-tritium shell enclosed by a coating of Li.

In the hammer model the beam mass (M_b) is much larger than the mass of the fuel M_p. In this case Eqs. (33) and (34) together with the definition of the hydrodynamic efficiency

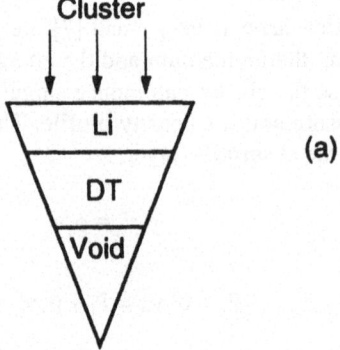

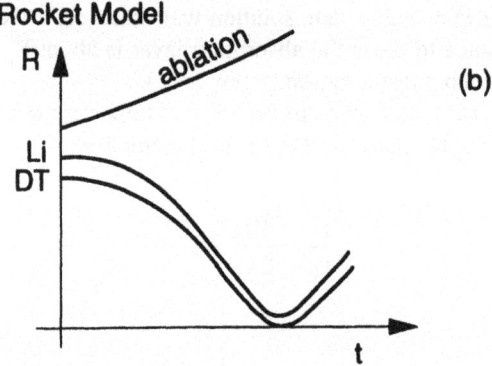

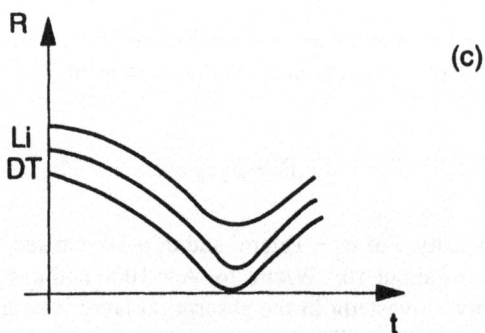

Figure 16. Schematic profiles of (a) cluster driven pellet, (b) radius-time diagram for the rocket model in comparison with (c) the hammer model.

$$\eta_H = \frac{\frac{1}{2} M_p v^2}{\frac{1}{2} M_b u^2} \tag{65}$$

give for $x = M_p/M_b$,

$$\eta_H = \frac{x}{(1+x)^2}.\tag{66}$$

η_H has a maximum at $x = 1$ yielding an efficiency of 25%. However, for the hammer model one has

$$x \ll 1, \eta_H \sim x \tag{67}$$

impliying that the hammer model is not efficient.

In order to get an understanding of the velocities involved in this case, it is useful to compare the kinetic energy E_{kin} of a deuterium-tritium beam with the internal energy E_{int} of the DT fuel.

$$E_{kin}(kJ/mg) = 5\left(\frac{v}{10^7 \, cm/sec}\right) \tag{68}$$

$$E_{int}(kJ/mg) = 120\left(\frac{T}{keV}\right) \tag{69}$$

Transforming η_H of the kinetic energy into internal energy (i.e. $E_{int} = \eta_H E_{kin}$) one gets for $\eta_H = 10\%$ that a speed of 3.5×10^8 cm/sec (in the center of mass system) is required in order to heat the fuel to 5 keV. This speed is ten times higher than the required speed of implosion of the fuel of an ICF target driven by ablative processes. It is now adequate to define a cluster beam "density" ρ_c by

$$\rho_c = \frac{2E}{S\tau_b v_0^3} = \frac{(Am_p)^{3/2} I}{\sqrt{2} \varepsilon_0^{3/2}} \tag{70}$$

where $\tau_b v_0$ is the cluster "length", S its cross section, τ_b, v_0 and ε_0 are the cluster pulse duration, velocity and energy respectively and a cluster has a mass of A nucleons. I is the irradiance (W/cm^2). For example, if $A = 10^3$, $I = 10^{15}$ W/cm^2 and $\varepsilon_0 = 1$ MeV one gets a cluster density of 0.23 g/cm^3, similar to the initial D-T fuel density. In the hammer model one can describe the "cluster beam"-target interaction as a collision satisfying the following equations:

$$\frac{\rho_{0c}}{\rho_{1c}} = \frac{\rho_0}{\rho_1} = \frac{\gamma - 1}{\gamma + 1} \tag{71}$$

$$P = \frac{\rho_{0c} v_0^2}{\left(1 - \frac{\rho_{0c}}{\rho_{1c}}\right)} \cdot \left[1 + \left(\frac{\rho_{0c}}{\rho_{1c}}\right)\frac{(1 + \frac{\rho_0}{\rho_1})}{(1 - \frac{\rho_{0c}}{\rho_{1c}})}\right]^{-2}. \tag{72}$$

ρ_{0c} (given by Eq. (70)) and ρ_{1c} are the cluster densities before and after the "collision", ρ_0 and ρ_1 are the densities before and after the collision in the fuel (the pellet) and P is the pressure associated with the collision. For a strong shock wave the right hand side of Eq.

(71) is justified. Taking $\gamma = 5/3$ for the fuel (as well as for the cluster beam) one gets a pressure of

$$P = \frac{4}{3}\rho_{0c}v_0^2 \cdot \left[1 + \frac{5}{3}\frac{\rho_{0c}}{\rho_0}\right]^{-2} \quad (73)$$

Using for example $\rho_0 = 0.2$ g/cm^3 and $\rho_{0c} \simeq \rho_0$ one gets for a cluster velocity of $v_0 = 5.10^8$ cm/s a pressure of about $2.5.10^4$ Mbar. Using Eqs. (71) and (72) one obtains a pressure which scales as

$$P \sim I\left(\frac{\varepsilon_0}{A}\right)^{-1/2} \quad (74)$$

This approach, which is equivalent to the impact model, is relevant if $M_t < M_b$, i.e.,

$$\rho_0 R_a \leq L\rho_{0c} \quad (75)$$

where R_a is the range that the clusters deposit their energy in and L is the beam length (= $v_0\tau_b$).

IX. A SPECIFIC CASE

In order to have a scenario illustrating the general problem let us take a typical numerical example.

Let us assume a total target mass of 2 mg (fuel + coating). In order to reach imploding speeds much higher than 10^7 cm/s, the energy required according to Eq. (67) will be of the order of 1 MJ. A beam of clusters of 2000 atomic units with a total kinetic energy of 1 MJ will require 0.6×10^{18} clusters of 10 MeV per cluster. The total mass of those clusters will be 2 mg, i.e., the same mass as the target. In this case, 50% of the driving energy will go to the ablation process and 50 % directly to kinetic energy. The mass of the compound body after the interaction will be lower than 4 mg because of the ablated mass. Moreover, the speed profile inside the compound body will not be uniform, because the lighter parts of the target (i.e., the fuel) will be accelerated to higher speeds. Summarizing this picture, it seems possible to accelerate a fuel mass of 1 mg to speeds much higher than 10^7 cm/s, which means that fusion conditions could be achieved at the end of the implosion phase. Nevertheless, the detailed performance of the implosion phase will play an important role in the fuel conditions obtained at the end of this phase.

The momentum equation can be written for the cluster-beam target interaction

$$M_t \frac{dv_t}{dt} = P_a + \Pi \quad (76)$$

where P_a is the ablation pressure produced by the plasma, Π is the beam direct pressure (as a consequence of momentum conservation) and M_t is the target mass with an average velocity v_t. The hot rocket model (Sec. VI) is available for $P_a \gg \Pi$ while the hammer (Sec. VIII) model is defined for $P_a \ll \Pi$. The intermediate case, $P_a \sim \Pi$ is also discussed (Sec. VII). For a cluster beam with a velocity v_o, the different regimes are defined by:

(a) rocket model (RM): ($M_t \gg M_b$)

$$P_a \gg \Pi \Leftrightarrow \frac{v_o}{c} \gg \frac{4}{f_\gamma} \qquad (77)$$

where c is the speed of sound and f_γ is defined below Eq. (40) and describes the equation of state effect. For $\gamma = 5/3$ one has $f_\gamma = 1$.

(b) hammer model (HM): ($M_t \ll M_b$)

$$P_a \ll \Pi \Leftrightarrow \frac{v_o}{c} \ll \frac{4}{f_\gamma} \qquad (78)$$

(c) intermediate model (IM): ($M_t \sim M_b$)

$$P_a \sim \Pi \Leftrightarrow \frac{v_o}{c} \sim \frac{4}{f_\gamma} \qquad (79)$$

In order to evaluate numerically the validity of Eqs. (77)-(79) it is necessary to estimate the plasma temperature in the corona. For this purpose we use

$$E_h = \frac{E_b M_t}{M_b + M_t}$$

where:
E_h = internal energy (heat) of the compound body
E_b = beam kinetic energy = $M_b u^2/2$
and assume that the energy converted into heat goes only to heating the plasma, i.e.

$$E_b \frac{M_t}{M_t + M_b} \approx \frac{3 M_t R_a kT}{A_t m_p L_t} \qquad (80)$$

where E_b is the beam energy, $A_t m_p$ (where m_p is the proton mass) is the target coating atomic mass. In Eq. (80) $\gamma = 5/3$ was taken and an equal temperature for electrons and ions is assumed. R_a is the energy absorption range and L_t is the target thickness. For example, for $M_b \simeq 0.1$mg, $M_t \simeq 0.5$mg, $A_t \simeq 6$, $E_b = 0.1$MJ, $R_a = 1\mu$m and $L_t \simeq 70\mu$m one gets a temperature of about 300 eV. (For this set of numbers a simulation was performed and a maximum temperature of 400 eV in the corona was achieved). For a beam cluster with A = 2000 and $\varepsilon_o = 10$ MeV (the energy of one cluster), a velocity $v_o \simeq 3.10^7$ cm/sec is obtained in comparison with a speed of sound c $\simeq 8.6.10^6$ cm/sec (see Eq. (80)). In this case Eq. (79) is satisfied so that we have an intermediate model. Eq. (79) (i.e. $P_a \sim \Pi$) was also consistent with our simulation. The analytical result gives $P_a \sim 2I/v_o \simeq 70$ Mb while in the simulation $P_a \sim 25$ Mb was achieved.

The hydrodynamic efficiency is defined by Eq. (53).
From the analysis of previous sections it turns out that

$$\eta_H(IM) > \eta_H(RM) > \eta_H(HM) \tag{81}$$

Therefore the intermediate model seems to be more effective for cluster driven ICF. It should be clear that for all of these models the irradiance I must be very high (typical irradiance is 10^{15} W/cm^2).

X. DIRECT DRIVE: A NUMERICAL SIMULATION

Elaborating upon the previous stopping analysis and the hydrodynamic framework of the driver-pellet interaction we can now proceed to a full fledged one dimensional Lagrangian simulation of an ICF compression of a thermonuclear pellet through intense cluster ion beams.

A. Target Design

Eq. (25) in Sec. III predicts that uniform illumination of an ICF capsule by heavy-ion cluster beams may generate high driving pressure at the target surface which may lie in the range 500–1000 Mbars, and thus can implode the capsule very efficiently. Moreover we note that, since Eq. (25) has been derived using a simplified picture of CIB target interaction, it may overestimate the driving pressure. In order to take into account this possibility, we have carried out target implosion simulations using 100 and 500 Mbars driving pressure. A self-consistent target will be presented later. Moreover, the cluster ions surrounding the target may act as a high-ρ, high-Z imploding tamper and the thickness of such a tamper will be determined by straggling of the cluster ions. It is to be noted that if we use a pure DT shell target, the cluster ions may mix into the fuel due to hydrodynamic instabilities that will be generated at the end of the compression phase due to the density difference at the fuel-cluster boundary. This can lead to reduction of thermonuclear energy output. To overcome this problem we have designed a target which has a low-ρ, low-Z lithium shell surrounding the DT fuel. This lithium shell can act as a stabilizer because the density across the fuel-lithium interface will be comparable at the end of the compression phase. The cluster ions may mix into the lithium region, but that will not effect the fuel. Such a design has already been used previously [24].

Fig. 17 shows the target used in the present study. The inner target radius R_i = 3.02 mm and it contains 4 mg solid DT fuel that is surrounded by a lithium shell. The inner radius and the fuel mass in this is the same as in the case of the HIBALL-I target [25] and the HIBALL-II target [24] respectively. A low-ρ DT gas is also filled inside the fuel shell. The initial gas density is about 3.26×10^{-3} kg/m^3. It has been shown [24] that, in a working reactor system there always will be some heat in the reactor chamber due to the previous implosion shot. When a new capsule will enter the reactor chamber, this heat will diffuse into the target and the target will be filled by DT vapor. The density of this vapor in case of HIBALL-I and the HIBALL-II reactor systems was found to be $\sim 3.26 \times 10^{-3}$ kg/m^3. In the present case we also take this value for the gas density.

The time variation of the driving pressure pulse is given in figure 18 by a box pulse. It is seen that the driving pressure is constant in time and the pulse duration is τ. In these calculations we use a driving pressure of 500 MB and τ = 5 nsec. Since the driving pressure is very high, the capsule is imploded very efficiently and a reasonably high energy output can be obtained even with an unshaped input pulse. According to Sec. III, the ablation pressure can be obtained by a CIB with E/A $\leq$ 20 keV/a.m.u.

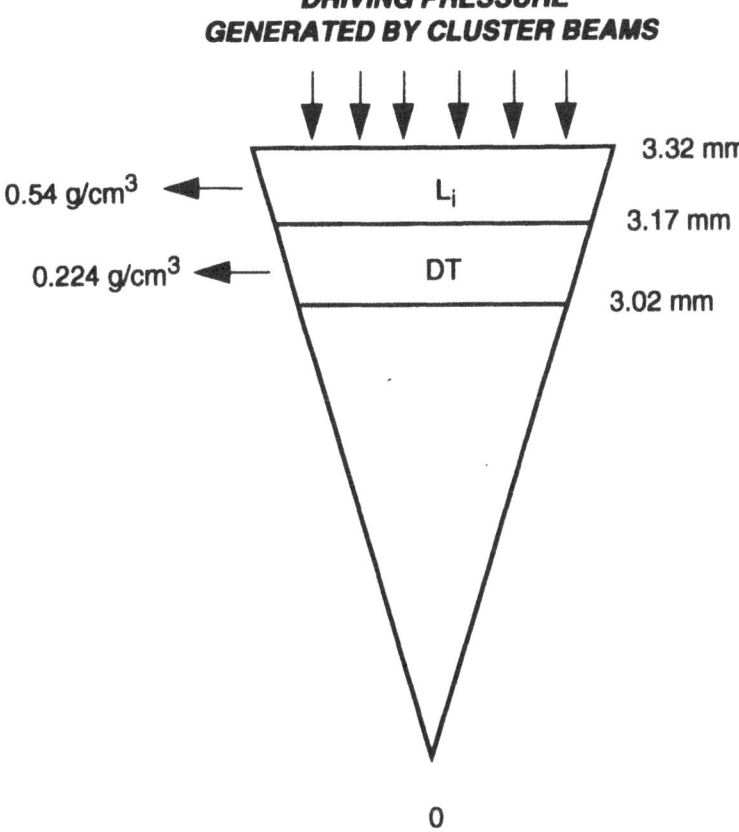

Figure 17. Target initial conditions.

We now present simulation results of the target implosion. These simulations have been carried out using a one-dimensional, three-temperature, Lagrangian inertial fusion computer code, MEDUSA-KAT [26].

B. Compression Phase

The driving pressure of 500 MB is applied at the outer target boundary at t = 0 and the implosion is set in. In Fig. 19(a) we plot the density, ρ, the pressure P, radiation temperature, T_r and matter temperature, T vs the target radius at t = 0.336 nsec. It is seen that a strong shock has been launched into the target and the lithium shell has been compressed to a density of 3×10^3 kg/m^3.

It is to be noted that in case of heavy-ion-beam driven, direct drive targets a significant mass of the capsule is heated by the driver ions. This is because the incident ions have a relatively long range and therefore they deposit their energy over a large thickness of the outer shell. The absorption region of the target is heated and it expands. In the present case, on the other hand, the range of the cluster is extremely small and therefore we can ignore the penetration depth of the cluster ions into the target. The target as a whole is compressed, as shown from Fig. 19(a). It is also seen that the matter temperature is ~ 3 ×

10^6 K in the compressed region and ahead of the compressed part it rapidly drops to a very low value. The radiation temperature, on the other hand, is lower in the compression region and is about 10^6 K. Ahead of the compressed part, $T_r > T$. This is due to the high conductivity of the photons.

Fig. 19(b) shows the same variables as Fig. 19(a), but at t = 2.29 nsec. It is seen that the shock has passed through the inner fuel boundary and the inner boundary is set in motion. At this time R_i, the inner fuel radius, has become ~ 1.58 mm from the initial value of 3.02 mm. It is also seen that the outer part of the solid DT shell has been compressed by the shock whereas the inner part has a density below the solid DT density due to the expansion of the inner fuel boundary. Moreover $T_r > T$ in the fuel region because the photons have a high conductivity. In the gas region, on the other hand, the ion temperature, T_i >> the electron temperature, T_e because of the strong shock heating. Also $T_e > Tr$. Since the gas density is low, the electron-ion and electron-radiation equilibration time scale is much longer compared to the hydrodynamic time scale.

C. Achievement of Ignition and Burn

The driving pressure leads the capsule to void closure. The fuel is compressed and strongly heated and the fuel is ignited. In Fig. 20(a) we plot matter temperature, T, radiation temperature, T_r, density, ρ and pressure P along target radius at t = 4.19 nsec. It is seen that the DT gas has been compressed to a density ~3 x 10^2 kg/m^3 whereas the solid DT shell has a much higher density. The radius of the gas region is about 170 μm. It is also seen that in the gas region and the inner half of the solid DT shell, T is ~7 x 10^7 K and

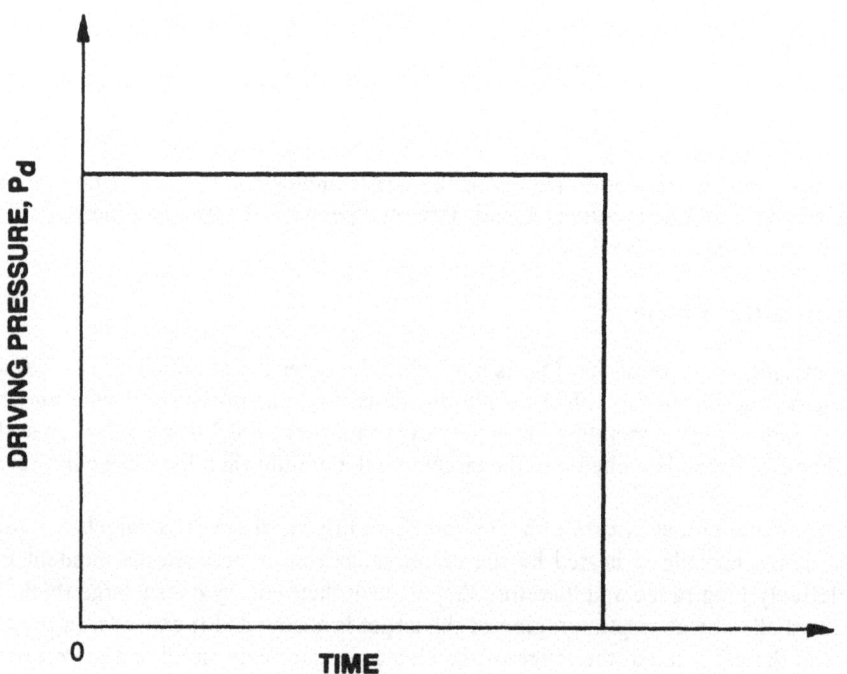

Figure 18. Time variation of input pressure pulse.

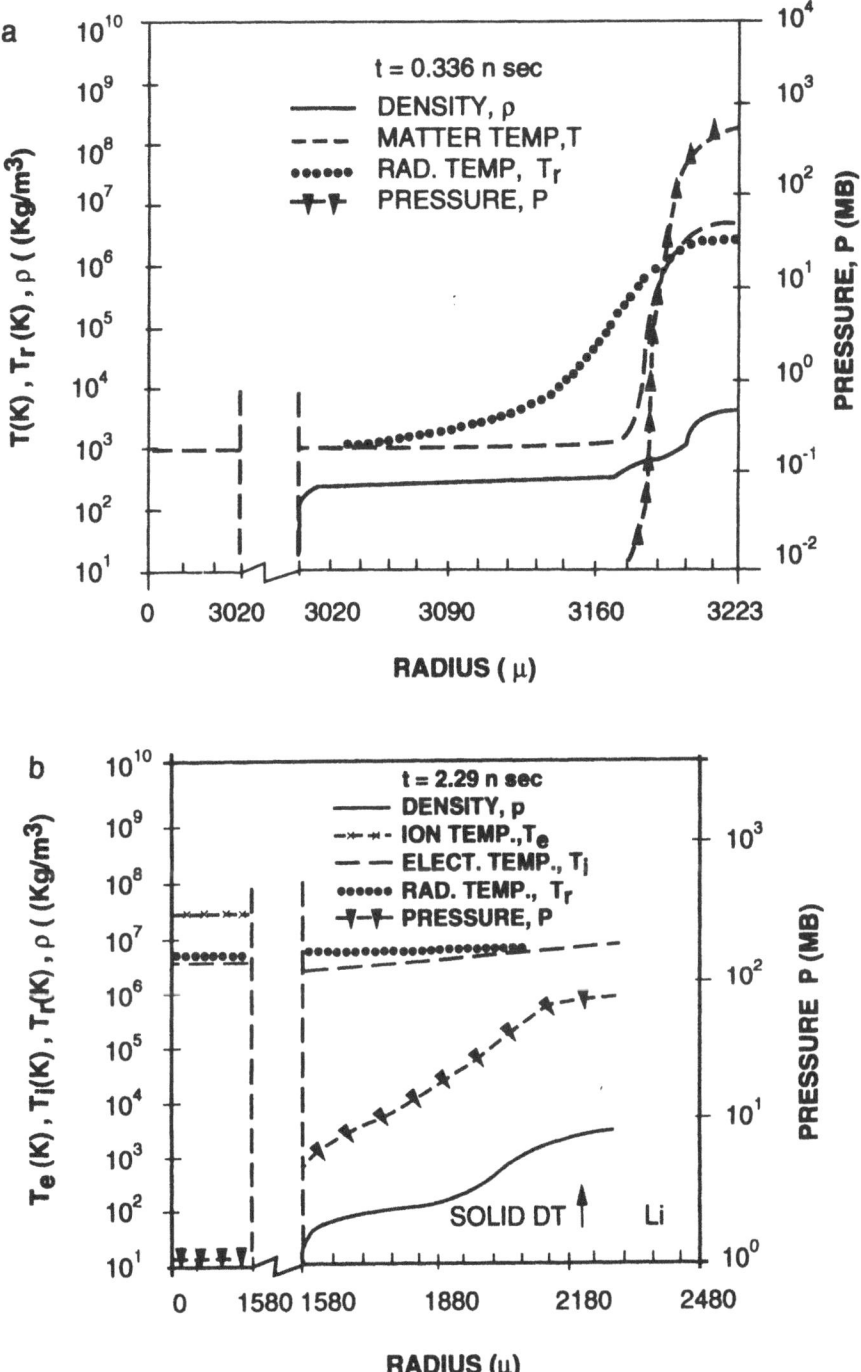

Figure 19. Matter temperature, T, radiation temperature, T_r, density, ρ, and pressure P versus target radius at, (a) t = 0.336 nsec., (b) t = 2.29 nsec.

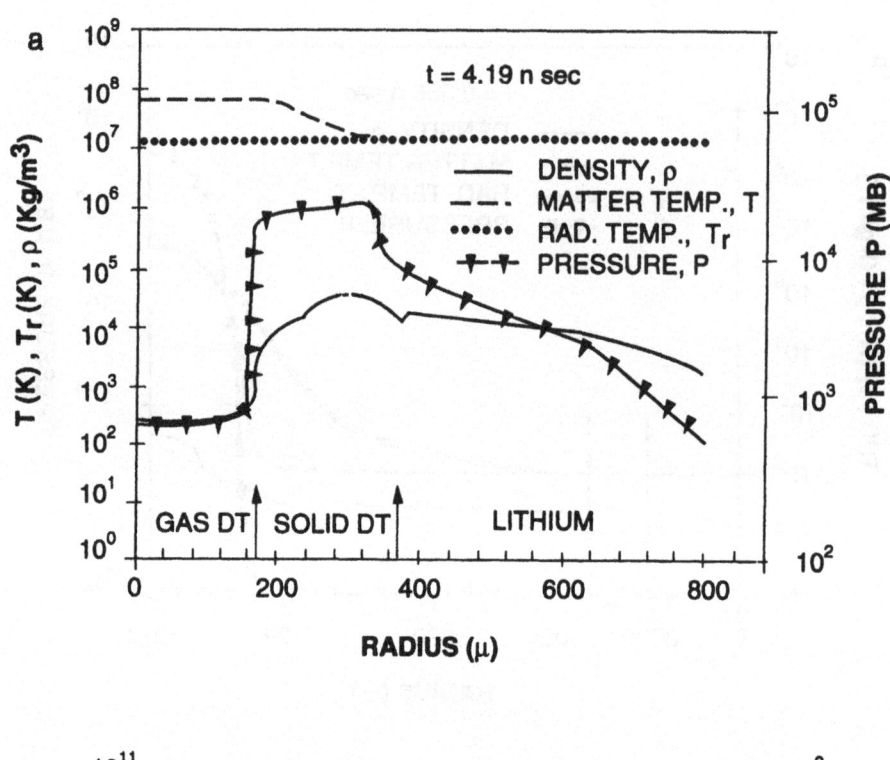

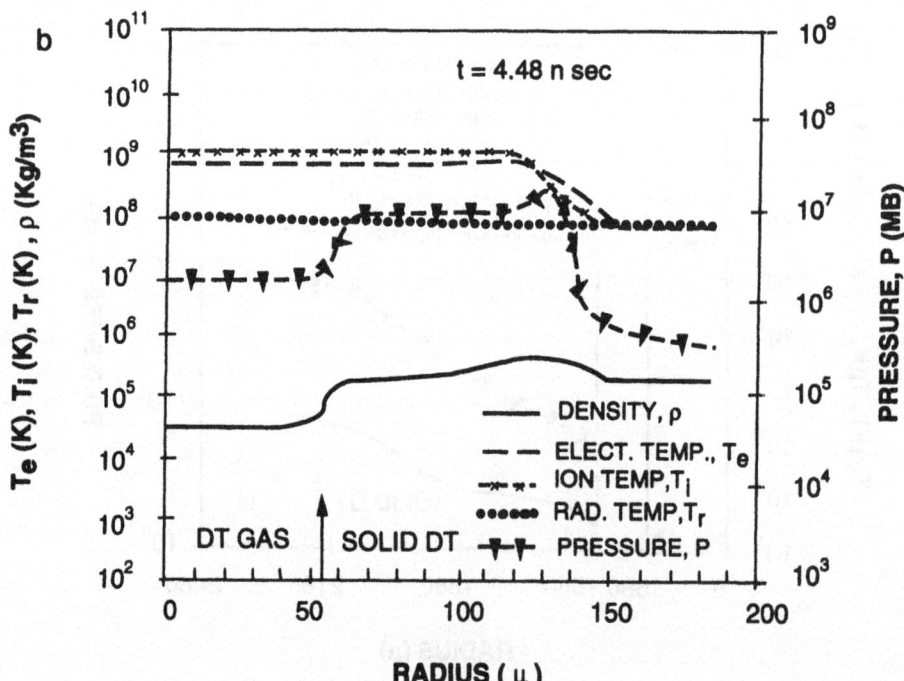

Figure 20. Same as in Fig. 19 at (a) t = 4.19 nsec, (b) t = 4.48 nsec, and (c) t = 4.49 nsec.

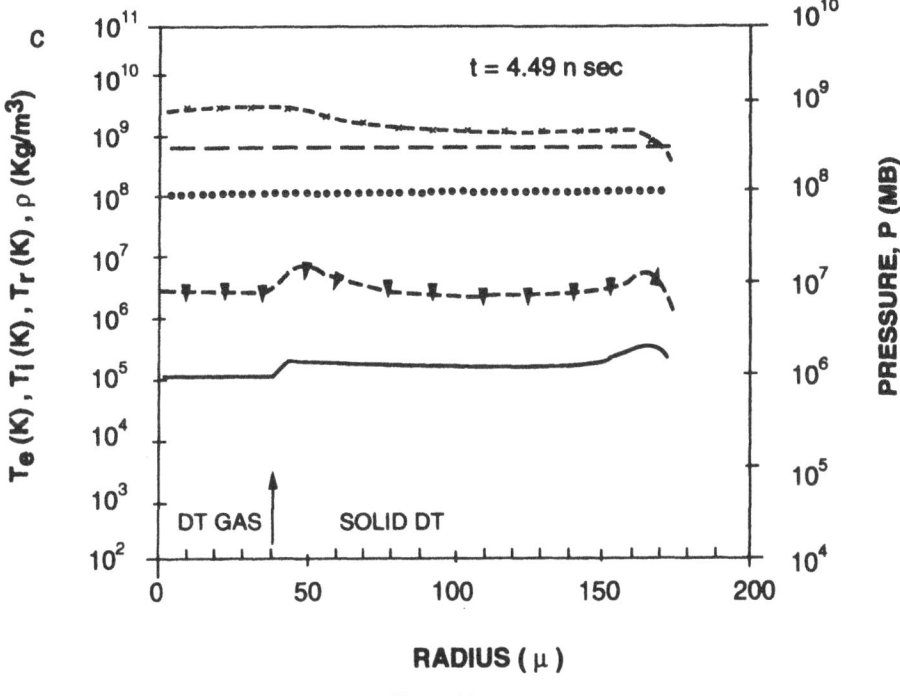

Figure 20c.

in the outer part of the capsule T is $\sim 10^7$ K. The radiation temperature, T_r, is uniform throughout the entire target. The nuclear reactions start in the fuel as soon as the temperature becomes ~ 1 keV, but below 5 keV, the radiation loss rate exceeds the α-particle energy deposition rate.

In Fig. 20(b) we plot the same variables as in Fig. 20(a), but at t = 4.48 nsec. It is observed that the entire fuel is burning efficiently. Inside the fuel region $T_i > T_e > T_r$. Also $T_i \sim 10^9$ K, $T_e \sim 6 \times 10^8$ K and $T_r \sim 10^8$ K. Ahead of the burning zone $T_e > T_i$ because thermal electrons conduct energy in this region. Fig. 20(c) is plotted at t = 4.49 nsec which shows that the entire fuel has been compressed to a density above 10^5 kg/m^3 and the target outer radius has been reduced to a value of 170 μm from an initial value of 3.32 mm.

D. Discussion of Results

In Fig. 21 we plot the outer target boundary, the outer fuel boundary and the inner fuel boundary as a function of time. It is seen that, because of the high driving pressure, the target is imploded in less than 5 nsec, whereas in case of the heavy-ion-driven numerically simulated HIBALL-I and HIBALL-II targets, which have the same dimensions as the present target, the implosion time was much longer, namely ~ 35 nsec. This is because in the latter two cases the driving pressure generated by the incident heavy ions was much lower.

The target implosion in the present calculations produces an energy output of the order of 630 MJ, which may be sufficient to run a reactor.

We also carried out an implosion simulation of the target shown in Fig. 17 using a lower driving pressure, namely, 100 MB. The pulse in this case is not a box pulse but a

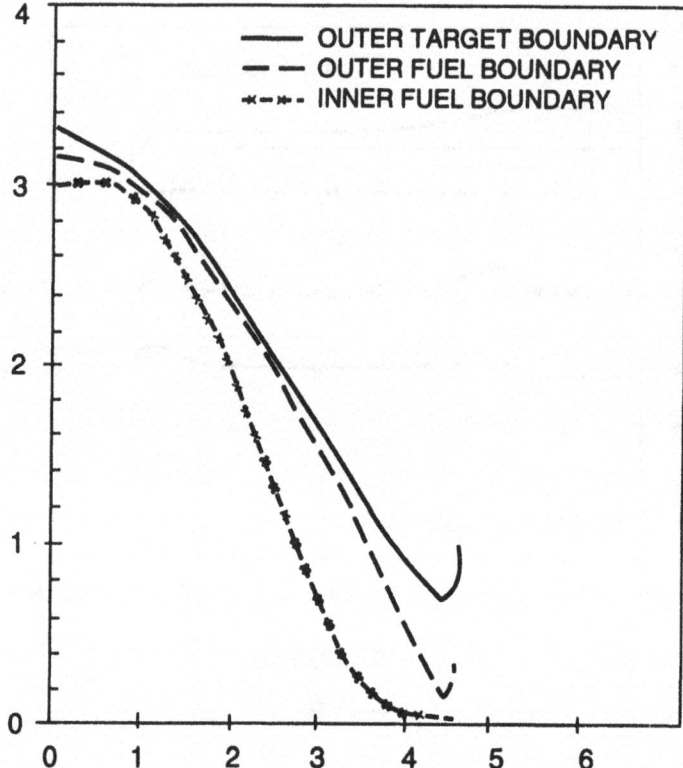

Figure 21. Coordinate interfaces as a function of time (in nsec).

two-step shaped pulse shown in figure 22. It starts with a prepulse having a driving pressure $P_1 = 10$ Mbars that increases linearly to a higher value, $P_2 = 100$ Mbars. The duration of the prepulse $\tau_1 = 7$ nsec, while that of the main pulse $\tau_2 = 3$ nsec. The implosion in this case yields an output energy of about 474 MJ, which may be sufficient to operate a CIB-operated ICF reactor system. One may also improve the energy output by using a more sophisticated pulse shape. The above calculations show that heavy ion cluster beams can efficiently implode reactor-sized ICF targets.

The target evolution emerging from Figs. 19 and 20 clearly demonstrates that inner-target and blowoff temperatures reach rapidly, in less than 2 nsec, the same plateau value.

It could be appreciated that target swift heating might require a hot-matter stopping calculation. This is likely to result in a shorter range than the cold ones considered here. However, being mostly motivated by the new qualitative features of a novel driving scheme, we postpone those refinements for further inquiry.

On the other hand, sticking closely to a cold-target approach might lead us to consider regimes beyond validity of the Bethe-like approach used in Sec. III. Nonetheless, despite being qualitatively different, the quantitative output is not expected to change enormously because the given driven stopping is close to its maximum value.

It is also instructive to address the final focusing issues that are so crucial for direct-drive heavy-ion-driven ICF. As is well known, in that case the final 100 cm of ion trajectory are likely to be perturbed by target radiation during implosion. In the present situation, thanks to CIB slow velocity, and much shorter compression time, only the final 2 mm are likely to be disturbed.

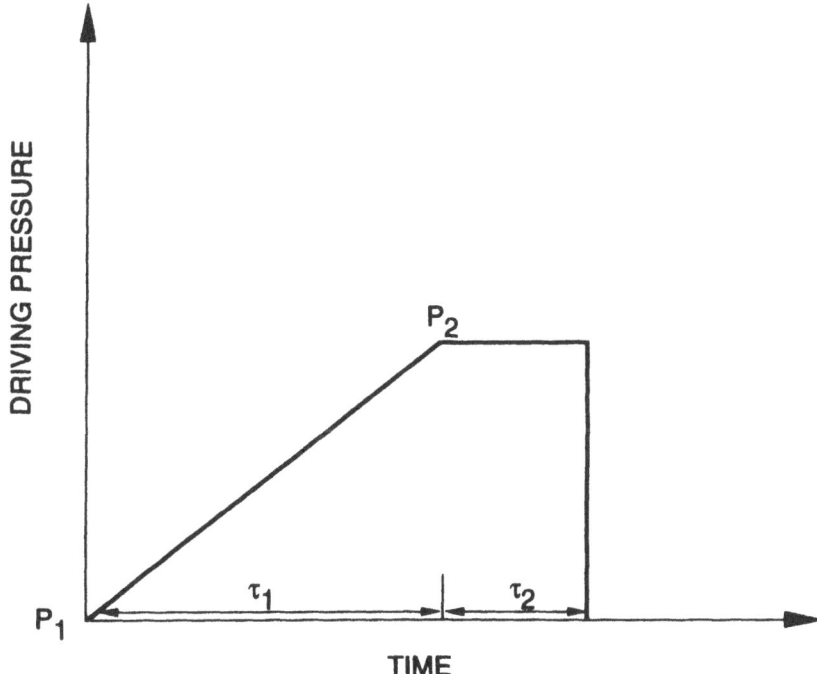

Figure 22. Time variation of driving pressure pulse with peak pressure 100 Mbar.

Finally, Figs. 19 and 20 show that good coupling efficiency is achieved already after the first nsec of compression time, thanks to a very rapid temperature rise.

XI. PROSPECTS FOR INDIRECT DRIVE

The potentialities of cluster ion beams for driving directly a thermonuclear pellet look therefore highly promising. Nevertheless, the most attractive option might still be provided by the so-called indirect drive approach. Such an appreciation is largely documented by the much shorter ranges, in otherwise similar stopping conditions, experienced by correlated ion fragments as compared to isolated ones, in cold and hot matter as well. This fortunate occurrence thus allows to pinpoint to a much thinner convector material. One may thus expect a more efficient confinement of the hard X rays bath within a more sphericalike hohlraum around the inner pellet containing a thermonuclear fuel.

Fig. 23 displays range-energy relationships for the correlated stopping calculated according to Sec. III for several densities of a gold convector. The given energy range $E/A \sim$ a few hundreds keV/a.m.u. is realistic in connection with presently existing linear accelerating facilities. Extensive simulations [27] are underway to implement those results in a two-dimensional simulation making use of the hohlraum targets features displayed on Figs. 24 and 25. The nearly spherical geometry featured by Fig. 24 implies that the convector material be removed in casing at suitable chosen locations. The target pictured on Fig. 25 allows for a maximum flexibility around the target chamber as far as the handling of intense cluster ion beams is concerned. Hohlraum temperatures in the keV range are likely to be produced by the above considered range-energy relationships.

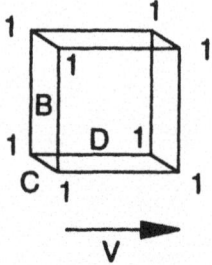

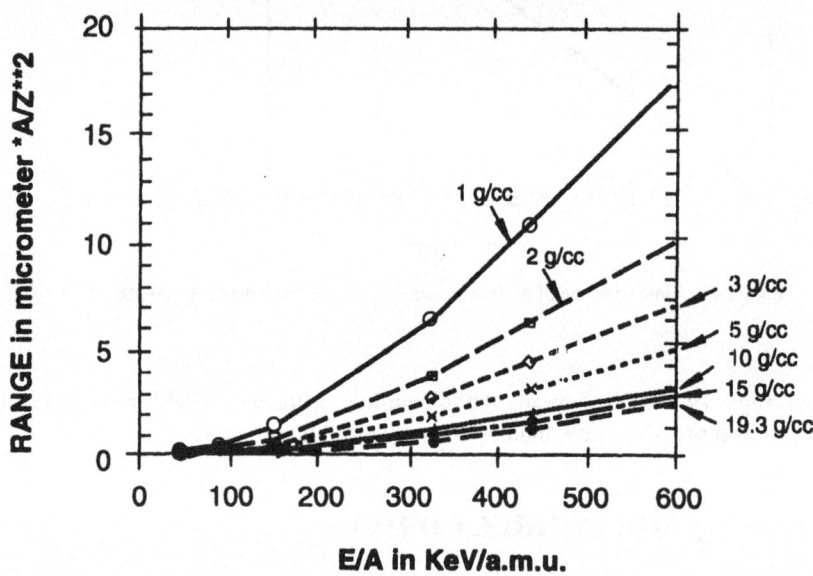

Figure 23. Range-energy relationships for eight correlated charges in a Au target at various densities.

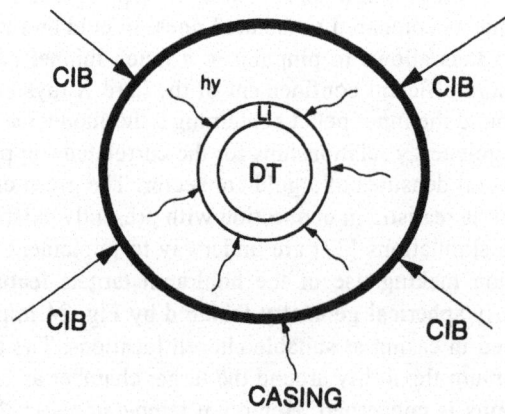

Figure 24. Hohlraum target with convertors removed in casing.

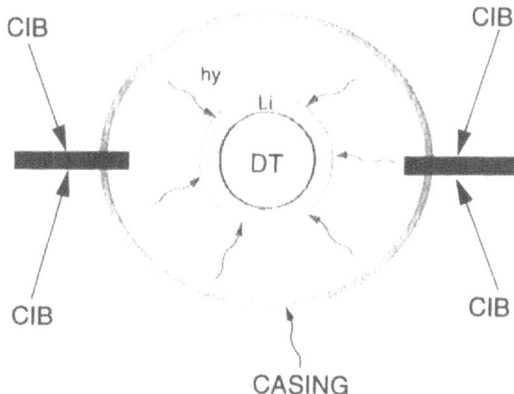

Figure 25. Hohlraum target with external CIB irradiation.

REFERENCES

1. F. Winterberg, *Z. Phys. A* **296**, 3(1980) and also T. Yabe and T. Mochizuki, *Jpn. J. Appl. Phys.* **22** L262 (1983).
2. A.W. Mashke, *Proceedings of HIF 84 International Symposium on Heavy Ion Fusion*, Tokyo, 1984, p. 168.
3. De Heer W.A., *Rev. Mod. Phys.* **65**, 611 (1993).
4. C. Deutsch, *Laser Part. Beam* **8**, 541 (1990).
5. M.A. Duncan and D.H. Rouvray *Sci. Am.* **261**, 60 (1989).
6. D. Beynon, *Philos. Trans. R. Soc. London Ser. A* **300**, 613 (1981).
7. C. Deutsch and N.A. Tahir, *Phys. Fluids* **B4**, 3735 (1992).
8. C. Deutsch, *Phys. Rev. E* **51**, 271 (1995).
9. C. Deutsch and P. Fromy, *Phys. Rev. E* **51**, 632 (1995).
10. P. Scheier, R. Robl, B. Schiestl and T.D. Märk, *Chem. Phys. Lett.* **20**, 141 (1994), and P. Scheier and T.D. Märk, *Phys. Rev. Lett.* **73**, 54 (1994).
11. P. Attal, S. Della-Negra, D. Gardès, J.D. Larson, Y. Le Beyec, R. Vienet-Lequé and B. Waast, *Nucl. Instr. Methods* **A238**, 293 (1993).
12. Z. Vager, R. Naaman and E.P. Kanter, *Science* **244**, 426 (1989).
13. A.Z. Mekjian, *Phys. Rev. C* **41**, 2103 (1990).
14. C. Deutsch, *Laser Part. Beam* **10**, 355 (1992).
15. A. Bret and C. Deutsch, *Fus. Eng. Des.* **32–33**, 517 (1996).
16. G. Basbas and R.H. Ritchie, *Phys. Rev. A* **25**, 1943 (1982). For a recent update of atomic ion and cluster ion stopping in cold matter see the Proceedings of the International Conference (St Malo) on Polyatomic ion impact on solids and related phenomena, *Nucl. Instr. Meth.* **B88**, 1 (1994).
17. A. Bret and C. Deutsch, *Phys. Rev. E* **47**, 127 (1993) and *E* **48**, 2989 (1993).
18. J. D'Avanzo, M. Lontano and P.F. Bortignon, *Phys. Rev. A* **45**, 6126 (1992) and *E* **47**, 3574 (1993).
19. B.D. Fried and S.D. Conte, *The Plasma Dispersion Function* (Academic Press, New York, 1961).
20. M. Farizon, N.V. De Castro-Garcia, B. Farizon-Maruy and M.J. Gaillard, *Phys. Rev.* **145**, 179 (1992).
21. S. Eliezer, J.M. Martinez-Val and C. Deutsch, *Laser Part. Beam* **13**, 43 (1995).
22. J.M. Martinez-Val and M. Piera, *Fusion Technology* **23**, 218 (1993).
23. C. Yamaka and S. Nakai, *Nature* **319**, 757 (1986).
24. K.A. Long and N.A. Tahir, *Phys. Rev.* **A35**, 2631 (1987).
25. N.A. Tahir and K.A. Long, *Nucl. Fusion* **23**, 887 (1983).
26. N.A. Tahir, K.A. Long and E.W. Laing, *J. Appl. Phys.* **60**, 898 (1986).
27. N.A. Tahir, D.H.H. Hoffmann, J.A Maruhn, and C. Deutsch, *Nucl. Instr. Methods* **B88**, 127 (1994).

34

INERTIAL FUSION ENERGY

An Approach to Low Maintenance and Cost of Electricity, and the Role of the National Ignition Facility Testing the Target Physics*

B. Grant Logan

Lawrence Livermore National Laboratory
Livermore, California 94551

Inertial Fusion Energy (IFE) is the primary fusion approach alternative to the Tokamak in the U.S., with a significant IFE driver development program on heavy-ion accelerators supported by DOE Energy Research, and with a large inertial confinement fusion target physics program supported by DOE Defense Programs, aimed at the goal of fusion ignition and energy gain in the laboratory with the National Ignition Facility (NIF) around 2005. The four basic development requirements for practical and economical IFE are:

1. Development of high gain targets (fusion yield/driver energy into the target).
2. Development of a driver with adequate efficiency, high-pulse-rate capability, high reliability, and acceptable cost.
3. Development of IFE fusion target chambers that can achieve long life with low maintenance at ~ 5 Hz pulse rates, and using low activation materials.
4. Development of low cost target fabrication techniques.

The heavy-ion accelerator driver development program, described in talks by Roger Bangerter, Tom Fessenden, and Bill Hermannsfeldt in this conference, is the prime IFE driver development in the U.S. and Europe to satisfy requirement 2. Using indirect-drive laser-driven targets, the NIF[1] can resolve most of the critical target physics issues required to design indirect-drive heavy-ion fusion targets for requirement 1. An IFE development facility, called the Engineering Test Facility (ETF), is required to provide integral testing of all four requirements for IFE[2].

---

* Work performed under the auspices of the U.S. Department of Energy by the Lawrence Livermore National Laboratory under Contract No. W-7405-ENG-48.

Given the same proper symmetry and flux of soft X-rays around the outer ablation layer of fuel capsules in indirect-drive targets, the dependence of capsule gain (capsule yield/X-ray energy absorbed) on capsule smoothness and aspect ratio (affecting hydrodynamic growth factors), implosion velocity, and convergence ratio, can be determined independent of whether the source of soft X-rays was from laser beam or heavy-ion deposition near the ends of the hohlraum. With a 1.8 MJ glass laser, the NIF is expected to demonstrate indirect-drive capsule gains of 30 to 100, sufficient to enable the design of heavy-ion targets for an overall target gain of 50 to 100 at 5 MJ of beam energy for an IFE power plant. Using other laser indirect drive targets including X-ray shine shields placed within the hohlraum cavity, the NIF can also be used to test the transport and symmetry of soft-X-rays in a heavy ion type hohlraum geometry, with greater fidelity.

A recently-completed study[3] called HYLIFE-II explores a liquid-wall approach to a long-lived fusion chamber for IFE power plants. An array of fixed and oscillating jets of Flibe molten salt (Li_2BeF_4) protect the vessel walls from direct neutrons, soft X-rays, and plasma debris from the targets. The oscillating Flibe jets form periodic cavities in the center for indirect drive targets to be illuminated by heavy-ion beams at two ends. The Flibe jets are injected into the chamber with a sufficient downward flow velocity to clear the chamber of target debris and liquid splash between shots at pulse rates of 6 Hz. The attenuation of target-induced shocks and 14 MeV neutrons through approximately half a meter thickness of Flibe jets reduces the peak stresses and neutron damage and activation to the outer 304 stainless steel vessel walls sufficiently that the HYLIFE-II fusion chamber can survive the life of the plant (> 30 full-power years of operation) without replacement. The elimination of the need to periodically replace blankets as in other fusion concepts allows HYLIFE-II plants to achieve higher availability (85% instead of the usual 75% assumed), and a lower maintenance cost, reducing the cost-of-electricity (CoE) another 10%. The greatly reduced activation allows the HYLIFE-II fusion chamber to be non-nuclear-grade construction, reducing the direct costs of the fusion chamber and other Flibe loop components another 50%.

The HYLIFE-II driver is based on a recirculating induction accelerator[4], which is estimated to reduce the heavy-ion driver direct costs by a factor of about two, from about $1 B direct ($2 B direct + indirect cost), to about $0.5 B direct ($1 B direct + indirect cost). The combined cost reductions due to the lower driver cost, lower maintenance and availability, results in an estimated CoE of 4.7 cents/kWehr at 1 GWe net power for HYLIFE-II, down to 3.5 cents/kWehr at 2 GWe plant size. The driver and fusion chamber advances incorporated in HYLIFE-II may thus allow fusion to finally reach a CoE competitive with future advanced fission or coal plants. At even larger plant sizes of 8 GWe appropriate for producing hydrogen synfuel for low-emission vehicles by water electrolysis, a HYLIFE-II plant using a target factory and a single driver with a beam switchyard to drive four fusion chambers is found capable of providing equal fuel cost per mile as for gasoline-powered vehicles[5].

REFERENCES

1. E. M. Campbell, *in the Proceedings of the 14th International Conference on Plasma Physics and Controlled Fusion Research*, Wurzberg, Germany, IAEA-CN-56/B-3-1-1, Vol. III, p. 113 (1992).
2. The U.S. National Energy Policy Act of 1992.
3. R. W. Moir, *et al.*, Fusion Technology **25**, pp. 5-25, (1994).
4. J. J. Barnard, *et al.*, Lawrence Livermore National Laboratory Report #UCRL-LR-108095
5. B. G. Logan, R. W. Moir, and M. A. Hoffman, Lawrence Livermore National Laboratory report #UCRL-JC-115787, (June 1994) submitted to *Fusion Technology*.

MAGNETIZED TARGET FUSION

An Ultrahigh Energy Approach in an Unexplored Parameter Space

Ronald C. Kirkpatrick, Irvin R. Lindemuth, Robert E. Reinovsky, and Peter T. Sheehey

Los Alamos National Laboratory
Los Alamos, New Mexico 87545

ABSTRACT

Magnetized target fusion is a concept that may lead to practical fusion applications in a variety of settings. However, the crucial first step is to demonstrate that fusion ignition can be achieved by this method. Among the possibilities for doing this is an ultrahigh energy approach to magnetized target fusion, one powered by explosive pulsed power generators that have become available for application to thermonuclear fusion research. In addition to providing a target implosion driver, explosive pulsed power sources can also be used to provide the preheated and magnetized plasma required within the target prior to implosion. In a collaborative effort between Los Alamos and the All-Russian Scientific Research Institute for Experimental Physics (VNIIEF) a very powerful helical generator with explosive power switching has been used to produce an energetic magnetized plasma. Several diagnostics have been fielded to ascertain the properties of this plasma. We are intensively studying the results of the experiments and calculationally analyzing the performance of this experiment.

INTRODUCTION

The physical characteristics of inertial confinement fusion and magnetic confinement fusion differ by about ten orders of magnitude. The approaches to creating the fusion plasma by these two methods are equally different. Because of the extreme differences and the vast parameter space that separates these two approaches, it is logical to consider the possibility of an intermediate approach to fusion. We believe that an idea suggested many years ago, and sporadically examined over the past 20 years may provide an approach to fusion energy that avoids the difficulties of the previous extremes. This intermediate approach to thermonuclear fusion is called magnetized target fusion (MTF). Los Alamos coined this phrase only about two years ago, but in the past the same concept has been called variously magnetized fuel, fast liner fusion, and magnetothermally insulated fusion.

We are now studying one attempt to accomplish the first step in the MTF process, the creation of a hot, magnetized plasma suitable for subsequent implosion. The MAGO experiment fielded jointly by Los Alamos and the All-Russian Scientific Research Institute for Experimental Physics (VNIIEF) promises to provide the necessary components for that first step. In fact, VNIIEF has fielded many experiments, and Los Alamos has joined the effort only lately, but has also brought to the effort advanced diagnostic and computational capabilities that should greatly improve our understanding of the plasma creation process and the properties of the plasma created.

Below we will present the salient facts for MTF and describe the US/Russian joint effort to take the first step toward realizing magnetized target fusion.

WHAT IS MAGNETIZED TARGET FUSION?

MTF is a relatively untried approach to fusion ignition, requiring two elements: a) a means of preheating and magnetizing the thermonuclear fuel and b) a compression system. The key aspects are 1) an embedded magnetic field is used to reduce the thermal conductivity of a hot plasma, not to confine the plasma, and 2) the plasma and field are subsequently compressed together until fusion ignition conditions are achieved.

For ICF, compression of the fusion fuel is called implosion and involves very strong shocks to accelerate a shell (or "pusher" that contains the fusion fuel) to high implosion velocity. This is because as fusion ignition conditions are approached the energy loss rates become high, and a high hydrodynamic work rate is necessary to overcome them until fusion energy production is sufficient to overcome them. The major energy loss process for typical ICF targets is thermal conduction, but for targets with high fuel density the bremsstrahlung radiative loss from the fusion fuel can be more important. Some details of ignition process in ICF targets were discussed in another paper in this symposium [1].

The strategy of MTF is to suppress the losses from the fusion fuel, first by reducing the density of the fuel to reduce radiative losses, making the conduction losses dominant, and then by using a magnetic field to suppress the conduction losses. We think that this reduction of thermal conductivity should be classical [2]. This strategy drastically reduces the overall losses, which means that the work rate necessary to overcome the reduced losses can be much smaller, greatly relaxing the need for high implosion velocity.

Because at high temperature the plasma electrical conductivity is high, the embedded magnetic field is effectively frozen into the plasma, so that the two move together, and as the plasma is compressed, so is the magnetic field. This means that the magnetic field will be amplified by a factor of about 100 in the course of about a thousand fold volumetric compression, which corresponds to a convergence of a factor of ten for a spherical target.

The MTF parameter space is shown in Figure 1. It is presented as an initial condition parameter space, that is, contours of unity gain plotted in the initial density, initial implosion velocity space. Three sets of contours are plotted in this initial condition space: B_o = 16, 40 and 100 KG. The plots are made for specified DT mass, initial temperature, and implosion energy [3].

It is clear that the MTF region in the lower left lies far from the ICF region in the upper right hand corner of the plot. This survey result suggests that MTF allows extremely low implosion velocities. Also, as the initial magnetic field is increased, the two fusion regions connect, so that there is a continuum between the two. The survey code used to produce Figure 1 did not allow for fusion energy deposition, so that fusion ignition was not possible. However, ignition is a possibility for MTF. What Figure 1 shows is that even without ignition MTF still provides a net gain in an accessible part of parameter space.

Because MTF does not need high implosion velocities, there are no strong shocks involved that serve to raise the temperature of the fusion fuel prior to continued compression by the pusher or liner. This means that it is necessary to heat the fusion fuel by an auxiliary means before compression. In Figure 1 the initial temperature of 50 ev was used. Ohmic heating is a convenient choice because it is also desired to have a magnetic field in the plasma. However, ohmic heating alone may not suffice, because as the plasma temperature rises, the resistivity falls. There is a limit to the temperature attainable by simple ohmic heating. It is also possible to employ an MHD means of shock heating the fusion fuel, which is done in the case of the MAGO experiment. This will be discussed later.

ACCESS TO THE MTF PARAMETER SPACE

Because the MTF target plasma is low density, for the same mass target, an MTF target must be larger than its ICF counterpart. This coupled with the fact that the MTF fusion region is accessible with lower implosion velocities means that the implosion times for MTF are much longer than for ICF. This means that the power and intensity on target for MTF are much lower than for ICF.

This is shown in Figures 2 and 3 as lines of constant values of power and intensity in the same initial condition space as Figure 1. It should be noted that the addition of a magnetic field also extends the conventional ICF region to lower density in this diagram than

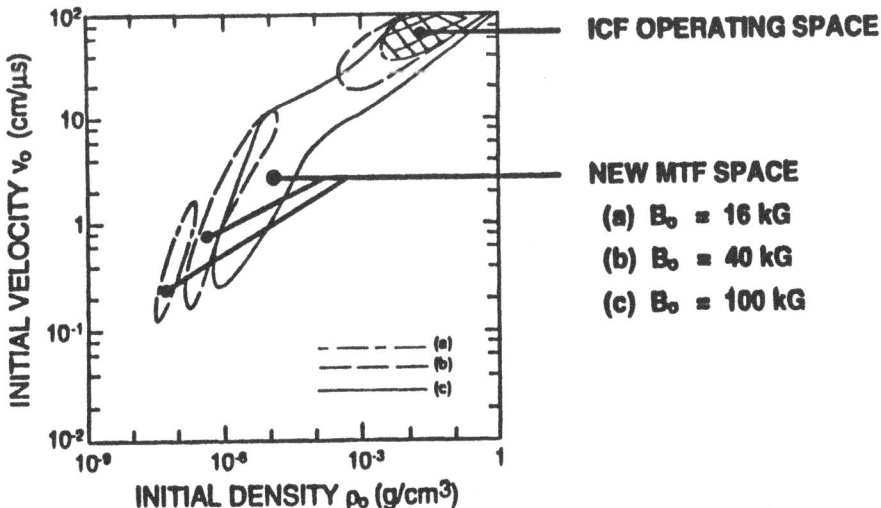

Figure 1. The initial condition space for MTF based on results derived from a "zero-dimensional survey code that follows the dynamics of an imploded magnetized fusion target [3]. The target plasma starts at 50 ev.

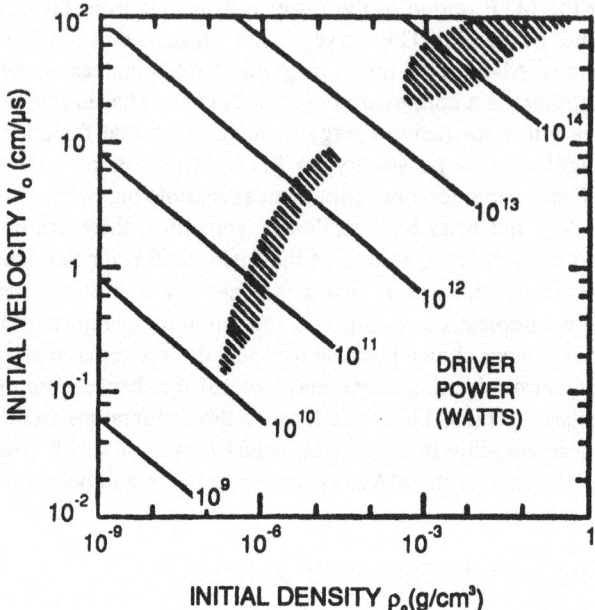

Figure 2. Lines of constant power in the MTF initial condition space.

would be the case for B = 0. In fact fully two or more orders of magnitude separate the extremes of the two regions in power, and four in intensity. These are significant factors that should make a substantial difference in the requirements for realizing fusion energy. In principle, the MTF fusion region is much more easily accessed than ICF. At the same time it avoids many of the classical magnetic confinement instabilities.

The MTF fusion region changes as the DT mass and implosion energy are increased. In initial condition space plots shown in figures 1 through 3 there is no DT alpha energy deposition allowed in the calculations (i.e., no "self heating" in the plasma). This means that for these survey calculations, ignition was not allowed to occur. Therefore, the gains calculated were a lower limits, and the sizes of the MTF fusion regions should be larger. The details of the survey calculations from which these figures were derived are given in the 1983 paper [3].

It should be clear that MTF is not a simple matter of just adding a magnetic field to a conventional ICF target. The magnetic field necessitates target plasma preparation, in a lower initial density fusion fuel, which means a larger target. This coupled with the possibility that the target can be imploded much slower means that the target should operate at much lower power and intensity on target. The lower implosion velocity means more massive pushers and longer dwell times near maximum compression, which matches the longer fusion burn time due to the lower density. No pulse shaping should be required, but there is a requirement for plasma preparation before implosion.

DT ALPHA ENERGY DEPOSITION

Because the magnetic field is compressed along with the fusion fuel, the gyro radius of the charged fusion reaction products decreases. For DT the one charged reaction prod-

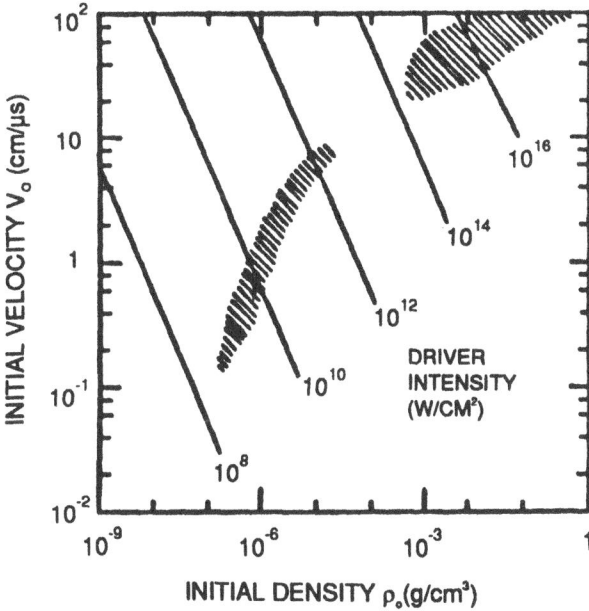

Figure 3. Lines of constant intensity in the MTF initial condition space.

uct is the DT alpha with an initial energy of 3.5 Mev, an atomic mass of 4 and a charge of 2. If the gyro radius is much smaller than the distance from where the DT alpha is born to the pusher, or if the mean free path is sufficiently short, then most of the fusion energy carried by that reaction product will be deposited in the fusion fuel. However, most of the fusion energy escapes with the DT 14 Mev neutron. If sufficient fusion energy is redeposited in the fusion fuel, it can ignite, meaning that the temperature will keep on increasing even after the implosion has reversed and expansion cooling now aids the energy loss processes in the fusion fuel.

An example of the complex trajectory of a DT alpha in a magnetized DT fusion fuel is shown in Figure 4. Figure 5 shows the region in the ρR temperature plane where DT ignition is possible for MTF. Notice that it is greatly extended to lower values of areal density ρR from the ICF region on the far right. This figure shows the fusion regions for several values of target plasma masses. For a magnetic field of 5 MG, the MTF ignition region disappears for DT mass below about 1 μg. This sets a lower limit on the energy necessary for MTF ignition.

PREVIOUS EXPERIENCE WITH MTF

The basic ideas for MTF seem to have had their origins both in the US and in Russia. Because both Edward Teller and Andrei Sakharov were engaged in nuclear weapons work, the genesis of the ideas are not completely documented. It is clear that the basic ideas came from the groups they worked with during those secret times. The first published work on this topic did not appear until the late 1970's, and probably independently of the secret sources of information. Sandia National Laboratory published an early report on

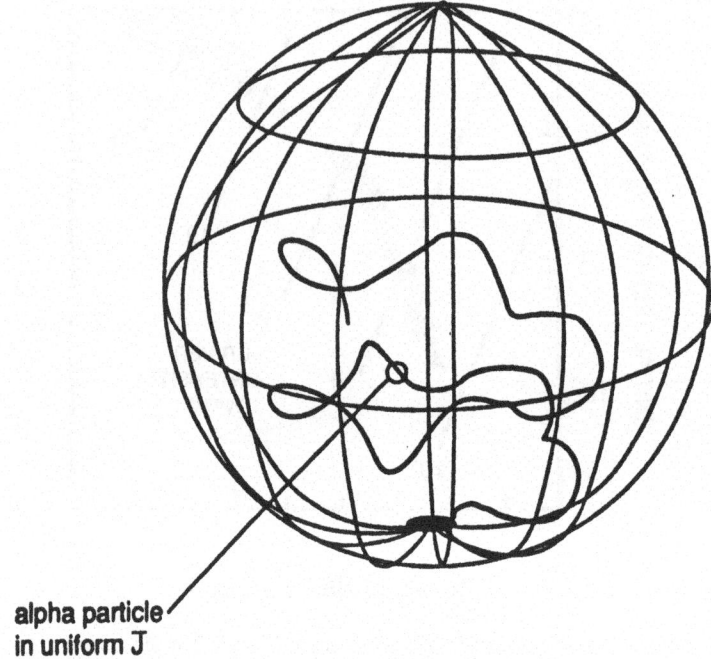

alpha particle in uniform J

Figure 4. Trajectory of a DT alpha particle in a magnetized plasma with uniform current density. During the implosion, the field in the target can be amplified, and possibly sufficiently to reduce the Larmor radius of the DT alpha and thereby significantly enhance the fusion energy "self-heating".

the Phi-target work in 1977 [4,5]. Two years later Mokhov and his colleagues published a very sketchy paper on a device intended to realize fusion with a magnetic implosion [6]. Since the late 1970's there have been several advances in our understanding, but the Sandia Phi-targets have remained the single best example of an integrated MTF experiment. Figure 6 shows various geometries for MTF, including the Phi-target. These drawings are not to scale. The Phi-target had a diameter of 3 mm. The MAGO experiment has a radius of 10 cm.

We believe that developments in the last decade warrant a renewed examination of MTF. First, modeling of Sandia Phi-target provided a better understanding of the essential physics. Second, survey code results put the MTF into a perspective against the back-drop of other fusion concepts [3]. In addition, there are now many new plasma diagnostics, pulsed power facilities, and improved computational tools. We think that one item that may be most important is the development of new methods for producing a hot, magnetized target plasma. The high density Z-pinch is one, and the Russian MAGO experiment is another. We are also considering other methods for creating a target plasma.

The Russian acronym MAGO is properly translated as "magnetic compression". It has become the acronym used for a variety of related activities which include explosive pulsed power and the creation of a hot, magnetized target plasma. VNIIEF has demonstrated 200 MJ high explosive pulsed power (HEPP) drivers, and has done a magnetically driven cylindrical implosion that had 25 MJ of kinetic energy. VNIIEF's unique target plasma formation scheme (which we refer to as the MAGO experiment), has produced up

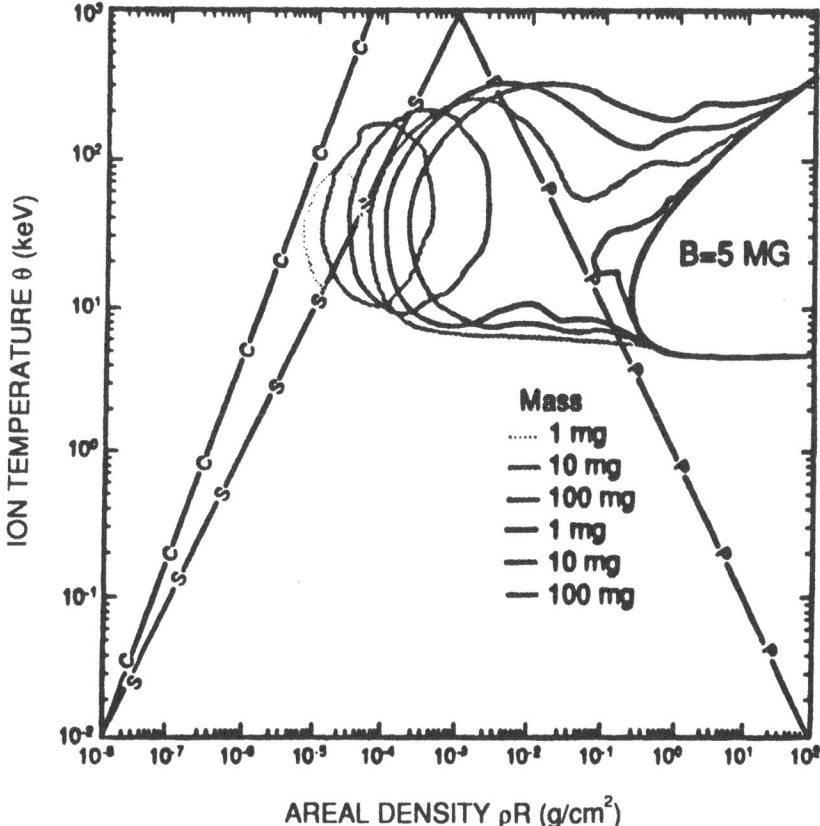

Figure 5. Mass dependence of the new MTF region in a Lindl-Widner diagram. The specific heat of DT is 100 J/µg.Kev, so about 5 MJ is required to to drive 1 mg of DT to fusion temperature.

to 5×10^{13} neutrons (without implosion). Now, US/Russian collaboration in ultrahigh magnetic fields and pulsed power is familiarizing US scientists with Russia's remarkable accomplishments.

In September of 1993 the first joint experiment involving the scientists from the US and Russian nuclear weapons design laboratories was performed at VNIIEF in Arzamas-16 (Sorova), Russia. The experiment was the first tangible result of an unprecedented scientific collaboration established through the support and encouragement of the highest levels of the governments of both nations. That first joint experiment was followed by a series of joint high magnetic field experiments in December 1993 and August 1994, and also a series of MAGO experiments in April 1994 and October 1994.

EXPLOSIVE PULSED POWER

The magnetic flux compression technique for converting the chemical energy of high explosives to intense electrical pulses that can provide intensely concentrated magnetic energy, is one of the legacies of Andrei Sakharov, father of the Soviet H-bomb and

SNLA φ-target

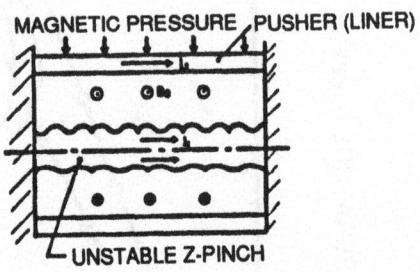
Z-pinch plasma/Z-pinch liner

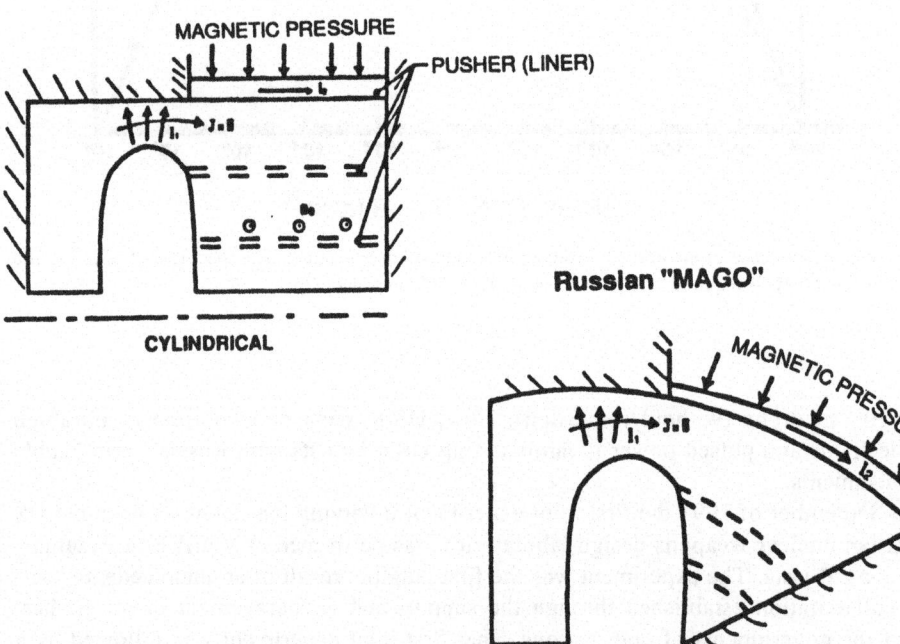

Figure 6. Examples of possible magnetized fusion targets. The scales range from a few millimeters to several centimeters, and the geometries from cylindrical to spherical.

winner of the Nobel Peace Prize. Beginning with his leadership, VNIIEF has created unmatched magnetic flux compression capability: 200 MJ at 200 MA, to produce fields in excess of 10 MG (1000 T). This advanced capability was not well known, so US researchers are just now pondering possible applications. Through the LANL/VNIIEF collaboration, VNIIEF is familiarizing US scientists with its unique capabilities.

The principle of magnetic flux compression is based on Lenz's Law. Electrical current in an inductive circuit stores energy in the form of a magnetic field in the inductance of the circuit. By decreasing the inductance in the circuit, the inductively stored energy can be amplified:

$$L_o \cdot I_o = L_f \cdot I_f$$
$$I_f = L_o \cdot I_o / L_f$$
$$1/2\, L_f I_f^2 = 1/2\, L_o \cdot I_o^2 \cdot L_o/L_f$$

This can also be viewed from the standpoint of flux conservation in a resistanceless current loop. The magnetic energy density is increased by decreasing the area of the current loop. This principle has been applied to two general types of magnetic flux compression generators.

The helical explosive magnetic generator (HEMG) is the one pursued most extensively in the US, but VNIIEF has also developed the disk explosive magnetic generator (DEMG), which has several advantages over former generators. Figure 7 illustrates a DEMG. These generators are modular. Each module consists of a shaped disk of high explosive sandwiched between two inductive cavities. The high explosive is initiated on axis and feeds a transmission line at the periphery. This makes a very low inductance transmission line possible. The modules can be stacked to increase the energy supplied without significantly effecting the delivery time. Systems using as many as 25 modules have been developed at VNIIEF, and a 3-module, 1-meter-diameter VNIIEF system has delivered 100 MJ at 256 MA into a 3.3 nH load. Scaling directly to a 10-module 1-meter-diameter system would produce 269 MJ at 196 MA into a 14 nH load.

LINER IMPLOSIONS

In September of 1993 the first joint US/Russian experiment involved implosion of a thin liner using explosive pulsed power. The 6-centimeter-diameter, 2-centimeter long liner load can be considered as the pusher of an MTF target (if a suitable plasma could be formed within it).

A DEMG was coupled to a liner load through an electro-exploded foil (fuse) opening switch. The experimental configuration and equivalent circuit are shown in Figure 8. The experiment consists of (1) a capacitor bank (not shown in the configuration, but shown in the circuit) that provides an initial current for (2) a helical generator (HEMG), which provides an initial current for (3) the DEMG, which is coupled through (4) switches to (5) the liner load at the top of the configuration. 20 MA was supplied to the load in less than 1 µs.

The HEMG, DEMG, fuse and closing switch operated as expected, but a transmission line insulation breakdown limited the current to the liner and led to some asymmetry in the implosion. Los Alamos optical current measurement confirmed the DEMG performance, and the joint experiment provided the first opportunity for VNIIEF to familiarize

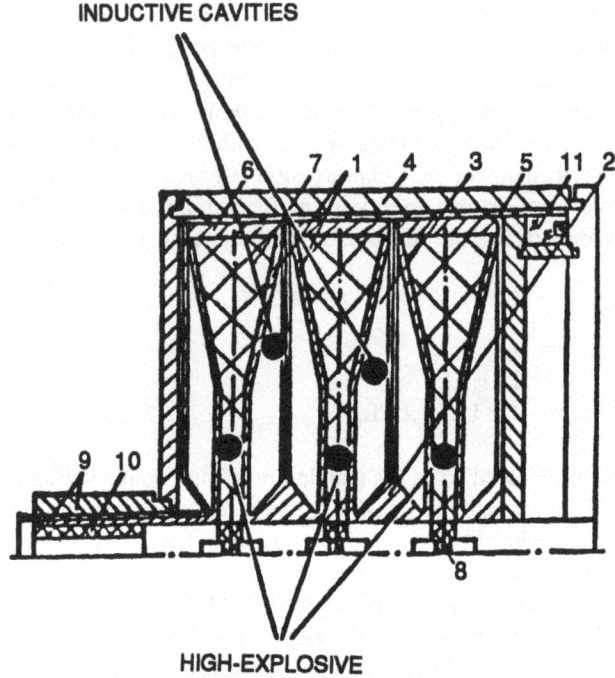

Figure 7. A 3-module disk explosive magnetic generator (DEMG).

Los Alamos personnel with their assembly and operating procedures. This experiment has led to significant improvement in the Los Alamos theoretical and computational models.

THE MAGO EXPERIMENT

The MAGO experiment is an approach to creating a hot, magnetized target plasma that has been under development at VNIIEF for several years. The geometry of this experiment is shown in Figure 9 together with a circuit diagram.

The experiment consists of an HEPP system that drives the MAGO experiment chamber through a complex explosive switching module. No DEMG is used for this experiment, since the energy and current needed to create the plasma are not very high. The HEPP system works thus: A capacitor is discharged through the entire system, including the experiment chamber. This puts a seed current in the helical generator. Then, the capacitor is isolated by closing switch S1. Next, the high explosives are initiated to operate the HEMG, thus amplifying the current through the chamber. Then, the switch S2 is operated to isolate the experiment chamber while the HEMG continues to operate, achieving a very much higher current (approaching 15 MA). Because the time constant for the isolated chamber is long, the current continues to flow there with little or no decay. At maximum current the explosively operated opening switch (indicated as a variable resistor in series with S2) is triggered. This suddenly introduces the very high current to the experiment

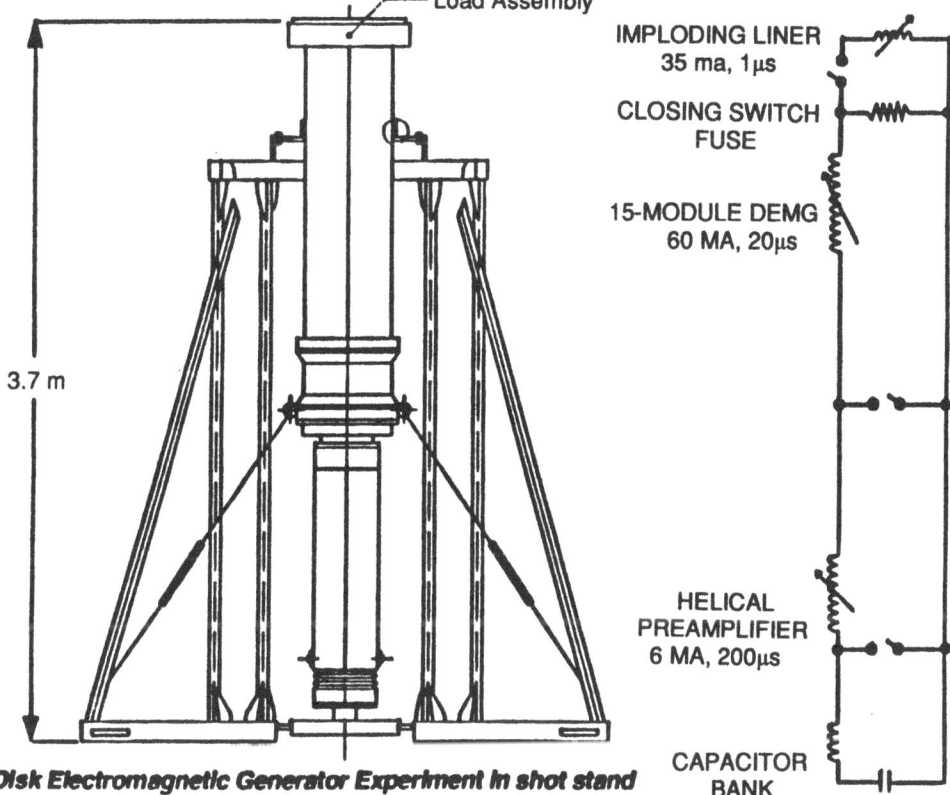

Figure 8. Configuration for the DEMG liner implosion experiment. An equivalent circuit for the system is shown as well. The capacitor bank shown in the equivalent was remotely located (not shown in the experiment configuration).[8]

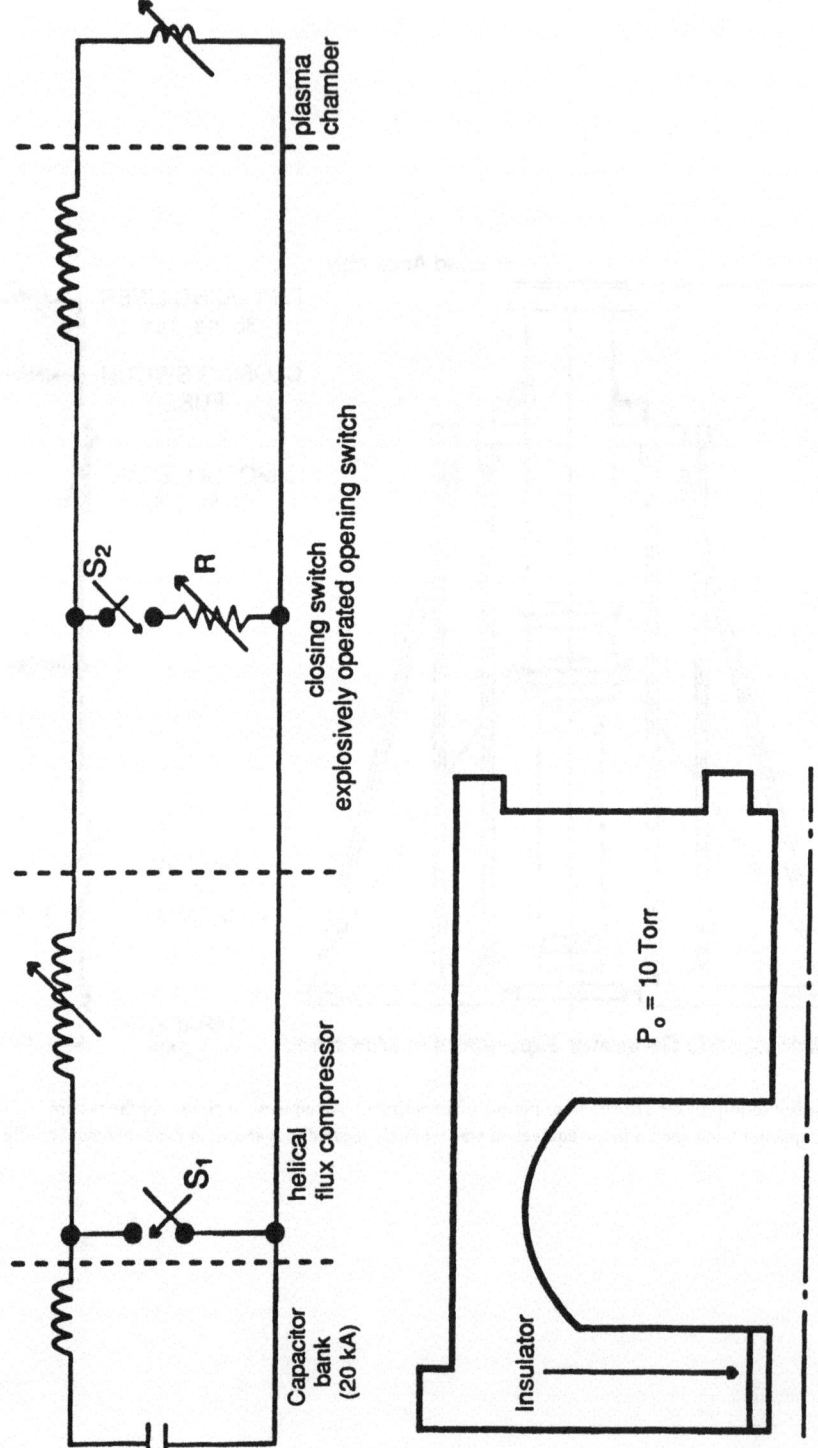

Figure 9. The MAGO target plasma creation experiment chamber with an equivalent circuit for the HEMG, switches, and chamber.

chamber, causing a breakdown and subsequent operation of the chamber. This will be described in the next paragraph. The sudden introduction of approximately ten times higher current to the experiment chamber produces a sufficient voltage across the insulator and in the gap between the chamber wall and the electrode to break down the DT gas inside. The dynamic behavior of the plasma in the chamber is important in determining the continued time history of the current.

The experiment chamber is divided into two volumes by a barrier mounted on a rod along the axis of the chamber. The early current through the rod and walls of the chamber sets up a magnetic field in the 10 torr DT gas in the chamber. No ionization occurs because the voltage across the insulator is not very high. When the high current is suddenly switched into the chamber, breakdowns occur in the chamber, first across the gap (or "nozzle") between the barrier and the outer wall. The current path broadens and the heated plasma expansion sends a shock into the second chamber. The process is not yet fully understood, but we see from the B-dot probes that the breakdown across the insulator occurs some time later. As the dynamics develop, the current across the nozzle provides a pondermotive force on the plasma in the nozzle, accelerating the plasma in the gap to high velocity. This high velocity plasma runs into the previously shocked (and thereby ionized) gas in the second chamber, producing a second shock. The temperature rises to high value in the region between these two shocks. The temperature of the bulk plasma is much lower. It is the hot plasma that is mainly responsible for the neutron output of up to 5×10^{13}.

The MAGO IIP experiment was a pure deuterium experiment, fielded in the same way as other MAGO experiments. Because there were no primary 14 Mev neutrons, this experiment afforded an opportunity to make a neutron ratio measurement that provided a rough lower limit for the bulk electron temperature in the plasma. That is because the tritons produced in the DD reaction are energetic, and while being slowed down in the plasma (mainly by electrons), they can undergo fusion reactions with the deuterium nuclei, thus producing a 14 Mev neutron. The ratio of these 14 Mev neutrons with the DD neutrons from the branch of the DD reaction that has one neutron and a light helium nucleus as its products is greater for higher electron temperatures because the range of the triton is longer for hotter electrons. It is a lower limit because the less reactions than otherwise will occur if the range is longer than the distance to the wall, or the ions substantially slow the tritons. A rough measurement was made by using activation detectors for the 14 Mev neutrons. A lower limit of 110 ev seemed to be indicated for the electron temperature.

In the MAGO I experiment we made time resolved neutron measurements, obtained neutron activation, did optical interferometry for electron density, obtained time resolved visible spectroscopy, did time integrated near-UV spectroscopy, and obtained many current and field measurements. Much of this data is still being analyzed, but we know that about 8×10^{12} neutrons were produced, the activation agreed with the time of flight measurements, the interferometry saw electron densities up to about 8×10^{16}, and the spectroscopy saw mostly continuum. While the analysis of the data is still in progress, the results so far are reasonably consistent with computations.

We obtained extensive diagnostics from MAGO II as well as insured the integrity of our common timing base, which will allow us to correctly correlate features in the various measurements. In the MAGO II experiment we also got a similar neutron yield to that of MAGO I (about 10^{13}). The visible spectroscopy in correlation with the interferometry and features in the B-dot probes looks very interesting. We now see more than continuum.

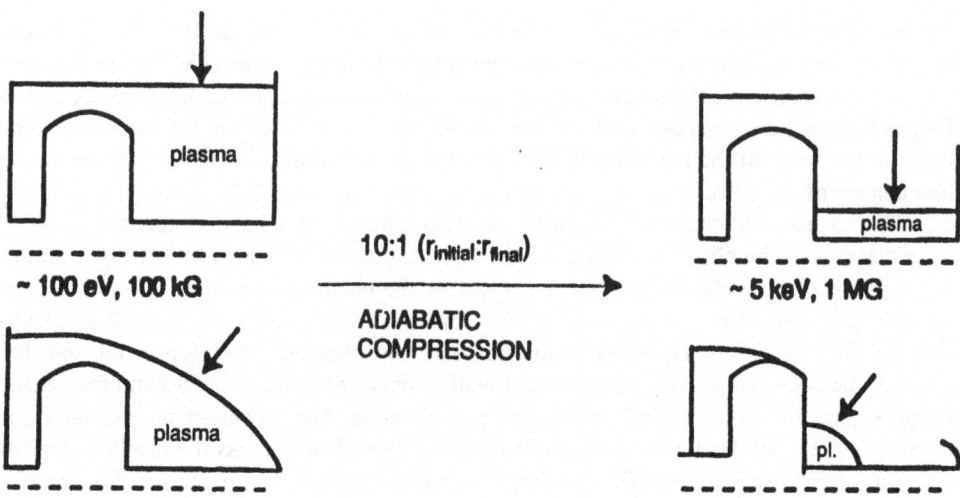

Figure 10. Adiabatic compression of the hot, magnetized plasma to fusion conditions requires only a modest convergence.

Time resolved, filtered X-ray measurements were made, which should contain spectral information on the emission from the plasma. The data reduction will take some time.

On the basis of computations done for our most recent experiment (MAGO II), it appears that the performance of the MAGO experiment is somewhat sensitive to the ratio of the early current to the later high current. In that experiment, the two currents had a smaller difference, and the neutron output was somewhat less than expected ($\sim 10^{13}$). The computed bulk plasma temperatures using the measured current histories were about 270 ev in MAGO I and 160 ev in MAGO II. Computations predict a plasma suitable for implosion and should settle into a Kadomtsev stable wall-confined Z-pinch profile.

A PROOF-OF-PRINCIPLE EXPERIMENT

The reasonably high bulk temperature in the MAGO experiments makes it interesting as a vehicle for a proof-of-principle experiment. To be suitable for MTF, the bulk plasma thermal time scale must be greater than the time required for compression. For MAGO with an implosion velocity on the order of 1 cm/μs, the time scale must be about 10 μs. Yakubov has pointed out that any impurities in the bulk plasma will reduce this time scale, especially as the plasma density increases [7].

VNIIEF has from the very beginning of this collaboration suggested methods of accomplishing a compression of the second chamber of the MAGO experiment. Two such methods are illustrated in Figure 10. Depending on dimensionality of the compression and the temperature of the bulk plasma, it may take only a modest convergence to reach fusion ignition temperatures. However, the temperature, field, and ρR of the MAGO plasma are too low for fusion ignition if cylindrical compression is used. If the plasma life time is in accord with the calculations, a cylindrical compression of the existing design of the MAGO second chamber should provide a very convincing proof-of-principle experiment. If the bulk plasma were compressed from about 6 cm down to about 1.5 cm, with the central rod unchanged in diameter, then neglecting losses, the plasma would be driven from 300 ev to 2.7 Kev, and the field would be pumped to tens of megagauss.

THE PROSPECT FOR FUSION IGNITION

Besides offering a vehicle for an MTF proof-of-principle experiment, the MAGO experiment may also provide a route to demonstrating fusion ignition. An attempt to achieve fusion ignition would require a redesign of the MAGO second chamber so that convergence along the axis as well as radially occurred. Assuming hemispherical compression, a convergence of ten would more than suffice for ignition if the initial temperature were 300 ev and the compression was reasonably adiabatic. We must emphasize that this discussion is very speculative at this stage, because neither the bulk plasma temperature or the life time of the bulk plasma has been established with certainty yet.

FUSION SYSTEM CONFIGURATION

MTF permits a complete rethinking of the entire driver/target configuration. Conventional target drivers (lasers, light ions, etc.) are optimized for conventional targets and are probably not appropriate for MTF in its extremes. For sufficiently strong magnetic fields, there is a continuum between conventional ICF target space and MTF space, suggesting a possible role for existing drivers. If the MAGO experiments confirm our calculations, the VNIEF 200 MJ HEPP systems should allow a fusion ignition experiment to be attempted in the near future with a minimal capital outlay.

AN MTF PROGRAM

An MTF program should focus on three major goals: We must have a convincing proof-of-principle demonstration of quasi-adiabatic heating ($T_f > 2\, T_o$) of a magnetized plasma by a magnetically driven liner. Achievement of fusion ignition temperatures (> 4 KeV) and high neutron yield (> 10^{14}) should be the next step. Finally, it is necessary to demonstrate thermonuclear fusion ignition and a significant burn-up fraction.

Each goal requires both a hot, magnetized target plasma and a target implosion system. The obvious target plasma formation candidates are the VNIIEF MAGO plasma for which characterization is in progress, and the Los Alamos cryogenic fiber Z-pinch for which only the early stages are characterized. There are two reasonable candidates for an implosion system: Both Los Alamos and VNIIEF have had extensive experience with cylindrical liners magnetically driven to velocities in the range of 0.3 to 1.0 cm/µs with convergences up to about 10. In addition, a quasi-spherical implosion at 0.5 cm/µs with a convergence of about 3 has been demonstrated by VNIIEF and Phillips Lab. This could satisfy conditions necessary for a proof-of-principle experiment. Each goal requires extensive computations to evaluate the trade-offs between initial temperature of the target plasma, the convergence required, the liner (or pusher) velocity, and the energy required.

We believe that a six year program costing about $ 3–6 M per year may suffice to demonstrate fusion ignition for MTF. For this program each of the goals would require about three years with a computational effort defining each of the necessary experiments and providing analysis of their results. The tasks for each goal can be overlapped to some degree, so that six years is a reasonable estimate of the time needed to accomplish all three goals. With more resources the time to demonstration could be shortened.

PROSPECTS FOR A FUSION REACTOR

What are the prospects for an MTF reactor? We recognize that the explosively driven magnetic flux compression generators cannot be used for economic power production. However, the most pressing need is to demonstrate fusion ignition. Ruby lasers first demonstrated lasing, but they are rarely used today. Nor do transoceanic airlines still use reciprocating engines. Based on our present knowledge, existing 200 MJ DEMG capabilities are more than adequate to achieve thermonuclear fusion ignition via MTF. MTF does not require a major capital equipment investment before the very first ignition experiment can even be attempted. Only when a large fusion burn-up fraction has been achieved can we truly say that controlled fusion is just an engineering problem. Just as the myriad of laser applications could not have been predicted long before lasing was demonstrated, the potential of MTF (or any other fusion concept) cannot be realistically evaluated until ignition is demonstrated.

SUMMARY

Magnetized target fusion has many attractive features. These include low implosion velocities, low convergence ratios, enhanced self-heating, the use of existing drivers with excess energy, and the possibility of demonstrating fusion ignition at low operating cost and no capital cost. All these features make MTF very attractive as a future research direction in our quest for controlled fusion. They also point toward the possibility of a very economical fusion power technology. However, the first major step is to demonstrate fusion ignition. Depending on the resources committed to this effort, that first major step could be accomplished at a much earlier date than is currently projected for the mainline fusion efforts and at a much lower cost.

ACKNOWLEDGMENTS

We wish to commend and thank the members of the Los Alamos and VNIIEF teams for their excellent performance in accomplishing the collaboration on the MAGO series of experiments.

REFERENCES

1. Kirkpatrick, R.C., Proceedings, Current Trends in International Fusion Research, (ed. E. Panarella, Plenum Press 1996), p. 319.
2. Dawson, J.M., Okuda, H., and Rosen, B., "Collective Transport in Plasmas", in Methods in Computational Physics 16 (1976) 281.
3. Lindemuth, I.R., and Kirkpatrick, R.C., "Parameter Space for Magnetized Fuel Targets in Inertial Confinement Fusion," Nucl. Fusion 23 (1983) 263.
4. Widner, M.M., "Neutron Production from Relativistic Electron Beam Targets", Bull. Am. Phys. Soc. 22 (1977) 1139.

5. Sweeny M.A., and Farnsworth, A.V., "High-Gain, Low-Intensity ICF Targets for a Charged-Particle Beam Fusion Driver", Nuclear Fusion 21 (1981) 41.
6. Mokhov, V.N., Chernychev, V.K., Yakubov, V.B., et al., Sov. Phys. Dokl. 24 (1979) 557.
7. Yakubov, V.B., Los Alamos National Laboratory Seminar, October (1994).
8. A. Buyko et al., "Investigation into the Possibility of Obtaining Thermonuclear Magnetized Plasma in a System with Magnetic Compression–Mago," Zababakhin Scientific Talks, Kishtym, Russia (January 1992).

36

CONCLUDING REMARKS

Emilio Panarella, Chairman of the Steering Committee

Advanced Laser and Fusion Technology, Inc.
189 Deveault St., No. 7, Hull, P.Q. J9Z 1S7 Canada, and
Department of Electrical Engineering
University of Tenneessee
Knoxville, Tennessee 37996–2100

It seems just yesterday when we started together here, in this same room, a journey of five eventful days that demonstrated how valuable our work is in plasma physics and fusion. I believe that there is no doubt in the mind of everyone that the works presented here are not only a reflection of our capabilities, but also of the love and perhaps passion that we put in that work. Each presentation was of such high calibre to be at same time a mirror of its author and of its value. In the past five days I did not miss any of your presentations, and I felt proud to be together with such outstanding scientists as you all are. The question and answer period after each presentation was profound for its contribution to the clarification of the already clear presentations, and I felt that it was a natural stimulating conclusion of your talks.

What we have done in the past five days cannot go unnoticed. With the demonstration of the individual and collective value of our work, we have left a strong mark of our presence within the plasma and fusion community. In the past five days we have proclaimed that plasma and fusion science has a value per se, independent of the ultimate goal, be it energy production or industrial applications. We have proclaimed that the basic value of the physics of plasmas has predominance over other considerations. We have also demonstrated that, while our desire to increase our knowledge in the physics of plasmas should be acknowledged and accepted, this does not distract us from the ambitious goal of reaching the objective of controlled fusion. We have sent a strong signal that the methodology of science, which can be expressed in a few simple words, namely 'one step at a time', is what guides us in our progression towards scientific and technological goals. We have also sent a signal that the selection criterion in 'big versus small science' should be used with exponentially increasing care when one moves from small to big.

As you have all witnessed, this symposium has procceded in a very smooth way, thus demonstrating not only the care with which the Steering Committee has done its work in the past six months or so, but also the care with which the speakers prepared themselves for their presentations. The members of the Steering Committee felt that they had an obligation towards the International Advisory Board, which counts such distiguished scientists as Bruno Coppi, George Miley, Hans Persson, Richard Post, Ravi Sudan, Liu Shenggang, Edward Teller, Guillermo Velarde, Chiyoe Yamanaka, and Masaji Yoshikawa, to make their best efforts in reaching excellence also on the organizational aspect of the symposium.

The assessment of our work in now in the hands of our Panel of Evaluators. Knowing that the Panel is composed of scientists as Edward Creutz, Arthur Kantrowitz, Joseph Lannutti, Hans Schneider-Muntau, Glenn Seaborg, Frederick Seitz, and William Thompson, I feel confident that our work will be assessed in positive terms. For the attention that the evaluators have paid in listening carefully our presentations, I would like to thank them warmly. For the difficullt task that lies ahead for them, in the next few months, I would like to say that they certainly deserve our gratitude.

In my opinion, this symposium has allowed us to make a quantum jump in what we intend to achieve, and the next step will be easier. Since we do not want to jeopardize the chances of a greater success in the next symposium, we want to assess carefully what we have achieved so far in order to plan carefully for next symposium. This process will start tonight, when the International Advisory Board and Steering Committee members will meet for the third time, and will continue in the next few months, when we will reach a deliberation on when to have the next symposium. I can only say now that we will have a 2nd International Symposium on 'Current Trends in International Fusion Research', that it will be held within the next two years or so, and that it will be held again here in Washington.

In conclusion, my thanks go to all of you. Clearly, *you* are the symposium, and *you* have made it the success story that we have all seen so clearly in the past five days.

37

REPORT OF THE EVALUATORS

Edward C. Creutz, Arthur R. Kantrowitz (Chairman), Joseph E. Lannutti,
Hans J. Schneider-Muntau, Glenn T. Seaborg, Frederick Seitz, and
William B. Thompson

The motivation of this conference is clear. Fusion represents one of the three possible permanent energy sources, the others being some convincingly safe development of conventional nuclear power, and some economical and practical way of using solar energy. Since this is fairly clear, it is far from obvious why more resources are not being devoted to the exploration of fusion, unless in the present anti-nuclear atmosphere even a reduction of the radioactive waste by a factor of 10^7 might not force career activists to declare fusion reactors "safe". It is also not clear why what funds there are, are concentrated in only a very few, not completely promising avenues of research. The conference explored methods of broadening that exploration.

The conference contained a description of current research in fusion, although that presentation was somewhat distorted. There was a strong representation from the ICF community which seeks to obtain useful power from mini-thermonuclear explosions produced by delivering an enormous flux of energy to the surface of a tiny fuel pellet; but a very limited presentation from the MCF group that hopes to burn a stationary plasma confined by a magnetic field. Tokamak research was the subject of a single survey paper, and there was no mention of stellarator systems.

What is clear, however, is that the major-flagship-programs in both ICF and MCF offer chiefly massive engineering, when the required scientific basis still has important uncertainties. Research progress is exceedingly expensive, and uncertainly promised practical reactor at least a generation in the future. In MCF, the major program is ITER, the international thermonuclear experimental reactor, which hopes to ignite a tokamak confined D.T. fuel, and control the burn for up to 1000 secs, long enough to meet many of the problems of a power producing reactor. A future development could include elements of a breeding blanket, hence all the elements of a reactor. But the physical scale is enormous, the engineering challenges formidable, the time scale for results at least a decade, the physics basis of its performance uncertain, and the price an astronomical $\sim\$2 \times 10^{10}$.

In the ICF program the situation is a little better. A National Ignition Facility is proposed: a monster laser system, which at a cost of $\$10^9$ will produce enough power to ignite a small pellet; though again there are many questions in basic physics that need to be addressed. Moreover, the driver cannot be scaled to produce power, since the repetition rate for firing the glass lasers is far too low. An alternative driver is needed, and most effort is centred on heavy ion beams, a large current of heavy, slightly charged ions accelerated to several G.V. focussed on a target of ~1 cm^2 area. The development and construction costs of the necessary accelerator have been assessed (by its proponents) at ~$\$10^9$, with a development time of ~10 years. This does, however, require ion current densities orders of magnitude greater than any so far produced, and projections are extremely uncertain. Moreover, to make a reactor would also require a target chamber which would confine the explosion, and the associated radiation, efficiently collect the energy produced, and still permit targeting the heavy ion beams. Finally, methods of producing very precisely designed targets at a few cents each need to be developed. There is no shortage of suggested methods of doing these, but development, probably at a high cost, is needed.

A major lesson from the lead programs is that fusion is very difficult—that a major disturbing role is played by ubiquitous instabilities which, if they do not destroy the system outright, lead to random departures from uniformity, to anomalous transport in MCF, and anomalous transmission, reflection and absorption in ICF. These fluctuations limit the usefulness of simple theoretical models and are extremely difficult to include in computations. Moreover phenomena change with scale; what is innocuous in a small experiment may become disastrous in a scaled-up version.

It is this turbulence-like fluctuating behaviour that makes the theory of plasma physics so difficult. Moreover, although vast sums have been spent on fusion research, the fraction devoted to well-conceived experimental demonstrations of the basic physics supporting that research has been derisory. Lacking a well-based and useable theory, projections have been based on uncertain empirical scaling laws: uncertain, because in contrast to hydrodynamics, plasma behaviour depends on many processes, and the number of dimensionless combinations of basic parameters is too great to permit simple scaling.

The conference papers could be grouped into sets. There were a few that dealt with the general problem of fusion research without defining specific areas (Teller, Rostoker, Perkins). A number presented reports on the mainline programs, and suggested slightly different pathways, modifying or replacing existing programs, or representing possible future developments (Coppi, Pegoraro, Logan, Kilkenny, Lindman, Martinez-Val, Yamanaka, Bangerter). There was a further set that described lines of research that had been followed quite seriously until cast aside in the tokamak frenzy, with suggestions as to how these could profitably be resumed (Post, Siemon, Phillips, Slough, Brzosko), while a number of others suggested magnetic confinement schemes that had either not been studied experimentally, or at best considered only superficially (Mauel, Dolan, Miley, Koloc, Roth). Several papers explored the gap between ICF where there is essentially no confinement of the fuel explosion, and MCF where the plasma is permanently confined. These impulsing fusion systems used pulsed power sources and magnetic confinement (Kirkpatrick, Smirnov, Rostoker, Degan, Dolan, Schlachter, Panarella, Bystriskii). Several papers were concerned with non-thermonuclear reactions between ion beams (Maglich, Rostoker, Miley), and two surveyed the present position and future promise of μ catalyzed fusion (Jones, Bystriskii).

Many of the presentations started with promising ideas but soon showed the deadly intrusion of Empire Building into research. EB forces investigators either to compete in making near-term promises, or risk being marginalized. EB encourages proposals for large scale experiments when smaller scale would produce information more quickly.

However, the mainline programs are on such a scale, both in money, effort and time, that even rather unlikely alternatives are worth a close scrutiny, and in most cases, more theoretical and experimental investigation.

THE CONFERENCE

I. General Papers

Teller described the importance of fusion and the need for more rapid progress than the main programs provided. He also observed that the popular fear of low level radiation would be a problem for the development of fusion, as well as fission, and must be taken seriously.

Rostoker gave a brief history of fusion research in which projects tended to increase in size rather than produce results, and were often stopped before important information had been obtained. He sketched the mainline programs to indicate the need for alternative approaches; and suggested that those promising near-term results might be able to get industrial funding. [*He did not mention that the unkept promises of near term results had already discredited fusion research.*]

Perkins described an assessment program for alternative fusion concepts, a proposal to provide a critical impartial evaluation of the feasibility and scalability of new fusion proposals, making use of the considerable numerical modelling facilities of LLNL. [*His illustrations were a little disturbing, showing engineering scaling of the FRC and Spheromac systems; with, however, little concern for the instabilities that will probably dog both systems.*]

II. Mainline Research

1. MCF-Tokamaks

Coppi gave a critical survey of the tokamak program, which has produced modest amounts of fusion power with both JET and TFTR. He pointed out the anomalous behaviour of these systems: transport coefficients are 10 to 1000 times classical; fluctuations in density, temperature, and current are observed and the discharges often end with a fairly violent disruption. [*He did not, however, describe the transition from low to high confinement modes, and the role played by edge localized modes in determining transport.*]

He briefly described the ITER program; a large (30m x 30m) expensive (2×10^{10}) Tokamak, which is intended to produce ignition of a D.T. fuel, i.e., to reach a state where the fuel temperature is preserved by its own reaction products and to maintain the burn for ~1000 sec. To do this, most of the elements of a reactor-confinement, burn refuelling, ash removal, wall loading energy removal, and radiation bombardment—must be satisfied. He was, however, more interested in IGNITOR, a development of the high field high density Alcator tokamak series to a scale where a few seconds of TD ignition could be obtained, a much more modest proposal than ITER.

Pegoraro described the IGNITOR project: a high field, high current, high plasma density, ohmicly heated Tokamak, operating at low β, so that there is a large margin of stability, which could demonstrate ignition and a few second burn of DT for 1/10 the price of ITER and (if funded) in 1/10 the time. Design studies, using a reasonable experiment based model of transport, have been completed, and the engineering of the modules needed to make up the system demonstrated by constructing modular elements; field coils; structural clamps and 1/3 of the vacuum vessel. The complete system costs 10^9 and could be completed in 5 years, if adequately funded. [*However, the observation of the H and VH modes in tokamaks indicate that supplementary heating is not as dangerous as it once seemed; hence a purely ohmic system seems less necessary.*]

2. ICF.

a. Target Design and Objectives.

Logan (LLNL): observed that present experiments and theoretical understanding permitted extrapolation (factor of ~5) to NIF which should reach ignition in a laser heated pellet, then described the further development needed to produce an ICF reactor: High gain targets—gains 50–150 (these should be developed with the aid of NIF); efficient high repetition rate drivers: $\sim 10^{15}$ W for $\sim 10^{-8}$ sec, with a rep rate of several shots/second; long lasting target chambers, $\sim 10^8$ shots and low cost targets, ~$.30 each.

Target chambers might be protected by liquid jets. Heavy ion beams seem the most promising driver candidate. Estimated cost for developing 1000 MW power plant ~$3 x 10^9.

[*But: 1) Target chamber design must protect walls, and also permit transmission and focussing of beams (so far this seems to require high vacuum), 2) High current, high energy accelerators have got to be developed, though induction accelerators look promising, 3) Rayleigh-Taylor instabilities put severe restrictions on lack of symmetry in target design!*]

Kilkenny (LLNL) discussed the Nova program and the consequent NIF design.

  i. Numerical modelling of targets gives good agreement with experiment and permits extrapolation to NIF.
  ii. Indirect drive has produced Hohlraum temperature T = 275 eV, in a ~3 mm radius cavity with 0.5% asymmetry, given implosion velocities of 3.5 x 10^7 cm/sec, convergence factors of 21 and densities ~170 g/cm^3.

But the Hohlraum is not free of beam imprints; gross asymmetries can appear if incident beams are not properly focussed on walls; and plasma effects—stimulated Raman and Brillouin scattering—affect transmission, symmetry, and especially reflection; thus the Holhraum does not avoid the plasma problems of direct drive. The NIF laser requires 192 beamlets, with a total energy of 1.8 MJ, and a maximum power of 500 TW, with a pointing accuracy of 50 µ, at a wavelength of .35 µ, reached by 2nd and 3rd harmonic conversion at an efficiency of 80%.

Initial targets have a shell of Br. doped polythene with a radius of 1.11 mm and a cryogenic DT shell enclosing a .87 mm hollow filled with DT gas at 0.3 mg/cc. The surface finish tolerance is 0.08 µ (800Å), and with a convergence factor of 36 yields a density of 1200 g/cm^3!

Experimentally, X-rays produced in the Hohlraum agree with 2D simulations; and collapse of doped plastic shells demonstrate Rayleigh-Taylor instabilities with saturated growth, which determines the surface finish requirements of the pellet.

Lindman (LLNL) discussed target physics research at LANL. Experiments on large and thin walled Hohlraums show the conditions needed for adequate X-ray symmetry; and experiments on re-remission from gas balloons show plasma effects on reflection; while transmission experiments on collapsing cylinders show the growth of RT modes. Reasonably detailed agreement has been obtained between experimental results and numerical modelling. [*Predictive ability of computational schemes?*].

Martinez-Val (Madrid) on target physics: Distinguished between compression modes in which the pusher is burnt off before the fuel collapse is completed, which reduces confinement times but avoids dilution of fuel, and stagnation in which pusher remains at the collapse point—this may introduce mixing of pusher with fuel, although it increases reaction times. He also proposed an asymmetric guided collapse involving two pellets being fired at one another through converging one-shot channels.

Yamanaka (Osaka) described the Japanese ICF programs.

ICF ignition requires T ~ 5keV: $\rho/\rho_{solid} \approx 500-1000$; ρR ~ 0.3 g cm^{-2}. Progress towards this using laser drive requires inhibition of anomalous transmission through plasma clouds. One promising route is to induce short phase coherence in the incident beams by (a) random phase plates [phase becomes random in space, but correlated in time]; (b) multimode films [increases $\Delta\omega$]; (c) temporal incoherence [random in time, longitudinal incoherence]; (d) indirect heating [*this doesn't seem to avoid the problem*].

Experimentally, targets have been either polystyrene D shells with a 50 atmosphere DT filling or cryogenic DT foam lined shells.

Results: Gekko XII, which operates in the green with 5% random phase and an energy of 10 kJ in 2µsec has produced densities ρ ~120 g cm^3. With high radiation symmetry, high target symmetry; maximum density ρ ~ 600 g cm^{-3}, $\rho R \cong$ 0.6 g.cm^2. Again Rayleigh-Taylor instabilities have been observed at saturated growth. Holhraum experiments have produced T ~ 200 keV and convergence factors of 18, with densities of 110 g/cm^3.

b. ICF Drivers.

Bangerter (LLNL): Reviewed again the advantages and requirements of ICF and especially the driver requirements: Drivers must be reliable, durable, efficient [$\eta G > 10$; η = efficiency, G = target gain], have an adequate rep rate (1 to 10 shots/sec), and not be too costly. The energy/pulse needs to be 1 ~ 10 MJ, in ~ 10^{-8} sec, with a focus of < 5 mm, and the energy must be deposited in ~ 0.1 g/cm^2. A promising driver is an ion beam. There are two possibilities here. Light ions at E ~10–100 MeV; with currents of 10^6–10^7 amps; or heavy ions with energies of 1–10 GeV and currents of 1,000 to 10,000 Amps. In high vacuum at small charge/mass ratios, it might be possible to focus H.I. beams by classical ballistic methods; L.I. beams however need to be electromagnetically self-focussing.

Accelerators: The induction linac, in which a transformer produces the required accelerating electric field, shows promise. Such a device has accelerated ~10^4 Amps of electrons. *To get adequate beam quality at high currents, the Coulomb repulsion must be compensated, which introduces a new element into accelerator design.*

Estimated cost of an accelerator system, using electrostatic acceleration for the first few MeV, followed by an induction Linac, with a fair amount of beam blending (and little emittance growth) is 10^9. L. I. beams with electron neutralization might be lower cost; but adequate focus has not been demonstrated.

[Is HIF consistent with target chamber design? Ballistic focussing requires low Z/M; but a background gas could strip the beam ions, and drive the transmission problem back toward LIF].

Fessenden (LBNL) gave a description of the induction Linac program.

Present status (MB E-4): Has 4 10 mA beam lines accelerating ions from 0.2–0.9 MeV. Current studies are on combining the 4 beams into 1, without emittance growth. Ion sources are Zeolite alkali metal sources, Cs$^+$ and K$^+$. Also, a driver scale 2 MeV injector has been developed yielding ~1 A at 100 keV.

Recirculation experiments using K$^+$ accelerated from 8–300 keV in a circular system with a radius 14.4 meters have been successful.

Next step: ILSE—an induction linac with driver scale beams to be used for studies of beam manipulation, focussing and superposition. ELISE—the electrostatically focused low E component was begun October 1995. Full scale driver should be ready by 2010, for a fusion demo in 2020! *[Note time scale.]*

Deutsch (Orsay) suggested use of ion clusters $M = 50,000$, $Q \simeq 50$, which are easily produced and accelerated as very high M/Z ions. He reported on stopping of clusters in solid target, and decrease of range produced by ion pair correlations. *[Misguided; problems of obtaining high currents $10^6 A$ at 100 MeV, more serious, and deposition most likely in blow-off plasma.]*

Cheeseman (LLNL)—Suggested production of particle beams at spherical surface, which have intensity and energy vs. time profiles reversing the adiabatic expansion of highly compressed pellet, producing fusion at the core. *(Why not just use similar, but less intense beam to drive a target (LIF)]?*

III. Magnetic Confinement (Well-Studied Areas)

Post (LLNL)—open-ended systems

1. History of the mirror program; a) magnetic confinement by μ, J conservation demonstrated; b) confinement of plasmas spoiled by interchange instabilities; c) problem solved by Joffe bars and containment in magnetic wells; d) microinstabilities—loss cone modes—produce anomalous losses; e) 2 XII B—High energy neutral beams in a seed plasma gave promising results; f) Tandem mirror, trapping wells solved loss cone problem, but at cost of higher conduction losses and low Q.
2. Defense of low Q systems. Mirror systems permit efficient energy extraction and re-injection. If cycling efficiency is high, low Q still may permit economical and feasible power extraction (c.f. jet engine).

[Probably true that termination of the mirror program was badly timed, but the empire and its costs had got out of hand!]

Siemon (LLNL)—Field Reversed Configurations

1. Parameters $n \simeq 10^{15}$ cm^{-3}, T or 1 → 1.0 keV, B ≃ 5.7 kg, τ ≃ .3 × 10^{-4}, nτ ≃ 10^{11};
2. Compression increases β by factor 7, n 10x, T 6x [nτ unaltered].
3. Stability: No field shear; bad curvature; means MHD unstable [close walls inhibit all but tilt modes]. However, kinetic effects due to finite ion Larmor radius (ρ_i) reduce growth rates; for s = r_s/ρ_i [separatrix radius/ion Larmor radius] <3. Could stability be maintained at large s by introducing energetic ions, or αparticles?

Slough (U. of Washington)—Stabilization

1. Higher modes stabilized by magnetic multipole (since system rotates)
2. Tilt modes stable at small s (r_s/ρ_i)
3. Changing shape of separatrix (from elliptical to race track) improves stability.

[Note that instability not detected in early systems; destructive instability may not be easily recognized in a pulsed system.]

Phillips (LANL)—Future of the Reversed Field Pinch

1. Attractive as a reactor ($\beta \sim 20\%$ and the plasma produced $B_\theta \gg$ coil produced B_ϕ), but confinement times $\sim 10^{-3}$ sec. Theoretically, a nearby conducting wall should stabilize plasma but experiments to date have been in systems with many wall gaps. Perhaps results would be improved by confining shells. Suggested modest experiment $I_\phi \sim 150$ kA, $\tau \approx 30$ msec using standard iron core transformer cooling. *[Is this Zeta redux?]*

Mauel (Columbia)—D-He^3 Dipole Fusion Reactor

1. Consider a plasma confined in a given stationary magnetic field. Particle motions are then determined by the adiabatic motions μ (associated with the gyro frequency), J (with the bounce frequency of parallel motion) and Ψ, the flux function conserved by the magnetic drifts. Then if $\partial F/\partial \Psi|_{\mu,J} = 0$, the system is stable against low frequency modes. In a dipole field the flux volume $V \sim L^4$ where L is the equatarial radius of a field line (McIlwane's parameter) hence for fixed $\partial F/\partial \Psi = 0$ $n \sim L^{-4}$, $nT \sim L^{-6.4}$

He suggested using a superconducting ring to produce a dipole field and placing boundary sufficiently far away. *[Reactor size ~ ITER! Problems maintaining F in face of collisions, development of electric fields; high frequency instabilities?]*

IV. Magneto-Electrostatic Confinement

1. Electrostatic.

Dolan (INEL)—Research on electrostatic and magnetic confinement

Basic idea: converging ion beams attracted to either a spherical grid or a trapped electron cloud, for high density and beam recycling.

Lavent'ev—focussed current of 12–25 mA.cm^{-2}, using a 4 kV grid; ratio $I_{circulation}/I_{grid} \approx 7$; density $n \approx 10^{11}$.

Hirsch—Injected electrons to form a central charge (grid +), then six energetic ion beams; found anomalous flux of D-T neutrons, and suggested multiple potential wells had been formed. However, theory suggested 1) wells could only be shallow; 2) neutron production not anomalous. Effects have not been found in repetitions of experiment.

Miley—Ionized a gas puff near outer walls, and let ions fall toward negative spherical grid. Fluxes of $\sim 10^5$ neutrons/sec.

Possible developments: magnetically shielded grids. *[Note extensive experience with magnetic shielding in magnetically confined systems (e.g. octupole)]*

2. Magneto-Electrostatic.

This is mostly concerned with the use of electrostatic fields to reduce losses from cusp confining magnetic fields. Cusp confinement takes two forms. At low β the trapping of most particles is due to the increase in B near the cusp and associated mirror trapping; however, in many cusp systems, there are low field regions where the adiabatic invariants are lost and a much higher loss rate possible. At high β, the dielectric currents exclude the magnetic field from the plasma, except at the edges, and losses are by hydrodynamic flow through the cusp. The losses are then determined by the effective cusp radius [in the early NYU analyses, this was given as a geometric mean of the Larmor radii ($r = \sqrt{\rho_e \rho_i}$)].

Electrostatic plugging puts an anode and a cathode at the cusp, the anode near the plasma to reflect ions and the cathode further out to reflect the electrons that stay much closer to the field lines. Problems are that the plasma screens the electrode potentials, and streaming particles produce high frequency instabilities near the loss cone. Geometries range from the simple spindle cusp which has 1 ring and 2 point cusps to Surmac, where the plasma surface is covered with many small field cusps. Two variants are the Polywell, in which electrons trapped in a multipole magnetic cusp form an electrostatic well to trap and focus energetic beams of ions, and the Penning trap, which has a similar goal, and is designed as a multipole in which the trapped density can exceed the Brillouin limit n_B [$n_B mc^2 = B^2/2$].

Experiments:

a. Early spindle cusp losses were dominated by electron neutral collisions
b. Electrostatic plugging could reduce ion losses by 85%
c. Jupiter 1A (a spindle cusp at Kharkov) showed a central electron density spike and a loss radius of $r \sim \rho_e$
d. Stansfield (U. of Quebec) in a spindle cusp obtained densities $n = 10^{11}$ at $T \approx 1$ keV, but millisecond confinement
e. Jupiter 1M (again at Kharkov) had $B \simeq 13$ T, a confining electrostatic potential $\phi \sim 6$ keV and obtained densities of $n = 10^{12}$ cm^{-3}, $T \simeq 100$ eV and $\tau \simeq 10^{-3}$ sec. Injected electrons however were lost in $\sim 10^{-5}$ sec, and the plasma decayed in $\sim 10^{-3}$ sec without the beam. However the plasma confinement time was 1/5 x classical.
f. Jupiter 2M (at Kharkov) has 7 ring cusps and is designed for $B_{cusp} \simeq 4$ T and $B_{ring} = 2$ T and has been run with $B_{ring} = 0.5$ T, yielding central densities of $n = 10^{11}$ and confinement times of $\sim 2 \times 10^{-3}$ sec.

Reece Roth (Tennessee)—*Electrostatic biasing of toroidal systems.* Used emitting electrodes to apply electrostatic potentials to a bumpy torus, and found the ion transport modified by the field; the ion flux was parallel to E, and could be against the density gradient. Loss was dominated by the electrodes. [*Compare with experiments of R. Taylor (UCLA) on biasing a Tokamak.*]

Barnes et al. (LLNL)—*PFX experiment.* These authors suggest that a multipole trap could confine a single species plasma at densities greater than n_B, the limit for a simple Penning trap, and that this could be a target for ion beams. Detailed design studies presented but no experiments.

Miley—*Electrostatic neutron sources.* Using ions produced near the wall of a spherical vessel and accelerated toward a spherical grid has produced D-T neutron fluxes of $\sim 10^6$/sec^{-1}, limited by the heat load to the grid. Believes, with aid of a DT moderator, and modifications to grid, this could be scaled to $\sim 10^{12}$ thermal neutrons, and be a cheap and useful source for medical purposes.

3. Pulsed Magnetic Systems.

Brzosko et al.—*Plasma Focus.* System: Plasma is produced and magnetically accelerated, either in a short cylinder, or a large disk; and collapses on axis to a dense, highly unstable line discharge, which in D or DT produces an intense burst of neutrons. Experimentally, the neutron production Y scales as the square of the input energy, $Y \sim W^2$ [this holds for $\sim W \simeq 100$ kJ]; and scales to thermonuclear breakeven at 100 MJ [a scale factor of 10^6!] However, it could be a useful radiation source at intermediate energies. To progress, experiments need to be done with plasma production away from magnetically protected insulators (gas puffs!)

Degnan (Phillips A.F.L.)—Described experiments on production, compression, and acceleration of plasma toroids between conical containing walls. Proposes to target these on a gas puff target *[a variation of plasma focus, without the current channel.]*

Scudder (LANL)—*Dense Z-pinch*. System: A large capacitor bank is discharged through a low inductance line [matched] into a small solid D_2 fiber. Currents of $\sim 10^6$ amps are produced in fractions of a microsecond $dI/dt \sim 2 \times 10^{13}$ amps/sec, and neutron fluxes produced. Sausage mode (or Rayleigh-Taylor?) instabilities appear in 5×10^{-8} sec, but enhance the ion heating rate. Stability may be modified by non-ideal effects. *[No evidence and scaling is uncertain.]*

Rostoker, Wessel et al.—*Thermonuclear fusion in a staged Z-pinch*. Modified the Z-pinch by adding a gas ring around the central conductor, then dI/dt in the core goes up to 5×10^{15} amps/sec.

Experiments at 10 TW (Angara 5) have produced $I = 2$ MA in $\sim 10^{-7}$ sec, and central densities of $n \simeq 10^{21} - 10^{22}$ with $10^9 - 10^{10}$ neutrons/shot.

Theory has considered one dimensional models of the motion of the liner and the core as the current rises, and predicts breakeven at $I \approx 10$ MA. *[Stability problems are mentioned, but not analyzed.]*

Bystritski et al.—*Staged Z-pinch as a super-radiant X-ray source*. Adds to the staged Z pinch configuration an ion beam between core and liner. On compression, inner shell electrons are excited and X-ray superradiance possible.

Panarella—*Spherical shocks in plasma*. Has modified the θ pinch geometry to produce a more or less spherical converging shock, which produces a hot spike at the center. Multiple shocks produce further increase in temperature, and process is numerically scaled to a thermonuclear reactor. Experimentally $\sim 10^7$ neutrons produced and the system makes a convenient neutron sources. *[The author does not invoke the Guderly singularity, but this is a version of ICF and not surprisingly the breakeven input energy is well above that for adiabatic compression.]*

Koloc (and Reece Roth)—*Plasmas and Ball lightning*. Invoke a model of ball lightning. [Lightning channel viewed as a Bennett pinch is wriggle unstable, and a loop may be cut off to form an isolated current ring, surrounded by a plasma mantle.] *[Is this in pressure equilibrium with the surrounding atmosphere?]* Suggests that a similar configuration is stable and can be produced in a spherical pressure vessel (cf. spheromak). If the vessel can stand adequate pressure, then thermonuclear energy might be produced in a convenient configuration. [Stable if a Taylor state with uniform pressure; at 1,000 atmospheres with $T = 10$ keV [4×10^5 normal temp] and $n \simeq 2 \times 10^{16}$ and $\tau \simeq 0.01$ sec to satisfy the Lawson criteria.] Process might be enhanced by spherical shocks. *[Microstability and transport problems need to be faced.]*

Kirkpatrick (LANL) — *Magnetized Target Fusion*. Modify an ICF target by including an internal closed magnetic field that is amplified by compression. Then confinement is determined by the mass of the pusher, or tamper: energy gain becomes possible for smaller fuel masses, and confinement time may be 10^{-7} sec [rather than 10^{-9}]. [*Note: impulse/unit area* $\approx p\tau$ *is defined by ignition conditions*].

Fusion is possible if 1) magnetic flux is compressed and 2) energy is transported across the field lines classically. Then minimum mass ~ 1 μg, B ~ 5 x 10^6 g and convergence factors of only 15–20 are needed.

Sweeney (Sandia) — *Phi-target experiment*. The φ-target is a gas filled microballoon with a conducting filament crossing its center. A pulsed electron beam drives a high current through the filament, producing a magnetic field of ~ 10 kG. The target is then collapsed by a 5 TW light ion beam, in 10^{-7} sec. Scaling to ignition requires a factor of 10 in ion beam power. [*But, the B field destroys symmetry, and modifies hydrodynamics. Is adequate compression possible in the scaled up system?*]

Smirnov—*Staged Z-pinch as a target driver*. Modifies staged Z-pinch by replacing center filament by a hollow metal cylinder. When collapse occurs cavity becomes an X-ray filled Holhraum and could be used to drive a fuel pellet. Simulations suggest that an input energy of 350 KJ could produce a current of 15 MA, and a radiation flux of 500 TW/cm^2. Experiments at 5–10 KJ produced flux of 3 TW/cm^2, with a 30% conversion efficiency.

Lindemuth (LANL)—*Review of Magnetic Compression*. Russian (Inst. for Experimental Physics) and LANL (providing diagnostics) experiments on compressing magnetic fields have produced 200 MJ with explosive drivers, and 0.5 MJ with current driven multiplying shocks. Fluxes at 5 x 10^{13} DT neutrons have been produced. Although not a potential reactor, this might be a short cut to ignition.

V. Ion Beam Fusion

Maglich

1. Describes Migma concept. [Trapped circulating ion beams, colliding at a focus, to produce fusion]
2. New developments
   a. Strong focussing magnetic fields. Radial gradients modified to give focussing and defocussing of beam, leading to net improved focus.
   b. Neutralize beam charge by electrons oscillating along B field
   c. Results: Max ion density increased from 10^8 to 10^{10} effective $n\tau \approx 10^{11}$.
   d. Computation: Show possible scaling to reactor [1000 x $n\tau$]

Rostoker (Irvine)

1. Reviewed evidence that ion beams (large gyroradius) behave classically, a) Maglich results; b) D.C.X.; c) beams injected into Tokamaks, where beam ions slow down classically, in spite of anomalous behaviour of plasma.
2. New suggestion
   a. Inject ionized beams into a reversed field configuration and neutralize with cold electrons. [*Stability? Production?*]
   b. Computations suggest a promising reactor.

Post (LLNL)—Suggests injecting D-He^3 beams in mirror geometry. In a long enough system [several km], fusion reactions between primary beams could produce fusion energy without neutrons.

VI. μ-Catalyzed Fusion

Bystritski (Dubna)—History: Initial estimates of fusion ~1 reaction/captured $\mu [\chi = 1]$

Surprises: 1) Resonant formation of DD μ molecule [$\tau \approx 10^{-9}$ sec instead of 10^{-6} (at solid density); 2) Resonant formation of DT μ molecule; 3) Small value of ω_s (probabilities of μ-He^4 capture in reactions). Initial estimates 1%, revised theory 0.6%; experiment ~ 0.4%; 4) Reasonably large value (0.3) for recycling He bound μ.

Present situation: Many cross sections are known, but phenomena depend on density and temperature of the trapping gas, and low temp. with 1.5 x solid density targets $\chi \simeq 170$. This is marginal for a fusion reactor; but could fuel a hybrid system (fusion and fission).

Jones (Utah)—Projected experiment $\rho \simeq 2.4 \times$ solid density, may yield $\chi = 1000$ (extrapolation of observed density dependence).

Need low temperature for DT μ formation, but high temperature for He μ stripping and μ recycling. Suggests solid DT pellets in a warm DT plasma? [*Experimental evidence needed to back this up*].

RECOMMENDATIONS

We are evaluating fusion proposals at a sad time when drastic budget cuts are imminent. It seems to us that this is the primary consideration in judging what kinds of projects should be saved. We must prepare for the day when the need for new sources of energy will again be perceived as important enough to justify substantial increases in federal funding. The strategy to be used depends on prophecy. If that day is seen as only a few years off then a holding pattern might make sense. A very different strategy would be indicated in preparation for a new start some decades ahead. What would we like to know in that case which would make a new start for fusion as attractive as possible?

Regarding ITER we endorse the position taken by Thomas H. Stix and his 8 cosigners in a letter to SCIENCE (Vol. 271, p. 891, 16 February 1996). They say "At present it is not known how to construct a fusion reactor economically" (The entire letter is reproduced in the Appendix). Therefore our *first priority should be to pursue alternate approaches until* we know how to construct an economical fusion reactor.

The most important need from the tokamak program is to use their large scale high temperature plasmas in an effort to gain better understanding of their unique physics. For example, we must understand the collapse or disruption phenomena discovered 33 years ago (Gorbunov and Razumova, Atomnaya Energaya *15*, 363 , 1963) and still not understood. We should use the big tokamaks to provide information precious for a new start rather than continuing to regard them as a step on an engineering road to a practical reactor.

In ICF, the NIF program should have highest priority and supporting work needed to produce ignition followed up, i.e., the Nova program, and studies on Holhraum dynamics and on target collapse, both at LLNL and LANL (as well as the Rochester program).

In MCF, the tokamak program should be supported until ignition is reached or a better understanding of their physics leads to more economical approaches. This means support not only for the ITER design study, but support for those many smaller Tokamak experiments (D.III.D is an example) that are studying the complex physics of magnetic confinement, have the flexibility to explore many aspects of Tokamak physics, show promise of improving Tokamak design, and as such are essential aspects of the MCF program, and although not discussed at this conference need to be maintained and amplified.

IGNITOR seems a rapid and cheap short cut to ignition and certainly merits support—perhaps in parallel with ITER—even though it does not tackle many of the reactor engineering problems faced by the larger device. ITER, of course, is designed to answer questions beyond that of ignition and may be the quickest approach to an actual reactor, and thus justify its cost.

In ICF, the next order of priority (after NIF) should probably be the development of the Induction Linac accelerator. Logically, this would be pushed after the demonstration of the laser produced ignition, but the time scale is such that overlap is worthwhile.

A further development is that of light ion beams, which using pulsed power technology might be cheaper and more practical than HIF. An important study here is magnetized targets. If, in spite of their inherent lack of symmetry, they can be scaled to ignition and make LIF beams useable, the field of ICF would be transformed. The modest experimental basis for this encourages exploration, and in ICF, we rate this third in order of priority.

The mirror program had reached a high level of sophistication, and although losses through the electron channel were never successfully suppressed, a modest program on low Q mirror systems looking for efficient energy transfer schemes would be worth supporting.

There were a number of suggestions for magnetically confined geometry in which most of the current is carried by the plasma. A major problem here is stability. Theoretical studies, mostly in the MHD limit, suggest that many of these are grossly unstable, and strong experimental and theoretical arguments must be made if these are to receive much support. Non-ideal effects—finite Larmor radius -collisions, etc., may reduce instability growth but no convincing analysis has addressed reactor conditions. One class of system is MHD stable: the minimum energy states; but this implies uniform pressure. Reactor conditions might be maintained with high pressure on the walls, but the temperature and density gradients required will lead to microinstabilities and anomalous transport. Very detailed theoretical studies supported by well designed small scale experiments are essential before large resources can be devoted to this area.

There are a number of suggestions that employ pulsed power either to compress a pellet or to produce a transient dense plasma as the site of a thermonuclear burn. Most of these are many orders of magnitude away from reactor conditions, but do produce useful bursts of radiation. Some of these—plasma focus, the fast Z pinch, the staged Z pinch, the spherical pinch, or the electrostatically confined ion beam-have been or could be developed as useful pulsed radiation sources. Further study at this level might explore the stability of these systems, and some may eventually be scaled to fusion levels. A major engineering problem will be the high output power. At present, however, it is as intense radiation sources that they merit support, and their value as such sources should determine the support level.

Cusp systems have high MHD stability, but losses are too high. Appropriate small scale experiments should be devoted to this study, both a low and high β. Once that problem has been understood, such systems might be developed on a reactor scale, but at present only small scale experiments merit support.

Ion beams as fusion sources still have far to go, but intense ion beams are important for ICF, as are storage rings, and perhaps neutralized beams; and it is probably in this context that ion beam studies are apt to prove most useful. Eventually an ion beam reactor might be possible, but since the ion beams would need to be neutralized, and since the system of beams and electrons is far from thermodynamic equilibrium, some kinds of instability are likely; and a strong experimental and theoretical case has to be made for extrapolation. Critical experiments can perhaps be best done at very low energies, where disturbing effects can be most easily seen. Again, theory and experiment need to be developed before investing in a large program.

μ catalyzed fusion is in a class by itself. The problems here are so different from those of other approaches to fusion and the surprises so unexpected and important that theoretical and experimental studies of the underlying physics should be supported. If χ can be experimentally shown to reach 1000, reactor concepts can be considered, but only then!

Finally, very high priority should be given to the program suggested by Perkins *et al* of providing an impartial review of the physics, engineering, reactor prospects and development path for promising new ideas.

APPENDIX

Science Magazine Letter which appeared in Volume 271, p. 8911 (16 February 1996).

Fusion Prospects

This letter relates to the article by Andrew Lawler about the International Thermonuclear Experimental Reactor (ITER) initiative (News & Comment. 19 Jan., p. 282). We have long believed that fusion has an inestimable potential as a long-term energy source. Eight of us have devoted much of our working lives to fusion and to the physics of tokamaks in particular. Through personal participation, we have each gained a warm respect for the high level of international cooperation that has been a hallmark of fusion research.

It is important to distinguish three different aspects of ITER:

1. The ITER organization, with policy and direction set by the ITER Council;
2. The ITER Engineering Design Activities (EDA) effort, a 6 year international program established in July 1992 to perform the detailed physics and engineering design for an engineering test reactor based on the ITER Conceptual Design Activities Final Report (1991); and
3. The decision to be made by the four parties to the ITER agreement about whether or not to actually *construct* this machine. (The decision is scheduled to be made after July 1998).

To achieve its stated goal of ignition and long-term burn, the ITER Conceptual Design report proposed an enormous machine that was a large extrapolation of existing tokamaks. The EDA has since performed a useful but expensive exercise in detailing and costing this machine, which is currently referred to as the "Interim Design".

At present, it is not known how to construct a fusion reactor economically. To enter directly into the actual construction of the Interim Design—a single machine of grandiose scale and cost in time, human effort, and money—suggests otherwise and would establish a commitment to a highly specific direction of development from which it would prove increasingly difficult and embarrassing to depart.

The Interim Design would take four giant steps simultaneously, each of them untested: Huge size, ignition, long pulse with fusion-grade plasmas, and very large volume superconducting magnets operating at their ultimate magnetic fields. To focus the world's fusion research efforts on the construction of this machine, which would not produce its first plasma until 2008 or even later, would constitute a high-risk choice.

The stated target for the Interim Design was to achieve ignition and remain in a state of equilibrium burn for 1000 seconds while producing 1500 megawatts of fusion power. Experimental data and theoretical understanding do not support the achievement of this performance under the proposed operating regime. Separate and serious concerns have been raised that the actual plasma and fusion performance in this machine would not even reach the fallback target, significant long-term power amplification. Such a failure of an internationally sponsored machine would be a humiliating setback for the entire fusion world.

The EDA has already defined the Interim Design sufficiently, and a broad-based independent assessment could be made now, rather than after July 1998.

The ITER Council and the EDA leadership have not given serious attention to strategic alternatives for international collaboration, other than moving along doggedly into the construction of the Interim Design.

The learning curve in fusion science is steep. For example, great advances in experiment, theory, and computation have occurred just in the past 5 years. With support, this rate of progress may be expected to continue. The speed of major device construction should be scaled to the appearance of new ideas and fundamental improvements (and we should use history as a guide).

We place the highest priority in fusion research on the maintenance of robust national programs (inluding the multinational Joint European Torus (JET)). Scientific advances of enormous potential value to economical fusion power have grown out of these programs. Efforts that use existing facilities, or those that can be built in the future, would be our best use of time and money. We suggest that the ITER council start immediately to identify new strategies to give us well-targeted and cost-effective ways to advance the world's quest for this enormously desirable energy source.

Thomas H. Stix*
Department of Astrophysical Sciences
Princeton University
Princeton, NJ 08543 USA

*Cosigners: **Ira B. Bernstein**, Department of Mechanical Engineering, Yale University, New Haven, CT 06520, USA; **Bruno Coppi**, Department of Physics, Massachusetts Institute of Technology, Cambridge, MA 02139, USA; **John M. Dawson**, Department of Physics, University of California, Los Angeles, CA 90095, USA; **Harold P. Furth**, Department of Astrophysical Sciences, Princeton University; **Chuan-Sheng Liu**, Department of Physics, University of Maryland, College Park, MD 20742, USA; **Roald Sagdeev**, Department of Physics, University of Maryland; **Andrew Sessler**, Lawrence Berkeley Laboratory, University of California, Berkeley, CA 94720, USA; **Ravi Sudan**, Laboratory for Plasma Studies, Cornell University, Ithaca, NY 14853, USA.

BIOGRAPHIES OF EVALUATORS

Dr. Edward C. Creutz
Dr. Arthur R. Kantrowitz, Chairman
Dr. Joseph E. Lannutti
Dr. Hans J. Schneider-Muntau
Dr. Glenn T. Seaborg
Dr. Frederick Seitz
Dr. William B. Thompson

EDWARD CHESTER CREUTZ

Edward Chester Creutz: Education: B.S. Mathematics and Physics, University of Wisconsin, 1936; Ph.D. Physics, U. of Wisconsin, 1939; Thesis: Resonance Scattering of Protons by Lithium.

Professional Experience: 1977–1984, Director, Bishop Museum, Honolulu, HI; 1976–1977, Acting Deputy Director, National Science Foundation, Washington, DC; 1975–1977, Assistant Director for Mathematical and Physical Sciences, and Engineering, National Science Foundation; 1970–1975, Assistant Director for Research, National Science Foundation (Presidential appointee); 1955–1970, Vice President, Research and Development, General Atomic, San Diego, CA; 1955–1956, Scientist at large, Controlled Thermonuclear Program, Atomic Energy Commission, Washington, DC; 1948–1955, Professor and Head, Department of Physics, and Director, Nuclear Research Center, Carnegie Institute of Technology, Pittsburgh, PA; 1946–1948, Associate Professor of Physics, Carnegie Institute of Technology; 1944–1946, Group Leader, Los Alamos, NM; 1942–1944, Group Leader, Manhattan Project, Chicago, IL; 1939–1942, Instructor of Physics, Princeton University, Princeton, NJ.

1945 ff, Consultant to AEC, NASA, Industry.

1960 ff, Editorial Advisory Board: American Nuclear Society, Annual Reviews, Handbuch der Physik, Interdisciplinary Science Reviews, Handbook of Chemistry and Physics.

Publications: 65 in fields of Physics, Metallurgy, Mathematics, Botany, and Science Policy.

Patents: 18 Nuclear Energy Applications.

Honors: Phi Beta Kappa; Tau Beta Pi; Sigma Xi; National Science Foundation Distinguished Service Award; University of Wisconsin, College of Engineering, Distinguished Service Citation; American Nuclear Society, Pioneer Award.

Memberships: National Academy of Sciences; AAAS, Fellow; American Physical Society, Fellow; American Association of Physics Teachers; American Nuclear Society.

ARTHUR R. KANTROWITZ

Arthur R. Kantrowitz, Professor of Engineering at the Thayer School of Engineering of Dartmouth College, earned his B. S., M.A., and Ph.D. degrees in physics at Columbia University. He taught aeronautical engineering and engineering physics at Cornell for ten years, and then founded and was CEO of the Avco Everett Research Laboratory. He is a member of the National Academy of Sciences and the National Academy of Engineering, and a fellow of the American Academy of Arts and Sciences, the American Physical Society, the American Institute of Aeronautics and Astronautics, the American Association for the Advancement of Sciences, and the American Astronautical Society. He was a Fulbright and Guggenheim Fellow, and recipient of the Roosevelt Medal of Honour for Distinguished Service in Science. He is an honorary trustee of the University of Rochester, an honorary life member of the Board of Governors of the Technion, and an honorary professor of the Hauzhong Institute of Technology in Wuhan, China. He holds 21 patents, and has published extensively. He is a director of the Hertz Foundation, and a member of the advisory board to television's popular "Nova" program. He has served our government on advisory boards to the Ford White House, the Department of Commerce, NASA, the General Accounting Office, and the National Science Foundation.

JOSEPH E. LANNUTTI

Joseph E. Lannutti: Born May 4, 1926, Malvern, PA; married with three children. High Energy Experimental Particle Physicist, Ph.D. 1957, University of California at Berkeley.

Served in the U.S. Army, 28th Infantry Division, European Theatre, in WWII 1944–46; Worked in Motive Power Division of the Pennsylvania Railroad Company 1943–44 and 1946–47.

Worked as a theoretical physicist designing guided missile autonavigation systems, at the North American Aviation, Inc., Downey, CA 1952–53; and as a research assistant at the University of California Radiation Laboratory 1953–57.

Came to Florida State University (FSU) as Assistant Professor in September 1957; promoted to Professor of Physics in 1965; appointed Associate Vice President for Academic Affairs in 1984.

Established laboratory for High Energy Particle Physics research at FSU and has been principal investigator since 1957. Continuous research funding from U.S. Department of Energy with FY93 budget over $1M; present personnel approximately 30.

Published more than 150 scientific articles and abstracts.

Established FAMU/FSU College of Engineering in 1982 and co-directed it until September 1984. Enrollment in 1992–93 approximately 2,000 students.

Established Supercomputer Computations Research Institute (SCRI) in 1984 and was Director 1984–1993. SCRI is the first federal/state/industry computations research institute of its kind in the United States and has a FY93 budget of approximately $10M; present personnel approximately 70.

Appointed Associate Vice President for Research 1992.

National Organization Memberships

Past

- High Energy Physics Advisory Panel, U.S. Department of Energy
- Oak Ridge Associated Universities, Board of Directors, Chairman of Council
- University Research Association, Board of Trustees, Chairman of Physics Committee
- Southeastern University Research Association, Board of Trustees
- Users Executive Committee, Fermi National Accelerator Laboratory

Present

Member of User Organizations at

- Brookhaven National Accelerator Laboratory, Long Island, NY
- Fermi National Laboratory, Batavia, IL
- Stanford Linear Accelerator Laboratory, Palo Alto, CA
- Center for Nuclear Research for Europe, Geneva, Switzerland
- University of Research Association Fermilab Review Committee

HANS J. SCHNEIDER-MUNTAU

Professional Interests: Advancement of magnet technology, development of state-of-the-art magnet systems, laboratory management

Education: Ph.D., Electrical Engineering, University of Munich, 1967
M.S., Electrical Engineering, University of Stuttgart, 1962
B.S., Electrical Engineering, University of Stuttgart, 1958

Professional Experience:

1991 - Present — Deputy Director, National High Magnetic Field Laboratory. Director, Magnet Development and Technology Group, NHMFL. Professor of Mechanical Engineering.

1972–1991 — Chief Engineer, High-Field Magnet-Laboratory, Grenoble, of the Max-Planck-Institute für Festkörperforschung, Stuttgart. Responsible for magnet development and administration, worked on development of resistive, pulsed and hybrid magnets and facility improvements.

1967–1972 — Head of the Development Laboratory, European Space Research Institute, Frascati. Worked on space simulation experiments, development of high-voltage ns discharges, capacitor banks and pulsed laser sources.

1962–1967 — Scientist, Institut für Plasmaphysik, Garching, of the Max-Planck- Gesellschaft, Munich. Worked on fusion technology, developed pulsed neutron sources, and fast high voltage discharges.

GLENN T. SEABORG

Glenn T. Seaborg is currently University Professor of Chemistry (the most distinguished titled bestowed by the Regents), Associate Director-at-Large of the Lawrence Berkeley Laboratory, and Chairman of the Lawrence Hall of Science at the University of California, Berkeley.

He received his A. B. in Chemistry from UCLA in 1934 and his Ph.D. in Chemistry from Berkeley in 1937. He has served on the faculty of the Berkeley campus since 1939 and was Chancellor of that campus 1958–1961. In 1961 Dr. Seaborg was appointed Chairman of the Atomic Energy Commission by President John F. Kennedy. He was subsequently reappointed by both Presidents Johnson and Nixon, serving in that position until 1971.

Winner of the 1951 Nobel Prize in Chemistry (with E. M. McMillan) for his work on the chemistry of the transuranium elements, Glenn Seaborg is one of the discoverers of plutonium (element 94). During World War II he headed the group at the University of Chicago's Metallurgical Laboratory which devised the chemical extraction processes used in the production of plutonium for the Manhattan Project. He and his coworkers have since discovered none more transuranium elements: americium (element 95), curium (96), berkelium (97), californium (98), einsteinium (99), fermium (100), mendelevium (101), nobelium (102), and Element 106. He has been honoured by the recommendation by the co-discoverers of Element 106 that it be named "seaborgium," with the symbol Sc. He holds over 40 patents, including those on elements americium and curium (making him the only person ever to hold a patent on a chemical element).

In 1944, Dr. Seaborg formulated the actinide concept of heavy element electronic structure which accurately predicted that the heaviest naturally occurring elements together with synthetic transuranium elements would form a transition series of actinide elements in a manner analogous to the rare earth series of lanthanide elements. This concept, one of the most significant changes in the periodic table since Mendeleev's 19th century design, shows how the transuranium elements fit into the periodic table and thus demonstrates their relationships to other elements.

His co-discoveries include many isotopes which have practical applications in research medicine and industry (such as iodine-131, technetium-99m, cobalt-57, cobalt-60, iron-55, iron-59, zink-65, cesium-137, manganese-54, antimony-124, californium-252, americium-241, plutonium-238), as well as the fissile isotopes plutonium-239 and uranium-233.

Dr. Seaborg continues to work as an active research scientist, with a research group in the search for new isotopes and new elements at the upper end of the periodic table, including a search for the "superheavy' elements. The group is also investigating the mechanism of the reactions of heavy ions with heavy element target nuclei. Another aspect of the research program is concerned with the determination of the chemical properties of the heaviest chemical elements.

Seaborg is the author of numerous books. During the spring of 1994, he published three books: *Chancellor at Berkeley*, a memoir about developments on the Berkeley campus during his chancellorship, *Modern Alchemy: The Selected Papers of Glenn T. Seaborg*, and *The Plutonium Story: The Journals of Professor Glenn T. Seaborg 1939–1946*, which describes his discovery of plutonium and work on the Manhattan Project. *Elements Beyond Uranium* (1990) is a comprehensive summary of all aspects of transuranium elements. He has written a trilogy about his service as chairman of the Atomic Energy Commission: *Kennedy, Khrushchev, and the Test Ban* (1981), *Stemming the Tide:*

Arms Control in the Johnson Years (1987), and *The Atomic Energy Commission under Nixon: Adjusting to Troubled Times* (1993); the first two books focus on arms control issues, while the Nixon book is more general. He has also authored over 500 scientific articles and guided the graduate studies of more than 65 successful Ph.D. candidates. In addition to the Nobel Prize and a great many other awards for his work in chemistry, science education and community service, Dr. Seaborg has been awarded 50 honorary doctoral degrees.

Among his many interests are international cooperation in science, history of science (documenting the early history of nuclear science), nuclear arms control (advocating a comprehensive test ban treaty), and hiking. A member of the National Commission on Excellence in Education which published the much-publicized report *A Nation At Risk* in 1983 and Chairman of the Lawrence Hall of Science, Dr. Seaborg is recognized as a national spokesman on education, addressing in particular the crisis in the mathematics and science education.

FREDERICK SEITZ

Frederick Seitz was born in San Francisco on July 4, 1911. He received his bachelor's degree from Stanford in 1932 and his Ph.D. from Princeton in 1934. He has written some classic works in physics including *Modern Theory of Solids* (1940), was co-editor of the series *Solid State Physics* (started in 1954), and examined the evolution of science in *The Science Matrix* (1992).

Seitz's early career included positions at the University of Pennsylvania, the Carnegie Institute of Technology, and General Electric. During World War II, he worked for the National Defense Research Committee, the Manhattan District, and as a consultant to the Secretary of War. From 1946 to 1947 he was director of the training program on peaceful uses of atomic energy at Oak Ridge National Laboratory. Appointed professor of physics at the University of Illinois in 1949, Seitz became department chair in 1957 and dean and vice president for research in 1964. He joined the Rockefeller University as its president in 1968.

Dr. Seitz was elected to the National Academy of Sciences in 1951, serving as part-time president for three years before assuming full-time responsibility in 1956. He has served as advisor to NATO, the President's Science Advisory Committee, the Office of Naval Research, the National Cancer Advisory Board, the Smithsonian Institution, and other national and international agencies. He has been honored with the Franklin Medal (1965), the Compton Medal — the highest award of the American Institute of Physics (1970), the National Medal of Science (1973), two NASA Public Service Awards (1969 and 1979), the National Science Foundation's Vannevar Bush Award (1983), as well as honorary degrees from 32 universities worldwide. In 1993, the University of Illinois renamed its Materials Research Laboratory in Dr. Seitz's honor. Stanford University has honored him with the Hoover Medal and Princeton University with the Madison Medal.

WILLIAM BELL THOMPSON

William B. Thompson first came to the University of California, San Diego in September 1961 as a Visiting Professor from the Culham Laboratory where he was head of the Theoretical Physics Division in 1961–63. From 1963 to 1965 he served as Professor of

Plasma Physics at Oxford University and returned to UCSD as Professor of Physics in November 1965, serving as department chairman 1969–72.

He received his B.A. degree in physics and mathematics in 1944 and his M.A. in physics in 1947 from the University of British Columbia, Vancouver. In 1950 he received his Ph.D. in applied mathematics from the University of Toronto, Ontario. He then went as a Senior Fellow to the Theoretical Physics Division of Harwell, the research laboratory of the United Kingdom Atomic Energy Authority, where he stayed for 10 years moving to Culham when that laboratory was founded.

Dr. Thompson's career has been devoted to the development of the theory of plasma physics and its applications in Controlled Thermonuclear Research. He was the theoretical member of the small group that started this research in England in 1950— a group that finally developed into the Culham Laboratory for Controlled Thermonuclear Fusion.

Professionally, he has served as Consultant to several private industries and government bodies; as a council member for the Plasma Physics Division of the American Physical Society, as Associate Editor of the *Journal of Plasma Physics*, joint editor of *Advances in Plasma Physics*, and *Plasma Physics and Controlled Thermonuclear Fusion*, and on the editorial board of *Interscience Tracts on Physics and Astronautics*.

He has published one monograph on plasma physics, and many articles on plasma theory and controlled thermonuclear fusion.

PARTICIPANTS

Dr. Richard Aamodt
Lodestar Research Corp.
P.O. Box 4545
Boulder, CO 80306
U.S.A.

Dr. Enan Ahmed
2600 Virginia Avenue, # 800
Washington, DC 20037
U.S.A.

Dr. D. A. Baker
P-4 MS E554
Los Alamos National Laboratory
Los Alamos, NM 87545
U.S.A.

Dr. R. O. Bangerter
Lawrence Berkeley Laboratory
Accelerator and Fusion
Research Division
1 Cyclotron Rd.
Berkeley, CA 94720
U.S.A.

Dr. Daniel Charles Barnes
Grp. T15
MS B217
Los Alamos National Laboratory
Los Alamos, NM 87545
U.S.A.

Dr. Jeff Beckstead
Interscience, Inc.
105 Jordon Road
Troy, NY 12180
U.S.A.

Dr. L. Bilbao
INFIP
Universidad de Buenos Aires
Ciudad Universitaria - Pab. I
1428 Buenos Aires
ARGENTINA

Dr. John Blewett
7405 Sweetbriar Dr.
College Park, MD 20740
U.S.A

Dr. J. S. Brzosko
Department of Physics
Stevens Institute of Technology
Hoboken, NJ 07030
U.S.A.

Dr. J. H. Brownell
Group-X-5
Los Alamos National Laboratory
MS-259
Los Alamos, NM 87545
U.S.A.

Dr. Viacheslav M. Bystriskii
Joint Institute of Nuclear Researches
141980 Dubna, Moscow Region
RUSSIA

Dr. Vitaly M. Bystriskii
Department of Physics
University of California
Irvine, CA 92717-4575
U.S.A.

Mr. André Cadieux
Cadieux & Chevalier
983 De Salière
Saint-LUC, P.Q. J2w 1A4
CANADA

Dr. Edward Michael Campbell
L-481
Lawrence Livermore National Laboratory
P.O. Box 5508
Livermore, CA 94551
U.S.A.

Dr. Peter Cheeseman
2557 Webster Street
Palo Alto, CA 94301
U.S.A.

Dr. Vladimir K. Chernyshev
All-Russian Scientific Research Institute
of Experimental Physics
Nizhegorodsky region, RF
607200 Arzamas - 16
RUSSIA

Dr. Chan-Kyoo Choi
School of Nuclear Engineering
Purdue University
West Lafayette, IN 47907
U.S.A.

Mr. Apichai Chvajarernpun
Royal Thai Embassy
Scientific Attaché
4301 Conneticut Ave., NW
Washington, DC 20008
U.S.A.

Dr. R.L. Coldwell
Department of Physics
University of Florida
Gainesville, FL 32611-2085
U.S.A.

Dr. Bruno Coppi
Department of Physics
Massachusetts Institute of Technology
Rm. 26-217
Cambridge, MA 02139
U.S.A.

Dr. Edward C. Creutz
P.O. Box 2757
Rancho Santa FE, CA 92067
U.S.A.

Dr. Ben Cross
Economic Development Division
Fusion/ITER Program
Bldg. 773-41A
Westinghouse Savannah River Co.
Aiken, SC 29808
U.S.A.

Dr. Randolph R. Davis
44 Redding Ridge Dr.
North Potomac, MD 20878
U.S.A.

Dr. James H. Degnan
USA Phillips Laboratory, PL/WSP
High Energy Plasma Division
3550 Aberdeen SW
Kirtland AFB, NM 87117-5776
U.S.A.

Dr. C. Deutsch
L.P.B. P. (Associé au CNRS)
Bâtiment 212
Université-Paris XI
91405 ORSAY
FRANCE

Dr. Vineet Dhyani
Institute of Plasma Physics
Czech Academy of Sciences
Za Slovankou 3, P.O. Box 17
18211 Prague-8
CZECH REPUBLIC

Participants

Dr. T. J. Dolan
INEL-Idaho National Engineering
 Laboratory
P.O. Box 1625
Idaho Falls, ID 83415
U.S.A.

Dr. Donald Ernst
Termacore, Inc.
780 Eden Road
Lancaster, PA 17601
U.S.A.

Dr. T. J. Fessenden
Lawrence Berkeley Laboratory
1 Cyclotron Rd.
Bldg. 47/112
Berkely, CA 94720
U.S.A.

Dr. Nikolai Filippov
I.Y. Kurchatov Institute of Atomic
 Energy
123182 Moscow
RUSSIA

Dr. Lydie Gerard
CEV-M
B.P. No. 7
77181 Courtry
FRANCE

Dr. Francesco Giammanco
Dipartimento Di Fisica
Universitá di Pisa
56100 Pisa
ITALY

Dr. John Gilleland
Bechtel National, Inc.
50 Beale Street
San Francisco, CA 94105-3965
U.S.A.

Dr. George Goldenbaum
Office of the Dean
College of Computer, Mathematical
and Physical Sciences
University of Maryland
College Park, MD 20742
U.S.A.

Dr. W. B. Hermannsfeldt
Stanford Linear Accelerator Center
Stanford University
P.O. Box 4349
Stanford, CA 94309
U.S.A.

Dr. Julio Herrera
Instituto de Ciencias Nucleares
Universidad Nacional Autonoma
Circuito exterior C.U.
A.P. 70-543
04510 Mexico DF
MEXICO

Dr. Yukihiro Hirano
JAERI Washington Office
1825 K Street NW
Suite 1203
Washington, DC 20008
U.S.A.

Dr. Arthur Kantrowitz
4 Downing Road
Hanover, NH 03755
U.S.A.

Dr. Predhiman K. Kaw
Institute for Plasma Research
Bhat Gandhinagar
Gujarat 382 424
INDIA

Dr. J. D. Kilkenny
Lawrence Livermore Laboratory
L473
Livermore, CA 94550
U.S.A.

Dr. R. Kirkpatrick
Los Alamos National Laboratory
MS-B229
Los Alamos, NM 87545
U.S.A.

Mr. Paul M. Koloc
President
Phaser Corporation
Bx 1037
College Park, MD 20740-1037
U.S.A.

Mr. Oh-Kab Kwon
Embassy of Korea
Scientific Attaché
2600 Virginia Avenue, NW
Washington, DC 20008
U.S.A.

Dr. John Landis
Stone and Webster Engineering Corp.
245 Summer Street
P.O. Box 2325
Boston, MA 02107
U.S.A.

Dr. Joseph Lannutti
217E Westcott Bldg.
Florida State University
Tallahassee, FL 32306-4093
U.S.A.

Dr. M. Laroussi
Department of Electrical and
Computer Engineering
The University of Tennessee
102 Ferris Hall
Knoxville, TN 37996-2100
U.S.A.

Dr. Herik Lerner
20 Pine Knoll Dr.
Lawrenceville, NJ 08648
U.S.A.

Dr. I. R. Lindemuth
Mail Stop 645
Los Alamos National Laboratory
Los Alamos, NM 87545
U.S.A.

Dr. Erick L. Lindman
Los Alamos National Laboratory
MS-F645, X-1, LANL
Los Alamos, NM 87545
U.S.A.

Mr. Viacheslav Lisitsin
Embassy of the Russian Federation
Scientific Attaché
1125 16th Steet, NW
Washington, DC 20036
U.S.A

Dr. B. Grant Logan
Lawrence Livermore National Laboratory
P.O. Box 5508, L-481
Livermore, CA 94550
U.S.A.

Dr. Bogdan C. Maglich
Advanced Physics Corporation
4199 Campus Dr., Suite 690
Irvine, CA 92715
U.S.A.

Dr. J. M. Martinez-Val
Institute of Nuclear Fusion,
    DENIM, E.T.S.I.I.
Madrid Polytechnic University
Jose Gutierrez Abascal, 2
28006 Madrid
SPAIN

Dr. Giorgio Mattiello
Addetto Scientifico
Embassy of Italy
1601 Fuller St. NW
Washington, DC 20009
U.S.A.

Participants

Dr. Michael E. Mauel
Department of Applied Physics
Columbia University
New York, NY 10027
U.S.A.

Dr. George H. Miley
216 Nuclear Engineering Laboratory
University of Illinois
103 S. Goodwin Ave.
Urbana, IL 61801
U.S.A.

Mr. Ronald Militello
Starborne Corporation
14625 Lake Forest Drive
Lutz, FL 33549
U.S.A.

Dr. V. N. Mokhov
All Russian Scientific Research
Institute of Experimental Physics
 (VNIIEF)
Nizhay Novgorod Region
607200, Arzamas-16
RUSSIA

Dr. H. Momota
National Institute for Fusion Science
Nagoya University
Nagoya 464
JAPAN

Dr. Hendrik J. Monkhorst
Quantum Theory Project
361 Williamson Hall
University Of Florida
Gainesville, FL 32611-8435
U.S.A.

Mr. Lev Mukhin
Embassy of the Russian Federation
Scientific Attaché
1125 16th Street, NW
Washington, DC 20036
U.S.A.

Dr. Ghulam Murtaza
Head, Fusion Laboratory
Department of Physics
Quaid-I-Azam University
Islamabad
PAKISTAN

Dr. John H. Nuckolls
Lawrence Livermore National
 Laboratory, L-1
P.O. Box 808
Livermore, CA 94550
U.S.A.

Dr. Emilio Panarella
Advanced Laser and Fusion
 Technology, Inc.
189 Deveault St., No. 7
Hull, P.Q. J8z 1S7
CANADA
and
Dept. of Electr. and Comp. Engineering
University of Tennessee
514 South Stadium Hall
Knoxville, TN 37996-2100
U.S.A.

Dr. Francesco Pegoraro
Dipartimento di Fisica Teorica
Universitá di Torino
via P. Giuria, 1
10125 Torino
ITALY

Dr. Martin Peng
Oak Ridge National Laboratory
P.O. Box Y
Bldg. 9204-1 MS-15
Oak Ridge, TN 37831-8072
U.S.A.

Dr. L. John Perkins
L-644
Lawrence Livermore National Laboratory
P.O. Box 5511
Livermore, CA 94550
U.S.A.

Dr. Hans Persson
Dept. of Fusion Plasma Physics
Alfvén Laboratory
Royal Institute of Technology
S-100 44 Stockolm
SWEDEN

Dr. James A. Phillips
P-4 MS E554
Los Alamos National Laboratory
Los Alamos, NM 87545
U.S.A.

Dr. Richard F. Post
Lawrence Livermore National Laboratory
P.O. Box 808, L-640
Livermore, CA 94550
U.S.A.

Dr. Jorge Pouzo
Head Of Plasma Group
Instituto De Fisica Arroyo Seco (IFAS)
UNC, Pinto 399
7000 Tandil
ARGENTINA

Dr. Ravi Prakash
Embassy of India
Scientific Attache
2536 Massachussetts Ave., NW
Washington, DC 20008
U.S.A.

Dr. Hafiz-Ur Rahman
Institute of Geophysics and
  Planetary Physics
University of California
Riverside, CA 92521-0412
U.S.A.

Dr. P. Ravetto
Politecnio di Torino
Dipartimento di Energetica
Corso Duca degli Abruzzi, 24
10129 Torino
ITALY

Dr. Charles W. Roberson
Physics Department
Code 412
Office of Naval Research
800 N. Quincy Street
Arlington, VA 22217
U.S.A.

Dr. Benjamin Robouch
ENEA/EURATOM
Laboratori Nazionali di Frascati
Via Enrico Fermi, 27
I-00044 Frascati (Roma)
ITALY

Dr. Norman Rostoker
Department of Physics
University of California
Irvine, CA 92697-4575
U.S.A.

Dr. J. Reece Roth
Department of Electrical and
  Computer Engineering
The University of Tennessee
316 Ferris Hall
Knoxville, TN 37996-2100
U.S.A.

Dr. Robin Roy
Project Director
Office of Technology Assessment
Congress of the United States
Washington, DC 20510-8025
U.S.A.

Dr. Nikos A. Salingaros
Division of Mathematics,
Computer Science and Statistics
The University of Texas
6900 North Loop 1604 West
San Antonio, TX 78249-0664
U.S.A.

Participants

Dr. H. Schneider-Muntau
National High Magnetic Field Laboratory
Florida State University
1800 E. Paul Dirac Drive
Tallahassee, FL 32306-4005
U.S.A.

Dr. David Woodward Scudder
P-1 E-526
Los Alamos National Laboratory
Los Alamos, NM 87545
U.S.A.

Dr. Glenn T. Seaborg
Bldg. 70/A Rm. 3307
Lawrence Berkeley Laboratory
1 Cyclotron Road
Berkeley, CA 94720
U.S.A.

Dr. Frederick Seitz
The Rockeffeller University
1230 York Ave.
New York, NY 10021-6399
U.S.A.

Dr. Liu Shenggang
The University of Electronic Science
and Technology of China
Chegdu, Sichuan Province, 610054
PEOPLE'S REPUBLIC OF CHINA

Dr. J. Shlachter
High Energy Density Physics
P-1, ES26
Los Alamos National Laboratory
Los Alamos, NM 87545
U.S.A.

Dr. Richard E. Siemon
MS H854
Los Alamos National Laboratory
P.O. Box 1663
Los Alamos, NM 87545
U.S.A.

Dr. D. P. Singh
Institute of Atomic and Molecular Physics
National Research Council
56127 Pisa
ITALY

Dr. Elio Sindoni
Istituto di Fisica
Via Celoria, 16
I-20133 Milano
ITALY

Mr. Carlos D. Souto
Embassy of Portugal
Scientific Attache
2125 Kalorama Road, NW
Washington, DC 20008
U.S.A.

Dr. John Slough
Radmond Plasma Physics Laboratory
University of Washington
14700 NE 95th St.
Seattle, WA 98052
U.S.A.

Mr. Michael Stephens
Canadian Embassy
Scientific Attaché
501 Pennsylvania Ave, NW
Washington, DC 20001
U.S.A.

Dr. Charles Striffler
Department of Electrical Engineering
University of Maryland
College Park, MD 20742
U.S.A.

Dr. M. A. Sweeney
Sandia National Laboratory
Mail Stop 1186
P.O. Box 5800
Albuquerque, NM 87185-1186
U.S.A.

Hon. Dick Swett
Member, House of Representatives
Room 230
Cannon House Office Building
Washington, DC 20510-0230
U.S.A.

Dr. A. Tayyib
2600 Virginia Avenue, #800
Washington, DC 20037
U.S.A.

Dr. Edward Teller
Lawrence Livermore National Laboratory
P.O. Box 808, L-0
Livermore, CA 94551
U.S.A.

Dr. W. B. Thompson
Department of Physics
University of California at San Diego
LA Jolla, CA 92093
U.S.A.

Dr. Guillermo Velarde
Institute of Nuclear Fusion
Polytechnic University of Madrid
José Gutierrez Abascal, 2
E-28006 Madrid
SPAIN

Dr. Frank J. Wessel
Department of Physics
University of California
Irvine, CA 92697-4575
U.S.A.

Dr. C. Yamanaka
Institute of Laser Technology
2-6 Yamada-oka
Suita, Osaka 565
JAPAN

Dr. Bruce A. Zeitlin
IGC Advanced Superconductors, Inc.
1875 Thomaston Avenue
Waterbury, CT 06704
U.S.A.

INDEX

Absorption, 249
Accelerators, 279
  induction accelerators, 279
    ILSE, 279
    SBTE, 279
    recirculation in, 280
    MBE-4, 280
Adiabatic compression, 485–487
Adiabatic invariants, 155, 156
Advanced confinement, 119
Advanced plasma regimes, 132
Alpha particle confinement, 125
Alpha particles, 214, 215, 221, 225
Alternate fusion concepts, 365, 370, 489–495
AMBAL, 160
Aneutronic, 485, 486
Angular momentum, 203
Anomalous transport, 491
Antiproton catalyzed fusion, 372
Auxiliary heating, 126
Average beta, 121

Ball lightning, 459, 475–479, 487
  characteristics, 461, 571
  diameter, 461, 462
  duration, 461
  energy content, 461, 463
  energy density, 461
Beam plasma–target fusion, 135–147; *see also* IEC
Beam smoothing, 300
Blow-by, 188
Boron neutron capture therapy (BNCT), 147
Bremsstrahlung, 38
Bumpy torus, 427

Cannonball, 303
Cannonball target, 272
Capital costs, 366
CIB, 498, 499, 509
Cluster-driven targets, 519
  hydrodynamic regime, 519
Cluster ion beam, 497
Cluster ion beam direct drive
  compression phase, 531

Cluster ion beam direct drive (*cont.*)
  ignition and burn, 532
  target design, 530
Cluster ion fragmentation, 500
Colliding beams, 33
Compact fusion reactor, 378
Compact machine, 125
Compact toroid, 179, 181
Compressed density, 268
Compression
  magnetic, 500
  magnetic field, 544
  magnetic flux, 551
  plasma, 544
Conduction losses, 211, 216, 217, 242
Conductivity
  electrical, 544
Cone angle, 188
Confinement, 119, 122, 442
Confinement, advanced, 119
Confinement of high temperature plasma, 375
Confinement time, 52
Conical-coaxial compression, 185
Conical targets, 53
Controlled fusion, 5; *see also* Fusion
Convergence
  factor of, 544
Coulomb barrier reduction, 371

DEMG, 551, 552
Deuterium-tritium, 213; *see also* DT
D-^{3}He, 121
D-^{3}He, 154, 162, 167
  open-ended systems, 155
  mirror machine, 155
D-^{3}He pilot plant, 142; *see also* IEC
D-^{3}He pilot plant design, 142
D-^{3}He reactor, 33, 149, 154, 164
Diamagnetic current, 121
Diamagnetism, 203
Dipole fusion, *see also* D-^{3}He reactor
  anomalous plasma trasport in, 149
  concept, 149, 569
  reactor, 149

Direct drive, 45, 264, 312, 498, 530
Divergence jets, 477–479
DPF, 11, 12
Drift pumping, 174
Driver, 259, 328
DT, 8, 16, 17, 125, 126, 211, 409, 410, 413, 490

Economics of fusion, 366
Economy of scale, 367
ECS, 508, 511, 512, 515, 516, 517
EFBT, 421, 427, 440, 441, 446, 447, 452, 453, 455, 456
Electric field bumpy torus, 421, 427
Electrode plasma, 182, 190
Electromagnetic trap, 197
Electrostatic plugging, 197
End losses, 163
Energy confinement time, 153, 238, 239, 240, 243
Energy transport, 252
Engineering test facility, 541
Enhanced correlated stopping, 508
ETF, 541
Explosive pulsed power, 551

Fermi-degenerated, 308
Field, magnetic, 551
    embedded magnetic, 544
Field reversed, 33
Field reversed confinement, 121
Field reversed configurations, 121, 370, 371, 387, 568; *see also* FRC
Fireball, 78, 80, 82
Fireball expansion, 77
Fission, 489
Fluctuation-induced transport, 452
Fossil fuels, 490
FRC, 161, 370, 371, 387, 568
    stability, 122, 381
Fuel
    thermonuclear, 544
Fusion, 153, 489
Fusion, controlled thermonuclear, 5, 375, 376
    burn-up fraction, 558
Fusion breakeven conditions, 211, 215, 216, 220
Fusion energy, 279, 459
Fusion fuel, 39
Fusion products, 40
Fusion reactor, 378, 470, 558

Gamma 10 experiment, 160
GEKKO XII, 297, 299, 316, 567
Gas dynamics trap, 174
    axi-symmetric magnetic field, 175
    MHD stabilization, 175
Grids for, 139–140

Halo, 138
Hammer model, 525
Heating, 446

Heating time, 51
Heavy Ion Inertial Fusion, 279
    accelerator-driver, 284
    environmental aspects, 284
    economics of, 283
Helical plasma, 477
HEMG, 552, 553, 554
HEPP, 553
HEPS, 135
High beta, 34
High explosive power, 550
High gain target, 309
High magnetic field, 125
High voltage, 197
Hohlraum, 295, 362, 542, 566
Homogeneous targets, 59
Hot rocket model, 520, 523
Hybrid magnetic, 143–145
Hydrodynamic instabilities, 40
HYLIFE-II, 542

ICF, 8, 245, 247, 295, 320, 497, 499, 563, 564, 567, 572, 574
IEC, Inertial Electrostatic Confinement, 135–147, 570
    concepts for, 136
    grids for, 139–140
    history of, 135
    hybrid magnetic, 143–145
    modes of operation, 138
        halo, 138
        STAR, 138
    neutrons from, 138–139
    simulation of, 140
    scale-up, 140–143, 145–146
    scaling laws, 142
    Penning trap, 145
    Polywell, 144
IFE, 541, 542
Ignition, 119, 211, 253, 257, 295, 312, 321, 323, 532, 544, 556
Ignition, National Ignition Facility, 9; *see also* NIF
Ignition physics, 125
IGNITOR, 125, 565, 574
Implosion, 48, 295
    computer simulation, 507
    Coulomb explosion, 502
    liner, 551
    velocity, 544, 546
    x-ray driven, 295
Implosion and Ignition, 253
Impurities, 381
Inflight Coulomb stability, 502
Indirect drive, 46, 269, 303, 537
Inductance
    of the circuit, 551
Inductance growth, 185
Inductively stored energy, 551
Inertial-electrostatic plasma confinement, 197

Inertial confinement fusion, 8, 88, 89, 245, 279, 320, 371, 497; see also ICF
Inertial fusion energy, 43, 541; see also ICF
Instability, 257
Instability growth, 179, 194
Instabilities of Z-pinch, 375
Insulator, 555
Ioffe bars, 157
Ion fragments stopping
  in cold matter, 508
  in dense plasma, 513
Ionization energy, 76
Irradiation uniformity, 298
ITER, 207, 246, 475, 490, 493, 563, 565, 573, 575, 576

Kernel, 475, 477
Knockons, 483
Koloc model, 464, 466, 467

Laser plasma interaction, 249
Lawson criteria, 211, 212, 232, 233, 237, 242, 320
LHART, 308
Lightning, 460
Linear collider, 164, 165, 166
Liner implosion, 551
Loss
  energy, 544
  radiative, 544
Loss-cone angle, 175
Low frequency fluctuations, 380, 381
Low-Q fusion systems, 154
  direct converters, 162
  linear collider, 164
  D-Helium^3 fusion, 164
  velocity-modulated beams, 170

Macrotor, 425
Magnetic confinement, 7; see also MCF
  open ended, 153
    adiabatic invariants, 155
    magnetic moment, 156
    magnetic wells, 158
    multiple mirror, 159
    tandem mirror, 159
Magnetic cusp, 197
Magnetic-electrostatic confinement, 569
Magnetic field errors, 380, 381
Magnetized plasma ring, 179, 181
Magnetized plasmas, 179, 191
Magneto-electric confinement, 421, 570
Magnetoplasmoid, 477
Magnetic confinement fusion, 370
Magnetic shielding, 197
Magnetized plasmas, 179, 191
Magnetized target fusion,
  aneutronic fusion with MTF, 329
  applications, 330
  concept, 319, 543, 544

Magnetized target fusion (cont.)
  critical issues, 329
  driver requirement, 328
  DT alpha energy deposition, 547
  Lindl–Widner diagram, 323, 324
  MTF target, 320
  Phi-target experiment, 325, 327, 549, 572
  parameter space, 546
  requirement for fusion ignition, 321, 323
Mantle, 466, 475–480
MAGO experiment, 549, 550, 552, 555
MCF, 563, 564, 573
Measurements
  currents, 555
  field, 555
  filtered x-rays, 555
  neutron, 555
  time-of-flight, 555
Megajoule, 181, 185, 187, 188, 190
MHD, 175, 181, 188, 189, 190, 192
MHD
  plasma instabilities, 149, 158
  stability, 125, 176, 388, 574
MIGMA, 145; see also IEC
Minimum particle density, 243
Mirror ratio, 156
Mixing, 307
Modes of operation, 138
MTF, 179, 191, 192, 571
Multifragmentation, 500
Muon catalysis, 372
Muon catalyzed fusion, 389, 401, 564, 573, 575
  cold fusion, 389
  cycling rate, 390, 401
  density effect, 391, 392
  expected fusion yield, 397
  hydrogen, 401
  hyperfine structure, 403
  in mixtures of hydrogen isotopes, 401
  isotopes, 401
  measurement of alpha-muon sticking, 394
  muon, 401
  muonic atom, 402
  muon-capture losses, 392
  sticking coefficient, 394, 396, 403
  sticking in high C targets, 396
  temperature effects, 398
  transition rate, 408
  trition, 403
  under barrier, 402
  µ-molecular, 401
  µ-atomic, 401
µ-meson, 8

National Ignition Facility, 541, 563; see also NIF
Neutron, 213
Neutron, 147; see also IEC
Neutron emission, 68
Neutron generation, 74

Neutron source, 7, 69
Neutron source, 135–147; *see also* IEC
Neutron yield, 81, 83, 85, 87, 10
Neutrons from, 138–140
NIF, 9, 541, 573
Non-adiabatic ions, 33
Non-Maxwelliam plasma, 135–147; *see also* IEC
Non-neutral plasma, 135–147; *see also* IEC
NOVA laser, 295, 566

Ohmic heating, 125
Open-ended magnetic confinement, 153, 568
Open magnetic confinement, 197
Optical spectroscopy, 182, 184

Particle and energy confinement, 125
$p^{11}B$ (protium boron), 485
Penning trap, 145
PFX experiment, 570
Piezoelectric probe, 192, 193
Piston field, 188
Plasma flow switches, 194
Plasma focus, 11, 12, 570; *see also* DPF
    application and near term payoff, 11
    breakeven, 11, 16
    fusion energy, 29
    neutron source, 25
    plasma parameters, 20
    pulse-power, 20
    safe fission reactor, 29
Plasma mantle: *see* Mantle
Plasmak, 475, 476, 477, 478, 483, 485, 486, 487
PMK, 477, 479, 482, 483, 486, 487
Pollution, 490
Poloidal flux, 122
Polywell, 144
Positron, 147; *see also* IEC
Positron emission tomography (PET), 147
P.P.R.I., 494
Propulsion design, 145–146
Propulsion, space, 145; *see also* IEC
Pulsed power, 194, 543, 549, 551

Radiation energy, 76
Radiation transport, 253
Radioactivity, 5, 6, 7
Radiography, 190
Ramsauer gas, 468
Reactor conditions, 211, 212
Reactor plant equipment, 366
Reconnection, 181
Reversed-field mirror, 153, 161
Reversed-Field Pinch, *see also* RFP
    behavior of, 380, 381
    future recommendations for, 381, 382
    history of, 376, 378, 379
    reactor potential, 378, 569
RFP, 375, 376, 378, 380, 381
Rigid rotor, 34

Ring cusps, 197
Rocket model, 256
Rotating magnetic field, 123
Rotational kink, 37

Sand boxes, 493
Scale up, 140–143, 145–146
Scaling law
    spherical pinch, 142
    ZT-40M, 380, 381
Self-colliding beams, 33
Shape enhanced fusion, 372
Shiva Star, 181
Shock implosion, 84
Shock wave, converging, 78
Shock wave, explosion, 74, 76
Shock waves, 74
    simulations of, 140
Soft X-ray, microlithography, 69
Solid liner, 179, 180
S-parameter, 122
Spark ignition, 48
Spherator, 149
Spherical liner, 190–194
Spherical pinch, 67
    analytical study, 73
    imploding shock waves, 112, 571
    inductive, 69
    industrial applications, 106
    neutron yield, 82
    numerical study, 88
    radiation source, 106
    resistive, 69
    scaling laws, 69, 71, 73, 83, 85, 88
    X-ray generator, 109
Spheromak, 371, 475, 485
Staged Z-pinch, 333
    as super radiant X-ray source, 347, 571
    driving hohlraum ICF target, 362
    field reversal, 358
    hardware requirement, 360
    implosion dynamics, 352
    inertial confinement with, 333, 347, 358, 571
    inner shell vacancy generation, 350
    magnetic flux, 333, 358
    modelling, 336
    radiation, 333, 358
    stability, 333, 342, 358
Stagnation, 3207
STAR, 138
Stellarator, 370

Tandem mirror, 159, 161
Target
    magnetized, 543
    Phi-, 549
Target design, 53, 530, 566
Target physics, 43, 566
Therapy, Boron, 147; *see also* IEC
Thermal conductivity, 217, 218, 219

Index

Thermal transport coefficients, 131
Thermonuclear fusion, 125
Tilt mode, 36
TMX, 160
Tokamak, 7, 119, 489, 563, 565
Tokamak biasing, 425
Tokamak plasma, 425
Transport coefficients, 131

2XIIB, 159, 568

Uncertainties in fusion development, 371
U.S. Congress, 493

VNIIEF, 543, 550, 551, 552, 557

Volume ignition, 49

Woltjier–Taylor equilibrium, 181
Working fluid, 179, 190, 191, 192

Z-Pinch
  dense Z-pinch, 385, 571
  fibre pinch, 386
  history of, 375, 376
  instabilities, 375, 385
  simple, 375
  unstable, 375
ZT-40M
  behaviour of, 380
  scaling of, 380

The manufacturer's authorised representative in the EU is Springer Nature Customer Service Centre GmbH, Europaplatz 3, 69115 Heidelberg, Germany. If you have any concerns regarding our products, please contact ProductSafety@springernature.com

Printed and bound by CPI Group (UK) Ltd, Croydon, CR0 4YY

02/01/2026

02028263-0012